T0211685

Communications in Computer and Information Science　1141

Commenced Publication in 2007
Founding and Former Series Editors:
Phoebe Chen, Alfredo Cuzzocrea, Xiaoyong Du, Orhun Kara, Ting Liu,
Krishna M. Sivalingam, Dominik Ślęzak, Takashi Washio, Xiaokang Yang,
and Junsong Yuan

More information about this series at http://www.springer.com/series/7899

Vladimir M. Vishnevskiy ·
Konstantin E. Samouylov ·
Dmitry V. Kozyrev (Eds.)

Distributed Computer and Communication Networks

22nd International Conference, DCCN 2019
Moscow, Russia, September 23–27, 2019
Revised Selected Papers

 Springer

Editors
Vladimir M. Vishnevskiy
V. A. Trapeznikov Institute
of Control Sciences
Moscow, Russia

Konstantin E. Samouylov ⓘ
Peoples' Friendship University of Russia
Moscow, Russia

Dmitry V. Kozyrev ⓘ
V. A. Trapeznikov Institute
of Control Sciences
Moscow, Russia

Peoples' Friendship University of Russia
Moscow, Russia

ISSN 1865-0929   ISSN 1865-0937 (electronic)
Communications in Computer and Information Science
ISBN 978-3-030-36624-7   ISBN 978-3-030-36625-4 (eBook)
https://doi.org/10.1007/978-3-030-36625-4

This Springer imprint is published by the registered company Springer Nature Switzerland AG
The registered company address is: Gewerbestrasse 11, 6330 Cham, Switzerland

Preface

This volume contains a collection of revised selected full-text papers presented at the 22nd International Conference on Distributed Computer and Communication Networks (DCCN 2019), held in Moscow, Russia, September 23–27, 2019.

DCCN 2019 is an IEEE (Region 8 + Russia Section) technically cosponsored international conference. It is a continuation of traditional international conferences of the DCCN series, which took place in Sofia, Bulgaria (1995, 2005, 2006, 2008, 2009, 2014); Tel Aviv, Israel (1996, 1997, 1999, 2001); and Moscow, Russia (1998, 2000, 2003, 2007, 2010, 2011, 2013, 2015, 2016, 2017, 2018) in the last 22 years. The main idea of the conference is to provide a platform and forum for researchers and developers from academia and industry from various countries working in the area of theory and applications of distributed computer and communication networks, mathematical modeling, methods of control, and optimization of distributed systems, by offering them a unique opportunity to share their views as well as discuss the perspective developments and pursue collaboration in this area. The content of this volume is related to the following subjects:

1. Communication networks algorithms and protocols
2. Wireless and mobile networks
3. Computer and telecommunication networks control and management
4. Performance analysis, QoS/QoE evaluation, and network efficiency
5. Analytical modeling and simulation of communication systems
6. Evolution of wireless networks toward 5G
7. Internet of Things and Fog Computing
8. Cloud computing, distributed and parallel systems
9. Probabilistic and statistical models in information systems
10. Queuing theory and reliability theory applications
11. High-altitude telecommunications platforms
12. Security in infocommunication systems

The DCCN 2019 conference gathered 174 submissions from authors from 26 different countries. From these, 132 high quality papers in English were accepted and presented during the conference. The current volume contains 52 extended mostly application-oriented papers which were recommended by session chairs and selected by the Program Committee for the Springer post-proceedings.

All the papers selected for the post-proceedings volume are given in the form presented by the authors. These papers are of interest to everyone working in the field of computer and communication networks.

We thank all the authors for their interest in DCCN, the members of the Program Committee for their contributions, and the reviewers for their peer-reviewing efforts.

September 2019

Vladimir Vishnevskiy
Konstantin Samouylov

Organization

DCCN 2019 was jointly organized by the Russian Academy of Sciences (RAS), the V.A. Trapeznikov Institute of Control Sciences of RAS (ICS RAS), the Peoples' Friendship University of Russia (RUDN University), the National Research Tomsk State University, and the Institute of Information and Communication Technologies of Bulgarian Academy of Sciences (IICT BAS).

International Program Committee

V. M. Vishnevskiy (Chair)	ICS RAS, Russia
K. E. Samouylov (Co-chair)	RUDN University, Russia
Ye. A. Koucheryavy (Co-chair)	Tampere University of Technology, Finland
S. M. Abramov	Program Systems Institute of RAS, Russia
S. D. Andreev	Tampere University of Technology, Finland
A. M. Andronov	Riga Technical University, Latvia
N. Balakrishnan	McMaster University, Canada
A. S. Bugaev	Moscow Institute of Physics and Technology, Russia
S. R. Chakravarthy	Kettering University, USA
T. Czachorski	Institute of Computer Science of Polish Academy of Sciences, Poland
A. N. Dudin	Belarusian State University, Belarus
D. Deng	National Changhua University of Education, Taiwan
A. V. Dvorkovich	Moscow Institute of Physics and Technology, Russia
Yu. V. Gaidamaka	RUDN University, Russia
P. Gaj	Silesian University of Technology, Poland
D. Grace	York University, UK
Yu. V. Gulyaev	Kotelnikov Institute of Radio-engineering and Electronics of RAS, Russia
J. Hosek	Brno University of Technology, Czech Republic
V. C. Joshua	CMS College, India
H. Karatza	Aristotle University of Thessaloniki, Greece
N. Kolev	University of São Paulo, Brazil
J. Kolodziej	Cracow University of Technology, Poland
G. Kotsis	Johannes Kepler University Linz, Austria
T. Kozlova Madsen	Aalborg University, Denmark
U. Krieger	University of Bamberg, Germany
A. Krishnamoorthy	Cochin University of Science and Technology, India
A. E. Koucheryavy	Bonch-Bruevich Saint-Petersburg State University of Telecommunications, Russia
Ye. A. Koucheryavy	Tampere University of Technology, Finland

N. A. Kuznetsov	Moscow Institute of Physics and Technology, Russia
L. Lakatos	Budapest University, Hungary
E. Levner	Holon Institute of Technology, Israel
S. D. Margenov	Institute of Information and Communication Technologies of Bulgarian Academy of Sciences, Bulgaria
N. Markovich	ICS RAS, Russia
A. Melikov	Institute of Cybernetics of the Azerbaijan National Academy of Sciences, Azerbaijan
G. K. Miscoi	Academy of sciences of Moldova, Moldavia
E. V. Morozov	Institute of Applied Mathematical Research of the Karelian Research Centre RAS, Russia
V. A. Naumov	Service Innovation Research Institute (PIKE), Finland
A. A. Nazarov	Tomsk State University, Russia
I. V. Nikiforov	Université de Technologie de Troyes, France
P. Nikitin	University of Washington, USA
S. A. Nikitov	Institute of Radio-engineering and Electronics of RAS, Russia
D. A. Novikov	ICS RAS, Russia
M. Pagano	Pisa University, Italy
E. Petersons	Riga Technical University, Latvia
V. V. Rykov	Gubkin Russian State University of Oil and Gas, Russia
L. A. Sevastianov	RUDN University, Russia
M. A. Sneps-Sneppe	Ventspils University College, Latvia
P. Stanchev	Kettering University, USA
S. N. Stepanov	Moscow Technical University of Communication and Informatics, Russia
S. P. Suschenko	Tomsk State University, Russia
J. Sztrik	University of Debrecen, Hungary
H. Tijms	Vrije Universiteit Amsterdam, The Netherlands
S. N. Vasiliev	ICS RAS, Russia
M. Xie	City University of Hong Kong, Hong Kong, China
Yu. P. Zaychenko	Kyiv polytechnic institute, Ukraine

Organizing Committee

V. M. Vishnevskiy (Chair)	ICS RAS, Russia
K. E. Samouylov (Vice Chair)	RUDN University, Russia
D. V. Kozyrev	RUDN University and ICS RAS, Russia
A. A. Larionov	ICS RAS, Russia
S. N. Kupriyakhina	ICS RAS, Russia
S. P. Moiseeva	Tomsk State University, Russia
T. Atanasova	IICT BAS, Bulgaria
I. A. Gudkova	RUDN University, Russia

S. I. Salpagarov   RUDN University
D. Yu. Ostrikova   RUDN University

Organizers and Partners

Organizers

Russian Academy of Sciences
RUDN University
V.A. Trapeznikov Institute of Control Sciences of RAS
National Research Tomsk State University
Institute of Information and Communication Technologies of Bulgarian Academy of Sciences
Research and Development Company "Information and Networking Technologies"

Support

Information support is provided by the IEEE (Region 8 + Russia Section) and the Russian Academy of Sciences. The conference has been organized with the support of the "RUDN University Program 5-100."

Contents

Computer and Communication
Networks and Technologies

Modeling of a Two-Way Communication System with a Special Searching for Customers

Attila Kuki[✉], Tamás Bérczes, János Sztrik, and Ádám Tóth

Faculty of Informatics, University of Debrecen, Debrecen, Hungary
{kuki.attila,berczes.tamas,sztrik.janos,
toth.adam}@inf.unideb.hu

Abstract. In this paper a special system with two-way communication is modelled by a finite and an infinite sources queueing system with retrial. The system is unreliable, the server may subject to random breakdowns. Customers from the finite source are the first order or regular customers, while the customers from the infinite source are the second order or the invited customers. The novelty of this paper is to investigate and model this unreliable system with different idle and busy breakdown intensity and different service rates for first and second order customers. In case of a busy server for first order customers, they can retry their requests. In case of an idle server, the second order customers are called for service. The effect of the breakdown and repair intensities are also investigated. The system balance equations are formulated, and the steady state probabilities can be obtained. Here the MOSEL-2 tool is used for these calculations. By the help of these probabilities the common performance measures are calculated and displayed.

Keywords: Retrial queues · Two-way communication · Unreliable system · Searching for customers

1 Introduction

Queueing theory has been investigated since decades for modeling of various problems of computer science, telecommunication systems, etc. As the complexity of the considered systems has been increased rapidly, developing new approaches of queueing models were necessary. Mainly the telephone switching centers motivated a new model. It was the retrial queueing systems. In case of busy lines or operators, the incoming call is not lost, but it is redirected to a virtual waiting facility, to the orbit, and it can retry the call again. These types of models were investigated by Falin, Templeton, Artalejo and more authors [3,4,10,13,14,21]. Real-life situations require, that in the models the customers generate their calls or request from a finite number of population. These demands lead to study the finite source models [3,12]. Furthermore, the considered real-life systems are unfortunately unreliable, that is the server or other parts of the

© Springer Nature Switzerland AG 2019
V. M. Vishnevskiy et al. (Eds.): DCCN 2019, CCIS 1141, pp. 3–14, 2019.
https://doi.org/10.1007/978-3-030-36625-4_1

systems can lose their efficiency or may breakdown. These types of unreliable systems were investigated e.g. in [9,22,23,25].

An other new general model was developed for not to lose the customers, who are not able or not disposed to wait the service (in the queue or in the orbit). Demands based on real applications were the motivation for developing the two-way communication systems. The key assumption is, that the idle server makes an outgoing call for the customers. One of the first paper on the retrial queueing system with two-way communication was presented by Falin [11]. So far several authors have investigated this type of models [7,8,15–17,19,20].

In business and economic application fields (e.g. trade and IT companies) where the agent can promote their new services, products, discounts, etc. it is very important to increase the performance and the utilization of the core facility (server) of the system. See, e.g. in [1,2,6,14,18,24].

This paper deals with a special case of searching for the customers, and in the background an unreliable server with breakdowns and repairs is supposed. Two types of sources are considered. The organization has a finite number of goodwill customers. They are the first order customers, making primary calls towards the organization (server). These clients are served according to the common retrial queueing discipline. The idle periods of the server is utilized for making outgoing calls towards the customers in the second, infinite source. The clients in this infinite source (second order customers) will contact the organization with some special interest. In case of a busy server (meanwhile another regular customer arrived), this special second order customer is treated as a non-preemptive priority client. In this model there is no distinction made between the service times for the two types of calls. The server is non-reliable, it is subject to random breakdowns. Different cases for busy time breakdown and repair is considered. The remaining parts of this work contain the followings. In Sect. 2 the model definition, the underlying Markovian process with 2 dimensions and the applied parameters are described. In Sect. 3 the steady-state probabilities are considered, and some performance measures (utilization, response times, etc.) are provided by the help of MOSEL-2 tool. At the end of the paper the results are summarized in a Conclusion.

2   Description of the Model

The considered system is modelled by a finite and infinite source retrial queueing system with a single server. The functionality of the model is displayed on Fig. 1.

The model has two sources. The first one is finite, the number of customers is N. They are the first order customers. These customers generate a job towards the server with an exponentially distributed inter-request time. The generation rate of a single first order customer is λ_1. If the server is idle, the service starts immediately. After the service, the job goes back to the source. The service time is again exponentially distributed with parameter μ_1. When the server is busy, the job is transferred to the orbit. The maximum size of the orbit is N. From the orbit the jobs after a random (exponential, with parameter ν) time keep retrying their request to the server until they are served.

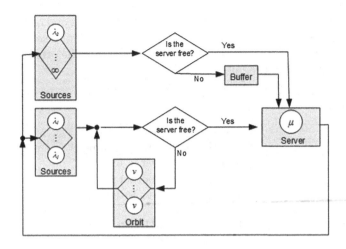

Fig. 1. The system model

The system has an infinite number of sources, as well. They are the second order customers. The idle server makes a call towards this infinite source, and the jobs in the source generate a request. The distribution of the inter-generation times are exponential, with parameter λ_2. Here λ_2 is the generation rate from the infinite source. In case, when the server is idle at the time of arrival of a second order customer, the service starts immediately. The service times are exponentially distributed with parameter of μ_2. When a second order customer finds the server busy, several working modes can be considered.

- The second order job is transferred back to the infinite source,
- The second order job takes place in a priority buffer. When the server becomes idle, the service of this job will start.

In this model the single server is an unreliable server, it may subject to breakdown. When the server is up, it will breakdown after a random time with exponentially distribution. The breakdown intensities are γ_0 for the idle server and γ_1 for the busy server. In case of a breakdown, a repair process starts immediately. The repair time is exponentially distributed with parameter γ_2. When a first order customer finds the server down, it will be transferred to the orbit. A second order customer also may arrive. The idle server makes a call for the customers, and during the request generation time (with parameter λ_2) a breakdown might occur. In this situation different cases can be investigated.

- The second order job is transferred back to the infinite source,
- The second order job takes place in a priority buffer. When the server becomes up, the service of this job will start.

The server may breakdown in a busy state, as well. A first order or a second order customer is under service at the time of breakdown. The first order customers can be transferred to the orbit or to the source, or the jobs may remain at the server. The service will continue after the repair. The second order customers also may remain at the server or may sent back to the infinite source.

Let us denote $O(t)$ and $S(t)$ the number of requests in the orbit and the state of the server at a given time point of t.

Let us define the state of the server by $S(t)$, that is

$$
S(t) = \begin{cases} 0, \text{ when the server is idle} \\ 1, \text{ when the server is busy} \\ \quad \text{with a first order customer} \\ 2, \text{ when the server is busy} \\ \quad \text{with a second order customer} \\ 3, \text{ when the server is down.} \end{cases}
$$

It is easy to see, that the maximum size of the orbit is N. From here, the state space representation of the Markovian-process $(S(t), O(t))$ can be described as a set of $\{0, 1, 2, 3\} \times \{0, 1, 2, ..., N\}$ elements. Although, the system has an infinite source, the maximum number of the customers in the system is $(N + 1)$ (N in the orbit and one second order customer under service), there is no stability problems regarding the system. The state space is finite.

All of the times, time intervals considered in the model, are exponentially distributed and totally independent from each other.

Let us consider the non-buffered model, when a second order customer under service is sent back to the source in case of breakdown. For this case the system balance equations for the steady-state system probabilities can be formulated as follows:

$$
p_{i,j} = \lim_{t \to \infty} P(S(t) = i, O(t) = j),
$$
$$
i = 0, 1, 2, 3 \text{ and } j = 0, 1, ..N
$$

$$
[(N - j)\lambda_1 + \lambda_2 + j\nu + \gamma_0]\, p_{0,j} = \mu_1 p_{1,j} + \mu_2 p_{2,j} + \gamma_2 p_{3,j}
$$

$$
[(N - j - 1)\lambda_1 + \mu_1 + \gamma_1]\, p_{1,j}
$$
$$
= (N - j)\lambda_1 p_{0,j} + (j + 1)\nu p_{0,j+1}
$$

$$
[(N - j)\lambda_1 + \mu_2 + \gamma_1]\, p_{2,j} = \lambda_2 p_{0,j}
$$

$$
[(N - j)\lambda_1 + \gamma_2]\, p_{3,j} = \gamma_0 p_{0,j} + \gamma_1 p_{1,j-1} + \gamma_1 p_{2,j}
$$

with $p_{1,-1} = p_{0,N+1} = 0$.

Similarly, consider the case in the non-buffered model, when a first order customer under service remains at the server in case of breakdown. The second order customer is sent back to the source. Because the exponentially distributed service time, the restarted or the continued services have the same characteristics. For this case the system balance equations for the steady-state probabilities can be formulated as follows:

$$p_{i,j} = \lim_{t \to \infty} P(S(t) = i, O(t) = j),$$

$$i = 0, 1, 2, 3 \text{ and } j = 0, 1, ..N$$

$$[(N - j)\lambda_1 + \lambda_2 + j\nu + \gamma_0] \, p_{0,j} = \mu_1 p_{1,j} + \mu_2 p_{2,j} + \gamma_2 p_{3,j}$$

$$[(N - j - 1)\lambda_1 + \mu_1 + \gamma_1] \, p_{1,j}$$
$$= (N - j)\lambda_1 p_{0,j} + (j + 1)\nu p_{0,j+1}$$

$$[(N - j)\lambda_1 + \mu_2 + \gamma_1] \, p_{2,j} = \lambda_2 p_{0,j}$$

$$[(N - j)\lambda_1] \, p_{3,j} = \gamma_0 p_{0,j} + \gamma_1 p_{1,j} + \gamma_1 p_{2,j}$$

with $p_{0,N+1} = 0$.

The system balance equations for the steady-state system probabilities in the other cases can be obtained by similar way.

Solving manually these balance equations is rather difficult. There exist several effective tools performing the background calculations. In this paper the MOSEL-2 tool was used. When the steady-state probabilities are calculated, this tool provides the well known performance characteristics. These measures are obtained using the following formulas.

– *Utilization 1*

$$U_1 = \sum_{o=0}^{N} p_{1,o}$$

– *Utilization 2*

$$U_2 = \sum_{o=0}^{N} p_{2,o}$$

– *Average number of jobs in the orbit*

$$\overline{O} = \sum_{s=0}^{3} \sum_{o=0}^{N} o p_{s,o}$$

– *Average number of active primary users*

$$\overline{M} = N - \overline{O} - U_1$$

– *Average generation rate of primary users*

$$\overline{\lambda_1} = \lambda_1 \overline{M}$$

– *Mean time spent in orbit by using Little-formula*

$$\overline{W} = \frac{\overline{O}}{\overline{\lambda_1}}$$

3   Numerical Results

The most important goal of these types of stochastic systems is to obtain the performance measures and system characteristics. Usually the throughput, utilization, response times, waiting times, queue length are considered. Here the utilization and waiting time in the orbit are focused.

Table 1. Numerical values of model parameters

Case studies

No.	N	λ_1	λ_2	μ	ν	γ_0	γ_1	γ_2
Fig. 2	100	x-$axes$	2	3	0.05	0.01	0.01	1
Fig. 3	100	x-$axes$	2	3	0.05	0.1	0.01	1
Fig. 4	100	x-$axes$	2	3	0.05	0.01, 0.1	0.01, 0.1	1
Fig. 5	100	x-$axes$	2	3	0.05	0.01, 0.1	0.01, 0.1	1
Fig. 6	100	x-$axes$	2	3	0.05	0.01	0.01	1
Fig. 7	100	x-$axes$	2	3	0.05	0.1	0.1	1
Fig. 8	100	0.2	2	3	0.05	x-$axes$	x-$axes$	1
Fig. 9	100	0.2	2	3	0.05	x-$axes$	x-$axes$	1
Fig. 10	100	0.2	2	3	0.05	0.1	0.1	x-$axes$

There exist several methods to calculate the system measures. Solving directly the balance equations is rather difficult in most cases. Effective software tools can be used to get the steady-state system probabilities. From these probabilities the performance measures can be computed directly or by the help of the considered tool. In this paper the MOSEL-2 tool is used. This is not a simulation tool. The system equations are build up and solved by one of the utilities developed for MOSEL-2. Here the SPNP (Stochastic Petri Net Program) is used (see in [5]). The following figures illustrates the most interesting numerical results. The numerical values of the applied parameters in the model are listed in Table 1. Most figure compares to different cases:

- In case of busy state breakdown, the first order and second order customers are interrupted. The first order customers are sent back to the orbit, the second order customers are sent back to the source. On figure these cases are denoted with blue lines dotted by diamonds.
- The service of both types of customers are interrupted. The customers are left at the server. After the repair their service will continue or restart. Because of the exponentially distributed service time, this difference - restart or continue - has no effect to the system characteristics. On figure these cases are denoted with orange lines dotted by squares.

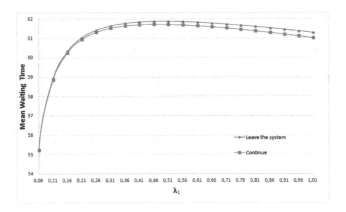

Fig. 2. Mean waiting time vs. λ_1 (Color figure online)

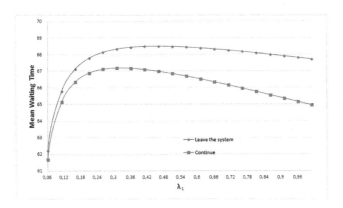

Fig. 3. Mean waiting time vs. λ_1 (Color figure online)

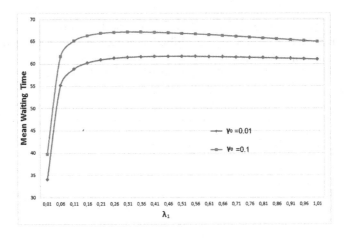

Fig. 4. Mean waiting time vs. λ_1 (Color figure online)

On Fig. 2 the running parameter (values of x-axes) is the first order generation rate λ_1. The failure rate is small for this figure. There are not so significant differences between lines. The waiting time of the 'leave the system' case is greater, because the first order jobs goes to the orbit and they have to try again.

Figure 3 displays the same situation with ten times greater failure rate, which will cause a much more significant deviance between the cases. The interruption is more frequent and the first order customer are sent back to the orbit more frequently, which results higher waiting times. The two considered failure rates are compared on Figs. 4 and 5 for 'Continue' and for 'Leave the system' scenarios, respectively. The expected results can be seen, the waiting times are higher for greater values of failure rates.

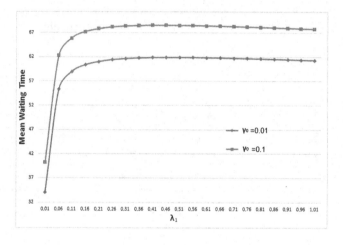

Fig. 5. Mean waiting time vs. λ_1 (Color figure online)

Figure 6 shows the utilization in function of the first order generation rate. The failure rate here is small, so the differences between the two scenarios are also small. The utilization is greater for the 'Continue' case, because after the repair the server state will be busy, immediately. While for the other scenario the server will be idle, and an exponential retrial, first or second order generation will take place. For Fig. 7 the parameters are the same, but the Failure rate, which is again ten times greater than on the Fig. 6. Consequently, the differences in utilization are more significant.

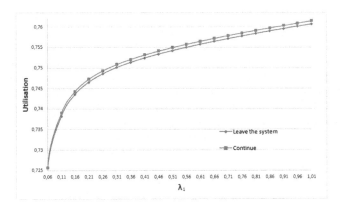

Fig. 6. Utilization vs. λ_1 (Color figure online)

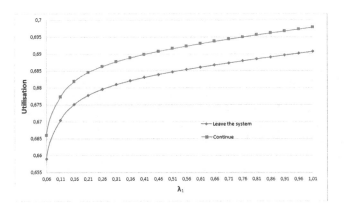

Fig. 7. Utilization vs. λ_1 (Color figure online)

On Figs. 8 and 9 the failure rate, γ_0 and γ_1 is the running parameter. The two parameters move together. On Fig. 8 the mean waiting time is considered. Here the parameter is modified in wider range than on Figs. 2 and 3, but the tendencies are the same. The higher the failure rate is, the higher the waiting times are. Additionally, waiting times for 'Leave the system' scenario is higher, as well.

On Fig. 9 the failure rate, γ_0 (and γ_1 with the same way) is the running parameter. With higher failure rate the utilization will decrease, and comparing the two scenarios, utilization is higher for the 'Continue' scenario.

Figure 10 investigates the effect of repair rate. Since the repair rate and the average repair time are reciprocal values, higher repair rate means shorter repair time. According this, it can be seen, that for higher repair rate the waiting times will decrease. Comparing the two cases, 'Leave the system' has greater waiting times.

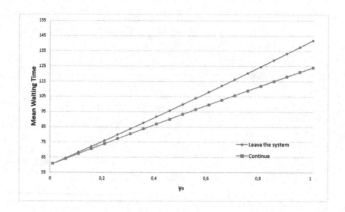

Fig. 8. Mean waiting time vs. γ_0 and γ_1 (Color figure online)

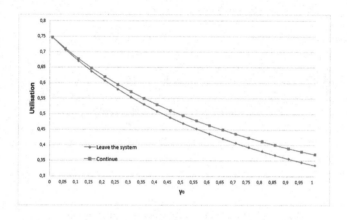

Fig. 9. Utilization vs. γ_0 and γ_1 (Color figure online)

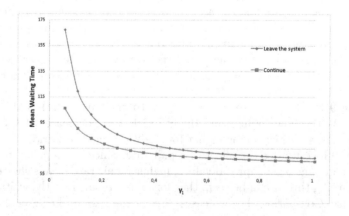

Fig. 10. Mean waiting time vs. γ_2 (Color figure online)

4   Conclusion

In the present paper a special two-way communication system was investigated. First order customers come from a finite source, while in case of an idle server, second order customers are able to reach the system via a direct call. Different cases can be considered. For simplicity, the service rates for the first and the second order customers were supposed to be different. Similarly, different failure rates are considered for idle server and busy server.

The main focus was to compare the 'Continue' and the 'Leave the system' scenarios. Based on the results displayed on the figures above, it can be stated, that the system performance (in waiting times and utilization) is better for the 'Continue' case. For the results, the buffered case of the second order customers was considered. It is closer to the real life situation. When a customer is called for service from the outside world, and in the meantime the server becomes busy, give the chance for the called customer to be served.

Acknowledgment. The research work of János Sztrik and Ádám Tóth was supported by the Austrian-Hungarian Bilateral Cooperation in Science and Technology project 2017-2.2.4-TeT-AT-2017-00010.

The research work of Attila Kuki and Tamás Bérczes was supported by the construction EFOP-3.6.3-VEKOP-16-2017-00002. The project was supported by the European Union, co-financed by the European Social Fund.

References

1. Aguir, S., Karaesmen, F., Akşin, O.Z., Chauvet, F.: The impact of retrials on call center performance. OR Spectrum **26**(3), 353–376 (2004)
2. Aksin, Z., Armony, M., Mehrotra, V.: The modern call center: a multi-disciplinary perspective on operations management research. Prod. Oper. Manag. **16**(6), 665–688 (2007)
3. Artalejo, J.: Retrial queues with a finite number of sources. J. Korean Math. Soc. **35**, 503–525 (1998)
4. Artalejo, J., Corral, A.G.: Retrial Queueing Systems: A Computational Approach. Springer, Heidelberg (2008). https://doi.org/10.1007/978-3-540-78725-9
5. Begain, K., Bolch, G., Herold, H.: Practical Performance Modeling, Application of the MOSEL Language. Kluwer Academic Publisher, Boston (2001)
6. Brown, L., et al.: Statistical analysis of a telephone call center: a queueing-science perspective. J. Am. Stat. Assoc. **100**(469), 36–50 (2005)
7. Dimitriou, I.: A retrial queue to model a two-relay cooperative wireless system with simultaneous packet reception. In: Wittevrongel, S., Phung-Duc, T. (eds.) ASMTA 2016. LNCS, vol. 9845, pp. 123–139. Springer, Cham (2016). https://doi.org/10.1007/978-3-319-43904-4_9
8. Dragieva, V., Phung-Duc, T.: Two-way communication M/M/1 retrial queue with server-orbit interaction. In: Proceedings of the 11th International Conference on Queueing Theory and Network Applications, p. 11. ACM (2016)
9. Dragieva, V.I.: Number of retrials in a finite source retrial queue with unreliable server. Asia-Pac. J. Oper. Res. **31**(2), 23 (2014). https://doi.org/10.1142/S0217595914400053

10. Falin, G.I.: Waiting time in a single-channel queuing system with repeated calls. Mosc. Univ. Comput. Math. Cybern. **4**, 66–69 (1977)
11. Falin, G.: Model of coupled switching in presence of recurrent calls. Eng. Cybern. **17**(1), 53–59 (1979)
12. Falin, G., Artalejo, J.: A finite source retrial queue. Eur. J. Oper. Res. **108**, 409–424 (1998)
13. Falin, G., Templeton, J.G.C.: Retrial Queues. Chapman and Hall, London (1997)
14. Gans, N., Koole, G., Mandelbaum, A.: Telephone call centers: tutorial, review, and research prospects. Manuf. Serv. Oper. Manag. **5**(2), 79–141 (2003)
15. Nazarov, A., Phung-Duc, T., Paul, S.: Heavy outgoing call asymptotics for MMPP/M/1/1 retrial queue with two-way communication. In: Dudin, A., Nazarov, A., Kirpichnikov, A. (eds.) ITMM 2017. CCIS, vol. 800, pp. 28–41. Springer, Cham (2017). https://doi.org/10.1007/978-3-319-68069-9_3
16. Nazarov, A.A., Paul, S., Gudkova, I., et al.: Asymptotic analysis of Markovian retrial queue with two-way communication under low rate of retrials condition. In: Proceedings 31st European Conference on Modelling and Simulation, pp. 687–693 (2017)
17. Phung-Duc, T., Rogiest, W.: Two way communication retrial queues with balanced call blending. In: Al-Begain, K., Fiems, D., Vincent, J.-M. (eds.) ASMTA 2012. LNCS, vol. 7314, pp. 16–31. Springer, Heidelberg (2012). https://doi.org/10.1007/978-3-642-30782-9_2
18. Pustova, S.: Investigation of call centers as retrial queuing systems. Cybern. Syst. Anal. **46**(3), 494–499 (2010)
19. Sakurai, H., Phung-Duc, T.: Two-way communication retrial queues with multiple types of outgoing calls. Top **23**(2), 466–492 (2015)
20. Sakurai, H., Phung-Duc, T.: Scaling limits for single server retrial queues with two-way communication. Ann. Oper. Res. **247**(1), 229–256 (2016)
21. Templeton, J.: Retrial queues. Top **7**(2), 351–353 (1999). https://doi.org/10.1007/BF02564732
22. Wang, J., Zhao, L., Zhang, F.: Performance analysis of the finite source retrial queue with server breakdowns and repairs. In: Proceedings of the 5th International Conference on Queueing Theory and Network Applications, pp. 169–176. ACM (2010)
23. Wang, J., Zhao, L., Zhang, F.: Analysis of the finite source retrial queues with server breakdowns and repairs. J. Ind. Manag. Optim. **7**(3), 655–676 (2011). https://doi.org/10.3934/jimo.2011.7.655
24. Wolf, T.: System and method for improving call center communications. US Patent App. 15/604,068, 30 November 2017
25. Zhang, F., Wang, J.: Performance analysis of the retrial queues with finite number of sources and service interruptions. J. Korean Stat. Soc. **42**(1), 117–131 (2013). https://doi.org/10.1016/j.jkss.2012.06.002

Accurate and Interval Estimates of the Probability of Network Service Availability for Communication Networks

Yu. Zaychenko[ID], V. Vasyliev, D. Vishtal[ID], and N. Lyubashenko[✉][ID]

National Technical University of Ukraine "Igor Sikorsky Kyiv Polytechnic Institute",
37, pr. Peremogy, Kyiv, Ukraine
zaychenkoyuri@ukr.net, ndp1992@bigmir.net

Abstract. One of the important and universal characteristics of the network performance for both a user and a network owner is the probability of the network service availability at any time. To obtain an accurate estimate of the probability of a particular network service availability in the class of binary stochastic models, effective methods of structure function decomposition are used. The paper discusses the issues of obtaining accurate estimates of the probability of network service availability for arbitrary pairs of network nodes. Lower bound and upper bound estimates for the probability of network service availability are constructed for large dimension networks with a complex structure.

Keywords: Binary stochastic model · Availability of network services · Alternating renewal process · Boolean algebra · Isomorphism of properties and classes algebra

1 Introduction

Over recent decades, problems of network structures reliability belong to the priority areas of research in the reliability theory.

Formally, the communication network is interpreted as a weighted graph without loops. The graph can be both unoriented and oriented. Elements of the graph are weighted by weight parameters. Weight values are usually assigned to the edges, assuming that the nodes are absolutely reliable. Weights can be also assigned to nodes.

For networks with renewal, the role of weight parameters is played by the factors of network elements availability. Failures of elements are caused by technical problems and external influences.

For practice, it is important to know that any pair of network subscribers can get a connection despite the failure of the elements (network service availability).

The availability of a network service means the ability to provide communication for any user at any time.

Educational Scientific Complex "Institute for Applied System Analysis".

V. M. Vishnevskiy et al. (Eds.): DCCN 2019, CCIS 1141, pp. 15–26, 2019.
https://doi.org/10.1007/978-3-030-36625-4_2

A few formal definitions of the availability of network services are given below:

1. The network has the (s-t) availability property at time t, if there is at least one simple path from node s to node t.
2. The network has the property of full availability at time t, if there is at least one simple path for any pair of network nodes.
3. The network has the property of full availability at time t, if there is at least one network spanning tree.
4. The network has the property of full availability at time t, if at least one available element exists in each of the cutting sets of the network.

Of course, the definitions (2–4) are equivalent, but they represent the structure of the connectivity property of a network in different ways.

The probability of network service availability allows to objectively define which of the compared networks is better suited to its purpose. Monitoring of this characteristic makes it possible to make timely necessary adjustments to the operation and development of the network.

The task of assessing the network service availability belongs to the class of so-called "computationally hard" problems of combinatorial logic of properties and classes. For this reason, it is not always possible to obtain an accurate estimate of the availability of a network service for very large networks, and upper bound and lower bound estimates are used instead of it.

Availability of networks can be characterized by assessment methodologies [1] such as Reliability Block Diagram (RBD), Fault Tree Analysis (FTA) [2] and so on.

Typical algorithms for computing network availability include the state enumeration method [3], sum of disjoint products method [4], factorization method [5], minimal cuts method [6] and cellular automata [7,8].

In this paper formalization of the problem of assessing the probability of network service availability in the form suitable for machine implementation is based on the isomorphism of the Boolean properties algebra (predicates) and the corresponding Boolean classes algebra. The predicate of the network service availability is represented by the corresponding structure function.

We consider the problem of estimating the stationary probability of a pair network connection. It is important to note that the concept of the network (s-t) connectivity is closely related to transport flow tasks. The stationary probability of the network (s-t) connectivity can be considered as one of the upper bound estimates for the stationary probability of a full network connectivity [9]. The stationary probability of a full network connectivity can be considered as a guaranteed lower bound for the probability of connectivity of any pairs of nodes.

2 Binary Stochastic System Model

1. Let $C = \{1, 2, \ldots, n\}$ be an indexed finite set of the system elements.
   The number $|C|$ is called the order of the system.
   Binarity means that the elements and the whole system take values in the set $B_2 = \{1, 2\}$.

The evolution of the i-th element in time is modeled by the corresponding alternating renewal process $x_i(t), i = 1, 2, \ldots, n$.

$$x_i(t) = \begin{cases} 1, \text{ if at the moment } t \text{ } i\text{-th element is operable} \\ 0, \text{ otherwise} \end{cases}.$$

We need some results of the renewal theory in a convenient interpretation. For the i-th element, the probability to be in an accessible state at time t is called the non-stationary availability factor and is denoted by

$$P\{x_i(t) = 1\} = P_i(t), i = 1, 2, \ldots, n.$$

As t increases, the nonstationary availability factor tends to a constant value - the stationary availability factor of the i-th element. Its value is defined as the ratio of the average length of the availability interval of the i-th element to the average length of the cycle of the i-th alternating process:

$$\lim_{t \to \infty} P\{x_i(t) = 1\} = \frac{M[\theta_{x_i}]}{M[\theta_{x_i}] + M[\xi_{x_i}]} = P_i, i = 1, 2, \ldots, n.$$

2. Let the vector of the elements state

$$X(t) = (x_1(t), x_2(t), \ldots, x_n(t))$$

uniquely determine the state of the system. The corresponding function (two-valued predicate) is called the structure function of the system:

$$\varphi(x_1(t), \ldots, x_n(t)) = \begin{cases} 1, \text{ if at the time } t \text{ the system is operable} \\ 0, \text{ otherwise} \end{cases}.$$

Formally, the role of a truth (false) set can be performed by any function of the logic algebra, with the exception of function-constants.

Most systems have the monotonicity property, so we consider only the case when the structure function $\varphi(x_1(t), \ldots, x_n(t))$ is monotone.

3. Deterministic properties of structure functions are as follows:
   (a)

$$\bigwedge_{i=1}^{n} x_i \leq \varphi_s(x_1(t), \ldots, x_n(t)) \leq \bigvee_{i=1}^{n} x_i$$

   (all variables are significant).
   (b) Reservation scale theorem:

$$\varphi_s(x_1 \vee \tilde{x}_1, x_2 \vee \tilde{x}_2, \ldots, x_n \vee \tilde{x}_n) \geq \varphi_s(x_1, x_2, \ldots, x_n) \vee \varphi_s(\tilde{x}_1, \tilde{x}_2, \ldots, \tilde{x}_n),$$
$$\varphi_s(x_1 \wedge \tilde{x}_1, x_2 \wedge \tilde{x}_2, \ldots, x_n \wedge \tilde{x}_n) \leq \varphi_s(x_1, x_2, \ldots, x_n) \wedge \varphi_s(\tilde{x}_1, \tilde{x}_2, \ldots, \tilde{x}_n).$$

   (c) Any monotone structure function is uniquely representable in the following form:

$$\bigvee_{I} f_i(x_1, \ldots, x_n) \equiv \varphi_s(x_1, \ldots, x_n) \equiv \bigwedge_{J} \psi_j^d(x_1, \ldots, x_n),$$

where:

$I = \{f_1(x_1, \ldots, x_n), \ldots, f_m(x_1, \ldots, x_n)\}$ - full set of first implicants of the function $\varphi_s(x_1, \ldots, x_n)$,

$J = \{\psi_1(x_1, \ldots, x_n), \ldots, \psi_r(x_1, \ldots, x_n)\}$ - full set of first implicants of the function $\varphi_s^d(x_1, \ldots, x_n)$.

$$\varphi_s^d(x_1, \ldots, x_n) = \bar{\varphi}(\bar{x}_1, \ldots, \bar{x}_n).$$
$$\psi_s^d(x_1, \ldots, x_n) = \bar{\psi}(\bar{x}_1, \ldots, \bar{x}_n).$$

From property (c) and the properties of two dual automorphisms

$$H : U \to U | \forall Q \in U, H(Q) = \bar{Q},$$
$$D : U \to U | \forall Q \in U, D(Q) = Q^d$$

we obtain immediate results:

$$\bigvee_J \psi_j(x_1, \ldots, x_n) \equiv \varphi_s^d(x_1, \ldots, x_n) \equiv \bigwedge_I f_i^d(x_1, \ldots, x_n), \tag{1}$$

$$\bigvee_J \psi_j(\bar{x}_1, \ldots, \bar{x}_n) \equiv \bar{\varphi}_s(x_1, \ldots, x_n) \equiv \bigwedge_I f_i^d(\bar{x}_1, \ldots, \bar{x}_n), \tag{2}$$

$$\bigvee_I f_i(\bar{x}_1, \ldots, \bar{x}_n) \equiv \bar{\varphi}_s^d(x_1, \ldots, x_n) \equiv \bigwedge_J \psi_j^d(\bar{x}_1, \ldots, \bar{x}_n), \tag{3}$$

$$\bigvee_{I_1 \subset I} f_i(x_1, \ldots, x_n) < \varphi_s(x_1, \ldots, x_n) < \bigwedge_{J_1 \subset J} \psi_j^d(x_1, \ldots, x_n), \tag{4}$$

$$\bigvee_{J_1 \subset J} \psi_j(\bar{x}_1, \ldots, \bar{x}_n) < \bar{\varphi}_s(x_1, \ldots, x_n) < \bigwedge_{I_1 \subset I} f_i^d(\bar{x}_1, \ldots, \bar{x}_n). \tag{5}$$

Using these properties makes it possible to construct all possible lower bounds and upper bounds estimates for the network connectivity probability. It is sufficient to use Boolean lattice inequalities from properties (4) and (5).

The axioms of Boolean algebra are consistent with the laws of intuitive logic for sets, propositions, and object properties (predicates).

The process of system functioning is a discrete-continuous process of random walk along a Boolean vector lattice. The time of transition from this state to the neighboring state is neglected.

As a characteristic of system functioning quality, we consider the probability of the network service availability

$$P\{\varphi(x_1, \ldots, x_n) = 1\}.$$

The formalization of the problem of estimating the stationary probability of pair connectivity in the form suitable for machine implementation is based on the isomorphism of Boolean algebra of functions representing properties

$$U_\Phi = \langle x_1, x_2, \ldots, x_n; \wedge; \vee; ^- \rangle$$

and the corresponding Boolean class algebra

$$U_K = \langle P(B_2^n); \cap; \cup; ' \rangle,$$

where:

$\{x_1, \ldots, x_n\}$ - generators of Boolean algebra of functions;
B_2^n - Boolean vector lattice (vector space of network elements);
$|B_2^n| = 2^n; dim(B_2^n) = n$.

3   Problem Formulation

Given:

$G(Y, X)$ - network structure;
$Y = \{y_1, \ldots, y_l\}$ - set of network nodes;
$X = \{x_1, \ldots, x_n\}$ - set of network channels;
$s, t \in Y$ - two arbitrary vertices of the network;
θ_{x_i} - random time intervals of availability of the i-th network channel;
ξ_{x_i} - random time intervals of unavailability of the i-th network channel;
p_i - factor of stationary availability of the i-th channel of the network:

$$p_i = \frac{M[\theta_{x_i}]}{M[\theta_{x_i}] + M[\xi_{x_i}]}, \quad \bar{p}_i = 1 - p_i = q_i.$$

For the sake of simplicity, we will consider all nodes to be absolutely reliable. Assumptions:

1. Dependence between random variables $\{\theta_{x_i}\}$ and $\{\xi_{x_i}\}$, $i = 1, \ldots, n$, can be neglected.

$$\exists M[\theta_{x_i}] < \infty \text{ and } \exists M[\xi_{x_i}] < \infty.$$

2. The vector of the network elements state uniquely determines the (s-t)-connectivity ((s-t)-availability) of the network.
3. The network is viewed as a renewable system with fully available renewal.
4. For a given mode of information load on the network (small, medium, large), a stationary mode is set up.
5. The network status relatively to $\varphi_{s,t}$-accessibility (unavailability) is determined by the resource state of its elements at a given load on the network.

Required to find the stationary probability of the network (s-t)-connectivity:

$$P\{\varphi_{s,t}(x_1, \ldots, x_n) = 1\}.$$

To solve the problem, the following algorithms are proposed:

- algorithm for finding all shortest (s-t)-paths of the network;
- algorithm for finding all minimal (s-t)-cuts of the network;

– algorithm for orthogonalization of the connectedness function in its most compact representation.

As a result of orthogonalization, we obtain a computational scheme for estimating the probability of (s-t)-connection, which is minimal from the point of view of the computational complexity.

By assigning values p_i to variables x_i in the orthogonalized minimal disjunctive normal form (MDNF)

$$\varphi_{s,t}(x_1,\ldots,x_n) = 1,$$

and q_i to variables with negations $\bar{x}_i$, we get the representation of the (s-t)-connection probability as a function of the availability of the network elements.

4   Algorithm for Finding the Representation of (s-t)-Connection Relation on a Graph G(Y,X)

A formal representation of (s-t)-connection relation ($s, t \in Y$):

$$\varphi_{s,t}(x_1,\ldots,x_n) = \bigvee_{\{f_i\}\in I} \left(\bigwedge_{k_i\in\{f_i\}} x_{k_i} \right),$$

where $\{f_i\}$ is a list of indices of the variables forming a simple (s-t)-path (the minimal carrier of the connectedness property).

$I = \{\{f_1\}, \{f_2\}, \ldots, \{f_m\}\}$ - complete list of all (s-t)-paths.

Let us consider an example of algorithms for finding (s-t)-paths and (s-t)-cuts in the network graph (see Fig. 1).

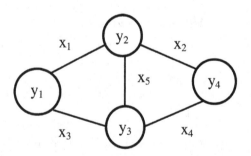

Fig. 1. Graph G(Y,X).

4.1   Algorithm for Finding Network (s-t)-Paths

Input information: the logical adjacency matrix of the graph M_G (see Fig. 2).

$N_v = 4$ - number of vertices;
$Y = \{y_1, \ldots, y_l\} = \{y_1, y_2, y_3, y_4\}$ - set of vertices;
$N_a = 5$ - number of arcs;
$X = \{x_1, \ldots, x_n\} = \{x_1, x_2, x_3, x_4, x_5\}$ - set of arcs.

Find: (s-t)-paths from y_1 to y_4 ($s = y_1$, $t = y_4$).

$$
M_G = \begin{array}{c} \\ y_1 \\ y_2 \\ y_3 \\ y_4 \end{array}
\begin{array}{cccc} y_1 & y_2 & y_3 & y_4 \\ 0 & 1 & 3 & 0 \\ 1 & 0 & 5 & 2 \\ 3 & 5 & 0 & 4 \\ 0 & 2 & 4 & 0 \end{array}
=
\begin{array}{c} \\ y_1 \\ y_2 \\ y_3 \\ y_4 \end{array}
\begin{array}{cccc} y_1 & y_2 & y_3 & y_4 \\ 0 & x_1 & x_3 & 0 \\ x_1 & 0 & x_5 & x_2 \\ x_3 & x_5 & 0 & x_4 \\ 0 & x_2 & x_4 & 0 \end{array}
$$

Fig. 2. Adjacency matrix for graph G(Y,X).

1. If the value of (s, t)-element in the logical adjacency matrix M_G is not equal 0, then there is a path of length 1 from vertex s to vertex t, we store it into memory. In this example, $(y_1, y_4) = 0$. Thus, paths of length 1 do not exist.
2. Iterations start with $N = 1$, where N is an iteration variable.
   $y_s^{(1)}$ equals to the s-th row of the matrix M_G.
3. We multiply (logically) row $y_s^{(N)}$ by M_G and get a new row.
4. If the number of iterations N is less than the number of vertices minus one $(N_v - 1)$, then go to step 3, otherwise - the end of the algorithm.

For our example first iterations are as follows.
We multiply row $y_s^{(1)}$ by M_G:

$$
(0\ 1\ 3\ 0) \begin{pmatrix} 0\ 1\ 3\ 0 \\ 1\ 0\ 5\ 2 \\ 3\ 5\ 0\ 4 \\ 0\ 2\ 4\ 0 \end{pmatrix} = (0;\ 3 \wedge 5;\ 1 \wedge 5;\ (1 \wedge 2) \vee (3 \wedge 4)).
$$

The result is stored in row $y_s^{(2)}$. Further, we analyze the value of the t-th element of $y_s^{(2)}$.

In our case, it is $(1 \wedge 2) \vee (3 \wedge 4)$ - two shortest paths of length 2.
The value of the t-th element of $y_s^{(2)}$ equals zero:

$$
y_s^{(2)} = (0;\ 3 \wedge 5;\ 1 \wedge 5;\ 0).
$$

Output information:
For this example, 4 shortest paths were found

$$\varphi_{14}(x_1, x_2, x_3, x_4, x_5) = (x_1 \wedge x_2) \vee (x_3 \wedge x_4) \vee (x_1 \wedge x_4 \wedge x_5) \vee (x_2 \wedge x_3 \wedge x_5).$$

List of (s-t)-paths:

$$\{f_1\} = \{1, 2\}; \ \{f_2\} = \{3, 4\}; \ \{f_3\} = \{1, 4, 5\}; \{f_4\} = \{2, 3, 5\}.$$

4.2 Algorithm for Finding (s-t)-Cuts of the Network in General Case

Input information:
  MDNF representation $\varphi_{s,t}(x_1, \ldots, x_n)$.
  Find: (s-t)-cuts of network.
  The algorithm consists in finding the dual function

$$\varphi_{s,t}^d(x_1, \ldots, x_n)$$

for $\varphi_{s,t}(x_1, \ldots, x_n)$, represented in MDNF.
  The indices of the variables forming the first implicants of the dual function $\varphi_{s,t}^d(x_1, \ldots, x_n)$ are minimal (s-t)-cuts.

$$\varphi_s^d(x_1, \ldots, x_n) = \bar{\varphi}(\bar{x}_1, \ldots, \bar{x}_n).$$

$J = \{\{\psi_1\}, \{\psi_2\}, \ldots \{\psi_r\}\}$ - a complete list of all (s-t)-cuts,
where $\{\psi_j\}$ - j-th (s-t)-cut, $j = 1, \ldots, r$.
  For the given graph $G(Y, X)$ in Fig. 1 it is necessary to find a dual function $\varphi_{y_1, y_4}^d(x_1, \ldots, x_5)$ and to represent it in the minimal conjunctive normal form (MCNF). Using the result of the previous problem, we obtain a complete list of (s-t)-cuts:

$$\varphi_{y_1, y_4}^d(x_1, \ldots, x_5) = (x_1 \vee x_2) \wedge (x_3 \vee x_4) \wedge (x_1 \vee x_4 \vee x_5) \wedge (x_2 \vee x_3 \vee x_5)$$
$$= (x_1 \vee x_2 x_4 \vee x_2 x_5) \wedge (x_3 \wedge x_2 x_4 \vee x_4 x_5) = x_2 x_4 \vee x_1 x_3 \vee x_1 x_4 x_5 \vee x_2 x_3 x_5$$
$$\varphi_{y_1, y_4}(x_1, \ldots, x_5) = \varphi_{y_1, y_4}^{dd}(x_1, \ldots, x_5)$$
$$= \bigwedge_{\{\psi_j\} \in J} (\bigvee_{r_j \in \{\psi_j\}} x_{r_j}).$$

List of (s-t)-cuts:

$$\{\psi_1\} = \{2, 4\}; \ \{\psi_2\} = \{1, 3\} \ \{\psi_3\} = \{1, 4, 5\}; \ \{\psi_4\} = \{2, 3, 5\}.$$

4.3 Algorithm for Orthogonalization of the Connectedness Function in General Case

Input information: MDNF (MCNF) representation

$$\varphi_{s,t}(x_1, \ldots, x_n)$$

and

$$\varphi_{s,t}^d(x_1, \ldots, x_n).$$

Find: Orthogonal MDNF (MCNF) $\varphi_{s,t}$ and $\varphi_{s,t}^d$.

1. Choose one of the representations

$$\varphi_{s,t}(x_1, \ldots, x_n)$$

or

$$\varphi^d_{s,t}(x_1, \ldots, x_n),$$

which contains the minimal number of characters (the most compact). The selected representation is orthogonalized.

2. The generalized De Morgan formula is applied to the chosen function: for function $\varphi_{s,t}(x_1, \ldots, x_n)$ represented in MDNF:

$$\varphi_{s,t}(x_1, \ldots, x_n) = f_1 \vee \cdots \vee f_m = f_1 + \bar{f}_1 f_2 + \bar{f}_1 \bar{f}_2 f_3 + \cdots + \bar{f}_1 \ldots \bar{f}_{m-1} f_m,$$

where: f_i - i-th network (s-t)-path; m - the number of network (s-t)-paths.

You can also apply the generalized De Morgan formula for orthogonalization of the MCNF representation:

$$\varphi_{s,t}(x_1, \ldots, x_n) = f_1^d \wedge \cdots \wedge f_m^d = f_1^d - f_1^d \bar{f}_2^d - f_1^d f_2^d \bar{f}_3^d - \cdots - f_1^d \ldots f_{m-1}^d \bar{f}_m^d.$$

For a function $\varphi^d_{s,t}(x_1, \ldots, x_n)$, presented in MDNF:

$$\varphi^d_{s,t}(x_1, \ldots, x_n) = \psi_1 \vee \cdots \vee \psi_r = \psi_1 + \bar{\psi}_1 \psi_2 + \bar{\psi}_1 \bar{\psi}_2 \psi_3 + \cdots + \bar{\psi}_1 \ldots \bar{\psi}_{r-1} \psi_r,$$

where:

ψ_i - i-th (s-t)-cut of the network;
r - the number of (s-t)-cuts.

5 Estimation of (s-t)-Connection Probability of Communication Network

Task definition: we have the communication network and nodes s and t (see Fig. 3).

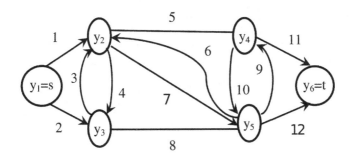

Fig. 3. Graph of network.

For each channel, an availability factor is

$$p_i = 0.8, i = 1, \ldots, 12.$$

Required to find:

1. List of all shortest (s-t)-paths.
2. List of all shortest (s-t)-cuts.
3. Orthogonal representation form.
4. The probability of the (s-t)-connection.

To solve this problem it is necessary to construct a matrix of logical adjacency for a given graph structure (see Fig. 4).

$$M_G \quad \begin{array}{c|cccccc} & y_1 & y_2 & y_3 & y_4 & y_5 & y_6 \\ y_1 & 0 & 1 & 2 & 0 & 0 & 0 \\ y_2 & 0 & 0 & 4 & 5 & 6 & 0 \\ y_3 & 0 & 3 & 0 & 0 & 8 & 0 \\ y_4 & 0 & 5 & 0 & 0 & 10 & 11 \\ y_5 & 0 & 7 & 8 & 9 & 0 & 12 \\ y_6 & 0 & 0 & 0 & 0 & 0 & 0 \end{array}$$

Fig. 4. Matrix of logical adjacency.

The adjacency matrix is the input for a corresponding computer program, which we have used to find paths and cuts. We obtain the following result (Tables 1 and 2).

Table 1. List of (s-t)-paths.

$|\{f_i\}| = 3$	$|\{f_i\}| = 4$	$|\{f_i\}| = 5$
$\{1, 5, 11\}$	$\{1, 4, 8, 12\}$	$\{1, 4, 8, 9, 11\}$
$\{1, 6, 12\}$	$\{1, 5, 10, 12\}$	$\{2, 3, 5, 10, 12\}$
$\{2, 8, 12\}$	$\{1, 6, 9, 11\}$	$\{2, 3, 6, 9, 11\}$
	$\{2, 3, 5, 11\}$	$\{2, 5, 7, 8, 11\}$
	$\{2, 3, 6, 12\}$	
	$\{2, 8, 9, 11\}$	

Table 2. List of (s-t)-cut sets.

$|\{\psi_j\}| = 2$	$|\{\psi_j\}| = 3$	$|\{\psi_j\}| = 4$	$|\{\psi_j\}| = 5$
$\{1,2\}$	$\{1,3,8\}$	$\{2,4,5,6\}$	$\{1,3,7,9,12\}$
$\{11,12\}$	$\{5,6,8\}$	$\{6,8,10,11\}$	$\{2,4,6,10,11\}$
	$\{5,9,12\}$		

For orthogonalization we choose a function

$$\varphi_{s,t}^d(x_1,\ldots,x_{12}),$$

since the representation

$$L(\varphi_{s,t}^d(x_1,\ldots,x_{12}))$$

is simpler than

$$L(\varphi_{s,t}(x_1,\ldots,x_{12})).$$

The calculation formula:

$$\begin{aligned}
h(p_1,\ldots,p_{12}) = 1 &- q_1q_2 - (1-q_1q_2)q_{11}q_{12}\\
&-p_2(1-q_{11}q_{12})q_1q_3q_8\\
-(1-q_{11}q_{12})(1&-q_1q_2-p_2q_1q_3)q_5q_6q_8\\
&-p_1p_5p_8p_{12}q_2q_4q_6q_{10}q_{11}\\
&-p_2p_5p_8p_{11}q_1q_3q_7q_9q_{12}\\
&-p_{11}(1-q_1q_2-p_2q_1q_3q_8\\
-(1-q_1q_2&-p_2q_1q_3)q_6q_8)q_5q_9q_{12}\\
-p_1p_8(1&-q_{11}q_{12}-p_{11}q_9q_{12})q_2q_4q_5q_6\\
-p_5p_{12}(1&-q_1q_2-p_2q_1q_3)q_6q_8q_{10}q_{11}.
\end{aligned} \tag{6}$$

To find the probability of (s-t)-connectivity, we substitute the values of availability factors in formula received (6).

Finally, the probability of (s-t)-connectivity:

$$h(p_1,\ldots,p_{12}) = 0{,}9003.$$

To assess and analyze the probability of availability of network services in large-scale networks, one can use known lower and upper bound estimates. These estimates can be constructed on the basis of the deterministic properties of the monotonic structures.

6   Conclusion

The approach for estimating the probability of (s-t)-accessibility is proposed. It is based on the isomorphism of properties algebra and classes algebra, also some order considerations.

The idea is also suitable for estimating the probability of full connectivity of large-dimensional communication networks.

The found probability of pairing and full connectivity allows you to analyze the availability properties with changing network load.

For a comparative analysis of the network structures reliability it is necessary to be able to calculate the probabilities of their full connectivity. The structure with higher probability is more preferable.

In the case of excessively large network dimension, when an accurate estimate is impossible, lower and upper bound estimates are constructed formally using inequalities (4), (5). The error in the estimate is determined by the difference between the boundary estimates (upper and lower).

This approach is suitable for future scientific search:

- to solve problems of designing networks with specified properties;
- to estimate an average time, when a network is steady and a service is available;
- to assess the importance of network elements [10].

The importance of an element shows the degree of its influencing the network availability ratio when providing a network service.

References

1. Zio, E.: Reliability engineering: old problems and new challenges. Reliab. Eng. Syst. Saf. **6**, 128–133 (2008)
2. Volkanovski, A., Cepin, M., Mavko, B.: Application of the fault tree analysis for assessment of power system reliability. Reliab. Eng. Syst. Saf. **94**, 1116–1127 (2009)
3. Shier, D.R.: Network Reliability and Algebraic Structures. Clarendon Press, Oxford (1991)
4. Wilson, J.M.: An improved minimizing algorithm for sum of disjoint products. IEEE Trans. Reliab. **39**, 42–45 (1990)
5. Wood, R.K.: Factoring algorithms for computing k-terminal network reliability. IEEE Trans. Reliab. **35**, 269–278 (1986)
6. Lin, Y.K.: Using minimal cuts to evaluate the system reliability of a stochastic-flow network with failures at nodes and arcs. Reliab. Eng. Syst. Saf. **75**, 41–46 (2002)
7. Rocco, C.M., Zio, E.: Solving advanced network reliability problems by means of cellular automata and Monte Carlo sampling. Reliab. Eng. Syst. Saf. **89**, 219–226 (2005)
8. Zio, E., Podofillini, L., Zille, V.: A combination of Monte Carlo simulation and cellular automata for computing the availability of complex network systems. Reliab. Eng. Syst. Saf. **91**, 181–190 (2006)
9. Zaychenko, Yu., Vasyliev, V., Vishtal, D., Khotyachuk, R.: Analysis of the viability of the DTS at the design stage. In: International Scientific and Practical Conference on Decision Making Systems and Information Technologies, Chernivtsi, pp. 48–49 (2004)
10. Beichelt, F., Franken, P.: Zuverlässigkeit und Instandhaltung. Mathematische Methoden. Verlag Technik, Berlin (1983)

SDN Load Prediction Algorithm Based on Artificial Intelligence

Artem Volkov[1], Konstantin Proshutinskiy[1], Abuzar B. M. Adam[2],
Abdelhamied A. Ateya[1,3], Ammar Muthanna[1,4(✉)], and Andrey Koucheryavy[1]

[1] The Bonch-Bruevich State University of Telecommunications,
Saint Petersburg, Russia
artem.n@5glab.ru, kotpro95@mail.ru, akouch@mail.ru
[2] School of Communications and Information Engineering, Chongqing University
of Posts and Telecommunications, Chongqing, People's Republic of China
L201810010@stu.cqupt.edu.cn
[3] Electronics and Communications Engineering, Zagazig University,
Zagazig 44519, Ash Sharqia Governorate, Egypt
a.ashraf@zu.edu.eg
[4] Peoples' Friendship University of Russia (RUDN University),
6 Miklukho-Maklaya Street, Moscow 117198, Russia
ammarexpress@gmail.com

Abstract. 5G/IMT-2020 networks have to provide new technical
requirements for realizing new services such as Tactile Internet, medical services and others. 5G infrastructure will be based on Software-
Defined networking and Network Function Virtualization for providing
new quality level. In general, a significant number of the available Internet services and applications require exact value of network parameters
such as latency, jitter, RTT and bandwidth. The SDN-based technologies
should be able to control and manage dynamic QoS for different new services, which are a time constraint. For this reason, SDN-controller, like
the main element of network infrastructure, must be stable and protected from external different threats. There are many works were on
this task. Most of these works are goaled on stress tests of hardware
and software parts, also one of the de-facto tests for each controllers is -
generating OpenFlow "packetin" message from special traffic generator.
Nevertheless, in "life mode" controller can be loaded differently, for example, uneven service load. We cannot build in advance various theoretical
models of the controller load. In this regards, there is a need to develop
a new approach for monitoring and prediction algorithm for build predicted models of OpenFlow activities. Also, this algorithm has to be
independent of the hardware features of the controller and another technical integration peculiarities. In this paper proposed a novel approach
for SDN load prediction based on artificial intelligence algorithms and
totally monitoring of OpenFlow channels activities. Also in this paper,
the possibility justification for predicting the load on hardware part, with
the help of OpenFlow thread analytics was given.

Keywords: 5G · Artificial intelligence · SDN · Prediction

© Springer Nature Switzerland AG 2019
V. M. Vishnevskiy et al. (Eds.): DCCN 2019, CCIS 1141, pp. 27–40, 2019.
https://doi.org/10.1007/978-3-030-36625-4_3

1 Introduction

By 2020 it is expected to begin a new era of mobile communication system with great and efficient capabilities by announcing the fifth generation of mobile communication systems (5G) [1]. 5G networks are expected to offer the opportunity to launch, efficiently and cost-effectively, numerous new services thus, creating an ecosystem for technical and business innovation. There are some challenges associated with the design and realization of the 5G cellular system introduced in details in [1–3]. One of these challenges is the ultra-low latency required for the 5G system, which is of order of 1 ms [4]. However, it is worth noting that 5G networks it' not only new technologies and infrastructure approaches for cellular network part.

Moreover, for time constraint's IMT-2020 services we have to provide new quality requirements level. Latency sensitive services (e.g., Tactile Internet, Medical networks, real-time IoT applications, streaming video, autonomous driving, and machine-to-machine communication) are of wide interest [2]. Such services promise to improve the quality of life in many fields of human society [5]. For the time being, the traffic generated by smart devices becomes a significant part of Internet [6]. Potentially, SDN-based technologies 5G will be used in Internet community for managing all types of services and therefore, SDN is to be tasked to manage these kinds of demands as well. In general, a significant number of the available Internet services and applications require exact value of network parameters such as latency, jitter, RTT and bandwidth [7]. The SDN-based technologies should be able to control and manage dynamic network quality parameters for different time constraint applications [8].

To implement the high quality requirements, it is necessary to ensure the stability of 5g/IMT-2020 control systems. One of these systems is the controller of software-defined networks. During the development of the concept of software-defined networks, a lot of research work has been carried out (including companies already implementing SDN solutions) on various solutions testing, in particular-stress tests. In most cases, the purpose of research is to find the limit of the controller's operation on parallel-served threads, evaluate the advantages of one controller architecture over another, find the fastest controller, the most functional, etc. Thus, knowledge of the workload of the controller allows you to assess the performance of the network as a whole. Given this fact, predicting the behavior of a software-configured network controller is one of the primary tasks [9].

One of the approaches to this task is to develop a special application for the operating system on which the SDN controller is deployed. This application asks the system for the values of the hardware parameters (CPU, RAM, ROM, Cores activities, system flows and etc.) [10]. The values of the requested hardware parameters can be transferred further to the analytical system. However, this approach has several disadvantages such as: dependence on the hardware, type and version of the operating system. Including increases the likelihood of the controller failing due to incorrect operation of this application with the controller's operating system.

Considering the above fact, it is required to analyze and develop a new approach in analytics and forecasting the load on the controller of a software-defined network. This article proposes a new approach to monitoring and predictive analytics of the load on the controller, based on analytics of only the total service flows from the switches to the SDN controller. This approach will solve the problem of dependence on the hardware and also the operating system on which the controller of the SDN is deployed. Considering the particularities of the obtained statistical data, the analysis of existing mathematical methods for forecasting data was carried out, a multiparametric analysis was also conducted, the result of which shows the dependence of the activity of service flows and the load on the controller's hardware. In the process, a server for monitoring and predictive analysis of flows in a software-configured network was also developed. The obtained results confirm the feasibility of the proposed method for predicting the load of the SDN controller in 5G/IMT-2020 communication networks.

2   Justification. Multiparametric Analyse

Part 1. Investigated Model. General Architecture

At the moment, there are many works aimed at the research and development of various methods of testing the controller, including stress testing. As noted above, this paper discusses the possibility of implementing monitoring of the hardware load of the controller using monitoring and intelligent analytics from a group of service flows – OpenFlow. To check the efficiency of the proposed approach, a model software-defined network with a running Video-on-Demand service at the data Plane level was used.

Analytical (hereinafter - the network application), part of which is the module for predicting the activity of service flows OpenFlow, was developed in the Python programming language as a WEB server operating on the basis of the MVC model. This server runs on top of the Northbound API controller and orchestrator model network lab. As a controller of the SDN network, this model network uses – OpenDaylight Beryllium SR4, as an Orchestrator of the infrastructure – OpenStack [11]. The Data Plane level is based on the Mikrotik switches with OpenFlow Protocol support [12]. The general architecture of the segment is shown in Fig. 1. It is worth noting that the figure also shows, in addition to the main elements of the model infrastructure [6], the objects of study and their parameters.

To form the data Sets, the application sent requests every second to the controller of the software-configured network. After that, the process of filtering the resulting OpenFlow tables occurred in such a way as to further transfer the values of the OpenFlow service flows to the processing. The next step, the obtained Data Set was processed to form a Data Set to build analytical models, the calculation of which is given below in the article.

Based on the data analyzing, displayed in two global parts of the table: Match Field and Actions, we can concluded that they can compose a metamodel of threads. The reduced structure of the switch flow table is shown in

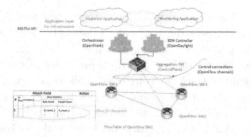

Fig. 1. General architecture

Fig. 1. One of the important features of this data is that, based on the "Byte Count" and "Packet Count" counters, it is impossible to accurately determine the exact packet length in the flow. Since at one time the counters can be equal to: "Byte Count" - 1500, "Packet Count" - 3. Accordingly, it is not possible to determine the exact length of each packet registered in the stream over a period of time based on this data: $\delta T = 1$ [s].

It is also worth noting that the counters display the total value of the [«Byte Count», «Packet Count»] parameters. However, in addition to these counters, in Flow Table there is another parameter «Time Stamp», which allows you to estimate at each time instantaneous value [ByteCount-delta] and [PacketCount-delta]. Thus, for an arbitrary period of time ΔT, having counts of values [Byte Count], [Packet Count], [TimeStamp] it is possible to make a Data Set with an established data structure, where each count displays the instantaneous value [ByteCount-delta] and [PacketCount-delta]. Counts of values [Byte Count], [Packet Count], [TimeStamp] – are formed by every second requests to the SDN controller through the RESTful API program interface. The structure formed on the basis of the DataSet-RQ queries with "raw" data (1) and the formula (2) of its conversion to the required DataSet-ML format with instantaneous values (3) are given below.

If, PacketCount-delta – PC-delta, ByteCount-delta – BC-delta, a TimeStamp-deltas – TS = 1 [sec.] = const, then:

$$DataSet_{RQ} = \begin{bmatrix} [TimeStamp] & [ByteCount] & [PacketCount] \\ TimeStamp_{11} & ByteCount_{12} & PacketCount_{13} \\ TimeStamp_{21} & ByteCount_{22} & PacketCount_{23} \\ ... & ... & ... \\ TimeStamp_{N1} & ByteCount_{N2} & PacketCount_{N3} \end{bmatrix} \quad (1)$$

$$\begin{cases} BCdelta_{N2} = ByteCount_{N2} - ByteCount_{(N-1)2}, & if\ N \geq 1 \\ PCdelta_{N2} = PacketCount_{N2} - PacketCount_{(N-1)2}, & if\ N \geq 1 \end{cases} \quad (2)$$

$$DataSet_{ML} = \begin{bmatrix} [TimeStamp] & [ByteCount] & [PacketCount] \\ TS & BC_{delta12} & PC_{delta13} \\ TS & BC_{delta22} & PC_{delta23} \\ ... & ... & ... \\ TS & BC_{deltaN2} & PC_{deltaN3} \end{bmatrix} \tag{3}$$

In so doing, the calculation of the total values of the parameters for a specified period of time is made by the following formulas (4, 5):

$$ByteCount_{\triangle T} = \sum_{N=1}^{N=\triangle T/TS} BC_{deltaN2} \tag{4}$$

$$PacketCount_{\triangle T} = \sum_{N=1}^{N=\triangle T/TS} PC_{deltaN2} \tag{5}$$

Part 2. Justification. Multiparametric Analyze

The primary requirement was to check the existence of a relationship between the activity of OpenFlow flows and the change in the load of the hardware of the software-configured network controller. Figure 2 shows the structure of the experimental stand.

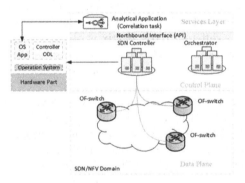

Fig. 2. Experimental stand

To test the dependence between the activity of the OpenFlow service flows and the hardware load on the SDN controller, two applications were developed «OS App» and «Analytical Application». «OS App» - is a server application for the Linux operating system on which the opendaylight SDN controller is deployed. This application requests the values of the hardware load parameters («CPU Value», «RAM Value») from the operating system. Values of the requested metrics are available through the REST API of this application to third-party applications over the network. «Analytical Application» - this application is also developed as a server, but already running on top of the Northern controller interface (REST API). The application also requests parameter values from the «OS App» through its API over the network, and eventually generates a Data Set of synchronized parameters to evaluate from the dependency.

To assess the dependence between the parameters, after analyzing the existing mathematical methods, it was proposed to use multiparameter correlation analysis. To carry out the analysis, a series of data is determined for comparison and a multidimensional data matrix (X) is formed, which is subsequently reduced to a typical (U). The rows of such matrix (not given) correspond to the results of registration of all observed parameters of objects (OpenFlow stream and SDN-controller hardware) in one experiment, and the columns contain the results of observations of one parameter (factor) in all experiments. To determine the series, we denote the number of parameters by m, where $(m > 1)$, and the number of observations by n. In the matrix, the element x_{ij} corresponds to the value of the j parameter in the i observation. In this case, it is allowed to have empty values of some elements, for example, due to omissions in the registration of parameter values. In the context of this research project – parameters are recorded every second. However, in a multivariate analysis, it is desirable to eliminate missing values. To do this, there are two approaches: striking out the corresponding rows of the matrix or entering the average values instead of the missing ones. In this article, the input data series are subjected to normalization before the direct correlation analysis based on the mechanism of entering the average values instead of the missing ones. Further methods of matrix X processing are based on the following assumption: if the object of study is subjected to a new survey and a different data matrix is obtained, then after its processing using the same methods, results close to the results of the first matrix will be obtained. This assumption is based on the statistical hypothesis of matrix formation.

Thus, the object of study in multidimensional analysis is a multidimensional random variable represented by a finite volume sample. It is also worth noting that the parameters characterizing the object of study have a different physical meaning, and the data matrix changes significantly if the scales in which the selected parameters are measured change. Accordingly, the data matrix is reduced to a standard form, that is, the values of the parameters are standardized (option). As noted above, the standardized matrix will be denoted by U. The matrix X, according to a number of parameters, constructed on the basis of the DataSet-ML matrix, is as follows:

$$X = \begin{bmatrix} ByteCount & PacketCount & CPU & RAM \\ x_{11} & x_{12} & x_{13} & x_{14} \\ x_{21} & x_{22} & x_{23} & x_{24} \\ x_{31} & x_{32} & x_{33} & x_{34} \\ x_{i1} & x_{i2} & x_{i3} & x_{i4} \end{bmatrix} \tag{6}$$

The transformation of the matrix X, formed from the data of the selected parameters (1) to the standardized matrix U, is as follows: For each studied parameter $j = 1, 2, ..., 4$ weighted estimates are calculated using the following formula

$$u_{ij} = \frac{x_{ij} - \mu(x_j)}{\delta(x_j)}, where\ i = 1, 2, ..., n; j = 1, 2, ..., 4; \tag{7}$$

The formula (2) uses the following two parameters, namely the mathematical expectation $\mu(x_j)$ and the variance $\delta(x_j)$, which are calculated by the following formulas:

$$\mu(x_j) = \frac{1}{n} \times \sum_{i=1}^{n} x_{ij} \tag{8}$$

$$\mu(x_j) = \delta^2(x_j) = \frac{1}{n} \times \sum_{i=1}^{n} (x_{ij} - \mu(x_j))^2 \tag{9}$$

Thus, calculating the elements "U_{ij}" is composed of a matrix U, which becomes the subsequent object of processing. The impact of general factors, the presence of objective laws in the behavior of the studied objects lead only to the appearance of the so-called static dependence. Static called dependence, in which a change in one of the values entails a change in the distribution of others, and these values are known to take some values with certain probabilities. There is also a special case of static dependence, which is more suitable for the model considered in this article, namely the correlation dependence, which in turn characterizes the relationship between the values of some random variables with the average value of others. It is worth noting that the correlation dependence usually describes a causal relationship between the values of the used parameters of the analyzed analytical model. The correlation dependence is determined by various parameters, among which the most common are the indicators characterizing the relationship of two random variables (pair indicators): the correlation moment, the correlation coefficient. The estimation of the correlation moment (co variance coefficient) of the two variants x_j and x_k is calculated from the initial matrix X:

$$\xi_{jk} = \frac{1}{n} \times \sum_{i-1}^{n} (x_{ij} - \mu_1(x_k)) \tag{10}$$

The co variance coefficient ρ-jk of normalized random variables is called the correlation coefficient and its estimate is calculated by the following formula:

$$P_{jk} = \frac{1}{n} \times \sum_{n}^{1} U_{ij}U_{ik} \tag{11}$$

It is worth noting that the correlation coefficient does not depend on the values of random variables, but on their variations, so if the value of the magnitude increases by an order of magnitude, then the coefficient will not change. so, the value of the correlation coefficient lies within from -1 to $+1$. If the random variables U_j and U_k are independent, then the coefficient ρ_{jk} is necessarily equal to zero, but the converse is incorrectly. The correlation coefficient ρ_{jk} characterizes the importance of the linear relationship between random variables (parameters):

where $\rho_{jk} = 1$, the values of u_{ij} and u_{ik} are fully coincide, i.e. the values of the parameters take the same values. In other words, there is a functional dependence, that is, knowing the value of one parameter, you can clearly specify the value of another parameter;

where $\rho_{jk} = -1$, the values of u_{ij} and u_{ik} take opposite values. And in this case there is also a functional dependence;

where $\rho_j k = 0$, the values of u_{ij} and u_{ik} practically are unrelated by a linear relation. However, in this case, does not mean the absence of any other (for example, non-linear) connections between the parameters;

where $|\rho_{jk}| > 0$, $|\rho_{jk}| < 1$, there is no single-valued linear relation of u_{ij} and u_{ik}. And the less the absolute value of the correlation coefficient, the less the values of one parameter can predict the value of another.

The so-called functional dependence was mentioned above, which essentially reflects the relation of the considered quantity with one or many other quantities, if this quantity depends only on this set of factors. However, in this case it is necessary to take into account the fact that functional relations are more mathematical abstractions, as in real situations there is an infinitely large number of properties of the object and the external environment that affect each other. Therefore, speaking about the possible functional dependence, it is necessary to take into account the fact that the selected number of parameters, on which the direct correlation analysis is limited. In so doing, a number of parameters are determined in each specific practical task, for example, as in the current research work, the justified in this article the number of parameters selected for analysis is only limited to four.

Thus, in this task (the study of the relation between the activity of service flows and the load of the SDN controller), the Analytical Application calculates the following indicators.

We adopt the following conventions: ByteCount – BtCt;

PacketCount – PcCt;
CPU value – Cpu;
RAM value – Ram;

$$\xi_{(1\times4)x} = \begin{bmatrix} BtCt|PcCt & BtCp|Cpu & BtCp|Ram & PcCt|Ram \\ \xi_{1x1} & \xi_{1x2} & \xi_{1x3} & \xi_{1x4} \end{bmatrix} \tag{12}$$

$$\rho_{(1\times4)U} = \begin{bmatrix} BtCt|PcCt & BtCp|Cpu & BtCp|Ram & PcCt|Ram \\ \rho_{1x1} & \rho_{1x2} & \rho_{1x3} & \rho_{1x4} \end{bmatrix} \tag{13}$$

In matrices (10, 12) the first line shows the attitude of the parameters between which the corresponding indicator is calculated (correlation moment or coefficient).

Part 2. Neural Network

Today, artificial neural network is widely used to solve various problems in real life. Such tasks as speech recognition, text recognition, predicting complex models are now more solved with the help of neural networks, achieving high results.

At the moment, there is a large variety of neural networks. This problem uses a neural network to predict the load. The architecture of the developed neural network is shown in Fig. 3. The data stream is fed to the input layer of neurons (placeholders) from the formed DataSet-ML. Placeholders are connected to the first layer of the neural network by a fully connected structure. The output is two neurons that generate a series of predicted values. The Neural network receives a fixed length data entry, so the data is divided into 200-line segments or 10 s. Activity tags are converted to the unitary code. Data is divided into training and practice sets in the ratio of 8:2. The network model contains 4 fully connected layers, each of which contains 10 hidden nodes. The hyperparameters of learning: Optimizer:Adam.

Number of epochs: 20
Number of samples per iteration: 1024
Learning speed: 0.0025

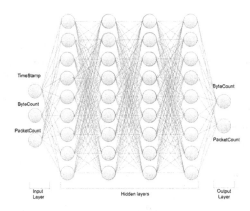

Fig. 3. Architecture of neural network

3 Test Results

As already shown above, in this article the practical tests of the proposed method of monitoring and forecasting the load on the controller of software-configured network is divided into two stages. The first step is to conduct a study to confirm the established hypothesis about the direct impact of the activity of the Open-Flow service message flow on the controller and, accordingly, on its hardware. The second part of the practical tests, with the obtained positive results of the first part, is the development and training of a neural network for monitoring and forecasting the activity of OpenFlow service flows.

For practical tests, a stand was assembled, the structure and description of which are given in paragraph 2.1 of this article. Soft was developed to calculate indicators for the dependence between the studied parameters. This software first generates a Data Set, the structure of which is displayed in paragraph 2.2 of this article, and then calculates the corresponding indicators ξ_X and ρ_U.

However, as already noted in justifying the choice of this mathematical method, the correlation moment has a certain feature, the correlation moment has a certain feature, namely, it depends on the units of measurement of independent quantities, which does not fully assess the level of correlation of the two quantities. For this, the original matrix is reduced to the form normalized by the formulas specified above. On the basis of this matrix, a point graph was also built for the distribution of weighted estimates relative to each other. The resulting graph was based on a sample of 100 values from the total Data Set and is displayed in Fig. 4.

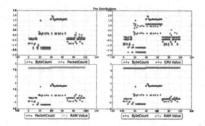

Fig. 4. Scatter plot of weighted estimates

On each of the subgraphs is displayed in the form of signatures corresponding data series of the matrix, the distribution of which are given. To assess the full level of correlation of the studied values, the correlation coefficient is calculated.

To analyze the data obtained, we turn to the properties of the coefficient of correlation coefficient, which were displayed in paragraph 4.2 of this article. All obtained values of the correlation coefficients satisfy the following property $|\rho_{jk}| > 0$, $|\rho_{jk}| < 1$. On the basis of this property it is possible to draw the following conclusion: there is a relation between the studied parameters, while the more value of the correlation coefficient, the more it is possible to predict the value of another parameter. From the obtained values in this paper we are interested in the values of the correlation coefficient between BtCt|Cpu and PcCt|Ram, equal to 0,89 and 0,92 respectively. The values of these indicators allow us to conclude that there is a relationship between the OpenFlow service flow and the hardware load of the SDN controller. In this case, there is no explicit linear relationship between them. However, it should be noted that the values of the correlation coefficient between the PcCt|Ram indicators are close to the value 1, which allows us to conclude that the proposed method of monitoring and forecasting the load on the controller can work effectively. And accordingly, to assess the operation of the controller to a certain extent, it is sufficient to build a predictive analytical model of OpenFlow flows, where the total PacketCount of a group of flows is a predictable parameter. Also, to visualize the obtained indicators, a bar chart was built, which displays the corresponding values of the correlation coefficients of the pairs of the studied values. The diagram is shown in Fig. 5.

For the second part of the practical tests, the stand has been upgraded with regarding to the first part of the practical tests. The structure of the stand and its description was given above. The second part of the tests was to train the developed neural network, the architecture and parameters of which are given in the second paragraph of this article. According to the obtained DataSet-ML a scatter plot values was constructed. The diagram shown in Fig. 6. In this diagram, it is visually possible to select several areas (clusters) of the distribution points, in the area of which corresponding the average value of the packet length in the flow prevails.

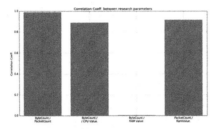

Fig. 5. Bar graph of the obtained correlation coefficients

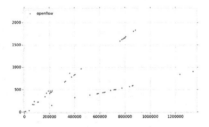

Fig. 6. Scatter plot of DataSet values

During neural network training (estimation of neural network prediction accuracy), During the neural network training (assessing the accuracy of the neural network prediction), the MSE parameter - Mean Square Error was observed as a performance evaluation parameter. The change in the MSE parameter value is shown in Fig. 7.

Figure 7 shows two graphs. The red dashed line shows the change of the MSE parameter when the network is running on the training DataSet, the green line shows the change of the MSE parameter on the working (real) Data Set. The graph shows that the selected architecture and parameters of the neural network are optimal for generating load prediction. Ultimately also, in the process of neural network training, the process of its training for the formation of forecasts was monitored. For clarity of the process, a graph of real and predicted values of the neural network was constructed. Figure 8 shows some screenshots of training the neural network to show the progress. In each of the areas shown in Fig. 8, there are two graphs showing real and predicted values, respectively, blue and green graphs. The left graph on the first line with the caption (Epoch 0, Batch 0) displays the real and first forecast. The graph shows that the predicted values are very different from the real ones. That is, the green graph is much lower than the main one – the blue graph. The right graph on the first line with the caption (Epoch 0, Batch 50) shows the next forecast of the neural network after the first one. The graph shows (green) that now the predicted values exceed the real ones, while the graph is overstated.

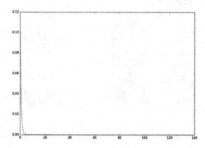

Fig. 7. Mean square error (Color figure online)

The bottom two graphs, with captions (Epoch 19, Batch 250), (Epoch 19, Batch 300) show the last steps of network training. According to the graphs, it is clear that the neural network has learned and now can correctly predict the activity of OpenFlow service flows in a real software-configured network. It can be seen that in the last (lower right graph), the real graph coincides with the predicted.

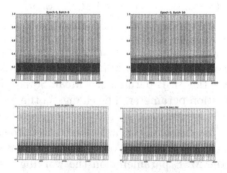

Fig. 8. Neural learning process. Real and predicted OpenFlow activity (Color figure online)

After the neural network has learned, the program saves the state of its architecture and parameters so that it is possible to load its state from another software module of the «Analytical Application» server. Further, the software module that loaded the state of the neural network generates predictive data on the data test that the neural network receives on the input neurons (placeholders).

4 Conclusion

This article proposed an intelligent monitoring method for one of the control systems in 5G/IMT-2020 communication networks, namely the controller of a software-configured network. The intelligence of the method and, accordingly, the uniqueness lies in the ability to predict the load on the controller of a software-configured network based only on the use of activity data from OpenFlow service flows. The results obtained allow us to summarize that it is possible to realize the forecast of the load on the controller's hardware can be implemented in the form of a network application that analyzes remotely the activity of incoming OpenFlow flows. Accordingly, the problem is solved depending on the type of controller hardware, including the operating system on which the controller is deployed.

Also, it is worth noting that this application (monitoring method) can be used not only for the purpose of predicting the load on the controller, but also for the tasks of managing flows in the network on the Data Plane. For example, another network application (or another service) has the possibility to request the controller's load values from a given application and take it into account in the decision-making process to manage flows at the DataPlane level.

In future works, which will be based on this work, it is planned to implement the work of this application in real time, including the output of the predicted Analytics in the WEB-interface of the developed server. Also, implement internal programming interfaces so that other applications that perform data Plane level monitoring tasks can comprehensively assess the development of a software-defined network as a unified system. The formation of such a complex system will allow a more balanced decision to be made on the management and scaling of a software-configured network. Thus, by closing various issues of monitoring and management, it is possible to arrive at a common, intelligent core of the network, which will be adapt and changed for the corresponding traffic. Taking into account the features of Smart Services traffic and strict quality criteria for their implementation, only a smart, intelligent network can provide them and at the same time provide a high guarantee of infrastructure operation.

Acknowledgment. The publication was prepared with the support of the "RUDN University Program 5-100".

References

1. 5G PPP architecture working group white paper, view on 5G architecture, July 2016
2. Ateya, A.A., Muthanna, A., Makolkina, M., Koucheryavy, A.: Study of 5G services standardization: specifications and requirements. In: 2018 10th International Congress on Ultra Modern Telecommunications and Control Systems and Workshops (ICUMT), pp. 1–6. IEEE, November 2018
3. Jiang, D., Liu, G.: An overview of 5G requirements. In: Xiang, W., Zheng, K., Shen, X.S. (eds.) 5G Mobile Communications, pp. 3–26. Springer, Cham (2017). https://doi.org/10.1007/978-3-319-34208-5_1
4. Ateya, A.A., Muthanna, A., Gudkova, I., Abuarqoub, A., Vybornova, A., Koucheryavy, A.: Development of intelligent core network for tactile internet and future smart systems. J. Sens. Actuator Netw. **7**(1), 1 (2018)
5. Feasibility Study on New Services and Markets Technology Enablers, document 3GPP TR 22.891, ver. 14.2.0, September 2016
6. Volkov, A., Khakimov, A., Muthanna, A., Kirichek, R., Vladyko, A., Koucheryavy, A.: Interaction of the IoT traffic generated by a smart city segment with SDN core network. In: Koucheryavy, Y., Mamatas, L., Matta, I., Ometov, A., Papadimitriou, P. (eds.) WWIC 2017. LNCS, vol. 10372, pp. 115–126. Springer, Cham (2017). https://doi.org/10.1007/978-3-319-61382-6_10
7. Ateya, A., Al-Bahri, M., Muthanna, A., Koucheryavy, A.: End-to-end system structure for latency sensitive applications of 5G, vol. 6, pp. 56–61 (2018)
8. Farris, I., Taleb, T., Khettab, Y., Song, J.: A survey on emerging SDN and NFV security mechanisms for IoT systems. IEEE Commun. Surv. Tutor. **21**(1), 812–837 (2018)
9. Muhizi, S., Shamshin, G., Muthanna, A., Kirichek, R., Vladyko, A., Koucheryavy, A.: Analysis and performance evaluation of SDN queue model. In: Koucheryavy, Y., Mamatas, L., Matta, I., Ometov, A., Papadimitriou, P. (eds.) WWIC 2017. LNCS, vol. 10372, pp. 26–37. Springer, Cham (2017). https://doi.org/10.1007/978-3-319-61382-6_3
10. Muhizi, S., Ateya, A.A., Muthanna, A., Kirichek, R., Koucheryavy, A.: A novel slice-oriented network model. In: Vishnevskiy, V.M., Kozyrev, D.V. (eds.) DCCN 2018. CCIS, vol. 919, pp. 421–431. Springer, Cham (2018). https://doi.org/10.1007/978-3-319-99447-5_36
11. https://www.opendaylight.org/ . Accessed May 2019
12. Muthanna, A., et al.: Secure and reliable IoT networks using fog computing with software-defined networking and blockchain. J. Sens. Actuator Netw. **8**(1), 15 (2019)

QoS Based Method for Energy Optimization in ZigBee Wireless Sensor Networks

A. Alexandrov$^{(\boxtimes)}$ (ID), V. Monov, R. Andreev, and J. Doshev

Institute of Information and Communication Technologies,
Bulgarian Academy of Sciences, Sofia, Bulgaria
{akalexandrov,vmonov}@iit.bas.bg, rumen@isdip.bas.bg, jodo@abv.bg
http://www.iict.bas.bg

Abstract. ZigBee based Wireless Sensor Networks (WSNs) are a reliable and adaptive solution for environment level monitoring. In this paper, we use QoS technique for energy optimization of the existing Zig-Bee communication protocol. The proposed new method and algorithm uses a combination of the embedded in the ZigBee wireless nodes Link Quality Indicator (LQI) and Received Signal Strength Indicator (RSSI) as critical parameters. The proposed algorithm is specifically tailored to be a stateless, localized algorithm with minimal control overhead. A real wireless node transmission experiments and analysis are provided to validate our claims.

Keywords: WSN · ZigBee · LQI · RSSI · Power optimization · Energy efficiency · 802.15.4

1 Introduction

The Wireless sensor networks - WSNs are one of the most important technologies in the 21st of Century. They are an essential part of today's life due to their wide range of applications. The WSNs are used mainly for IoT and industrial applications such as factory automation and control, environmental monitoring, etc.

These type of applications have a relatively low data rate but high demand for Quality of Service (QoS) in terms of reliability and timeliness along with energy efficiency. The WSN's lifetime depends mainly on the level of the sensor nodes' energy consumption.

One possible way to save energy is the management of the transmission power of the sensor node in the function of the inter-node distance and the RF link quality. The energy management can be realized by the embedded communication protocol and/or by a custom design software algorithm.

This paper is supported by the National Scientific Program "Information and Communication Technologies for a Single Digital Market in Science, Education and Security (ICTinSES)", financed by the Ministry of Education and Science.

© Springer Nature Switzerland AG 2019
V. M. Vishnevskiy et al. (Eds.): DCCN 2019, CCIS 1141, pp. 41–52, 2019.
https://doi.org/10.1007/978-3-030-36625-4_4

The typical WSN is built by a large number of battery-powered multi-functional wireless sensor nodes. The proper communication protocol related to the specific needs of an industrial or IoT based WSN is critical for the reliability of the sensor network.

One of the possible implementations of WSNs is based on the IEEE 802.15.4 technical standard families. The IEEE 802.15.4 based standards define physical, and media access MAC/PHY layers on low-speed wireless networks. They are supported by the IEEE 802.15 Task Force and were defined in 2003 and later extended till 802.15.4v in 2017. The 802.15.4 standard family is the basis for ZigBee, 6LowPan, ISA100.11a, WirelessHart, and MiWi communication protocols, each of which is an upgrade of the standard and builds layers that are not defined by 802.15.4 [1,2].

ZigBee Protocol. The ZigBee protocol was created and ratified by ZigBee Alliance companies. This communication protocol is designed to provide easy-to-use secure and reliable wireless architecture. The ZigBee protocol is based on the MAC/PHY layers of the IEEE 802.15.4 standard and operates in unlicensed radio bands, including 2.4 GHz, 900 MHz, and 868 MHz radio frequencies [3,4]. The protocol allows touch devices to communicate in a variety of network topologies (star, tree, a cluster tree, and mesh) and allows the design of devices with the autonomy of up to 2–3 years without a battery change.

The main advantages of the ZigBee protocol include:

- allows to build a scalable WSNs with fast and straightforward sensor nodes add/remove functionality;
- allows a programmable long duty cycle low power consumption;
- relatively low latency;
- Direct Sequence Spread Spectrum (DSSS) [5];
- supports up to 65,000 nodes per network;
- 128-bit AES packet encryption with built-in avoidance procedure packet collisions.

The IEEE 802.15.4 standard defines a wireless communications mechanism characterized by low data rate and low power consumption supporting only star topologies and peer-to-peer. Technically, the IEEE 802.15.4 standard mainly focuses on the development of PHY and MAC layers that are primarily suited for wireless communication rather than for building large-scale networks [6–9].

The ZigBee Specification was created in 2004 by the ZigBee Alliance to define specifications for building large wireless networks upgrading the existing IEEE 802.15.4 standard that specifies only PHY and MAC layers suitable for relatively small networking such as LR-WPAN (Low Rate - Wireless Personal Area Network) [10–13]. As an 802.15.4 standard extension, the ZigBee specification defines a stack consisting of network and security layer (NWK).

Mesh Topology. The mesh topology has a similar tree structure but is considerably more flexible. With this topology, all sensor devices have the right to

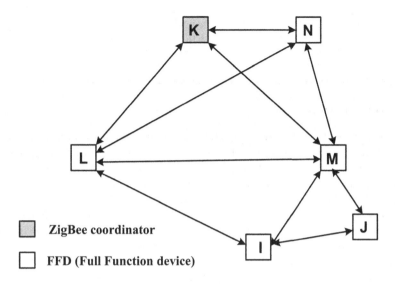

Fig. 1. ZigBee mesh topology

communicate with each other without the need to send a message first to the ZigBee coordinator device (Fig. 1).

For example, if device J has to send a message to device K, the possible routes may be $J \rightarrow I \rightarrow L \rightarrow K$ or $J \rightarrow M \rightarrow K$ or $J \rightarrow I \rightarrow M \rightarrow N \rightarrow K$ or $J \rightarrow I \rightarrow L \rightarrow N \rightarrow K$ and so on. If a device fails and a route fails, the message may reach its endpoint on alternative routes [8].

2   Related Works

Several QoS based protocols have been proposed in the literature for power optimization in battery powered WSNs. In [14], the authors propose a comparative analysis and a new algorithm for transmission power adjustment based on RSSI measurements. In [15] is proposed a couple of power control mechanisms based on Signal Noise Ratio (SNR) estimation - the multiplicative-increase additive-decrease power control (MIAD-PC), and the packet error rate power control (PER-PC). In [16] a new QoS and Energy Aware Cooperative Routing Protocol is proposed. The protocol is based on RSSI and node energy consumption in a competitive context (RSSI/energy-CC). In [17] a transmission power control (ATCP) is proposed to achieve energy efficiency and improve the communication between neighbor sensor nodes. The main disadvantage of the proposed mechanism is that the ATCP uses only RSSI to calculate the level of the link quality.

In general, most of the power optimization techniques presented in the literature use RSSI or PRR based calculations as a QoS metric based on the IEEE 802.15.4 standard. However, in the case of a high level of RF noise and RF interference from neighbor WSNs, the RSSI parameter is not a viable indicator to fully determine the QoS. Other disadvantages of [14–17] were that the proposed protocols and algorithms were evaluated by using only a simulation.

3   QoS Based Method for ZigBee Energy Optimization

3.1   Problem Definition

The approach of using the RSSI as the only parameter for power transmission optimization of a sensor node is not reliable in many cases. The main goal of the RSS indication is to measure the received RF power in a selected radio channel.

The problem is that the RSSI can not make a differentiation between the signal level of the useful radio link data and the level of the RF noise with the same frequency in the channel. Therefore, the new generation of the sensor devices and their transmission modules have an embedded RF Link Quality Indication (LQI) functionality. LQI is an estimate of "how easily a received signal can be demodulated by accumulating the magnitude of the error between ideal constellations and the received signal over the 64 symbols immediately following the sync word" [4].

3.2   Problem Solving

The proposed QoS method for energy optimized inter-node communication uses a combination of the embedded in the sensor devices Link Quality Indicator (LQI) and Received Signal Strength Indicator (RSSI). The LQI and RSSI are a standard option for 802.15.4 and ZigBee compatible RF front-ends (RF transmitting modules). The proposed procedure and method of the QoS based sensor node energy optimization inter-node is similar to procedures in 802.15.4 related Radio Channel Assessment and Active scanning procedures.

Energy Detection ED (Energy Detection). The energy detection function enables the system to determine the energy level of specific channels. Each radio signal available in the frequency range of the selected channel increases its energy level. The primary task of this feature is to find all potential interfering sources.

Active and Passive Scan. In 802.15.4 standard and in the related ZigBee stack the active and passive scan functions are designed to help the system to detect the presence of similar wireless networks in the area. During the process of initialization of a ZigBee based WSN, it must perform at least one active scan.

The proposed method uses the active scanning function of the ZigBee compatible sensor node in the WSN and sends a special beacon request including a QoS request and Personal Operating Space (POS) code and which is used for the sensor nodes synchronization. After that, the sensor node analyzes the received answers (beacon frames), including the LQI and RSSI parameters of the neighbor nodes in the RF transmission range. The diagram on Fig. 2 shows a typical procedure for QoS based inter-node communication.

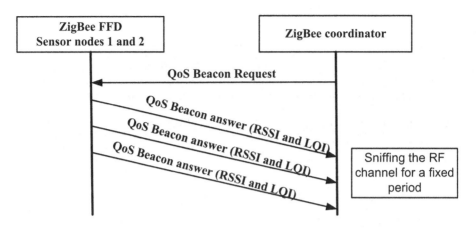

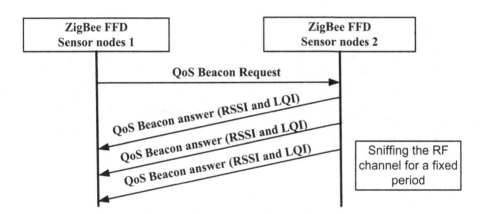

Fig. 2. QoS beacon scanning

The transmit/receive beacon procedure of LQI and RSSI related data from the sensor nodes in the transmission range is repeated after a fixed time intervals. The average data values of the neighbor sensor node LQI (Q_{LQI}) and RSSI Q_{RSSI} parameters, which belongs to the specified sensor node RF link are the basis for a calculation of the inter-node QoS parameter Q_{Qos}.

$$Q_{iQoS} = AQ_{iRSS} + BQ_{iLQ} \tag{1}$$

Where:

Q_{LQ} - represents the average value of the received LQI;
Q_{RSS} - represents the average value of the received RSSI;
i - represents the number of nodes in the transmission range;
A and B are weight coefficients related to the expected level of QoS.

The Fig. 3 below represents the QoS based procedure for sensor inter-node communication and power optimization.

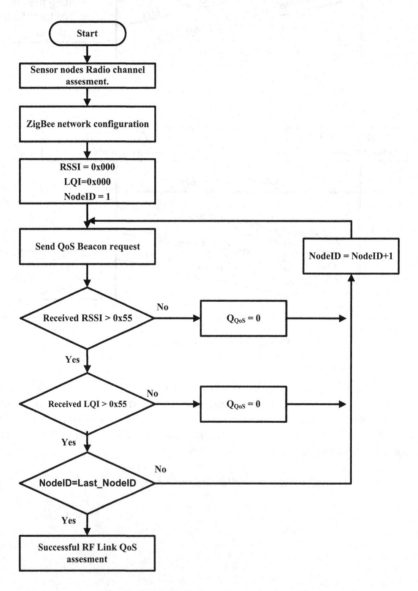

Fig. 3. RF link QoS assessment algorithm

3.3   RSSI and LQI Monitoring Levels

In the TI CC2531 chip the LQI value is an unsigned 8-bit integer ranging from 0x00 to 0xFF (or 0xF0 in stacks released prior to Z-stack 3.0.2), with the maximum value representing the best possible link quality. The relationship between LQI and the "link weight" for purposes of routing support in the stack is such that LQI values of 0x55 map to the lowest costs of 3, while LQIs below 0x55 represent links with high error rates, so the worst-case cost of 7 is assigned. In our case the real measured LQI value of 0x55 represents approximately 70% reliability of receiving the packet intact. It is important to note that in the TI CC2531 chip the LQI measurement is based on the chip error rate of the every packet which is received from the neighbor sensor node of the inbound route.

Therefore it provides a link quality information specific to the current link connection and depends how the neighboring device transmits the current packet to the local device. This LQI data is used to determine the connectivity level between neighboring sensor nodes and to select parent devices when joining the ZigBee network.

The RSSI value in TI CC2531 RF front-end hardware is a signed 8-bit integer with values from approximately -100 to $+127$, with each value representing the energy level (in dBm) at the radio's receiver. It is important to be noted that the RSSI measurement is based on the peak (maximum) energy level detected by the radio receiver on the current frequency over the first 8 symbol periods (about $128\,\mu s$) of the current packet being received. That signal energy at a given frequency which in our case is $2.405\,GHz$ equal to channel 11 can come from any transmitter/interferer on that frequency, whether it is another nearby node on the same network, a device from a different ZigBee coordinator on the same channel, or a non-ZigBee, non-802.15.4 interferer, such as a WiFi transmitter or microwave with big harmonics for example.

The measured RSSI level does not concern to any specific link with another device, it can also be assessed outside of the incoming message handler context.

3.4   Experimental Data

The experimental ZigBee based lab hardware consists of three ZigBee CC2531 USB dongle development boards model CC2531EMK. The ZigBee development boards were configured as one ZigBee coordinator and two FFD. All the USB ZigBee CC2531 RF development boards are mounted on NUCs model Gigabyte GB-BACE-3150 with installed Linux, TI CC13x2 SDK, Wireshark packet sniffer and software power meter for battery consumption simulation and measurement. Additionally, a separate emulator of RF noise was created by Software Defined Radio (SDR) hardware based on one NUCs model Gigabyte GB-BACE-3150 with mounted modified as RF generator (transmitter) USB SDR RTL2832U RF dongle and antenna. The custom design RF generator can transmit RF noise in band $1.8\,GHz$–$3\,GHz$ including harmonics.

The used in the current experiments ZigBee communication protocol is build on the basis of the TI Z-stack 3.0.2 software implementation. The proposed new QoS assessment optimization algorithm is implemented as a modification and an upgrade of the existing Z-stack source code in GitHub https://github.com/zstackio/zstack.

All the experiment was realized in ZigBee channel 11 (equal to 2.405 GHz). The lab hardware and configuration diagram are shown on Fig. 4 and the photos of the lab hardware - on Figs. 5 and 6.

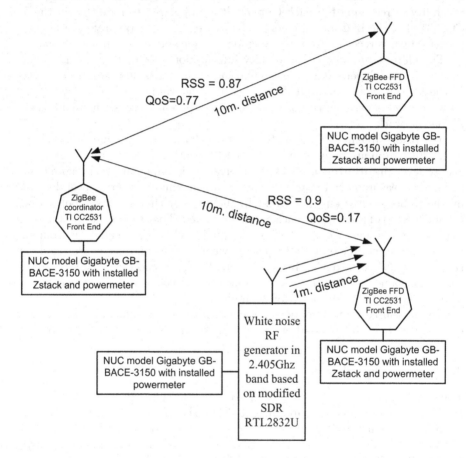

Fig. 4. ZigBee CC2531 transmission power levels

Fig. 5. ZigBee CC2531 transmission power levels

Fig. 6. ZigBee CC2531 transmission power levels

The results from the experiments by real ZigBee inter-nodes communication, based on Texas Instruments (TI) RF chip CC2531 are shown in Table 1.

Table 1. Transmission power levels provided by TI CC2531 and power consumption analysis

Transmission power (dBm)	CC2531 Current consumption mA	Battery capacity of regular ZigBee WSN node in hours	Battery capacity of ZigBee node with QoS management in hours
−28	23,0	331	382
−20	24,0	216	263
−10	25,5	120	202
−6	26,0	102	191
−3	27,0	82	174
0.5	28,0	79	165
+1	29,0	68	141
+2.5	31,0	48	141

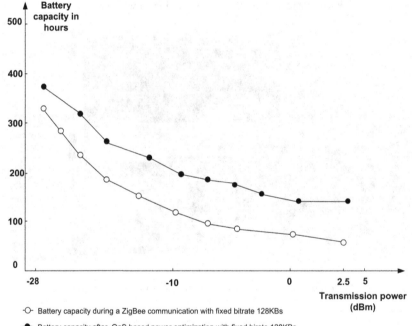

-○- Battery capacity during a ZigBee communication with fixed bitrate 128KBs

-●- Battery capacity after QoS based power optimization with fixed birate 128KBs

Fig. 7. ZigBee CC2531 transmission power levels

The experimental results visualized on Fig. 7 illustrate that the proposed QoS method can improve sensitively, in some cases with around 15% the sensor node battery shelf live.

4   Conclusion

The proposed new QoS based method and algorithm for sensor node energy optimization are developed with the main objective to provide a reliable and energy efficient inter-node communication. In addition to the existing routing protocols which use the RSSI indicator as to the only criteria for RF link quality the proposed method uses the combination of LQI and RSSI as much more reliable QoS parameter. The implemented power optimization can achieve up to 15% better sensor node shelf live as a result of lower level RF transmission power, better RF channel collision avoidance and minimized packet retransmission because of the relatively less data loss.

References

1. Ahn, G.S., Campbell, A.T., Veres, A., Sun, L.H.: Service differentiation in stateless wireless ad hoc networks. In: Proceedings of IEEE INFOCOM 2002, June 2002
2. Hill, J., Szewczyk, R., Woo, A., Hollar, S., Culler, D., Pister, K.: System architecture directions for network sensors. In: ASPLOS (2000)
3. Tashev, T.D., Hristov, H.R.: Modeling of synthesis of information processes with generalized nets. J. Cybern. Inform. Technol. 3(2), 92–104 (2003). SOA, Academic Publishing House "Prof. Marin Drinov"
4. Texas Instrument: Calculation and usage of LQI and RSSI. http://e2e.ti.com/support/lowpowerrf/w/designnotes/calculation-and-usage-of-lqi-and-rssi.aspx
5. Kurose, J.F., Ross, K.W.: Computer Networking a Top-Down Approach Featuring the Internet. Addison Wesley Longman Inc., Boston (2000). ISBN 0-201-47711-4
6. Atanasova, T.: Modelling of complex objects in distance learning systems. In: Proceedings of the First International Conference - Innovative Teaching Methodology, 25–26 October 2014, Tbilisi, Georgia, pp. 180–190. ISBN 978-9941-9348-7-2
7. Ding, W., Koubaa, A., Cunha, A., Alves, M., Tovar, E.: A time division beacon scheduling mechanism for IEEE 802.15.4/ZigBee cluster- tree wireless sensor networks. Research Group, Polytechnic Institute of Porto, RuaAlatonio Bernardino de Almedia, Porto, Portugal, vol. 431, pp. 4200–4072
8. Othman, F., Bouabdallah, N., Boutaba, R.: Load-balancing routing scheme for energy-efficient wireless sensor networks. In: IEEE GLOBCOM 2008, New Orleans, LA USA, December 2008
9. Bhadoria, R.S., Sahu, D., Dixit, M.: Proficient routing in wireless sensor networks through grid based protocol. Int. J. Commun. Syst. Netw. (IJCSN) 1(2), 104–109 (2012)
10. Shuaib, K., Alnuaimi, M., Boulmalf, M., Jawhar, I., Sallabi, F., Lakas, A.: Performance evaluation of IEEE 802.15.4: experimental and simulation results. J. Commun. 4, 29–37 (2007)
11. Ergen, S.C., Varaiya, P.: Energy efficient routing with delay guarantee for sensor networks. ACM Wirel. Netw. J. 13(5), 679–690 (2007)
12. Al-Karaki, J.N., Kamal, A.E.: Routing techniques in wireless sensor networks. IEEE J. Wirel. Commun. 11(6), 6–28 (2004)
13. Balabanov, T., Zankinski, I., Barova, M.: Strategy for individuals distribution by incident nodes participation in star topology of distributed evolutionary algorithms. Cybern. Inf. Technol. 16(1), 80–88 (2016). Print ISSN: 1311–9702, Online ISSN: 1314–4081

14. Jeong, J., Culler, D.E.: Empirical analysis of transmission power control algorithms for wireless sensor networks. In: Proceedings of International Conference Networked Sensing Systems (INSS 2007), Braunschweig, Germany, pp. 27–34 (2007)
15. Ares, B.Z., Park, P.G., Fischione, C., et al.: On power control for wireless sensor networks: system model, middleware component and experimental evaluation. In: Proceedings (ECC 2007), Kos, Greece, pp. 4293–4300 (2007)
16. Maalej, M., Cherif, S., Besbes, H.: QoS and energy aware cooperative routing protocol for wildfire monitoring wireless sensor networks. Sci. World J. **2013**, 11 (2013). https://doi.org/10.1155/2013/437926. Article ID 437926
17. Lin, S., et al.: ATPC: adaptive transmission power control for wireless sensor networks. ACM Trans. Sens. Netw. (TOSN) **12**(1), 1–31 (2016)

A Problem of Optimal Location of Given Set of Base Stations in Wireless Networks with Linear Topology

Roman Ivanov[1], Amir Mukhtarov[1,2(✉)], and Oleg Pershin[2]

[1] V.A. Trapeznikov Institute of Control Sciences of Russian Academy of Sciences,
65 Profsoyuznaya Street, Moscow 117997, Russia
iromcorp@gmail.com, mukhtarov.amir.a@gmail.com
[2] Gubkin Russian State University of Oil and Gas (National Research University),
Moscow, Russia
pershino@mail.ru

Abstract. The paper describes a special case of a base station placement in wireless network with linear topology. Each station is equipped with access point (for example IEEE 802.11) that is used by the objects to send their data via the network, and with a relay equipment that allows the station to connect to the neighboring stations. Each station is described by the coverage radius and the communication radius to other stations. The goal is to determine such placement of the stations which will maximize the total coverage on the linear section by the given set of stations. Two formal problem statements are described and analyzed: the statement in the extremal combinatorial form and the formulation of the problem as a mixed-integer linear programming model. The main result is the development of a special branch and bound algorithm for solving the problem represented in the combinatorial form. The results of a comparative computational experiment by three approaches: by the brute force method, by the branch and bound algorithm and solving the problem in the form of the mixed-integer linear programming are given.

Keywords: Wireless network · Optimal base stations placement · Combinatorial model · Branch and bound method

1 Introduction

At the present time, the problem to supervise and control the area where industrial and civil facilities, technological installations, moving vehicles, etc., are placed, is a very important.

The problem of optimal equipment placement on a given redundant set of possible placement locations comes to light in the process of creating the distributed control systems. In this paper we consider the infrastructure consisting

The publication was supported in part by Russian Foundation for Basic Research (RFBR) according to the research project No. 19-07-00919.

© Springer Nature Switzerland AG 2019
V. M. Vishnevskiy et al. (Eds.): DCCN 2019, CCIS 1141, pp. 53–64, 2019.
https://doi.org/10.1007/978-3-030-36625-4_5

of multiple base stations that interact to each other to form the broadband communication network.

We review the particular case of a such problem, when the territory to be controlled is a one-dimensional linear route. It could be a section of a road transportation network, a linear part of a main pipeline, or a field communications line. For example, a backbone network of road side units (RSUs) is an essential part of modern intelligent transportation systems, where the roadside units are used to collect and distribute information among the mobile users (see Fig. 1).

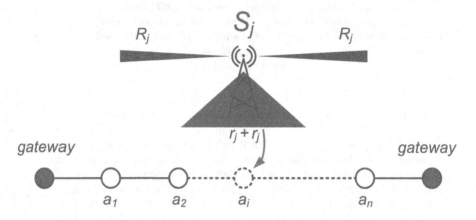

Fig. 1. Optimal station placement problem parameters including places and stations description.

Similar problems were discussed in a number of publications. Brahim et al. [1] describes the problem of station deployment which will maximize the coverage area while restricted by a total cost. The input data includes the set of potential places for stations deployment and the preliminarily collected statistics of the users' traffic. Cavalcante et al. suggested the model of maximum coverage with delay restrictions and analyzed it with the genetic algorithm [2]. The problem of correct choice of road side units placement scheme for maximizing coverage of territory are presented in [3]. Liu et al. [4] formulated road side units (RSUs) placement problem which will maximize the probability of the average connectivity in vehicular ad-hoc network as a combinatorial optimization problem. The problem is solved by the Expansion and Coloration Algorithm (ECA), with an addition of taking into consideration the road traffic characteristics. In articles [5, 6] the authors consider the extremal problem of choosing base station types and their placement along the line road. They formulated and discussed the problem in a form of the mixed-integer programming model.

Within the framework of a wide class of optimal placement of capacities problems, the problem considered in this article in addition to the specificity of one-dimensional structure of the territory to be controlled, also has the substantial feature caused by distance limit presence between base stations.

The main result of our work is the development of a special branch and bound algorithm for solving the problem of optimal location for the given set of base stations in wireless network with linear topology represented in a combinatorial form. The paper also describes the problem statement in the form of a mixed-integer linear programming model. The results of a comparative computational experiments to solve the problem: (a) by the brute force method, (b) by the branch and bound method and (c) in the form of the mixed-integer linear programming model are presented.

2   The Placement Problem in the Combinatorial Form

Let be we have a line segment α of length L with the ends points a_0 and a_{n+1}. Inside of the segment $\alpha = [a_0, a_{n+1}]$ a finite set of arranged points $A = \{a_i\}_{i=1}^n$, $a_{i+1} > a_i$ is given; these points correspond to the set of vacant places where the stations can be placed. Each point a_i is defined with its one-dimensional coordinate l_i. Let's also denote a set of stations as $S = \{s_j\}_{j=1}^m$. Each station has two parameters: a coverage radius r_j and a communication radius R_j. Then we can also define a station $s_j \in S$ as a set of parameters: $s_j = \{r_j, R_j\}$.

There are special stations s_0 and s_{n+1} which are gateways. These stations are already placed at the ends a_0 and a_{n+1} of the segment α correspondingly. For those stations $r_0 = r_{n+1} = 0$ and $R_0 = R_{m+1} = 0$.

A *feasible stations placement* $P = \{a_i, s_j\}$ is a sequence of pairs for which the following constraints are satisfied:

1. left connectivity: $\forall (a_i, s_j)$, $1 \leq i \leq n$, either $\exists (a_k, s_q) : l_k < l_i$ and $l_i - l_k \leq \min\{R_j, R_q\}$, or $l_i - l_0 \leq R_j$;
2. right connectivity: $\forall (a_i, s_j)$, $1 \leq i \leq n$, either $\exists (a_t, s_g) : l_t > l_i$ and $l_t - l_i \leq \min\{R_j, R_g\}$, or $l_{n+1} - l_i \leq R_j$;
3. $|P| = m$.

The first and the second requirements guarantee that all stations are connected in a chain and this chain ends in the gateways. The third requirement guarantee that all stations have been placed.

Let's define the set of all feasible placements as G.

The coverage value $z(P)$ is corresponded to each placement P. This function is defined as the length of disjoint union of segments τ, $\tau \subset \alpha$ such that each segment is included in the coverage only if it is contained in the coverage area of some station from placement P, and each point $a \in \alpha$ belongs to a single segment τ.

We introduce the concept of "non-coverage" of the segment α:

$$f(P) = L - z(P)$$

Now we can formulate the optimal placement problem as the extremal combinatorial problem in the combinatorial form.

Problem 1. It is required to find a permissible placement P^*, such that

$$P^* = \operatorname*{argmin}_{P \in G} f(P)$$

Let us denote the set of all combination of m stations on n places (not only feasible placements) as Γ. The number of elements $\gamma \in \Gamma$ is

$$\gamma = C_n^m \times m!$$

3   The Brute-Force Method

To solve the problem described above, we will first consider the brute-force method. For it to be implemented we define an algorithm to build a binary search tree, also referred to as a branch tree. This algorithm will be also useful for us to develop the branch and bound algorithm that will be described further.

The algorithm builds a tree by splitting a set G. This approach involves a use of a well-known method based on a variation of a binary variable. The variable π_{ij} that is used for splitting is defined as follows:

- $\pi_{ij} = 1$ if the station s_j is placed into point a_i;
- $\pi_{ij} = 0$ otherwise.

The procedure is an iterative one: on ν iteration we fix the value of the variable π_{ij} to 0 or 1 that will cause the splitting the entire set G_ν into two child subsets G_ν^1 and G_ν^2, which correspond to child vertexes of the parent vertex corresponding to the set G_ν on the branch tree. Let's assume that G_ν^1 is obtained by setting $\pi_{ij} = 1$ and G_ν^2 is obtained by setting $\pi_{ij} = 0$.

Now we will describe how we choose such variable among the set of all variables $\Pi = \{\pi_{ij}\}$. Based on previously chosen variables and its values on some iteration, the given set forms a disjoint union of three sets: Π^+ - the set of variable equals to 1 (allowed variable), Π^- - the set of variables equals to 0 (forbidden variable), and Π^f - undefined variables.

To split the set G_ν the variable π_{ij} are chosen among Π^f with the lowest possible i and j. It means that we decide: either locate the station s_j on the place a_i or not.

We describe movement on the branch tree.

After splitting the set G_ν in two subsets G_ν^1 and G_ν^2, these subsets on the branch tree are assigned the indices $G_{\nu+1}$ and $G_{\nu+2}$, respectively.

When forming a branch tree, two types of steps are defined: the "direct" step and the "back" step. The direct step is the movement "in depth" along the same branch of the tree to execute the partition of the current subset G_ν. The back step is the step which performs the transition from the set G_ν to one of the previously formed subsets. The back step is executed in cases when either the set G_ν consists of one variant of station location or the set G_ν is empty. In these cases, the corresponding vertex of the tree is called "closed".

For movement on a tree we will use the LIFO rule. Direct steps will be performed until a vertex is obtained that must be closed. This corresponds to a movement along the same tree branch. The subset G_ν^1 will be examine first of the subsets G_ν^1 and G_ν^2.

If the vertex will be not be closed as a result of the examination, then further movement on the same branch will continue (the execution of a direct step). If the vertex will be closed, then the back step will be executed. The back step is the transferred to the unclosed vertex which is the last formed vertex among unexamined vertexes.

The process stops after set of Π^f becomes empty.

4   The Branch and Bound Algorithm

We have developed the branches and bound algorithm to reduce the search on the branch tree.

In order to develop the branch and bound algorithm for solving the placement problem using the branch tree described in the previous section, we need only to develop methods for investigating vertices for the possibility of closure.

In accordance with the branch and bound technique the vertex will be closed in following three cases.

Case 1. Set G_ν is empty, i.e. there is no feasible placement for given sets of allowed and forbidden variables.

Case 2. It has been proved that in set G_ν there is no feasible placement with the value of objective function $f(P)$ which is less than already found $f(\widehat{P})$ named as "current best value". The initial value of $f(\widehat{P})$ is set higher then possible optimal value (e.g. $f(\widehat{P})$ is L).

Case 3. An optimal solution for set G_ν has been found.

4.1   Case 1

Here we should proof that conductivity or completeness constraints described above are not fulfilled.

To prove completeness infeasibility, it is sufficient to show that for given allowed and forbidden π_{ij} there is no vacant place for some unplaced station.

Obviously, such a check is algorithmically easy to implement: it suffices to show that there exists k such that for a station s_k for which $\pi_{ik} \neq 1$ for all i there is no a_i for which simultaneously $\pi_{ij} \neq 1$ for some j, $j \neq k$, and $\pi_{ik} \neq 0$.

To prove connectivity infeasibility, we first should check the following requirement for the initial set $G = G_0$: the distance between two neighboring places should not be greater than the second largest communication radius R_j of stations in S. If the requirement is not satisfied, then the set of all feasible placements G is empty and the problem has not the solution.

We will describe now the check procedure for G_ν, $\nu > 0$. We will assume that G_ν is obtained by splitting the parent set by variable $\pi_{kt} = 1$ and the set contains more than one placement P. The algorithm consists of three steps.

Step 1. Check that each distance R_t and R_h, where h is an index of a station placed into position a_d just before a_k, is greater than $l_k - l_d$. If the closest place from the left is a_0 than it checks R_t only.

Step 2. Check that both R_t and the largest R_j between unplaced stations is not less than the distance between a_k and place a_i, the closest to it from the right. If such place is a_{n+1} the check is applied only to R_t and the distance $l_{n+1} - l_k$. If all stations are placed, then G_ν consists of the single placement and this case will be analyzed further.

Step 3. If the number of unplaced stations is more than 1, then we check that the distance between the two neighboring points is not greater than the second largest communication radius among unplaced stations, and the distance between a_{n+1} and a_n is not greater than the largest one.

If only one unallocated station remains, then we check that among the still unoccupied points to the right of the point a_k there is at least one such point that the distances from this point to the point a_k and simultaneously from this point to the point a_{n+1} is no greater than the communication radius R_j of the station remaining unallocated.

If on at least one step is failed, then G_ν is empty and the vertex must be closed. In this case, the back step is performed.

If the set G_ν is obtained by setting $\pi_{kt} = 0$, then one should check that the distance between the place with the greatest index among those where stations are placed (considering a_0) and a_k is not exceeded the maximum connectivity range of unplaced stations. If this is not true, then the set G_ν is empty and the back step is performed.

4.2   Case 2

Let us name as a *partial non-coverage* the value

$$\Delta(k, d, p, t) = max\{(a_k - a_d) - (r_p - r_t), 0\}. \tag{1}$$

This value is defined for any two places a_d and a_k, $k > d$, where the stations s_p and s_t are placed in situations where there is no other station between them.

It is obvious that non-coverage $f(P)$ for every placement P is calculated as the sum of all $\Delta(k, d, p, t)$ between the places where stations are deployed including the ends of α.

Let us build the lower bound $W(G_\nu)$ for $f(P)$, i.e.

$$W(G_\nu) \leq f(P), P \in G_\nu.$$

If it can be shown that $W(G_\nu) \geq f(\widehat{P})$, then G_ν does not contain the placement better than already found $\widehat{P}$ and the vertex should be closed.

First we shall build $W(G_0)$.

Since all the stations must be placed, then the maximum coverage of α is obtained in situations when it would be possible to place all the stations without intersections of their coverage radiuses. Each station s_j covers the segment of length $2r_j$, therefore, a total non-coverage cannot exceed the value.

$$W\left(G_0\right) = \max\left\{L - \sum_{j=1}^m 2r_j, 0\right\}.$$

Now we will define $W(G_\nu)$, $\nu > 0$ as the sum of partial sums w_1 and w_2.

$$W\left(G_\nu\right) = w_1 + w_2.$$

Suppose G_ν is obtained by splitting a parent set by setting the variable $\pi_{kt} = 1$. Then w_1 is "non-coverage" of the segment from a_0 to a_k while w_2 is non-coverage of the segment from a_k to a_{n+1}.

The sum w_1 is calculated by formula (1) by summing non-coverages between the places where stations have been placed already.

The sum w_2 is calculated as

$$w_2 = max\left\{(l_{n+1} - l_k) - \left(r_t + \sum_{j \in S_v} 2r_j\right), 0\right\}, \tag{2}$$

where S_ν is the set of unplaced stations.

The formula (2) is analogous to the formula (1) except the fact, that it is applicable to the part of segment α which is to the right from the last place where any station has been placed.

If $\pi_{kt} = 0$ the estimation $W(G_\nu)$ remains unchanged after splitting.

4.3   Case 3

In this case we review only sets G_ν, which consist of a single placement P, and $f(P)$ is obtained as the sum of non-coverages $\Delta(k, d, p, t)$ between known places where stations are placed.

If for the given placement the inequality $f(P) < f(\widehat{P})$ takes place, then the placement P becomes a new current best value $\widehat{P}$ and backward step is applied.

The branch and bound algorithm will stop when all the vertexes of the searching tree will be closed.

The solution is

$$P^* = \widehat{P}, \widehat{f}(P^*) = f(\widehat{P}).$$

5   Example 1

Input Data
The segment α with $L = 50$, end points a_0 and a_4 with coordinates $l_0 = 0$ and $l_4 = 50$ is given. There are internal points a_1, a_2, a_3 with coordinates $l_1 = 20$, $l_2 = 30$, $l_3 = 40$.

The set of stations $S = \{s_j\}$, $j = 1,2$ is given, where station s_j has parameters: $r_1 = 20$, $R_1 = 40$; $r_2 = 5$, $R_2 = 20$. There are special stations s_0 and s_4, on the end points with $r_0 = R_0 = r_4 = R_4 = 0$.

We have to find a feasible placement P^*, such that

$$P^* = \min_{P \in G} f(P)$$

The process of solving the problem is presented in the form of a binary search tree (see Fig. 2).

The initial value of $f(\widehat{P}) = L = 50$, $G_0 = G$.

The vertices of the tree indicated by $\emptyset$ correspond to the sets G_ν for which there are no feasible placements.

Two placements P_1 and P_2 were obtained as the current best solutions with $f(P_1) = 15$ and $f(P_2) = 5$.

The optimal solution is $P^* = P_2, f(P^*) = f(P_2) = 5$.

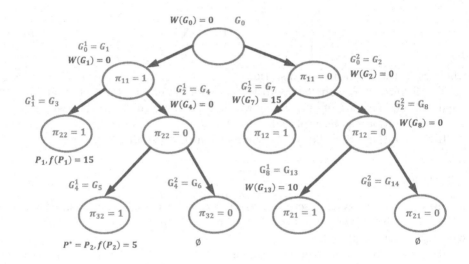

Fig. 2. The search tree of branch and bound algorithm.

6 The Statement of Problem as a Mixed-Integer Programming Model

Here we will formulate our placement problem in the form of a mixed-integer programming model. The description of problem is given in Sect. 2.

Let us introduce binary variables x_{ij} where $x_{ij} = 1$ if a station s_j is placed at point a_i and $x_{ij} = 0$ otherwise.

Let us introduce binary variables e_i where $e_i = 1$ if any station is placed at point a_i and $e_i = 0$ otherwise.

By definition

$$e_i = \sum_j^m x_{ij}, \ i = 1, 2, \ldots, n.$$

We have $e_0 = 1$ and $e_{n+1} = 1$ for the end points.
Let us formulate the following system of the problem constraints.
Each station must be placed in one and only one place

$$\sum_{i=1}^n x_{ij} = 1, \ j = 1, 2, \ldots, m.$$

At any point there can be no more than one station

$$\sum_{j=1}^m x_{ij} \le 1, \ i = 1, 2, \ldots, m.$$

We will introduce non-negative variables y_i^+ and y_i^- for points a_i, $i = 0, 1, 2, \ldots, n$, $n + 1$. Variables y_i^+ and y_i^- are the area sizes (right and left from point a_i) which are covered by station placed at point a_i.
Values of variables y_0^+, y_0^-, y_{n+1}^+, y_{n+1}^- equal 0.
The values of coverages are not greater than the coverage radius of the station located at a_i, and equal to 0 if is no station at a_i:

$$y_i^+ \le \sum_{j=1}^m x_{ij} r_j, \ i = 1, 2, \ldots, n,$$

$$y_i^- \le \sum_{j=1}^m x_{ij} r_j, \ i = 1, 2, \ldots, n.$$

The total coverage area between any two points a_i and a_k on which the stations are located cannot exceed the distance between these points.
For $i = 1, \ldots, n$:

$$y_i^+ + y_k^- \le \frac{l_k - l_i}{2} (e_i + e_k) + (2 - e_i - e_k) \ L, \ k = i + 1, \ldots, n + 1,$$

$$y_i^- + y_k^+ \le \frac{l_i - l_k}{2} (e_i + e_k) + (2 - e_i - e_k) \ L, \ k = i - 1, \ldots, 0.$$

This condition excludes the effect from intersections of station coverages when calculating the total coverage value for the entire segment α.
According to the conditions of the problem, the station located at a_i must be connected with at least one station on the left and one station on the right, including stations at the end points a_0 and a_{n+1}.
We will introduce variables $z_{ijk}, i = 1, 2, \ldots, n; \ j = 1, 2, \ldots, m; \ k = 1, 2, \ldots, n; \ k \ne i$ to formulate this requirement where:

- $z_{ijk} = 1$ if a station s_j is located at point a_i and connected with a station which is located at point a_k;
- $z_{ijk} = 0$ otherwise.

We will also introduce variables z_{ij0} and z_{ijn+1} where $z_{ij0} = 1$ if a station s_j is located at point a_i and connected with a station s_0 which is located at point a_0 and $z_{ij0} = 0$ otherwise; $z_{ijn+1} = 1$ if a station s_j is located at point a_i and connected with a station s_{n+1} which is located at point a_{n+1} and $z_{ijn+1} = 0$ otherwise.

Stations must be at both points so that they can be connected:

$$z_{ijk} \leq e_i, \forall i, j, k,$$

$$z_{ijk} \leq e_k, \forall i, j, k.$$

The station s_j which is located at a_i must be connected with at least one station which is located right from a_i and at least one station which is located left from a_i

$$\sum_{k=i+1}^{n+1} z_{ijk} \geq x_{ij}, \forall i, j,$$

$$\sum_{k=0}^{i-1} z_{ijk} \geq x_{ij}, \forall i, j.$$

The communication radius R_j of the station located at the point a_i, must be no less than the distance to the point a_k, where there is a station with which it is connected:

$$z_{ijk} \left(R_j - (a_i - a_k) \right) \geq 0, \ k = i - 1, \ldots, 0, \ j = 1, 2, \ldots, m,$$

$$z_{ijk} \left(R_j - (a_k - a_i) \right) \geq 0, \ k = i + 1, \ldots, n + 1, \ j = 1, 2, \ldots, m.$$

Objective function

$$f = \sum_{i=1}^{n} \left(y_i^+ + y_i^- \right) \rightarrow max$$

7   Numerical Results

The algorithms Branch and Bound (BnB) and brute-force algorithm (BF) were implemented using Python. Table 1 shows the results of solving several problems for a different number of locations and a different number of stations using the B and B algorithm, the BF algorithm and the standard program for solving mixed-integer problem in the MATLAB package. We compare the number of vertices in the search trees so that the execution parameters of the algorithms do not depend on the speed of the machine and/or the quality the computer program. For each set of stations and set of placements 10 examples were computed with different numerical input data. For B and B and the MATLAB package the table shows the average execution parameters of the number of vertices in the search tree for each of the 10 examples.

Table 1. The results of solving problems.

Places	Stations	BF	BnB	MILP
7	5	17550	933	753
9	5	71090	6478	2669
10	5	126180	1041	8551
12	6	No	8294	38569
13	6	No	18485	30369

The numerical results. "No" means the problem was not solved after 3 h.

8   Conclusion

In this paper the problem of finding optimal location for the given set of base stations in wireless network with linear topology was analyzed. The problem has been formulated as an extremal combinatorial problem and also as mixed-integer linear programming model. The branch and bound algorithm for solving the problem in combinatorial form was developed. The results of the computer experiment show that the branch and bound algorithm is more effective than the brute-force algorithm and using of the branch and bound algorithm also more effective than to solve the problem represented as a mixed-integer programming model.

References

1. Brahim, M., Drira, W., Filali, F.: Roadside units placement within city scaled area in vehicular ad-hoc networks. In: 3rd International Conference on Connected Vehicles and Expo (ICCYE 2014) (2014)
2. Cavalcante, E., Aquino, A., Pappa, G., Loureiro, A.: Roadside unit deployment for information dissemination in a VANET: an evolutionary approach. In: 14th Annual Conference Companion on Genetic and Evolutionary Computation (aECCO 2012), New York, USA, pp. 27–34 (2012)
3. Lee, J., Kim, C.M.: A roadside unit placement scheme for vehicular telematics networks. In: Kim, T., Adeli, H. (eds.) ACN/AST/ISA/UCMA - 2010. LNCS, vol. 6059, pp. 196–202. Springer, Heidelberg (2010). https://doi.org/10.1007/978-3-642-13577-4_17

4. Liu, H., Ding, S., Yang, L., Yang, T.: A connectivity-based strategy for roadside units placement in vehicular ad hoc networks. Int. J. Hybrid Inf. Technol. **7**, 91–108 (2014)
5. Vishnevsky, V.M., Larionov, A., Smolnikov, R.V.: Optimization of topological structure of broadband wireless networks along the long traffic routes. In: Vishnevsky, V., Kozyrev, D. (eds.) DCCN 2015. CCIS, vol. 601, pp. 30–39. Springer, Cham (2016). https://doi.org/10.1007/978-3-319-30843-2_4
6. Ivanov, R., Pershin, O., Larionov, A., Vishnevsky, V.: On a problem of base stations optimal placement in wireless networks with linear topology. In: Vishnevskiy, V.M., Kozyrev, D.V. (eds.) DCCN 2018. CCIS, vol. 919, pp. 505–513. Springer, Cham (2018). https://doi.org/10.1007/978-3-319-99447-5_43

Optimal Quantization Methods in Problems of Visual Data Multichannel Wavelet-Compression

Gleb Verba and Kirill Bystrov$^{(\boxtimes)}$

Moscow Institute of Physics and Technology, Institutsky lane 9, Dolgoprudny, Russia
{verba,kirill.bystrov}@phystech.edu

Abstract. Telecommunication systems and multimedia technologies advance is closely related to inevitable increase of transferred media data volumes and sets higher requirements to data coding systems efficiency, including video compression algorithms.

Using multichannel scheme of the discrete wavelet transform (DWT) performs a new promising approach to video coding presumed to increase compression ratio while saving restored image quality.

Coefficients quantization leads to inevitable image quality loss, however, is necessary for data compression. A rigorous analysis of statistical properties of frequency subbands of decomposition coefficients, considering inherent features of wavelet coefficients, may help to develop a new quantization method that would provide maximum compression of the output data while maintaining image quality, thereby allowing more efficient use of the DWT features.

Thus, the aim of this study was to find methods of optimal and quasi-optimal quantization of DWT coefficients.

A few innovative quantization methods were proposed within the framework of the paper and their efficiency was investigated. Some of the proposed techniques give a stable efficiency gain compared to standard quantization schemes.

Keywords: Statistical properties of discrete orthogonal transforms · Wavelet transform · Video coding · Quantization · Data compression

1 Introduction

The discrete wavelet transform is widely used in video coding. For example, image compression standard JPEG2000 [1] and video codecs Dirac [2] and Motion JPEG 2000 [3] use two-channel signal decomposition based on lowpass and highpass FIR filters. Using multichannel transform scheme may increase the compression ratio while maintaining the quality of the restored image [4].

Quantization is one of the most important stages of video coding process that defines the volume of the output data and the quality of the restored signal. DWT is based on division the image into spatio-frequency domains and allows

© Springer Nature Switzerland AG 2019
V. M. Vishnevskiy et al. (Eds.): DCCN 2019, CCIS 1141, pp. 65–76, 2019.
https://doi.org/10.1007/978-3-030-36625-4_6

more efficient video compression by means of reducing psycho-visual redundancy in images [5]. Namely, it applies more flexible quantization of individual signal components taking into account the features of human perception of visual information. In this connection the choice of quantization methods remains an actual issue of video coding.

Thus, the purpose of this paper is to study the statistical properties of the frequency domains of wavelet decomposition of images, determine improved methods for wavelet coefficients quantization on the basis of these properties and verify their applicability in problems of video compression.

The novelty of the work lies in the methods proposed in the paper, as well as the result they give in comparison to standard quantizing schemes, which consists in a guaranteed increase of the ratio of the restored signal quality and the output data volume.

2   Quantization of Wavelet Coefficients

As a rule, linear quantization is applied, however, other approaches also find a use. Boundaries of Q quantizing zones $(t_j, j = \overline{0, Q})$ and Q recovery levels $(d_j \in (t_j, t_{j+1}), j = \overline{0, Q-1})$ are assumed to be adjustable parameters within the general formulation of the problem. If the value of the coefficient $x_n, n = \overline{1, N}$ lays in range $[t_j, t_{j+1})$, after quantizing it is approximated with a value d_j, i.e.

$$\hat{x}_n = q(x_n),$$
$$q(x) = \{d_j, x \in [t_j, t_{j+1}), j = \overline{0, Q-1}\}. \tag{1}$$

The linear quantization has two parameters: the position of the central recovery level d_0, usually set to zero, and the quantizing step Δd. Then parameters t_j, d_j are defined in the following way:

$$d_j = d_0 + j\Delta d, \tag{2}$$

$$t_j = d_0 + \left(j - \frac{1}{2}\right)\Delta d. \tag{3}$$

Among the methods of non-linear quantization the Lloyd-Max method [6–8] deserves special attention. It is based on the criterion of minimizing the mean-square deviation (which is equivalent to Peak Signal-to-Noise Ratio (PSNR) maximizing) arising from quantization of the set of continuous quantities with a finite set of discrete levels, whose number Q is assumed to be prespecified. Most of the algorithms following the idea of deviation minimization have iterative implementations and their application leads to quantization grid thickening the near the distribution maxima and rarefying at the distribution minima.

As the papers [9,10] show, application this method to I-frames in wavelet video coding gives a slight result, noticeable mainly at low bit rates. It is worth mentioning that even though the above-described method maximizes PSNR, however, it doesn't take into account the entropy change arising from grid distortion. As a result, the method leads not only to PSNR growth, but as well to output entropy rise, often bringing the profit to naught.

A few proposals for improvement of linear quantization were formulated and tested in the framework of this article.

2.1   Constant Shift of the Quantization Grid

Consider the distribution of the original image (Fig. 1a) and of its DWT subband containing the lowest frequencies across the height and the width of the same image — the Low-Low zone (LL) (Fig. 1b). As the figures show, the distribution of the LL coefficients replicates the shape of the original one.

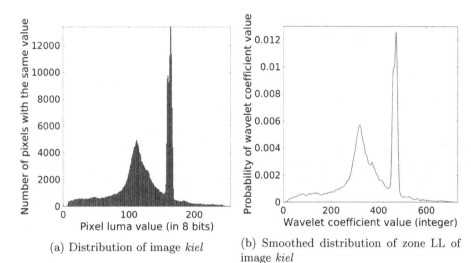

(a) Distribution of image *kiel*

(b) Smoothed distribution of zone LL of image *kiel*

Fig. 1. Distribution of the original image pixels and of the LL coefficients of its wavelet decomposition

The distribution has two spaced maxima and the right one is very narrow. As alteration the quantizing step leads to quantizing grid scaling, a significant amount of values that corresponds to the distribution mode, laying far from zero, changes the accuracy of approximation by their recovery levels simultaneously. This may lead to an unpleasant effect of unpredictable effectiveness loss at individual quantizing step values. An example of such phenomenon is depicted in Fig. 2.

Shifting the quantizing grid origin prior to quantization execution is expected to suppress the arising oscillations. The dependency of the output data volume and the reconstructed image quality characteristics on the grid shift value at a fixed quantizing step is shown in Fig. 3. The following computation is required to determine the optimal shift value.

Consider the quantization error distribution. Let ε_n denote the deviation of the nth coefficient from its original value:

$$\varepsilon_n = x_n - \hat{x}_n = x_n - q(x_n), n = \overline{1, N}, \tag{4}$$

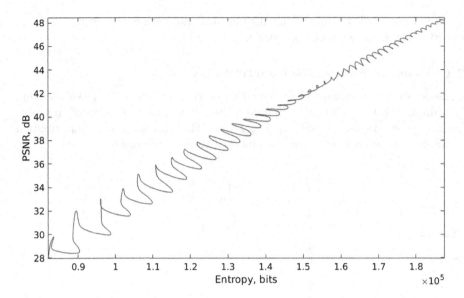

Fig. 2. Curve entropy-PSNR for LL band of the DWT coefficients of the image *kiel* when applying the standard quantizing technique. Other bands were neither subjected to quantization nor taken into account when computing the entropy

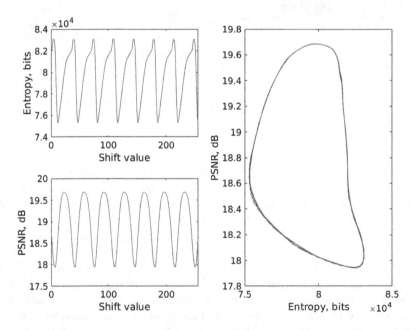

Fig. 3. Alteration of the metrics of the image *kiel* when varying the grid shift value at the quantizing step 100. The shift value is given in terms of the original image pixels units (range from 0 to 255). Entropy computation takes the LL band into account. Other bands are neither quantized nor considered

Then the error distribution can be written as follows:

$$f_{\Delta X}(\varepsilon) = \sum_{n=1}^{N} \delta(\varepsilon - \varepsilon_n). \tag{5}$$

It may be shown that in the case of linear quantization it can be expressed in the following way:

$$f_{\Delta X}(\varepsilon) = \sum_{j=-\infty}^{+\infty} f_X(\varepsilon + d_0 + j\Delta d), \varepsilon \in \left[-\frac{\Delta d}{2}, \frac{\Delta d}{2}\right). \tag{6}$$

Let us express the error mean square via the error distribution:

$$\sigma^2 = \sum_{n=1}^{N} \Delta x_i{}^2 = \int_{-\frac{\Delta d}{2}}^{+\frac{\Delta d}{2}} \varepsilon^2 f_{\Delta X}(\varepsilon) d\varepsilon \tag{7}$$

It is evident that the more distant from zero the error distribution peak is, the more it contributes to the expression (7), therefore its relocation towards the origin is presumed to cause rise of PSNR. It emerges from formula (6), that for single-mode images the image mode is assumed to initiate the child mode in error distribution, if the mode is narrow enough and the quantization grid is not too dense. Consequently, shifting the distribution by the mode value may lead to relocation the error peak towards zero.

It is necessary to calculate the smoothed distribution in order to determine the mode of a DWT band, that turns out to be labour-consuming operation. However, taking into account the similarity of the LL band distribution to the original one allows to replace it with the original distribution mode search. This approach is more convenient as the original distribution has a discrete structure. Using the arithmetic average and the median is convenient because of their low computation complexity.

Thus, one of the ideas of this study proposes shifting the origin of the quantizing grid applied to the DWT coefficients by a value depending on the image statistical properties.

2.2   Block-Wise Shift of the Quantization Grid

The previous idea development requires considering separate blocks of the image. The frequency distribution may vary sharply from one block to another one. Figure 5 shows the distributions for the blocks marked in Fig. 4. A proposition can be made that independent quantizing of individual blocks may allow to take into account local distribution features and therefore minimize the quantization error as well as reduce the spread of the quantized coefficients. The latter becomes a result of extraction an array of shifts determined separately for each block, thus the quantized coefficients give data about local features inside a block.

The arithmetic average and the median value utilization can be extended to the proposed method, meanwhile the mode computation appears to be complex, as it can't be found through the original image mode calculation because of low compliance between the blocks of the image and the ones of the band, therefore it requires a smoothed distribution evaluation within each block. The shift value can also be determined by exhaustive search in a specified set of values. This method got an improved modification based on FFT computation which reduces the algorithmic complexity with approximate keeping the compression effectiveness.

2.3   Null-Zone Extension

Another interesting idea of modifying the quantizing algorithm is the extension of the null-zone, i.e. the central zone of the quantizing grid, relatively to the other zones in all bands but LL [11,12]. An example of a grid with the central zone extended 1.5 times is depicted in the Fig. 6.

Fig. 4. Blocks of the image *kiel* containing ships (the blue square) and sky (the red square) (Color figure online)

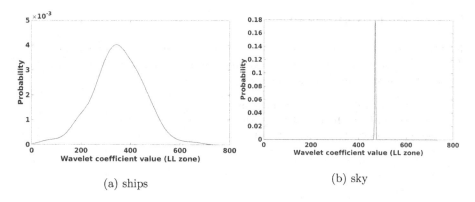

(a) ships

(b) sky

Fig. 5. Frequency distribution of two blocks of the image *kiel*

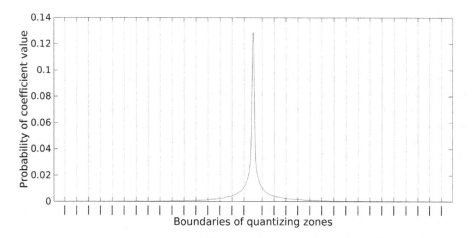

Fig. 6. Quantizing zones boundaries disposition imposed upon distribution of ML zone coefficients of image *kiel* at quantizing step 12 and extension coefficient 1.5

This idea may appear to contradict the Lloyd-Max algorithm, that contracts the zones close to the zero point, where the distribution maximum is reached. As it was said, the Lloyd-Max algorithm implements optimization on PSNR without regard to entropy change. While constricting the zones beside the maximum, it makes the distribution between the levels more uniform, raising the entropy. On the contrary, the central zone extension increases the part of values corresponding to one recovery level and, as a consequence, decreases the uncertainty. Meanwhile the recovery error is expected to rise slightly. The other quantization zones don't change their width though they are shifted, therefore the arising distortions should be concerned with the central zone.

3   Practical Implementation

Results of application the following quantizing techniques to coefficients arrays of each frequency zone of wavelet decomposition for an individual image were investigated in the frame of the research:

- simple constant shift of the quantizing grid;
- block-wise shift of the quantizing grid;
- null-zone extension of the quantizing grid.

Three-channel filterbank 23/23/23 — 13/13/13 applied in [4] was chosen for wavelet decomposition. The investigated quantization techniques were tested on an array of images of different resolutions (from 4CIF to Full HD) [13] using MATLAB modelling environment. Metric "entropy-PSNR", which characterizes the ratio of the output data volume and the quality of the reconstructed signal, was applied to estimate the effectiveness of the proposed methods. Testing was carried out for the luma component.

4   Results

Using simple shift of the quantization grid outside the LL zone doesn't lead to profit, as the other frequency subbands have distributions condensed near the zero point that makes quantizing grid shift unreasonable. On the contrary, the method is adequate for LL zone quantization, and the most significant result is reached when using the distribution main mode as the shift value. Nevertheless, division the LL band into blocks and individual shift value selection for each block lead to even better result, as various areas mainly have completely different distributions.

For the block-wise shift technique for LL zone quantization the best results are obtained when using division into square blocks of small size (5×5) and choosing the shift value equal to the median of the distribution inside the block. The conforming results are depicted in Fig. 7. Division into blocks of larger size gives profit as well, though it is not as significant. The result of shift by the value of the block main mode is practically identical to the one of shift by the value of the median, however, the computation of the latter is simpler, so its application is preferable. It is also necessary to take into account that the considered method should be applied to deepest levels of wavelet decomposition (the given examples use three of them), namely their LL part, as an additional level of the DWT gives a better result.

All zones (except LL) can be compressed by expansion the central quantization zone on the first and second transform levels. Figure 8 demonstrates the results arising owing to extension the null-zone by various factors. The best results are mostly obtained with the extension parameter set equal to 1.5. Null-zone extension is meaningless for deeper levels of the transform.

Issues of effective combination of the techniques proposed in this study are of the greatest interest. A typical example of testing results for the techniques presented in the article is depicted in Fig. 9.

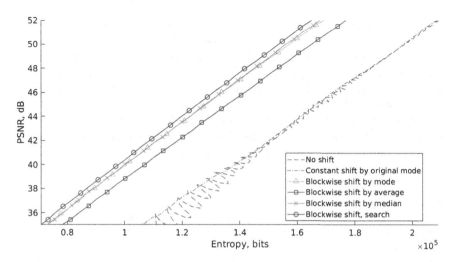

Fig. 7. Curve entropy-PSNR at quantizing the LL band of DWT coefficients of the image *kiel* when shifting the coefficient values by their various statistics in blocks 5×5. Only LL zone coefficients were taken into account when computing the entropy

As the results of this research show, the block-wise shift can be used for compressing the low-frequency image components (Fig. 7), while the null-zone extension can be used to compress the high-frequency image components (Fig. 8). Nonetheless, the reckless combination of the proposed ideas may lead to unexpected results: the application of some described methods may stop giving a gain

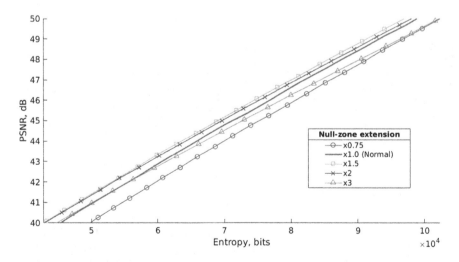

Fig. 8. Curve entropy-PSNR at quantizing the ML band of DWT coefficients of the image *kiel* when extending the null-zone. Other bands were neither subjected to quantization nor taken into account when computing the entropy

in terms of "entropy - PSNR" or even lead to losses by taking into account all decomposition components and quantizing them in the same way (as it shown in Fig. 9, where a separate null-zone extension at second level of image decomposition lead to loss in "entropy - PSNR" metric in comparison with simple linear quantization without it). This is due to the fact that the extension of the null zone decreases the PSNR of the restored image, although it decreases the entropy of the output stream. Whereas the "entropy - PSNR" curve (as in Figs. 7 and 8) after null-zone extension in a particular frequency subband remains to the left of the original one (without null-zone extension), the quality of the high-frequency component decreases with respect to the LL zone of the same decomposition level and to subbands of other levels for each curve point corresponding to a selected quantization step.

Thus, with the extension of the null-zone of quantization grid for high-frequency components at one transform level it becomes necessary to amend the quantization step for high-frequency subbands at other levels to achieve optimality in terms of the "entropy - PSNR" ratio. As it was mentioned in [11], if linear quantization is performed for all frequency subranges at the same transform level then it is optimal to apply quantization with an equal step for these subranges; therefore, it makes no sense to introduce individual corrections for each wavelet subband at the same level. Meanwhile, a better joint implementation of the set of proposed methods may require introducing individual amendment coefficients to quantizing steps of different levels of the transform.

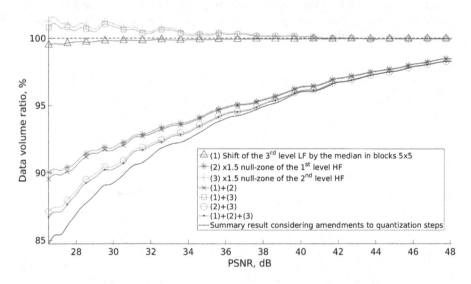

Fig. 9. The volume of the quantized data for the whole set of wavelet subbands for image *kiel* when applying individual techniques of quantization improvement for three-channel decomposition. The last curve introduces improving amendments to ratios of quantizing steps of different levels of the DWT. Quantization and consideration are applied to all frequency zones. The data volume when applying linear quantization uniformly across all the subbands is taken as 100%. LF - low frequency image components; HF - high frequencies image components.

Results of such implementation is shown in Fig. 9 - it gives better results in terms of "entropy - PSNR" than simple combination of all proposed techniques (without quantization step amendments on different levels of image decomposition). Nonetheless, it should be considered that the evaluation of individual amendment coefficients to quantizing steps of each levels of the transform requires more complex calculations.

5 Conclusion

The relationship between quantization parameters for wavelet coefficients and their statistical properties was investigated within the framework of this paper. The study also proposed new techniques for wavelet coefficients quantization, including simple and block-wise shift of the quantization grid, as well as the idea of null-zone extension for the quantization grid.

The obtained results indicate the possibility of effective combination of the proposed techniques allowing to increase the compression ratio for wavelet decomposition of the image. The best result can be obtained by application the three-level DWT together with the block-wise shift technique in the most low-frequency zone (LL zone of the third level) and null-zone extension in all (high-frequency) subbands at the first and second levels of the transform.

A promising direction of development of the proposed methods for quantizing wavelet coefficients is presented by the study of the dependence of the approximating curves parameters that associate the required level of image reconstruction quality with the quantization parameters on the statistical characteristics of this image.

References

1. Taubman, D.S., Marcellin, M.W.: JPEG2000: standard for interactive imaging. Proc. IEEE **90**, 1336–1357 (2002)
2. Onthriar, K., Loo, K.K., Xue, Z.: Performance comparison of emerging Dirac video codec with H.264/AVC. In: International Conference on Digital Telecommunications (ICDT 2006), p. 22 (2006)
3. ISO/IEC 15444-3: Information technology - JPEG 2000 image coding system - Part 3: Motion JPEG 2000, 2002 (2002)
4. Bystrov, K., Dvorkovich, A., Dvorkovich, V., Gryzov, G.: Usage of video codec based on multichannel wavelet decomposition in video streaming telecommunication systems. In: Vishnevskiy, V.M., Samouylov, K.E., Kozyrev, D.V. (eds.) DCCN 2017. CCIS, vol. 700, pp. 108–119. Springer, Cham (2017). https://doi.org/10.1007/978-3-319-66836-9_10
5. Sonal, D.K.: A study of various image compression techniques. RIMT-IET, Hisar COIT **8**, 91–102 (2007)
6. Lloyd, S.P.: Least squares quantization in PCM. IEEE Trans. Inf. Theory **IT–2**(2), 129–137 (1982)
7. Max, J.: Quantizing for minimum distortion. IRE Trans. Inf. Theory **IT–6**, 7–12 (1960)

8. Andries, B., Lemeire, J., Munteanu, A.: Optimized quantization of wavelet subbands for high quality real-time texture compression. In: 2014 IEEE International Conference on Image Processing (ICIP), pp. 5616–5620, Paris (2014). https://doi.org/10.1109/ICIP.2014.7026136

9. Trong, N.D.: Optimization of quantization method for a wavelet-based video codec (in Russian). In: DSPA, vol. 3, pp. 44–48 (2018)

10. Nam, D.T., Gryzov, G.Y., Dvorkovich, A.V., Dvorkovich, V.P.: Nonlinear quantization method for wavelet-based video codec. In: 2018 Engineering and Telecommunication (EnTMIPT), vol. 2018, pp. 25–29, Moscow, Russia (2018). https://doi.org/10.1109/EnT-MIPT.2018.00013

11. Vorobiev, V., Gribunin, V.: Theory and practice of wavelet transform (in Russian). Publishing House of the Military University of Communications, Saint-Petersburg (1999)

12. Martínez-Rach, M.O., Granado, O.L., Peral, P.P., Malumbres, M.P.: Impact of dead zone size on the rate/distortion performance of wavelet-based perceptual image encoders. In: 2016 5th International Conference on Multimedia Computing and Systems (ICMCS), pp. 65–70, Marrakech (2016). https://doi.org/10.1109/ICMCS.2016.7905671

13. Tested images. https://github.com/tovoidcast/dwt_test_benchmark

The Use of Intra Prediction Method
in Wavelet-Based Video Coding Systems

Gleb Verba, Kirill Bystrov$^{(\boxtimes)}$, Viktor Dvorkovich, and Gennady Gryzov

Moscow Institute of Physics and Technology, Institutsky lane 9, Dolgoprudny, Russia
{verba,kirill.bystrov}@phystech.edu, v.dvorkovich@mail.ru,
gryzov@gmail.com

Abstract. Modern image and video compression technologies include both lossless compression methods, such as entropy coding, Inter-frame and Intra-frame coding, and lossy compression methods, such as discrete orthogonal transforms with quantization. All these techniques are actively applied in video codecs based on the H.264 and H.265 standards.

Discrete wavelet transform (DWT) is one of the most perspective versions of discrete orthogonal transforms. Both Intra-frame coding and wavelet decomposition of images allow to reduce the volume of transmitted data, but they are not used together in video coding systems. The combined usage of these methods seems to be a promising approach in terms of visual data compression.

Thereby the aim of this research is to develop a new technique based on the combination of Intra-frame coding and the DWT and to test the applicability of the proposed method in the image compression tasks.

The effectiveness of various implementations of the proposed algorithm, including those based on contexts and using several levels of wavelet decomposition of images, was evaluated in the study.

Keywords: Intra Prediction · Intra-frame coding · Discrete wavelet transform · Video coding · Data compression · Context-based coding

1 Introduction

Intra Prediction is widely used in video coding. It is a transformation that allows to reduce the data entropy at the input of entropy encoder by coding only residual values obtained using the prediction schemes. H.264 [1] and H.265 [2] are the examples of standards which include this technique.

Discrete wavelet transform (DWT) is another efficient approach to image coding. It is based on spatial-frequency analysis of input data and is also applied in video coding standards [3–5].

Combining these two methods opens up new prospects for video compression systems and offers a novelty because, although a number of related studies have been carried out [6,7], no video coding system performing such concept has been implemented yet.

© Springer Nature Switzerland AG 2019
V. M. Vishnevskiy et al. (Eds.): DCCN 2019, CCIS 1141, pp. 77–88, 2019.
https://doi.org/10.1007/978-3-030-36625-4_7

Thus, the purpose of this study is to attempt to combine Intra Prediction and DWT techniques and to test the applicability of the proposed method in the image compression tasks.

2   Intra-frame Prediction in the Space of Wavelet Coefficients

Intra Prediction method matches each coefficient $s(i;j)$ and its predicted value $\hat{s}(i;j)$, which is determined by a prediction scheme using some spatially nearby reference data. In practice, a frame is divided into blocks, for which common reference values are used and which occupy the line above and/or the column to the left of the current block. The array of prediction errors $e(i;j)$ is fed to the input of entropy encoder. This approach [8–10], which is used in H.264 [1] and H.265 [2] standards, is common for most video codecs. Block-based intra-frame prediction technique was significantly improved during last decade by reducing its computational complexity [11,12] and by increasing the compression ratio for prediction modes [13,14].

The forward and inverse prediction processes for wavelet coefficients are described by the following system:

$$\begin{cases} e(i;j) = s(i;j) - \hat{s}(i;j) \\ s(i;j) = e(i;j) + \hat{s}(i;j) \end{cases} \tag{1}$$

Wavelet coefficients have a special correlation pattern: in zones, which are low-frequency in one of dimensions (refer to them as *semi-low-frequency*), the lines with horizontal or vertical orientation (depending on the specific frequency zone) are observed near the peculiarities of the original image—these are lines of the same sign with similar moduli (Fig. 1). This leads to the assumption that prediction with appropriate orientation will also be effective for *semi-low-frequency* subbands. The low-frequency subband is similar to the original image—that allows to suppose that Intra Prediction may be effectively applied to it.

2.1   Intra Prediction for Absolute Values of Wavelet Coefficients

As oscillations are characteristic of the high-frequency zones, it is more plausible to expect approximate constancy of the magnitude of these oscillations rather than their values proper, thus this paper proposes a new technique carrying out division the initial data into a field of moduli and a field of signs (the latter of is defined only for non-zero values in order to avoid redundancy introduction) prior to prediction. Note that using this technique requires taking into account the field of sign as well.

The frame of sign of DWT coefficients after quantization is depicted in the Fig. 2. The values oscillate by zero and there are horizontal or vertical lines of constant sign in *semi-low-frequency* subbands.

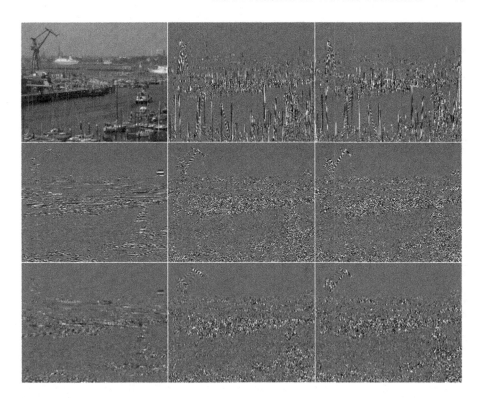

Fig. 1. Wavelet decomposition of image *kiel* by three-channel filter bank

2.2 Context-Based Coding for Residual Frame of Intra-predicted Wavelet Coefficients

Context-based coding is an entropy coding technique which is based on computing conditional probabilities of symbols depending on preceding symbol sequences, which allows to improve encoding efficiency. Since the frequency subbands of wavelet-decomposed images have specific correlation between their coefficients, their statistical properties can be used for forming contexts of wavelet coefficients. This approach finds an application in a range of papers on compression, including wavelet-based compression [15].

Each wavelet coefficient is assigned a specific context defined by a set of neighboring coefficients. A table of conditional probabilities of symbols appearance is constructed for each context and is applied in subsequent entropy coding. Thus, each coefficient is coded using one of the probability tables depending on the context of the current coefficient. Compression is achieved because the conditional entropy in the general case is not greater than unconditional one and strictly less due to the presence of correlation. The complexity of the context is determined by the number of adjacent elements used, as well as the mapping of the set of their values to the set of contexts.

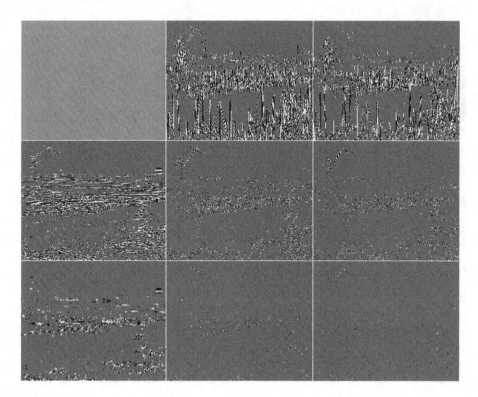

Fig. 2. Field of sign of DWT decomposition of luma component of image *kiel* at quantization step 33. Grey points correspond to zero quantized values, white points—to positive ones, black points—to negative ones

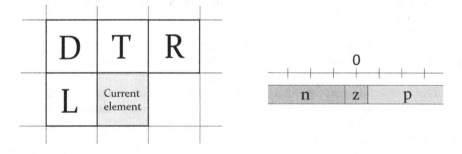

Fig. 3. Context model used in study: four neighboring coefficients (left L, top-left D, top T and top-right R) and their values range: negative n, zero z and positive p values.

In the framework of this paper it is suggested to apply the following model for context formation (Fig. 3).

3   Practical Implementation

Prediction was applied to frames of wavelet coefficients after their quantization. The coder scheme is depicted in Fig. 4.

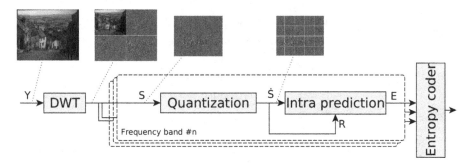

Fig. 4. Coder scheme: Y - luma component of original image, S - coefficients of frequency subband of wavelet-decomposed image, $\dot{S}$ - quantized wavelet coefficients of subband, E - array of prediction errors, R - reference data (the row above or/and the column to the left) for Intra Prediction in subband

Prediction process is preceded by picking out reference values taking place in the upper row or/and the left column. The remaining part is divided into blocks of size 4×4, 8×8, 16×16 or 32×32.

The following prediction modes, defined in the H.264 recommendation, were considered in the framework of the paper (Table 1): the 8 directed modes and the DC-mode, that had been defined for luma blocks of sizes 4×4 and 8×8, and the planar prediction mode, that had been defined for luma and chroma blocks of size 16×16. Also adative mode was use for chosing optimal mode for each block. All the modes were extended to all working block sizes.

The efficiency of applying intra prediction to a block was estimated using a certain distortion metric. If applying intra prediction didn't lead to the metric value growth, the block was skipped without prediction. Either a mode was applied to the entire frame, or the adaptive mode was used: each block passed each prediction mode alternately, whereupon the mode minimizing the chosen metric was determined. Using adaptive mode requires transmission of the vector of values denoting the modes chosen for each block. The used metrics are: SAD (sum of absolute differences), SSD (sum of squared differences), SATD (SAD for Hadamard transform of the prediction remainder), *SAHD (SATD modification: SAD for Haar transform of the prediction remainder), kNN [16].

Table 1. Tested intra prediction modes.

Mode index	Prediction mode
0	Vertical
1	Horizontal
2	DC
3	Diagonal down-left
4	Diagonal down-right
5	Vertical-right
6	Horizontal-down
7	Vertical-left
8	Horizontal-up
9	Planar
10	Adaptive

Three-channel filterbank 23/23/23—13/13/13 applied in [17] was chosen for wavelet transform implementation. The investigated quantization techniques were tested on an array of images of different resolutions (from 4CIF to Full HD) [18] using MATLAB modelling environment. Metric "entropy-PSNR", which characterizes the ratio of the output data volume and the quality of the reconstructed signal, was applied to estimate the effectiveness of the proposed methods. The entropy calculation did not take into account the data necessary to represent the contexts.

4   Results

The use of intra prediction in the domain of wavelet coefficients of LL subband provides guaranteed and noticeable gain in "entropy – PSNR" (Fig. 5). Adaptive mode has a significant advantage over single modes and the smallest block size is optimal. Nonetheless, the execution of an additional wavelet decomposition step of the same subband is more efficient (Fig. 6).

At the same time the benefit of intra prediction usage for *semi-low-frequency* subbands is approximately a few percent in terms of initial entropy of the subband, and only for the corresponding directed mode and adaptive mode (Fig. 7). Use of intra prediction in high-frequency subbands only decreases the compression ratio (Fig. 8).

Division into modulus and sign causes almost no change of the ratio "entropy–PSNR" in any frequency band of the tested images (Fig. 9).

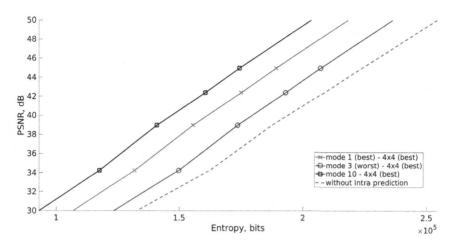

Fig. 5. Entropy-PSNR ratio of LL subband of *kiel* image in case of DWT with intra prediction. The size of the block that provides the best compression is chosen as a parameter for each curve and is included in the legend

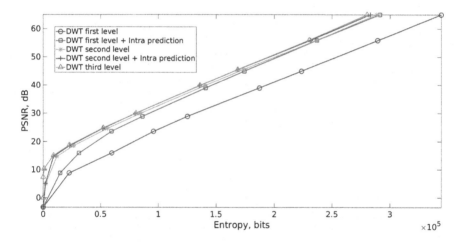

Fig. 6. Entropy-PSNR ratio of LL subband of *kiel* image in case of DWT with intra prediction. The block size is 4×4

The best results for the proposed method were obtained with a preliminary recalculation of the entropy of the frequency subbands, taking into account the contexts of the wavelet coefficients: the use of intra prediction becomes more efficient for LL zone (Fig. 10), while for other subbands the entropy decreases significantly, both with and without the use of prediction (Fig. 11). Contexts consideration increases the compression ratio by up to 25%, depending on the frequency subband and the image.

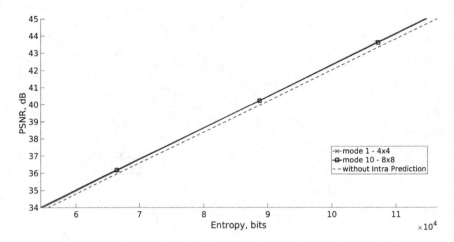

Fig. 7. Entropy-PSNR ratio of LM subband of *kiel* image in case of DWT with intra prediction. The size of the block that provides the best compression is chosen as a parameter for each curve and is included in the legend

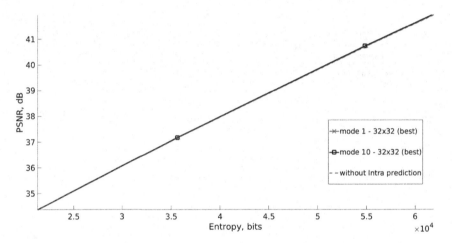

Fig. 8. Entropy-PSNR ratio of MM subband of *kiel* image in case of DWT with intra prediction. The size of the block that provides the best compression is chosen as a parameter for each curve and is included in the legend

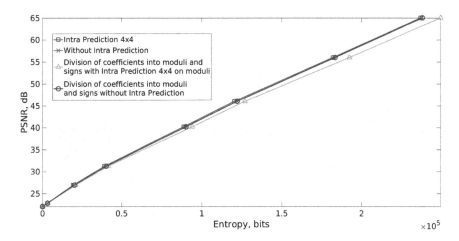

Fig. 9. Curves "entropy–PSNR" when using no transforms, after simple prediction, after division into moduli and signs and after prediction in the domain of the derived moduli for LM band of image *kiel*. The curve for the case of moduli extraction coincides with the one for the case of no transforms

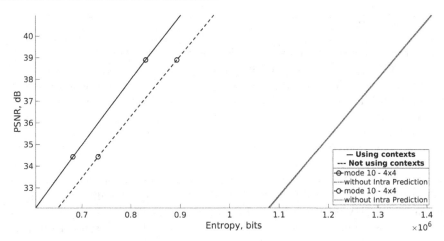

Fig. 10. Entropy-PSNR ratio of LL subband of *kiel* image in case of DWT intra prediction with context usage and without it. Only best prediction modes are shown in graph. The block size (4×4) providing the best compression is chosen

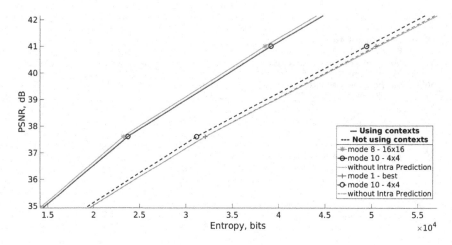

Fig. 11. Entropy-PSNR ratio of LH subband of *kiel* image in case of DWT intra prediction with context usage and without it. Only best prediction modes are shown in graph. The block size providing the best compression is chosen

5   Conclusion

Application of Intra Prediction modes in the way they are used in the H.264/H.265 standards to the result of the wavelet decomposition of the image is unjustified: in spite of entropy reduction, it is still more efficient to execute additional DWT for low-frequency coefficients (Fig. 6)—both in terms of the ratio of compression and quality of the reconstructed image and in terms of computational complexity; and for other frequency domains it does not guarantee an improvement in compression (Fig. 7and 8). This is explained by the fact that H.264/H.265 Intra Prediction modes are adapted to be applied to the components of YUV image, meanwhile the correlation between the wavelet coefficients for all frequency subbands (except LL) is completely different than the correlation in YUV. Conversely, the LL zone represents the approximation of the original image, which allows to reduce the entropy with such type of the Intra Prediction.

Modeling (Fig. 9) demonstrates that extracting the moduli of the wavelet coefficients without prediction gives a result, identical to the absence of transformations, while the prediction for absolute values of coefficients gives a negative result. The first observation may be explained simply: if the distribution is initially approximately symmetrical about zero, then the uncertainty of the value is reduced by 1 bit when turning to absolute values, since the probability to meet each value becomes twice as high. This difference of 1 bit is compensated for by introducing the sign that is positive or negative with approximately equal probability. The second observation demonstrates the fallacy of the assumption that the coefficient moduli are correlated.

The results of this study show evidently that the use of contexts of wavelet coefficients not only reduces the initial entropy of a frequency subband of the wavelet-decomposed image (Fig. 11), but also improves the Intra Prediction results in LL zone (Fig. 10). Thereby it is proved that the use of contexts in wavelet video coding is a promising approach, which, however, requires more thorough research. One of the possible ways to improve the efficiency of the proposed method is to select a context model simultaneously with the choice of prediction mode, as well as taking into account the implementation features of the entropy coder.

References

1. Wiegand, T., Sullivan, G.J., Bjontegaard, G., et al.: Overview of the H.264/AVC video coding standard. IEEE Trans. Circ. Syst. Video Technol. **13**(7), 560–576 (2003)
2. Sullivan, G.J., Ohm, J.-R., Han, W.-J., et al.: Overview of the high efficiency video coding (HEVC) standard. IEEE Trans. Circ. Syst. Video Technol. **22**(12), 1649–1668 (2012)
3. Taubman, D.S., Marcellin, M.W.: JPEG2000: standard for interactive imaging. Proc. IEEE **90**, 1336–1357 (2002)
4. ISO/IEC 15444–3:2002, Information technology - JPEG 2000 image coding system - Part 3: Motion JPEG 2000 (2002)
5. Onthriar, K., Loo, K.K., Xue, Z.: Performance comparison of emerging Dirac video codec with H.264/AVC. In: International Conference on Digital Telecommunications (ICDT06), p. 22 (2006)
6. Mai, Z., Nasiopoulos, P., Ward, R.: A wavelet-based intra-prediction lossless image compression scheme. In: 2009 Digest of Technical Papers International Conference on Consumer Electronics, pp. 1–2, Las Vegas (2009)
7. Elarabi, T., Sammoud, A., Abdelgawad, A., Li, X., Bayoumi, M.: Hybrid wavelet - DCT intra prediction for H.264/AVC interactive encoder. In: 2014 IEEE China Summit and International Conference on Signal and Information Processing (ChinaSIP), pp. 281–285, Xi'an (2014)
8. Nan, Z., Baocai, Y., Dehui, K., Wenying, Y.: Spatial prediction based intra-coding [video coding]. In: 2004 IEEE International Conference on Multimedia and Expo (ICME) (IEEE Cat. No.04TH8763), vol. 1, pp. 97–100, Taipei (2004). https://doi.org/10.1109/ICME.2004.1394134
9. Joint Video Team (JVT) of ISO/IEC MPEG and ITU-T VCEG: New Intra Prediction using Intra-Macroblock Motion Compensation, JVT-C151 (2002)
10. Lainema, J., Bossen, F., Han, W., Min, J., Ugur, K.: Intra coding of the HEVC standard. IEEE Trans. Circ. Syst. Video Technol. **22**(12), 1792–1801 (2012). https://doi.org/10.1109/TCSVT.2012.2221525
11. Yi, H., Qin, H.: The optimization of HEVC intra prediction mode selection. In: 2017 4th International Conference on Information Science and Control Engineering (ICISCE), pp. 1743–1748, Changsha (2017). https://doi.org/10.1109/ICISCE.2017.364
12. Adireddy, R., Palanisamy, N.K.: Effective approach to reduce complexity for HEVC intra prediction in inter frames. In: 2014 Twentieth National Conference on Communications (NCC), pp. 1–5, Kanpur (2014). https://doi.org/10.1109/NCC.2014.6811337

13. Sanchez, G., Fernandes, R., Agostini, L., Marcon, C.: DCDM-intra: dynamically configurable 3D-HEVC depth maps intra-frame prediction algorithm. In: 2018 25th IEEE International Conference on Image Processing (ICIP), pp. 1782–1786, Athens (2018). https://doi.org/10.1109/ICIP.2018.8451620
14. Agarwal, P., Jiang, M., Ling, N., Zheng, J., Zhang, P.: Enhanced intra prediction mode coding by using reference samples. In: IEEE International Workshop on Signal Processing Systems (SiPS), 296–299, Cape Town (2018). https://doi.org/10.1109/SiPS.2018.8598421
15. Jiang, X., Song, B., Zhuang, X.: An enhanced wavelet image codec: SLCCA PLUS. In: International Conference on Audio, Language and Image Processing (ICALIP) (2018)
16. Boltz, S., Wolsztynski, E., Debreuve, E., et al.: A minimum-entropy procedure for robust motion estimation. In: International Conference on Image Processing (2006)
17. Bystrov, K., Dvorkovich, A., Dvorkovich, V., Gryzov, G.: Usage of video codec based on multichannel wavelet decomposition in video streaming telecommunication systems. In: Communications in Computer and Information Science, Distributed Computer and Communication Networks. DCCN 2017, vol. 700, pp 108–119 (2017). https://doi.org/10.1007/978-3-319-66836-9_10
18. Tested images. https://github.com/tovoidcast/dwt_test_benchmark

Performance of Transport Connection with Selective Failure Mode When Competing for Throughput of Data Transmission Path

Pavel Mikheev$^{(\boxtimes)}$, Pavel Pristupa, and Sergey Suschenko

National Research Tomsk State University, Lenina str., 36, 634050 Tomsk, Russia
doka.patrick@gmail.com, ssp.inf.tsu@gmail.com

Abstract. An indicator model of transport connection is proposed for selective failure mode when competing among different subscribers connections for throughput of transmission path. The indicator of competition is the queue of competitive data flows in transit nodes of transport connection with specified parameters. The analysis of available throughput in different conditions when competing is carried out.

Keywords: Transport protocol · Selective failure mode · Competition for resources · Throughput · Protocol parameters · Round-trip delay · Mathematical model · Markov chain

1 Introduction

The most important operational characteristic of a subscriber connection controlled by a computer network transport protocol is its throughput. This indicator is largely determined by the intensity of external flows relative to this connection, which has at least a part of common route with it. The main indicator of "external" load on the path in which the studied transport connection is laid is the size of the queues ahead of protocol data blocks of analyzed connection in transit nodes. Monitoring such an indicator allows us to evaluate the distribution of queue lengths in transit nodes from external network streams in regard to analyzed connection and to use when calculating the operational characteristics of the connection and the choice of protocol parameters for the communication time between a given pair of subscribers. Known models of asynchronous control procedures of a separate data link and transport protocol [1–7] do not allow us to take into account the load on shared network resources which is provided by the neighboring with other virtual connections, aggregated on different sections of the path in separate links of the route of a given subscriber connection, and present itself as "External" queues in transit nodes. The study of data transmission process in a loaded transport connection [8,9] was carried

Performed under state assignment No. 2.1718.2017/4.6.

out with significant restrictions on the values of the protocol parameters and characteristics of data transmission path. The paper proposes a mathematical model of transport connection controlled by transport protocol in the selective failure mode, which takes into account, apart from the distortion factor in the forward and reverse data transmission paths and retransmission mechanisms, caused by distortions and the timeout of non-reception of response from the recipient of the information flow, and also non-zero queues lengths from "external" inter-subscriber connections for end-to-end timeout durations with interval and below restrictions.

2   Indicator Data Transmission Path Model

Let us consider the exchange between subscribers connected by a multi-link data path. Assume that the following assumptions are true: The nodes of the path are connected by duplex communication channels having the same speed in both directions. The length of the tract, expressed in the number of hop is equal to D_n. The return channel, on which confirmation is delivered to the sender about the validity of the reception of sequence of data segments, has a length D_o. The probabilities of segment distortion in the communication channel are specified for the forward $R_n(d)$, $d = \overline{1, D_n}$ and reverse—$R_o(d)$, $d = \overline{1, D_o}$ the transmission directions of each segment of the hop is given. Then the reliability of transmission of data segments along the path from the source to the addressee and back will be $F_n = \prod_{d=1}^{D_n}(1 - R_n(d))$; $F_o = \prod_{d=1}^{D_o}(1 - R_o(d))$. The processing time of segments in path nodes is the same. Interacting subscribers have an unlimited flow of segments for transmission, and the exchange is carried out by segments of the same length. The recipient's confirmation of the validity of received data is transferred in the segments of the counter flow. We believe that the retransmission of segments is organized in accordance with the selective rejection procedure [1]. We also assume that the loss of segments due to the absence of buffer memory at the nodes of the path does not occur. Probability function is given b_n, $n = \overline{0, N}$ that each segment from the flow of analyzed connection in transit nodes will meet a queue of size $n \leq N$, where N is the maximum queue size determined by the capacity of buffer pools of transit nodes. We will call cycle time t necessary for output of a segment to a line. The cycle is determined by the sum of the segment output time to the line, the signal distribution time in the communication channel and processing time of the segment by the receiving node. The timeout S, expressed in duration t, runs before the start of transmission of the first segment of sequence and is fixed for all segments within the window width. We assume that the size of the controlled protocol window is determined by the value of W, and $S > W$—sets the duration of the timeout for waiting for confirmation of validity of data delivery. It is obvious that the sum of lengths of forward and reverse data paths $D = D_n + D_o$ can be interpreted as the duration of the round-trip delay the unloaded path, expressed in cycles t. After next segment is transferred, protocol copies it to the queue of transmitted but not confirmed data and starts a timeout. As soon as the queue size becomes equal to the width of the window W, the control protocol suspends transmission while waiting for the acknowledgement or the expiration of timeout S for confirmation. Upon receiving the confirmation, the segments that reached the addressee without distortion are removed from the queue. When S timeout expires, the corresponding segment is retransmitted and timeout starts again. Then the time of confirmation by the sender of end-to-end acknowledgement is distributed according to the geometric law with the parameter F_o and the duration of the sampling cycle t. The operation of virtual connection controlled by transport protocol in a loaded multilink data transmission path with segment queues before sending data or confirmations can be described by a markovized process of the dynamics of a queue of transmitted but not confirmed segments in which the queue size ahead of the forward or reverse data flow of the test connection is additional variable of Markov process. In the state of Markov chain (i, n) the source sent a sequence of size $i - n$ segments, which in the process of transfer in one of the links met a queue with length of n segments. The coordinates $i = \overline{0, W + n}$, $n = \overline{0, N}$ of the states of Markov chain correspond to the number of segments transmitted but not confirmed by the recipient and the time from the beginning of the transmission of the sequence, while the values $i = \overline{W + n + 1, S - 1}$, $n = \overline{0, N}$ correspond to the time during which the sender is not active and is waiting for confirmation of a receipt of the validity of transmitted sequence from the W segments. We define by $P(i, n)$, $i = \overline{0, S - 1}$, $n = \overline{0, N}$, the probabilities of the states of the Markov chain. Then the sequence of transmitted, but not confirmed data segments of considered virtual connection with a zero-length queue grows to the state of a Markov chain with coordinates $(D - 1, 0)$ with probability b_0. Further increase in size of this sequence occurs with probability $b_0(1 - F_o)$. In states (i, n), $i = \overline{D - 1 + n, S - 1}$, $n = \overline{0, N}$, it is possible for the sender to receive acknowledgement and depending on the delivery results, the sender transmits new segments (with a positive acknowledgement), or repeatedly—distorted. Since transmitted sequence of segments of the virtual connection under study may encounter a queue of non-zero length at any moment of transferring process (on the path of the sequence to the addressee or when transferring confirmation to the sender of information flow), the transition from state $(i, 0)$, $i = \overline{0, S - 2}$ to state (i, n), $i = \overline{0, S - 2}$, $n = \overline{1, N}$ occurs with probability b_n.

3   State Probabilities of Markov Chain

Let us define π_{in}^{jm} the transition probabilities of Markov chain, where (i, n) are the coordinates of the initial one, and (j, m) are the altered states of the chain. Then the dynamics of the process of transmitting information flow in the selective failure mode in loaded data transmission path can be set with the following values of transition probabilities:

$$
\pi_{in}^{jm} = \begin{cases}
b_0, & j = i+1, m = 0; \ i = \overline{0, D-2}, n = 0; \\
b_0(1 - F_o), & j = i+1, m = 0; \ i = \overline{D-1, S-2}, n = 0; \\
b_m, & j = i, m = \overline{1, N}; \ i = \overline{0, S-2}, n = 0; \\
b_0 F_o, & j = D-1, m = 0; \ i = \overline{D-1, W-1}, n = 0; \\
b_0 F_o, & j = W+D-2-i, m = 0; \ i = \overline{W, W+D-2}, n = 0; \\
b_0 F_o, & j = 0, m = 0; \ i = \overline{W+D-1, S-2}, n = 0; \\
1, & j = 0, m = 0; \ i = S-1, n = \overline{0, N}; \\
1, & j = i+1, m = n; \ i = \overline{0, D-2+n}, n = \overline{1, N}; \\
1 - F_o, & j = i+1, m = n; \ i = \overline{D-1+n, S-2}, n = \overline{1, N}; \\
F_o, & j = D-1, m = 0; \ i = \overline{D-1+n, W-1+n}, n = \overline{1, N}; \\
F_o, & j = W+n+D-2-i, m = 0; \\
& \quad i = \overline{W+n, W+n+D-2}, n = \overline{1, N}; \\
F_o, & j = 0, m = 0; \ i = \overline{W+n+D-1, S-2}, n = \overline{1, N}.
\end{cases}
\tag{1}
$$

The variety of solutions of system of equilibrium equations for Markov chain state probabilities is determined by relations between the protocol parameters W, S, the total path length D and the maximum length of queues N. Since timeout length must exceed the window width, and be no shorter than the round-trip delay ($S \geq D$), exceeding the waiting time in queues from protocol data blocks of corresponding traffic prior to transmission in transit nodes a wide variety of solutions for different areas of change in the values of the Protocol parameters and queue lengths are distinguished. Analysis of the transmission process in analytical form for arbitrary values of protocol parameters in the conditions of competition for network resources is possible only under the assumption that the "external" queues have a non-zero length ($b_0 = 0$).

4   Analysis of Transmission Process with Lower Restrictions on Duration of Timeout

Consider the transfer process for protocol parameters related to the total path length and the maximum queue size of the form inequalities $W \geq D$, $S \geq D + W + N - 1$. The system of equilibrium equations is written as follows:

$$
P(0,0) = F_o \sum_{n=1}^{N} \sum_{i=D+W+n-2}^{S-2} P(i,n) + \sum_{n=0}^{N} P(S-1,n);
\tag{2}
$$

$$
P(i,0) = F_o \sum_{n=1}^{N} P(D+W+n-2-i, n), \ i = \overline{1, D-2};
\tag{3}
$$

$$
P(D-1,0) = F_o \sum_{n=1}^{N} \sum_{i=D+n-1}^{W+n-1} P(i,0);
\tag{4}
$$

$$P(0,n) = b_n P(0,0), \; n = \overline{1,N}; \tag{5}$$

$$P(i,n) = b_n P(i,0) + P(i-1,0), \; i = \overline{1, D-1}, \; n = \overline{1,N}; \tag{6}$$

$$P(i,n) = P(i-1,n), \; i = \overline{D, D+n-1}, \; n = \overline{1,N}; \tag{7}$$

$$P(i,n) = (1 - F_o)P(i-1,n), \; i = \overline{D+n, S-1}, \; n = \overline{1,N}. \tag{8}$$

Let us find a solution to this system of equations. According to Eq. (7), we obtain: $P(i,n) = P(D-1,n)$, $i = \overline{D, D+n-1}$, $n = \overline{1,N}$, and from (8) we have: $P(i,n) = (1 - F_o)^{i-D-n+1}P(D-1,n)$, $i = \overline{D+n, S-1}$, $n = \overline{1,N}$. Taking into account these relations from (3), (4) for $i = \overline{1, D-1}$ we find:

$$P(i,0) = F_o(1 - F_o)^W \sum_{m=1}^{N} P(D-1,m), \; i = \overline{1, D-2},$$

$$P(D-1,0) = \left(1 - (1 - F_o)^{W-D+1}\right) \sum_{m=1}^{N} P(D-1,m).$$

Substituting the relations found in (6) with (5) taken into account, we obtain

$$P(i,n) = b_n \left[P(0,0) + (1 - F_o)^{W-i-1}\left(1 - (1 - F_o)^i\right) \sum_{m=1}^{N} P(D-1,m) \right],$$

$$i = \overline{1, D-2}, \; n = \overline{1,N},$$

$$P(D-1,n) = b_n \left[P(0,0) + \left(1 - (1 - F_o)^{W-1}\right) \sum_{m=1}^{N} P(D-1,m) \right], \; n = \overline{1,N}.$$

Hence, we successively express for arbitrary $n = \overline{1,N}$ through $P(D-1,n)$ the probabilities of states $P(D-1,m)$ $m = \overline{n+1, N}$:

$$P(D-1,n) = \frac{b_n}{1 - \left(1 - (1 - F_o)^{W-1}\right) \sum\limits_{m=1}^{N} b_m} \left[P(0,0) + \right.$$

$$\left. + \left(1 - (1 - F_o)^{W-1}\right) \sum_{m=n+1}^{N} P(D-1,m) \right], \; n = \overline{1,N}. \tag{9}$$

When $n = N$ from here we come to: $P(D-1,N) = \frac{b_N P(0,0)}{1-F_o^{W-1}}$. Substituting this relation into (9) for values of n from $N-1$ to 1, we recursively find functional expressions for the probabilities of states $P(D-1,n)$ through $P(0,0)$: $P(D-1,n) = \frac{b_n P(0,0)}{1-F_o^{W-1}}$, $n = \overline{1,N}$. Hence, from previously found relations, we finally obtain the probability distribution of states of Markov chain

$$P(i,0) = F_o \frac{P(0,0)}{(1-F_o)^i}, \; i = \overline{1, D-2};$$

$$P(D-1,0) = \frac{\left(1 - (1 - F_o)^{W-D+1}\right) P(0,0)}{(1-F_o)^{W-1}};$$

$$P(i,n) = \frac{b_n P(0,0)}{(1-F_o)^i}, \quad i = \overline{0, D-2}, \ n = \overline{1, N};$$

$$P(i,n) = \frac{b_n P(0,0)}{(1-F_o)^{W-1}}, \quad i = \overline{D-1, D+n-1}, \ n = \overline{1, N};$$

$$P(i,n) = \frac{b_n(1-F_o)^{i-D-n+1} P(0,0)}{(1-F_o)^{W-1}}, \quad i = \overline{D+n-1, S-1}, \ n = \overline{1, N},$$

and from the normalization condition we find the probability of the initial state

$$P(0,0) = \frac{F_o(1-F_o)^{W-1}}{1 + F_o(1+\bar{N}) + (1-F_o)^{W-D+1} - (1-F_o)^W - \sum\limits_{m=1}^{N} b_m(1-F_o)^{S-D+1-m}},$$

where $\bar{N} = \sum_{n=1}^{N} n b_n$.

Let us consider the solution found in a number of special cases. For deterministic return path $(F_o = 1)$, the space of significant states (i, n) forms a plane of an isosceles along coordinates i and n triangle $i = \overline{D-1, D-1+n}, \ n = \overline{0, N}$:

$$P(D-1, 0) = \frac{1}{2+\bar{N}};$$

$$P(i, 0) = \frac{b_n}{2+\bar{N}}, \quad i = \overline{D-1, D-1+n}, \ n = \overline{1, N}.$$

With an unlimited width of the window $(w = \infty)$ the states (i, n), $i = \overline{0, D-2}$, $n = \overline{0, N}$ are non-recurrent $(P(i, n) = 0)$ and probabilities of the state of Markov chain take the form

$$P(D-1, 0) = \frac{F_o}{1 + F_o(1+\bar{N})};$$

$$P(i, n) = \frac{b_n F_o}{1 + F_o(1+\bar{N})}, \quad i = \overline{D-1, D-1+n}, \ n = \overline{1, N};$$

$$P(i, n) = \frac{b_n(1-F_o)^{i-D-n+1} P(0,0)}{1 + F_o(1+\bar{N})}, \quad i \geq D-1+n, \ n = \overline{1, N}.$$

Let us consider the process of data transfer in conditions where the width of the window does not exceed the duration of the round-trip delay $(W \leq D)$, and the size of the timeout is limited from below $(S \geq D + W + N - 1)$. According to (1) the system of equilibrium equations given above will change as follows. Equations (2), (5), (8) will remain unchanged, (3)—true for $i = \overline{1, W-1}$, Eq. (4) will take the form $P(D-1, 0) = 0$, Eq. (6)—true for $i = \overline{1, W-1}$, $n = \overline{1, N}$, Eq. (7)—for $i = \overline{W, D-1+n}, \ n = \overline{1, N}$. The solution of the system of equilibrium equations is as follows

$$P(i, 0) = \frac{F_o P(0,0)}{(1-F_o)^i}, \quad i = \overline{1, W-1};$$

$$P(i, n) = \frac{b_n P(0,0)}{(1-F_o)^i}, \quad i = \overline{0, W-1}, \ n = \overline{1, N};$$

$$P(i,n) = \frac{b_n P(0,0)}{(1-F_o)^{W-1}}, \quad i = \overline{W-1, D+n-1}, \quad n = \overline{1, N};$$

$$P(i,n) = \frac{b_n(1-F_o)^{i-D-n+1} P(0,0)}{(1-F_o)^{W-1}}, \quad i = \overline{D+n-1, S-1}, \quad n = \overline{1, N},$$

and from the normalization condition we obtain the probability of the initial state

$$P(0,0) = \frac{F_o(1-F_o)^{W-1}}{2 + F_o(D-W+\bar{N}) + (1-F_o)^W - \sum_{m=1}^{N} b_m(1-F_o)^{S-D+1-m}}.$$

If $F_o = 1$ only states are significant

$$P(W-1,0) = \frac{F_o^2}{D-W+N+2};$$

$$P(i,n) = \frac{b_n F_o^2}{D-W+N+2}, \quad i = \overline{W-1, D-1+n}, \quad n = \overline{1, N}.$$

The unlimited duration of timeout leads to probability of the initial state of the following form:

$$P(0,0) = \frac{F_o(1-F_o)^{W-1}}{2 + F_o(D-W+\bar{N}) - (1-F_o)^W}.$$

For start-stop protocol $(W = 1)$ we obtain

$$P(0,0) = \frac{F_o}{1 + F_o(D+\bar{N}) - \sum_{m=1}^{N} b_m(1-F_o)^{S-D+1-m}}.$$

5   Analysis of Transfer Process with Interval Limits on the Duration of Timeout

Let us consider the operation of a transport connection with interval restrictions on protocol parameters and the maximum queue size of the form $W \geq D$, $D + W - 1 \leq S \leq D + W + N - 1$, $1 \leq N \leq D - 2$. Under these restrictions, Eqs. (2), (3) of the original system of equilibrium Eqs. (2)–(8) are converted to

$$P(0,0) = F_o \sum_{n=1}^{S-D-W+1} \sum_{i=D+W+n-2}^{S-2} P(i,n) + \sum_{n=0}^{N} P(S-1,n);$$

$$P(i,0) = F_o \sum_{n=1}^{S-D-W+i} P(D+W+n-2-i,n), \quad i = \overline{1, D+W+N-S-1}; \quad (10)$$

$$P(i,0) = F_o \sum_{n=1}^{N} P(D+W+n-2-i,n), \quad i = \overline{D+W+N-S, D-2}. \quad (11)$$

From Eqs. (7), (8), (10), (11), (4) we find:

$$P(i,n) = P(D-1,n), \quad i = \overline{D, D+n-1}, \quad n = \overline{1,N};$$

$$P(i,n) = (1-F_o)^{i-D-n+1} P(D-1,n), \quad i = \overline{D+n, S-1}, \quad n = \overline{1,N};$$

$$P(i,0) = F_o(1-F_o)^{W-i-1} \sum_{m=1}^{S-D-W+i} P(D-1,m), \quad i = \overline{1, D+W+N-S-1};$$

$$P(i,0) = F_o(1-F_o)^{W-i-1} \sum_{m=1}^{N} P(D-1,m), \quad i = \overline{D+W+N-S, D-2};$$

$$P(D-1,0) = \left(1-(1-F_o)^{W-D+1}\right) \sum_{m=1}^{N} P(D-1,m).$$

Substituting these dependencies in (6) with (5) we obtain

$$P(i,n) = b_n \left[P(0,0) + (1-F_o)^{W-i-1} \sum_{m=1}^{S-D-W+i} P(D-1,m) - (1-F_o)^{W-1} \times \right.$$
$$\times \sum_{m=1}^{S-D-W} P(D-1,m) - \sum_{m=S-D-W+1}^{S-D-W+i} P(D-1,m)(1-F_o)^{S-D-m} \right],$$
$$i = \overline{1, D+W+N-S-1}, \quad n = \overline{1,N};$$

$$P(i,n) = b_n \left[P(0,0) + (1-F_o)^{W-i-1} \sum_{m=1}^{N} P(D-1,m) - (1-F_o)^{W-1} \times \right.$$
$$\times \sum_{m=1}^{S-D-W} P(D-1,m) - \sum_{m=S-D-W+1}^{N} P(D-1,m)(1-F_o)^{S-D-m} \right],$$
$$i = \overline{D+W+N-S, D-2}, \quad n = \overline{1,N};$$

$$P(D-1,n) = b_n \left[P(0,0) + \left(1-(1-F_o)^{W-i}\right) \sum_{m=1}^{S-D-W} P(D-1,m) - \right.$$
$$\left. - \sum_{m=S-D-W+1}^{N} P(D-1,m)\left(1-(1-F_o)^{S-D-m}\right) \right], \quad n = \overline{1,N}.$$

Then, successively for an arbitrary $n = \overline{1, S-D-W}$ we express $P(D-1,n)$ in terms of the probabilities of states $P(D-1,m)$, $m = \overline{n+1, N}$ and rewrite this equation in the form:

$$P(D-1,n) = b_n \left[P(0,0) + \sum_{m=n+1}^{N} P(D-1,m) - (1-F_o)^{W-i} \sum_{m=n+1}^{S-D-W} P(D-1,m) - \right.$$
$$\left. - \sum_{m=S-D-W+1}^{N} P(D-1,m)(1-F_o)^{S-D-m} \right] \bigg/ \left[1 - \left(1-(1-F_o)^{W-1}\right) \sum_{m=1}^{n} b_m \right], \quad n = \overline{1,N}. \quad (12)$$

If $n = S - D - W$ from here we come to:

$$P(D-1, S-D-W) = \frac{b_{S-D-W}}{1-(1-F_o)^{W-1} \sum\limits_{m=1}^{S-D-W} b_m} \left[P(0,0) + \right.$$

$$\left. + \sum_{m=S-D-W+1}^{N} P(D-1,m)\left(1-(1-F_o)^{S-D-W}\right) \right].$$

Substituting this relation in (12), then for the values of n from $S - D - W - 1$ to 1 we find the functional expressions for the probabilities of the states $S - D - W - 1$ through $P(0,0)$ and $P(D-1,m)$, $m = \overline{S-D-W+1, N}$ and we simplify Eq. (12) to:

$$P(D-1,n) = \frac{b_n\left[P(0,0) + \sum\limits_{m=S-D-W+1}^{N} P(D-1,m)\left(1-(1-F_o)^{S-D-m}\right)\right]}{1-\left(1-(1-F_o)^{W-1}\right)\sum\limits_{m=1}^{S-D-W} b_m}, \quad n = \overline{1,N}. \quad (13)$$

From here it is consistent for $n = \overline{S-D-W+1, N}$ we obtain

$$P(D-1,n) = b_n\left[P(0,0) + \sum_{m=n+1}^{N} P(D-1,m)\left(1-(1-F_o)^{S-D-m}\right)\right] \Big/ \left[1 - \sum_{m=1}^{n} b_m + \right.$$

$$\left. + (1-F_o)^{W-1} \sum_{m=1}^{S-D-W} b_m + \sum_{m=S-D-W+1}^{n} b_m(1-F_o)^{S-D-mW-1} \right].$$

From this relation, we consistently express for n from N to $S - D - W + 1$ taking into account (13) and we finally express $P(D-1,n)$ it through the probability of the initial state $P(0,0)$ and according to dependences found earlier, we obtain the probabilities of the states of Markov chain:

$$P(i,0) = \frac{P(0,0)F_o(1-F_o)^{W-i-1} \sum\limits_{m=1}^{S-D-W+i} b_m}{(1-F_o)^{W-1} \sum\limits_{m=1}^{S-D-W} b_m + \sum\limits_{m=S-D-W+1}^{N} b_m(1-F_o)^{S-D-m}}, \quad (14)$$

$$i = \overline{1, D+W+N-S-1};$$

$$P(i,0) = \frac{P(0,0)F_o(1-F_o)^{W-i-1}}{(1-F_o)^{W-1} \sum\limits_{m=1}^{S-D-W} b_m + \sum\limits_{m=S-D-W+1}^{N} b_m(1-F_o)^{S-D-m}}, \quad (15)$$

$$i = \overline{D+W+N-S, D-2};$$

$$P(D-1,0) = \frac{P(0,0)\left(1-(1-F_o)^{W-D+1}\right)}{(1-F_o)^{W-1} \sum\limits_{m=1}^{S-D-W} b_m + \sum\limits_{m=S-D-W+1}^{N} b_m(1-F_o)^{S-D-m}};$$

$$P(0,n) = P(0,0)b_n, \ n = \overline{1,N}; \tag{16}$$

$$P(i,n) = \frac{P(0,0)b_n \left[(1-F_o)^{W-i-1} \sum\limits_{m=1}^{S-D-W+i} b_m + \sum\limits_{m=S-D-W+1}^{N} b_m(1-F_o)^{S-D-m}\right]}{(1-F_o)^{W-1} \sum\limits_{m=1}^{S-D-W} b_m + \sum\limits_{m=S-D-W+1}^{N} b_m(1-F_o)^{S-D-m}}, \tag{17}$$

$$i = \overline{1, D+W+N-S-1}, \ n = \overline{1,N};$$

$$P(i,n) = \frac{P(0,0)b_n(1-F_o)^{W-i-1}}{(1-F_o)^{W-1} \sum\limits_{m=1}^{S-D-W} b_m + \sum\limits_{m=S-D-W+1}^{N} b_m(1-F_o)^{S-D-m}}, \tag{18}$$

$$i = \overline{D+W+N-S, D-2}, \ n = \overline{1,N};$$

$$P(i,n) = \frac{P(0,0)b_n}{(1-F_o)^{W-1} \sum\limits_{m=1}^{S-D-W} b_m + \sum\limits_{m=S-D-W+1}^{N} b_m(1-F_o)^{S-D-m}}, \tag{19}$$

$$i = \overline{D-1, D+n-1}, \ n = \overline{1,N};$$

$$P(i,n) = \frac{P(0,0)b_n(1-F_o)^{i-D-n+1}}{(1-F_o)^{W-1} \sum\limits_{m=1}^{S-D-W} b_m + \sum\limits_{m=S-D-W+1}^{N} b_m(1-F_o)^{S-D-m}}, \tag{20}$$

$$i = \overline{D+n, S-1}, \ n = \overline{1,N}.$$

The probability of the initial state found from the normalization condition has the form:

$$P(0,0) = \left((1-F_o)^{W-1} \sum\limits_{m=1}^{S-D-W} b_m + \sum\limits_{m=S-D-W+1}^{N} b_m(1-F_o)^{S-D-m}\right) \Bigg/ \Bigg[1+$$

$$+ F_o(1+\bar{N}) + (1-F_o)^{W-D+1} - \sum\limits_{m=1}^{N} b_m(1-F_o)^{S-D+1-m} - (1-F_o)^W \sum\limits_{m=1}^{S-D-W} b_m -$$

$$- \left(1 - F_o(D+W+N-S)\right) \sum\limits_{m=S-D-W+1}^{N} b_m(1-F_o)^{S-D-m} \Bigg].$$

It is easy to verify that this distribution is cross-linked with the previously obtained distribution for restrictions on the duration of the time-out at the $S = D+W+N-1$.

Let us analyze the process of information transfer in the transport connection with the size of the sliding window not exceeding the duration of the round trip delay ($W \leq D$), and the interval limits on the duration of time-out and maximum size of the queue type $D+W-1 \leq S \leq D+W+N-1$, $1 \leq N \leq W-2$. Under these conditions, the Eq. (2) of the initial system of equilibrium Eqs. (2)–(8) is converted to

$$P(0,0) = F_o \sum\limits_{n=1}^{S-D-W+1} \sum\limits_{i=D+W+n-2}^{S-2} P(i,n) + \sum\limits_{n=0}^{N} P(S-1,n).$$

Equation (3) is redefined by relations (10) and (11). In this case, Eq. (11) is valid for a set of indices $i = \overline{D + W + N - S, W - 1}$. Equation (4) takes the form $P(D - 1, 0) = 0$, Eq. (6) is satisfied for $i = \overline{1, W - 1}$, $n = \overline{1, N}$, Eq. (7)—for $i = \overline{W, D + n - 1}$, $n = \overline{1, N}$. The stationary probabilities of the states of Markov chain described by these equations, up to the probability of the initial state, take the form (14) – (20), but expressions (15) and (18) are valid for indices $i = \overline{D + W + N - S, W - 1}$, and (19) for $i = \overline{W - 1, D + n - 1}$. According to the normalization condition, the initial state is determined by the relation

$$P(0,0) = F_o \left((1 - F_o)^{W-1} \sum_{m=1}^{S-D-W} b_m + \sum_{m=S-D-W+1}^{N} b_m (1 - F_o)^{S-D-m} \right) \Big/ \Big[2 +$$

$$+ F_o(D - W + \bar{N}) - \sum_{m=1}^{N} b_m (1 - F_o)^{S-D+1-m} - (1 - F_o)^{W} \sum_{m=1}^{S-D-W} b_m -$$

$$- \sum_{m=S-D-W+1}^{N} b_m (1 - F_o)^{S-D-m+1} \Big].$$

6 Available Throughput of the Transport Connection

The capacity of a transport connection under the conditions of competition of flows of different corresponding subscribers for the throughput of data transmission path is defined as the relation of the average amount of data transmitted between two consecutive acknowledgement to the average time acknowledgement [4,5]. Contribution to the speed of the virtual connection is given by those states of Markov chain for which it is possible to obtain acknowledgement. Normalized per unit throughput of virtual connection in loaded path is determined by the relation of the average number of data segments transmitted by the sender between two consecutive acknowledgement to the average time between acknowledgement expressed in the number of intervals of duration $t : Z(W, S) = \overline{V}/\overline{T}$. Since acknowledgements are transferred in each segment independently and arrive to the sender every cycle t, provided that they are not distorted in the path of length D from the recipient to the sender of the information flow the average time between acknowledgement is distributed according to the geometric law with the parameter F_o and will be: $\overline{T} = 1/F_o$. The average volume of data transmitted between acknowledgements taking into account the fact that each segment of the test connection with the probability b_n, $n = \overline{0, N}$ meets the size of the queue n and contributes to the amount of information transmitted inversely proportional to the value $n + 1$, is determined by generalizing the relation given in [4]

$$\overline{V} = \sum_{n=0}^{N} \frac{1}{n+1} \left[\sum_{l=2D-1+n}^{W+2D-2+n} \bar{l} P(l, n) + \sum_{l=W+2D-1+n}^{S-1} \overline{W} P(l, n) \right].$$

Values $\bar{l}$ and $\bar{W}$ are determined by the average number of segments that reached the addressee in selective procedure for repeating distorted segments:

$$\bar{l} = (l - 2D - n + 2)F_n, \quad \bar{W} = WF_n.$$

Then dependence of throughput of the virtual connection on the protocol parameters (W, S), characteristics of transmitting path (D, F_n, F_o) and load parameters $(b_n, n = \overline{1, N})$ will take the form:

$$Z(W, S) = F_n F_o \sum_{n=0}^{N} \frac{1}{n+1} \left[\sum_{l=2D-1+n}^{W+2D-2+n} (l - 2D + 2 - n)P(l, n) + W \sum_{l=W+2D-1+n}^{S-1} P(l, n) \right].$$

Hence, for an arbitrary width of the window when $S \geq D + W + N - 1$ we finally get

$$Z_c(W, S) = \begin{cases} F_n \dfrac{\sum\limits_{n=1}^{N} \frac{b_n}{n+1} \left[1 - (1-F_o)^W - WF_o(1-F_o)^{S-D-n+1} \right]}{2 + F_o(D - W + \bar{N}) - (1-F_o)^W - \sum\limits_{n=1}^{N} b_n(1-F_o)^{S-D-n+1}}, & W < D; \\[3em] F_n \dfrac{F_o^2 \left(1 - (1-F_o)^{W-D+1} \right) + \sum\limits_{n=1}^{N} \frac{b_n}{n+1} \left[1 - (1-F_o)^W - WF_o(1-F_o)^{S-D-n+1} \right]}{1 + F_o(\bar{N}+1) + (1-F_o)^{W-D+1} - (1-F_o)^W - \sum\limits_{n=1}^{N} b_n(1-F_o)^{S-D-n+1}}, & W \geq D. \end{cases}$$

For interval limits on the duration of the timeout and the queue size of competitors $1 \leq N \leq D - 2$ the speed of transport connection in a competitive data transmission environment will be

$$Z_c(W, S) = F_n \left\{ F_o^2 \left(1 - (1 - F_o)^{W-D+1} \right) + \sum_{n=1}^{N} b_n - \sum_{n=1}^{S-D-W+1} \frac{b_n}{n+1} \left[(1 - F_o)^W + \right. \right.$$

$$+ WF_o(1 - F_o)^{S-D-n+1} \right] - \sum_{n=S-D-W+2}^{N} \frac{b_n}{n+1} (1 - F_o)^{S-D-n+1} \left[1 + \right.$$

$$\left. + F_o(S - D - n + 1) \right] \right\} \Big/ \left\{ 1 + F_o(1 + \bar{N}) + (1 - F_o)^{W-D+1} - (1 - F_o)^W - \right.$$

$$\left. - \sum_{n=1}^{N} b_n(1 - F_o)^{S-D-n+1} \right\}.$$

In the case of an absolutely reliable return channel $(F_o = 1)$, available throughput of transport connection $W \leq D$ is largely determined by the proximity of window width to the duration of round-trip delay

$$Z_c(W, S) = \frac{F_n}{2 + D - W + \bar{N}} \sum_{n=1}^{N} \frac{b_n}{n+1},$$

and for $W \geq D$—is invariant to D

$$Z_c(W, S) = \frac{F_n}{2 + \bar{N}} \left[1 + \sum_{n=1}^{N} \frac{b_n}{n+1} \right].$$

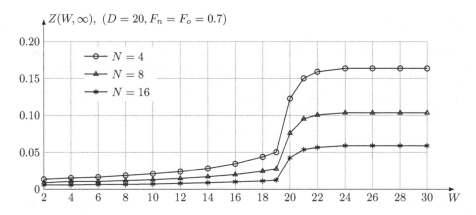

Fig. 1. The dependence of the available bandwidth on the window size with a uniform distribution of the queue length for different values of N

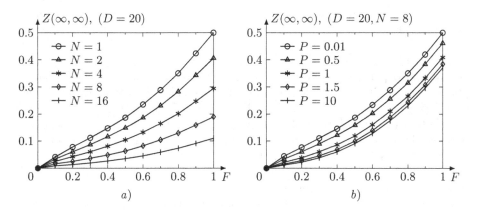

Fig. 2. The dependence of the available bandwidth on the reliability of data transmission $F = F_n = F_o$; *(a)* with a uniform distribution of the queue length for different values of N; *(b)* for different values of the truncated geometric distribution parameter of the queue length P

The unlimited duration of the timeout $(S \to \infty)$ when $W < D$ leads to the dependence of the form

$$Z_c(W, \infty) = \frac{F_n\left(1 - (1 - F_o)^W\right) \sum_{n=1}^{N} \frac{b_n}{n+1}}{2 + F_o(D - W + \bar{N}) - (1 - F_o)^W},$$

and for unlimited increasing width of the window we obtain

$$Z_c(\infty, \infty) = \frac{F_n}{1 + F_o(\bar{N} + 1)}\left[F_o^2 + \sum_{n=1}^{N} \frac{b_n}{n+1}\right].$$

Numerical analysis shows that the available throughput for the transport connection $W \geq D$ is practically invariant to the duration of round-trip delay, significantly decreasing from the saturation range at $W = D$ and $F_o < 1$ (see Fig. 1). In case $W < D$ available throughput is underloaded and the effective data transmission rate is significantly reduced. With increasing competition between subscribers for the throughput of the transmission path, the queue size increases, and the speed of information transfer decreases rapidly (see Fig. 2).

7 Conclusion

The analysis of competitor process of information flows of various inter-subscriber connections for the throughput on shared sections of the path has been carried out. An indicator model of transport connection, competing for the throughput of individual sections of the route, in the form of Markov chain with discrete time, describing the dynamics of queue of sent but not confirmed protocol data blocks, is proposed. The distribution of states of Markov chain under various operating conditions of transport connection is obtained. Analytical dependencies of transport connection speed are found for different ratios between parameters of transport protocol, the characteristics of network channels and load parameters. Numerical studies of available throughput of transport connection in selective re-transmission mode showed that the transmission rate between subscribers is determined by the reliability of data transmission, distribution of queue length of protocol units in transit nodes, and the ratio between duration of round-trip delay and the window width. The direction of further research is to single out the task of analyzing the available throughput of transport connections with interval restrictions on the size of the queues of competitive flows and duration of the end-to-end timeout of transport protocol. It is important to analyze the efficiency of application of forward error correction procedures at transport protocol level with exclusive and competitive use of network communication channels.

References

1. Boguslavskii, L.B.: Data Flow Control in Computer Networks. Energoatomizdat, Moscow (1984)
2. Gelenbe, E., Labetoulle, J., Pugolle, G.: Performance evaluation of the HDLC protocol. Comput. Netw. **2**, 409–415 (1978)
3. Borodikhin, E.A., Korotaev, I.A.: Analysis of functioning and optimization of HDLC protocol. Avtomatica i vichislitelnaya technica (Autom. Comput. Tech.) (2), 47–51 (1993)
4. Sushchenko, S.P.: Analytical models of asynchronous procedures for data link control. Avtomatica i vichislitelnaya technica (Autom. Comput. Tech.) **22**(2), 32–40 (1988)
5. Kokshenev, V.V., Sushchenko, S.P.: Performance analysis of asynchronous procedure of data link control. Vichislitelnie technologii (Comput. Technol.) **13**(5), 61–65 (2008)

6. Ewald, N.L., Kemp, A.H.: Analytical model of TCP NewReno through a CTMC. In: Bradley, J.T. (ed.) EPEW 2009. LNCS, vol. 5652, pp. 183–196. Springer, Heidelberg (2009). https://doi.org/10.1007/978-3-642-02924-0_15
7. Padhye, J., Firoiu, V., Towsley, D.F., Kurose, J.F.: Modeling TCP Reno performance: a simple model and its empirical validation. IEEE/ACM Trans. Netw. **8**(2), 133–145 (2000)
8. Kokshenev, V.V., Mikheev, P.A., Sushchenko, S.P.: Analysis of the selective mode of the transport protocol in loaded data path. Vestnik TSU. Seriya upravlenie, vichislitelnaya technika i informatika (Series of Control, Comp. Eng. Inf.) **3**(24), 78–94 (2013)
9. Herrero, R.: Modeling and comparative analysis of forward error correction in the context of multipath redundancy. Telecommun. Syst. Model. Anal. Des. Manag. **65**(4). 783–794 (2017)

Adaptation of the Frame Synchronization Algorithm of the Second Generation Satellite Broadcasting Standard DVB-S2 for Communication System

Liubov Antiufrieva$^{(\boxtimes)}$ and Aleksandr Ivchenko

Moscow Institute of Physics and Technology (National Research University),
Moscow, Russian Federation
{antyufrieva,ivchenko.av}@mipt.ru

Abstract. The possibility of adapting the frame synchronization algorithm of second generation satellite broadcasting standard DVB-S2 for communication system is considered. The issue of frame synchronization is being studied in conditions when the base station determines the parameters of the physical layer and the time and frequency slot. Frame synchronization algorithms used in DVB-S2 are analyzed. An adaptation of the frame synchronization algorithm for communication system that solves the problem of frame synchronization from the first frame is proposed.

Keywords: Communication system · Broadcasting system · Satellite communication · Satellite modem · DVB-S2 · Physical layer frame · Synchronization algorithm · Frame synchronization

1 Introduction

The second generation standard of satellite broadcasting DVB-S2 and its extension DVB-S2X have high spectral efficiency [1–4]: Quasi Error Free operation near (0.7 to 1 dB) Shannon limit. This is due to the powerful FEC (Forward Error Correction) based on LDPC (Low density parity-check) codes concatenated with BCH (Bose – Chaudhuri – Hocquenghem) codes. This is due to the large LDPC code block length (length of normal block is 64 800 bits, medium block - 32 400 bits, short block - 16 200 bits), large number of iterations on LDPC and the concatenation LDPC with BCH. In addition, the physical layer frame structure provides high PL framing efficiency (up to 99.72%). For this reason, the ability to build a communication system based on signal-code structures of the DVB-S2/S2X standard seems promising.

Frame synchronization is one of the synchronization subsystems on the signal receiver. Block diagram of the synchronization system DVB-S2 receiver is shown in Fig. 1. It runs after matched filtering and symbol synchronization. Serves to extract information about frame boundaries [3,5–7]. This information is necessary to decode the block code and data-aided carrier recovery.

© Springer Nature Switzerland AG 2019
V. M. Vishnevskiy et al. (Eds.): DCCN 2019, CCIS 1141, pp. 104–114, 2019.
https://doi.org/10.1007/978-3-030-36625-4_9

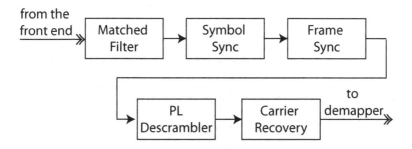

Fig. 1. Block diagram of the synchronization system DVB-S2 receiver

One of the differences between a communication system and a broadcasting system is the need to ensure synchronization from the first frame. Frame synchronization algorithms, commonly used in the DVB-S2 standard [3, 8–11], do not provide synchronization from the first frame at low values of the signal-to-noise ratio. This leads to a decrease in the efficiency of satellite channel at the beginning of communication session.

There is a technique in which a dummy frame [12] transmitted at the beginning of a communication session is used for synchronization, but this leads to a decrease in the efficiency of using a satellite channel.

Frame synchronization in the standard consists in finding the beginning of the frame and decoding the information necessary for demapping and decoding. The formation of PLFRAME is is shown in the Fig. 2. A data block in the format of a bit sequence is supplemented with service information up to BBFRAME (BaseBend frame), after that Base-Bend scrambling is applied on BBFRAME, Forward Error Correction (FEC), including Low-density parity-check (LDPC), Bose – Chaudhuri – Hocquenghem (BCH) coding and bit interleaving (if it is necessary), form FECFRAME. Mapping it into constellation forms an XFECFRAME. It is split into 90 character slots. For the pilot mode, every 16 slots are set 36 pilot symbols identified by $I = (1/\sqrt{2})$, $Q = (1/\sqrt{2})$. XFECFRAME with pilots go to physical layer scrambling. At the beginning of the frame, the PLHEADER (Physical Layer header) is placed. It contains a SOF (Start Of Frame) sequence (26 bits), and is protected by the Reed-Muller code information about frame structure on the PLSCODE (Physical Layer Signaling cod) including information about the channel coding rate, the modulation type, the frame length, the presence or absence of pilot symbols - total 7 bits [1]. The PLSCODE length is 64 bits. The coding process is shown in the Fig. 4. Each even bit is a copy or inversion of the previous one. The header of the physical layer frame is modulated by BPSK $\pi/2$. The structure of the physical layer frame of the DVB-S2 standard is shown in Fig. 3.

This article considers the problem of providing frame synchronization from the first frame under the conditions of a communication system in which the central station collects information about the energy of channels, determines the communication session schedule and signal parameters (modulation type, code rate, frame size, absence or presence of pilot symbols). Thus, the PLSCODE

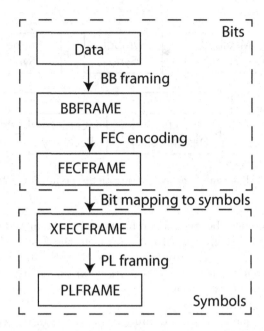

Fig. 2. Frame formation DVB-S2 [12]

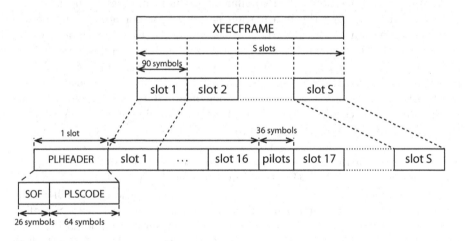

Fig. 3. Physical layer frame structure DVB-S2 [1]

sequence is known to both the transmitter and the receiver. In addition, it is proposed that at the end of the communication session once again transmit PLHEADER.

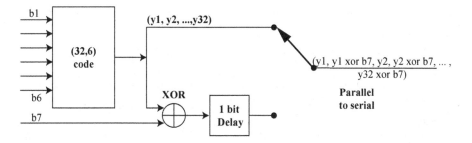

Fig. 4. PLSCODE generation DVB-S2 [1]

The Frame Synchronization Methods section further describes methods for synchronizing the DVB-S2 standard using correlation of signal differential coefficients and signal correlation. Numerical indicators of the methods and their capabilities are given, schedules of their work are given. The disadvantages of existing methods are described.

The Proposed Method section describes in detail the approach and algorithms used in other developments. A graph is presented with correlation values and a calculation of the dependence of the probability of false detection and the absence of detection for various threshold values of the algorithm.

2   Frame Synchronization Methods

In conditions of broadcasting with adaptive coding and modulation (ACM), the receiver has no information about PLSCODE. For this reason, synchronization is performed using the SOF sequence known at the transmitter and receiver and information about the structure and possible values of PLSCODE.

The signal at the receiver after symbol synchronization can be described by the formula [13]:

$$s(k) = a(k) \cdot e^{j(2\pi k F_d T + \theta)} + n(k), \tag{1}$$

where $a(k)$ is the constellation point characterizing the signal symbol formed on the transmitter, F_d - frequency shift (Hz), T - symbol period (s), θ - phase shift and $n(k)$ is a complex white Gaussian noise with variance σ_n^2. The product of F_d and T is called the normalized frequency shift. $F_d T$ shows frequency shift F_d to bandwidth B_n ratio:

$$B_n T = 1. \tag{2}$$

2.1   Synchronization by the Covariance of the Pair-Wise Difference

Figure 5 illustrates a frame synchronization algorithm using pair-wise difference [3,8]. The received signal s is fed to the shift register, at the beginning of which the differentials c_k are calculated:

$$c_k = s(k)s^*(k+1), \tag{3}$$

where $s^*(k+1)$ is the complex conjugation of $s(k+1)$. From (1) and (3), we obtain:

$$c_k = a(k)a^*(k+1)e^{-j2\pi F_d T} + n(k)a^*(k+1)e^{-j(2\pi(k+1)F_d T + \theta)} + \tag{4}$$

$$+ n(k+1)^* a(k)e^{j(2\pi k F_d T + \theta)} + n(k)n(k+1)^*.$$

With a high signal-to-noise ratio, white Gaussian noisewith variance $\sigma_n^2 \to 0$. In this case:

$$c_k \approx a(k)a^*(k+1)e^{-j2\pi F_d T}. \tag{5}$$

Differential coefficients have a constant phase shift equal to $2\pi F_d T$. The covariance cov for differential coefficients will be calculated as:

$$cov = \sum_{k=1}^{N-1} c_k a^*(k)a(k+1). \tag{6}$$

The correlation $corr$ will be calculated as:

$$corr = \frac{cov}{\sigma_c \sigma_{aa^*}} = \frac{cov}{\sqrt{\sum_{k=1}^{N-1} c_k c_k^* \sum_{k=1}^{N-1} a(k)a(k+1)a^*(k)a^*(k+1)}} \tag{7}$$

At a high SNR, the frequency shift will produce a constant phase shift of the covariance and correlation calculated by the differential coefficients. With noise comparable in power to the signal, the noise terms in (4) will be comparable to the first term. For this reason, this algorithm is weakly resistant to noise.

The shift register is divided into two parts (Fig. 5): the right side detects the SOF signal, the left one detects the PLSCODE signal. This requires the knowledge of SOF and PLSCODE structure [3,8].

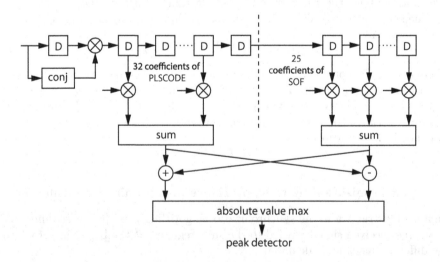

Fig. 5. Frame synchronization algorithm with differentials [3]

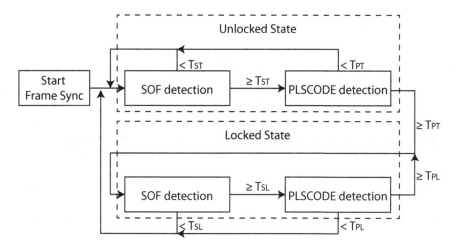

Fig. 6. Signal correlation synchronization algorithm

This algorithm allows frame synchronization at a frequency shift of up to 50% of the bandwidth. In this case, the differentials are unstable with respect to noise. To increase the probability of synchronization, an analysis of several peaks of covariance is used [9] and an energy correction is introduced [10]. However, at low signal-to-noise ratios (up to $-2\,$dB), analysis of several frames (from 4 to several dozen depending on the modification of the algorithm) frames for synchronization is necessary [8–11].

2.2   Signal Correlation Synchronization

An alternative method of the physical layer frame synchronization is the calculation of correlation with the threshold solution [3]. The block diagram of the algorithm is shown in the Fig. 6. The algorithm works in two modes: search mode and tracking mode.

At the beginning of synchronization, the algorithm is in the search state. For the received signal, the normalized correlation function is calculated with the SOF sequence. The obtained values are compared with the T_{ST} threshold for SOF in the search stage. If the threshold is exceeded, a correlation with 128 possible PLSCODEs is calculated for the next 64 characters. The obtained values are compared with the threshold value T_{PT} for PLSCODE in the search state. When a PLSCODE with correlation value higher than threshold is found, the beginning of the frame is considered to be detected and the algorithm goes into tracking mode.

In tracking mode, the presence of SOF and PLSCODE at the expected position is checked (the frame length is calculated from the information encoded in PLSCODE). Accordingly, the correlation values are compared with the T_{SL} and T_{PL} thresholds. If the correlation does not exceed the threshold value, the algorithm goes into the search mode.

The correlation value *corr* is calculated as:

$$corr = \frac{\sum\limits_{k=1}^{N} s(k)a^*(k)}{\sqrt{\sum\limits_{k=1}^{N} s(k)s^*(k) \sum\limits_{k=1}^{N} a(k)a^*(k)}} \tag{8}$$

Substituting (1) into (8), we obtain a smaller dependence of the correlation value on the noise level than in the first algorithm. In this case, the correlation value considerably depends on the magnitude of the frequency shift.

Signal correlation is more robust to noise, but less resistant to frequency shift. With a frequency shift of more than 0.4% of the bandwidth, the algorithm stops working. For narrowband channels, the frequency shift may exceed this value.

3   Proposed Method of Frame Synchronization

As a basis for the proposed method, the synchronization algorithm for the correlation of the signal is taken as being more resistant to noise (Fig. 7). Since PLSCODE is known on the receiver, the correlation is counted immediately across the entire PLHEADER.

The correlation value for the second algorithm is highly dependent on the frequency shift (Fig. 8). When $F_d T = 0.004$, the absolute value of the signal correlation takes 0.8 from its maximum value (see Table 1).

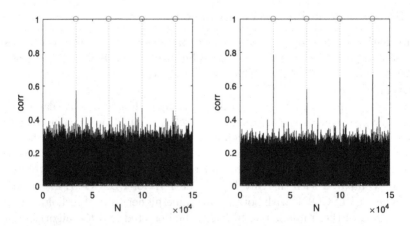

Fig. 7. Correlation values (solid line), calculated by the differentials (a) and signal (b) with a signal-to-noise ratio of $-2\,\mathrm{dB}$. The dotted line indicates the position of the correlation peaks (beginning of the frame) without noise and distortion

Table 1. The dependence of the absolute value of the correlation PLHEADER on the normalized frequency shift calculated for the signal

F_dT	$|corr|$	F_dT	$|corr|$	F_dT	$|corr|$	F_dT	$|corr|$
0	1	0.01	0.109	0.02	0.104	0.03	0.096
0.001	0.987	0.011	0.101	0.021	0.057	0.031	0.07
0.002	0.948	0.012	0.073	0.022	0.01	0.032	0.04
0.003	0.884	0.013	0.139	0.023	0.034	0.033	0.01
0.004	0.8	0.014	0.184	0.024	0.071	0.034	0.02
0.005	0.699	0.015	0.21	0.025	0.1	0.035	0.046
0.006	0.585	0.016	0.217	0.026	0.119	0.036	0.067
0.007	0.464	0.017	0.207	0.027	0.128	0.037	0.083
0.008	0.341	0.018	0.183	0.028	0.126	0.038	0.09
0.009	0.221	0.019	0.147	0.029	0.115	0.039	0.091

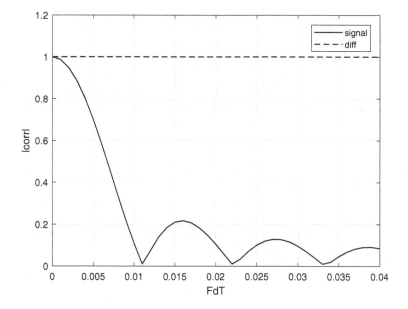

Fig. 8. The dependence of the absolute value of the correlation PLHEADER on the normalized frequency shift calculated for differentials (dashed line) and for the signal (solid line)

The frequency shift instability for narrow channels is compensated by the construction of a frequency grid $\{0, 0.004, -0.004, 0.008, ...\}$ from the bandwidth, where the maximum shift is determined by the characteristics of the equipment. Correlation is considered to be frequency shifted by PLHEADER.

The grid spacing is chosen to ensure stable synchronization at a frequency shift of more than 0.4% of the bandwidth.

The search for the threshold value T is a compromise between the probability of falsely exceeding the threshold p_f and the probability of not detecting the beginning of the frame p_n (see Table 2). To increase the probability of symbolic synchronization from the first frame, it was decided to lower the threshold in the search mode and equate it to the threshold in the tracking mode.

Table 2. Dependence of the probability of false detection of p_f and the probability of not detecting the beginning of frame p_n of the threshold value T at a signal-to-noise ratio of $-2\,\text{dB}$ without checking the next PLSCODE.

T	p_f	p_n	T	p_f	p_n
0.35	$103 \cdot 10^{-6}$	$0.105 \cdot 10^{-8}$	0.4	$5 \cdot 10^{-6}$	$7.5 \cdot 10^{-8}$
0.36	$58 \cdot 10^{-6}$	$0.32 \cdot 10^{-8}$	0.41	$2.5 \cdot 10^{-6}$	$14.3 \cdot 10^{-8}$
0.37	$32 \cdot 10^{-6}$	$0.64 \cdot 10^{-8}$	0.42	$1.3 \cdot 10^{-6}$	$23 \cdot 10^{-8}$
0.38	$17.6 \cdot 10^{-6}$	$1.5 \cdot 10^{-8}$	0.43	$0.77 \cdot 10^{-6}$	$46 \cdot 10^{-8}$
0.39	$9.6 \cdot 10^{-6}$	$3.7 \cdot 10^{-8}$	0.44	$0.36 \cdot 10^{-6}$	$98 \cdot 10^{-8}$

This increases the probability of false detection of the beginning of the frame. To reduce the probability of false detection, the presence of the next PLHEADER (tracking mode) is checked. If there is no detection of the next PLHEADER in the right place, the algorithm continues the search, starting with the character following the false start of the frame (frames are not skipped). The values of the additive white Gaussian noise can be considered as independent random variables. The probability of false detection after checking is p_f^2. Probability of no detection after verification is $2p_n$.

The threshold value $T = 0.4$ is selected for the algorithm. For the algorithm with the next PLSCODE check, with a signal-to-noise ratio of $-2\,\text{dB}$, this threshold value gives the probability of a false detection $2.5 \cdot 10^{-11}$ and the probability of not detecting the start of frame $1.5 \cdot 10^{-7}$.

4   Conclusion

In this article, frame synchronization algorithms commonly used in the second generation DVB-S2 satellite broadcasting standard were considered, their shortcomings were identified, the frame synchronization algorithm was adapted for use in a communication system, the optimal parameters of the algorithm were selected, and the probabilities of incorrect operation were calculated. The proposed algorithm solves the problem of frame synchronization from the first frame in the conditions of a communication system in which the central station determines the schedule of communication sessions and signal parameters.

The results of the work can be used to build a satellite communications system. In the future, it is planned to work on the adaptation of the DVB-S2 frequency synchronization algorithms for use in a communication satellite system and the optimization of the proposed algorithm for FPGA prototyping.

5 Discussion

The development is carried out with the aim of providing satellite communications in the regions at extreme frequencies, in which synchronization speed is important and there is a struggle for every decibel. The problems of the DVB-S2 family are the need to maintain the previous fleet of devices and outdated standards. In our work, we were free to deviate from this paradigm and are engaged in the development of a satellite communications standard, focusing on its efficiency and work in conditions of high noise. In the future, it will be necessary to develop a network management system on a satellite hub that can intelligently distribute available satellite resources and adaptively configure satellite modems in near real-time mode.

Acknowledgments. We express our gratitude to the staff of the Laboratory of Multimedia Systems and Technologies of the Moscow Institute of Physics and Technology and Alexander Dvorkovich for their support and contribution to satellite communications research.

References

1. ETSI EN 302 307-1 V1.4.1 (2014–11) Digital Video Broadcasting (DVB). Second generation framing structure, channel coding and modulation systems for Broadcasting, Interactive Services, News Gathering and other broadband satellite applications. Part 1: DVB-S2. ETSI (2014)
2. Draft ETSI EN 302 307-2 V1.1.1 (2014–10) Digital Video Broadcasting (DVB). Second generation framing structure, channel coding and modulation systems for Broadcasting, Interactive Services, News Gathering and other broadband satellite applications. Part 2: DVB-S2 Extensions (DVB-S2X). ETSI (2014)
3. ETSI TR 102 376-1 V1.2.1 (2015–11) Digital Video Broadcasting (DVB). Implementation guidelines for the second generation system for Broadcasting, Interactive Services, News Gathering and other broadband satellite applications. Part 1: DVB-S2. ETSI (2015)
4. ETSI TR 102 376-2 V1.1.1 (2015–11) Digital Video Broadcasting (DVB). Implementation guidelines for the second generation system for Broadcasting, Interactive Services, News Gathering and other broadband satellite applications; Part 2: S2 Extensions (DVB-S2X). ETSI (2015)
5. Casini, E., Gaudenzi, R., Ginesi, A.: DVB-S2 modem algorithms design and performance over typical satellite channels. Int. J. Satell. Commun. Netw. **22**(3), 281–318 (2004). https://doi.org/10.1002/sat.791
6. Lima, E.R., et al.: A detailed DVB-S2 receiver implementation: FPGA prototyping and preliminary ASIC resource estimation. In: 2014 IEEE Latin-America Conference on Communications (LATINCOM) (2014). https://doi.org/10.1109/LATINCOM.2014.7041856

7. Savvopoulos, P., Papandreou, N., Antonakopoulos, T.: Architecture and DSP implementation of a DVB-S2 baseband demodulator. In: 2009 12th Euromicro Conference on Digital System Design, Architectures, Methods and Tools (2009). https://doi.org/10.1109/DSD.2009.228

8. Sun, F.W., Jiang, Y., Lee, L.N.: Frame synchronization and pilot structure for second generation DVB via satellites. Int. J. Satell. Commun. Netw. **22**(3), 319–339 (2004). https://doi.org/10.1002/sat.793

9. Qing, L., Xiaoyang, Z., Chuan, W., Yulong, Z., Yunsong, D., Han, J.: Optimal frame synchronization for DVB-S2. In: IEEE International Symposium on Circuits and Systems (2008). https://doi.org/10.1109/ISCAS.2008.4541578

10. Lee, D., Kim, P., Sung, W.: Robust frame synchronization for low signal-to-noise ratio channels using energy-corrected differential correlation. EURASIP J. Wirel. Commun. Netw. **2009**(1), 345989 (2009). https://doi.org/10.1155/2009/345989

11. Miyashiro, H., Boutillon, E., Roland, C., Roland, J., Vilca, J., Diaz, D.: Improved multiplierless architecture for header detection in DVB-S2 standard. In: 2016 IEEE International Workshop on Signal Processing Systems (SiPS) (2016). https://doi.org/10.1109/SiPS.2016.51

12. Giraud, X., Lesthievent, G., Meric, H.: Receiver synchronisation based on a single dummy frame for DVB-S2/S2X beam hopping systems. In: 2018 25th International Conference on Telecommunications (ICT) (2018). https://doi.org/10.1109/ICT.2018.8464847

13. Wu, N., Wang, H., Kuang, J., Fei, Z., Fan, G.: A modified carrier frequency estimator for DVB-S2 system. In: 2007 IEEE Wireless Communications and Networking Conference (2007). https://doi.org/10.1109/WCNC.2007.452

The Performance Evaluation of IEEE 802.11 for Different Sized DCF WLAN Under Unsaturated Condition

Leonid Abrosimov(ID) and Margarita Rudenkova$^{(\boxtimes)}$(ID)

National Research University "Moscow Power Engineering Institute",
Krasnokazarmennaya 14, Moscow 111250, Russia
{AbrosimovLI,RudenkovaMA}@mpei.ru
https://mpei.ru

Abstract. This paper proposed the analytical model to estimate the transmission efficiency of wireless LAN (WLAN) depending on the load of wireless network. The load of wireless network is defined as the intensity of the user application data and the number of connected wireless users. The computation is made for the fundamental protocol to access a wireless media Distributed Coordination Function (DCF). This model is able to evaluate the transmission condition in real environment. The numerical and experimental results are presented in the article.

Keywords: 802.11 · DCF · WLAN · Performance · Efficiency

1 Introduction

The wireless LAN (WLAN) is one of the most popular method to gain access to corporate resources, Internet or hotspot services. The throughput of the WLAN media grows rapidly, today we have value of maximum data rate up to several gigabits. The traffic characteristic of the user applications and the throughput requirements vary significant. Some applications consume a lot of resources of the media and some require less. There is no methodology to determine or predict the resource needs for the network which load is changing rapidly. The WLAN is an example of the network which changes during every working hour. Some user applications experience the transmission problems even if the high date rate media is used. The problems source is the media access control protocol which consumes additional time for data transmission. This data transmission time depends on the load of wireless media which depends on an amount of users connected to WLAN and the data rate of user applications. Therefore, with wireless media load increasing, the transmission time will increases significantly. The goal of this paper is the analytical model which is able to estimate the transmission efficiency in real WLAN and provide the comparison method of the different medium control protocol. A lot of researchers studied performance and efficiency of WLAN. Authors [1] made an emulation environment and a

V. M. Vishnevskiy et al. (Eds.): DCCN 2019, CCIS 1141, pp. 115–126, 2019.
https://doi.org/10.1007/978-3-030-36625-4_10

prototype testbed for performance evaluation which may be useful to estimate performance of WLAN with different media access protocols. But these results cannot be applied to constantly changing real environment. In [2,3] the authors estimate the performance of WLAN with the different media access protocols using simulation, in this paper the analytical comparison solution of WLAN with the different intensity of the user application data and the number of connected wireless users is provided. Since [4] the researches have developed many accurate analytic models of the WLAN medium access control [5–8]. As example, in [8] authors improved model [4] to obtain a saturation delay. However, the analytical models in [4–8] developed only for saturated conduction which is not suitable in real environment. This paper considers the method of the transmission efficiency evaluation for real environment condition. The authors [9–11] studied the unsaturation performance of DCF. However this results is unsuitable in case of real environment with different intensity of the user application data and therefore this paper provides another model to estimate the transmission efficiency in such case. Authors in [12] provide new analytical model using two-dimensional continuous-time Markov chain and suggested a few macroscopic state of backoff stages for this. But this model is not able to evaluate the transmission efficiency. In this paper transmission efficiency criterion is suggested as performance of wireless media which determines the average response time of wireless media for transmission of user application data packet. This paper provides the analytical model to estimate the performance of different sized DCF WLAN under unsaturated condition. This model is developed to determine transmission conditions and evaluate transmission efficiency in real environment.

1.1   The Model Formulation

This paper propose an analytical model which is used to estimate the efficiency of using WLAN. The efficiency of using WLAN depends on the load of wireless network which depends on the number of connected users and the rate of user application data. In this paper we consider the WLAN (see Fig. 1) which contains the access point AP, the server Server (emulates the LAN services and Internet) and $N-1$ wireless users STA. The STAs are connected using IEEE 802.11 wireless media.

The throughput of wireless media is C. The throughput C determines τ time which is needed to transmit one bit:

$$\tau = \frac{1}{C} \tag{1}$$

The average size of the packets which is transmitted through wireless media is l. Let μ^1 be the service rate of wireless media is needed to service only one user:

$$\mu^1 = \frac{C}{l} \tag{2}$$

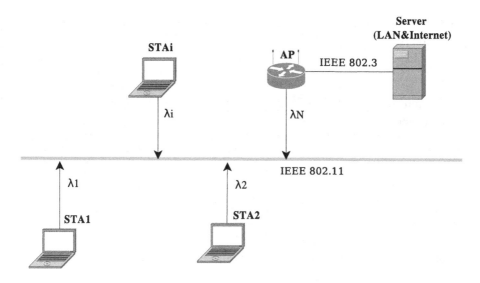

Fig. 1. The WLAN

In this case (2) the collision does not occur. The service rate of the wireless media μ^1 determines the average service time t^0:

$$t^0 = \frac{1}{\mu^1} \qquad (3)$$

Let apply $M/M/1/\infty$ queue to describe the wireless media which is used to service the packets for only one i-th user, λ_i ($i = \overline{1, N}$) is the intensity of the user application data. Then the response time T^1 for one user is:

$$T^1 = \frac{1}{(\mu^1 - \lambda_i)} \qquad (4)$$

Let introduce criterion to estimate transmission efficiency of wireless media without collisions as the performance of wireless media. The performance of the wireless media u^1 which is needed to service one user without using any collision resolution technique is:

$$u^1 = \mu^1 - \lambda_i \qquad (5)$$

The load of wireless media for one user is determined as:

$$\rho^1 = \frac{\lambda_i}{\mu^1} \qquad (6)$$

However, if we consider more than one wireless user, we need the specific protocol which provides the specific collision resolution technique. But, first of all, this technique increases t^0 by value Δt^a which is needed to resolve the conflicts, and the second one, this technique increases λ_i by value $\Delta \lambda^a$, because we have the retransmission probability of the "corrupted packets". So we have new actual value of the actual service time t^a and new actual intensity of the user application data λ^a:

$$t^a = t^0 + \Delta t^a \tag{7}$$

$$\lambda^a = \lambda_i + \Delta \lambda_i^a \tag{8}$$

Therefore, the actual service rate of wireless media μ^a for more than one user is determined as:

$$\mu^a = \frac{1}{t^0 + \Delta t^a} \tag{9}$$

Thus, the response time of wireless media T^a for more than one user is:

$$T^a = \frac{1}{\mu^a - \sum_{i=1}^{N} \lambda_i^a} \tag{10}$$

The performance of wireless media u^a which is needed to service more than one user and takes the account of using the specific collision resolution technique is determined as:

$$u^a = \mu^a - \sum_{i=1}^{N} \lambda_i^a \tag{11}$$

1.2   The Distributed Coordination Function Protocol

Now, it is possible to estimate the performance of wireless media by computation of Δt^a and $\Delta \lambda^a$ for specific collision resolution technique. In this paper we consider the basic medium access control protocol of IEEE 802.11 WLAN Distributed coordination function (DCF). The Fig. 2 describes the DCF packet transmission technique.

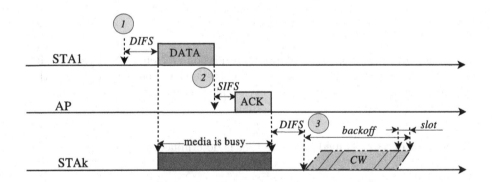

Fig. 2. The Distributed coordination function

- The STA1 senses that the wireless media is idle and waits for a period of time which is equal t_{DIFS} (DCF interframespace). When the STA1 sends the data packet. If the STAk senses that the media is busy it must wait for a time until it is sensed idle.
- The AP receives the data packet from the STA1 and transmits ACK frame after a period of time t_{SIFS} (short inter frame space).
- Then the STAk have sensed idle of wireless media after t_{DIFS} and at this point it wait for a randomly selected backoff interval t_b which is decremented every time STAk senses the channel to be idle. The STAk starts transmission when the backoff timer reaches zero.

The random backoff interval is random number b of the time slot σ:

$$t_b = b \cdot \sigma \tag{12}$$

where

$$b = Random\,(0, w - 1) \tag{13}$$

the w is the contention window.

Contention window w is depend on the number of retransmission occurs in the unsuccessful transmission attempt and determined as:

$$\text{for } n_r < m : w = W_i = W_0 \cdot 2^{n_r} \tag{14}$$

$$\text{for } n_r \geq m : w = W_i = W_m \tag{15}$$

where W_0 - the minimum contention window,
$W_m = CW_{max}$ - the maximum contention window,
n_r - the current number of transmission attempts,
m - the number of transmission attempts, when the current contention window w reaches the value of maximum contention window.

Thus, the service time of one packet t^{DCF} for wireless media using DCF collision resolution technique is:

$$t^{DCF} = t_{DATA} + t_{DIFS} + t_{SIFS} + t_{ACK} \tag{16}$$

where t_{DATA} is the transmission time of one data packet,
t_{ACK} is the transmission time of the ACK frame.
Therefore, the service rate of DCF wireless media μ_{DCF} is:

$$\mu^{DCF} = (t_{DATA} + t_{DIFS} + t_{SIFS} + t_{ACK})^{-1} \tag{17}$$

The DCF collision technique is able to decrease the collision probability, however, the collision probability exists and the new version of corrupted packets will be sent again. Then the collision of more than two packets occurs at following stages:

(1) the STA senses that the wireless media is idle;

(2) the competing STA generates the similar random number b and get the similar backoff interval.

The probability of idle wireless media ρ_{wm} is determined by load of wireless media:

$$p_{wm} = 1 - \rho_{wm} = \frac{\sum_{i=1}^{N} \lambda_i}{\mu^{DCF}} \tag{18}$$

Let the probability of similar backoff interval (two STA get the similar backoff interval) is:

$$p_w = \frac{1}{(CW_{max} - W_0)^2} \tag{19}$$

Let the probability of STA i-th collision ($i = \overline{1, N}$) is determined as the number of combinations C_i^2 :

$$p_i^{DCF} = \frac{C_i^2}{(CW_{max} - W_0)^2} \tag{20}$$

The backoff interval extends the transmission time by Δt_i^{DCF} .

$$\Delta t_i^{DCF} = n_{ri} \cdot \sum_{j=1}^{n_{ri}} t_{bji} \tag{21}$$

Thus, the intensity of user application data increase for $\Delta \lambda_i^{DCF}$:

$$\Delta \lambda_i^{DCF} = p_{wm} \cdot p_i^{DCF} \cdot \sum_{i=1}^{N} (\lambda_i + \frac{1}{\Delta t_i^{DCF}}) \tag{22}$$

Hence, the response time of wireless media that uses DCF collision resolution technique is:

$$T^{DCF} = \frac{1}{\mu^{DCF} - (\sum_{i=1}^{N} \lambda_i + p_{wm} \cdot p_i^{DCF} \cdot \sum_{i=1}^{N} \lambda_i)} \tag{23}$$

and the performance of wireless media wireless media that uses DCF collision resolution technique is:

$$u^{DCF} = \mu^{DCF} - (\sum_{i=1}^{N} \lambda_i + p_{wm} \cdot p_i^{DCF} \cdot \sum_{i=1}^{N} \lambda_i) \tag{24}$$

2   The Model Validation

To validate the model we use the results of the real experiment with two wireless stations, one server and the access point Cisco Aironet 1232ag. The WLAN parameters used to obtain numerical results are specified for the Orthogonal frequency division multiplexing (OFDM) PHY and $C = 54\,\text{Mbps}$ (see Table 1).

Table 1. The WLAN parameters

σ	$9us$	t_{DIFS}	$34us$
CW_{max}	1023	l_{ACK}	$14\,Bytes$
W_0	15	l_{CTS}	$14\,Bytes$
t_{SIFS}	$16\,us$		

The analytical model determines the maximum achievable service rate under such conditions. We also determine the number of packets in service n to establish congestion conditions.

$$n = \frac{\Lambda}{u^{DCF} - \Lambda} \tag{25}$$

where Λ is the summary rate of two STA and Server application data ($N = 3$) and determined as:

$$\Lambda = \sum_{i=1}^{N} \lambda_i \tag{26}$$

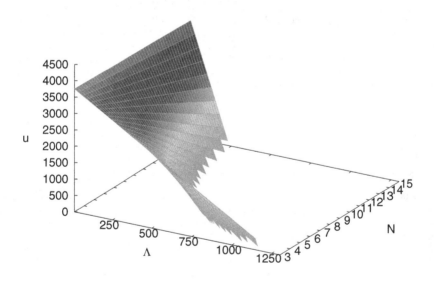

Fig. 3. The Performance u

The Fig. 3 shows wireless media additional resources for DCF protocol. These analytical results can be used to compare different WLAN medium access protocols.

To obtain result the several stage of experiment is used. Each stage is a time period $t_s = 120\,s$ with specific rate of user application date. In the first experiment two wireless stations and the server generates similar rate of user application data (Table 2).

Table 2. The rate of user application data (Experiment 1)

s	Λ	λ_1	λ_2	λ_3
1	30	10	10	10
2	150	50	50	50
3	300	100	100	100
4	450	150	150	150
5	600	200	200	200
6	900	300	300	300
7	1200	400	400	400
8	1500	500	500	500
9	1800	600	600	600
10	2100	700	700	700
11	2400	800	800	800
12	2700	900	900	900
13	3000	1000	1000	1000
14	3300	1100	1100	1100
15	3600	1200	1200	1200
16	3900	1300	1300	1300

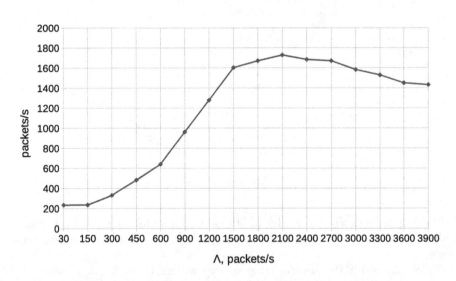

Fig. 4. The real service rate versus summary rate of data applications (Experiment 1)

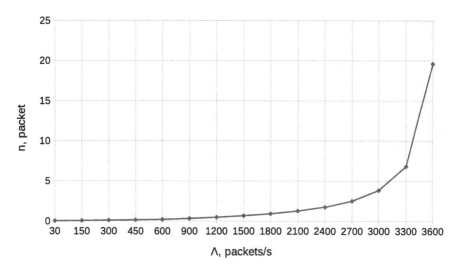

Fig. 5. Packets in service versus summary rate of data applications (Experiment 1)

Table 3. The rate of user application data (Experiment 2)

s	Λ	λ_1	λ_2	λ_3
1	40	10	10	20
2	120	10	50	60
3	220	10	100	110
4	420	10	200	210
5	820	10	400	410
6	1220	10	600	610
7	1620	10	800	810
8	2020	10	1000	1010
9	2420	10	1200	1210
10	3020	10	1500	1510
11	3220	10	1600	1610
12	3420	10	1700	1710
13	3620	10	1800	1810

The values of the Fig. 4 were gained using AP wireless interface statistic table. The Fig. 4 shows that service rate increases Λ until <2100 packet per second and the transmission efficiency decrease. The n curve Fig. 5 shows that the packets in service significantly increase at $\Lambda > 2100$ packet per second. It shows that the transmission problems occurs at stage number 10.

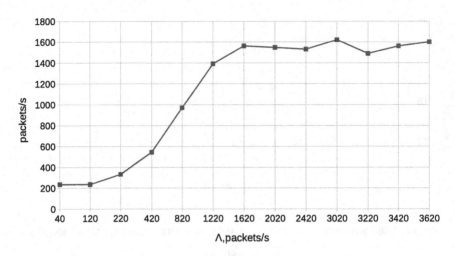

Fig. 6. The real service rate versus summary rate of data applications (Experiment 2)

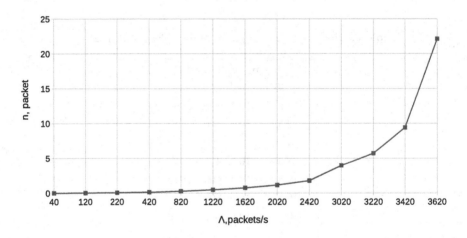

Fig. 7. Packets in service versus summary rate of data applications (Experiment 2)

The Fig. 6 shows the dependence of real wireless media services rate from summary rate of application data λ. The Fig. 6 shows that service rate increases Λ until <1620 packet per second and the transmission efficiency decrease.

In the second experiment (Table 3) the wireless stations and the server generates different rate of user application data, STA1 and Server: small constant rate of user application data, STA2 increases rate of user application data).

The n curve Fig. 7 shows that the packets in service start increase at Λ >1620 packet per second. It shows that the wireless media has transmission problems at stage number 7.

We have shown how to determine transmission conditions and transmission problems. We also have done the performance comparison of the WLAN under the different conditions. This comparison shows efficiency of current WLAN using. However, this comparison method is also suitable to estimate WLAN transmission efficiency when WLAN has different configuration settings.

3 Conclusion

In this paper the analytical model to estimate transmission efficiency of WLAN and to compute the performance of wireless medium depending on the load of wireless network is presented. The performance characterizes an achievable service rate of application data. The performance value is applicable to comparison of the different WLAN. The model evaluates the transmission conditions in real environment and proposed for adaptive switching of the protocol to access a wireless media for matching resources of wireless media and wireless network load in real environment. The model will be improved in future works using statistics obtained in experiments.

References

1. Chen, Z., Fu, D., Gao, Y., Hei, X.: Performance evaluation for WiFi DCF networks from theory to testbed. In: 2017 IEEE International Symposium on Parallel and Distributed Processing with Applications and 2017 IEEE International Conference on Ubiquitous Computing and Communications (ISPA/IUCC), Guangzhou, pp. 1364–1371 (2017)
2. Shaaban, S., El Badawy, H.M., Hashad, A.: Performance evaluation of the IEEE 802.11 wireless LAN standards. In: World Congress on Engineering, vol. 1 (2008)
3. Ali, Q.: Performance evaluation of WLAN internet sharing using DCF & PCF modes. Int. Arab J. e-Technol. **1**(1), 38–45 (2009)
4. Bianchi, G.: Performance analysis of the IEEE 802.11 distributed coordination function. IEEE J. Sel. Areas Commun. **18**, 535–547 (2000)
5. Wu, H., Peng, Y., Long, K., Cheng, S., Ma, J.: Performance of reliable transport protocol over IEEE 802.11 wireless LANs: analysis and enhancement. In: Proceedings of IEEE Information Communications (INFOCOM), pp. 599–607 (2002)
6. Hadzi-Velkov, Z., Spasenovski, B.: Saturation throughput delay analysis of IEEE 802.11 DCF in fading channel. In: Proceedings of IEEE International Conference on Communications (ICC 2003), pp. 121–126 (May 2003)
7. Weng, C.-E., Chen, H.-C.: The performance evaluation of IEEE 802.11 DCF using Markov chain model for wireless LANs. Comput. Stand. Interfaces. **44**, 144–149 (2016)
8. Ziouva, E., Antonakopoulos, T.: CSMA/CA performance under high traffic conditions: throughput and delay analysis. Comput. Commun. **25**, 313–321 (2002)
9. Liaw, Y.S., Dadej, A., Jayasuriya, A.: Performance analysis of IEEE 802.11 DCF under limited load. In: Proceedings of the Asia-Pacific Conference on Communications, Perth, Western Australia, vol. 1, pp. 759–763 (2005)
10. Malone, D., Duffy, K., Leith, D.: Modeling the 802.11 distributed coordination function in nonsaturated heterogeneous conditions. IEEE/ACM Trans. Netw. **15**(1), 159–172 (2007)

11. Senthilkumar, T.D., Krishnan, A., Kumar, P.: Nonsaturation throughput analysis of IEEE 802.11 distributed coordination function. In: Proceedings of the 5th International Conference on Electrical and Computer Engineering, ICECE 2008, Dhaka, Bangladesh, pp. 472–477 (2008)
12. Li, X., Narita, Y., Gotoh, Y., Shioda, S.: Performance analysis of IEEE 802.11 DCF based on a macroscopic state description. IEICE Trans. Commun. (2018)
13. IEEE Standard for Information technology-Telecommunications and information exchange between systems local and metropolitan area networks-specific requirements - Part 11: Wireless LAN Medium Access Control (MAC) and Physical Layer (PHY) Specifications. In: IEEE Std 802.11-2016 (Revision of IEEE Std 802.11-2012), pp. 1–3534, 14 December 2016

Energy Estimation for VANET Performance Based Robust Neural Networks Learning

Ali R. Abdellah[1,2(✉)], Ammar Muthanna[2,3], and Andrey Koucheryavy[2]

[1] Electronics and Communications Engineering, Electrical Engineering Department, Al-Azhar University, Qena 83513, Egypt
alirefaee@azhar.edu.eg
[2] The Bonch-Bruevich Saint-Petersburg State University of Telecommunications, Pr. Bolshevikov, 22, St. Petersburg 193232, Russia
akouch@mail.ru
[3] Applied Probability and Informatics, Peoples' Friendship University of Russia (RUDN University), Moscow 117198, Russia
ammarexpress@gmail.com

Abstract. Abstract - Vehicular ad-hoc networks (VANETs) technology has emerged recently as an important research area. They have today been established as reliable networks that vehicles use for communication purpose on highways or urban environments. VANET is a network that is formed when vehicles with wireless transceiver have the need to communicate with each other. It is composed of models based communication among vehicles and vehicle with a high mobility feature. The power consumption by wireless communications might become a major concern in VANET design. Artificial neural networks (ANNs) are one of the most popular and promising areas of artificial intelligence (AI) research. In this paper, we study the performance estimation of VANET in terms of energy consumption and the throughput; we propose the robust neural networks learning that are based on a family of robust statistics estimators, commonly known as M-estimators to replace the traditional MSE performance function order to robustify the neural networks learning in the case of high-quality clean data (noise free). Comparative study between the robust and the traditional performance functions was established in this paper using VANET performance estimation.

Keywords: VANET · Vehicle to Vehicle (V2V) · Neural networks · M-estimators · Robust statistics

1 Introduction

Nowadays the increasing of demands for wireless communication technologies have a great interest in research on self-organizing, self-healing networks beyond the interference of any centralized or pre-established infrastructure. The networks without any centralized or pre-established infrastructure are known as Ad hoc networks. Ad-hoc Networks are the category of wireless networks that uses

© Springer Nature Switzerland AG 2019
V. M. Vishnevskiy et al. (Eds.): DCCN 2019, CCIS 1141, pp. 127–138, 2019.
https://doi.org/10.1007/978-3-030-36625-4_11

multi-hop radio relaying. Vehicular Ad hoc Networks (VANETs) is an application of Mobile Ad Hoc Networks (MANETs). VANETs is the most advanced technology that has equipped to offer an intelligent transportation system by using the wireless communication capabilities between vehicles to vehicles and roadside units (RSUs) to vehicles according to the IEEE 802.11p standard. VANET provides a wide range of safety and non-safety applications. Safety application provides safety to passengers such as lane change warning, collision detection, etc [1]. In addition, it provides comfort and commercial applications to road users such as electronic toll collection, audio/video exchanging, electronic payments, route guidance, weather information, mobile E-commerce, internet access, etc.

The main aim of VANET architecture is to enable the connection between vehicles or between vehicles and fixed roadside units leading to the following three possibilities [2]:

* Vehicle-to-Vehicle (V2V) ad-hoc network allows the direct vehicular communication without depending on the fixed infrastructure support and can be mainly employed for security, safety and dissemination applications.
* Vehicle-to-Infrastructure (V2I) network allows a vehicle to communicate with the roadside unit essentially for information and data gathering applications.
* Hybrid architecture combines both V2I communications and V2V. In this scenario, a vehicle can communicate with the roadside infrastructure either in a multi-hop or single hop fashion, depending on the distance, i.e. if it can or not access easily the roadside unit. It facilitates long-distance connection to the internet or to vehicles that are far.

In VANET nodes are communicate with RSU (Roadside unit) and other nodes wirelessly. So they will easily get data concerning road traffic, blockage or work on the road, road accident, etc. By using these information drivers can able to take smart decisions according to road conditions like changing the routes to the destination, slowing down their speed, etc. They can additionally ask queries to RSU about the route to a destination, availability of parking, hotels, petrol station, hospitals in an unknown area [3]. Even though a VANET is a type of MANET it differs from MANET. Figure 1 shows the architecture of VANET.

* A VANET is characterized by a rapidly changing but somewhat predictable topology.
* network topology changes very fast so fragmentation regularly happens.
* Due to the high speed of nodes diameter of a network in VANET is relatively small.
* Redundancy is limited both temporally and functionally.
* Do not need any infrastructure.
* Predictable topology (using digital map).
* No problem with power [4,5].

Neural networks (NNs) are parallel computing devices, which are basically an attempt to make a computer model of the brain by massively distributed parallel processing consisting of simple processing units. These units are computational elements called neurons or nodes that have a neurological characteristic, it stores

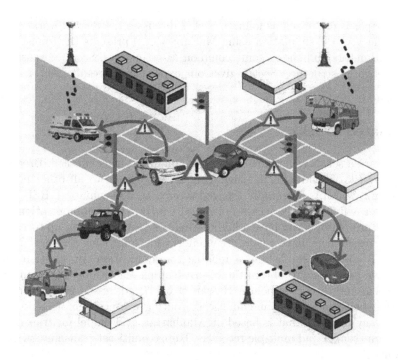

Fig. 1. The architecture of VANET

practical knowledge and empirical information to make it available to the user by adjusting the weights. The main objective is to develop a system to perform various computational tasks faster than the traditional systems [6].

The most popular feed-forward neural networks scheme makes use of the backpropagation (BP) strategy and a minimization of the mean squared error (MSE). Recently, various robust BP learning algorithms have been proposed. Commonly, they take advantage of the idea of robust estimators. This approach was adopted to the neural networks learning algorithms by replacing the MSE with a performance function of such a shape that the impact of outliers may be, in certain conditions, reduced or even removed [13].

In our work and for first time we investigate the performance of VANET in terms of energy consumption and the throughput using the robust neural network learning which is trained using backpropagation learning algorithm that uses a family of robust statistics estimators called M-estimators as cost functions (performance functions) instead of the most famous traditional MSE – cost function in order to robustify the neural networks learning for getting the best performance in the case of high-quality clean data (noise free). In addition, this is done with a view to see which algorithm will produce better results and has faster training for the application under consideration. We compare the performance of neural networks in terms of Root Mean Square Error (RMSE) as a merit function and the training speed in seconds.

The paper is organized as follows: Sect. 2 discusses the related works; Sect. 3 explain the VANET simulation using Matlab; Sect. 4 presents the robust statistics M-estimators and shows some common M-estimators; Sect. 5 discusses the neural networks structure; Sect. 6 gives our experimental result; Sect. 7 conclusion.

2 Related Works

VANETs are a set of vehicles with wireless communication enabled. Broadcasting is the task of sending a message from a source node to all other nodes in the network, which is repeatedly referred to as data dissemination. RSU is considered as a wireless LAN access point and can provide communications with infrastructure. Also, it can have a higher range of communication up to 1,000 m.

Many researchers have made great efforts for improving the performance of VANET and for establishing a reliable communication. In addition, many efforts have been done to obtain the robust leaning algorithms, we summarized few of them and discussed below. Hassan et al. [7] addressed the performance evaluation between unicast and multicast routing protocols implemented in a vehicular environment that is based on Manhattan grid model for transmission between one sender and multiple receivers. Unlike multicast transmission in geocast routing, the multiple receivers for the paper scenario are not located in a specific geographic region. They evaluated the performance evaluated in term of average end-to-end delay, throughput, packet delivery ratio and routing overhead. The results reveal a consistent performance for multicast protocols as the number of receiving nodes increases during the transmission.

Rehman et al. [8] he studied performance evaluation of VANETs based on routing protocols in different scenarios. He focused and inspected various routing protocols including AODV, DSR and DSDV for the purpose to find out protocols best suited for all scenarios. The comparison and evaluation of various routing protocols is done on the basis of different performance metric criteria like data throughput, PDR, end to end delay or latency and network stability etc.

Li et al. [9] proposed a connectivity-sensed routing protocol (CSR) for VANETs in urban scenario. CSR utilizes vehicle distribution information collected by intersection infrastructure to help vehicles select a road not only with progress to destination but also with better network connectivity. Moreover, simulation results demonstrate that the CSR protocol achieves much lower end-to-end delay, higher delivery rate, and higher throughput than traditional routing protocols.

El-Melegy et al. [10] presented several methods in order to robustify neural network training algorithms. First, they exploited a family of robust statistics estimators, commonly known as M-estimators, in order to robustify the learning process of the back-propagation learning algorithm. The performance of the trained NN using backpropagation learning algorithm, that uses M-estimators, was evaluated for the task of function approximation and dynamical model identification. As these M-estimators sometimes do not have adequate insensitivity to outliers, so they used the statistically more robust estimator of the least median of squares(LMedS) and developed a stochastic algorithm to minimize a related cost function.

Mohamed et al. [11] studied the performance of the multilayer feed-forward neural networks they presented M-estimators as performance functions alternatives of Mse Performance function in the case of using high quality clean data (noise free). He compared between Mse and M-estimators in two applications crab classification, and function approximation.

Mohamed et al. [12] addressed the problem of fitting a functional model to data corrupted with outliers using a multilayered feed-forward neural network (MFNN). He proposed new activation functions that are based on M-estimators to replace the traditional activation functions. The proposed activation function was evaluated on synthetic data, contaminated with varying degrees of outliers, and compared them to existing neural network training algorithms.

Ellah et al. [13] studied the robust backpropagation learning algorithm study for feed forward Neural networks. He investigated of the robustness of neural networks learning in presence of data corrupted with outliers. He used many of the methods for improving the robust learning process. He proposed the robust statistics M-estimators as performance function instead of the traditional MSE performance function. In addition, he implemented a new approach by selecting a new activation functions that are based on M-estimators to replace the traditional activation functions. He used a lot of application to test the performance of neural networks.

Ellah et al. [14] studied the performance of four different artificial neural network (ANN) training algorithms, which are conjugate gradient with Fletcher-Reeves updates, conjugate gradient with Polak - Ribiére updates, resilient backpropagation, and conjugate gradient with Powell - peal restart. He compared their performance in terms of Root Mean Square Error as a merit function and the training speed in seconds. The examined neural networks trained by aforementioned backpropagation learning algorithms, that used the robust M-estimators performance functions instead of MSE one, in order to get robust learning in the presence of outliers. He noticed that Traincgf is the best algorithm in terms of RMSE, while the Traincgp is the best in terms of training speed.

3   VANET Simulation Using Matlab

In this section, the VANET in the Urban City simulation is implemented as follow: We used Matlab for generating a realistic mobility model for VANETs. The routing protocols AODV have been implemented over the generated realistic mobility model to analyze and evaluate their behavior and performance.

In order to generate a mobility map, firstly the road network needs to be created. It has three main modules: City Size, Nodes and RSU's. Here one has to specify the city size, where the nodes will be travelling in random directions for the AODV implementation. Maximum the city size and large number nodes will be required to show the working of the AODV. Similarly, more number of

RSU's need to be installed. If the city size is larger than the simulation time extends automatically. So according to the work, we assumed the city size is 100 metres and both in respect to X-axis and Y-axis.

The urban city will be designed so that the desired nodes could travel random directions on the dedicated paths. The module in Fig. 2 illustrates the nodes (vehicles) with dot and the RSU's locations along with their ID numbers which is assigned to them by the network architecture or topology designer. The source node in this model indicated by node number 20 and the destination node number 70.

This module is responsible for defining number of nodes, flow of nodes that will specify the groups of nodes movements flow on the simulation and turning ratio that will define the probability of directions on each junction. Simulation module is used to visualize the configured topology, also specify the beginning, and end time of simulation.

The positioning the RSU's (Road Side Unit) in the urban city map for assistance to the nodes moving in random direction for communication with other nodes which are far from other nodes range. An RSU can be attached to an infrastructure network, which in turn can be connected to the Internet. RSUs can also communicate to each other directly or via multi-hop. Figure 2 shows VANET Simulation in an Urban city.

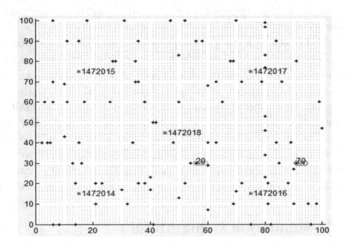

Fig. 2. VANET Simulation in an Urban city

4   Robust Statistics M-Estimators

M-estimators have gained great popularity in the neural networks community. The term M-estimator is a broad class of estimators of maximum likelihood type, which play an important role in robust statistics. It is the simplest approach both computationally and theoretically.

Recently many researches exploited M-estimators as performance function in order to robustify the NN learning process. M-estimators use some performance functions which increase less than that of least square estimators as the residual departs from zero. When the residual error exceeds a threshold, the M-estimator suppresses the response instead. Therefore, M-estimator performance function is more robust for the presence of the outliers than MSE performance function [13]. The authors in [11] introduced a family of robust statics M-estimators instead of traditional performance functions of MSE. It is well known that this family provided high reliability for robust NN training in the presence of contaminated data. Therefore, they recommended the use of this family of estimators as a good alternative of MSE performance function, in the presence of clean and contaminated data [11]. The traditional MSE performance function will be replaced by M-estimators based performance functions in order to improve the learning process and hence the robustness of neural networks learning.

Assumes that x_i the residual of the ith datum, i.e. the difference between the ith observation and its fitted value. The M-estimators estimate the parameters by solving the nonlinear minimization problem. M-estimators try to minimize the error by replacing the squared residuals with another function of the residuals, yielding.

$$\min \sum_{i=1}^{n} \rho(r_i) \tag{1}$$

That is, the estimator must yield the smallest value of squared of residuals computed for the entire data set. Where $\rho(.)$ is function with the following properties:

* $\rho(x) \geq 0$ for all x and has a minimum at 0.
* $\rho(x) = \rho(-x)$ for all x.
* $\rho(x)$ increases as x increases from 0, but doesnt get too large as x increases.

Table 1. Some commonly used M-estimators.

Type	$\rho(x)$	$\psi(x)$	$\omega(x)$
L2	$x^2/2$	x	1
L1	$\lvert x \rvert$	$sgn(x)$	$1/\lvert x \rvert$
Fair	$c^2[\frac{\lvert x \rvert}{c} - log(1 + \frac{\lvert x \rvert}{c})]$	$x/(1 + \lvert x \rvert c)$	$1/(1 + \lvert x \rvert c)$
Huber $\begin{cases} if\ \lvert x \rvert \leq k \\ if\ \lvert x \rvert \geq k \end{cases}$	$\begin{cases} \frac{x^2}{2} \\ k(\lvert x \rvert - k2) \end{cases}$	$\begin{cases} x \\ ksgn(x) \end{cases}$	$\begin{cases} 1 \\ \frac{k}{\lvert x \rvert} \end{cases}$
Cauchy	$\frac{c^2}{2}log(1 + (x(c))^2)$	$1/(1 + (x(c))^2)$	$1/(1 + (x(c))^2)$
GM	$\frac{x^2}{2}/(1 + x^2)$	$x/(1 + x^2)^2$	$1/(1 + x^2)^2$
LMLS	$log(1 + 1/2x^2)$	$x/(1 + 1/2x^2)$	$1/(1 + 1/2x^2)$

5   Neural Network Structure

In this paper, we used a multilayer feed-forward neural network structure with one hidden layer having 20 hidden neurons. The network is trained with the traditional backpropagation algorithm using the M-estimators mentioned above. In this work, we have used the Matlab neural network toolbox with the following settings:

* A sigmoidal function (Tansig) was chosen to be the activation function for all neurons in the hidden layers and a "Purlin" activation function was chosen to be the activation function for the neuron in output layer.
* The network training function which is used here network training function is Traincgf (conjugate gradient backpropagation with Fletcher-Reeves updates)
* The maximum number of epochs to train is 1000 epochs.
* The goal (Minimum Performance Value) is 0.001. Were used in this study due to its high speed and accuracy.

Where we have the input/output training patterns for the examined performance, the supervised learning mode was considered. The training set data which had been studied was supplied as input for the ANN, where the throughput of VANET act as the input data, while the desired output of the ANN was shown by energy consumption.

Input values need to be normalized in the range $[-1, 1]$, which corresponds to the min-max actual values. Subsequently, testing the ANN requires a new independent set (test sets) in order to validate the generalization capacity of the estimation model.

Prior to the training phase, the data that had been gathered were analyzed and pre-processed. The datasets were divided into inputs and outputs, being subsequently randomly split into three subcategories: training set (70%), testing set (15%) and validation set (15%).

6   Simulation Results

In this section the proper performances of all aforementioned performance functions, both proposed and traditional performance function will try to find the optimum VANET performance.

To compare the performance of robust and traditional performance functions we use root mean square error (RMSE) of each model,

$$RMSE = \sqrt{\frac{\sum_{i=1}^{n} (t_i - y_i)^2}{N}} \qquad (2)$$

Where the target t_i is the actual value of the function at x_i and y_i is the output of the network given x_i as its input. The results presented below are the average response of training. This was done to take into account the different initial values of weights and bias at the beginning of each training. Table 2 shows RMSE values and processing time for all mentioned performance functions.

Table 2. The performance of networks trained with the robust and traditional estimators.

VANET Performance based estimated energy consumption		
Performance Function	RMSE	Processing Time
MSE	6.1706e−004	2.6875
Cauchy	5.0916e−004	2.4844
GM	4.4933e−004	2.7500
Fair	4.8975e−004	2.5000
L1	4.8023e−004	6.0469
LMLS	4.8023e−004	2.5000
Huber	5.0048e−004	2.4531

Table 2 displays the performances of neural networks in term of RMSE and processing time in order to investigate the best performance, which use traditional MSE or M-estimators as performance function.

It is clear from tabulated results that, GM estimator has the best performance with RMSE value 4.4933e−004, in comparison to other estimators, and the L1 and LMLS estimators have the same RMSE value 4.8023e−004, which is semi-equal to the GM estimator. In addition, the Fair, Huber and Cauchy estimators have approximately semi-equal RMSE values in comparison to the others. On the other hand, the traditional MSE performance function provides so poor performance with RMSE value 6.1706e−004, in comparison with others.

It is clear from tabulated results that, the neural network trained using Huber estimators has the shortest training time = 2.4531 s, and hence it the fastest training between its peers.

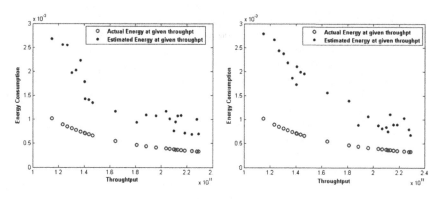

Fig. 3. (a) Model predicted using traditional MSE estimator, (b) Model predicted using robust Fair estimator.

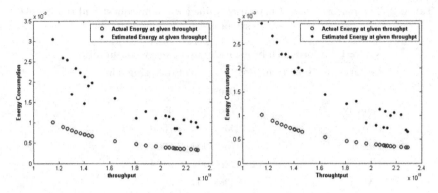

Fig. 4. (a) Model predicted using traditional L1 estimator, (b) Model predicted using robust GM estimator.

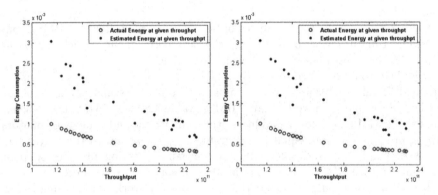

Fig. 5. (a) Model predicted using traditional Cauchy estimator, (b) Model predicted using robust LMLS estimator.

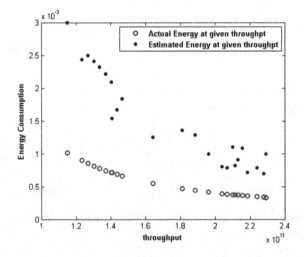

Fig. 6. Model predicted using robust Huber estimator

Looking at Figs. 3, 4, 5 and 6 all responses of neural networks trained using the traditional MSE performance function and robust M-estimators performance function in terms of VANET throughput and energy consumption.

As shown in Figs. 3, 4, 5 and 6 all responses are semi similar to actual model except the model predicted by traditional MSE performance function that slightly deviate from the actual model.

7   Conclusion

In this paper, ANN has been proposed to find the optimum performance for VANET, in case of the throughput and energy consumption as input and target for neural networks respectively. The performance of neural network learning process has been estimated in terms of RMSE and processing time. The ANN can be used very efficiently as an estimator, especially in an VANET with high mobility feature, when nature of the network is complex and highly nonlinear. We proposed a family of robust statistics M-estimators as robust performance functions for training neural networks. It is well known that this family provided high reliability for robust NN training in many applications. Based on the mentioned above result we noticed that the GM estimator has the best performance in term RMSE value and the Huber estimator is the best in the term of speed of training. We recommend this family of estimators as a good performance function for robust training neural networks in a comparison to traditional MSE performance function.

References

1. Nagaraj, U., Kadam, N.N.: Study of statistical models for route prediction algorithms in VANET. J. Inf. Eng. Appl. 1(4) (2011)
2. Yasser, A., Zorkany, M., Abdel Kader, N.: VANET routing protocol for V2V implementation: a suitable solution for developing countries. Cogent Eng. 4(1), 1362802 (2017). https://doi.org/10.1080/23311916.2017.1362802
3. Mármol, F.G., Pérez, G.M.: TRIP, a trust and reputation infrastructure-based proposal for vehicular adhoc networks. J. Netw. Comput. Appl. 35(3), 934–941 (2012)
4. Gadkari, M.Y., Sambre, N.B.: VANET: routing protocols, security issues and simulation tools. IOSR J. Comput. Eng. (IOSRJCE) 3(3), 28–38 (2012). ISSN 2278-0661
5. Balon, N.: Increasing Broadcast Reliability in Vehicular Adhoc Networks. The University of Michigan, Ann Arbor (2006)
6. Eluyode, O.S., Akomolafe, D.T.: Comparative study of biological and artificial neural networks. Eur. J. Appl. Eng. Sci. Res. 2(1), 36–46 (2013)
7. Hassan, A., Ahmed, M.H., Rahman, M.A.: Performance evaluation for multicast transmissions in VANET. In: 2011 24th Canadian Conference on Electrical and Computer Engineering (CCECE), Niagara Falls, pp. 001105–001108 (2011)
8. Rehman, M.U., Ahmed, S., Khan, S.U., Begum, S., Ahmed, S.H.: Performance and execution evaluation of VANETs routing protocols in different scenarios. EAI Endorsed Trans. Energy Web Inf. Technol. 5(17), 1–5 (2018). https://doi.org/10.4108/eai.10-4-2018.154458

9. Li, C., Wang, M., Zhu, L.: Connectivity-sensed routing protocol for vehicular ad hoc networks: analysis and design. Int. J. Distrib. Sens. Netw. **11**(8), 1–11 (2015). https://doi.org/10.1155/2015/649037
10. El-Melegy, M., Essai, M., Ali, A.: Robust training of artificial feedforward neural networks, vol. 1, pp. 217–242. Springer, Berlin (June 2009). https://doi.org/10.1007/978-3-642-01082-8_9
11. Zahra, M.M., Essai, M.H., Ellah, A.R.A.: Performance functions alternatives of MSE for neural networks learning. Int. J. Eng. Res. Technol. (IJERT) **3**(1), 967–970 (2014)
12. Essai, M.H., Ellah, A.R.A.: M-estimators based activation functions for robust neural network learning. In: The IEEE 10th International Computer Engineering Conference (ICENCO2014), pp. 76–81, December 29–30, Cairo (2014)
13. Ellah, A.R.A., Essai, M.H., Yahya, A.: Robust backpropagation learning algorithm study for feed forward neural networks. Thesis, Al-Azhar University, Faculty of Engineering (2016)
14. Ellah, A.R.A., Essai, M.H., Yahya, A.: Comparison of different backpropagation training algorithms using robust M-estimators performance functions. In: The IEEE 2015 Tenth International Conference on Computer Engineering & Systems (ICCES), pp. 384–388, December 23–24, Cairo (2015)

Mathematical Model of Four-Dimensional Parametric Systems Based on Block Diagonal Matrix with 2 × 2 Blocks

K. Vytovtov$^{(\boxtimes)}$ ⓘ and E. Barabanova ⓘ

Astrakhan State Technical University, Tatischeva 16 street, Astrakhan, Russia
vytovtov_konstan@mail.ru

Abstract. In this work the accurate analytical model for the wide class of four-dimensional linear deterministic dynamic systems with arbitrary piecewise constant parameters is presented for the first time. Here the fundamental solution system is found as the 4×4 block diagonal matrix with the 2×2 blocks at main diagonal. It is important that the solution is obtained without transformation into another basis. The presented model is valid if eigennumbers of initial differential equation systems are inverse in pair. The numerical example demonstrating the practical application of the developed model is also presented.

Keywords: Mathematical model · Fourth order differential equations · Dynamic system

1 Introduction

Mathematical modeling of physical processes is one of the most relevant for understanding these processes [1–9]. And, in particular, the very important scientific problem is studying dynamic systems with two degrees of freedom. Such systems can be mechanical [1–3], electrical [1–3], biological [3,4], electromagnetic [5,6], optic [7] quantum [8], economic [9] nature.

In mechanics it can be a system of matched oscillators, the electrical analogue is a system of matched electrical RLC-oscillatory circuits. A wave analogue is an electromagnetic wave in an anisotropic medium. A quantum one is a two-dimensional crystal lattice. A biological one is the process of spreading infection or interaction of populations [10]. In [4] the author has proposed a model of an isolated beating heart.

One-dimensional systems with piecewise constant parameters have been studied theoretically in [11]. In that work it has been considered the cases of one or two jumps of the functions and the accurate theory of these systems has been presented. Two-dimensional systems have been the subjects of investigations

The reported study was founded by RFBR according to the research project 18-37-00059/18.

ⓒ Springer Nature Switzerland AG 2019
V. M. Vishnevskiy et al. (Eds.): DCCN 2019, CCIS 1141, pp. 139–151, 2019.
https://doi.org/10.1007/978-3-030-36625-4_12

in [5,12]. The authors of [12], for example, have considered systems with periodic coefficients. Three-dimensional ones have been researched in [2], and four-dimensional ones have been under consideration, for example, in [6,13]. The theories of linear [14] and nonlinear [15] systems have also been described in the scientific literature.

For deterministic fourth-order dynamical systems with constant parameters the analysis has been carried out quite deeply and widely presented in the scientific literature. But for systems with variable parameters there is no general method of analysis for today. As rule numerical [8,9,12] methods have been used to solve such problems. For systems with piecewise constant parameters the method based on the fundamental solution matrix has been used [5,6]. For four-dimensional systems this is a 4 × 4 matrix. In this case, a resulting matrix must be obtained as a product of interval matrices with constant parameters. However for systems with a large number of intervals, a resulting expression is unwieldy and it cannot be a subject of future analysis. In fact the matrix of fundamental solutions has been written only for two intervals. Moreover, in a number of practical problems, it is necessary to satisfy the conditions of continuity for the value proportional to the first derivative but not for the first derivative.

Solving this problem is of great importance to create linear communication devices based on anisotropic materials. So, in the general case, an elliptically polarized wave is propagating in an anisotropic medium [6,16–18], and in the classical theory such a wave is decomposed into waves of right- and left-circular polarization. But as rule it is necessary to carry out analysis of waves of linear polarization when developing specific devices, i.e. it is necessary to consider field components in two orthogonal planes. Earlier this problem is solved for homogeneous media only. However real linear devices of communication systems as rule include inhomogeneous, in particular stratified, anisotropic structures. And for this case, there is no solution for today. Thus the solution of this problem is of great practical importance.

In problems of satellite communication systems, when waves propagate in upper atmosphere, a medium can be considered as inhomogeneous anisotropic. Moreover a wave is elliptically polarized in this case. Therefore investigation of polarization plane rotation is very important for such a system.

Then the analytical model of wave behaviour will solve a number of important problems in area of communication systems. There are scattering, reflection, attenuation of waves among these problems.

In this paper the new analytical method that allows us to investigate a four-dimensional linear parametric system with arbitrary piecewise constant parameters is presented. The important feature of the considered systems is the fact that the characteristic equation of the initial differential equation system for each interval with constant parameters must be bi-square. Such systems correspond, for example, to the case of harmonic wave propagation under an oblique angle within a longitudinally magnetized gyrotropic stratified medium. The results are also important for analyzing the behavior of parametric amplifiers, control Bragg filters, studying the interaction of various biological populations and social systems. Despite the fact that the majority of real dynamic systems are non-linear, the results obtained by the authors are of great practical importance, since they are the first approximation in the solution for non-linear systems. Of course, there can be no jump-like changes in the parameters of nature systems, but such an approximation is used quite often [5,6] in the scientific literature.

2   Statement of the Problem

In this paper we consider a system described by the fourth order linear ordinary homogeneous differential equations with arbitrary piecewise-constant coefficients [16]

$$c_{11}(t)\frac{\partial^2 U(t)}{\partial t^2} + c_{12}(t)\frac{\partial^2 V(t)}{\partial t^2} + \omega_1^2(t)U(t) + \alpha_1(t)V(t) = 0$$
$$c_{21}(t)\frac{\partial^2 U(t)}{\partial t^2} + c_{22}(t)\frac{\partial^2 V(t)}{\partial t^2} + \omega_2^2(t)V(t) + \alpha_2(t)U(t) = 0 \tag{1}$$

where in a general case the coefficients are complex functions of the independent variable t:

$$\omega_k = \begin{cases} \omega_{k,1}, & 0 < t < t_1 \\ \omega_{k,2}, & t_1 < t < t_2 \\ \cdots & \cdots \\ \omega_{k,n}, & t_{n-1} < t < T \end{cases} \tag{2}$$

$$\alpha_k = \begin{cases} \alpha_{k,1}, & 0 < t < t_1 \\ \alpha_{k,2}, & t_1 < t < t_2 \\ \cdots & \cdots \\ \alpha_{k,n}, & t_{n-1} < t < T \end{cases} \tag{3}$$

$$c_k = \begin{cases} c_{k,1}, & 0 < t < t_1 \\ c_{k,2}, & t_1 < t < t_2 \\ \cdots & \cdots \\ c_{k,n}, & t_{n-1} < t < T \end{cases} \tag{4}$$

Here i is the number of an interval with constant parameters. The example of parameters for the twelve intervals is shown in Fig. 1. Each of the parameters can jump for any value of the independent variable t.

As it is noted above, system the (1) describes the large class of problems in the mechanic and electromagnetic wave theory, quantum physics, etc. At the same time in a number of applied problems it is necessary to satisfy the continuity conditions not for the functions and for their first derivatives but for the functions and values proportional to their first derivatives $h_1(z) = \partial U/\partial t$, $h_2(z) = \partial V/\partial t$ at boundaries of intervals with constant parameters.

Additionally note that to correct using the presented below method there is the main requirement for the considered systems: the characteristic equation of

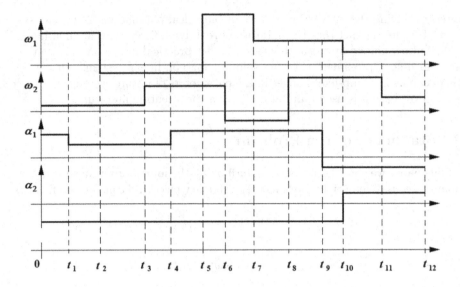

Fig. 1. The possible coefficients of the system (1)

the initial differential Eq. (1) must be necessarily bi-square. It can be the system of four first-order system

$$\frac{\partial U_1}{\partial t} = a_{11}(t)V_1 + a_{12}(t)V_2$$

$$\frac{\partial U_2}{\partial t} = a_{21}(t)V_1 + a_{22}(t)V_2$$

$$\frac{\partial V_1}{\partial t} = b_{11}(t)U_1 + b_{12}(t)U_2 \tag{5}$$

$$\frac{\partial V_1}{\partial t} = b_{21}(t)U_1 + b_{22}(t)U_2$$

that can be reduced to the system (1) without problems. Let us assume that all four functions U_1, U_2, V_1, V_2 must be continuity at interval boundaries.

Development of the accurate analytical method, convenient to practical uses in applied sciences, is our main aim here. First of all, representation of a fundamental solution matrix as a block diagonal matrix with 2×2 blocks would be preferable in most applied problems. However such a matrix has not been presented in the scientific literature for today. To solve this problem it is necessary to present fundamental solution matrices of each interval with constant coefficients in a block form, since a fundamental matrix of a system with piecewise-constant parameters can be obtained as an interval matrix product [5,6,16,19]. Here we also take into account the fact that the 4×4 matrix for the system with constant coefficients in the analytical form has been found [6,16]. The important solution requirement is that a block matrix and an initial fundamental solution matrix must be written in the same basis. Indeed, for example, in problems of layered media electrodynamics it is necessary to satisfy the continuity conditions for tangential components of the electromagnetic field but not for derivatives. Therefore, the transformation to other coordinate system is impractical in our case. Similar requirements are imposed to hydrodynamics problem solutions. Thus the fourth-order differential equations system on the i-th interval with constant coefficients

$$\frac{\partial U_1}{\partial t} = a_{11}V_1 + a_{12}V_2$$

$$\frac{\partial U_2}{\partial t} = a_{21}V_1 + a_{22}V_2$$

$$\frac{\partial V_1}{\partial t} = b_{11}U_1 + b_{12}U_2 \tag{6}$$

$$\frac{\partial V_1}{\partial t} = b_{21}U_1 + b_{22}U_2$$

has a 4×4-matrix of fundamental solutions

$$\mathbf{L}_i = \begin{pmatrix} m_{11}, m_{12}, m_{13}, m_{14} \\ m_{21}, m_{22}, m_{23}, m_{24} \\ m_{31}, m_{32}, m_{33}, m_{34} \\ m_{41}, m_{42}, m_{43}, m_{44} \end{pmatrix} \tag{7}$$

The elements of (7) has been written in the analytical form in the elementary functions [6]. The fundamental solution matrix of a system with piecewise constant parameters is hereby found as a product of interval matrices. However, working with 4×4 matrices in solving applied problems is not convenient, since expressions are unwieldy. We propose the transformation of this matrix to a block diagonal matrix with 2×2 blocks:

$$\mathbf{L}_i = \begin{pmatrix} \mathbf{M}_i, & 0 \\ 0, & \mathbf{N}_i \end{pmatrix} \tag{8}$$

The resulting matrix for N intervals with constant parameters under the continuity function condition can be also found as a product of the interval matrices (8).

3   Method

To solve the (8) we use the substation

$$\begin{aligned} U_1 &= U_{11} + U_{12} \\ U_2 &= \xi_1 U_{11} + \xi_2 U_{12} \\ V_1 &= \zeta_1 V_{11} + \zeta_2 V_{12} \\ V_2 &= V_{11} + V_{12} \end{aligned} \tag{9}$$

where U_{11} and V_{11} are the components of one of the two eigenmodes, U_{12} and V_{12} are the components of the second eigenmode. Then the Eq. (6) can be reduced to the four first order equations

$$
\begin{aligned}
\frac{\partial U_{11}}{\partial t} &= \gamma_1 U_{11} \\
\frac{\partial U_{12}}{\partial t} &= \gamma_2 U_{12} \\
\frac{\partial V_{11}}{\partial t} &= \chi_1 V_{11} \\
\frac{\partial V_{12}}{\partial t} &= \chi_2 V_{12}
\end{aligned}
\tag{10}
$$

where

$$
\begin{aligned}
\gamma_1 &= \frac{a_{11}\xi_2 + a_{12}\zeta_1\xi_2 - a_{21} - a_{22}\zeta_1}{\xi_2 - \xi_1} \\
\gamma_2 &= \frac{a_{11}\xi_1 + a_{12}\zeta_2\xi_1 - a_{21} - a_{22}\zeta_2}{\xi_2 - \xi_1} \\
\chi_1 &= \frac{b_{21}\zeta_2 + b_{22}\xi_1\zeta_2 - b_{11} - b_{12}\xi_1}{\xi_2 - \xi_1} \\
\chi_2 &= \frac{b_{21}\zeta_1 + b_{22}\xi_2\zeta_1 - b_{11} - b_{12}\xi_2}{\xi_2 - \xi_1}
\end{aligned}
\tag{11}
$$

Here the influence coefficients are defined by the expressions

$$
\xi_{1,2} = \frac{a_{22}b_{22} - a_{11}b_{11} + a_{21}b_{12} - a_{12}b_{21}}{2(a_{11}b_{12} + a_{12}b_{22})}
$$
$$
\times \left[1 \pm \sqrt{1 + \frac{4(a_{22}b_{21} + a_{21}b_{11})(a_{11}b_{12} + a_{12}b_{22})}{(a_{11}b_{11} - a_{22}b_{22} + a_{12}b_{21} - a_{21}b_{12})^2}} \right]
\tag{12}
$$

$$
\zeta_{1,2} = -\frac{b_{12}\xi_{1,2} + b_{11}}{b_{22}\xi_{1,2} + b_{21}}
\tag{13}
$$

Then we obtain the differential equations for eigenmodes

$$
\frac{\partial^2 U_{11}}{\partial t^2} - \gamma_1 \chi_1 U_{11} = 0
$$
$$
\frac{\partial^2 U_{12}}{\partial t^2} - \gamma_2 \chi_2 U_{12} = 0
\tag{14}
$$

from (10). Taking into account the obtained results the matrix of fundamental solutions of the Eq. (6) for the eigenmodes on an interval with constant parameters can be written in the form:

$$
\mathbf{M}_{1,2} = \begin{pmatrix} \cosh \Omega_{1,2}t & \dfrac{\gamma_{1,2}}{\Omega_{1,2}} \sinh \Omega_{1,2}t \\ \dfrac{\Omega_{1,2}}{\gamma_{1,2}} \sinh \Omega_{1,2}t & \cosh \Omega_{1,2}t \end{pmatrix}
\tag{15}
$$

where from (14)

$$\Omega_{1,2} = \sqrt{\gamma_{1,2}\chi_{1,2}} \tag{16}$$

are frequencies of the system eigenmodes. The matrices (15) are obtained by well-known method [17] for the initial conditions

$$\begin{pmatrix} U_{11}(0) \\ V_{11}(0) \end{pmatrix} = \begin{pmatrix} 1 \\ 0 \end{pmatrix} \quad \begin{pmatrix} U_{11}(0) \\ V_{11}(0) \end{pmatrix} = \begin{pmatrix} 0 \\ 1 \end{pmatrix}$$

$$\begin{pmatrix} U_{12}(0) \\ V_{12}(0) \end{pmatrix} = \begin{pmatrix} 1 \\ 0 \end{pmatrix} \quad \begin{pmatrix} U_{12}(0) \\ V_{12}(0) \end{pmatrix} = \begin{pmatrix} 0 \\ 1 \end{pmatrix} \tag{17}$$

Here $U_{11}(0)$, $U_{12}(0)$, $V_{11}(0)$, $V_{12}(0)$ are the mode components at $t = 0$. Therefore

$$\begin{pmatrix} U_1(0) \\ V_2(0) \end{pmatrix} = \begin{pmatrix} U_{11}(0) + U_{12}(0) \\ V_{11}(0) + V_{12}(0) \end{pmatrix}$$

$$= \begin{pmatrix} 1 & 0 \\ 0 & 1 \end{pmatrix} \begin{pmatrix} U_{11}(0) \\ V_{11}(0) \end{pmatrix} + \begin{pmatrix} 1 & 0 \\ 0 & 1 \end{pmatrix} \begin{pmatrix} U_{12}(0) \\ V_{12}(0) \end{pmatrix}$$

$$= \mathbf{M}_1 \begin{pmatrix} U_{11}(0) \\ V_{11}(0) \end{pmatrix} + \mathbf{M}_1 \begin{pmatrix} U_{11}(0) \\ V_{11}(0) \end{pmatrix} \tag{18}$$

where $U_1(0)$, $V_2(0)$ are the resulting functions at $t = 0$. Analogously at an arbitrary t

$$\begin{pmatrix} U_1(t) \\ V_2(t) \end{pmatrix} = \begin{pmatrix} U_{11}(t) \\ V_{11}(t) \end{pmatrix} + \begin{pmatrix} U_{12}(t) \\ V_{12}(t) \end{pmatrix}$$

$$= \mathbf{M}_1(t) \begin{pmatrix} U_{11}(0) \\ V_{11}(0) \end{pmatrix} + \mathbf{M}_2(t) \begin{pmatrix} U_{12}(0) \\ V_{12}(0) \end{pmatrix} \tag{19}$$

Note that the expressions (18) and (19) are key ones in our transformations. Let us consider the expressions (18) and (19) in detail. It is followed from (18) and (19) that the matrices $\mathbf{M}_1(t)$ and $\mathbf{M}_2(t)$ relate the functions $U_1(t)$ and $V_2(t)$ at the arbitrary point t to these components at the point $t = 0$. However, for utilizing these relations we must know $U_{11}(0)$, $U_{12}(0)$, $V_{11}(0)$, $V_{12}(0)$ at $t = 0$ that are determined by boundary conditions, and therefore by the parameters of a previous interval with constant parameters. Moreover boundary conditions at interfaces of intervals are written for the components $U_1(t)$ and $V_2(t)$ but not for the eigenmode components $U_{11}(t)$, $U_{12}(t)$, $V_{11}(t)$, $V_{12}(t)$. Thus the eigenmode components can be discontinuous at interval interfaces and continuity conditions must be applied to $U_1(t)$ and $V_2(z)$ only. Therefore let us write

$$\mathbf{M}(t) \begin{pmatrix} U_1(0) \\ V_2(0) \end{pmatrix} = \mathbf{M}_1(t) \begin{pmatrix} U_{11}(0) \\ V_{11}(0) \end{pmatrix} + \mathbf{M}_2(t) \begin{pmatrix} U_{12}(0) \\ V_{12}(0) \end{pmatrix} \tag{20}$$

or in the scalar form

$$M_{11}U_1(0) + M_{12}U_2(0)$$
$$= m_{11}^{(1)}U_{11}(0) + m_{12}^{(1)}V_{11}(0) + m_{11}^{(2)}U_{12}(0) + m_{12}^{(2)}V_{12}(0) \tag{21}$$
$$M_{21}U_1(0) + M_{22}U_2(0)$$
$$= m_{21}^{(1)}U_{11}(0) + m_{22}^{(1)}V_{11}(0) + m_{21}^{(2)}U_{12}(0) + m_{22}^{(2)}V_{12}(0)$$

where $m_{11}^{(1)}$, $m_{12}^{(1)}$, $m_{21}^{(1)}$, $m_{22}^{(1)}$ are the elements of the matrix $\mathbf{M}_1(t)$; $m_{11}^{(2)}$, $m_{12}^{(2)}$, $m_{21}^{(2)}$, $m_{22}^{(2)}$ are the elements of the matrix $\mathbf{M}_2(t)$; M_{11}, M_{12}, M_{21}, M_{22} are the elements of the matrix $\mathbf{M}(t)$. Then taking into account (9) we obtain

$$M_{11}U_{11}(0) + M_{11}U_{12}(0)$$
$$+ M_{12}V_{11}(0) + M_{12}V_{12}(0) = m_{11}^{(1)}U_{11}(0)$$
$$+ m_{12}^{(1)}V_{11}(0) + m_{11}^{(2)}U_{12}(0) + m_{12}^{(2)}V_{12}(0)$$
$$M_{21}U_{11}(0) + M_{21}U_{12}(0) \tag{22}$$
$$+ M_{22}V_{11}(0) + M_{22}V_{12}(0) = m_{21}^{(1)}U_{11}(0)$$
$$+ m_{22}^{(1)}V_{11}(0) + m_{21}^{(2)}U_{12}(0) + m_{22}^{(2)}V_{12}(0)$$

Now form (10) we have

$$V_{1,2}(t) = \frac{\Omega_{1,2}}{\gamma_{1,2}}U_{1,2}(t) = -\sqrt{\frac{\chi_{1,2}}{\gamma_{1,2}}}U_{1,2}(t) \tag{23}$$

Then substituting (23) in the first Eq. (22) we have

$$M_{11}U_{11}(0) + M_{11}U_{12}(0) + \sqrt{\frac{\chi_1}{\gamma_1}}M_{12}V_{11}(0)$$
$$+ \sqrt{\frac{\chi_1}{\gamma_1}}M_{12}V_{12}(0) = m_{11}^{(1)}U_{11}(0) \tag{24}$$
$$+ \sqrt{\frac{\chi_1}{\gamma_1}}m_{12}^{(1)}U_{11}(0) + m_{11}^{(2)}U_{12}(0) + \sqrt{\frac{\chi_2}{\gamma_2}}m_{12}^{(2)}U_{12}(0)$$

This equality must be satisfied for any $U_{11}(0)$ and $U_{12}(0)$. Therefore we obtain

$$M_{11} + \sqrt{\frac{\chi_1}{\gamma_1}}M_{12} = m_{11}^{(1)} + \sqrt{\frac{\chi_1}{\gamma_1}}m_{12}^{(1)}$$
$$M_{11} + \sqrt{\frac{\chi_2}{\gamma_2}}M_{12} = m_{11}^{(2)} + \sqrt{\frac{\chi_2}{\gamma_2}}m_{12}^{(2)} \tag{25}$$

Solving (25) we find

$$M_{11} = m_{11}^{(2)} + \sqrt{\frac{\chi_2}{\gamma_2}} m_{12}^{(2)}$$
$$- \sqrt{\frac{\chi_2}{\gamma_2}} \frac{m_{11}^{(1)} - m_{11}^{(2)} + \sqrt{\frac{\chi_1}{\gamma_1}} m_{12}^{(1)} - \sqrt{\frac{\chi_2}{\gamma_2}} m_{12}^{(2)}}{\sqrt{\frac{\chi_1}{\gamma_1}} - \sqrt{\frac{\chi_2}{\gamma_2}}}$$
$$M_{12} = \frac{m_{11}^{(1)} - m_{11}^{(2)} + \sqrt{\frac{\chi_1}{\gamma_1}} m_{12}^{(1)} - \sqrt{\frac{\chi_2}{\gamma_2}} m_{12}^{(2)}}{\sqrt{\frac{\chi_1}{\gamma_1}} - \sqrt{\frac{\chi_2}{\gamma_2}}}$$

(26)

Analogously from the second equation of (22) we can write

$$M_{21} = m_{21}^{(2)} + \sqrt{\frac{\chi_2}{\gamma_2}} m_{22}^{(2)}$$
$$- \sqrt{\frac{\chi_2}{\gamma_2}} \frac{m_{21}^{(1)} - m_{21}^{(2)} + \sqrt{\frac{\chi_1}{\gamma_1}} m_{22}^{(1)} - \sqrt{\frac{\chi_2}{\gamma_2}} m_{22}^{(2)}}{\sqrt{\frac{\chi_1}{\gamma_1}} - \sqrt{\frac{\chi_2}{\gamma_2}}}$$
$$M_{22} = \frac{m_{21}^{(1)} - m_{21}^{(2)} + \sqrt{\frac{\chi_1}{\gamma_1}} m_{22}^{(1)} - \sqrt{\frac{\chi_2}{\gamma_2}} m_{22}^{(2)}}{\sqrt{\frac{\chi_1}{\gamma_1}} - \sqrt{\frac{\chi_2}{\gamma_2}}}$$

(27)

Substituting the elements of the matrix (15) in (26) and (27) we obtain

$$\mathbf{M}(t) = \begin{pmatrix} -\dfrac{\sqrt{\chi_2\gamma_1}\exp(\Omega_1 t) - \sqrt{\chi_1\gamma_2}\exp(\Omega_2 t)}{\sqrt{\chi_1\gamma_2} - \sqrt{\chi_2\gamma_1}} & -\sqrt{\chi_1\chi_2}\dfrac{\exp(\Omega_1 t) - \exp(\Omega_2 t)}{\sqrt{\chi_1\gamma_2} - \sqrt{\chi_2\gamma_1}} \\[2ex] \sqrt{\gamma_1\gamma_2}\dfrac{\exp(j\Omega_1 t) - \exp(\Omega_2 t)}{\sqrt{\chi_1\gamma_2} - \sqrt{\chi_2\gamma_1}} & \dfrac{\sqrt{\chi_1\gamma_2}\exp(\Omega_1 t) - \sqrt{\chi_2\gamma_1}\exp(\Omega_2 t)}{\sqrt{\chi_1\gamma_2} - \sqrt{\chi_2\gamma_1}} \end{pmatrix}$$

(28)

Analogously it can be found the matrix $\mathbf{N}(t)$ that describes the components $U_2(t)$, $V_1(t)$

$$\mathbf{N}(t) = \begin{pmatrix} -\dfrac{\zeta_2\xi_1\sqrt{\chi_2\gamma_1}\exp(\Omega_1 t) - \zeta_1\xi_2\sqrt{\chi_1\gamma_2}\exp(\Omega_2 t)}{\zeta_1\xi_2\sqrt{\chi_1\gamma_2} - \zeta_2\xi_1\sqrt{\chi_2\gamma_1}} & -\zeta_1\zeta_2\sqrt{\chi_1\chi_2}\dfrac{\exp(\Omega_1 t) - \exp(\Omega_2 t)}{\zeta_1 x i_2\sqrt{\chi_1\gamma_2} - \zeta_2 x i_1\sqrt{\chi_2\gamma_1}} \\[2ex] \xi_1\xi_2\sqrt{\gamma_1\gamma_2}\dfrac{\exp(j\Omega_1 t) - \exp(\Omega_2 t)}{\zeta_1\xi_2\sqrt{\chi_1\gamma_2} - \zeta_2\xi_1\sqrt{\chi_2\gamma_1}} & \dfrac{\zeta_1\xi_2\sqrt{\chi_1\gamma_2}\exp(\Omega_1 t) - \zeta_2\xi_1\sqrt{\chi_2\gamma_1}\exp(\Omega_2 t)}{\zeta_1\xi_2\sqrt{\chi_1\gamma_2} - \zeta_2\xi_1\sqrt{\chi_2\gamma_1}} \end{pmatrix}$$

(29)

Then the fundamental solution matrix in the form (8) can be written instead the 4×4-matrix [17,18].

The resulting matrix of a system with piecewise parameters in accordance to well-known theory must be found as product of interval matrices. In the considered case it is following

$$
\begin{aligned}
\mathbf{L}_{\Sigma} &= \begin{pmatrix} \mathbf{M}(T) & 0 \\ 0 & \mathbf{N}(T) \end{pmatrix} = \\
&\begin{pmatrix} \prod_{i=P}^{1} \mathbf{M}(t_i) & 0 \\ 0 & \prod_{i=P}^{1} \mathbf{N}(t_i) \end{pmatrix}
\end{aligned}
\tag{30}
$$

Thus in this work the fundamental solution matrix of the fourth order differential equation system in the kind of the diagonal block matrix with 2×2-blocks is found.

4   System with Periodic Coefficients

One of the important scientific problems is investigating a periodic two-dimensional system with periodic piecewise parameters. Moreover, the problem of determining boundaries of solution stability regions is one of the main problems for systems with periodic parameters. Such a problem is important, for example, for determining the conditions of parametric resonance of mechanical systems, excitation conditions of parametric generators, bandgaps of Bragg filters. These boundaries correspond to periodic solutions and in accordance with Lyapunov's theory [19] are determined by eigenvalues of the fundamental solution matrix. Thus, if all eigenvalues are less or equal to one in absolute value then the solutions are stable. For the block matrix (8), four eigenvalues are defined as solutions of its fourth-order characteristic equation. In this case the characteristic equation is return [19] and it can be decomposed into two square ones:

$$
\begin{aligned}
\lambda^2 + \mathrm{tr}\mathbf{M}(T)\lambda + \det \mathbf{M}(T) = 0 \\
\lambda^2 + \mathrm{tr}\mathbf{N}(T)\lambda + \det \mathbf{N}(T) = 0
\end{aligned}
\tag{31}
$$

Therefore the stability conditions of the considered system are

$$
\begin{aligned}
-\frac{\mathrm{tr}\mathbf{M}(T)}{2} + \sqrt{\frac{\mathrm{tr}^2\mathbf{M}(T)}{4} - \det\mathbf{M}(T)} \leq |1| \\
-\frac{\mathrm{tr}\mathbf{M}(T)}{2} - \sqrt{\frac{\mathrm{tr}^2\mathbf{M}(T)}{4} - \det\mathbf{M}(T)} \leq |1| \\
-\frac{\mathrm{tr}\mathbf{N}(T)}{2} + \sqrt{\frac{\mathrm{tr}^2\mathbf{N}(T)}{4} - \det\mathbf{N}(T)} \leq |1| \\
-\frac{\mathrm{tr}\mathbf{N}(T)}{2} - \sqrt{\frac{\mathrm{tr}^2\mathbf{N}(T)}{4} - \det\mathbf{N}(T)} \leq |1|
\end{aligned}
\tag{32}
$$

Thus the presented approach greatly simplifies the solution of the stability problem for a two-dimension parametric system with piecewise constant parameters, since now we are talking about analyzing 2×2 matrices instead of 4×4 matrices.

5   Numerical Example

In this section we present the numerical example that demonstrate important practical application of the development mathematical model.

Here it is considered the Bragg filter based on the stratified ferrite structure. The filter contains the double-layered periods described by the scalar permittivity ϵ and the dyadic magnetic permeability

$$\mu = \begin{pmatrix} \mu_{xx} & -j\mu_{xy} & 0 \\ j\mu_{xy} & \mu_{xx} & 0 \\ 0 & 0 & \mu_{zz} \end{pmatrix} \tag{33}$$

were $j = \sqrt{-1}$. The geometry of the problem is shown in Fig. 2. In the figure: z is the anisotropy axis, z' is the incident wave direction. Here it is assumed that structure is infinite in x and y directions, and it is stratified along z axis. In general case an elliptically polarized wave propagates within the structure. We decomposed such a wave into two linear polarized waves. In the considered example the wave frequency is 4 GHz, the incidence angle is 50°. The elements of the dyadic (33) for the first layer of the period is $\mu_{xx} = 2.2$, $\mu_{xy} = 0.5$, $\mu_{zz} = 0.999$. The elements of (33) for the second layer of the period is $\mu_{xx} = 3.5$, $\mu_{xy} = 1.7$, $\mu_{zz} = 0.999$.

The presented method allows us to study bandgap structure of the filter. Indeed, in accordance to Liapunov theory a system is stable if all eigennumbers of a fundamental matrix are less or equal to unite. This case corresponds to passband of the Bragg filter. In other case a system is unstable, and this corresponds to stopband of the filter. The results of the numerical calculations of the fundamental matrix eigennumbers are presented in Fig. 3. Here we can see

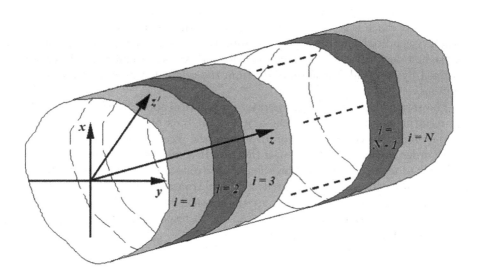

Fig. 2. The geometry of the problem

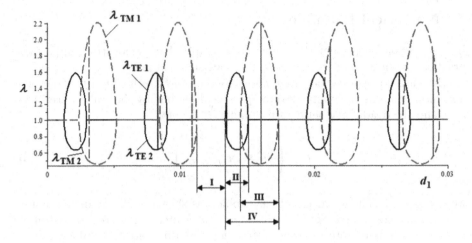

Fig. 3. The bandgap structure of the filter

the dependence of the eigennumber modules on the first layer thickness if the thickness of the second layer is 1 mm.

The regions with $\lambda = 1$ correspond to passband of the Bragg filter. And we have stopbands if $\lambda \neq 1$. Therefore the region I is passband and IV is stopband.

6   Conclusion

In this paper we propose the transformation of the 4×4 matrix of fundamental solutions to the block diagonal matrix with 2×2 blocks for the first time. The block matrix is found in the same coordinate system as the original 4×4 matrix. Such the approach is convenient for the cases when it is necessary to satisfy the boundary conditions for the functions proportional to the first derivative but not for the first derivatives of the functions. All expressions are analytic and they are obtained by identical transformations from the original system of fundamental solutions.

The obtained model is very important for practical applications. For example, the resulting matrix simplifies the stability analysis of the considered fourth order system. Indeed, now we can investigate the second order Eq. (31) instead fourth order equation in the classical theories. In some cases, in particular in the problems of reflection and transmission of electromagnetic waves in anisotropic materials, the expressions for the reflection and transmission coefficients are also much simpler, and now we must not use the classical 2×2 reflection or translation matrices for anisotropic media but we can the corresponding simple expressions for isotropic media.

References

1. Torok, J.S.: Analytical Mechanics: With an Introduction to Dynamical Systems. Wiley, New York (1999)
2. Amerongen, J.: Dynamical Systems for Creating Technology. Controllab Products B.V., Enschede (2010)
3. Hirsch, M.W., Smale, S., Devaney, R.L.: Differential Equations, Dynamical Systems, and Introduction to Chaos. Elsevier Academic press, New York (2004)
4. Fruchter, G., Ben-Haim, S.: Stability analysis of one-dimensional dynamical systems applied to an isolated beating heating. J. Theor. Biol. 148, 175–192 (1991)
5. Vytovtov, K.A.: Analytical investigation of stratified isotropic media. J. Opt. Soc. Am. A 22(4), 689–696 (2005)
6. Vytovtov, K.: An analytical method for investigating periodic stratified media with uniaxial bianisotropy. Radiotekhnika i electronika 46(2), 159–165 (2001)
7. Vytovtov, K., Barabanova, E., Zouhdi, S.: Optical switching cell based on metamaterials and ferrite films. In: 12th International Congress on Artificial Materials for Novel Wave Phenomena - Metamaterials, Espoo, Finland (2008)
8. Khalili, F.Ya.: Quantum oscillation systems. Physical department of MSU, Moscow (2008). (in Russian)
9. Majumdar, M., Gidea, M., Niculescu, C.P.: Chaotic dynamical systems: an introduction. Econ. Theory 4(5), 641–648 (1994)
10. Bratus, A.S., Novojilov, A.S., Platonov, F.P.: Dynamic Systems and Models in Biology. Fizmatlit, Moscow (2010). (in Russian)
11. Tumwesigye, A.B.: On one-dimensional systems and commuting elements in noncommutative algebras. Malardalen University Press (2016)
12. Rossetto, B., Zhang, Y.: Two-dimensional dynamical system with periodic coefficients. J. Bifurc. Chaos 19(11), 3777–3790 (2009)
13. Gilchrist, O.: The free oscillations of conservative quasi-linear systems with two degrees of freedom. Int. J. Mech. Sci. 3, 286–311 (1961)
14. Chen, C.T.: Linear System Theory and Design. Oxford University Press Inc., Oxford (1998)
15. Khalil, H.K.: Nonlinear Systems, 3rd edn. Prentice Hall, Upper Saddle River (2002)
16. Vytovtov, K.A., Tarasenko, Yu.S.: Analytical investigation of one-dimensional magneto electric photonic crystals. The 2x2 matrix approach. J. Opt. Soc. Am. A: 24(11), 3564–3572 (2007)
17. Passler, N.C., Paarmann, A.: Generalized 4×4 matrix formalism for light propagation in anisotropic stratified media: study of surface phonon polaritons in polar dielectric heterostructures. J. Opt. Soc. Am. B 34(10), 2128–2139 (2017)
18. Mounier, D., Kouyate, M., Pezeril, T., Vaudel, G., Ruello, P., Picart, P., Breteau, J., Gusev, V.: 4×4 matrix algebra in the theory of optical detection of picosecond acoustic pulses in anisotropic media. Chin. J. Phys. 49(1), 191–200 (2011)
19. Gantmacher, F.R.: Applications of the Theory of Matrices, 336P. Dover Publications, New York (2005)

PD-NOMA Power Coefficients
Calculation While Using QAM Signals

Ya. V. Kryukov$^{(\boxtimes)}$ [ID], D. A. Pokamestov[ID], and E. V. Rogozhnikov[ID]

Tomsk State University of Control Systems and Radioelectronics, Tomsk, Russia
kryukov.tusur@gmail.com
https://tusur.ru

Abstract. Non-orthogonal multiple access method with power channel division is one of the most perspective multiple access methods for the next generation mobile communication systems. One of the actual issues is the power calculation and distribution method development between user channels. The current paper provides a power coefficients calculation method for channel multiplexing with symmetric QAM modulation, while using apriori information about the radio link estimation at each receiving side. The proposed method effectiveness is shown while using the mathematical simulation.

Keywords: NOMA · PD-NOMA · Multiple access · Power allocation · Channel multiplexing · Non-orthogonal multiple access

1 Introduction

The spectral efficiency increasing problem in wireless mobile communication systems is one of the most important, cause the frequency resource is limited. Thus, new solutions will be developed to improve the bandwidth usage efficiency. Today there are several ways to improve spectral efficiency. One of them is the new multiple access method usage.

The PD (Power Domain) multiple access method which implies the channel multiplexing is one of the NOMA (Non-Orthogonal Multiple Access) methods proposed to be used in the fifth-generation wireless broadband communication systems [1–6]. PD-NOMA method implies the user channels (or layers) to be multiplexed by power in a single frequency band [7–12]. Thus, each user channel frequency band increases and each channel power coefficient is calculated based on each user's transmission channel estimate. The user with a low signal-to-noise ratio in the channel is provided with the largest power share and vice versa. It becomes possible to distribute the communication resource between users in three domains at once, namely time, frequency and power while using PD-NOMA and OFDMA together. PD-NOMA method is non-orthogonal, since the user channels interfere with each other, however, this interference is controlled while multiplexing. From an theoretic perspective, it is known that non-orthogonal channel multiplexing using superposition coding at the transmitter

© Springer Nature Switzerland AG 2019
V. M. Vishnevskiy et al. (Eds.): DCCN 2019, CCIS 1141, pp. 152–162, 2019.
https://doi.org/10.1007/978-3-030-36625-4_13

side and successive interference cancellation (SIC) at the receiver side not only outperforms orthogonal multiplexing, but also is optimal in the sense of achieving the capacity region of the downlink broadcast channel [13,14]. Note that NOMA can also be applied to uplink (multiple access channel) with SIC being applied at the base station side [15].

International researchers group confirmed that additional PD-NOMA use can provide spectral efficiency gain within the conditions where the user telecommunication channels characteristics are not the same and have a significant difference.

2   PD-NOMA System Model

The key idea of PD-NOMA is the using of the power domain for multiple access to simultaneously serve users in the same time-frequency resource element. Thus, the users have different values of signal power. According to PD-NOMA, a user with a weak communication channel gets the largest power coefficient, and a user with a strong channel gets the least power coefficient. In PD-NOMA, the transmitter applies superposition of the signals from all users and then allocates more power to the user with lower signal-to-interference and noise ratio SINR (worse channel conditions), which is far from the transmitter, and allocates less power to the user with higher SINR (better channel conditions), which is closer to the transmitter.

Let us consider a scenario, where several user devices are found in the coverage area of base station (eNB) and K is the number of users (UE). Each user channel is allocated partial power p_k, which determines the channel noise immunity and its capacity (Fig. 1) The power calculation of a specified user channel depends on the requirement to the capacity and channel condition $SINR_k$. Then common transport signal $S(i)$ is the superposition of channel symbols $x_k(i)$ with weight coefficients p_k:

$$S(i) = \sum_{k=1}^{K} \sqrt{p_k} x_k(i)$$

The structure of the PD-NOMA channel is shown in Fig. 1. The signal Z_k at the k-th receiver is a common transport signal S, passed through the k-th own channel of propagation:

$$Z_k = H_k \otimes S + N_k,$$

where Z_k is input signal, H_k is channel response and N_k is additive noise of the k-th receiver.

At the receiver terminals, each user uses SIC to demodulate their own signal, since all users receive the same superimposed signal $S(i)$. The SIC method consists of sequential signal demodulation with the highest power value with its subsequent regeneration and removing from the received group signal [15–17]. Then it becomes possible to demodulate the signal with the second high power value, and so on, until the weakest signal is demodulated. The structure of the SIC process is shown in Fig. 2. Error correcting encoding/decoding operations are

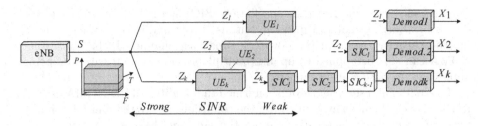

Fig. 1. The structure of the PD-NOMA channel

not required during regeneration, however, they allow correction bit errors and more accurately recover channel symbols. If SIC demodulation occurs without encoding/decoding, method is called SL-SIC (Symbol Level - SIC). Otherwise, the interference cancellation occurs at the codeword level and this method is called CL-SIC (Code Level - SIC) [18,19].

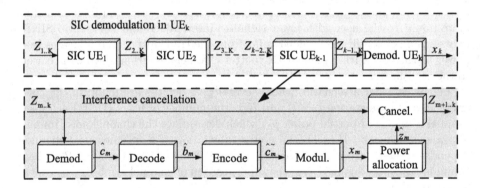

Fig. 2. The structure of the SIC process

This process is performed K times until the weakest user demodulates his own signal. The user located near the transmitter removes the signals of the other users as they act as interference. On the other hand, the farthest user, which has the highest allocated power and, therefore, the maximum contribution to the received superimposed signal, will initially demodulate his signal. The main challenge for the transmitter is to correctly assign the power coefficients p_k for each user according to the channel conditions SINR_k, as this affects the performance of the SIC.

The SINR at the demodulator input of the k-th PD-NOMA channel is calculated on the basis of the additive interference power of the k-th channel and the system interference of the remaining nondemodulated channels located in the same time-frequency resource element:

$$SINR_k = \begin{cases} \dfrac{\alpha_k p_k}{\alpha_k \sum\limits_{i=k+1}^{K} p_i + N_k} & 1 \le k < K \\[3mm] \dfrac{\alpha_k p_k}{N_k} & k = K \end{cases}$$

where N_k is the interference power of the kth channel, α_k is the channel attenuation coefficient, p_k is the power coefficient of the kth channel, p_i is the power coefficients of channels $i < k$. The sum of partial powers $\sum\limits_{i=k+1}^{K} p_i$ of channels $i < k$ is a systematic interference for channel k.

3 Issue Statement

The channel power allocation is one of the acute PD-NOMA issues [20] and is a key for effective PD-NOMA using. Since NOMA is based on the SIC order, the served users achieve unequal rates and this could be critical for scenarios with strict fairness constraints. The works in [21–26] analyze the performance of the several NOMA schemes in terms of outage probability and achievable sum-rate for a fixed power allocation without discussing potential fairness issues. Fairness can be supported through appropriate power allocation of the superimposed transmitted data flows. However, it is required to calculate the power coefficients of each user channel in any NOMA scheme. In [25] authors investigates the maximization of the secrecy sum rate of a SISO NOMA system, where each user has a predefined quality of service requirement.

The proposed algorithms of power coefficients calculation are based on the Shannon theorem [27]. However, none method considers the real signal constructions features, which impose a hard limit on the allowable channel power proportions between the channels.

A detection area is defined for each reference QAM constellation symbol. Exceeding the allowable power proportions may result in the detection areas intersection, which results in the demodulation errors increase. This is extremely critical for PD-NOMA due to the fact the channels are dependent (non-orthogonal) on each other, and the SIC demodulation result of each next channel depends on the previous one demodulation result.

Detection areas intersection examples during two QPSK channels compaction are shown in Fig. 2. Here the dotted line marks the area beyond which the signal sample hitting probability under the given power normal interference influence is close to zero. The channel powers permissible ratio scenario is provided in Fig. 1a. In this case, the lower (marked in blue) and upper layer (marked in red) symbols will be demodulated accurately.

The symbols within one quadrant are located too close to each other (Fig. 2c) in the event of insufficient upper layer power, which results in the symbol error probability increase during the upper layer signal demodulation. On the other hand, in the case of the upper layer overcapacity, part of adjacent quadrants symbols will also be too close to each other (Fig. 2b), which results in the symbol error probability increase when demodulating PD-NOMA channels lower layer signal. Therefore, it is required to calculate such upper and lower layers channel powers values during the compaction, at which reliable demodulation will be ensured and there will be no detection areas overlap.

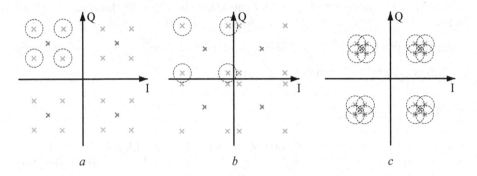

Fig. 3. Two QPSK channels power domain multiplexing: a - admissible channel power; b, c - unacceptable channel power

4   The Proposed Power Allocation Method for Two QAM Channels

The paper proposes a method for the allowable channel power ratio range calculation when multiplexing two PD-NOMA channels. It is theoretically possible to multiplex an infinite channels number, however, this will result in a multiple linear growth of the group signal formation and processing clearing complexity. It is inappropriate to produce a seal by power of over 2 channels for mobile communication systems since this results in a significant increase in the transmission channel and computational power quality requirements with a slight spectral efficiency increase.

The proposed method is based on the QAM symbols confident detection areas calculation for a given erroneous detection probability ρ_{er} and the known signal-to-noise ratio at the demodulator input. Let p_1 be the lower (most powerful) layer channel power while p_2 – the upper (least powerful) layer, and $\gamma = p_1/p_2$ the power ratio. The following restrictions are imposed on the p admissible values:

$$\begin{cases} p_{1,2} > 0 \\ p_1 > p_2 \\ p_2 = P - p_1 \end{cases} \qquad (1)$$

where P is the group signal power.

Let a famous apriori information on the mean square error (MSE) estimates of the normal interference $\sigma_{1,2}$ and the transmission coefficients $\alpha_{1,2}$, which can be received at the user receivers and transmitted to the base station base via the feedback channel. Let's calculate Z symmetric section (circle) radius relative to the constellation reference, which will get the received symbol under the normal interference influence with $(1 - \rho_{er})$ probability:

$$Z = \sigma * F_{inv}\left(\frac{(2 - \rho_{er})}{2}\right) \tag{2}$$

where $F_{inv}(t)$ is the $F(x)$ probability integral inverse function of the following form:

$$F(x) = \frac{2}{\sqrt{\pi}} \int_0^x e^{-t^2} dt \tag{3}$$

Calculate p_2 value, required to ensure the upper-level characters erroneous detection probability in the second user receiver to be no more than Rosh. Therefore, D_2 distance between the QAM constellation points must be at least $2Z_2$. Then the successful (with an error probability no more than ρ_{er}) inner constellation demodulation condition can be written the following way:

$$D_2/2 - Z_2 \geq 0 \tag{4}$$

D_k calculation expression in the k-th layer:

$$D_k = 2\sqrt{\frac{\alpha_k p_k}{M_k}} \tag{5}$$

where p_k is the k-th layer channel power and M_k is the k-th layer QAM normalization factor with Q_k index, which can be calculated:

$$M_k = \frac{2}{3}(Q_k - 1) \tag{6}$$

p_2 value, which implies the condition (3) fulfilling, can be calculated by substituting (2, 4, 5) in (3) and having resolved the square inequality with respect to p_2. Root of inequality (3) which satisfies the condition (1):

$$p_2^d \geq \frac{Z_2^2 \cdot M_2}{\alpha_2} \tag{7}$$

Thus, if the condition (6) is being fulfilled, the upper layer QAM-signal in the second user demodulator will be correctly demodulated with a symbol error probability of no more than ρ_{er}. Then the power ratio doubled (6):

$$\gamma_d \geq \frac{p_2^d}{P - p_2^d} \tag{8}$$

Failure (7) will result in the situation shown in the Fig. 1c. Then we'll get the outer constellation (lower layer) detection condition with the symbol error probability of no more than ρ_{er}. For this, the distance between D_1 outer constellation

points must be greater than the distances sum between the inner constellation
extreme points and $2Z_1$:

$$\frac{D_1}{2} - (\sqrt{Q_2} - 1)D_2 - Z_1 \geq 0 \tag{9}$$

So, we obtain a square inequality with respect to p_2 having substituted (2,
4, 5) in (8) and expressing $p_1 = P - p_2$. Root satisfying condition (1):

$$p_2^u \leq \frac{\alpha_1 P M_2 + G Z_1}{\alpha_1 V} - \frac{2G(M_2 Z_1 - \sqrt{(\alpha_1 P V - G Z_1)} + \sqrt{Q_2(\alpha_1 P V - G Z_1)})}{\alpha_1 V^2} \tag{10}$$

where the following replacements are introduced in order to simplify:

$$\begin{cases} G = 4M_1 M_2 Z_1 \\ V = 4M_1(1 - \sqrt{Q_2})^2 + M_2 \end{cases}$$

When performing (9), the lower layer QAM-signal will be demodulated with
the error probability of no more than ρ_{er}. Then the power ratio, which satisfies (8):

$$\gamma_u \geq \frac{p_2^u}{P - p_2^u} \tag{11}$$

So, in order to ensure successful group signal demodulation with the symbol
error probability of no more than ρ_{er}, the upper and lower layers power ratio
must lie within:

$$\gamma_d \leq \gamma \leq \gamma_u \tag{12}$$

The following steps can be taken if it is impossible to simultaneously fulfill
conditions (11) and (1) for any γ value:

(1) Increase group signal total power P.
(2) Increase error probability ρ_{er}.
(3) Reduce QAM modulation index of one of the layers.

The final γ value choice depends on which layer (upper or lower) requires
an additional noise immunity. The decision algorithm must consider the error
magnitude in transmission channel estimation and channel fluctuations degree.
For example, it's better to provide additional noise immunity to the second
channel if one of the channels is stable while the second is changeable over time.
The channel powers ratio can be obtained as arithmetic mean while having equal
channels priority:

$$\gamma = \frac{\gamma_u + \gamma_d}{2} \tag{13}$$

5   Simulation

The simulation goal is to evaluate the proposed power calculation method effectiveness while compacting two user channels, as well as comparing PD-NOMA and OFDMA channels noise immunity. PD-NOMA and OFDMA channels formation and processing occurred while being influenced by normal interference during the iteration simulation.

The network state is constantly changing in real communication systems due to both distribution channels, the users number and throughput changes etc. It was shown in the works [3–6] that PD-NOMA method is expedient to use when compacting user channels with the most different signal-to-noise ratio SINR values (for example, on the cell edge and in the cell center). Therefore, such simulation scenario is used in the work.

Simulation scenario. There are two user devices with time-varying amplitude-frequency transmission channel characteristic (Fig. 4) in the cell. OFDMA and PD-NOMA communication channels are organized with the required capacity for two users on the base station side. Both the estimated signal attenuation in $\alpha_{1,2}$ channels as well as the intrinsic thermal noise power level $N_{1,2}$ of both receivers are available when the channel is being formed. QAM modulation indexes $Q_{1,2}$ are selected based on the required bandwidth. The group signal power $P = 1\,W$. Error correction coding is not used, channel demultiplexing and bit errors number calculation SINR are estimated at the receiving side in each channel.

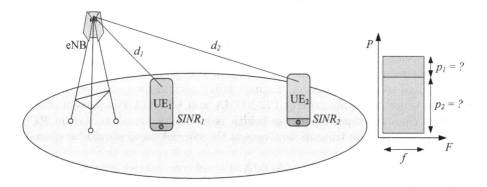

Fig. 4. Simulation scenario

PD-NOMA Multiplexing. The boundary values γ_d and γ_u are calculated with $\rho_{er} = 10^{-6}$ detection error probability while using (1–10) each time while considering $Q_{1,2}$ values. The calculation occurs at an increased ρ_{er} value if ρ_{er} can't be provided in the given propagation channel conditions. γ and $p_{1,2}$ channel powers value are calculated according to (12). Next, the user channels are multiplexing in the power domain in a single frequency band with $p_{1,2}$ portions and the group signal is transmitted to the radiochannel.

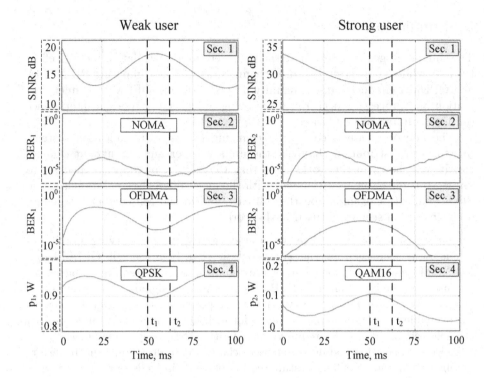

Fig. 5. Simulation result

OFDMA Multiplexing. OFDMA channel compaction is performed in the frequency domain, and the frequency band between users is divided equally, but both user channels use all of the available radiation power P in their own band. It is required to ensure the same channels bandwidth in the total system bandwidth in order to ensure correct PD-NOMA and OFDMA comparison. Each OFDMA channel transmission bandwidth is in 2 times smaller than in PD-NOMA, therefore the transmission rate in the current band should be several times higher.

The simulation result for PD-NOMA channels compaction with quadrature modulation indices $Q_1 = 4$ and $Q_2 = 16$ (respectively, $Q_1 = 16$, $Q_2 = 256$ for OFDMA) is provided in the Fig. 4. The time dependencies are provided for each user: Sect. 1 – SINR estimate at the demodulator input; Sect. 2 - BER in PD-NOMA channels; Sect. 3 - BER in OFDMA channels; Sect. 4 – channel power values (for PD-NOMA);

The dependencies, shown in the Fig. 4 confirm the users channel power calculation (Sect. 4) occurs adaptively to the transmission channel state (Sect. 1). BER value (Sects. 2 and 3) of both PD-NOMA channels changes proportionally, which confirms compliance with the priority equality.

Earlier it was said that PD-NOMA method has got certain benefit if comparing with OFDMA if the user transmission channels SINR is different. For example, it's possible to notice that BER value in both PD-NOMA channels is less than in OFDMA while analyzing the time interval $t_1...t_2$. PD-NOMA channels are more noise-proof and it is advantageous to use the PD-NOMA method from the spectral efficiency point of view at a given point in time. The base station, analyzing each user's transmission channels current state, decides whether to use the PD-NOMA.

6 Conclusion

Channel power calculation method for PD-NOMA two user channels seal while using symmetric QAM modulation (with indices 4,16,64, etc.). The current method provides an opportunity to calculate the upper and lower layers channel power, based on a priori information about the user transmission channels state assessment. It is possible to carry out power channel coefficients adaptive calculation relatively to the transmission channel while using the proposed method. The proposed method effectiveness is confirmed by simulation.

Acknowledgements. The work is supported by the Russian Federation President grant to ensure young Russian scientists research state support. Grant number MK-1126.2019.9.

References

1. Marcus, M.J.: 5G and IMT for 2020 and beyond (spectrum policy and regulatory issues). IEEE Wirel. Commun. **22**(4), 2–3 (2015)
2. Soldani, D., Manzalini, A.: Horizon 2020 and beyond: on the 5G operating system for a true digital society. IEEE Veh. Technol. Mag. **10**(1), 32–42 (2015)
3. Marsch, P., Da Silva, I., El Ayoubi, S.E., Boldi, O.M., et al.: 5G RAN architecture and functional design. METIS White Paper (2016)
4. Alliance, N.: 5G white paper. Next generation mobile networks, white paper (2015)
5. 5G PPP Architecture Working Group. View on 5G Architecture, White Paper (2016)
6. 5G Network Architecture Design, IMT-2020 White Paper (2016)
7. Kryukov, Ya.V.: Generation and demodulation signals of multi-channel communication systems with channel multiplexing by power. Ph.D. dissertation, Tomsk State University of Control Systems and Radioelectronics (2015)
8. Benjebbour, A., Saito, K., Li, A.: Non-orthogonal multiple access (NOMA): concept, performance evaluation and experimental trials. In: 2015 International Conference on Wireless Networks and Mobile Communications (WINCOM), 14 January 2015, pp. 1–6, Marrakech, Morocco (2015)
9. Saito, Y., Benjebbour, A., Kishiyama, Yo., et al.: System-level performance evaluation of downlink non-orthogonal multiple access (NOMA). In: 2013 IEEE 24th International Symposium on Personal, Indoor and Mobile Radio Communications (PIMRC), pp. 611–615 (2013)
10. Saito, Y.: Non-orthogonal multiple access (NOMA) for cellular future radio access. In: 2013 IEEE 77th Vehicular Technology Conference (VTC Spring) (2013)

11. Islam, S.M.R., et al.: Power-domain non-orthogonal multiple access (NOMA) in 5G systems: potentials and challenges. IEEE Commun. Surv. Tutor. **19**(2), 721–742 (2016)
12. Al Rabee, F., Davaslioglu, K., Gitlin, R.: The optimum received power levels of uplink non-orthogonal multiple access (NOMA) signals. In: 2017 IEEE 18th Wireless and Microwave Technology Conference (WAMICON). IEEE, pp. 1–4 (2017)
13. Tse, D., Viswanath, P.: Fundamentals of Wireless Communication. Cambridge University Press, Cambridge (2005)
14. Caire, G., Shamai, S.: On the achievable throughput of a multi-antenna Gaussian broadcast channel. IEEE Trans. Inf. Theory **49**(7), 1692–1706 (2003)
15. Higuchi, K., Benjebbour, A.: Non-orthogonal Multiple Access (NOMA) with successive interference cancellation for future radio access. IEICE Trans. Commun. **98**(3), 403–414 (2015)
16. Patel, P., Holtzman, J.: Analysis of a simple successive interference cancellation scheme in a DS/CDMA system. IEEE J. Sel. Areas Commun. **12**(5), 796–807 (1994)
17. Ling, B., et al.: Multiple decision aided successive interference cancellation receiver for NOMA systems. IEEE Wirel. Commun. Lett. **6**(4), 498–501 (2017)
18. Saito, K., et al.: Performance and design of SIC receiver for downlink NOMA with open-loop SU-MIMO. In: 2015 IEEE International Conference on Communication Workshop (ICCW). IEEE, pp. 1161–1165 (2015)
19. Saito, K., et al.: Link-level performance of downlink NOMA with SIC receiver considering error vector magnitude. In: 2015 IEEE 81st Vehicular Technology Conference (VTC Spring). IEEE, pp. 1–5 (2015)
20. Ding, Z., Yang, Zh, Fan, P., et al.: On the performance of non-orthogonal multiple access in 5G systems with randomly deployed users. IEEE Signal Process. Lett. **21**(12), 1501–1505 (2014)
21. Dai, L., Wang, B., Yuan, Y., et al.: Non-orthogonal multiple access for 5G: solutions, challenges, opportunities, and future research trends. IEEE Commun. Mag. **53**(9), 74–81 (2015)
22. Timotheou, S., Krikidis, I.: Fairness for non-orthogonal multiple access in 5G systems. IEEE Signal Process. Lett. **22**(10), 1647–1651 (2015)
23. Kryukov, Ya.V., Demidov, A.Ya., Pokamestov, D.A.: Power calculation algorithm in non-orthogonal multiple access. Proc. TUSUR **19**(4), 91–94 (2016)
24. Oviedo, J.A., Sadjadpour, H.R.: A fair power allocation approach to NOMA in multiuser SISO systems. IEEE Trans. Veh. Technol. **66**(9), 7974–7985 (2017)
25. Zhang, Y., Wang, H.-M., Yang, Q., et al.: Secrecy sum rate maximization in non-orthogonal multiple access. IEEE Commun. Lett. **20**(5), 930–933 (2016)
26. Yang, Z., et al.: A general power allocation scheme to guarantee quality of service in downlink and uplink NOMA systems. IEEE Trans. Wirel. Commun. **15**(11), 7244–7257 (2016)
27. Shannon, C.E.: A mathematical theory of communication. Bell Syst. Tech. J. **27**(3), 379–423 (1948)

Distributed Data Compression Algorithm for Low-Power Wide-Area Networks

S. V. Dushin and S. A. Frolov$^{(\boxtimes)}$

Institute of Control Sciences of RAS, 65 Profsoyuznaya Street, Moscow, Russia
s.dushin@inbox.ru, sergey@frolov.ru

Abstract. A distributed data compression algorithm for low-power wide-aria networks is proposed. The algorithm is based on prediction of observed process on server side with controlling prediction error on the end device side. The prediction is completed by recursive linear prediction algorithm using Levinson-Durbin recursion. The algorithm provides obtaining the estimated values in real time on server side with no sending any data through the network until prediction error exceeded threshold. It allows to use wireless transceiver less intensively saving the battery budget this way, and to reduce number of packets and its length saving the wireless channel capacity. The efficiency of algorithm is investigated on end device and server models using the real data collected by CO_2, humidity and light sensors.

Keywords: Low-power wide-area network · Edge computing · Distributed computing · Linear prediction · Levinson-Durbin recursion

1 Introduction

The concept of Internet of Things (IoT) involves a lot of smart sensors (end devices) to measure different physical values and sending data to application servers. In many use cases the sensors cannot be connected to the power system, so it is powered by non-rechargeable batteries.

In general the autonomous powering imposes many restrictions on sensors hardware and software design. The actual engineering approach to achieve acceptable energy efficiency assumes several specific steps: management of peripheral device powering by CPU, avoid any interfaces with pull-up resistors, use very low quiescent current ICs, use special low-power wide supply range microcontrollers and its sleep mode features and, of course, use the low-power wide-area networks (LPWAN) to communicate with server. Designed this way autonomous smart sensors are used to measure temperature, pressure, humidity, different gas concentration (CO, CO_2, ammonia), etc. As usual the sensors interconnect with server with period from 5 min to several hours.

Since application of IoT systems is continuously growing, the industry faces new issues concern to autonomous sensor network deployment.

The most actual one is the growing of number of connected sensors and deployed networks (usually in limited non-licensed spectra). The growing data

V. M. Vishnevskiy et al. (Eds.): DCCN 2019, CCIS 1141, pp. 163–173, 2019.
https://doi.org/10.1007/978-3-030-36625-4_14

stream generated by sensors should be transmitted through modern LPWANs, which have low payload capacity. So increase of LPWANs bandwidth and effective compression of network payload are strongly needed to normal progress of IoT systems.

The second important issue is a lifetime of the sensors. As many sensors places in hard-to-reach areas and its amount continuously growing, the battery replacement become more and more tedious for operators. So the lifetime provided by modern sensors (typically couple of years) is not enough for convenient usage in applications that require large-scale deployment.

A lot of promising applications demand very intensive peak measurement and delivery data to server with low latency. For example, such requirements can be faced in following applications: controlling of micro-climate in buildings (heat transfer, air flow controlling), energy disaggregation, identification of heating system topology. In this cases the sensor network should provide burst mode to obtain large amount of data from dedicated sensors in short time.

The conception of distributed computing can be used to increase the payload capacity of sensor network. In this work we propose the distributed data compression algorithm that is developed especially for LPWANs. The algorithm provides to increase the payload capacity without additional latency for the data and to increase the energy efficiency of sensors.

2 Requirements to Data Compression in LPWANs

2.1 Decomposition of Power Consumption

To develop efficient data compression algorithm, which provides both high data compression ratio and increasing of energy efficiency, the specific conditions of LPWANs should be taken in account.

First of all, the power consumption profile of end device should be considered as the energy efficiency is critical. As mentioned above, the sensor's CPU controls power mode of peripherals devices activating it only in properly time. Usually the control is performed by sleep mode of ICs or external MOSFET switches. Between data transmission cycles the microcontroller also turns to sleep mode. So the different schematic parts have a different active on-time and power consumption profiles. Taking it in account let analyze the power consumption of the sensor in typical application to highlight most power consuming operations among sensor activity (measuring, interconnect with server, etc.).

The LoRaWAN class A temperature sensor are considered below as example. The sensor is built on following components: microcontroller STM32L151 [1], transceiver SX1272 [2], temperature sensor B57861. The battery voltage is +3.6 V dropped on low-dropout regulator (LDO) to +3.3 V to supply the circuit. The output power of transmitter is +17 dBm.

To obtain next sample of measured value the microcontroller and sensitive element should be engaged. During this operation the energy consumes on microcontroller, sensor circuit and LDO. The calculated energy that is spent to obtain one sample is presented of Fig. 1.

To send the sample to server we need to engage microcontroller and LoRaWAN transceiver. After transmission, it is necessary to keep the receiver turned on in two time windows due to the requirements of the LoRaWAN MAC protocol. During data transmission the energy consumes on microcontroller, transmitter and LDO. During receive cycle the energy consumes on microcontroller, receiver and LDO. The calculated energy that is spent on this operation is also presented on Fig. 1. The values are calculated according datasheets. The receive operation isn't shown, because the class A device is considered.

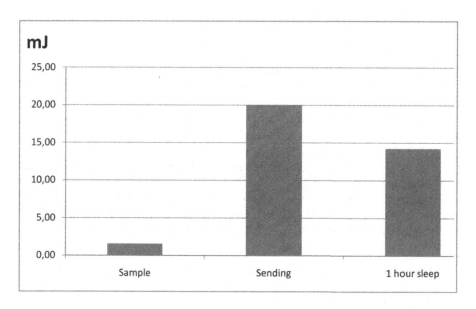

Fig. 1. Energy consumption for basic operations of LoRaWAN temperature sensor

The most power-consuming operation in considered case is the sending data to server. This is typical situation for many end devices, however some type of sensitive elements demands significant more energy to be work in compare to considered one (for example NDIR CO_2 sensors). It will be take in account in the study below.

Anyway the transceiver activity must be keep as low as possible to improve energy efficiency, as this is most important issue to increase power efficiency. On other hand the algorithm should avoid any significant computational complex operation on the end device side, because it can load microcontroller and dramatically increase its power consumption. So the rational usage of transceiver and low computational complexity on the end device side are required to data compression algorithm.

2.2   Network Payload

Other important issue is the network payload. To reach high compression ratio, the algorithm should be developed taking into account the type of information to be compressed and its statistical characteristics.

As mention above, the modern battery powered sensors usually used for measure different physical processes (temperature, pressure, humidity, etc), so follow we will consider these types of payload, but modifications of proposed below algorithm can be implemented to transmit voice/sounds and even pictures through the LPWANs as well.

It is easy to see, the measured values are slow-changes processes with strong inner dependencies. So it can be considered as an auto-correlated stochastic process. If the observed process is wide-sense stationary (WSS) process on dedicated time interval, the next sample can be defined as linear combination of previous samples and predicted by the filter predictor. Such technique called linear prediction and used for: audio and video signal compression, prediction of the signals and adaptive filtering. However usually data compression algorithms perform all computational complex operations to compress the data on end device side. In considered case it is not allow, because of power efficiency requirement. The other problem is significant delay introduced by the designed this way algorithms.

Consequently the classical structure of data compression algorithms isn't suitable for battery-powered sensor networks connected by LPWANs, and the another way to compress the data need to be developed.

3   Distributed Data Compression Algorithm for LPWANs

3.1   Algorithm Description

The main idea of proposed algorithm is to calculate the filter predictor coefficients on server side. It allows to keep computational complexity of end device part of algorithm as low as possible. The coefficients calculated by server can be used on server and end device sides to estimate next data sample synchronously. The end device knows both actual measured value and its estimation, so it can control the predefined allowable error and send the updates to server, when estimation error exceeds the threshold. The calculated coefficients are sent to the end device using receive channel, which usually already exists (in particular, in LoRaWAN protocol).

The diagram illustrates one iteration of algorithm is shown on Fig. 2. After the packet from end device has received, the server calculates and updates filter predictor coefficients. Updated coefficients send to end device (for LoRaWAN Class A it can be done in receive window after transmission). Optionally server can send maximum allowable error. After that both server and end device start to estimate the next samples of the signal synchronously (the tight synchronization is not required), wherein the end device controls estimation error and number of completed iterations. If the estimation error or number of iterations without update have exceeded predefined thresholds, the end device sends packet with actual data to the server.

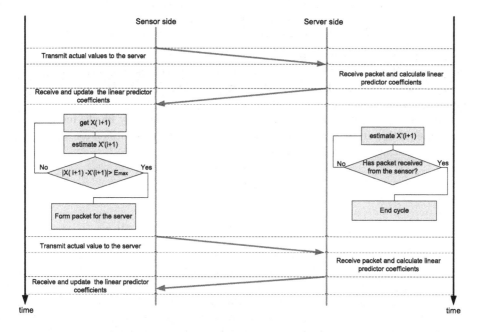

Fig. 2. Distributed data compression algorithm

So the proposed algorithm allows to predict observed process on server side and control the error in real time.

3.2 Linear Prediction Technique

To describe the math side of algorithm let consider linear prediction technique. The idea of linear prediction is to represent current sample as a linear combination of past samples (1).

$$\hat{x}(i) = \sum_{n=1}^{N} a_n * x(i-n) \qquad (1)$$

The formula (1) describes the convolution of N past samples of observed process and coefficients a_n. The coefficients can be considered as taps of finite impulse response filter, which is called filter-predictor.

The optimal coefficients of predictor in terms of minimizing MSE (means squire error) can be found to solve Yule-Walker Eq. (2).

$$\begin{pmatrix} R_0 & R_1 & \cdots R_{N-1} \\ R_2 & R_0 & \cdots R_{N-2} \\ \vdots & \vdots & \ddots & \vdots \\ R_{N-1} & R_{N-2} & \cdots & R_0 \end{pmatrix} * \begin{pmatrix} a_1 \\ a_2 \\ \vdots \\ a_N \end{pmatrix} = - \begin{pmatrix} R_0 \\ R_1 \\ \vdots \\ R_{N-1} \end{pmatrix} \qquad (2)$$

The traditional methods to solve Yule-Walker equation have computational complexity $O(N^3)$. However the autocorrelation matrix is Toeplitz matrix, so the fast computational algorithm called Levinson-Durbin recursion can be used. This algorithm provides computational complexity $O(N^2)$. The main idea of Levinson-Durbin recursion is to solve Yule-Walker equation by blocks increasing of order equation from 1 to N. The generalized form of the recursion for arbitrary order $n + 1$ described by formulas (3)–(6).

$$k_{n+1} = \frac{\alpha_n}{e_n}, \qquad (3)$$

$$A_{n+1}(z) = A_n(z) + k_{n+1} * z^{-1} * [z^{-n} * A_n(z)], \qquad (4)$$

$$e_{n+1} = [1 - |k_{n+1}|^2] * e_n, \qquad (5)$$

$$\alpha_n = R_{n+1} + a_{n,1} * R_n + a_{n,2} * R_{n-1} + \ldots + a_{n,n} * R_1, \qquad (6)$$

where e_n - error prediction, k_{n+1} - reflection coefficient, $A_n(z)$ - filter predictor response.

To do Levinson-Dusrbin recursion we are need to estimate the autocorrelation function. It can be calculated recursively updating previous values by new samples to reduce the computational complexity [6]. Of course, a possible long term non-stationarity of estimated process should be taken in account. There are two main ways to do this: introducing the forgetting factor or calculate the autocorrelation function within a window. The second one is used for proposed algorithm.

3.3   Data Transmitted to the Server

To keep the network payload and transmitter active time as small as possible, it is necessary to minimize the data size within packets, which sends by end device to server.

Since the estimated samples is accepted as well-predicted while end device keeps silence, the server can use these estimated values to update the filter predictor coefficients on the next iteration. Taken this in account only last measured sample (poorly predicted) can be transmitted to server. Moreover the end device can transmit only prediction error, as the actual value can be calculated by the server using simple formula (7).

$$x(i) = \hat{x}(i) + e_i \qquad (7)$$

Summarize, the algorithm consists of logical steps as shown on Fig. 2 with using following formulas:

– The samples of observed process are estimated using formula (2) on both server and device sides;
– The filter-predictor coefficients calculate according to (3)–(6);
– The actual samples on server side calculates according to (7).

The main parameters of algorithm are: predictor order (P), error threshold (D), maximum samples to predict (L), sampling frequency (F), window to estimate the autocorrelation function (W).

On the end device side the proposed algorithm measures the real values and compare it with estimated one. This operation demand only P multiplications and P sums in fixed point. So the algorithm provides low computational complexity on the end device side as was targeted above.

4   Simulation Results

To estimate the proposed algorithm efficiency, the Matlab/Simulink models of server side and sensor side is built. The real data, which is collected by sensor network deployed in Institute of control science of RAS, is used during experiments. In particular, it is data from intrabuilding CO_2, humidity and light sensors collected during two weeks. In additional to this real signals, the autoregression driven by the Gaussian noise and sinusoidal signal are considered. These signals is used to demonstrate the compression capability for fast-changed signals with different statistical characteristics.

The studies of the model with real data set allows precisely estimate data compression ratio and to calculate expected power economy in the conditions in which data are obtained.

The data compression ratio expresses as number of measured samples divided to number of sent samples through wireless channel (8).

$$CR = \frac{N_{sample}}{N_{trans}}, \tag{8}$$

where N_{sample} - number of measured samples, N_{trans} - number of transmitted samples through wireless channel.

The energy economy ratio can be expressed as energy, which is spent to work without proposed algorithm, divided to energy, which is spent if the proposed algorithm is used (9).

$$ER = \frac{N * (J_{sample} + J_{sending}) + J_{sleep}}{N * J_{sample} + N * J_{predictor} + K * J_{sending} + J_{sleep}}, \tag{9}$$

where N - number of measured samples, K - number of sent packets, $J_{predictor}$ - the energy spent to perform algorithm code, J_{sample} - the energy spent to get a sample, $J_{sending}$ - the energy spent to send packet, J_{sleep} - the energy spent in sleep mode.

To keep the study methodology simple, the $J_{predictor}$ can be taken equal to zero as the end device part of algorithm is very simple. So the formula (9) can be rewritten as follows:

$$ER = \frac{N * (J_{sample} + J_{sending}) + J_{sleep}}{N * J_{sample} + K * J_{sending} + J_{sleep}}, \tag{10}$$

The results of study are presented in Table 1. The error is defined in percent from full scale of data (signals). The energy ratio is estimated according to datasheets on components, which are used in humidity, light and NDIR CO_2 sensors.

The actual collected data and its estimation on server side for humidity sensor are presented on Fig. 3. The same for light sensor is presented of Fig. 4. On the figures, the original data is drawn by solid lines, the estimation one is drawn by dotted lines.

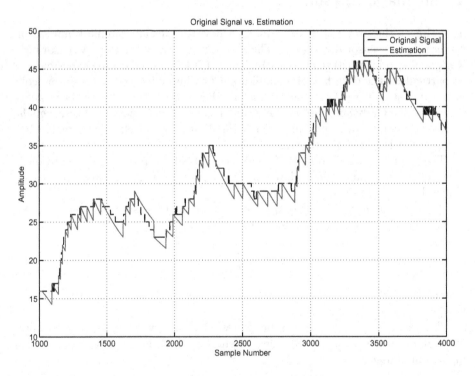

Fig. 3. Actual data of humidity sensor and its estimation on server side

Table 1. Results of study of algorithm efficiency in different applications

Data	Predictor order	Window	Error%	Comp ratio (CR)	En ratio (ER)
Humidity	5	100	5	8.4	4.4
Humidity	10	100	5	8.6	4.5
Humidity	10	10	5	1.2	1.2
Humidity	10	1000	5	64.7	7.4
Humidity	10	1000	1	8.5	4.4
Humidity	10	1000	10	124.1	7.8
Light	5	100	5	6.5	3.9
Light	10	100	5	6.8	4.0
Light	10	10	5	5.0	3.3
Light	10	1000	5	6.8	4.0
Light	10	1000	1	2.5	2.1
Light	10	1000	10	15.9	5.6
CO_2	5	100	5	6.5	1.9
CO_2	10	100	5	6.5	1.9
CO_2	10	10	5	1.1	1.1
CO_2	10	1000	5	29.2	2.1
CO_2	10	1000	1	6.0	1.8
CO_2	10	1000	10	56.2	2.2
AR	5	100	5	1.7	NA
AR	10	100	5	1.8	NA
AR	10	10	5	1.8	NA
AR	10	1000	5	1.9	NA
AR	10	1000	1	1.2	NA
AR	10	1000	10	2.9	NA
Sin	5	100	5	4.9	NA
Sin	10	100	5	5.2	NA
Sin	10	10	5	1.2	NA
Sin	10	1000	5	14.0	NA
Sin	10	1000	1	3.6	NA
Sin	10	1000	10	22.7	NA

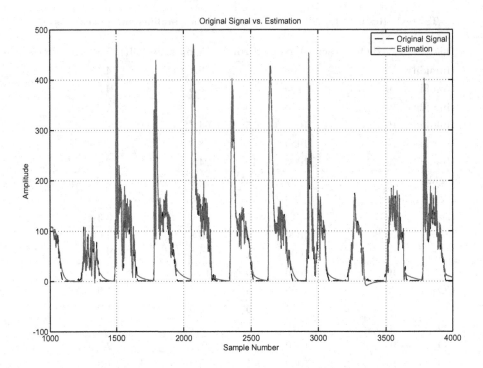

Fig. 4. Actual data of light sensor and its estimation on server side

5   Conclusion

The proposed algorithm provides significant data compression ratio and economy of battery budget for all considered signals. The maximum efficiency is achieved when the typical for IoT applications slow-changed physical values (CO_2 concentration, humidity and light) are estimated.

The efficiency of algorithm mainly depends of statistical characteristics of the data to be estimated, filter predictor order, window to calculate of autocorrelation function and allowable error.

The considered signals can be predicted with low-order filter predictor (sufficient predictor order is 5–10). However, many samples are required to estimate autocorrelation function, which is estimated recursively on server side. These feature avoids significant load of end devices and server hardware.

The algorithm is applied to LoRaWAN technology, but it can be adapted to other LPWANs.

References

1. Official site of STMicroelectronics, May 2019. https://www.st.com/resource/en/datasheet/cd00277537.pdf
2. Official site of Semtech Corporation, May 2019. https://www.semtech.com/uploads/documents/SX1272_DS_V4.pdf
3. Vaidyanathan, P.P.: The Theory of Linear Prediction. Morgan & Claypool Publishers (2008)
4. Jackson, L.B.: Digital Filters and Signal Processing, 2nd edn. Kluwer Academic Publishers (1989)
5. Ramirez, M.A.: A Levinson algorithm based on an isometric transformation of Durbin's. IEEE Signal Process. Lett. **15**, 99–102 (2008)
6. Haykin, S.: Adaptive Filter Theory, 5th edn. Prentice Hall (2014)
7. Hannan, E.J., Quinn, B.G.: The determination of the order an autoregression. J. Roy. Stat. Soc. B **41**, 190–195 (1979)
8. Mekki, K., Bajic, E., Chaxel, F., Meyer, F.: Overview of cellular LPWAN technologies for IoT deployment: Sigfox, LoRaWAN, and NB-IoT. In: 2nd IEEE International Workshop on Mobile and Pervasive Internet of Things, Athens (2018)
9. Chen, T., Ljung, L.: Implementation of algorithms for tuning parameters in regularized least squares problems in system identification. Automatica **50**, 2216–2220 (2013)
10. Hasan, A.H., Grachev, A.N.: On-line parameters estimation using fast genetic algorithm. J. Electr. Control Eng. (JECE) **4**(2), 16–21 (2014)

PPPXoE - Adaptive Data Link Protocole for High-Speed Packet Networks

Aleksandr Ivchenko⬥, Liubov Antiufrieva⬥, Pavel Kononyuk⬥,
and Alexander Dvorkovich$^{(\boxtimes)}$⬥

Moscow Institute of Physics and Technology, 9 Institutskiy per., Dolgoprudny, Russia
{ivchenko.av,antyufrieva,kononyuk.pa,dvorkovich.av}@mipt.ru

Abstract. The paper is devoted to the problems of guaranteed data transmission with a dynamically changing amount of packet loss ranging from 0 to 23% in channels with a bandwidth of more than 0.5 Mbps. The problems arising due to standard TCP loss protection mechanisms, such as retransmit flows and buffer overflows are considered. Mechanisms for recovering data and service information, tuning for the current state of the channel are proposed, the implementation of the adaptive protocol of the data link layer is described, and the results of its operation are presented. It also describes the operation of a separate protocol subsystem for low-speed channels starting from 1200 bps.

Keywords: Network protocols · Data link protocol · Adaptive internet protocol · Packet loss · Bit errors · Error correction codes · Forward error correction · Interleaving · Reed-Solomon codes · PPP · PPPoE

1 Introduction

As a rule, trunk wired communication lines have stable characteristics and a low amount of packet loss. The situation is different in the channels of the "last mile" connecting the end equipment with the access points of the telecom operator. Such channels have dynamically changing characteristics, such as high delays and delay variations, a significant amount of packet loss. Often such channels are low-speed, provided by outdated equipment. In addition, the paper concerns heterogeneous networks, individual sections of channels can be implemented using copper cables or microwave links.

A similar situation is in the provision of satellite channels or radio channels communications. The cause of the dynamic characteristics is the analog nature of the signal, the errors are associated with interference, attenuation in the line and various additive noises. In such channels, one can distinguish both packet losses and bit losses [1].

1.1 Article Structure

The rest of this section describes why we choose the link layer of the network model and the main problems of the TCP protocol that lead to significant loss of bandwidth and the complete collapse of the network.

© Springer Nature Switzerland AG 2019
V. M. Vishnevskiy et al. (Eds.): DCCN 2019, CCIS 1141, pp. 174–187, 2019.
https://doi.org/10.1007/978-3-030-36625-4_15

Section State of the art gives a brief overview of the subject area, some of which will be included in the next revisions of our protocol.

The Methods and algorithms section describes the error correction and interleaving, formats of the structures and also the problems that we encountered during development. The mechanism of adaptive settings of the protocol parameters depending on the state of the channel is described.

Section Low speed mode describes how data is transmitted at ultra-low speeds from 1.2 Kbps to 0.5 Mbps.

The section Adaptive Parameters gives an overview of the experimental results and gives some numerical parameters and their combinations.

The Results section contains graphs of the protocol, demonstrates its recovery ability and speed. Some comments are given to researchers on the features of testing similar developments.

1.2   Layer of Work in OSI Model

In the seven-level OSI model [2], we can consider the loss of segments and datagrams at the transport level, packets on the network, frames on the channel, and frames or bits on the physical level. The lower the level, the more opportunities for recovering lost data. At the same time, physical layer protocols, as a rule, possess to one degree or another recovery mechanisms - forward error correction (FEC) codes or cascades of such codes [3].

Application of recovery mechanisms is most effectively at the physical level. At this level there is all the service information, the headings of the protocols of higher levels, to which mathematical methods of recovery can be applied. The problem is that the solution for the physical layer is not universal, cannot be applied to already released equipment (except for solutions based on SDR - software defined radio [4]). Therefore, the PPPXoE protocol is implemented at the channel level, as the closest to the physical (Fig. 1). With software implementation at the data link level, the solution can be applied on heterogeneous equipment that uses a standard IP protocol stack and Linux operating system.

One of the development tenets is the development of the IP protocol stack and support for working with other standard protocols without changing the existing infrastructure. From the point of view of the IP stack, part of the libraries implementing the standard Point-to-Point [5] and Point-to-Point over Ethernet [6] protocols is being replaced, without changing the other communication mechanisms.

1.3   The Basic Mechanism for Guaranteed Data Delivery

TCP assumes [7] that any loss of data packets is caused by network congestion (buffer overflows of network devices). If there is missing part of the packet in the stream, a re-request of the lost or corrupted data is performed. This mechanism is the main reason of a sharp decrease of the transmission speed and can lead to the complete impossibility of communication in network losses of 2–5%

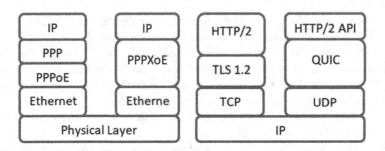

Fig. 1. PPPXoE and QUIC in the protocol stack

packets. This causes a number of problems. In the absence of just one packet, the buffer cannot transmit data and waits a re-requested packet. This leads to delays, buffer overflows and discards packets that have successfully arrived. The channel increases the flow of service data (requests for lost data). The more failed and dropped packets, the more retransmissions. On intermediate routers, data queues for sending grow and overflow [8]. Installing routers with a large amount of memory only pushes back the avalanche-like network drop, but does not prevents it.

Another reason is the basic TCP congestion control mechanism based on controlling the size of the congestion window - the amount of data sent to the network. In the initial phase, TCP tests the bandwidth available for the connection. In this phase, the slow start algorithm works, in which the size of the congestion window exponentially grows from the initial value equal to the size of one maximum TCP segment to the set threshold value. Then, in the overload prevention phase, the window size grows linearly. If, during a certain interval (timeout), the transmitting device has not received confirmation of packet reception, the new threshold value is set to half (or less) the current congestion window, the congestion window itself is reduced to the initial value and the algorithm starts again.

If packet loss does not occur, the size of the congestion window grows to the size of the window declared by the recipient, and the sender begins to transmit the amount of data corresponding to the capabilities of the recipient to the network. The window state is saved until the next packet loss. Thus, the more often the basic anti-congestion mechanism is triggered, the more dips in the transmission speed are and the less efficiently the bandwidth available for connection is used.

Unlike wired networks, in wireless communication channels short periods of packet loss or a sharp increase in delivery delays often are not caused by network congestion. They can be associated with short-term changes in the conditions in the radio channel, such as fading, interference, loss of radiovisibility of the receiving side. Repeat packet transmissions by link layer protocols play a role, as well as traffic prioritization procedures at intermediate nodes, as a result of which other services may experience delay variation, etc. These single events lead to the inclusion of a fairly conservative basic TCP congestion management mechanism, which entails a noticeable decrease in its efficiency and negatively affects the actual throughput: the time for opening data transfer sessions, downloading data, etc. increases, and the radio network bandwidth is not used completely.

Thus, there is a task of constructing such a communication protocol, which would ensure the restoration of lost and damaged data with minimizing re-requests, at the same time, proactively responding to changes in the communication channel.

2 State of the Art

There are a number of studies in the field of guaranteed data delivery and usage of FEC, but there are very few complex solutions in the form of new protocols for the IP stack. Most inventions are disparate in nature and have not yet included standards or network protocols.

Some papers study data transmission over multiple routes, including data duplication, balance of load across channels or optimization of the transmission route. So, in 2018, Salkuyeh and Abolhassani developed an algorithm for distributing video packets over channels of various quality in compliance with QoS [9].

Neural networks also have application. The team led by B. Mao developed the Tensor-based Deep Belief Architectures (TDBAs) for route prediction, trained on the edge routers' traffic patterns based on the extension of the Deep Belief Architecture (DBA) [10].

Tensors are great mathematical model for a multidimensional description of routes, network packets, and channel parameters. The algorithm finds one, the most optimal traffic route, while optimizing for average hop delays and packet loss. Studies are at the initial stage and already demonstrate a great potential, it is planned to include more parameters in the analysis, in particular, packet size and service time, QoS satisfaction metrics [11].

Unfortunately, at least one route with good channel characteristics does not always exist, especially in the case of the "last mile".

There are a number of FEC applications in the video transmission. The use of original Erasure codes to recover lost packets on the receiver significantly improves the quality of 4K and 8K ultrahigh definition television (UHDTV) over IP-transmission system for live program production. Kawamoto and Kurakake also note a significant difference in restoring ability with burst loss and random (byte) loss.

Forward error correction using an erasure code is a key technology to recover lost packets at the receiver in unidirectional transmission. In FEC using erasure code, special packets only for error correction (FEC packets) are generated by mathematical calculation, and both video and FEC packets are transmitted in the same transmission path. On the receiver side, lost packets are recovered using the received video/FEC packets.

FEC packets are created using the XOR operation. The obvious drawback of this approach is the impossibility of data recovery in case of loss of the FEC packet. The algorithm used does not guarantee the delivery of data, in case of an accidental (byte) loss, the probability of data recovery is high, in the case of a burst loss, if an FEC packet is lost, recovery is not possible. To combat burst losses, researchers distribute data between several FEC packets, thereby increasing their recovery ability. Researchers note that the developed method processes data faster than the Reed-Solomon and Raptor codes [12].

The paper of Wu is close to our. It presents the FEC coding scheme dubbed PATON (Priority-Aware and TCP-Oriented codiNg), aimed at delivering mobile HD video using TCP over wireless networks. The parameters of FEC redundancy and packet size similar to the PPPXoE were selected heuristically. In the paper, the fight against packet forwarding (retransmission) and Congestion control is being waged. PATON has the following solution procedures: (1) perform priority-aware frame scheduling to maximize video quality; (2) analyze TCP connection state to model the end-to-end delay; (3) adapt FEC redundancy level and packet size to minimize the effective data loss rate. Without the frame selection module, substantial amount of prioritized video frames [e.g., the I (Intra) frames] may be lost during bandwidth shrink and network congestion. It is an application-layer solution and can be implemented in real-time video communication systems without modifications on underlying protocols.

The system is not aimed at guaranteed data delivery, but reduces the amount of packet loss. PATON changes the code rate and packet size according to the pathChirp algorithm, the network status monitor on the recipient side periodically evaluates the network status and sends this information to the sender. The packet loss pattern is modeled based on Gilbert model and continuous time Markov chain. Since PPPXoE aims to completely recover losses, such a mechanism is too slow for us. PATON feedback is at least 2 times slower because it evaluates RTT (Round Trip Time), in our system channels are evaluated simultaneously and independently on both sides of the connection.

The system operates on losses up to 10% and improves the average video PSNR by up to 6.5 (23.6%), 10.2 (33.2%), and 14.3 (43.9%) dB compared to the CLOSET, Rateless and EEP (Equal Error Protection) schemes. Another disadvantage of the system is the optimization metric. The system's objective is to improve the quality of the video, but PSNR is used to assess the quality, which does not meet both the assessment of user-oriented QoE quality and the QoS estimation methods used in image quality comparison tasks [14].

The experimental QUIC (Quick UDP Internet Connections) protocol developed by Google is a transport protocol that runs on top of UDP (Fig. 1). A key feature is the acceleration of opening connections and negotiating TLS parameters (HTTPs). Unlike TCP, which uses the principle of "triple handshake", in QUIC handshake occurs in one step with a server that is already familiar and in two steps with a server that the client has not worked with before. The second stage is needed to open a secure communication channel and exchange cryptographic keys. As a result, QUIC has a lower connection and transmission delay than TCP. When transmitting data over a long distance (for example, from one continent to another) using a mobile device, the difference in the speed of establishing a connection between TCP with TLS and the QUIC packet can reach 300 ms [14,15].

QUIC no longer has a set of parameters associated with the IP addresses and ports of the server and client. Instead, the protocol works with the connection identifier UUID. This allows you to switch between Wi-Fi and the mobile network, each time without re-creating the connection (UUID is saved).

The protocol has its own cryptographic functions, multiplexing and setting up streaming, so only the truncated part of the HTTP/2 API used to communicate with remote servers remains on top of it. Initially, the protocol had non-adaptive redundancy of 10%. However, this data only duplicated parts of other packages; error correction codes were not applied. This approach has been shown to be ineffective; there is no FEC in the current edition, but research is underway in this direction. Also, the protocol is not compatible with networks that use NAT, Anycast, or ECMP technologies. They work with TCP connections and will not be able to recognize and regulate QUIC traffic [16].

3   Methods and Algorithms

This paper is a further development of an adaptive protocol for special-purpose networks [17]. It describes the mechanisms of error recovery, tuning of the channel, changes in the frame format.

The protocol integrates into the existing IP stack and replaces the PPPoE (Point-to-Point over Ethernet) protocol. RAW sockets are used, allowing you to bypass the TCP/UDP transport layer and independently create the desired frame format [18]. RAW sockets accept and transmit raw datagrams without adding any system headers (Ethernet traffic) to them. The mechanisms of transport and the connection establishment of the standard PPP protocol are used, the final data structure corresponds to no more than 1472 bytes for the purpose of further correct transmission over the channel.

3.1   Methods

Data recovery is based on the use of Reed-Solomon codes with character lengths of 8, 10 and 12 bits and interleaving. In the formula (1), one of the generating polynomials of the code is given. Initial sequences as well as an interleaving algorithm are known at the receiving and transmitting sides. The protocol settings for reception and transmission are independent and may vary.

$$g(x) = x^{10} + x^3 + 1 \qquad (1)$$

The developed PPPXoE protocol (eXtended) uses its own encapsulation, shown in Fig. 2. The original data is converted into a Data Frame: it is divided into DataBlock blocks, redundant information is added to the RS-code blocks. Before them is inserted the header for decoding. It was necessary to reduce the share of overhead information in personnel, leaving the minimum sufficient. The fields remained: frame type, number of data blocks sent, data frame length, FEC header code and check sum.

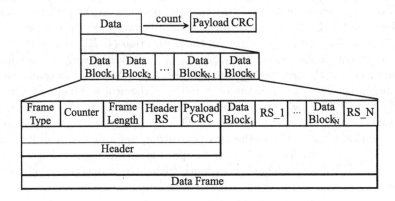

Fig. 2. Data frame

Super Frame is assembled from several Data Frames and service Control Frame (optional). The interleaving procedure is performed (Fig. 3) at the data byte level, partitioning into equal superblocks. After that, individual superblocks with headers (synchronization sequence, protocol version and revision, super-frame number, adaptation parameter number, CRC and FEC-protection of the header, and a number of others) are transmitted over the network (Fig. 4) using the PPP protocol transport, which adds 28 byte Ethernet header.

3.2 Channel Tunning

The protocol has feedback. Parameters are configured in both directions independently (one-way transfer is possible), the receiving side, in turn, sends the data back and signals the reception quality.

The problem of the initial adaptation algorithm was its inertness. When the session was initialized, the parameters were averaged (code rate 1/8, block size 64 bytes), a Control Frame overhead frame with operation statistics was inserted into the superframe every second, based on which new parameters for the code rate, block sizes and superframe were selected according to the current number of losses packages. The problem was that the settings did not change fast enough, packet loss occurred when switching modes. This approach led to the sequential switching of protection modes, one at a time.

In order to achieve positive feedback, a Pulse-frame was introduced (Fig. 5), which follows with each transmitted super-frame, which allowed rebuilding in fractions of a second from the 20% packet loss mode to 0%. The detuning speed depends on the frequency of incoming superframes. Information on bit rate, numbers of the last undelivered and unrestored frames, the number of frames in the send queue, and the current number of errors in the data are transmitted in the Pulse frame. At this second revision of PPPXoE Control Frame includes every 10 s.

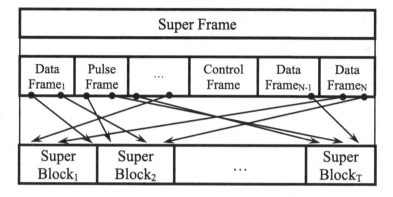

Fig. 3. SuperFrame forming

SYNC	VER	REV	SF number	Adapt Param	SFblock number	Enum Field	Hamming Code	Header CRC	Header RS	Super Block$_T$

Fig. 4. Data link structure, Ethernet header will be added further

Bytrate	Undelivered frames num	Unrecovered SF num	Queued RX frames	RX RS errors

Fig. 5. Pulse frame

3.3 Interleaving

In the new algorithm, the number of blocks is tightly related to the size of the blocks. The interleaver evenly decomposes into each block one byte from the other blocks. So, when transmitting 704 blocks, their size will be 704 bytes, if one of them is lost, it will be necessary to restore 1 byte in blocks. The percentage of errors in each block increases evenly up to the maximum possible on this parameter.

Previously, the interleaving method was taken from the field of digital signal processing and tuned to the random occurrence of errors on the radio channel in an interference environment. Data is written in rows, subtracted in columns. Also in each row a cyclic shift is performed. This approach works well when transmitting OFDM signals, when part of the sub-frequencies are incorrectly decoded. In our country, it was destructive due to the uneven distribution of errors. At 20% loss on the channel, in some blocks there were 46% and 90%, but these errors were random in nature and poorly tracked. Reed-Solomon codes could not fix so many errors.

3.4 Multiplexing

A different task was to maintain the total size of the SuperFrame structure along with the Ethernet header to match the standard MTU size of 1,500 bytes. Some modern network equipment is capable of processing larger MTUs, but none of the devices encountered has this function. The MTU size could be made smaller by changing the configuration of network devices, and the crushing procedure should be given to the GSO mechanism [19], but we went a different way. The data is split into ECC blocks (data groups together with their FEC code) in such a way as to be transmitted at a time, i.e. match the final MTU.

The size of the superframe is strictly related to the CR and DL parameters, but if the data is empty, we decided to duplicate part of the ECC blocks to fill the MTU. The minimum size of such a block is 32 bytes (16 bytes of data and 16 bytes of FEC), thus increasing the size by a multiple of 32 bytes. The size of the superframe and incoming blocks are both transmitted. The final superframe size and the multiplication coefficient are calculated from this information. ECC blocks are duplicated evenly, so it is possible to recover data if it is damaged. Thus, the reliability of data transmission is further enhanced.

4 Low Speed Mode

With a bandwidth below 0.5 Mbps, error correction codes begin to lose their effectiveness due to increasing delays. There was developed a separate protocol for these types of channels.

At the start of a communication session, the channel is tested and the protocol operation mode is selected - high-speed or low-speed. At low speed, data backup showed the best results. Data blocks are divided into smaller blocks with a multiplicity of 8 bytes. These blocks are duplicated with the reservation coefficient selected according to the channel status. So, if the coefficient is 2, then 1% loss is guaranteed to be corrected. The reserve ratio was tested up to 24, however this parameter can also be set to 100, so the protocol will work on almost hopeless channels.

Upon receipt, successfully received blocks fill the original block, and duplicate blocks are discarded. Also, as in fast speed channel mode, to check the block for bit errors, a CRC checksum is used for the header and data.

5   Adaptive Parameters

Previously, hyperparameters were passed separately with each DataFrame. Now hyperparameters: coding rate (CR - code rate), block size (Data Lenth) and super-frame size (SF Size) are tightly interconnected in proportional equations for various maximum errors (MER - maximum error rate). Combinations turned out to be 255, but during the testing only 25 modes were left for integer recovery abilities. Modes with large block sizes proved to be ineffective (blocks larger than 1.5 MB in length are processed by the Reed-Solomon code for a long time). Some combinations of hyperparameters showed instability of the recovery ability.

Code rates vary from CR15 equal to 1/16 to CR00 equal to 1 (for the case of a lossless channel, which means the absence of RS codes). So, CR00 encodes a code rate of 1, and CR05 11/16. Block lengths vary from 16 to 128 bytes. The size of the error correction code (ECC) block (data + FEC) is from 32 bytes with minimum protection up to 2048 (128 bytes data and 1920 FEC). The sizes of superframes are from 256 (16 × 16) to 4194304 (4096 × 1024) bytes.

Table 1 shows the correspondence of a number of block lengths in bytes to code rates

Table 1. Correspondence of ECC blocks size and code rates

	DL0	DL1	DL2	DL3	DL4	DL5	DL6	DL7
$CR01$	0	4	4	4	4	4	4	4
$CR05$	8	12	12	16	16	20	20	20
$CR12$	48	60	72	84	96	108	120	132
	DL8	DL9	DL10	DL11	DL12	DL13	DL14	DL15
$CR01$	4	8	8	8	8	8	8	12
$CR05$	24	28	32	36	40	44	52	60
$CR12$	144	168	192	216	240	288	366	384

The Table 2 shows a number of combinations of hyperparameters - adaptation protocol operation modes, including AP001 mode for error-free channels and AP228 - highly protected mode with a 15/16 code rate.

Table 2. Adaptive modes of protocol

Mode	SF Size, byte	CR	DL	Mode	SF Size, byte	CR	DL
$AP001$	2304	CR00	DL08	$AP100$	11664	CR09	DL01
$AP002$	400	CR01	DL00	$AP200$	82944	CR12	DL11
$AP003$	576	CR01	DL01	$AP212$	123904	CR14	DL07
$AP004$	784	CR01	DL02	$AP220$	262144	CR13	DL13
$AP005$	1024	CR01	DL03	$AP228$	589824	CR15	DL13

6   Results

Testing was carried out on a specially assembled stand. In the laboratory, a software modem was written that evenly punched packets. As the receiving and transmitting sides were personal computers of the following configuration:

- Sender: Intel Core i7-4770K 3,5 GHz 8 MB Cache, 8 GB DDR3 DualCh 1600 MHz;
- Receiver: Intel Core i5 2.0 GHz 2 MB Cache, 4 GB DDR3 1333 MHz.

Like the original PPPoE, the developed PPPXoE protocol occupies a maximum bandwidth of 896 Mbps. In the case of a channel without packet loss, data is transmitted at a speed of about 109,000 Kbps. Due to the redundancy in the headers, PulseFrame and ControlFrame, there is a slight loss of bandwidth. The original PPPoE protocol transmits at 110,000 Kbps. With an increase in packet loss, the code rate increases, the character of the bandwidth loss is linear, with a sharp drawdown during the transition from 0% to 1% of channel losses (Fig. 6). In Fig. 7 shows the recovery ability of the algorithms.

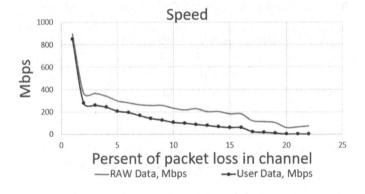

Fig. 6. Speed of protocol PPPXoE

In Fig. 8 is the result of comparing the work of the standard protocol and our. A complete comparison of the characteristics of PPPoE and PPPXoE was not carried out due to the need for further optimization of program codes. We parallelized the operation of the protocol by the number of processor cores, fixed errors associated with memory leaks - now it is allocated once for the entire operation of the protocol. We also redesigned the Reed-Solomon module of the Linux kernel. The standard module used the same memory area, which caused the kernel to freeze.

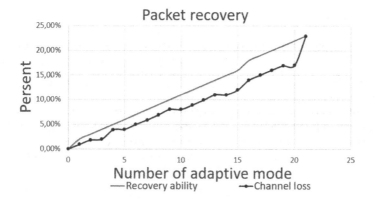

Fig. 7. Recovery ability of protocol PPPXoE

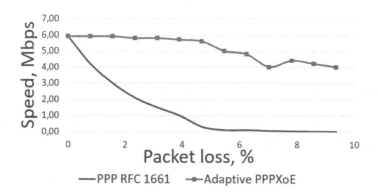

Fig. 8. Comparement of PPPoE and PPPXoE

6.1 Testing Notes

Carrying out any development without proper metrological support is impossible. So, impulse losses in packet loss emulation systems were noticed. With the declared 10%, there could be no consecutive 100 packets, but over a long period of time this is 10%. As a result, our own software was written that transparently deleted packages by normal and uniform distributions, but always out of 100 packages no more than the declared number was lost.

And one of the ready-made test benches, while limiting the bandwidth, did not take into account that the protocol increases the flow by adding redundant information - FEC codes. That is, restrictions were imposed on the transmitted source information of the transport layer, and not the channel. The issue was resolved by evaluating the additional traffic and changing the stand settings.

Unfortunately, this work was not intended to test the data of firmware environments, therefore this information is a warning to researchers.

7 Conclusion

The developed protocol provides guaranteed data delivery without re-requests with packet losses from 0 to 23%. The protocol recovers all missing and corrupted data. There may be small losses when switching from lossless mode, but they were not observed during testing.

It is required to optimize program codes, since not all algorithms have been redesigned for parallel operation. Also, the protocol does not correspond to the nature of the errors on the radio links, it is required to ensure stable operation on the radio channel, as well as to consider the possibility of applying these developments at the physical level to the 5G standards group.

To ensure work on heterogeneous IP networks, the possibility of developing a neural network with reinforced learning is being considered.

It is also required to transfer the protocol from the kernel level to the user level. This will allow you not to depend on the specific implementation of the kernel and, in particular, work on the old, third core. In addition, at the kernel level, vector processor operations do not work, which can significantly speed up the processing of error correction codes.

References

1. Zogovic, N., Dimic, G., Bajic, D.: PHY-MAC cross-layer approach to energy-efficiency and packet-loss trade-off in low-power, low-rate wireless communications. IEEE Commun. Lett. **17**, 661–664 (2013)
2. ISO/IEC 7498–1: ISO/IEC International Standard, Information Technology-open Systems Interconnection-Basic Reference Model: The Basic Model, 2nd edn (1994). http://standards.iso.org/ittf/PubliclyAvailableStandards/s020269_ISO_IEC_7498-1_1994(E).zip
3. Alay, O., Korakis, T., Wang, Y., Panwar, S.: Is physical layer error correction sufficient for video multicast over IEEE 802.11g networks? In: Proceedings of 6th IEEE Consumer Communications and Networking Conference, pp. 1–5. Las Vegas (2009). https://doi.org/10.1109/ccnc.2009.4784785
4. Perruisseau, J.: Carrier versatile reconfiguration of radiation patterns, frequency and polarization: a discussion on the potential of controllable reflectarrays for software-defined and cognitive radio systems. In: 2010 IEEE International Microwave Workshop Series on RF Front-ends for Software Defined and Cognitive Radio Solutions. https://doi.org/10.1109/IMWS.2010.5440982
5. RFC 1661, Internet Standard, The Point-to-Point Protocol (PPP) (1994). https://tools.ietf.org/html/rfc1661
6. RFC 2516, Internet Standard, A Method for Transmitting PPP Over Ethernet (PPPoE) (1999). https://tools.ietf.org/html/rfc2516
7. RFC 793, Internet Standard, Transmission control protocol (1981). https://tools.ietf.org/html/rfc793
8. Cai, Y., Liu, Y., Gong, W., Wolf, T.: Impact of arrival burstiness on queue length: an infinitesimal perturbation analysis. In: Proceedings of the 48th IEEE Conference on Decision and Control (CDC) held jointly with 2009 28th Chinese Control Conference, pp. 7068–7073 (2009). https://doi.org/10.1109/CDC.2009.539953

9. Salkuyeh, M.A., Abolhassani, B.: Optimal video packet distribution in multipath routing for urban VANETs. J. Commun. Netw. **20**, 198–206 (2018)
10. Kato, N., et al.: The deep learning vision for heterogeneous network traffic control - proposal, challenges, and future perspective. IEEE Wirel. Commun. Mag. **24**(3), 146–153 (2016)
11. Mao, B., et al.: A tensor based deep learning technique for intelligent packet routing. In: GLOBECOM 2017–2017 IEEE Global Communications Conference, pp. 1–6 (2017). https://doi.org/10.1109/GLOCOM.2017.8254036
12. Kawamoto, J., Kurakake, T.: XOR-based FEC to improve burst-loss tolerance for 8k ultra-high definition TV over IP transmission. In: GLOBECOM 2017–2017 IEEE Global Communications Conference, pp. 1–6 (2017). https://doi.org/10.1109/GLOCOM.2017.8254125
13. Wu, J., Cheng, B., Wang, M., Chen, J.: Priority-aware FEC coding for high-definition mobile video delivery using TCP. IEEE Trans. Mob. Comput. **16**, 1090–1106 (2017)
14. Langley, A., Riddoch, A., Wilk, A.: The QUIC transport protocol: design and internet-scale deployment. In: SIGCOMM 2017 Proceedings of the Conference of the ACM Special Interest Group on Data Communication, pp. 183–196 (2017). https://doi.org/10.1145/3098822.3098842
15. Rüth, J., Poese, I., Dietzel, C., Hohlfeld, O.: A first look at QUIC in the wild. In: Beverly, R., Smaragdakis, G., Feldmann, A. (eds.) PAM 2018. LNCS, vol. 10771, pp. 255–268. Springer, Cham (2018). https://doi.org/10.1007/978-3-319-76481-8_19
16. Michel, F., De Coninck, Q., Bonaventure, O.: QUIC-FEC: bringing the benefits of forward erasure correction to QUIC. In: IFIP Networking. IFIP (2019). https://doi.org/10.23919/IFIPNetworking.2019.8816838
17. Ivchenko, A.V., Erchenko, A.V., Sinolits, V.V., Suhotepliy, A.P., Durigin, V.V.: Adaptive data link protocole for special-purpose networks. Telecommun. Radio Eng. **2**, 12–19 (2019)
18. Phang, S.Y., Lee, H., Lim, H.: Design and implementation of V6SNIFF: an efficient IPv6 packet sniffer. In: 2008 Third International Conference on Convergence and Hybrid Information Technology, pp. 44–49. https://doi.org/10.1109/ICCIT.2008.279
19. de Bruijn, W., Dumazet, E.: Optimizing UDP for content delivery: GSO, pacing and zerocopy. In: Linux Plumbers Conference (2018)

Model of Optical Non-blocking Information Processing System for Next-Generation Telecommunication Networks

E. Barabanova[1]([✉]) [iD], K. Vytovtov[1] [iD], V. M. Vishnevskiy[2] [iD], and V. Podlazov[2] [iD]

[1] Astrakhan State Technical University,
Tatischeva 16 str., 414004 Astrakhan, Russia
elizavetaalexb@yandex.ru
[2] V.A. Trapeznikov Institute of Control Sciences of RAS,
Profsoyuznaya 65 str., 117997 Moscow, Russia

Abstract. The authors propose the new principle of optical information processing systems construction and the new method for parallel transmission of information and control signals, which can improve the performance of next generation telecommunication networks and all-optical supercomputing systems. The advantages of the new systems are a decentralized control of switching process and non-blocking switching scheme [1,2]. Here the authors present the set-theoretical model of the 16 × 16 switch for the first time, describe the structure and the algorithm of the system and present the bipartite directed graph of the switch. The proposed system has low complexity in compared with well-known ones. To prove this important property of the system we have carried out numerical calculations of our schemes and well-known tunable one of a kind non-blocking Clos scheme complexity. The calculation results showed that the complexity of the developed schemes is 1.5–3 times less than the Clos schemes.

Keywords: Optical switch · Decentralized control · Algorithm · Control · Complexity

1 Introduction

The IoT and $5G$ networks are predicted to create interconnection of big number of devices and so required of high performance data processing nodes development for transferring real-time data. To increase the performance of the next generation networks not only fast data processing algorithms are needed, but also fundamentally new devices must be proposed which allow increasing the speed

The reported study was founded by RFBR according to the research project 18-37-00059/18.

© Springer Nature Switzerland AG 2019
V. M. Vishnevskiy et al. (Eds.): DCCN 2019, CCIS 1141, pp. 188–198, 2019.
https://doi.org/10.1007/978-3-030-36625-4_16

of information interconnection an order of magnitude [3, 4]. One of such devices is an all optical processor, which has been actively developed by both domestic and foreign scientists [4–7]. For the information exchange between optical processors, various completely optical switching schemes must be used [1, 2, 7–12]. That schemes must have low complexity and satisfy two basic requirements: high speed and non-blocking. In [7], the possibility of connecting optical processors through well-known network topologies as the $3D$ torus, the GN hypercube, and dragonfly (DF) was investigated. The disadvantage of these schemes is switch blocking when arbitrary switching commands must be processed. From switching theory [9, 10], we know that there is only one rearrangeable non-blocking scheme known as Clos network, it has relatively low complexity but uses a central control device to implement a complex rearrangeable algorithm. In [12] the networks with the topology of a quasi-complete graph and or digraph and their invariant extensions have been proposed. They are designed to use small switches and have high switching complexity if there is a large number of subscribers in the network.

In [1, 2], for the information exchange between optical processors, the authors propose to use the 4×4 next generation optical decentralized control switching systems which process the control information and payload consistently and the control information is transmitted in the packet header. In this paper, for the first time, the authors propose a method for parallel transmission of control signals and data at the level of individual bits. The control and information signals have different wavelengths. A feature of our switching systems is a new method of non-blocking information processing. This method provides the increasing of the duty cycle of optical pulses, and combines switch and bus methods of conflicts prevention.

We have proposed the method of expansion of the next generation optical switching systems based on 4×4 switches for constructing 16×16, 256×256, 4096×4096 systems [1, 2]. However, in [1, 2], a mathematical model of the proposed systems and a theoretical foundation of their non-blocking have not been presented. In this paper, the set-theoretic model of a 16×16 switch and the corresponding bipartite graph are presented for the first time. Based on these models, the authors have developed the optical switching system algorithm that ensures the absence of collisions. In this paper for the first time we also present our scheme complexity numerical calculating method and the comparison results with similar-sized well-known Clos scheme complexity.

2   Set-Theoretical Switch 16×16 Model

The 16×16 switch scheme is presented in the Fig. 1 [1]. The scheme consists of input and output stages of four 4×4 switches (Fig. 1). The bipartite directed graph of the switch is presented in Fig. 2 for the first time.

The bipartite directed graph of the switch $G(A, B, C, D, Y, Z)$ is presented in Fig. 2 for the first time. The graph is connected because every pair of the distinct vertices is joined by the path between sets $A(G)$ and $B(G)$.

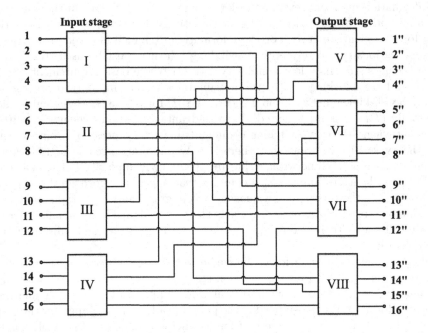

Fig. 1. The next generation 16 × 16 optical switching system scheme

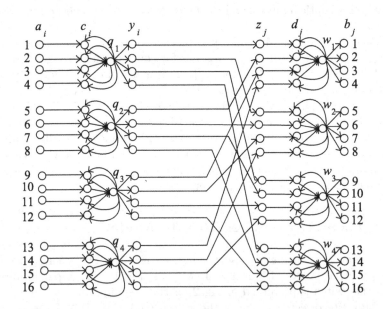

Fig. 2. The bipartite directed graph of the next generation optical 16 × 16 switch

Let $A = \{a_1, a_2, \ldots, a_{16}\}$ is the set of switch inputs; $B = \{b_1, b_2, \ldots, b_{16}\}$ is the set of switch outputs; $Y = \{y_1, y_2, \ldots, y_{16}\}$ is the set of outputs of the input stage switches; $Z = \{z_1, z_2, \ldots, z_{16}\}$ is the set of inputs of the output stage switches; $C = \{c_1, c_2, \ldots, c_{16}\}$ - is the set of delay lines of the input stage switch; $D = \{d_1, d_2, \ldots, d_{16}\}$ is the set of delay lines of the output stage switch; $Q = \{1, 0\}$ is the set of states of the optical integrated device 4×4 of the switch of the input stage (1-busy; 0-free); $W = \{1, 0\}$ -set of states of the optical integrated device 4×4 of the switch of the output stage (1-busy; 0-free) [1]. An optical integrated device is used for multiplexing signals and for control of delay lines.

Note that the degree of vertices $a_1 \ldots a_{16}$ and $b_1 \ldots b_{16}$ is equal to 1; the degree of vertices $c_1 \ldots c_{16}$ and $d_1 \ldots d_{16}$ is equal to 3 and vertices $y_1 \ldots y_{16}$ and $z_1 \ldots z_{16}$ is equal to 2 and the degree of vertices $q_1 \ldots c_{16}$ and $w_1 \ldots w_{16}$ is equal to 12.

The bipartite graph G consists of eight subgraphs G'. Let us denote the vertices of the subgraph g as $a_1 \ldots a_4$; $c_1 \ldots c_4$; q_1; $y_1 \ldots y_4$ and describe one of the subgraphs of the graph G by the adjacency matrix (Table 1).

Table 1. The adjacency matrix of subgraph G'

	a_1	a_2	a_3	a_4	c_1	c_2	c_3	c_4	q_1	y_1	c_2	c_3	c_4
a_1	0	0	0	0	1	0	0	0	1	0	0	0	0
a_2	0	0	0	0	0	1	0	0	1	0	0	0	0
a_3	0	0	0	0	0	0	1	0	1	0	0	0	0
a_4	0	0	0	0	0	0	0	0	1	0	0	0	0
c_1	1	0	0	0	0	0	0	0	2	0	0	0	0
c_2	0	1	0	0	0	0	0	0	2	0	0	0	0
c_3	0	0	1	0	0	0	0	0	2	0	0	0	0
c_4	0	0	0	1	1	0	0	0	2	0	0	0	0
q_1	0	0	0	0	1	1	1	1	0	1	1	1	1
y_1	0	0	0	0	0	0	0	0	1	0	0	0	0
y_2	0	0	0	0	0	0	0	0	1	0	0	0	0
y_3	0	0	0	0	0	0	0	0	1	0	0	0	0
y_4	0	0	0	0	0	0	0	0	1	0	0	0	0

The bipartite directed graph G includes feedbacks between sets of vertices C and Q and sets of vertices D and W. The graph G is named the bipartite directed graph with feedbacks and it describes the space-time divided of optical signals. For example, let us consider the paths between two parallel connections: the first input and the sixteen output and the fourth input and the first output. The first path includes the vertices a_1, c_1, q_1, y_4, z_{12}, d_{12}, w_4 and b_{16} and the second path includes the vertices a_4, c_4, q_1, y_1, z_1, d_1, w_1 and b_1. As you can see there is an interaction of connection lines in vertex q_1. As the vertex q_1 has additional feedback edges between each of vertices c_1, c_2, c_3, c_4 of subgraph G' there is no blocking in the system.

The formalized model describing the structure of the switch is defined by the following expression:

$$T = S_{c1} \cup S_{c2} \tag{1}$$

where the sets S_{C1} and S_{C2} describe the mathematical models of the input and output stages respectively. These sets can be represented in the following form:

$$S_{c1} = \bigcup_{i=1,4} (A_i \times Y_i) \tag{2}$$

$$S_{c2} = \bigcup_{j=1,4} (Z_j \times B_j) \tag{3}$$

where Cartesian products $A_i \times Y_i$ and $Z_j \times B_j$ describe i-switching block and j-switching block of input and output stages accordingly. We additionally define the set of communication lines between the input and output stages as M_L. We have divided the sets Y and Z into four subsets: $R_y = \{Y_k : Y_k \subset M_L; k = \overline{1,4}\}$; $R_z = \{Z_k : Z_k \subset M_L; k = \overline{1,4}\}$, where any subset Z_k contains only one element of each subset Y_k and vice versa. So the connection lines between input and output switching blocks can be described through the adjacency matrix (Table 2, Fig. 1).

Table 2. The adjacency matrix of 16×16 switching system graph

	I	II	III	IV	V	VI	VII	VIII
I	0	0	0	0	1	0	0	0
II	0	0	0	0	1	0	0	0
III	0	0	0	0	1	0	0	0
IV	0	0	0	0	1	0	0	0
V	0	0	0	0	1	0	0	0
VI	0	0	0	0	1	0	0	0
VII	0	0	0	0	1	0	0	0
VIII	0	0	0	0	1	0	0	0

The set of input variables $X = \{x_k\}$, $x_k \subset \{\lambda_i^1, \lambda_j^1, \lambda_i^2, \lambda_j^2, \lambda_p\}$ is arrived at the inputs of the system and it can be distributed among the outputs in accordance with the switch operation algorithm. The block diagram of the algorithm is presented in Fig. 3. The operators of occupying of the optical integrated device of the input $p_1(q_i)$ and output $p_2(w_i)$ 4×4 switches stages are defined as follows:

$$p_1(q_i) = \begin{cases} 1, & \text{if there is a communication channel between } a_i \text{ and } y_i \text{ through } q_i \\ 0, & \text{in other cases} \end{cases} \tag{4}$$

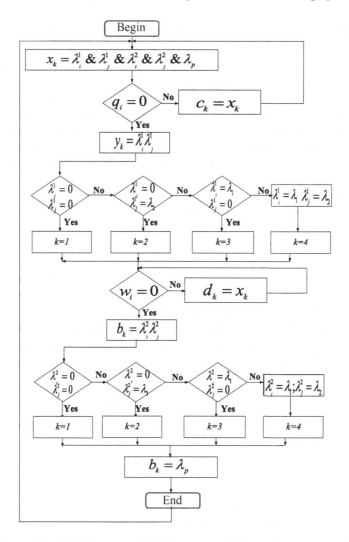

Fig. 3. The block diagram of the 16×16 switching system algorithm

$$p_2(w_i) = \begin{cases} 1, \text{ if there is the communication channel between } z_j \text{ and } b_j \text{ through } w_i \\ 0, \text{ in other cases} \end{cases}$$

$$(5)$$

Let $\nu_1(c_i)$ is the distribution operator of the variables x_k along the delay lines c_i.

$$\nu_1(c_i) = \begin{cases} x_k, \text{ if } x_k \text{ is at the input } a_i \text{ and } p_1(q_i) = 1 \\ 0, \text{ in other cases} \end{cases} \qquad (6)$$

and let $\nu_2(d_i)$ is the distribution operator of the variables x_k along the delay lines d_i.

$$\nu_1(d_i) = \begin{cases} x_k, \text{ if } x_k \text{ is at the input } a_i \text{ and } p_2(w_i) = 1 \\ 0, \text{ in other cases} \end{cases} \tag{7}$$

The problem of distributing of the set of input variables $X = \{x_k\}$ over the outputs is solved by two steps. At the first step of the algorithm, the communication channel is established through the switching unit of the input stage ($IS1$). For this, in accordance with a combination of control signals λ_i^1 and λ_j^1, the output number of one of the input stage switching blocks y_k is determined. So the first output corresponds to a combination of control signals $\{0; 0\}(\lambda_i^1 = 0, \lambda_j^1 = 0)$, the second one corresponds to $\{0; \lambda_2\}$ ($\lambda_i^1 = 0, \lambda_j^1 = \lambda_2$), the third one corresponds to $\{\lambda_1; 0\}$ ($\lambda_1^1 = \lambda_1, \lambda_j^1 = 0$), the fourth one corresponds to $\{\lambda_3; \lambda_4\}$ ($\lambda_i^2 = \lambda_3, \lambda_j^2 = \lambda_4$) (see Fig. 3 and Table 1). At the second step the communication channel is established through the switching unit of the output stage ($OS2$). For this, in accordance with a combination of control signals, the output number of one of the switching blocks of the output stage z_k is determined. So, the first output corresponds to the combination $\{0; 0\}$ ($\lambda_i^2 = 0, \lambda_j^2 = 0$), the second one corresponds to $\{0; \lambda_4\}$ ($\lambda_i^2 = 0, \lambda_j^2 = \lambda_4$), the third one corresponds to $\{\lambda_3; 0\}$ ($\lambda_i^2 = \lambda_3, \lambda_j^2 = 0$), the fourth one corresponds to $\{\lambda_3; \lambda_4\}$ ($\lambda_i^2 = \lambda_3, \lambda_j^2 = \lambda_4$) (see Fig. 1, Table 1) (Table 3).

Table 3. The combinations of the control signals corresponding to the output numbers (N) of the 16×16 switching system

N	Control signals	N	Control signals
1	0 0 0 0	9	λ_1 0 0 0
2	0 0 0 λ_4	10	λ_1 0 0 λ_4
3	0 0 λ_3 0	11	λ_1 0 λ_3 0
4	0 0 λ_3 λ_4	12	λ_1 0 λ_3 λ_4
5	0 λ_2 0 0	13	λ_1 λ_2 0 0
6	0 λ_2 0 λ_4	14	λ_1 λ_2 0 λ_4
7	0 λ_2 λ_3 0	15	λ_1 λ_2 λ_3 0
8	0 λ_2 λ_3 λ_4	16	λ_1 λ_2 λ_3 λ_4

Thus, we can introduce the operator of the existence of a communication channel between input a_i and output b_j as

$$S(a_i b_j) = \begin{cases} 1, \text{ if } p_1(q_i) = 0 \text{ and } p_2(w_i) = 0 \\ 0, \text{ in other cases} \end{cases} \tag{8}$$

As a result, the necessary condition for the absence of blocking and collisions during the information transmission can be written in the form

$$S(a_i b_i) = F\left[x_k(a_i), p(q_i), p_2(w_i)\right] =$$

$$x_k \& \overline{p_1(q_i)} \& \overline{p_2(w_i)}$$

(9)

Timing diagrams of control and information signals are presented in Fig. 4.

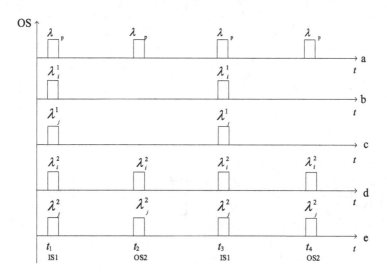

Fig. 4. Timing diagrams of control and information signals

At the moment t_1 one of the inputs of the switching system receives a bit of an information signal λ_p (Fig. 4a) and four control signals λ_i^1, λ_j^1, λ_i^2 and λ_j^2 (Fig. 4b, c, d, e). Two control signals are used to configure the input stage 4×4 switch (IS). At the moment t_2, one of the inputs of the output stage switch (OS) receives a bit of the information signal (Fig. 4a, d, e) and two control signals λ_i^2 and λ_j^2, which are used to configure the output stage 4×4 switch.

At the moment t_3, there is a transmission of information and control signals in accordance with the output number. Thus, the switch configuration process is repeated. On the timing diagram the control and information signals transmission from only one of several inputs is presented. It should be noted that there is no blocking in the system.

3   Numerical Calculations of Scheme Complexity

In order to evaluate the advantages of the proposed schemes, we have compared their complexity S and the required signal period T with the well-known non-blocking Clos scheme (Fig. 5).

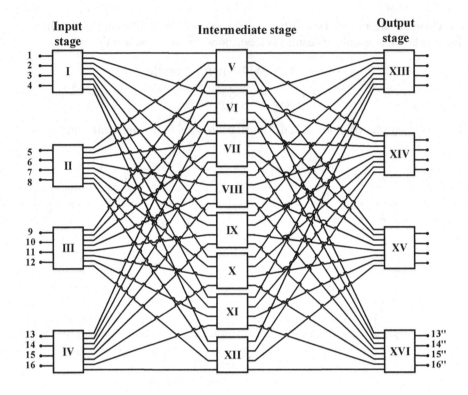

Fig. 5. The 3-stage Clos scheme

The complexity of three-stage Clos switching system is generally determined as following expression: $S_1 = 3m^3 = 3N_1^{3/2}$, where N_1 is the number of subscribers, and m is the number of inputs and outputs of the input stage switch. But Clos switching system is blocked in the case of parallel information transmission from arbitrary input to arbitrary output [...].

We have considered the non-blocking Clos scheme, which contains the first stage $m \times 2m$ switches, the intermediate stage $m \times m$ switches, and the output stage $2m \times m$ switches. Then the number of subscribers N_1 and complexity S_1 can be calculated as following expressions: $N_1 = m^2$, $S_1 = 6m^3 = 6N_1^{3/2}$. For example, for $m = 4$ (Fig. 5), we have $N_1 = 16$ and $S_1 = 6 \cdot 64 = 384 = N_1^{2.15}$ for $m = 16$ we have $N_1 = 256$ and $S_1 = 6 \cdot 4096 = 24576 = N_1^{1.86}$. The period of optical pulses T_1 equal to 1. The disadvantage of this scheme is external control by special rearrangeable algorithm to provide non-blocking property of the system.

For the proposed photon 16×16 switch (Fig. 1), the complexity and period of optical pulses can be calculated as following expressions: $S_1 = N_1^2 = 256$ and a $T_1 = 4$, the complexity S_2 of the photon 256×256 switch as $S_2 = 8192 = N_2^{1.62}$ with a period of optical pulses $T_2 = 49$. Thus, the complexity of the proposed 16×16 switch is 1.5 times less than the complexity of the Clos switch, and the complexity of the 256 × 256 switch is three times less than the complexity of the analogously Clos switch. The disadvantage of the presented systems is the larger size of the optical signals period required to provide non-blocking scheme with decentralized control. As the number of inputs increases, the number of control signals and the period of optical signals increases.

4 Conclussions

For the development of all optical switching systems that provide switching speed about several ps, various multistage schemes such as Benes, Spanke, Shpanke-Benes and Clos have been proposed early [9–11]. All these schemes are non-blocking only for a limited set of input-output pairs. Only the rearrangeable Close scheme is non-blocking but it must be controlled by external processor that uses a complex switching algorithm. Therefore speed of optical switching systems based on such type of schemes is really limited by ns. The strictly non-blocking schemes have been offered by authors recently [1,2]. However in those works the authors did not present the accurate proof of non-blocking property and self-turning functioning algorithm of new strictly non-blocking switching systems. Despite it is very important point in design of all-optical switches.

In this paper the accurate set-theoretical model and the bipartite directed graph of the 16 × 16 optical switching system that can be used in next generation communication networks are presented for the first time. The model allows us to formulate the non-blocking condition for the new scheme types for the first time too. Additionally we developed and presented the functioning algorithm of the 16 × 16 optical switching system [1] using the set-theoretical description and graph theory. The algorithm is based on the fundamentally new method of parallel processing of control and information signals that we described here for the first time. We also present the research results of dependencies of the proposed scheme complexity and signal periods on the number of inputs. The comparison of obtained results with the analogous parameters of the Clos scheme is carried out, and we can say that our schemes have smaller complexity but longer signal period to providing strictly non-blocking of all optical switching. The main advantage of presented switching method is that connection establishment control without external electron processor is realized for the first time. For this purpose processing of control and information signals is carried out inside the switching scheme. As a result, in our point of view, the developed schemes are perspective for all-optical next generation telecommunication networks and it can be used in real-time systems.

References

1. Vytovtov, K.A., Barabanova, E.A., Podlazov, V.S.: Model of next-generation optical switching system. Commun. Comput. Inf. Sci. **918**, 377–386 (2018)
2. Vytovtov, K.A., Barabanova, E.A., Barabanov I.O.: Next-generation switching system based on 8 × 8 self-turning optical cell. In: Proceedings of International Conference on Actual Problems of Electron Devices Engineering, pp. 306–310, Saratov (2018)
3. Vishnevskiy, V., Semenova, O.: Queueing system with alternating service rates for free space optics-radio hybrid channel. In: Vinel, A., Bellalta, B., Sacchi, C., Lyakhov, A., Telek, M., Oliver, M. (eds.) MACOM 2010. LNCS, vol. 6235, pp. 79–90. Springer, Heidelberg (2010). https://doi.org/10.1007/978-3-642-15428-7_9
4. Djordjevic, I.B.: Advanced Optical and Wireless Communications Systems. Springer, Heidelberg (2018)
5. Stepanenko, S.A.: Photon computer: structure and algorithms, parameter estimates. Photonika **7**(67), 72–83 (2017). (in Russian)
6. Stepanenko, S.A.: Multiprocessor environments of supercomputers. Efficiency scaling, Moscow, Fizmatlit (2016). (in Russian)
7. Kaliaev, I.A., Levin, I.I., Semernikov, E.A., Shmoilov, V.I.: Reconfigurable Multipipeline Computing Structures. Nova Science Publishers, Inc. USA (2012). ISBN 978-1-61942-854-6
8. El Bawab, T.S.: Optical Switching. Springer, Heidelberg (2006). https://doi.org/10.1007/0-387-29159-8
9. Kabacinski, W.: Nonblocking Electronic and Photonic Switching Fabrics. Springer, Heidelberg (2005). https://doi.org/10.1007/b137691
10. Hwang, F.K.: A survey of nonblocking multicast three-stage Clos networks. IEEE Commun. Mag. **41**, 34–37 (2003). 50th anniversary of Clos networks
11. Podlazov, V.S.: A comparison of system area networks: generalized extended multiring vs flattened butterfly. Autom. Remote Control **79**(3), 571–580 (2018)
12. Kutuzov, D., Stukach O.: Algorithms of parallel switching for multistage schemes. In: International Siberian Conference on Control and Communications (SIBCON), Proceedings. Siberian Federal University, Krasnoyarsk, Russia, 12–13 September (2013)

Accurate Mathematical Model of Two-Dimensional Parametric Systems Based on 2 × 2 Matrix

K. Vytovtov[1(✉)] [iD], E. Barabanova[1] [iD], and V. M. Vishnevskiy[2] [iD]

[1] Astrakhan State Technical University, Tatischeva 16 str., Astrakhan, Russia
vytovtov_konstan@mail.ru
[2] V.A. Trapeznikov Institute of Control Sciences of RAS,
Profsoyuznaya 65 str., 117997 Moscow, Russia

Abstract. In this paper we consider a two dimensional dynamical system with arbitrary piecewise constant parameters described by a linear homogeneous differential equations system with discontinuous coefficients. For such a system the fundamental solution matrix in the analytical form in elementary functions is found. The theorem saying that this matrix is the finite sum of the unimodular matrices with the certain influence coefficients is proved. The results allow us to carry out qualitative analysis of the corresponding dynamical systems, solve inverse problems, investigate the conditions for the oscillation stabilities. Obviously, the results can be used in theory of inhomogeneous and non-linear systems.

Keywords: Mathematical model · Two order differential equations · Linear dynamic system

1 Introduction

Studying two-dimensional dynamic systems is a classic problem in mechanics, microwave theory, optics, telecommunications, hydrodynamics, biology etc. During the past two centuries linear and non-linear systems have been widely described in scientific literature [1–10].

The analytical [1–4,8–10] and numerical [4–7] methods have been utilized to research such systems. Accurate analytical solutions for systems with constant parameters have been presented, for example, in [1,8,9]. Approximate analytical methods of small parameters have been used [4]. However, numerical methods have been applied as rule [5–7].

For today the fundamental matrix method [1–4,8–10] can be considered as most applicable for linear systems with piecewise constant parameters. Indeed, this matrix relates a state of a system at an arbitrary moment t to a state at

The reported study was founded by RFBR according to the research project 18-37-00059/18.

an initial moment $t = 0$. For example, solving stability problem in mechanics by using this method is simple and elegant [8,9]. In microwave theory and optics the fundamental matrix method allows us to find reflection and transmitted coefficients of stratified structures. The 2×2 matrix for two dimensional mechanical system has been presented in detail in [1–4]. In the case of four dimensional electromagnetic system the method has been applied in [11,12].

In this paper we present the fundamental matrix of an linear homogeneous two-dimensional dynamic system with arbitrary piecewise constant parameters for the first time. The solution is obtained in the analytical form in elementary functions. It also is proved that the fundamental matrix for the arbitrary intervals with constant parameters can be presented as the finite sum of the unimodular matrices. It is very important result as it is well-known [13] that a unimodular matrix describes a undamped oscillation. Thus we obtain the finite spectrum of a resulting oscillation.

2   Statement of the Problem

In this paper we consider a system described by the two-order linear ordinary homogeneous differential equations with arbitrary piecewise-constant coefficients [1–4].

$$\begin{cases} \dfrac{dy(t)}{dt} = a_{11}(t)x(t) + a_{12}(t)y(t) \\ \dfrac{dx(t)}{dt} = a_{21}(t)x(t) + a_{22}(t)y(t) \end{cases} \tag{1}$$

where in a general case the coefficients a_{11}, a_{12}, a_{21}, a_{22} are complex and piecewise constant:

$$a_{11}(t) = \begin{cases} a_{11}^{(1)}, \ 0 < t < t_1 \\ a_{11}^{(2)}, \ t_1 < t < t_2 \\ \cdots \quad \cdots \\ a_{11}^{(N)}, t_{N-1} < t < T \end{cases} \tag{2}$$

$$a_{12}(t) = \begin{cases} a_{12}^{(1)}, \ 0 < t < t_1 \\ a_{12}^{(2)}, \ t_1 < t < t_2 \\ \cdots \quad \cdots \\ a_{12}^{(N)}, t_{N-1} < t < T \end{cases} \tag{3}$$

$$a_{21}(t) = \begin{cases} a_{21}^{(1)}, \ 0 < t < t_1 \\ a_{21}^{(2)}, \ t_1 < t < t_2 \\ \cdots \quad \cdots \\ a_{21}^{(N)}, t_{N-1} < t < T \end{cases} \tag{4}$$

$$a_{12}(t) = \begin{cases} a_{12}^{(1)}, \ 0 < t < t_1 \\ a_{12}^{(2)}, \ t_1 < t < t_2 \\ \cdots \quad \cdots \\ a_{12}^{(N)}, t_{N-1} < t < T \end{cases} \tag{5}$$

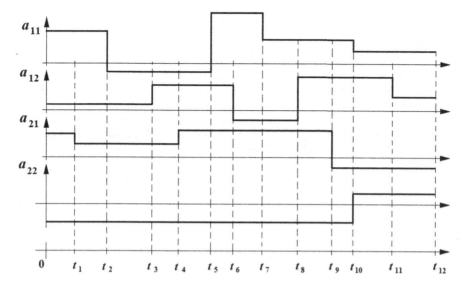

Fig. 1. The possible coefficients of the system (1)

Here the superscript is the number of an interval with constant parameters. The possible example of parameter is presented in Fig. 1. Each of the parameters can jump for any value at any moment t.

3   Method

3.1   The Fundamental Matrix of an i-th Interval with Constant Parameters

First of all we find the eigennumbers of the systems (1). In accordance to well-known method [1,2,4] we can write

$$k_{1,2} = \frac{a_{12} + a_{21}}{2} \pm \sqrt{\frac{(a_{12} + a_{21})^2}{4} + (a_{11}a_{22} - a_{12}a_{21})} \qquad (6)$$

It also is well-known [1–4,14,15] that the general solution of the system (1) can be written as the sum of two exponents

$$\begin{aligned} x(t) &= A\exp(k_1 t) + B\exp(k_2 t) \\ y(t) &= (k_1 - a_{21})A\exp(k_1 t) + (k_2 - a_{21})B\exp(k_2 t) \end{aligned} \qquad (7)$$

Now, the elements of the fundamental matrix must be found for the independent initial conditions $\{1;0\}^T$ and $\{0;1\}^T$ at $t = 0$ [1,8,9,14,15]. Here T is the transpose operation. In this way the coefficients A and B are

$$A = \frac{k_2 - a_{21}}{k_2 - k_1}$$

$$B = -\frac{k_1 - a_{21}}{k_2 - k_1}$$

(8)

for the initial condition $\{1; 0\}^T$. Now in accordance to [1, 2, 14, 15] the first column elements of the fundamental matrix are following

$$M_{11} = \frac{k_2 - a_{21}}{k_2 - k_1} \exp(k_1 t) - \frac{k_1 - a_{21}}{k_2 - k_1} \exp(k_2 t)$$

$$M_{21} = \frac{(k_1 - a_{21})(k_2 - a_{21})}{k_2 - k_1} [\exp(k_1 t) - \exp(k_2 t)]$$

(9)

Analogously taking into account the initial conditions $\{0; 1\}^T$ the constants A and B are obtained as

$$A = -\frac{1}{k_2 - k_1}$$

$$B = \frac{1}{k_2 - k_1}$$

(10)

and the second column elements of the matrix are

$$M_{12} = -\frac{1}{k_2 - k_1} \exp(k_1 t) + \frac{1}{k_2 - k_1} \exp(k_2 t)$$

$$M_{22} = -\frac{k_1 - a_{21}}{k_2 - k_1} \exp(k_1 t) + \frac{k_2 - a_{21}}{k_2 - k_1} \exp(k_2 t)$$

(11)

As a result the fundamental matrix for an i-th interval with constant parameters can be written in the form

$$\mathbf{M}_i = \frac{1}{k_2^{(i)} - k_1^{(i)}} \begin{pmatrix} \gamma_2^{(i)} \exp(k_1^{(i)} t^{(i)}) - \gamma_1^{(i)} \exp(k_2^{(i)} t^{(i)}) \\ \gamma_1^{(i)} \gamma_2^{(i)} \left[\exp(k_1^{(i)} t^{(i)}) - \exp(k_2^{(i)} t^{(i)}) \right] \end{pmatrix}$$

$$- \exp(k_1^{(i)} t^{(i)}) + \exp(k_2^{(i)} t^{(i)})$$

$$-\gamma_1^{(i)} \exp(k_1^{(i)} t^{(i)}) + \gamma_2^{(i)} \exp(k_2^{(i)} t^{(i)}) \Bigg)$$

(12)

where

$$\gamma_1^{(i)} = k_1^{(i)} - a_{21}^{(i)}$$
$$\gamma_2^{(i)} = k_2^{(i)} - a_{21}^{(i)}$$

(13)

Here the superscript indicates the interval number, and the subscript indicates the eigenmode number within the interval.

3.2   The Fundamental Matrix of an Interval with Piecewise Constant Parameters

In accordance to the method $[1,2,8,9,14,15]$ a fundamental matrix relates the dynamic variables $x(t)$ and $y(t)$ at an arbitrary point t to these variables $x(0)$ and $y(0)$ at $t = 0$. On other words, we have

$$\mathbf{U}(t_1) = \mathbf{M}_1\mathbf{U}(0) \tag{14}$$

for the first interval. Here $\mathbf{U}(t) = \{y(t), x(t)\}^T$, $\mathbf{M}_1$ is the fundamental matrix of the first interval (see (12)). Taking into account continuity conditions for dynamic variables $[1,2,8,9]$, we can write the analogous expression for two intervals with constant parameters.

$$\mathbf{U}(t_2) = \mathbf{M}_2\mathbf{U}(t_1) = \mathbf{M}_2\mathbf{M}_1\mathbf{U}(0) \tag{15}$$

Here $\mathbf{M}_2$ is the fundamental matrix of the second interval. Analogously it can be written

$$\mathbf{U}(T) = \mathbf{M}_N\mathbf{M}_{N-1}...\mathbf{M}_2\mathbf{M}_1\mathbf{U}(0) \tag{16}$$

for N intervals with constant parameters. Therefore a fundamental matrix for the arbitrary finite number N of intervals with constant parameters is the product of the interval matrices (12).

$$\mathbf{M} = \mathbf{M}_N\mathbf{M}_{N-1}...\mathbf{M}_2\mathbf{M}_1 = \prod_{i=N}^{1} \mathbf{M}_i \tag{17}$$

In classical scientific literature this formula is final and its further transformations have not been carried out. As a result there are no an expression of a fundamental matrix of a two dimensional system with arbitrary piecewise constant parameters in analytical form for today.

Our main purpose here is finding such a matrix. First of all scientific research of (17) is allowed us to formulate the following main theorem:

Theorem 1. *The fundamental matrix of the two-dimensional system (1) with arbitrary piecewise constant coefficients can be written as the finite sum of the unimodular matrices $\mathbf{M}_q$ with the certain influence coefficients ζ_q:*

$$\mathbf{M} = \sum_{q=1}^{2^N} \zeta_q\mathbf{M}_q \tag{18}$$

where N is the number of intervals with constant parameters,

$$\zeta_q = \frac{\prod_{i=1}^{N-1}(\gamma_{1+F_{q,i}}^{(i)} - \gamma_{2-F_{q,i}}^{(i+1)})}{\prod_{i=1}^{N}(k_2^{(i)} - k_1^{(i)})} \tag{19}$$

is the influence coefficients,

$$\mathbf{M}_q = \begin{pmatrix} (-1)^{\sum_{i=1}^{N} F_{q,i}} \gamma^{(1)}_{2-F_{q,i}} \exp\left[\sum_{i=1}^{2^N}(k^{(i)}_{1+F_{q,i}} t^{(i)})\right] \\ (-1)^{\sum_{i=1}^{N} F_{q,i}} \gamma^{(1)}_{2-F_{q,i}} \gamma^{(N)}_{1+F_{q,i}} \exp\left[\sum_{i=1}^{2^N}(k^{(i)}_{1+F_{q,i}} t^{(i)})\right] \\ (-1)^{\sum_{i=1}^{N} F_{q,i}} \gamma^{(1)}_{2-F_{q,i}} \gamma^{(N)}_{1+F_{q,i}} \exp\left[\sum_{i=1}^{2^N}(k^{(i)}_{1+F_{q,i}} t^{(i)})\right] \\ (-1)^{\sum_{i=1}^{N} F_{q,i}+1} \gamma^{(N)}_{2-F_{q,i}} \exp\left[\sum_{i=1}^{2^N}(k^{(i)}_{1+F_{q,i}} t^{(i)})\right] \end{pmatrix} \tag{20}$$

is the unimodular matrix.

Proof. The proof this theorem is given by using the mathematical induction method. At first, let us write the matrix for two intervals with constant parameters. The elements of the one obtained by using the interval matrix multiplication and the algebraic transformations are

$$M^{(2)}_{11} =$$
$$-\frac{\gamma^1_2(\gamma^1_1 - \gamma^2_2)}{(k^{(1)}_2 - k^{(1)}_1)(k^{(2)}_2 - k^{(2)}_1)} \exp\left(k^{(1)}_1 t^{(1)} + k^{(2)}_1 t^{(2)}\right)$$
$$+\frac{\gamma^1_2(\gamma^1_1 - \gamma^2_1)}{(k^{(1)}_2 - k^{(1)}_1)(k^{(2)}_2 - k^{(2)}_1)} \exp\left(k^{(1)}_1 t^{(1)} + k^{(2)}_2 t^{(2)}\right)$$
$$+\frac{\gamma^1_1(\gamma^1_2 - \gamma^2_2)}{(k^{(1)}_2 - k^{(1)}_1)(k^{(2)}_2 - k^{(2)}_1)} \exp\left(k^{(1)}_2 t^{(1)} + k^{(2)}_1 t^{(2)}\right)$$
$$-\frac{\gamma^1_1(\gamma^1_2 - \gamma^2_1)}{(k^{(1)}_2 - k^{(1)}_1)(k^{(2)}_2 - k^{(2)}_1)} \exp\left(k^{(1)}_2 t^{(1)} + k^{(2)}_2 t^{(2)}\right) \tag{21}$$

$$M^{(2)}_{12} =$$
$$\frac{(\gamma^1_1 - \gamma^2_2)}{(k^{(1)}_2 - k^{(1)}_1)(k^{(2)}_2 - k^{(2)}_1)} \exp\left(k^{(1)}_1 t^{(1)} + k^{(2)}_1 t^{(2)}\right)$$
$$-\frac{(\gamma^1_1 - \gamma^2_1)}{(k^{(1)}_2 - k^{(1)}_1)(k^{(2)}_2 - k^{(2)}_1)} \exp\left(k^{(1)}_1 t^{(1)} + k^{(2)}_2 t^{(2)}\right)$$
$$-\frac{(\gamma^1_2 - \gamma^2_2)}{(k^{(1)}_2 - k^{(1)}_1)(k^{(2)}_2 - k^{(2)}_1)} \exp\left(k^{(1)}_2 t^{(1)} + k^{(2)}_1 t^{(2)}\right)$$
$$+\frac{(\gamma^1_2 - \gamma^2_1)}{(k^{(1)}_2 - k^{(1)}_1)(k^{(2)}_2 - k^{(2)}_1)} \exp\left(k^{(1)}_2 t^{(1)} + k^{(2)}_2 t^{(2)}\right) \tag{22}$$

$$M^{(2)}_{21} =$$
$$-\frac{\gamma^1_2\gamma^2_1(\gamma^1_1 - \gamma^2_2)}{(k^{(1)}_2 - k^{(1)}_1)(k^{(2)}_2 - k^{(2)}_1)} \exp\left(k^{(1)}_1 t^{(1)} + k^{(2)}_1 t^{(2)}\right)$$
$$+\frac{\gamma^1_2\gamma^2_2(\gamma^1_1 - \gamma^2_2)}{(k^{(1)}_2 - k^{(1)}_1)(k^{(2)}_2 - k^{(2)}_1)} \exp\left(k^{(1)}_1 t^{(1)} + k^{(2)}_2 t^{(2)}\right)$$
$$+\frac{\gamma^1_1\gamma^2_1(\gamma^1_1 - \gamma^2_2)}{(k^{(1)}_2 - k^{(1)}_1)(k^{(2)}_2 - k^{(2)}_1)} \exp\left(k^{(1)}_2 t^{(1)} + k^{(2)}_1 t^{(2)}\right)$$
$$-\frac{\gamma^1_1\gamma^2_2(\gamma^1_1 - \gamma^2_2)}{(k^{(1)}_2 - k^{(1)}_1)(k^{(2)}_2 - k^{(2)}_1)} \exp\left(k^{(1)}_2 t^{(1)} + k^{(2)}_2 t^{(2)}\right) \tag{23}$$

$$M_{21}^{(2)} =$$
$$\frac{\gamma_1^2(\gamma_1^1 - \gamma_2^2)}{(k_2^{(1)} - k_1^{(1)})(k_2^{(2)} - k_1^{(2)})} \exp\left(k_1^{(1)}t^{(1)} + k_1^{(2)}t^{(2)}\right)$$
$$- \frac{\gamma_2^2(\gamma_1^1 - \gamma_2^2)}{(k_2^{(1)} - k_1^{(1)})(k_2^{(2)} - k_1^{(2)})} \exp\left(k_1^{(1)}t^{(1)} + k_2^{(2)}t^{(2)}\right)$$
$$- \frac{\gamma_1^2(\gamma_1^1 - \gamma_2^2)}{(k_2^{(1)} - k_1^{(1)})(k_2^{(2)} - k_1^{(2)})} \exp\left(k_2^{(1)}t^{(1)} + k_1^{(2)}t^{(2)}\right)$$
$$+ \frac{\gamma_2^2(\gamma_1^1 - \gamma_2^2)}{(k_2^{(1)} - k_1^{(1)})(k_2^{(2)} - k_1^{(2)})} \exp\left(k_2^{(1)}t^{(1)} + k_2^{(2)}t^{(2)}\right) \tag{24}$$

Expressions (21)–(24) already show the certain regularity in the elements of the matrix. However to fully understand this regularity, we write the matrix elements for three intervals with constant parameters. For example, the element $M_{11}^{(3)}$ after algebraic transformations is

$$M_{11}^{(3)} =$$
$$\frac{\gamma_2^1(\gamma_1^1 - \gamma_2^2)(\gamma_1^2 - \gamma_2^3)}{(k_2^{(1)} - k_1^{(1)})(k_2^{(2)} - k_1^{(2)})(k_2^{(3)} - k_1^{(3)})} \exp\left(k_1^{(1)}t^{(1)} + k_1^{(2)}t^{(2)} + (k_1^{(3)}t^{(3)})\right)$$
$$- \frac{\gamma_2^1(\gamma_1^1 - \gamma_2^2)(\gamma_1^2 - \gamma_1^3)}{(k_2^{(1)} - k_1^{(1)})(k_2^{(2)} - k_1^{(2)})(k_2^{(3)} - k_1^{(3)})} \exp\left(k_1^{(1)}t^{(1)} + k_1^{(2)}t^{(2)} + (k_2^{(3)}t^{(3)})\right)$$
$$- \frac{\gamma_2^1(\gamma_1^1 - \gamma_1^2)(\gamma_2^2 - \gamma_2^3)}{(k_2^{(1)} - k_1^{(1)})(k_2^{(2)} - k_1^{(2)})(k_2^{(3)} - k_1^{(3)})} \exp\left(k_1^{(1)}t^{(1)} + k_2^{(2)}t^{(2)} + (k_1^{(3)}t^{(3)})\right)$$
$$+ \frac{\gamma_2^1(\gamma_1^1 - \gamma_1^2)(\gamma_2^2 - \gamma_1^3)}{(k_2^{(1)} - k_1^{(1)})(k_2^{(2)} - k_1^{(2)})(k_2^{(3)} - k_1^{(3)})} \exp\left(k_1^{(1)}t^{(1)} + k_2^{(2)}t^{(2)} + (k_2^{(3)}t^{(3)})\right)$$
$$- \frac{\gamma_1^1(\gamma_2^1 - \gamma_2^2)(\gamma_1^2 - \gamma_2^3)}{(k_2^{(1)} - k_1^{(1)})(k_2^{(2)} - k_1^{(2)})(k_2^{(3)} - k_1^{(3)})} \exp\left(k_2^{(1)}t^{(1)} + k_1^{(2)}t^{(2)} + (k_1^{(3)}t^{(3)})\right)$$
$$+ \frac{\gamma_1^1(\gamma_2^1 - \gamma_2^2)(\gamma_1^2 - \gamma_1^3)}{(k_2^{(1)} - k_1^{(1)})(k_2^{(2)} - k_1^{(2)})(k_2^{(3)} - k_1^{(3)})} \exp\left(k_2^{(1)}t^{(1)} + k_1^{(2)}t^{(2)} + (k_2^{(3)}t^{(3)})\right)$$
$$+ \frac{\gamma_1^1(\gamma_2^1 - \gamma_1^2)(\gamma_2^2 - \gamma_2^3)}{(k_2^{(1)} - k_1^{(1)})(k_2^{(2)} - k_1^{(2)})(k_2^{(3)} - k_1^{(3)})} \exp\left(k_2^{(1)}t^{(1)} + k_2^{(2)}t^{(2)} + (k_1^{(3)}t^{(3)})\right)$$
$$- \frac{\gamma_1^1(\gamma_2^1 - \gamma_1^2)(\gamma_2^2 - \gamma_1^3)}{(k_2^{(1)} - k_1^{(1)})(k_2^{(2)} - k_1^{(2)})(k_2^{(3)} - k_1^{(3)})} \exp\left(k_2^{(1)}t^{(1)} + k_2^{(2)}t^{(2)} + (k_2^{(3)}t^{(3)})\right) \tag{25}$$

The element $M_{12}^{(3)}$ for tree intervals with constant parameters is

$$
\begin{aligned}
M_{12}^{(3)} = \\
-\frac{(\gamma_1^1 - \gamma_2^2)(\gamma_1^2 - \gamma_1^3)}{(k_2^{(1)} - k_1^{(1)})(k_2^{(2)} - k_1^{(2)})(k_2^{(3)} - k_1^{(3)})} \exp\left(k_1^{(1)}t^{(1)} + k_1^{(2)}t^{(2)} + (k_1^{(3)}t^{(3)}\right) \\
+\frac{(\gamma_1^1 - \gamma_2^2)(\gamma_1^2 - \gamma_1^3)}{(k_2^{(1)} - k_1^{(1)})(k_2^{(2)} - k_1^{(2)})(k_2^{(3)} - k_1^{(3)})} \exp\left(k_1^{(1)}t^{(1)} + k_1^{(2)}t^{(2)} + (k_2^{(3)}t^{(3)}\right) \\
+\frac{(\gamma_1^1 - \gamma_1^2)(\gamma_2^2 - \gamma_2^3)}{(k_2^{(1)} - k_1^{(1)})(k_2^{(2)} - k_1^{(2)})(k_2^{(3)} - k_1^{(3)})} \exp\left(k_1^{(1)}t^{(1)} + k_2^{(2)}t^{(2)} + (k_1^{(3)}t^{(3)}\right) \\
-\frac{(\gamma_1^1 - \gamma_1^2)(\gamma_2^2 - \gamma_1^3)}{(k_2^{(1)} - k_1^{(1)})(k_2^{(2)} - k_1^{(2)})(k_2^{(3)} - k_1^{(3)})} \exp\left(k_1^{(1)}t^{(1)} + k_2^{(2)}t^{(2)} + (k_2^{(3)}t^{(3)}\right) \\
+\frac{(\gamma_2^1 - \gamma_2^2)(\gamma_1^2 - \gamma_2^3)}{(k_2^{(1)} - k_1^{(1)})(k_2^{(2)} - k_1^{(2)})(k_2^{(3)} - k_1^{(3)})} \exp\left(k_2^{(1)}t^{(1)} + k_1^{(2)}t^{(2)} + (k_1^{(3)}t^{(3)}\right) \\
-\frac{(\gamma_2^1 - \gamma_2^2)(\gamma_1^2 - \gamma_1^3)}{(k_2^{(1)} - k_1^{(1)})(k_2^{(2)} - k_1^{(2)})(k_2^{(3)} - k_1^{(3)})} \exp\left(k_2^{(1)}t^{(1)} + k_1^{(2)}t^{(2)} + (k_2^{(3)}t^{(3)}\right) \\
-\frac{(\gamma_2^1 - \gamma_1^2)(\gamma_2^2 - \gamma_2^3)}{(k_2^{(1)} - k_1^{(1)})(k_2^{(2)} - k_1^{(2)})(k_2^{(3)} - k_1^{(3)})} \exp\left(k_2^{(1)}t^{(1)} + k_2^{(2)}t^{(2)} + (k_1^{(3)}t^{(3)}\right) \\
+\frac{(\gamma_2^1 - \gamma_1^2)(\gamma_2^2 - \gamma_1^3)}{(k_2^{(1)} - k_1^{(1)})(k_2^{(2)} - k_1^{(2)})(k_2^{(3)} - k_1^{(3)})} \exp\left(k_2^{(1)}t^{(1)} + k_2^{(2)}t^{(2)} + (k_2^{(3)}t^{(3)}\right)
\end{aligned}
\tag{26}
$$

The element $M_{21}^{(3)}$ for tree intervals with constant parameters is

$$
\begin{aligned}
M_{21}^{(3)} = \\
\frac{\gamma_2^1\gamma_1^3(\gamma_1^1 - \gamma_2^2)(\gamma_1^2 - \gamma_2^3)}{(k_2^{(1)} - k_1^{(1)})(k_2^{(2)} - k_1^{(2)})(k_2^{(3)} - k_1^{(3)})} \exp\left(k_1^{(1)}t^{(1)} + k_1^{(2)}t^{(2)} + (k_1^{(3)}t^{(3)}\right) \\
-\frac{\gamma_2^1\gamma_2^3(\gamma_1^1 - \gamma_2^2)(\gamma_1^2 - \gamma_1^3)}{(k_2^{(1)} - k_1^{(1)})(k_2^{(2)} - k_1^{(2)})(k_2^{(3)} - k_1^{(3)})} \exp\left(k_1^{(1)}t^{(1)} + k_1^{(2)}t^{(2)} + (k_2^{(3)}t^{(3)}\right) \\
-\frac{\gamma_2^1\gamma_1^3(\gamma_1^1 - \gamma_1^2)(\gamma_2^2 - \gamma_2^3)}{(k_2^{(1)} - k_1^{(1)})(k_2^{(2)} - k_1^{(2)})(k_2^{(3)} - k_1^{(3)})} \exp\left(k_1^{(1)}t^{(1)} + k_2^{(2)}t^{(2)} + (k_1^{(3)}t^{(3)}\right) \\
+\frac{\gamma_2^1\gamma_2^3(\gamma_1^1 - \gamma_1^2)(\gamma_2^2 - \gamma_1^3)}{(k_2^{(1)} - k_1^{(1)})(k_2^{(2)} - k_1^{(2)})(k_2^{(3)} - k_1^{(3)})} \exp\left(k_1^{(1)}t^{(1)} + k_2^{(2)}t^{(2)} + (k_2^{(3)}t^{(3)}\right) \\
-\frac{\gamma_1^1\gamma_1^3(\gamma_2^2 - \gamma_2^2)(\gamma_1^2 - \gamma_2^3)}{(k_2^{(1)} - k_1^{(1)})(k_2^{(2)} - k_1^{(2)})(k_2^{(3)} - k_1^{(3)})} \exp\left(k_2^{(1)}t^{(1)} + k_1^{(2)}t^{(2)} + (k_1^{(3)}t^{(3)}\right) \\
+\frac{\gamma_1^1\gamma_2^3(\gamma_2^1 - \gamma_2^2)(\gamma_1^2 - \gamma_1^3)}{(k_2^{(1)} - k_1^{(1)})(k_2^{(2)} - k_1^{(2)})(k_2^{(3)} - k_1^{(3)})} \exp\left(k_2^{(1)}t^{(1)} + k_1^{(2)}t^{(2)} + (k_2^{(3)}t^{(3)}\right) \\
+\frac{\gamma_1^1\gamma_1^3(\gamma_2^1 - \gamma_1^2)(\gamma_2^2 - \gamma_2^3)}{(k_2^{(1)} - k_1^{(1)})(k_2^{(2)} - k_1^{(2)})(k_2^{(3)} - k_1^{(3)})} \exp\left(k_2^{(1)}t^{(1)} + k_2^{(2)}t^{(2)} + (k_1^{(3)}t^{(3)}\right) \\
-\frac{\gamma_1^1\gamma_2^3(\gamma_2^1 - \gamma_1^2)(\gamma_2^2 - \gamma_1^3)}{(k_2^{(1)} - k_1^{(1)})(k_2^{(2)} - k_1^{(2)})(k_2^{(3)} - k_1^{(3)})} \exp\left(k_2^{(1)}t^{(1)} + k_2^{(2)}t^{(2)} + (k_2^{(3)}t^{(3)}\right)
\end{aligned}
\tag{27}
$$

The element $M_{22}^{(3)}$ for tree intervals with constant parameters is

$$
\begin{aligned}
M_{22}^{(3)} = \\
-\frac{\gamma_1^3(\gamma_1^1 - \gamma_2^2)(\gamma_1^2 - \gamma_2^3)}{(k_2^{(1)} - k_1^{(1)})(k_2^{(2)} - k_1^{(2)})(k_2^{(3)} - k_1^{(3)})} \exp\left(k_1^{(1)}t^{(1)} + k_1^{(2)}t^{(2)} + (k_1^{(3)}t^{(3)}\right) \\
+\frac{\gamma_2^3(\gamma_1^1 - \gamma_2^2)(\gamma_1^2 - \gamma_1^3)}{(k_2^{(1)} - k_1^{(1)})(k_2^{(2)} - k_1^{(2)})(k_2^{(3)} - k_1^{(3)})} \exp\left(k_1^{(1)}t^{(1)} + k_1^{(2)}t^{(2)} + (k_2^{(3)}t^{(3)}\right) \\
+\frac{\gamma_1^3(\gamma_1^1 - \gamma_1^2)(\gamma_2^2 - \gamma_2^3)}{(k_2^{(1)} - k_1^{(1)})(k_2^{(2)} - k_1^{(2)})(k_2^{(3)} - k_1^{(3)})} \exp\left(k_1^{(1)}t^{(1)} + k_2^{(2)}t^{(2)} + (k_1^{(3)}t^{(3)}\right) \\
-\frac{\gamma_2^3(\gamma_1^1 - \gamma_1^2)(\gamma_2^2 - \gamma_1^3)}{(k_2^{(1)} - k_1^{(1)})(k_2^{(2)} - k_1^{(2)})(k_2^{(3)} - k_1^{(3)})} \exp\left(k_1^{(1)}t^{(1)} + k_2^{(2)}t^{(2)} + (k_2^{(3)}t^{(3)}\right) \\
+\frac{\gamma_1^3(\gamma_2^1 - \gamma_2^2)(\gamma_1^2 - \gamma_2^3)}{(k_2^{(1)} - k_1^{(1)})(k_2^{(2)} - k_1^{(2)})(k_2^{(3)} - k_1^{(3)})} \exp\left(k_2^{(1)}t^{(1)} + k_1^{(2)}t^{(2)} + (k_1^{(3)}t^{(3)}\right) \\
-\frac{\gamma_2^3(\gamma_2^1 - \gamma_2^2)(\gamma_1^2 - \gamma_1^3)}{(k_2^{(1)} - k_1^{(1)})(k_2^{(2)} - k_1^{(2)})(k_2^{(3)} - k_1^{(3)})} \exp\left(k_2^{(1)}t^{(1)} + k_1^{(2)}t^{(2)} + (k_2^{(3)}t^{(3)}\right) \\
-\frac{\gamma_1^3(\gamma_2^1 - \gamma_1^2)(\gamma_2^2 - \gamma_2^3)}{(k_2^{(1)} - k_1^{(1)})(k_2^{(2)} - k_1^{(2)})(k_2^{(3)} - k_1^{(3)})} \exp\left(k_1^{(1)}t^{(1)} + k_2^{(2)}t^{(2)} + (k_1^{(3)}t^{(3)}\right) \\
+\frac{\gamma_2^3(\gamma_2^1 - \gamma_1^2)(\gamma_2^2 - \gamma_1^3)}{(k_2^{(1)} - k_1^{(1)})(k_2^{(2)} - k_1^{(2)})(k_2^{(3)} - k_1^{(3)})} \exp\left(k_2^{(1)}t^{(1)} + k_2^{(2)}t^{(2)} + (k_2^{(3)}t^{(3)}\right)
\end{aligned}
$$

$$(28)$$

Similar expressions can be written for four, five, six intervals. However in (25)–(28), the regularity in these expressions is clearly visible. First of all it is the regularity in the index change. This regularity can be taken into account by the new sign function [16]

$$
F_{q,i} = \frac{1}{2}\langle 1 + (-1)^i \mathsf{sign}\left\{\sin\left[\frac{\pi}{2^{N+1-i}}(2q-1)\right]\right\}\rangle \tag{29}
$$

Here i is the interval number, q is the number of the term in the sums (21)–(24) and (25)–(28). For example, the values of the sign function for three intervals with constant parameters are presented in Table 1.

Table 1. The sign function values for three intervals with constant parameters

q	$i=1, k_1=1$	$i=2, k_2=3$	$i=3, k_3=5$	$i=1, k_1=2$	$i=2, k_2=4$	$i=3, k_3=6$
1	1	1	1	0	0	0
2	1	1	0	0	0	1
3	1	0	1	0	1	0
4	1	0	0	0	1	1
5	0	1	1	1	0	0
6	0	1	0	1	0	1
7	0	0	1	1	1	0
8	0	0	0	1	1	1

It is easy to check that taking into account (29), the expression (25) can be written in the form

$$M_{11}^{(3)} = \frac{\gamma_{2-F_{q,i}}^{(1)}(\gamma_{1+F_{q,i}}^{(1)} - \gamma_{2-F_{q,i}}^{(2)})(\gamma_{1+F_{q,i}}^{(2)} - \gamma_{2-F_{q,i}}^{(3)})}{\prod_{i=1}^{3}(k_2^{(i)} - k_1^{(i)})}$$

$$\exp(k_{1+F_{q,i}}^{(1)}t^{(1)} + k_{1+F_{q,i}}^{(2)}t^{(2)} + k_{1+F_{q,i}}^{(3)}t^{(3)}) =$$

$$= \sum_{q=1}^{2^3}\left\{\gamma_{2-F_{q,1}}^{(1)}(-1)^{\sum_{i=1}^{N}F_{q,i}}\right.$$

$$\left.\frac{\prod_{i=1}^{2}(\gamma_{1+F_{q,i}}^{(i)} - \gamma_{2-F_{q,i}}^{(i+1)})}{\prod_{i=1}^{3}(k_2^{(i)} - k_1^{(i)})}\exp\left[\sum_{i=1}^{3}(k_{1+F_{q,i}}^{(i)}t^{(i)})\right]\right\}$$

(30)

Analogously, the expressions (26)–(28) can be presented in the forms

$$M_{12}^{(3)} = \frac{(\gamma_{1+F_{q,i}}^{(1)} - \gamma_{2-F_{q,i}}^{(2)})(\gamma_{1+F_{q,i}}^{(2)} - \gamma_{2-F_{q,i}}^{(3)})}{\prod_{i=1}^{3}(k_2^{(i)} - k_1^{(i)})}$$

$$\exp(k_{1+F_{q,i}}^{(1)}t^{(1)} + k_{1+F_{q,i}}^{(2)}t^{(2)} + k_{1+F_{q,i}}^{(3)}t^{(3)}) =$$

$$= \sum_{q=1}^{2^3}\left\{(-1)^{\sum_{i=1}^{N}(F_{q,i}+1)}\right.$$

$$\left.\frac{\prod_{i=1}^{2}(\gamma_{1+F_{q,i}}^{(i)} - \gamma_{2-F_{q,i}}^{(i+1)})}{\prod_{i=1}^{3}(k_2^{(i)} - k_1^{(i)})}\exp\left[\sum_{i=1}^{3}(k_{1+F_{q,i}}^{(i)}t^{(i)})\right]\right\}$$

(31)

$$M_{21}^{(3)} = \frac{\gamma_{2-F_{q,i}}^{(1)}\gamma_{1+F_{q,i}}^{(3)}(\gamma_{1+F_{q,i}}^{(1)} - \gamma_{2-F_{q,i}}^{(2)})(\gamma_{1+F_{q,i}}^{(2)} - \gamma_{2-F_{q,i}}^{(3)})}{\prod_{i=1}^{3}(k_2^{(i)} - k_1^{(i)})}$$

$$\exp(k_{1+F_{q,i}}^{(1)}t^{(1)} + k_{1+F_{q,i}}^{(2)}t^{(2)} + k_{1+F_{q,i}}^{(3)}t^{(3)}) =$$

$$= \sum_{q=1}^{2^3}\left\{\gamma_{2-F_{q,1}}^{(1)}\gamma_{1+F_{q,i}}^{(3)}(-1)^{\sum_{i=1}^{N}F_{q,i}}\right.$$

$$\left.\frac{\prod_{i=1}^{2}(\gamma_{1+F_{q,i}}^{(i)} - \gamma_{2-F_{q,i}}^{(i+1)})}{\prod_{i=1}^{3}(k_2^{(i)} - k_1^{(i)})}\exp\left[\sum_{i=1}^{3}(k_{1+F_{q,i}}^{(i)}t^{(i)})\right]\right\}$$

(32)

$$M_{22}^{(3)} = \frac{\gamma_{1+F_{q,i}}^{(3)}(\gamma_{1+F_{q,i}}^{(1)} - \gamma_{2-F_{q,i}}^{(2)})(\gamma_{1+F_{q,i}}^{(2)} - \gamma_{2-F_{q,i}}^{(3)})}{\prod_{i=1}^{3}(k_2^{(i)} - k_1^{(i)})}$$

$$\exp(k_{1+F_{q,i}}^{(1)}t^{(1)} + k_{1+F_{q,i}}^{(2)}t^{(2)} + k_{1+F_{q,i}}^{(3)}t^{(3)}) =$$

$$= \sum_{q=1}^{2^3}\left\{\gamma_{1+F_{q,i}}^{(3)}(-1)^{\sum_{i=1}^{N}(F_{q,i}+1)}\right.$$ (33)

$$\left.\frac{\prod_{i=1}^{2}(\gamma_{1+F_{q,i}}^{(i)} - \gamma_{2-F_{q,i}}^{(i+1)})}{\prod_{i=1}^{3}(k_2^{(i)} - k_1^{(i)})}\exp\left[\sum_{i=1}^{3}(k_{1+F_{q,i}}^{(i)}t^{(i)})\right]\right\}$$

Taking into account (30)–(33) we can assume that the elements of the fundamental matrix for the N constant parameter intervals can be written in the form

$$M_{11}^{(N)} = (-1)^{\sum_{i=1}^{N}F_{q,i}}\gamma_{2-F_{q,i}}^{(1)}$$
$$\frac{\prod_{i=1}^{N-1}(\gamma_{1+F_{q,i}}^{(i)} - \gamma_{2-F_{q,i}}^{(i+1)})}{\prod_{i=1}^{N}(k_2^{(i)} - k_1^{(i)})}\exp\left[\sum_{i=1}^{2^N}(k_{1+F_{q,i}}^{(i)}t^{(i)})\right]$$

$$M_{12}^{(N)} = (-1)^{\sum_{i=1}^{N}F_{q,i}+1}$$
$$\frac{\prod_{i=1}^{N-1}(\gamma_{1+F_{q,i}}^{(i)} - \gamma_{2-F_{q,i}}^{(i+1)})}{\prod_{i=1}^{N}(k_2^{(i)} - k_1^{(i)})}\exp\left[\sum_{i=1}^{2^N}(k_{1+F_{q,i}}^{(i)}t^{(i)})\right]$$ (34)

$$M_{21}^{(N)} = (-1)^{\sum_{i=1}^{N}F_{q,i}}\gamma_{2-F_{q,i}}^{(1)}\gamma_{1+F_{q,i}}^{(N)}$$
$$\frac{\prod_{i=1}^{N-1}(\gamma_{1+F_{q,i}}^{(i)} - \gamma_{2-F_{q,i}}^{(i+1)})}{\prod_{i=1}^{N}(k_2^{(i)} - k_1^{(i)})}\exp\left[\sum_{i=1}^{2^N}(k_{1+F_{q,i}}^{(i)}t^{(i)})\right]$$

$$M_{22}^{(N)} = (-1)^{\sum_{i=1}^{N}F_{q,i}+1}\gamma_{2-F_{q,i}}^{(N)}$$
$$\frac{\prod_{i=1}^{N-1}(\gamma_{1+F_{q,i}}^{(i)} - \gamma_{2-F_{q,i}}^{(i+1)})}{\prod_{i=1}^{N}(k_2^{(i)} - k_1^{(i)})}\exp\left[\sum_{i=1}^{2^N}(k_{1+F_{q,i}}^{(i)}t^{(i)})\right]$$

To prove the theorem, it is necessary to find the matrix for $N + 1$ intervals. Here we don't present these cumbersome algebraic transformations. However, after multiplying the corresponding matrices and replacing $M = N + 1$, we obtained the elements of the matrix in the form (34). Next, the solution (34) can be transformed in such a way as to distinguish unimodular matrices $\mathbf{M}_q$ in it. As a result, the fundamental matrix of the considered system is reduced to the form (20).

Thus, we obtained the fundamental matrix of a linear ordinary homogeneous two-dimensional dynamic system with arbitrary piecewise constant parameters for the first time.

4 Conclusions

In this paper a dynamical system with arbitrary piecewise constant parameters described by a linear homogeneous differential equations system with discontinuous coefficients is studied theoretically. For this case the matrix of fundamental solutions in the analytical form in elementary functions is found for the first time. The theorem saying that this matrix can be represented as a finite sum of unimodular matrices with certain contribution coefficients is proved. Despite the fact that the matrix looks quite complex, its application in specific problems gives us simple and convenient expressions. Moreover the regularities in the matrix noted in this work allow us to use it for programming this class problems.

The obtained results allow us to carry out analytical investigation of two-dimensional linear dynamical systems (homogeneous and inhomogeneous) [17, 18]. For example, this matrix gives us possibility to plot phase maps of systems and investigate the oscillation stability conditions. Indeed stability conditions of a linear two-dimension system are determined as the equality to two of this matrix trace.

We also can solve inverse problems for such class of systems. The obtained results can be used in the problems of designing optical devices based on stratified structures [8,10,19]. It can be Bragg filters, optical amplitude detectors, displacement devices. For example, absorbing and reflecting optical coatings have been analysed by using so-call translation matrix in [8], Bragg filters and periodic waveguides have been investigated by using so-call translation matrix in [10], the switching cell has been described in [19].

The result is very important in quantum physics for Schroedinger equation solving [20]. Such an equation, for example, describe behaviour of lattice of composite materials. For today, in scientific literature the analytical solution has been written for two intervals only. However it has been a very rough approximation of real problems. Thus numerical methods have been used for solving this problem as rule.

Obviously, the results can be also used in theory of inhomogeneous and non-linear systems. Indeed a solution of an inhomogeneous system is the sum of a general solution of a corresponding homogeneous system and a particular solution of an inhomogeneous system [17]. As for non-linear systems [4,5,17], the obtained result is a zero approximation written in the analytical form in elemental functions.

References

1. Arnold, V.I.: Ordinary Differential Equations. Springer-Verlag, Berlin (2006)
2. Arnold, V.I.: Mathematical Methods of Classical Mechanics. Graduate Texts in Mathematics. Springer-Verlag, New York (1989). https://doi.org/10.1007/978-1-4757-2063-1
3. Abeles, F.: La theorie generale des couches minces. Le Journal de Physique et le Radium 11, 307–310 (1950)

4. Nayfeh, A.H.: Introduction to Perturbation Methods. WILEY-VCH Verlag GmbH & Co. KGaA, Weinheim (2004)
5. Antman, S.S.: Nonlinear Problems of Elasticity. Applied Mathematical Sciences, 2nd edn. Springer-Verlag, New York (2005). https://doi.org/10.1007/0-387-27649-1
6. Vries, G., Hillen, T., Lewis, M., Muller, J., Schonfisch, B.: A Course in Mathematical Biology: Quantitative Modeling with Mathematical and Computational Methods. SIAM, Philadelphia (2006)
7. Cebeci, T., Cousteix, J.: Modeling and Computation of Boundary-Layer Flows. Springer, Heidelberg (2005). https://doi.org/10.1007/3-540-27624-6
8. Born, M., Wolf, E.: Principle of Optics, 7th edn. Pergamon, Oxford (1999)
9. Vytovtov, K.A.: Analytical investigation of stratified isotropic media. J. Opt. Soc. Am. A **22**(4), 689–696 (2005)
10. Yeh, P., Sari, S.: Optical properties of stratified media with exponentially graded refractive index. Appl. Opt. **22**(24), 4142–4145 (1983)
11. Berreman, D.W.: Optics in stratified and anisotropic media: 4×4-matrix formulation. J. Opt. Soc. Am. **62**, 502 (1972)
12. Mounier, D., et al.: 4×4 matrix algebra in the theory of optical detection of picosecond acoustic pulses in anisotropic media. Chin. J. Phys. **49**(1), 191–200 (2011)
13. Vytovtov, K.A., Tarasenko, Y.S.: Analytical investigation of one-dimensional magnetoelectric photonic crystals. The 2×2 matrix approach. J. Opt. Soc. Am. A **24**(11), 3564–3572 (2007)
14. Gantmacher, F.R.: The Theory of Matrices, vol. 1. Chelsea Publishing Co., New York (1959)
15. Gantmacher, F.R.: The Theory of Matrices, vol. 2. Chelsea Publishing Co., New York (1959)
16. Vytovtov, K.A.: An analytical method for investigating periodic stratified media with uniaxial bianisotropy. Telecommun. Radio Eng. (English translation of Elektrosvyaz and Radiotekhnika) **65**(14), 1307–1321 (2006)
17. Perko, L.: Differential Equations and Dynamical Systems. Texts in Applied Mathematics, 3rd edn. Springer, New York (1991). https://doi.org/10.1007/978-1-4684-0392-3
18. Kuznetsov, Y.A.: Elements of Applied Bifurcation Theory. Applied Mathematical Sciences, 3rd edn. Springer Verlag, New York (2004). https://doi.org/10.1007/978-1-4757-3978-7
19. Vytovtov, K., Barabanova, E., Zouhdi, S.: Optical switching cell based on metamaterials and ferrite films. In: Proceedings 12th International Congress on Artificial Materials for Novel Wave Phenomena - Metamaterials, Espoo, Finland, August 27th–September 1st (2018)
20. Pavelich, R.L., Marsiglio, F.: The Kronig-Penney model extended to arbitrary potentials via numerical matrix mechanics. Am. J. Phys. **83**, 773–781 (2015)

Analytical Modeling of Distributed Systems

On a Queueing-Inventory Problem in Passenger Transport System

Dhanya Shajin[1][iD], Jaison Jacob[2], V. M. Vishnevskiy[3],
and A. Krishnamoorthy[4(✉)]

[1] Department of Mathematics, Sree Narayana College, Chempazhanthy,
Thiruvananthapuram 695587, Kerala, India
`dhanya.shajin@gmail.com`
[2] Department of Mathematics, St. Aloysius College, Elthuruth,
Thrissur 680611, Kerala, India
`jaisjacobt@gmail.com`
[3] V. A. Trapeznikov Institute of Control Sciences of Russian Academy of Sciences,
65 Profsoyuznaya Street, Moscow 117997, Russia
`vishn@inbox.ru`
[4] Centre for Research in Mathematics, CMS College, Kottayam 686001, India
`achyuthacusat@gmail.com`

Abstract. We consider a queueing-inventory problem arising in transport of passengers (flight/train/bus) in which seats in the passenger vessel are assumed to be physically available inventory. Two types of customers – type 1 (high priority (HP)) and type 2 (low priority (LP)) arrive for service. High priority customers have a finite buffer to wait whose maximum capacity is $S + V$, where S is the capacity of the vessel and V is the number of overbookings permitted. Low priority customers wait in an infinite capacity queue. High priority customers have non-preemptive priority over low priority customers.

Arrival of customers form a marked Poisson process. Service time for each customer is exponentially distributed. Each customer asks for exactly one item from inventory which requires an exponentially distributed time for processing (reservation). The service time parameter varies with the "stage of common life time of items for reservation". Vehicle departure time is regarded as "realization of common life time (CLT) of seats in the vehicle". To be precise, inter departure time of vehicles is assumed to have Erlang distribution with K stages. Instantly the next vessel is scheduled. In addition to advanced reservation of seats (inventory), those customers who already reserved seats can "cancel their reservation", before CLT gets realized.

Depending on the number of overbookings, the vessel capacity for the scheduled departure is modified (for example, a larger vessel is employed if the number of overbooking at the time of departure is high enough; else the normal vessel is used).

We derive the stability condition for the system. Then we go about

A. Krishnamoorthy—Research supported by UGC No. F.6-6/2017-18/EMERITUS-2017-18-GEN-10822 (SA-II) and DST project INT/RUS/RSF/P-15.

V. M. Vishnevskiy et al. (Eds.): DCCN 2019, CCIS 1141, pp. 215–229, 2019.
https://doi.org/10.1007/978-3-030-36625-4_18

computing the system state distribution. From these we derive expressions for computing performance of the system. Finally we analyze an optimization problem associated with the model.

Keywords: Overbooking · Common life time · Schedule cancellation

1 Introduction

Queueing-inventory is introduced independently by Sigman and Simchi-Levi [10] and Melikov and Molchanov [8] in 1992. However, until the end of that decade very few findings were further reported (see Berman et al. [1], Berman and Kim [2]). However, from the beginning of this century about 150 papers appeared till date (see survey paper by Krishnamoorthy et al. [5]). A few among these discuss stochastic decomposition and product form solution.

Reservation and cancellation of items in inventory and the common life time (CLT) of stocked items are discussed for the first time in queueing-inventory literature in Krishnamoorthy et al. [6]. The same authors discussed a GI/M/1 type queueing-inventory in [7]. Subsequently Dhanya and Krishnamoorthy [3] extend their earlier work to overbooking for one time unit ahead. Dhanya et al. [4] discuss overbooking in customer transport in a very general context.

In this paper we aim at extending the findings in two of the last mentioned papers in the above paragraph to the case of vehicle schedule cancellation and deployment of larger vehicle for the current schedule based on the reservation status immediately before CLT realization (vehicle departure). We assume the CLT to be Erlang distributed (order K).

Two classes of customers arrive to the system for reservation of seats. They are labelled type 1 (high priority - HP) and type 2 (low priority - LP), respectively. The HP customers have a finite buffer of varying size (depending on seat availability) to wait. LP customers join in an infinite capacity queue waiting for their turn to be transferred to the buffer for service. HP customers have non-preemptive priority over LP customers. Service time of customers of both type follow independent exponential distributions with parameter depending on the stage in which the CLT is in.

Though at a glance the present work appears to be exclusively for customer transport, it can be suitably modified to deal with other types of queueing-inventory problems. This will be discussed in a followup paper.

The section wise breakup of this paper is as described: Sect. 2 deals with the mathematical modelling and analysis of the system under study. In this section we also discuss the system state distribution. Measures for evaluation of the performance of the system are provided in Sect. 3. In Sect. 4 we provide a few numerical illustrations for the performance. Further we discuss an optimization problem to decide on when to cancel a schedule or when to employ a larger vehicle for the current schedule.

2   Mathematical Formulation

Consider a single server queueing-inventory system with two types of customers, a finite buffer for type 1 customers of size depending on the availability of number of items and an infinite waiting space for type 2 customers. Customers arrive according to the Marked Poisson process of rates λ_1 and λ_2 for type 1 (high priority) and type 2 customers respectively. Type 1 customers have non-preemptive priority over LP customers. The inventoried items have a common life time (CLT) which means that they all perish together on realization of common life time. The distribution of the duration of this time is Erlang (K, θ). Service time of customers is exponentially distributed with parameter depending on the stage of CLT; that is, the service time parameter while CLT is in stage i is $\mu_i, \ 1 \leq i \leq K$ with $\mu_1 > \mu_2 > ... > \mu_K$. It is assumed that the CLT progresses as $K \to K - 1 \to ... \to 2 \to 1$. The system permits overbooking. We set an upper bound for overbooking as V. On realization of common life time the next schedule is announced and inventory level reaches its maximum S; consequently overbooked customers (if $< S/2$) are immediately served with these new items. Otherwise (overbooked customers $\geq S/2$), all the customers are served at the time of realization of CLT. Reservation of items and cancellation of sold items before CLT realization is permitted in the corresponding cycle. Cancellation takes place according to an exponentially distributed inter-occurrence time with parameter $i\eta$, when $(S+V-i)$ items are present in the inventory $0 \leq i \leq S+V-1$. At each epoch of CLT realization or when no customer is available in the buffer with atleast one space for customers, transfer of one LP customer occurs from infinite queue to the buffer. Also we assume that if the number of served items is less than $\dfrac{S}{4}$, then the schedule is cancelled; consequently these customers are provided that many items of the next schedule.

Define $N(t)$ as the number of customers in the infinite queue, $I(t)$, the number of items in the inventory, $B(t)$ is the number of customers in the buffer including the one in service and $E(t)$ is the stage of common life time. Then $\Omega = \{(N(t), I(t), B(t), E(t)), t \geq 0\}$ is a continuous time Markov chain on the state space $\{(0, i, n_2, j), 0 \leq i \leq S + V, 0 \leq n_2 \leq i, j = K, K - 1, ..., 1\} \bigcup \{(n_1, 0, 0, j), n_1 \geq 1, j = K, K - 1, ..., 1\} \bigcup \{(n_1, i, n_2, j), n_1 \geq 1, 1 \leq i \leq S + V, 1 \leq n_2 \leq i, j = K, K - 1, ..., 1\}$.

NOTE   Write $K_S = \lceil S/2 \rceil$ and $L_S = \lceil S/4 \rceil$.
   The transition rates are:

(a) Transitions due to arrival:

$(n_1, i, n_2, j) \to (n_1 + 1, i, n_2, j)$: rate λ_2 for $n_1 \geq 0, 1 \leq i \leq S + V,$
$1 \leq n_2 \leq i, j = K, K - 1, ..., 1$

$(n_1, 0, 0, j) \to (n_1 + 1, 0, 0, j)$: rate λ_2 for $n_1 \geq 0, j = K, K - 1, ..., 1$

$(n_1, i, 0, j) \to (n_1, i, 1, j)$: rate $\lambda_1 + \lambda_2$ for $n_1 \geq 0, 1 \leq i \leq S + V,$
$j = K, K - 1, ..., 1$

$(n_1, i, n_2, j) \to (n_1, i, n_2 + 1, j)$: rate λ_1 for $n_1 \geq 0, 2 \leq i \leq S + V,$
$1 \leq n_2 \leq i - 1, j = K, K - 1, ..., 1$

(b) Transitions due to service completions:

$(0, i, n_2, j) \to (0, i-1, n_2-1, j)$: rate μ_j for $1 \leq i \leq S+V, 1 \leq n_2 \leq i,$
$\qquad\qquad j = K, K-1, ..., 1$

$(n_1, i, n_2, j) \to (n_1, i-1, n_2-1, j)$: rate μ_j for $n_1 \geq 1, 2 \leq i \leq S+V,$
$\qquad\qquad 2 \leq n_2 \leq i, j = K, K-1, ..., 1$

$(n_1, 1, 1, j) \to (n_1, 0, 0, j)$: rate μ_j for $n_1 \geq 1, j = K, K-1, ..., 1$

$(n_1, i, 1, j) \to (n_1-1, i-1, 1, j)$: rate μ_j for $n_1 \geq 1, 2 \leq i \leq S+V,$
$\qquad\qquad j = K, K-1, ..., 1$

(c) Transitions due to common life time realization:

$(0, i, n_2, 1) \to (0, S+V, 0, K)$: rate θ for $0 \leq i \leq V - K_S, 0 \leq n_2 \leq i$

$(0, i, n_2, 1) \to (0, S+V, 0, K)$: rate θ for $L_S \leq i \leq S+V, 0 \leq n_2 \leq i$

$(0, i, n_2, 1) \to (0, S+i, 0, K)$: rate θ for $V - K_S + 1 \leq i \leq V, 0 \leq n_2 \leq i$

$(0, i, n_2, 1) \to (0, S+2V-i, 0, K)$: rate θ for $V+1 \leq i \leq V + L_S - 1, 0 \leq n_2 \leq i$

$(n_1, 0, 0, 1) \to (n_1-1, S+V, 1, K)$: rate θ for $n_1 \geq 1$

$(n_1, i, n_2, 1) \to (n_1-1, S+V, 1, K)$: rate θ for $n_1 \geq 1, 1 \leq i \leq V - K_S$
$\qquad\qquad 1 \leq n_2 \leq i$

$(n_1, i, n_2, 1) \to (n_1-1, S+V, 1, K)$: rate θ for $n_1 \geq 1, V+1 \leq i \leq S+V-L_S$
$\qquad\qquad 1 \leq n_2 \leq i$

$(n_1, i, n_2, 1) \to (n_1-1, S+i, 1, K)$: rate θ for $n_1 \geq 1, V - K_S + 1 \leq i \leq V$
$\qquad\qquad 1 \leq n_2 \leq i$

$(n_1, i, n_2, 1) \to (n_1-1, i, 1, K)$: rate θ for $n_1 \geq 1, S+V-L_S+1 \leq i \leq S+V$
$\qquad\qquad 1 \leq n_2 \leq i$

(d) Transition due to cancellation:

$(0, i, n_2, j) \to (0, i+1, n_2, j)$: rate $(S+V-i)\eta$ for $0 \leq i \leq S+V-1, 0 \leq n_2 \leq i$
$\qquad\qquad j = K, K-1, ..., 1$

$(n_1, 0, 0, j) \to (n_1-1, 1, 1, j)$: rate $(S+V)\eta$ for $n_1 \geq 1, j = K, K-1, ..., 1$

$(n_1, i, n_2, j) \to (n_1, i+1, n_2, j)$: rate $(S+V-i)\eta$ for $n_1 \geq 1, 1 \leq i \leq S+V-1,$
$\qquad\qquad 1 \leq n_2 \leq i, j = K, K-1, ..., 1$

(e) Transitions due to common life time:

$(0, i, n_2, j) \to (0, i, n_2, j-1)$: rate θ for $0 \leq i \leq S+V, 0 \leq n_2 \leq i,$
$\qquad\qquad j = K, K-1, ..., 2$

$(n_1, i, n_2, j) \to (n_1, i, n_2, j-1)$: rate θ for $n_1 \geq 1, 1 \leq i \leq S+V,$
$\qquad\qquad 1 \leq n_2 \leq i, j = K, K-1, ..., 2$

$(n_1, 0, 0, j) \to (n_1, 0, 0, j-1)$: rate θ for $n_1 \geq 1, j = K, K-1, ..., 2$

Thus the infinitesimal generator of Ω is of the form

$$Q = \begin{pmatrix} A_{00} & A_{01} & & \\ A_{10} & A_1 & A_0 & \\ & A_2 & A_1 & A_0 \\ & & \ddots & \ddots & \ddots \end{pmatrix}. \tag{1}$$

Each matrix A_0, A_1, A_2 are square matrix of order u and matrices A_{00}, A_{01}, A_{10} are of order $v \times v, v \times u, u \times v$ respectively where $u = K\left(1 + \dfrac{(S+V)(S+V+1)}{2}\right)$ and $v = \dfrac{(S+V+1)(S+V+2)K}{2}$.

2.1 Stability Condition

Let π be the steady state probability vector of $A = A_0 + A_1 + A_2$. Then

$$\pi A = 0, \quad \pi e = 1 \tag{2}$$

where

$$A_0 = \begin{pmatrix} L_0 & & \\ & \ddots & \\ & & L_{S+V} \end{pmatrix}, A_1 = \begin{pmatrix} H_0 & & & \\ M_1 & H_1 & C_1 & \\ & \ddots & \ddots & \ddots & \\ & & M_{S+V-1} & H_{S+V-1} & C_{S+V-1} \\ & & & M_{S+V} & H_{S+V} \end{pmatrix},$$

$$A_2 = \begin{pmatrix} C_0 & & & & & & & & B_0 \\ N_2 & & & & & & & & B_1 \\ & \ddots & & & & & & & B_2 \\ & & N_{V-K_S} & & & & & & \vdots \\ & & & N_{V-K_S+1} & & & B_{V-K_S+1} & & B_{V-K_S} \\ & & & & \ddots & & & \ddots & \\ & & & & & N_{V-1} & & & B_{V-1} \\ & & & & & & N_V & & B_V \\ & & & & & & N_{V+1} & & B_{V+1} \\ & & & & & & & \ddots & \vdots \\ & & & & & & N_d & & B_d \\ & & & & & & & N_{d+1} & B_{d+1} \\ & & & & & & & & \ddots & \ddots \\ & & & & & & & & & N_{V+S} & B_{V+S} \end{pmatrix}$$

with

$$d = S + V - L_S, L_0 = (\lambda_2 I), L_i = \begin{pmatrix} \lambda_2 I & & \\ & \ddots & \\ & & \lambda_2 I \end{pmatrix}_{i \times i}, 1 \leq i \leq S + V$$

$$N_i = \begin{pmatrix} U & & & \\ O & & & \\ & \ddots & \\ & & & O \end{pmatrix}_{i \times i - 1}, 2 \leq i \leq S + V, B_0 = \begin{pmatrix} T^0 \alpha \; \mathbf{0} \; \cdots \; \mathbf{0} \end{pmatrix},$$

$$B_i = \begin{pmatrix} T^0 \alpha \; \mathbf{0} \; \cdots \; \mathbf{0} \\ \vdots \\ T^0 \alpha \; \mathbf{0} \cdots \mathbf{0} \end{pmatrix}_{i \times S + V}, 1 \leq i \leq V - K_S \text{ and } V + 1 \leq i \leq S + V - L_S$$

$$B_i = \begin{pmatrix} T^0 \alpha \; \mathbf{0} \cdots \mathbf{0} \\ \vdots \\ T^0 \alpha \; \mathbf{0} \cdots \mathbf{0} \end{pmatrix}_{i \times S + i}, V - K_S + 1 \leq i \leq V$$

$$B_i = \begin{pmatrix} T^0 \alpha \; \mathbf{0} \cdots \; \mathbf{0} \\ \vdots \\ T^0 \alpha \; \mathbf{0} \cdots \; \mathbf{0} \end{pmatrix}_{i \times i}, S + V - L_S + 1 \leq i \leq S + V$$

$$M_1 = (U), M_i = \begin{pmatrix} O \cdots O \\ U \\ & \ddots \\ & & U \end{pmatrix}_{i \times i - 1}, 2 \leq i \leq S + V$$

$$C_0 = ((S + V)\eta I), C_i = \begin{pmatrix} (S + V - i)\eta I & & \\ & \ddots & \\ & & (S + V - i)\eta I \end{pmatrix}_{i \times i + 1}, 1 \leq i \leq S + V - 1$$

$$H_0 = (T - (\lambda_2 + (S + V)\eta)I), H_1 = (T - (\lambda_2 + (S + V - 1)\eta)I - U),$$

$$H_i = \begin{pmatrix} a_i\ \lambda_1 I & & & \\ & a_i\ \lambda_1 I & & \\ & & \ddots\ \ddots & \\ & & & a_i\ \lambda_1 I \\ & & & b_i \end{pmatrix}, 2 \le i \le S + V$$

$$a_i = T - (\lambda + (S + V - i)\eta)I - U, b_i = T - (\lambda_2 + (S + V - i)\eta)I - U,$$

$$U = \begin{pmatrix} \mu_K & & \\ & \ddots & \\ & & \mu_1 \end{pmatrix}, T = \begin{pmatrix} -\theta\ \theta & & \\ & \ddots\ \ddots & \\ & & -\theta\ \theta \\ & & -\theta \end{pmatrix}, T^0 = \begin{pmatrix} 0 \\ \vdots \\ 0 \\ \theta \end{pmatrix}, \alpha = (1\ 0\ \cdots\ 0)$$

with $I, O, \mathbf{0}, \mathbf{e}$ represent the identity matrix, zero matrix, zero vector, column vector of 1's respectively.

From (2) we have

$$\pi_0 E_0 + \pi_1 M_1 = 0,$$
$$\pi_{i-1}C_{i-1} + \pi_i E_i + \pi_{i+1}F_{i+1} = 0, 1 \le i \le S + V - K_S$$
$$\pi_i B_i + \pi_{S+i-1}C_{S+i-1} + \pi_{S+i}E_{S+i} + \pi_{S+i+1}F_{S+i+1} = 0, V - K_S + 1 \le i \le V - 1$$
$$\sum_{i=0}^{V-K_S} \pi_i B_i + \sum_{i=V}^{S+V-L_S} \pi_i B_i \pi_{S+V-1}C_{S+V-1} + \pi_{S+V}F_{S+V} = 0$$

where

$$E_i = L_i + H_i, \qquad 0 \le i \le S + V - L_S$$
$$E_i = L_i + H_i + B_i, S + V - L_S + 1 \le i \le S + V$$
$$F_i = M_i + N_i, \qquad 2 \le i \le S + V.$$

Solving the above system of equations we get

$$\pi_i = \pi_{i+1}\mathcal{U}_i,\ 0 \le i \le S + V - 1 \tag{3}$$

where

$$\mathcal{U}_i = \begin{cases} -M_1 [E_0]^{-1}, i = 0, \\ -F_{i+1}[\mathcal{U}_{i-1}C_{i-1} + E_i]^{-1}, 1 \le i \le S + V - K_S \\ -F_{i+1}[\mathcal{U}_{i-1}...\mathcal{U}_{i-S}B_{i-S} + \mathcal{U}_{i-1}C_{i-1} + E_i]^{-1}, S + V - K_S + 1 \le i \le S + V - 1. \end{cases}$$

Let $\mathcal{V}_i = \mathcal{U}_{S+V-i}$ for $1 \le i \le S + V$. From the normalizing condition $\pi\mathbf{e} = 1$ we get

$$\pi_{S+V}\left[I + \sum_{i=1}^{S+V}\prod_{j=1}^{i}\mathcal{V}_j\right]\mathbf{e} = 1. \tag{4}$$

Theorem 1. *The queueing-inventory system under study is stable if and only if*

$$\pi_{S+V}\mathcal{U}e < \pi_{S+V}\mathcal{V}e \tag{5}$$

Proof. The queueing-inventory system under study with the generator given in (1) is stable if and only if (see Neuts [9])

$$\pi A_0 \mathbf{e} < \pi A_2 \mathbf{e}. \tag{6}$$

Note that from the elements of A_0 and from A_2, we get

$$\pi A_0 \mathbf{e} = \pi_{S+V} \left[\sum_{i=0}^{S+V-1} \prod_{j=1}^{S+V-i} V_j L_i + L_{S+V} \right] \mathbf{e} \text{ and}$$

$$\pi A_2 \mathbf{e} = \pi_{S+V} \left[\sum_{i=2}^{S+V-1} \prod_{j=1}^{S+V-i} V_j N_i + N_{S+V} + \sum_{i=0}^{S+V-1} \prod_{j=1}^{S+V-i} V_j B_i + B_{S+V} + \prod_{j=1}^{S+V} V_j C_0 \right] \mathbf{e}.$$

Let $\mathcal{U} = \left[\sum_{i=0}^{S+V-1} \prod_{j=1}^{S+V-i} V_j L_i + L_{S+V} \right]$ and

$$\mathcal{V} = \left[\sum_{i=2}^{S+V-1} \prod_{j=1}^{S+V-i} V_j N_i + N_{S+V} + \sum_{i=0}^{S+V-1} \prod_{j=1}^{S+V-i} V_j B_i + B_{S+V} + \prod_{j=1}^{S+V} V_j C_0 \right].$$

Now using (6) we get the stated result.

2.2   Steady State Probability Vector

Let $\mathbf{x}$ be the steady state probability vector of $\mathcal{Q}$. Then $\mathbf{x}$ must satisfy the set of equations

$$\mathbf{x}\mathcal{Q} = 0, \mathbf{x}\mathbf{e} = 1. \tag{7}$$

Thus the above set of equations reduce to:

$$\begin{aligned}
\mathbf{x}_0 A_{00} + \mathbf{x}_1 A_{10} &= \mathbf{0}, \\
\mathbf{x}_0 A_{01} + \mathbf{x}_1 A_1 + \mathbf{x}_2 A_2 &= \mathbf{0}, \\
\mathbf{x}_{n-1} A_0 + \mathbf{x}_n A_1 + \mathbf{x}_{n+1} A_2 &= \mathbf{0}, \quad n \geq 2.
\end{aligned} \tag{8}$$

Under the assumption that the stability condition holds, we see that $\mathbf{x}$ is obtained as (see Neuts [9])

$$\mathbf{x}_n = \mathbf{x}_1 R^{n-1}, \quad n \geq 2 \tag{9}$$

where R is the minimal non-negative solution to the matrix quadratic equation:

$$R^2 A_2 + R A_1 + A_0 = O \tag{10}$$

and the boundary equations are given by

$$\begin{aligned}
\mathbf{x}_0 A_{00} + \mathbf{x}_1 A_{10} &= \mathbf{0}, \\
\mathbf{x}_0 A_{01} + \mathbf{x}_1 \left[A_1 + R A_2 \right] &= \mathbf{0}.
\end{aligned} \tag{11}$$

The normalizing condition (7) gives

$$\mathbf{x}_0 \left[I + \mathcal{K}(I - R)^{-1} \right] \mathbf{e} = 1 \tag{12}$$

where $\mathcal{K} = -A_{01} \left[A_1 + R A_2 \right]^{-1}$.

3   Some Important System Performance Measures

1. Expected number of customers in the queue: $E_N = \dfrac{K}{\theta} \left[\displaystyle\sum_{n_1=1}^{\infty} n_1 \mathbf{x}_{n_1} \mathbf{e} \right]$.

2. Expected   number   of   customers   in   the   buffer:
$$E_B = \frac{K}{\theta} \left[\sum_{n_1=1}^{\infty} \sum_{i=1}^{S+V} \sum_{n_2=1}^{i} n_2 \mathbf{x}_{n_1}(i, n_2) \mathbf{e} \right].$$

3. Expected number of items in the inventory:
$$E_I = $$
$$\frac{K}{\theta} \left[\sum_{i=V+1}^{V+S} \sum_{n_2=0}^{i} (i-V)\mathbf{x}_0(i, n_2)\mathbf{e} + \sum_{n_1=1}^{\infty} \sum_{i=V+1}^{V+S} \sum_{n_2=1}^{i} (i-V)\mathbf{x}_{n_1}(i, n_2)\mathbf{e} \right].$$

4. Expected rate of purchase: $E_{PR} = \dfrac{K}{\theta} \left[\displaystyle\sum_{n_1=0}^{\infty} \sum_{i=1}^{S+V} \sum_{n_2=1}^{i} \sum_{j=1}^{K} \mu_j x_{n_1}(i, n_2, j) \right]$.

5. Expected cancellation rate: $E_{CR} = \dfrac{K}{\theta} \left[\displaystyle\sum_{n_1=0}^{\infty} \sum_{i=0}^{S+V-1} (S + V - i)\eta \mathbf{x}_n(i)\mathbf{e} \right]$.

6. Expected loss rate of type 1 customers due to capacity restriction:
$$E_{LR} = \frac{K}{\theta}\lambda_1 \left[\sum_{n_1=0}^{\infty} \sum_{i=0}^{S+V} \mathbf{x}_{n_1}(i, i)\mathbf{e} \right].$$

7. Expected loss rate of items due to realization of common life time:
$$E_{LRI} = \frac{K}{\theta} \left[\sum_{n_1=1}^{\infty} \sum_{i=V+1}^{S+V} \sum_{n_2=1}^{i} x_{n_1}(i, n_2, 1) + \sum_{i=V+1}^{S+V} \sum_{n_2=0}^{i} x_{n_1}(i, n_2, 1) \right].$$

8. Expected rate of overbooking:
$$E_{OR} = \frac{K}{\theta} \left[\sum_{n_1=0}^{\infty} \sum_{i=1}^{V} \sum_{n_2=1}^{i} \mu_j \mathbf{x}_{n_1}(i, n_2)\mathbf{e} \right].$$

9. Probability that at any time inventory is at its maximum $(S + V)$ in the system: $P_{full} = \displaystyle\sum_{n_1=0}^{\infty} \mathbf{x}_{n_1}(S + V)\mathbf{e}$.

10. Probability of no item in the inventory: $P_{no}^{item} = \displaystyle\sum_{n_1=0}^{\infty} \sum_{i=0}^{V} \mathbf{x}_{n_1}(i)\mathbf{e}$.

11. Probability that the system is in maximum overbooked state:
$$P_{max} = \sum_{n_1=0}^{\infty} \mathbf{x}_{n_1}(0)\mathbf{e}.$$

3.1   Distribution of Number of Purchases Before Realization of Common Life Time

In this section we analyze the expected number of purchases in a cycle. Here a cycle is defined as the time elapsed from the beginning a scheduled is announced until its CLT is realized. First choose N such that

$$\sum_{n_1=0}^{N} \mathbf{x}_{n_1} \mathbf{e} > 1 - \epsilon \text{ for any pre-assigned } \epsilon \text{ with N depending on } \epsilon.$$

To get the distribution of the number of purchases we consider the Markov chain $\{(M(t), N(t), I(t), B(t), E(t)), t \geq 0\}$ where $M(t)$ represents the number of purchases, $N(t)$ is the number of customers in the queue, $I(t)$ is the number of items in the inventory, $B(t)$ is the number of customers in the buffer and $E(t)$ represents the stage of common life time. Thus the state space $\{(m, 0, i, n_2, j), m \geq 0, 0 \leq i \leq S + V, 0 \leq n_2 \leq i, j = K, K - 1, ..., 1\} \bigcup \{(m, n_1, 0, 0, j), m \geq 0, 1 \leq n_1 \leq N, j = K, K - 1, ..., 1\} \bigcup \{(m, n_1, i, n_2, j), m \geq 0, 1 \leq n_1 \leq N, 1 \leq i \leq S + V, 1 \leq n_2 \leq i, j = K, K - 1, ..., 1\} \bigcup \{\Delta\}$ where $\{\Delta\}$ is the absorbing state which means the realization of common life time. Thus its infinitesimal generator is of the form $\mathcal{N}_1 = \begin{pmatrix} 0 & 0 & 0 & \cdots \\ E & E_1 & E_0 & \\ E & & E_1 & E_0 \\ \vdots & & & \ddots & \ddots \end{pmatrix}$. Thus the transition rates are:

$(m, n_1, i, n_2, j) \rightarrow (m, n_1 + 1, i, n_2, j) : \lambda_2$	$m \geq 0, 0 \leq n_1 \leq N - 1, 1 \leq i \leq S + V,$ $1 \leq n_2 \leq i, j = K, K - 1, ..., 1$
$(m, n_1, 0, 0, j) \rightarrow (m, n_1 + 1, 0, 0, j) : \lambda_2$	$m \geq 0, 0 \leq n_1 \leq N - 1, j = K, K - 1, ..., 1$
$(m, n_1, i, 0, j) \rightarrow (m, n_1, i, 1, j) : \lambda_1 + \lambda_2$	$m \geq 0, 0 \leq n_1 \leq N, 1 \leq i \leq S + V,$ $j = K, K - 1, ..., 1$
$(m, n_1, i, n_2, j) \rightarrow (m, n_1, i, n_2 + 1, j) : \lambda_1$	$m \geq 0, 0 \leq n_1 \leq N, 2 \leq i \leq S + V,$ $1 \leq n_2 \leq i - 1, j = K, K - 1, ..., 1$
$(m, 0, i, n_2, j) \rightarrow (m + 1, 0, i - 1, n_2 - 1, j) : \mu_j$	$m \geq 0, 1 \leq i \leq S + V, 1 \leq n_2 \leq i,$ $j = K, K - 1, ..., 1$
$(m, n_1, i, n_2, j) \rightarrow (m + 1, n_1, i - 1, n_2 - 1, j) : \mu_j$	$m \geq 0, 1 \leq n_1 \leq N, 2 \leq i \leq S + V,$ $2 \leq n_2 \leq i, j = K, K - 1, ..., 1$
$(m, n_1, 1, 1, j) \rightarrow (m + 1, n_1, 0, 0, j) : \mu_j$	$m \geq 0, 1 \leq n_1 \leq N, j = K, K - 1, ..., 1$
$(m, n_1, i, 1, j) \rightarrow (m + 1, n_1 - 1, i - 1, 1, j) : \mu_j$	$m \geq 0, 1 \leq n_1 \leq N, 2 \leq i \leq S + V,$ $j = K, K - 1, ..., 1$
$(m, 0, i, n_2, 1) \rightarrow \{\Delta\} : \theta$	$m \geq 0, 0 \leq i \leq V - K_S, 0 \leq n_2 \leq i$
$(m, 0, i, n_2, 1) \rightarrow \{\Delta\} : \theta$	$m \geq 0, L_S \leq i \leq S + V, 0 \leq n_2 \leq i$
$(m, 0, i, n_2, 1) \rightarrow \{\Delta\} : \theta$	$m \geq 0, V - K_S + 1 \leq i \leq V, 0 \leq n_2 \leq i$
$(m, 0, i, n_2, 1) \rightarrow \{\Delta\} : \theta$	$m \geq 0, V + 1 \leq i \leq V + L_S - 1, 0 \leq n_2 \leq i$
$(m, n_1, 0, 0, 1) \rightarrow \{\Delta\} : \theta$	$m \geq 0, 1 \leq n_1 \leq N$
$(m, n_1, i, n_2, 1) \rightarrow \{\Delta\} : \theta$	$m \geq 0, 1 \leq n_1 \leq N, 1 \leq i \leq V - K_S$ $1 \leq n_2 \leq i$
$(m, n_1, i, n_2, 1) \rightarrow \{\Delta\} : \theta$	$m \geq 0, 1 \leq n_1 \leq N, V + L_S \leq i \leq S + V$ $1 \leq n_2 \leq i$
$(m, n_1, i, n_2, 1) \rightarrow \{\Delta\} : \theta$	$m \geq 0, 1 \leq n_1 \leq N, V - K_S + 1 \leq i \leq V$ $1 \leq n_2 \leq i$
$(m, n_1, i, n_2, 1) \rightarrow \{\Delta\} : \theta$	$m \geq 0, 1 \leq n_1 \leq N, V + 1 \leq i \leq V - L_S - 1$ $1 \leq n_2 \leq i$
$(m, 0, i, n_2, j) \rightarrow (m, 0, i + 1, n_2, j) : (S + V - i)\eta$	$m \geq 0, 0 \leq i \leq S + V - 1, 0 \leq n_2 \leq i$ $j = K, K - 1, ..., 1$
$(m, n_1, 0, 0, j) \rightarrow (m, n_1 - 1, 1, 1, j) : (S + V)\eta$	$m \geq 0, 1 \leq n_1 \leq N, j = K, K - 1, ..., 1$
$(m, n_1, i, n_2, j) \rightarrow (m, n_1, i + 1, n_2, j) : (S + V - i)\eta$	$m \geq 0, 1 \leq n_1 \leq N, 1 \leq i \leq S + V - 1,$ $1 \leq n_2 \leq i, j = K, K - 1, ..., 1$
$(m, 0, i, n_2, j) \rightarrow (m, 0, i, n_2, j - 1) : \theta$	$m \geq 0, 0 \leq i \leq S + V, 0 \leq n_2 \leq i,$ $j = K, K - 1, ..., 2$
$(m, n_1, i, n_2, j) \rightarrow (m, n_1, i, n_2, j - 1) : \theta$	$m \geq 0, 1 \leq n_1 \leq N, 1 \leq i \leq S + V,$ $1 \leq n_2 \leq i, j = K, K - 1, ..., 2$
$(m, n_1, 0, 0, j) \rightarrow (m, n_1, 0, 0, j - 1) : \theta$	$m \geq 0, 1 \leq n_1 \leq N, j = K, K - 1, ..., 2$

Let $\mathbf{y}_k$ be the probability that the number of purchases before realization of common life time. Then $\mathbf{y}_k$ be the probability that the absorption occurs from the kth level. Hence $\mathbf{y}_k$ is given by $\mathbf{y}_k = \zeta \left(-E_1^{-1} E_0\right)^k \left(-E_1^{-1} E\right)$, $k \geq 0$ where ζ is the initial probability vector is of the form $\zeta = \dfrac{\zeta_N}{\zeta_0}$ with $\zeta_N = (0, ..., 0, x_0(S + V - K_S + 1, 0, K), 0, ..., 0, x_0(S + V - K_S + 1, 1, K), 0, ..., 0, x_0(S + V, 0, K), 0, ..., 0, x_0(S + V, 1, K), 0, ..., 0, x_1(S + V - K_S + 1, 1, K), 0, ..., 0, x_1(S + V, 1, K), 0, ..., 0, x_N(S + V - K_S + 1, 1, K), 0, ..., 0, x_N(S + V, 1, K), 0, ..., 0)$ and

$$\zeta_0 = \sum_{i=S+V-K_S+1}^{S+V} [x_0(i,0,K) + x_0(i,1,K)] + \sum_{n_1=1}^{N} \sum_{i=S+V-K_S+1}^{S+V} x_{n_1}(i,1,K). \text{ Thus}$$

the expected number of customers getting service before realization of common

life time is $E_{\mathcal{N}_1} = \sum_{k=0}^{\infty} k\mathbf{y}_k$.

3.2 Distribution of Number of Cancellations Before Realization of Common Life Time

To derive the distribution of the number of cancellations in a cycle we consider the Markov chain $\{(M(t), N(t), I(t), B(t), E(t)), t \geq 0\}$ where $M(t)$ represents the number of cancellations and $I(t)$, $N(t)$, $B(t)$, $E(t)$ are defined in Sect. 3.1. The state space $\{(m,0,i,n_2,j), m \geq 0, 0 \leq i \leq S+V, 0 \leq n_2 \leq i, j = K, K-1, ..., 1\} \bigcup \{(m,n_1,0,0,j), m \geq 0, 1 \leq n_1 \leq N, j = K, K - 1, ..., 1\} \bigcup \{(m,n_1,i,n_2,j), m \geq 0, 1 \leq n_1 \leq N, 1 \leq i \leq S+V, 1 \leq n_2 \leq i, j = K, K-1, ..., 1\} \bigcup \{\Delta\}$ where $\{\Delta\}$ is the absorbing state which means the realization of common life time. Thus its infinitesimal generator is of the

form $\mathcal{N}_2 = \begin{pmatrix} 0 & 0 & 0 & \cdots \\ F & F_1 & F_0 & \\ F & & F_1 & F_0 \\ \vdots & & & \ddots & \ddots \end{pmatrix}$. Thus the transition rates are:

Transition	Rate	Conditions
$(m,n_1,i,n_2,j) \to (m,n_1+1,i,n_2,j)$	λ_2	$m \geq 0, 0 \leq n_1 \leq N-1, 1 \leq i \leq S+V,$ $1 \leq n_2 \leq i, j = K, K-1, ..., 1$
$(m,n_1,0,0,j) \to (m,n_1+1,0,0,j)$	λ_2	$m \geq 0, 0 \leq n_1 \leq N-1, j = K, K-1, ..., 1$
$(m,n_1,i,0,j) \to (m,n_1,i,1,j)$	$\lambda_1 + \lambda_2$	$m \geq 0, 0 \leq n_1 \leq N, 1 \leq i \leq S+V,$ $j = K, K-1, ..., 1$
$(m,n_1,i,n_2,j) \to (m,n_1,i,n_2+1,j)$	λ_1	$m \geq 0, 0 \leq n_1 \leq N, 2 \leq i \leq S+V,$ $1 \leq n_2 \leq i-1, j = K, K-1, ..., 1$
$(m,0,i,n_2,j) \to (m,0,i-1,n_2-1,j)$	μ_j	$m \geq 0, 1 \leq i \leq S+V, 1 \leq n_2 \leq i,$ $j = K, K-1, ..., 1$
$(m,n_1,i,n_2,j) \to (m,n_1,i-1,n_2-1,j)$	μ_j	$m \geq 0, 1 \leq n_1 \leq N, 2 \leq i \leq S+V,$ $2 \leq n_2 \leq i, j = K, K-1, ..., 1$
$(m,n_1,1,1,j) \to (m,n_1,0,0,j)$	μ_j	$m \geq 0, 1 \leq n_1 \leq N, j = K, K-1, ..., 1$
$(m,n_1,i,1,j) \to (m,n_1-1,i-1,1,j)$	μ_j	$m \geq 0, 1 \leq n_1 \leq N, 2 \leq i \leq S+V,$ $j = K, K-1, ..., 1$
$(m,0,i,n_2,1) \to \{\Delta\}$	θ	$m \geq 0, 0 \leq i \leq V-K_S, 0 \leq n_2 \leq i$
$(m,0,i,n_2,1) \to \{\Delta\}$	θ	$m \geq 0, L_S \leq i \leq S+V, 0 \leq n_2 \leq i$
$(m,0,i,n_2,1) \to \{\Delta\}$	θ	$m \geq 0, V-K_S+1 \leq i \leq V, 0 \leq n_2 \leq i$
$(m,0,i,n_2,1) \to \{\Delta\}$	θ	$m \geq 0, V+1 \leq i \leq V+L_S-1, 0 \leq n_2 \leq i$
$(m,n_1,0,0,1) \to \{\Delta\}$	θ	$m \geq 0, 1 \leq n_1 \leq N$
$(m,n_1,i,n_2,1) \to \{\Delta\}$	θ	$m \geq 0, 1 \leq n_1 \leq N, 1 \leq i \leq V-K_S$ $1 \leq n_2 \leq i$
$(m,n_1,i,n_2,1) \to \{\Delta\}$	θ	$m \geq 0, 1 \leq n_1 \leq N, V+L_S \leq i \leq S+V$ $1 \leq n_2 \leq i$
$(m,n_1,i,n_2,1) \to \{\Delta\}$	θ	$m \geq 0, 1 \leq n_1 \leq N, V-K_S+1 \leq i \leq V$ $1 \leq n_2 \leq i$
$(m,n_1,i,n_2,1) \to \{\Delta\}$	θ	$m \geq 0, 1 \leq n_1 \leq N, V+1 \leq i \leq V-L_S-1$ $1 \leq n_2 \leq i$
$(m,0,i,n_2,j) \to (m+1,0,i+1,n_2,j)$	$(S+V-i)\eta$	$m \geq 0, 0 \leq i \leq S+V-1, 0 \leq n_2 \leq i$ $j = K, K-1, ..., 1$
$(m,n_1,0,0,j) \to (m+1,n_1-1,1,1,j)$	$(S+V)\eta$	$m \geq 0, 1 \leq n_1 \leq N, j = K, K-1, ..., 1$
$(m,n_1,i,n_2,j) \to (m+1,n_1,i+1,n_2,j)$	$(S+V-i)\eta$	$m \geq 0, 1 \leq n_1 \leq N, 1 \leq i \leq S+V-1,$ $1 \leq n_2 \leq i, j = K, K-1, ..., 1$
$(m,0,i,n_2,j) \to (m,0,i,n_2,j-1)$	θ	$m \geq 0, 0 \leq i \leq S+V, 0 \leq n_2 \leq i,$ $j = K, K-1, ..., 2$
$(m,n_1,i,n_2,j) \to (m,n_1,i,n_2,j-1)$	θ	$m \geq 0, 1 \leq n_1 \leq N, 1 \leq i \leq S+V,$ $1 \leq n_2 \leq i, j = K, K-1, ..., 2$
$(m,n_1,0,0,j) \to (m,n_1,0,0,j-1)$	θ	$m \geq 0, 1 \leq n_1 \leq N, j = K, K-1, ..., 2$

Let $\mathbf{z}_k$ be the probability that the number of cancellations before realization of common life time. Hence $\mathbf{z}_k = \zeta \left(-F_1^{-1}F_0\right)^k \left(-F_1^{-1}F\right)$, $k \geq 0$ where ζ

is the initial probability vector (see Sect. 3.1). Hence the expected number of cancellations before realization of common life time is $E_{\mathcal{N}_2} = \sum_{k=0}^{\infty} k \mathbf{z}_k$.

3.3 Distribution of Number of Schedule Cancellations

In this section we compute the expected number of schedule cancellations over a given time interval. At the epoch when a common life time realization occurs, a random clock is started, realization time of which follows exponential distribution with parameter β. Consider the Markov chain $\{(M(t), N(t), I(t), B(t), E(t)), t \geq 0\}$ where $M(t)$ represents the number of schedule cancellations and $I(t)$, $N(t)$, $B(t), E(t)$ are defined in Sect. 3.1. The state space $\{(m, 0, i, n_2, j), m \geq 0, 0 \leq i \leq S + V, 0 \leq n_2 \leq i, j = K, K - 1, ..., 1\} \bigcup \{(m, n_1, 0, 0, j), m \geq 0, 1 \leq n_1 \leq N, j = K, K - 1, ..., 1\} \bigcup \{(m, n_1, i, n_2, j), m \geq 0, 1 \leq n_1 \leq N, 1 \leq i \leq S + V, 1 \leq n_2 \leq i, j = K, K - 1, ..., 1\} \bigcup \{\Delta\}$ where $\{\Delta\}$ is the absorbing state which means the realization of the random clock. Thus its infinitesimal generator is of the

form $\mathcal{N}_3 = \begin{pmatrix} 0 & 0 & 0 & \cdots \\ G & G_1 & G_0 & \\ G & & G_1 & G_0 \\ \vdots & & & \ddots & \ddots \end{pmatrix}$. Thus the transition rates are:

$(m, n_1, i, n_2, j) \rightarrow (m, n_1 + 1, i, n_2, j) : \lambda_2$ $\quad m \geq 0, 0 \leq n_1 \leq N - 1, 1 \leq i \leq S + V,$ $1 \leq n_2 \leq i, j = K, K - 1, ..., 1$

$(m, n_1, 0, 0, j) \rightarrow (m, n_1 + 1, 0, 0, j) : \lambda_2$ $\quad m \geq 0, 0 \leq n_1 \leq N - 1, j = K, K - 1, ..., 1$
$(m, n_1, i, 0, j) \rightarrow (m, n_1, i, 1, j) : \lambda_1 + \lambda_2$ $\quad m \geq 0, 0 \leq n_1 \leq N, 1 \leq i \leq S + V,$ $j = K, K - 1, ..., 1$

$(m, n_1, i, n_2, j) \rightarrow (m, n_1, i, n_2 + 1, j) : \lambda_1$ $\quad m \geq 0, 0 \leq n_1 \leq N, 2 \leq i \leq S + V,$ $1 \leq n_2 \leq i - 1, j = K, K - 1, ..., 1$

$(m, 0, i, n_2, j) \rightarrow (m, 0, i - 1, n_2 - 1, j) : \mu_j$ $\quad m \geq 0, 1 \leq i \leq S + V, 1 \leq n_2 \leq i,$ $j = K, K - 1, ..., 1$

$(m, n_1, i, n_2, j) \rightarrow (m, n_1, i - 1, n_2 - 1, j) : \mu_j$ $\quad m \geq 0, 1 \leq n_1 \leq N, 2 \leq i \leq S + V,$ $2 \leq n_2 \leq i, j = K, K - 1, ..., 1$

$(m, n_1, 1, 1, j) \rightarrow (m, n_1, 0, 0, j) : \mu_j$ $\quad m \geq 0, 1 \leq n_1 \leq N, j = K, K - 1, ..., 1$
$(m, n_1, i, 1, j) \rightarrow (m, n_1 - 1, i - 1, 1, j) : \mu_j$ $\quad m \geq 0, 1 \leq n_1 \leq N, 2 \leq i \leq S + V,$ $j = K, K - 1, ..., 1$

$(m, 0, i, n_2, 1) \rightarrow (m, 0, S + V, 0, K) : \theta$ $\quad m \geq 0, 0 \leq i \leq V - K_S, 0 \leq n_2 \leq i$
$(m, 0, i, n_2, 1) \rightarrow (m, 0, S + V, 0, K) : \theta$ $\quad m \geq 0, L_S \leq i \leq S + V, 0 \leq n_2 \leq i$
$(m, 0, i, n_2, 1) \rightarrow (m, 0, S + i, 0, K) : \theta$ $\quad m \geq 0, V - K_S + 1 \leq i \leq V, 0 \leq n_2 \leq i$
$(m, 0, i, n_2, 1) \rightarrow (m + 1, 0, S + V - i, 0, K) : \theta$ $\quad m \geq 0, V + 1 \leq i \leq V + L_S - 1, 0 \leq n_2 \leq i$
$(m, n_1, 0, 0, 1) \rightarrow (m, n_1 - 1, S + V, 1, K) : \theta$ $\quad m \geq 0, 1 \leq n_1 \leq N$
$(m, n_1, i, n_2, 1) \rightarrow (m, n_1 - 1, S + V, 1, K) : \theta$ $\quad m \geq 0, 1 \leq n_1 \leq N, 1 \leq i \leq V - K_S$ $1 \leq n_2 \leq i$

$(m, n_1, i, n_2, 1) \rightarrow (m, n_1 - 1, S + V, 1, K) : \theta$ $\quad m \geq 0, 1 \leq n_1 \leq N, V + L_S \leq i \leq S + V$ $1 \leq n_2 \leq i$

$(m, n_1, i, n_2, 1) \rightarrow (m, n_1 - 1, S + i, 1, K) : \theta$ $\quad m \geq 0, 1 \leq n_1 \leq N, V - K_S + 1 \leq i \leq V$ $1 \leq n_2 \leq i$

$(m, n_1, i, n_2, 1) \rightarrow (m + 1, n_1 - 1, S + 2V - i, 1, K) : \theta$ $\quad m \geq 0, 1 \leq n_1 \leq N, V + 1 \leq i \leq V - L_S - 1$ $1 \leq n_2 \leq i$

$(m, 0, i, n_2, j) \rightarrow (m, 0, i + 1, n_2, j) : (S + V - i)\eta$ $\quad m \geq 0, 0 \leq i \leq S + V - 1, 0 \leq n_2 \leq i$ $j = K, K - 1, ..., 1$

$(m, n_1, 0, 0, j) \rightarrow (m, n_1 - 1, 1, 1, j) : (S + V)\eta$ $\quad m \geq 0, 1 \leq n_1 \leq N, j = K, K - 1, ..., 1$
$(m, n_1, i, n_2, j) \rightarrow (m, n_1, i + 1, n_2, j) : (S + V - i)\eta$ $\quad m \geq 0, 1 \leq n_1 \leq N, 1 \leq i \leq S + V - 1,$ $1 \leq n_2 \leq i, j = K, K - 1, ..., 1$

$(m, 0, i, n_2, j) \rightarrow (m, 0, i, n_2, j - 1) : \theta$ $\quad m \geq 0, 0 \leq i \leq S + V, 0 \leq n_2 \leq i,$ $j = K, K - 1, ..., 2$

$(m, n_1, i, n_2, j) \rightarrow (m, n_1, i, n_2, j - 1) : \theta$ $\quad m \geq 0, 1 \leq n_1 \leq N, 1 \leq i \leq S + V,$ $1 \leq n_2 \leq i, j = K, K - 1, ..., 2$

$(m, n_1, 0, 0, j) \rightarrow (m, n_1, 0, 0, j - 1) : \theta$ $\quad m \geq 0, 1 \leq n_1 \leq N, j = K, K - 1, ..., 2$
$(m, 0, i, n_2, j) \rightarrow \{\Delta\} : \beta$ $\quad m \geq 0, 0 \leq i \leq S + V,$ $0 \leq n_2 \leq i, j = K, K - 1, ..., 1$

$(m, n_1, 0, 0, j) \rightarrow \{\Delta\} : \beta$ $\quad m \geq 0, 1 \leq n_1 \leq N,$ $j = K, K - 1, ..., 1$

$(m, n_1, i, n_2, j) \rightarrow \{\Delta\} : \beta$ $\quad m \geq 0, 1 \leq n_1 \leq N, 1 \leq i \leq S + V,$ $1 \leq n_2 \leq i, j = K, K - 1, ..., 1$

Let $\mathbf{v}_k$ be the probability that the number of schedule cancellations before realization of random clock. Hence $\mathbf{v}_k$ is given by

$$\mathbf{v}_k = \zeta \left(-G_1^{-1} G_0\right)^k \left(-G_1^{-1} G\right), \quad k \geq 0$$

where ζ is the initial probability vector (see Sect. 3.1). Hence the expected number of schedule cancellations before realization of random clock is $E_{\mathcal{N}_3} = \sum_{k=0}^{\infty} k\mathbf{v}_k$.

4   Numerical Illustrations

Next we proceed to a few numerical examples in order to bring out the system behaviour with respect to certain parameters. Here service time depends on stage of Erlang distribution with parameter $\mu_i, 1 \leq i \leq K$ and $\mu_1 \geq \mu_2 \geq \ldots \geq \mu_K$. We use $\mu_i = \mu(K - (i - 1))^{\gamma}$, $1 \leq i \leq K$ where $0 \leq \gamma \leq 1$. From Table 1, $E_{PR}, E_{CR}, E_{OR}, E_{LR}, E_{LRI}$ are seen increase with increasing value of K. However, value of θ increases the above mentioned measures are decrease (see Table 2).

Table 1. Effect of K: Fix $\lambda_1 = 0.5, \lambda_2 = 1, \eta = 1.5, \mu = 2, \theta = 3, S = 8, V = 10$

K	E_{PR}	E_{CR}	E_{OR}	E_{LR}	E_{LRI}
1	0.6664	0.2868	0.0184	0.1123	0.3256
2	1.3804	0.8051	0.0599	0.1836	0.6425
3	2.1238	1.3977	0.1115	0.2356	0.9564
4	2.8890	1.9819	0.1639	0.2799	1.2709
5	3.6716	2.5391	0.2139	0.3210	1.5870

Table 2. Effect of θ: Fix $\lambda_1 = 0.5, \lambda_2 = 1, \eta = 1.5, \mu = 2, K = 3, V = 10, S = 8$

θ	E_{PR}	E_{CR}	E_{OR}	E_{LR}	E_{LRI}
1	6.3705	4.3748	0.3549	0.4303	2.8645
1.5	4.2473	2.9881	0.2437	0.3425	1.9063
2	3.1855	2.2292	0.1815	0.2931	1.4297
2.5	2.5485	1.7403	0.1406	0.2601	1.1453
3	2.1238	1.3977	0.1115	0.2356	0.9564

Table 3 indicate that E_{LR} decreases with increasing value of η. However, E_{CR} and E_{PR} increase with increasing value of η. E_{LRI} first increases with increasing value of η and then decreases. However, E_{OR} first decreases and then increases with increasing value of η.

From Table 4, $E_{PR}, E_{CR}, E_{LR}, E_{LRI}$ are seen increase with increasing value of S. However, E_{OR} decreases with increasing value of S.

Table 5 is indication of the fact that E_{PR}, E_{CR}, E_{OR} increase with increase in value of V.

Table 3. Effect of η: Fix $\lambda_1 = 0.5, \lambda_2 = 1, \theta = 1.5, \mu = 2, K = 3, V = 10, S = 8$

η	E_{PR}	E_{CR}	E_{OR}	E_{LR}	E_{LRI}
0.5	4.2468	1.1753	0.2702	0.3837	1.8976
1	4.2472	2.0903	0.2420	0.3613	1.9079
1.5	4.2473	2.9881	0.2437	0.3425	1.9063
2	4.2473	3.9196	0.2512	0.3275	1.9027
2.5	4.2471	4.8913	0.2591	0.3154	1.8991

Table 4. Effect of S: Fix $\lambda_1 = 0.5, \lambda_2 = 1, \theta = 1.5, \eta = 1.5, \mu = 2, K = 3, V = 10$

S	E_{PR}	E_{CR}	E_{OR}	E_{LR}	E_{LRI}
4	12.7376	8.4969	3.2891	0.3378	5.0643
8	12.7383	8.1186	0.6558	0.6631	5.7516
12	12.7423	9.0624	0.3790	0.6397	5.8527
16	12.7443	9.9949	0.2521	0.6201	5.9012
20	12.7451	10.9211	0.1815	0.6026	5.9285

Table 5. Effect of V: Fix $\lambda_1 = 0.5, \lambda_2 = 1, \theta = 1.5, \eta = 1.5, \mu = 2, K = 3, S = 8$

V	E_{PR}	E_{CR}	E_{OR}	E_{LR}	E_{LRI}
6	12.7337	7.2303	0.4371	0.7211	5.8467
7	12.7352	7.4468	0.4961	0.7045	5.8216
8	12.7365	7.6673	0.5521	0.6895	5.7973
9	12.7375	7.8914	0.6053	0.6757	5.7740
10	12.7383	8.1186	0.6558	0.6631	5.7516

Table 6. Revenue function: Fix $\lambda_1 = 0.5, \lambda_2 = 1, \eta = 1.5, \mu = 2, \theta = 3$

S	$F(K,V,S)$ $V = 10, K = 3$	V	$F(K,V,S)$ $S = 8, K = 3$	K	$F(K,V,S)$ $S = 8, V = 10$
4	1063.1	6	831.0039	1	33.0326
8	870.8167	7	841.161	2	80.3756
12	840.825	8	851.1751	3	134.5606
16	832.986	9	861.0567	4	191.316
20	835.7836	10	870.8167	5	249.1483

4.1 Cost Analysis

To study the effect of overbooking and schedule cancellation, it is instructive to introduce operating costs and schedule cancellation cost:

If the number of overbookings $k_1 \geq \dfrac{S}{2}$, then the operating cost is V_h.

If the number of overbookings $k_2 < \dfrac{S}{2}$, then the operating cost is V_ℓ.

Thus we have the average total operating cost:

$$
\begin{aligned}
\mathcal{L} = \frac{K}{\theta} &\left[\sum_{i=0}^{V-K_S} \sum_{n_2=0}^{i} V_h x_0(i, n_2, 1) \quad \sum_{i=V-K_S+1}^{V} \sum_{n_2=0}^{i} V_\ell x_0(i, n_2, 1) \right. \\
&+ \sum_{n_1=1}^{\infty} \sum_{i=1}^{V-K_S} \sum_{n_2=1}^{i} V_h x_{n_1}(i, n_2, 1) + \sum_{n_1=1}^{\infty} V_h x_{n_1}(0, 0, 1) \\
&\left. + \sum_{n_1=1}^{\infty} \sum_{i=V-K_S+1}^{V} \sum_{n_2=1}^{i} V_\ell x_{n_1}(i, n_2, 1) \right].
\end{aligned}
$$

Next we consider the case of schedule cancellation which is the consequence of at most $S/4 - 1$ reservations at the time of the corresponding CLT realization. In this case the customers in that schedule are transferred to the next scheduled departure. The system has to pay for their accommodation and local hospitality. Expected amount of time these customers are held is $\dfrac{K}{\theta}$. Let the holding cost for each such customer per unit time be V_c. Thus the schedule cancellation cost

$$
\mathcal{C} = \frac{K}{\theta} V_c \sum_{n_1=0}^{\infty} \sum_{i=S+V-L_S+1}^{S+V} \mathbf{x}_{n_1}(i) \mathbf{e}.
$$

Based on the above system characteristics we define the following revenue (profit) function as: $F(K, V, S) = C_1 E_{PR} + C_2 E_{CR} + C_3 E_{OR} - C_4 E_I - C_5 E_{LR} - C_6 E_{LRI} - \mathcal{L} - \mathcal{C}$ where

- C_1: Revenue to the system due to per unit purchase
- C_2: Revenue to the system from each cancellation of purchase
- C_3: Revenue due to overbooking
- C_4: Holding cost per inventoried item per unit time
- C_5: Cost due to loss of priority customer due to capacity restriction per unit time
- C_6: Cost due to loss of items due to realization of common life time
- V_h, V_ℓ: varying operating costs

In order to study the variation in different parameters on profit function we first take the values $(C_1, C_2, C_3, C_4, C_5, C_6, V_h, V_\ell, V_c) = (\$80, \$30, \$20, \$2, \$5, \$40, \$75, \$50, \$35)$.

Table 6 represents the revenue to the system as a function of K, S and V.

Acknowledgment. The authors would like to thank the reviewers for their valuable comments and constructive suggestions that help to improve the presentation of this paper.

References

1. Berman, O., Kaplan, E.H., Shimshak, D.G.: Deterministic approximations for inventory management at service facilities. IIE Trans. **25**(5), 98–104 (1993)
2. Berman, O., Kim, E.: Stochastic models for inventory managements at service facilities. Commun. Statist. Stoch. Model. **15**(4), 695–718 (1999)
3. Shajin, D., Krishnamoorthy, A.: On a queueing-inventory systemwith impatient customers, advanced reservation, cancellation, overbooking and common life time. Oper. Res. (2019). https://doi.org/10.1007/s12351-019-00475-3
4. Shajin, D., Krishnamoorthy, A., Dudin, A.N., Joshua, V.C., Jacob, V.: On a queueing-inventory system with advanced reservation and cancellation for the next K time frames ahead: the case of overbooking. Queueing Syst. 1–35 (2019)
5. Krishnamoorthy, A., Shajin, D., Narayanan, V.C.: Inventory with positive service time: a survey. In: Anisimov, V., Limnios, N. (eds.) Advanced Trends in Queueing Theory: Series of Books "Mathematics and Statistics", Sciences. ISTE&Wiley, London (2019)
6. Krishnamoorthy, A., Shajin, D., Lakshmy, B.: On a queueing-inventory with reservation, cancellation, common life time and retrial. Ann. Oper. Res. **247**(1), 365–389 (2016)
7. Krishnamoorthy, A., Shajin, D., Lakshmy, B.: $GI/M/1$ type queueing-inventory systems with postponed work, reservation, cancellation and common life time. Indian J. Pure Appl. Math. **47**(2), 357–388 (2016)
8. Melikov, A.Z., Molchanov, A.A.: Stock optimization in transportation/storage systems. Cybern. Syst. Anal. **28**(3), 484–487 (1992)
9. Neuts, M.F.: Matrix-geometric solutions in stochastic models: an algorithmic approach. The Johns Hopkins University Press, Baltimore (1994 version is Dover Edition) (1981)
10. Sigman, K., Simchi-Levi, D.: Light traffic heuristic for an M/G/1 queue with limited inventory. Ann. Oper. Res. **40**, 371–380 (1992)

Queueing Inventory Model
for Crowdsourcing

Binitha Benny[1,2,3], V. M. Vishnevskiy[1,2,3]([✉]), and A. Krishnamoorthy[1,2,3]

[1] Department of Mathematics, St. Joseph's College for Women,
Alappuzha 688001, India
binithabennya@gmail.com
[2] V. A. Trapeznikov Institute of Control Sciences of Russian Academy of Sciences,
65 Profsoyuznaya Street, Moscow 117997, Russia
vishn@inbox.ru
[3] Centre for Research in Mathematics, CMS College, Kottayam 686001, India
achyuthacusat@gmail.com

Abstract. We consider a multi- server queueing inventory system with
two types of customers: Type I and Type II. Type II customers are
virtual ones. Arrival of both Type I and Type II customers follow two
independent Poisson processes. Type I are to be served by one of the
servers and service time is assumed to be exponentially distributed. Type
II customer may be served by a Type I customer having already been
served and ready to act as a server or by one of the servers with exponentially distributed service time. Type I customer has non preemptive
priority over Type II. Type II is served by a Type I only if inventory
is available after attaching inventory to the existing Type-I customers
available in the system. Type II is served by a Type I with probability
p on completion of latter's service and with complementary probability
$q = 1 - p$ served Type I leaves the system without serving a Type-II.
Fresh arrivals of both type of customers are permitted to join the system
only when excess inventory, which is defined as the difference between on
hand inventory and number of busy servers is positive. System capacity
for Type I is limited, where as Type II has unlimited waiting area. When
inventory level drops to $c + s$, an order for replenishment is placed to
bring the level to $c + S$. The ordered items are received after a random
amount of time which is exponentially distributed. A revenue function
is constructed and effect of probability p on the revenue function for
single server, two and three servers is numerically analyzed. Effects of
parameter β of the exponentially distributed lead time, on the loss rate
of customers and revenue function are numerically analyzed.

Keywords: Crowdsourcing · Queueing · Inventory

Research is supported by the University Grants Commission, Govt. Of India, under
Faculty Development Programme (Grant No. F.FIP/12th Plan/KLKE003TF05) in
Department of Mathematics, Cochin University of Science and Technology, Cochin-22.
B. Benny and A. Krishnamoorthy—Centre for Research in Mathematics, CMS College,
Kottayam.

V. M. Vishnevskiy et al. (Eds.): DCCN 2019, CCIS 1141, pp. 230–243, 2019.
https://doi.org/10.1007/978-3-030-36625-4_19

1   Introduction

Crowdsourcing is the process of getting work usually online from a crowd of people. It is a combination of the words 'crowd' and 'outsourcing'. The idea is to take work and outsource it to a crowd of workers. The principle of crowdsourcing is that more heads are better than one. By canvassing a large crowd of people for ideas, skills or participation, the quality of content and idea generation will be superior.

Wikipedia, the most comprehensive encyclopedia the world has ever seen, is a famous example of crowdsourcing. Instead of creating an encyclopedia on their own, they gave a crowd the responsibility to create the information. The concept of crowdsourcing is used by many industries such as food, consumer products, hotels, electronics and other large retailers. A number of examples of crowdsourcing can be found in [9].

According to Howe [10], "crowdsourcing represents the act of a company or institution taking a function once performed by employees and outsourcing it to a large network of people in the form of an open call. This can take the form of peer production(when the job is performed collaboratively), but is also often undertaken by sole individuals. The crucial prerequisite is the use of the open call format and the large network of potential labourers". The motivation for this paper is from Chakravarthy and Dudin [11] in which the authors use the crowdsourcing in the context of service sectors getting possible help from one group of customers who first receive service from the former and then opt to execute similar services to another group. They consider a multi-server queueing system with two type of customers, Type-I and Type-II. Type-I customers visit the store to procure items while Type-II customers place order over some medium such as internet, phone and expects them to be delivered. The store management use the customers visiting them as couriers to serve the other type of customers. Since not all in-store customers may be willing to act as servers, a probability is introduced for in-store customers to opt for serving the other type. They assumed that Type-I have non-preemptive priority over Type-II. This is the first reported work on crowdsourcing modelled in the queueing theory context. A multi-server priority queue with preemption in crowdsourcing is considered in Krishnamoorthy et al. [12] wherein the authors assume that arrival of a Type-I customer interrupts the ongoing service of any one of Type-II customers if any in service, and hence this preempted customer joins back as the head of the Type-II queue. The resources for service is assumed to be abundantly available in both papers. Limited number of work was reported related to queueing inventory models in crowdsourcing. In the present paper we assume finiteness of availability of item to be served. Thus, when the item is not available service cannot be provided.

The rest of the paper is arranged as follows. Mathematical formulation is taken up in Sect. 2. Section 3 provides the steady state analysis of the model. Some important performance measures are derived in this section. Numerical examples and an optimization problem are discussed in Sect. 4.

2 Model Description

Consider a queueing inventory system with c servers. There are two types of customers: Type I and Type II. Type II customers are virtual ones, ordering through phone or internet or through some other means. Type-I customers visit the store to procure the items. Arrival of Type I and Type-II customers follow two independent Poisson processes with parameter λ_1 and λ_2, respectively. Type I are to be served by one of the c servers with service time assumed to be exponentially distributed with parameter μ_1. Type II customer may be served by a Type I customer having already been served and ready to act as a server, or by one of the c servers. Type II when served by one of the servers, the service time is exponentially distributed with parameter μ_2. Type I customer has non-preemptive priority over Type II. That is, a Type-I customer can move ahead of all the Type-II customers waiting in the queue, but Type-II customers in service are not interrupted by Type-I customers. Type II is served by a Type I only if inventory is available after attaching the existing items to the priority customers already present. Type II is served by a Type I on completion of latter's service with probability $p, 0 \le p \le 1$ and with complementary probability $q = 1 - p$ served Type I will leave the system without serving Type-II. If a Type I customer serves a Type II customer, then that Type II customer will be removed from the system immediately on completion of the corresponding Type-I customer's service. Arrival of both type of customers is permitted only when excess inventory, which is defined as the difference between on hand inventory and number of busy servers, is positive. A finite waiting space L for Type I is assumed whereas Type II has unlimited waiting area. When inventory level drops to $c + s$, an order for replenishment is placed to bring it to $c + S$. We assume $c < s < L < S = c + 2L$. These items are obtained after a random amount of time exponentially distributed with parameter β.

Define

$N_1(t) =$ Number of Type II customers in queue (waiting for service) at time t,
$N_2(t) =$ Number of servers busy with Type II customers at time t,
$N_3(t) =$ Number of Type I customers in the system at time t,
$I(t) =$ Inventory level at time t.

Then
$$\{(N_1(t), N_2(t), N_3(t), I(t)) : t \ge 0\}$$
is a continuous time Markov chain whose state space is

$$\Omega = \bigcup_{i=0}^{\infty} \ell(i)$$

where $\ell(i)$ denotes level i. These levels are described as under:

$$\ell(0) = \{(0, j, k, l) : 0 \le j \le c, 0 \le k \le L, j \le l \le c + S\}$$

and

$$\ell(i) = \{(i,j,k,l) : i > 0, 0 \le j \le c, 0 \le k \le c - j - 1, j \le l \le j + k\}$$
$$\bigcup \{(i,j,k,l) : i > 0, 0 \le j \le c, c - j \le k \le L, j \le l \le c + S\},$$

where i, j, k, l denote number of Type-II customers in queue(waiting for service), number of servers busy with Type-II customers, number of Type-I customers in the system and the inventory level, respectively. The level 0, $\ell(0)$, can be further partitioned as

$$\ell(0) = \{(0,0), (0,1), (0,2), ..., (0,c)\}.$$

where the set of states $(0, j)$, $0 \le j \le c$ corresponds to the case when there is no Type II customer waiting in the queue and j of them are in service and each $\{(0,j) : 0 \le j \le c\}$ has $(L+1)(c+S-j+1)$ elements for $0 \le j \le c$. So $\ell(0)$ has $a = (L+1)\sum_{j=0}^{c}(c+S-j+1)$ elements. Similarly $\ell(i)$ can also be further partitioned as

$$\ell(i) = \{(i,0), (i,1), (i,2), ..., (i,c)\},$$

where the set of states (i, j) corresponds to the case when there are i Type II customer waiting in the queue and j are in service, each has $(1 + 2 + 3 + ... + c - j) + (L - (c - j - 1))(c + S - j + 1)$ elements for $0 \le j \le c$. Thus $\ell(i)$ has $b = \sum_{j=1}^{c}(j(j+1)/2 + \sum_{j=1}^{c+1}(L - (c - j))(c + S - j + 2)$ elements.

The transitions in the above Markov chain can be described as follows:

1. Transition due to arrival of customers:
   – Due to the arrival of Type I customer:
     - $(0,j,k,l) \to (0,j,k+1,l)$ with rate λ_1 if $j + k < l, 0 \le j \le c, 0 \le k \le c - j - 1, j \le l \le c + S$
     - $(i,j,k,l) \to (i,j,k+1,l)$ for $i \ge 1$ with rate λ_1 if $0 \le j \le c, c - j \le k \le L - 1, c + 1 \le l \le c + S$
   – Due to the arrival of Type II customer:
     - $(0,j,k,l) \to (0,j+1,k,l)$ with rate λ_2 if $j + k < l, 0 \le j \le c - 1, 0 \le k \le c - j - 1, j \le l \le c + S$
     - $(0,j,k,l) \to (1,j,k,l)$ with rate λ_2 if $j + k < l, j = c, c - j \le k \le L, c + 1 \le l \le c + S$
     - $(i,j,k,l) \to (i+1,j,k,l)$ with rate λ_2 for $0 \le j \le c, c - j \le k \le L, c + 1 \le l \le c + S$

2. Transitions due to service completions:
   – $(0,j,k,l) \to (0,j-1,k,l-1)$ with rate $j\mu_2$ for $1 \le j \le c, 0 \le k \le L, j \le l \le c + S$
   – $(0,j,k,l) \to (0,j,k-1,l-1)$ with rate $min(c - j,k,l - j)\mu_1$ for $0 \le j \le c, 1 \le k \le L, j \le l \le c + S$
   – $(i,j,k,l) \to (i,j,k-1,l-1)$ with rate $min(c - j,k,l - j)\mu_1$ for $0 \le j \le c, 1 \le k \le c - j, j \le l \le j + k$
   – $(i,j,k,l) \to (i-1,j,k-1,l-2)$ with rate $p(c-j)\mu_1$ for $0 \le j \le c-1, c-j \le k \le L, j + k + 1 \le l \le c + S$

- $(i,j,k,l) \rightarrow (i-1, j+1, k-1, l-1)$ with rate $q(c-j)\mu_1$ for $0 \le j \le c-1, k = c-j, c+1 \le l \le c+S$

3. Transition due to replenishment:

- $(0,j,k,l) \rightarrow (0,j,k,c+S)$ with rate β for $0 \le j \le c, 0 \le k \le L, 0 \le l \le c+s$
- $(i,j,k,l) \rightarrow (0, j+i, k, c+S)$ with rate β for $0 \le j \le c-i, 0 \le k \le c-j-i, 0 \le l \le j+k$
- $(i,j,k,l) \rightarrow (i,j,k,c+S)$ with rate β for $0 \le j \le c, c-j \le k \le L, j \le l \le c+s$

The infinitesimal generator of the above process is

$$
\mathcal{Q} = \begin{pmatrix}
A_{0,0} & A_{0,1} \\
A_{1,0} & A_{1,1} & A_{1,2} \\
A_{2,0} & A_{2,1} & A_{2,2} & A_{2,3} \\
A_{3,0} & A_{3,1} & A_{3,2} & A_{3,3} & A_{3,4} \\
\vdots \\
\vdots \\
A_{c-2,0}\ A_{c-2,1}\ A_{c-2,2} & \cdots & \cdots & \cdots\ A_{c-2,c-2}\ A_{c-2,c-1} \\
A_{c-1,0}\ A_{c-1,1}\ A_{c-1,2} & \cdots & \cdots & \cdots\ A_{c-1,c-2}\ A_{c-1,c-1}\ A_{c-1,c} \\
A_{c,0}\ \ A_{c,1}\ \ A_{c,2} & \cdots & \cdots\ \cdots\ A_{c,c-2}\ \ A_{c,c-1}\ \ A_{c,c}\ \ A_{c,c+1} \\
A_{c+1,1}\ A_{c+1,2} & \cdots & \cdots\ \cdots\ \cdots\ \cdots\ A_{c+1,c}\ A_{c+1,c+1} \\
A_{c+2,2}\ A_{c+2,3}\ \cdots\ \cdots \\
& \ddots \\
&& \ddots \\
&&& \ddots \\
&&&& \ddots \\
&&&&& A_{2c,c}\ \ A_{2c,c+1}\ \cdots \\
&&&&& \ \ A_{2c+1,c+1}\ \cdots \\
&&&&&& \ddots
\end{pmatrix}
$$

The matrices $A_{i,i-1}$ and $A_{i,i+1}$ denote the transitions from $\ell(i)$ to $\ell(i-1)$ and to $\ell(i+1)$ respectively and $A_{i,i}$ has as elements transition rates within $\ell(i)$. $A_{i,j}$ has as entries transition rates from $\ell(i)$ to $\ell(j)$ for $0 \le j \le i-2$ for $i \ge 2$. From the transitions described above we can see that $A_{i,i+1}$ are same for $i \ge 1$ and is denoted by A_0, $A_{i,i}$, for $i \ge 1$, are same and they are denoted by A_1, $A_{i,i-1}$, for $i \ge 1$, are same and they are denoted by A_2. Similarly, $A_{i,i-2}$ for $i \ge 3$, $A_{i,i-3}$ for $i \ge 4$, $A_{i,i-4}$ for $i \ge 5$, $\ldots$, $A_{i,i-(c-1)}$ for $i \ge c$ and $A_{i,i-c}$ for $i \ge c+1$ are same. They are denoted by $A_3, A_4, A_5, \ldots, A_c, A_{c+1}$ respectively. The model under study can be studied as a QBD process by combining the set of states as follows:

$$L(1) = \{\ell(1), \ell(2), \ell(3), \ldots, \ell(c)\}$$

$$L(2) = \{\ell(c+1), \ell(c+2), \ell(c+3), \ldots, \ell(2c)\}$$

$$L(3) = \{\ell(2c+1), \ell(2c+2), \ell(2c+3), \ldots, \ell(3c)\}$$

and so on. Thus, the new generator is

$$Q' = \begin{bmatrix} B_1 & B_0' & & & \\ A_2' & \tilde{A}_1 & \tilde{A}_0 & & \\ & \tilde{A}_2 & \tilde{A}_1 & \tilde{A}_0 & \\ & & \tilde{A}_2 & \tilde{A}_1 & \tilde{A}_0 \\ & & & \ddots & \ddots & \ddots \end{bmatrix}.$$

where the block entries appearing in Q' are obtained from those of Q as follows: $B_1 = A_{0,0}$,

$$B_0' = \begin{bmatrix} A_{0,1} & 0 & \cdots \end{bmatrix}, \quad A_2' = \begin{bmatrix} A_{1,0} \\ A_{2,0} \\ A_{3,0} \\ \vdots \\ A_{c,0} \end{bmatrix}$$

$$\tilde{A}_0 = \begin{bmatrix} 0 & & \cdots & & 0 \\ \vdots & & & & \vdots \\ 0 & & & & \\ A_0 & 0\ 0\ 0\ \cdots & & \cdots\ \cdots & 0 \end{bmatrix}$$

$$\tilde{A}_2 = \begin{bmatrix} A_{c+1} & A_c & & \cdots & & & A_2 \\ & A_{c+1} & A_c & & \cdots & & A_3 \\ & & A_{c+1} & A_c & & & A_4 \\ & & & \ddots & & & \\ & & & & \ddots & & \\ & & & & & \ddots & \\ & & & & & A_{c+1} & A_c \\ 0 & & \cdots & & \cdots & & A_{c+1} \end{bmatrix}$$

$$\tilde{A}_1 = \begin{bmatrix} A_1 & A_0 & & & & & \\ A_2 & A_1 & A_0 & & & & \\ A_3 & A_2 & A_1 & A_0 & & & \\ \vdots & & & \ddots & & & \\ & & & & \ddots & & \\ \vdots & & & & & \ddots & \\ & & & & & & A_1 & A_0 \\ A_c & A_{c-1} & A_{c-2} \cdots & \cdots & & & A_2 & A_1 \end{bmatrix}$$

3 Steady State Analysis

We proceed with the steady state analysis of the queueing-inventory system under study. The first step is to look for the condition for stability.

3.1 Stability Condition

Define $\tilde{A} = \tilde{A}_0 + \tilde{A}_1 + \tilde{A}_2$. Then it is the infinitesimal generator of the finite state continuous time Markov chain. Let $\tilde{\pi} = (\tilde{\pi}_1, \tilde{\pi}_2, \ldots \tilde{\pi}_c)$ be the steady state probability vector of this generator $\tilde{A}$. That is $\tilde{\pi}$ satisfies

$$\tilde{\pi}\tilde{A} = 0$$

and

$$\tilde{\pi}\mathbf{e} = 1$$

$\tilde{A}$ is a circulant matrix and so the vector $\tilde{\pi}$ is of the form $\tilde{\pi} = (\pi/c, \pi/c, \pi/c, \ldots \pi/c)$ where π satisfies

$$\pi A = 0$$

and

$$\pi\mathbf{e} = 1$$

with $A = A_0 + A_1 + A_2 + \ldots + A_{c+1}$, and $\pi = (\pi_0, \pi_1, \pi_2, \ldots \pi_c)$. The QBD type generator is stable if and only if

$$\tilde{\pi}\tilde{A}_0\mathbf{e} < \tilde{\pi}\tilde{A}_2\mathbf{e},$$

which on simplification yields

$$\pi/cA_0\mathbf{e} < \pi/c\{cA_{c+1}\mathbf{e} + (c-1)A_c\mathbf{e} + (c-2)A_{c-1}\mathbf{e} + \ldots 2A_3\mathbf{e} + A_2\mathbf{e}\} \quad (1)$$

i.e,

$\lambda_2 < Prob(\text{excess inventory level exceeds number of priority customer waiting})$

$Prob(r \text{ priority customers are in service})$

$r\,\mu_1\,p\,Prob(\text{atleast one low priority waiting})$

$+ Prob(l \text{ low priority customers in service})l\,\mu_2$

3.2   Steady State Probability Vector

Let $\mathbf{y} = (\mathbf{y_0}, \mathbf{y_1}, \mathbf{y_2}, \ldots)$ denote the steady state probability vector of $\mathcal{Q}'$. Then,

$$\mathbf{y}\mathcal{Q}' = 0, \mathbf{y}\mathbf{e} = 1.$$

Note that, $\mathbf{y_0} = \mathbf{x_0}$, and $\mathbf{y_1} = (\mathbf{x_1}, \mathbf{x_2}, \mathbf{x_3}, \ldots \mathbf{x_c})$, $\mathbf{y_2} = (\mathbf{x_{c+1}}, \mathbf{x_{c+2}}, \mathbf{x_{c+3}}, \ldots \mathbf{x_{2c}})$ and so on where $\mathbf{x} = (\mathbf{x_0}, \mathbf{x_1}, \mathbf{x_2}, \ldots)$ being the steady state probability vector of $\mathcal{Q}$. The component vectors are partitioned as

$$\mathbf{x_0} = \{x_0(j, k, l) : 0 \leq j \leq c, 0 \leq k \leq L, j \leq l \leq c + S\}$$

and

$$\mathbf{x_i} = \{x_i(j, k, l) : 0 \leq j \leq c, 0 \leq k \leq c - j - 1, j \leq l \leq j + k\}$$
$$\bigcup \{x_i(j, k, l) : 0 \leq j \leq c, c - j \leq k \leq L, j \leq l \leq c + S\}, \text{ for } i \geq 1$$

Under the stability condition (1), the steady state probability vector

$$\mathbf{y_i} = \mathbf{y_1} R^{i-1}, i \geq 2$$

where R is the minimal nonnegative solution to the matrix quadratic equation

$$R^2 \tilde{A}_2 + R \tilde{A}_1 + \tilde{A}_0 = 0,$$

and the vectors $\mathbf{y_0}$ and $\mathbf{y_1}$ are obtained by solving

$$\mathbf{y_0} B_1 + \mathbf{y_1} A_2' = 0$$

$$\mathbf{y_0} B_0' + \mathbf{y_1} [\tilde{A}_1 + R \tilde{A}_2] = 0$$

subject to the normalizing condition

$$\mathbf{y_0} + \mathbf{y_1} (I - R)^{-1} \mathbf{e} = 1$$

4   System Characteristics

1. Expected number of Type-II customers in the queue,

$$E_{TII} = \sum_{i=1}^{\infty} i \mathbf{x_i} \mathbf{e}.$$

2. Expected number of Type-I customers in system,

$$E_{TI} = \sum_{j=0}^{c} \sum_{k=1}^{L} k \sum_{l=j}^{c+S} x_0(j, k, l) + \sum_{i=1}^{\infty} \sum_{j=0}^{c} \sum_{k=1}^{c-j-1} k \sum_{l=j}^{j+k} x_i(j, k, l)$$

$$+ \sum_{i=1}^{\infty} \sum_{j=0}^{c} \sum_{k=c-j}^{L} k \sum_{l=j}^{c+S} x_i(j, k, l).$$

3. Rate at which Type-II customers leave with Type-I customers upon completion of latter's service,

$$R_{TII,TI} = \sum_{i=1}^{\infty} \sum_{j=0}^{c-1} p(c-j)\mu_1 \sum_{k=c-j}^{L} k \sum_{l=j+k+1}^{c+S} x_i(j,k,l).$$

4. Rate at which Type II customers served out by servers,

$$R_{TIIS} = \sum_{j=1}^{c} j\mu_2 \sum_{k=0}^{L} \sum_{l=j}^{c+S} x_0(j,k,l)$$

$$+ \sum_{i=1}^{\infty} \sum_{j=1}^{c} j\mu_2 \sum_{k=0}^{c-j-1} k \sum_{l=j}^{j+k} x_i(j,k,l)$$

$$+ \sum_{i=1}^{\infty} \sum_{j=1}^{c} j\mu_2 \sum_{k=c-j}^{L} \sum_{l=j}^{c+S} x_i(j,k,l).$$

5. Probability that a Type II customer leaves with a Type-I customer,

$$= \frac{1}{\lambda_2} R_{TII,TI}.$$

6. Probability that Type II customer leaves with service from one of c servers

$$= \frac{1}{\lambda_2} R_{TIIS}.$$

7. Probability that Type-I is lost due to no inventory,

$$P_{TInoinv} = \sum_{j=0}^{c} \sum_{k=0}^{c-j} \sum_{l=j}^{j+k} x_0(j,k,l) + \sum_{j=0}^{c} \sum_{k=c-j+1}^{L-1} \sum_{l=j}^{c} x_0(j,k,l)$$

$$+ \sum_{i=1}^{\infty} \sum_{j=0}^{c} \sum_{k=0}^{c-j} \sum_{l=j}^{j+k} x_i(j,k,l)$$

$$+ \sum_{i=1}^{\infty} \sum_{j=0}^{c} \sum_{k=c-j+1}^{L-1} \sum_{l=j}^{c} x_i(j,k,l).$$

8. Expected loss rate of Type-I customer due to no inventory,

$$E_{TIlossrate} = \lambda_1 P_{TInoinv}.$$

9. Expected loss rate of Type-II customer due to no inventory,

$$E_{TIIlossrate} = \lambda_2 P_{TIInoinv}.$$

10. Probability that an arriving Type-I customer is lost due to lack of space in buffer,

$$P_{nospace} = \sum_{i=0}^{\infty} \sum_{j=0}^{c} \sum_{l=j}^{c+S} x_i(j,L,l).$$

11. Probability that Type-II is lost due to no inventory,

$$P_{TIInoinv} = \sum_{j=0}^{c} \sum_{k=0}^{c-j} \sum_{l=j}^{j+k} x_0(j,k,l) + \sum_{j=0}^{c} \sum_{k=c-j+1}^{L} \sum_{l=j}^{c} x_0(j,k,l)$$

$$+ \sum_{i=1}^{\infty} \sum_{j=0}^{c} \sum_{k=0}^{c-j} \sum_{l=j}^{j+k} x_i(j,k,l)$$

$$+ \sum_{i=1}^{\infty} \sum_{j=0}^{c} \sum_{k=c-j+1}^{L} \sum_{l=j}^{c} x_i(j,k,l).$$

12. Probability that all servers are idle,

$$\sum_{l=0}^{c+S} x_0(0,0,l) + \sum_{i=1}^{\infty} \sum_{k=0}^{l} x_i(0,k,0).$$

13. Probability that all servers are busy,

$$\sum_{i=0}^{\infty} \sum_{j=0}^{c} \sum_{k=c}^{L} \sum_{l=c}^{c+S} x_i(j,k,l).$$

14. Probability that all servers are busy with Type-I,

$$\sum_{i=0}^{\infty} \sum_{k=c}^{L} \sum_{l=c}^{c+S} x_i(0,k,l).$$

15. Probability that all servers are busy with Type-II,

$$\sum_{i=0}^{\infty} \sum_{k=c}^{L} \sum_{l=c}^{c+S} x_i(c,k,l).$$

16. Probability that no server is busy with Type-I,

$$\sum_{j=0}^{c} \sum_{l=j}^{c+S} x_0(j,0,l) + \sum_{j=0}^{c} \sum_{k=1}^{L} x_0(j,k,j) + \sum_{k=1}^{L} \sum_{l=c+1}^{c+S} x_0(c,k,l)$$

$$+ \sum_{i=0}^{\infty} \sum_{j=0}^{c} \sum_{k=0}^{L} x_i(j,k,j) + \sum_{k=0}^{L} \sum_{l=c+1}^{c+S} x_i(c,k,l).$$

17. Probability that exactly 'm' servers are busy with Type-I,

$$= \sum_{i=0}^{\infty} \sum_{j=0}^{c-m} \sum_{k=m}^{L} x_i(j,k,j+1) + \sum_{j=0}^{c-m} \sum_{l=j+k+1}^{c+S} x_0(j,m,l).$$

18. Probability that no server is busy with Type-II,

$$\sum_{k=0}^{L} \sum_{l=0}^{c+S} x_0(0,k,l) + \sum_{i=1}^{\infty} \sum_{k=0}^{c} \sum_{l=0}^{j+k} x_i(0,k,l)$$

$$+ \sum_{k=c+1}^{L} \sum_{l=c+1}^{c+S} x_i(0,k,l).$$

19. Probability that exactly 'm' servers are busy with Type-II,

$$\sum_{k=0}^{L}\sum_{l=m}^{c+S} x_0(m,k,l) + \sum_{i=1}^{\infty}\left\{\sum_{k=0}^{c-m}\sum_{l=m}^{j+k} x_i(m,k,l) + \sum_{k=c}^{L}\sum_{l=m}^{c+S} x_i(m,k,l)\right\}.$$

20. Expected reorder rate,

$$E_R = k\mu_1 \sum_{k=1}^{c} x_0(0,k,c+s+1) + c\mu_1 \sum_{k=c+1}^{L} x_0(0,k,c+s+1)$$

$$+ \sum_{j=1}^{c} j\mu_2 \sum_{k=0}^{L} x_0(j,k,c+s+1)$$

$$+ \sum_{j=1}^{c}(c-j)\mu_1 \sum_{k=1}^{L} x_0(j,k,c+s+1)$$

$$+ \sum_{i=1}^{\infty}\sum_{j=0}^{c-1} q(c-j)mu_1 \sum_{k=c-j}^{L} x_i(j,k,c+s+1)$$

$$+ \sum_{i=1}^{\infty} c\mu_2 \sum_{k=0}^{L} x_i(j,k,c+s+1)$$

$$+ \sum_{i=1}^{\infty}\sum_{j=0}^{c-1} p(c-j)mu_1 \sum_{k=c-j}^{c+s+2-(j+1)} x_i(j,k,c+s+2).$$

21. Expected number of items in the inventory,

$$EI = \sum_{k=0}^{L}\sum_{l=1}^{c+S} lx_0(0,k,l) + \sum_{j=1}^{c}\sum_{k=0}^{L}\sum_{l=j}^{c+S} lx_0(j,k,l)$$

$$+ \sum_{i=1}^{\infty} x_i(0,1,1) + \sum_{k=2}^{c-1}\sum_{l=1}^{j+k} lx_i(0,k,l)$$

$$+ \sum_{i=1}^{\infty}\sum_{j=1}^{c}\sum_{k=0}^{c-j-1}\sum_{l=j}^{j+k} lx_i(j,k,l) + \sum_{i=1}^{\infty}\sum_{j=0}^{c}\sum_{k=c-j}^{L}\sum_{l=j}^{c+S} lx_i(j,k,l)$$

4.1   Optimization Problem

Based on the above performance measures we construct a revenue function. We define this revenue function as $\mathcal{RF}$ as

$$\mathcal{RF} = (C_1 - C_2 - C_3)R_{TII,TI} + (C_1 - C_2)R_{TIIS} - C_4 P_{nospace} - C_5 P_{noinv} - h_I EI$$

$$- C_2 E_R - h_{C_I} E_{TI} - h_{C_{II}} E_{TII}$$

where

- C_1 = Selling Cost per unit item
- C_2 = Purchase Cost per unit item
- C_3 = Incentive to Type-I for serving Type-II
- C_4 = Cost for loss due to lack of space in buffer
- C_5 = Cost for customer loss due to no inventory
- h_I = holding cost per unit time per unit item in the inventory
- h_{C_I} = holding cost per Type-I customer per unit time
- $h_{C_{II}}$ = holding cost per Type-II customer per unit time

In order to study the variation in different parameters on profit function we first fix the costs $C_1 = \$75, C_2 = \$50, C_3 = \$2, C_4 = \$10, C_5 = \$10, h_I = \$5, h_{CI} = \$5, h_{CI} = \$2$.

Effect of p on $\mathcal{RF}$

The effect of p on the revenue function for $c = 1, c = 2$ and $c = 3$ are given below:

Table 1. Value of revenue function for various p: Fix $c = 1, L = 8, S = 17, s = 5, \lambda_1 = 0.9, \lambda_2 = 0.8, \beta = 1, \mu_1 = 2, \mu_2 = 3$

p	0	0.25	0.5	0.75	1
$\mathcal{RF}$	−37.4033	−40.8083	−43.0853	−44.4609	−45.2468

Table 2. Value of revenue function for various p: Fix $c = 2, L = 8, S = 18, s = 5, \lambda_1 = 0.9, \lambda_2 = 0.8, \beta = 2, \mu_1 = 2, \mu_2 = 3$

p	0	0.25	0.5	0.75	1
$\mathcal{RF}$	−35.2785	−50.1243	−54.6316	−54.6847	−53.7205

As p increases the value of the revenue function is seen to decrease for $c = 1$ (Table 1) and $c = 3$ (Table 3). For $c = 2$ (Table 2), revenue function decreases first and then it shows a slight increase. This could be due to increase in incentives consequent to increase in number of low priority customers served by those of high priority.

Table 3. Value of revenue function for various p: Fix $c = 3, L = 8, S = 19, s = 5, \lambda_1 = 1, \lambda_2 = 1.1, \beta = 2, \mu_1 = 1.1, \mu_2 = 1.2$

p	0	0.25	0.5	0.75	1
$\mathcal{RF}$	515.4303	404.3992	339.7477	298.8594	267.5835

Effect of β on Loss Rates and Number of Items in Inventory

As β value increases loss rate of Type-I customers, Type-II customers due to no inventory is evaluated (Tables 4, 5 and 6) and we can see that loss rate of customers decreases, and as β increases expected number of items in the inventory increases.

Effect of β on Revenue Function

As β increases value of revenue function increases for $c = 2$ and $c = 3$ (Tables 8 and 9), whereas for $c = 1$ (Table 7) it decreases.

Table 4. Value of revenue function for various value of β: Fix $c = 1, L = 8, S = 17, s = 5, \lambda_1 = 0.9, \lambda_2 = 0.8, \mu_1 = 2, \mu_2 = 3, p = 0.5$

β	1	1.5	2	2.5	3
$E_{TIlossrate}$	0.0141	0.0049	0.0019	$9,1408 \times 10^{-4}$	4.8460×10^{-4}
$E_{TIIlossrate}$	0.0125	0.0042	0.0017	8.1252×10^{-4}	4.3075×10^{-4}
E_I	10.4024	10.7832	10.9746	11.0889	11.1644

Table 5. Value of revenue function for various values of β: Fix $c = 2, L = 8, S = 18, s = 5, \lambda_1 = 0.9, \lambda_2 = 0.8, \mu_1 = 2, \mu_2 = 3, p = 0.5$

β	1	1.5	2	2.5	3
$E_{TIlossrate}$	0.0081	0.0024	9.3577×10^{-4}	4.1559×10^{-4}	1.9997×10^{-4}
$E_{TIIlossrate}$	0.0072	0.0021	8.1711×10^{-4}	3.5942×10^{-4}	1.7045×10^{-4}
E_I	12.8770	13.2197	13.3877	13.4867	13.5519

Table 6. Value of revenue function for various values of β: Fix $c = 3, L = 8, S = 19, s = 5, \lambda_1 = 1, \lambda_2 = 1.1, \mu_1 = 1.1, \mu_2 = 1.2, p = 0.5$

β	1	1.5	2	2.5	3
$E_{TIlossrate}$	0.0749	0.0474	0.0342	0.0270	0.0224
$E_{TIIlossrate}$	0.0837	0.0535	0.0389	0.0307	0.0256
E_I	12.0157	13.1016	13.5791	13.8264	13.9762

Table 7. Value of revenue function for various values of β: Fix $c = 1, L = 8, S = 17, s = 5, \lambda_1 = 0.9, \lambda_2 = 0.8, \mu_1 = 2, \mu_2 = 3, p = 0.5$

β	1	1.5	2	2.5	3
$\mathcal{RF}$	-43.0853	-44.5015	-45.1581	-45.5057	-45.7053

Table 8. Value of revenue function for various values of β: Fix $c = 2, L = 8, S = 18, s = 5, \lambda_1 = 0.9, \lambda_2 = 0.8, \mu_1 = 2, \mu_2 = 3, p = 0.5$

β	1	1.5	2	2.5	3
$\mathcal{RF}$	-58.7539	-56.5816	-54.6316	-53.1138	-51.9845

Table 9. Value of revenue function for various values of β: Fix $c = 3, L = 8, S = 19, s = 5, \lambda_1 = 1, \lambda_2 = 1.1, \mu_1 = 1.1, \mu_2 = 1.2, p = 0.5$

β	1	1.5	2	2.5	3
$\mathcal{RF}$	195.6691	278.4610	339.7477	381.2734	410.2780

5   Conclusion

In this we considered a queueing inventory system useful in crowdsourcing. We investigated a multi-server queueing inventory model in which one type of customers are encouraged to serve another type of customers which improves the efficiency of the service facility. Here we assumed that resources to be provided to the customer on service completion to be finite. We assumed the arrival process to be Poisson and service times exponentially distributed. A revenue function is constructed and effect of probability p on the revenue function for single server, two server and three server is numerically analyzed. Effect of β on the loss rate of customers and revenue function is numerically analyzed. We propose to extend the above model where arrival is MAP and service time is Phase-type.

References

1. Balintfy, J.L.: On a basic class of inventory problems. Manag. Sci. **10**, 287–297 (1964)
2. Berman, O., Kaplan, E.H., Shimshak, D.G.: Deterministic approximations for inventory management at service facilities. IIE Trans. **25**(5), 98–104 (1993)
3. Berman, O., Kim, E.: Stochastic models for inventory managements at service facilities. Commun. Stat. Stoch. Models **15**(4), 695–718 (1999)
4. Bhat, U.N.: An Introduction to Queueing Theory: Modeling and Analysis in Applications. Springer, New York (2008). https://doi.org/10.1007/978-0-8176-8421-1
5. Bhat, U.N., Miller, G.K.: Elements of Applied Stochastic Processes. Wiley Science in Probability and Statistics, 3rd edn. Wiley, New York (2002)
6. Gross, D., Harris, C.M.: Fundamentals of Queueing Theory. Wiley, New York (1988)
7. Hadley, G., Whitin, T.M.: Analysis of Inventory Systems. Prentice Hall Inc., Englewood Cliffs (1963)
8. Neuts, M.F.: Matrix-Geometric Solutions in Stochastic Models: An Algorithmic Approach. The Johns Hopkins University Press, Baltimore (1981). [1994 version is Dover Edition]
9. What are the best examples of crowdsourcing. http://www.quora.com/
10. Howe, J.: Crowdsourcing: a definition, June 2006
11. Chakravarthy, S.R., Dudin, A.N.: A queueing model for crowdsourcing. J. Oper. Res. Soc. Jpn. **68**, 221–236 (2015)
12. Krishnamoorthy, A., Shajin, D., Manjunath, A.S.: On a multi-server priority queue with preemption in Crowdsourcing. In: First International Conference on Analytical and Computational Methods in Probability Theory and if Applications, ACMPT, RUDN University, Russia, pp. 145–157, October 2017

An MMAP/M/∞ Queueing System with an Offer Zone Working in a Random Environment

V. C. Joshua, Ambily P. Mathew$^{(\boxtimes)}$, and A. Krishnamoorthy

Department of Mathematics, CMS College, Kottayam 686001, Kerala, India
{vcjoshua,ambilypm,krishnamoorthy}@cmscollege.ac.in
http://www.cmscollege.ac.in

Abstract. We consider a tandem queueing network with two service stations without buffers. An infinite capacity station named as the main station provides usual paid service. Another station is a finite capacity station named as the offer zone. The offer zone is an intermediate station strategically designed to attract the maximum number of customers to the main station. The offer zone works under various random environments. Sojourn times of each random environment follows Phase Type distribution. Two types of customers arrive to the system according to an MMAP. Service times of customers at both the stations are exponentially distributed. The stationary probability distribution of the states of the Markov chain representing the proposed model is computed. Some operational and probabilistic characteristics of the system are determined. A control problem is discussed. A cost function is proposed. The effect of the maximum capacity of the offer zone on various performance measures are considered. Numerical as well as graphical illustrations are given.

Keywords: Main station · Offer zone · Tandem queue · Random environment

1 Introduction

In this paper we consider a queueing system with two service stations working in tandem. The theory of queues and tandem queueing networks are intertwined. Each of these can be used to supplement the other. Tandem queues or Multi-Stage queues serve as a link between these two. When we consider the tandem queue as a queueing network, it is one of the simplest queueing networks consisting of a number of sequential service counters or stations. A tandem queue with two stations and a finite queue in between the stations is considered in [3]. In [2], Gomez Corral considers a tandem queue with two stations in which an infinite queue is allowed before the first queue, but no queue in between the stations. In [6] Klimenok et al. consider a tandem queue with a finite number of stations. Tandem queue with cross-traffic in between the stations can be found in [7]. Studies in connection with tandem queues can be found in [1],[5] and [4].

© Springer Nature Switzerland AG 2019
V. M. Vishnevskiy et al. (Eds.): DCCN 2019, CCIS 1141, pp. 244–257, 2019.
https://doi.org/10.1007/978-3-030-36625-4_20

Algorithmic solutions to exponential tandem queues with blocking can be found in [8]. In the present paper, the tandem queue considered works with two stations. In most of the tandem queueing networks finite capacity service stations are considered. In the present model the main station works without buffers and it can accommodate an infinite number of customers. The service policy adopted at each station is self service policy. Two types of customers arrive to the system. Only one of the customers use the intermediate station for being served. In the present model, the first station of the tandem consists of an infinite number of servers and it is named as the main station. It provides usual paid services. The second station of the tandem queueing system is named as the offer zone and it consists of a maximum of N servers. 'Offer' is a strategy adopted by the operators of the service facility to attract more customers. The operator of the system aims at keeping more customers continue with the service. An 'offer' may be considered as rendering discounted service, trial service or rather free service. Thus 'offer' results in high cost burden to the system. Some customers may leave the system after enjoying discounted services. The offer zone acts as an intermediate station for attracting more customers to the main station. There are restrictions on the time a customer can spent being served at the offer zone. Two types of customers arrive to the system according to a Marked Markovian Arrival Process (MMAP). One type of customer directly enter the main station for being served. The second type of customer make a trial service at the offer zone and then decide whether to join main station or not. The environments of the offer zone are designed in such a way to attract more customers to the system. It aims at optimizing the cost effectiveness of the system. The following situation prevailing in the field of telecommunication motivated the designing of the model discussed.

In the field of telecommunication, the number of e-service providers, network operators and software vendors are increasing very fast. The scope, variety and range of various e-services provided by them are astonishing. As far as they are concerned, competition has become an unavoidable part of their corporate life. It has become not only a means of making more profit but a surety even for their existence. They adopt various strategies to attract more subscribers to their service. They design various strategies to attract subscribers of other networks to their service. To make the maximum number of customers to continue with their service, they announce various types of offers and free trials. It helps in minimizing the loss of subscribers of their network. The offers include free trial/special tariff packages and discounts on service packages. Usually the offers are provided for a short duration of time. There are establishment as well as operational costs associated with such strategic moves compared to the benefits acquired from the paid customers. So restrictions in time and number of subscriptions is mandatory in the case of services by means of offers. Cost analysis and optimization of control variables are very important in such a case. The present paper mathematically models a similar situation as a tandem queue without buffers. The arrival process in this tandem network is correlated and it is assumed to be Marked Markovian Arrival Process [MMAP]. We model the present queueing problem as a Continuous Time Markov Chain (CTMC) and analyze it by means of Matrix Analytic Methods [9].

2 Model Description

We consider a tandem queuing network with two service stations named as the main station and the offer zone. Both these stations provide immediate service for the customers upon arrival. The main station is of infinite capacity and provides usual paid service. The offer zone is of finite capacity N and it works under various random environments. Customers get the same type of required service from both the service stations. We consider each strategic plan to attract customers to the offer zone as an offer. We consider a package consisting of a combination of these offers as a particular environment. The environments of the offer zone include no offer environment too. The offer zone serves as an intermediate source for attracting those customers who have not yet decided whether to take the paid service from the main station or not.

We assume that there are a finite number of environments associated with the functioning of the offer zone. Let n be the number of these environments. Let $\{1, 2, \ldots, n-1\}$ denote the $n-1$ environments with one or more offers and n denote the no offer environment of the offer zone. The duration of each environment follows Phase type distribution. The generator matrix of the Markov process leading to the PH distribution depends on the current environment of the offer zone. Let p_r denote the probability that the offer zone is at environment r where $\{r = 1, 2, \ldots, n\}$. The duration of time the environment r works follow Phase type distribution with irreducible representation $PH(\beta_r, S_r)$ with M_r phases. The vector S_r^0 is given by $S_r^0 = -S_r e$. Let $\nu_r = \beta_r(-S_r)^{-1}e$ denote the mean duration of the environment r. We assume that all the customers in the offer zone are getting served in the same environment and so the offers given to those customers in service at the offer zone change with the change in the environment in which the offer zone works. Customers arriving to the system are categorized as type-A and type-B. A type-A customer do not try to take an offer and directly enters the main station for service. A type-B customer likes to have a trial service at the offer zone. After the service completion at the offer zone, a type-B customer decides whether to continue their service at the main station or to leave the system. The stochastic process corresponding to the arrival of both these types of customers are assumed to be an MMAP (Marked Markovian Arrival Process) with representation (D_0, D_1, D_2) where $D_1 = pD^*$, $D_2 = (1 - p)D^*$ for some D^* with $0 \leq p \leq 1$. After the service completion at the offer zone, we assume that a type-B customer moves to the main station to continue their service with probability η or leave the system forever with its complimentary probability $(1 - \eta)$. If the offer zone is full at the time of arrival, a type-B customer directly enters the main station with probability γ or leaves the system with probability $(1 - \gamma)$.

We assume that the service times at both the stations are exponentially distributed. Even though the service provided at both the main station and the offer zone are the same, we assume that the rates at which service is provided are different. Let the service time of a customer at the main station is exponentially distributed with rate μ and the distribution of the service time of a customer at the offer zone when it is working in environment r is exponential with rate μ_r where $\{r = 1, 2, \ldots, n\}$. The MMAP governing the arrival of type-A and type-B customers in the present model is described as follows:

Let the underlying Markov chain $\{\nu_t, t \geq 0\}$ be irreducible and let D be the generator of this Markov chain with state space $\{1, 2, 3, \ldots, m\}$. At the end of sojourn time in state i, which is exponentially distributed with a positive finite parameter $\lambda^{(i)}$ any of the following can happen. It can move to state j where $j \neq i$ without an arrival or it can move to state j with an arrival of a either a type A customer or the arrival of a type B customer. Let $D_0 = (d_{ij}(0))$ be the rate matrix corresponding to those transitions without an arrival and $D^* = (d_{ij}^*)$ be the rate matrix corresponding to an arrival of any customer to the system. Let $D_1 = (pd_{ij}^*(1))$ be the rate matrix corresponding to the arrival of type-A customer and let $D_2 = (1-p)d_{ij}^*(1)$be the rate matrix corresponding to the arrival of type-B customer where $0 \leq p \leq 1$. Then the MMAP under consideration is well be described by the parameter matrices (D_0, D_1, D_2) and $D = D_0 + D_1 + D_2$ is the infinitesimal generator of the markov chain corresponding to the MMAP. All the off-diagonal elements of D_0 and all the elements of D^* are non negative. We assume that the initial probability vector is the same as the stationary probability vector. That is our MMAP is a stationary MMAP.

The average total arrival intensity λ is defined by $\lambda = \theta D^* \mathbf{e}$, where θ is the invariant vector of the stationary distribution of the Markov chain $\{\nu_t, t \geq 0\}$. The vector θ is the unique solution of the system of equations $\theta D = \mathbf{0}$, $\theta \mathbf{e} = 1$ where $\mathbf{e}$ denotes a column vector of $1's$ and $\mathbf{0}$ is a row vector of $0's$. The average arrival intensity λ_A and λ_B of type-A and type-B customers respectively are defined by $\lambda_A = \theta D_1 \mathbf{e}$ and $\lambda_B = \theta D_2 \mathbf{e}$.

In the following sequel the following notations are used

- $\oplus$ represents Kronecker sum of matrices.
- $\otimes$ represents Kronecker product of matrices.
- I_M denotes an identity matrix of order M.
- $diag\{., ., ., \}$ represents a diagonal matrix of appropriate order with entries listed within the braces.

3   Matrix Analytic Solution

To formulate the above model mathematically, we introduce the necessary random variables as follows:

Let $N_1(t)$ be the number of customers in the main station, $N_2(t)$ the number of customers in the offer zone, $E(t)$ the environment of the offer zone and $S(t)$ the phase of the environment of the offer zone. Let $A(t)$ the phase of the arrival process. $E(t)$ can take any of the values $\{1, 2, \ldots, n\}$ depending on the ongoing environment of the offer zone. Then $\{N_1(t), N_2(t), E(t), S(t), A(t)\}$ is a Markov process and it describes the process under consideration.

We define the state space of the QBD under consideration and analyze the structure of its infinitesimal generator. The state space Ω consists of all elements of the form (i, j, r, s, t) where $i \geq 0, 0 \leq j \leq N; r = 1, 2, \ldots, n; s = 1, 2, \ldots, M_r$ for a fixed value of r and $t = 1, 2, \ldots, m$. Let the elements of Ω be ordered lexicographical. The infinitesimal generator Q of the Level Dependent QBD describing the $MMAP/M/\infty$ Queueing System with an Offer Zone working in a Random Environment is of the form

$$Q = \begin{pmatrix} A_1^0 & A_0 & & \\ A_2^1 & A_1^1 & A_0 & \\ & A_2^2 & A_1^2 & A_0 \\ & & \cdots\cdots\cdots \end{pmatrix} \tag{1}$$

where A_0, A_1^i, A_2^i are all square matrices whose entries are block matrices.

A_0 represents the arrival of a customer to the main station; that is transition from level $i \to i + 1$ where $i \geq 0$.

A_2^i represents departure of a customer after service completion at the main station when there are i customers in the main station; transitions from level $i \to i - 1$, for $i = 1, 2, \ldots$ and

A_1^i describes all transitions in which the level does not change (transitions within levels i).

The structure of the A_1^i for $i \geq 0$ are as follows:

$$A_1^i = \begin{pmatrix} E_1^0 & E_0 & & & \\ E_2^1 & E_1^1 & E_0 & & \\ & E_2^2 & E_1^2 & E_0 & \\ & & \cdots\cdots & \cdots & \\ & & & \cdots\cdots & \cdots \\ & & & E_2^N & E_1^N \end{pmatrix} \tag{2}$$

For $j = 0, 1, 2\ldots, N$, E_1^j is given by

$$E_1^j = \begin{pmatrix} C_1 & S_1^0 \otimes p_2\beta_2 \otimes I_m & S_1^0 \otimes p_3\beta_3 \otimes I_m & \cdots & S_1^0 \otimes p_n\beta_n \otimes I_m \\ S_2^0 \otimes p_1\beta_1 \otimes I_m & C_2 & S_2^0 \otimes p_3\beta_3 \otimes I_m & \cdots & S_2^0 \otimes p_n\beta_n \otimes I_m \\ S_3^0 \otimes p_1\beta_1 \otimes I_m & S_3^0 \otimes p_2\beta_2 \otimes I_m & C_3 & \cdots & S_3^0 \otimes p_n\beta_n \otimes I_m \\ \vdots & \vdots & \vdots & \ddots & \vdots \\ S_n^0 \otimes p_1\beta_1 \otimes I_m & S_n^0 \otimes p_2\beta_2 \otimes I_m & \cdots & \cdots & C_n \end{pmatrix} \tag{3}$$

For a given value of $j = 0, 1, 2, \ldots(N - 1)$, the matrices C_l is defined as

$$C_l = [(1 - p_l)S_l - (i\mu + j\mu_l)I] \oplus D_0$$

and for j=N

$$C_l = [(1 - p_l)S_l - (i\mu + j\mu_l)I] \oplus [D_0 + (1 - \gamma)D_2]$$

where $l = 1, 2, \ldots\ldots, n$.

E_0 is the matrix representation of the rates at which customers arrive to the offer zone and is given by

$$E_0 = diag\{I_{M_1} \otimes D_2, I_{M_2} \otimes D_2, \ldots, I_{M_n} \otimes D_2\} \tag{4}$$

E_2^j is the matrix representation of the rates at which those customers who have completed their service in the offer zone leave the system forever without going to the main station and is given by

$$E_2^j = diag\{I_{M_1} \otimes (1-\eta)j\mu_1 I_m, I_{M_2} \otimes (1-\eta)j\mu_2 I_m, \ldots, \ldots, I_{M_n} \otimes (1-\eta)j\mu_n I_m\} \tag{5}$$

$$A_0 = \begin{pmatrix} U_1 & & & & & \\ U_2^1 & U_1 & & & & \\ & U_2^2 & U_1 & & & \\ & & \cdots\cdots & & & \\ & & & \cdots & & \\ & & & U_2^{(N-1)} & U_1 & \\ & & & & U_2^N & U_1^N \end{pmatrix} \tag{6}$$

$$U_1 = diag\{I_{M_1} \otimes D_1, I_{M_2} \otimes D_1, \ldots, \ldots, I_{M_n} \otimes D_1\} \tag{7}$$

$$U_1^N = diag\{I_{M_1} \otimes [D_1 + \gamma D_2], I_{M_2} \otimes [D_1 + \gamma D_2], \ldots, \ldots, I_{M_n} \otimes [D_1 + \gamma D_2]\} \tag{8}$$

For $j = 1, 2, \ldots, N$ the matrices U_2^j are given by

$$U_2^j = diag\{I_{M_1} \otimes \eta j\mu_1 I_m, I_{M_2} \otimes \eta j\mu_2 I_m, \ldots, \ldots, I_{M_n} \otimes \eta j\mu_n I_m\} \tag{9}$$

- U_1 is the matrix representation of the rates at which the customers arrive to the main station when the offer zone is not fully occupied.
- U_1^N is the matrix representation of the rates at which the customers arrive to the main station when the offer zone is fully occupied.
- U_2^j is the matrix representation of the rates at which those customers who have completed their service in the offer zone enters the main station for continuing their service at the main station.

For $i = 1, 2, \ldots$, the matrices A_2^i are given by

$$A_2^i = I_{(N+1)} \otimes diag\{I_{M_1} \otimes i\mu I_m, I_{M_2} \otimes i\mu I_m, \ldots, \ldots, I_{M_n} \otimes i\mu I_m\} \tag{10}$$

The present model is a level dependent QBD and we apply Neuts-Rao truncation [10] for the analysis of the model. We assume that when the number of customers in the main station exceeds a certain limit, say K, service occurs at constant rates $K\mu$. In that situation the matrices A_2^i becomes A_2^K for $i \geq K$.

The infinitesimal generator Q^1 of the modified model becomes

$$Q^1 = \begin{pmatrix} A_1^0 & A_0 & & & \\ A_2^1 & A_1^1 & A_0 & & \\ & A_2^2 & A_1^2 & A_0 & \\ & & \cdots\cdots\cdots & & \\ & & & A_2 & A_1 & A_0 \\ & & & & \cdots\cdots\cdots & \end{pmatrix} \tag{11}$$

where
$$A_1 = A_1^K, A_2 = A_2^K.$$

In this model, there are infinite number of servers in the main station and therefore the system is always stable.

3.1   Stationary Distribution

The stationary distribution of the Markov process under consideration is obtained by solving the set of equations

$$\mathbf{x}Q^1 = 0; \mathbf{x}e = 1. \tag{12}$$

Let $\mathbf{x}$ be the steady- state probability vector of Q^1
Partition this vector in conformity with Q^1 as follows:

$$\mathbf{x} = (\mathbf{x_0}, \mathbf{x_1}, \mathbf{x_2}, \dots)$$

$$\mathbf{x_i} = (\mathbf{x_{i0}}, \mathbf{x_{i1}}, \dots, \mathbf{x_{iN}})$$

$$\mathbf{x}_{ij} = (\mathbf{x}_{ij1}, \mathbf{x}_{ij2}, \mathbf{x}_{ij3}, \dots \mathbf{x}_{ijn})$$

$$\mathbf{x}_{ijr} = (\mathbf{x}_{ijr1}, \mathbf{x}_{ijr2} \dots \mathbf{x}_{ijrM_r})$$

$$\mathbf{x}_{ijrs} = (x_{ijrs1}, x_{ijrs2} \dots, x_{ijrsm}).$$

x_{ijrst} is the probability of being in state (i, j, r, s, t) for $i \geq 0, j = 0, 1, ..., N$; $r = 1, 2, \dots n, s = 1, 2, \dots, M_r, t = 1, 2, \dots, m$. The steady-state probability vector is obtained as

$$\mathbf{x}_{(K-1)+i} = \mathbf{x}_{(K-1)} R^i, i \geq 0 \tag{13}$$

where R is the minimal non negative solution to the matrix quadratic equation

$$R^2 A_2 + R A_1 + A_0 = 0 \tag{14}$$

and the vectors $\mathbf{x_0}, \dots, \mathbf{x_{(K-1)}}$ are obtained by solving

$$\mathbf{x_0} A_1^0 + \mathbf{x_1} A_2^1 = 0 \tag{15}$$

$$\mathbf{x_{(i-1)}} A_0 + \mathbf{x_i} A_1^i + \mathbf{x_{(i+1)}} A_2^{(i+1)} = 0 \tag{16}$$

for $1 \leq i \leq (K-2)$

$$\mathbf{x_{(K-2)}} A_0 + \mathbf{x_{(K-1)}} \left[A_1^{(K-1)} + A_2 R \right] = 0 \tag{17}$$

subject to the normalizing condition

$$\sum_{i=0}^{(K-2)} \mathbf{x_i} + \mathbf{x_{(K-1)}}(I - R)^{-1}\mathbf{e} = 1. \tag{18}$$

4 Performance Measures of the System

With regard to the above system under consideration, we may be interested in evaluating a number of performance measures which may help us to optimize the capacity N and to minimize the expected loss of customers from the system. There are two types of loss of type-B customers from the system. The first of which namely type-I loss occurs due to the restriction on the maximum capacity of the offer zone, that is, when the offer zone is full with N number of customers. The second type of loss of type-B customers namely type-II loss, occurs from the offer zone after completing their service at the offer zone. We can identify the environment of the offer zone from which the maximum expected number of type-B customers are lost without joining the main station which in turn help us to redefine the offers so as to minimize type-II loss. Following are some performance measures of the system:

1. Expected number of customers in the main station

$$E[MS] = \sum_{i=0}^{\infty} i x_i \mathbf{e} \tag{19}$$

2. Expected number of customers in the offer zone

$$E[OZ] = \sum_{i=0}^{\infty} \sum_{j=0}^{N} \sum_{r=1}^{n} \sum_{s=1}^{M_r} \sum_{t=1}^{m} j x_{ijrst} \tag{20}$$

3. Expected number of customers in environment r of the offer zone

$$E[OZ(r)] = \sum_{i=0}^{\infty} \sum_{j=0}^{N} \sum_{s=1}^{M_r} \sum_{t=1}^{m} j x_{ijrst} \tag{21}$$

4. Probability of type-I loss

$$P[L_1] = \sum_{i=0}^{\infty} \sum_{r=1}^{n} \sum_{s=1}^{M_r} \sum_{t=1}^{m} (1 - \gamma) x_{iNrst} \tag{22}$$

5. Expected rate of type-I loss

$$ER[L_1] = \lambda_B P[L_1] \tag{23}$$

6. Expected rate at which type-B customers enter the main station from environment r of the offer zone

$$E[MS(r)] = \sum_{i=0}^{\infty} \sum_{j=0}^{N} \sum_{s=1}^{M_r} \sum_{t=1}^{m} j \mu_r \eta x_{ijrst} \tag{24}$$

for $r = 1, 2, ... n$

7. Expected rate at which type-B customers enter the main station after the service completion at the offer zone

$$E[OZMS] = \sum_{r=0}^{n} E[MS(r)] \tag{25}$$

8. Expected rate of type-II loss from environment r

$$ER[L_2(r)] = \sum_{i=0}^{\infty} \sum_{j=0}^{N} \sum_{s=1}^{M_r} \sum_{t=1}^{m} j\mu_r(1-\eta)x_{ijrst} \tag{26}$$

for $r = 1, 2, ...n$

9. Expected rate of type-II loss

$$ER[L_2] = \sum_{r=0}^{n} R[L_{E(r)}] \tag{27}$$

10. Fraction of time environment r operates

$$F(r) = \nu_r p_r / [\sum_{t=1}^{n} \nu_t p_t] \tag{28}$$

5   The Cost Function

For the cost analysis of the system we introduce the revenue and expenditure per customer as follows:

- Revenue R monetary units per unit time per customer undergoing service in the main station.
- Revenue R_r monetary units per unit time per customer undergoing service in the environment r of the offer zone where $r = 1, 2, ..., n$.
- Revenue R_d monetary units per type-A customer upon direct entry to main station.
- Operational cost of environment r per unit time, c_r monetary units.
- Holding cost h_r monetary units per customer in environment r

The Expected Total Profit (ETP) is given by

$$(\textbf{ETP}) = R * E[MS] + R_d * \lambda_A + (R_r) * \sum_{r=1}^{n} E[MS(r)]$$

$$- \sum_{r=1}^{n} [F_r * c_r] - \sum_{r=1}^{n} [E[OZ(r)] * h_r * F_r] \tag{29}$$

So the objective of the service providers or the operators of the system is to determine an optimal value of N for which the total expected profit (**ETP**) is maximum and to spot out the environment which contributes more customers to the main station under specified conditions.

6 Numerical Examples

6.1 Example-1

In example-1 we study the effect of parameters p and η on the expected number of customers in the main station. We assume that arrival to the main station and the offer zone occurs at exponential rates $p\lambda$ and $(1-p)\lambda$ respectively where $0 \leq p \leq 1$.

We consider the values of the variables as follows:

$$\lambda = 6, \mu = 8, \mu_1 = 10, \mu_2 = 12, \gamma = .5$$

$$M_1 = 2; M_2 = 2; n = 2$$

$$S_1 = \begin{pmatrix} -6 & 3 \\ 1 & -3 \end{pmatrix}; S_2 = \begin{pmatrix} -7 & 3 \\ 2 & -4 \end{pmatrix}$$

$$\beta_1 = [.5, .5]; \beta_2 = [.5, .5]$$

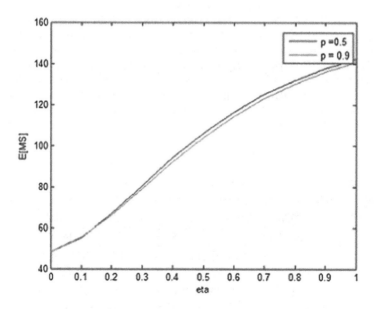

Fig. 1. Effect of η on $E[MS]$

From Tables 1 and 2, it is evident that for a given value of p, the number of customers in the main station increases with an increase in the value of η. Figure 1 shows that if the offers are designed in such a way to attract more customers to the main station, the number of customers continuing service at the main station can be increased. Figure 2 shows the combined effect of p and η on $E[MS]$.

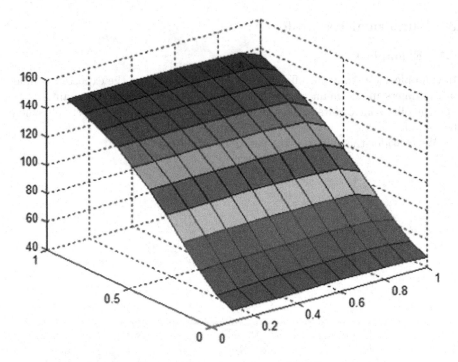

Fig. 2. Effect of p and η on $E[MS]$

Table 1. Effect of p and η on $E[MS]$

η	p				
	0.1	0.2	0.3	0.4	0.5
0	48.1111	48.1159	48.1538	48.2133	48.2832
0.1	53.5404	53.9409	54.3207	54.6705	54.978
0.2	65.3643	65.8186	66.2272	66.579	66.8581
0.3	79.5969	79.9629	80.2747	80.5199	80.6803
0.4	93.5508	93.8027	94.0012	94.1336	94.1811
0.5	105.9276	106.0867	106.197	106.247	106.2188
0.6	116.3643	116.4598	116.5131	116.514	116.446
0.7	124.9577	125.0135	125.0344	125.011	124.9288
0.8	131.9709	132.003	132.0077	131.976	131.895
0.9	137.6914	137.7093	137.7075	137.6767	137.6051
1	142.3762	142.3852	142.3823	142.3571	142.2987

6.2   Example-2

In example-2, we analyze the effect of N on various performance measures. We assume that arrival to the main station and the offer zone occurs at exponential

Table 2. Effect of p and η on $E[MS]$

η	p				
	0.6	0.7	0.8	0.9	1
0	48.3488	48.3886	48.3657	48.1928	47.1571
0.1	55.2253	55.3832	55.3959	55.118	53.0076
0.2	67.0402	67.0844	66.9083	66.2876	62.4921
0.3	80.7271	80.6106	80.2285	79.2927	73.9089
0.4	94.1136	93.8777	93.3619	92.244	85.755
0.5	106.0831	105.7876	105.2208	104.0545	97.0178
0.6	116.2823	115.9744	115.4174	114.2964	107.1701
0.7	124.7639	124.4729	123.9606	122.9367	116.032
0.8	131.7439	131.4846	131.0329	130.1277	123.6262
0.9	137.4749	137.253	136.8659	136.0825	130.0719
1	142.1916	142.0075	141.6827	141.0145	135.5218

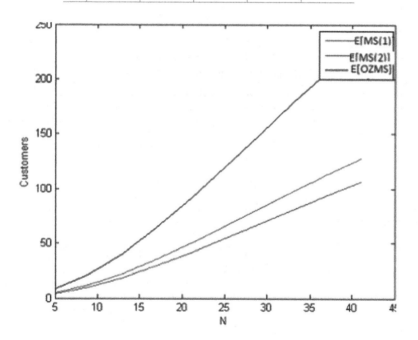

Fig. 3. Effect of N on the entry of type-B customers from the offer zone to the main station

rates $p\lambda$ and $(1-p)\lambda$ respectively where $0 \le p \le 1$. We fix the values of the variables as follows:

$$\lambda = 6, p = 0.3; \mu = 8, \mu_1 = 10, \mu_2 = 12, \gamma = 0.5$$

Table 3. Effect of capacity N

N	$E[MS]$	$E[OZ]$	$ER[L_1]$
5	58.7453	0.6671	0.0194
9	69.4407	1.7003	0.0142
13	78.0358	3.0920	0.0114
17	84.7510	4.7385	0.009
21	89.8669	6.5478	0.0081
25	93.6547	8.4447	0.0070
29	96.3461	10.3693	0.0062
33	98.1249	12.2737	0.0055
37	99.1312	14.1185	0.0049
41	99.4687	15.8698	0.0043

Table 4. Effect of capacity N

N	$E[MS(1)]$	$E[MS(2)]$	$E[OZMS]$
5	3.6707	4.4049	8.0756
9	9.8212	11.7855	21.6067
13	18.5529	22.2635	40.8163
17	29.2967	35.1561	64.4528
21	41.4415	49.7298	91.1714
25	54.4261	65.3113	119.7373
29	67.7752	81.3302	149.1054
33	81.1003	97.3203	178.4206
37	94.0833	112.9000	206.9833
41	106.4575	127.7489	234.2064

$$S_1 = \begin{pmatrix} -6 & 3 \\ 1 & -3 \end{pmatrix}; S_2 = \begin{pmatrix} -7 & 3 \\ 2 & -4 \end{pmatrix}$$

$\beta_1 = [0.5, 0.5]; \beta_2 = [0.4, 0.6]; p_1 = 0.5; p_2 = 0.5;$

$$M_1 = 2; M_2 = 2; n = 2$$

The data in Tables 3 and 4 shows the effect of N on various performance measures. Figure 3 shows that the number of type-B customers entering the main station from the offer zone increases with an increase in value of N. An optimum value of N can bedetermined based on the cost and revenues associated with the system.

7   Conclusion

The results in this paper may be extended to tandem queueing networks consisting of more than two service stations and to the case where the service time distributions are of so general say Phase type distributions. We plan to investigate such a general problem in future.

Acknowledgment. V.C Joshua and A. Krishnamoorthy thank the Department of Science and Technology, Government of India for the support given under the Indo-Russian Project $INT/RUS/RSF/P - 15$. A. Krishnamoorthy thanks the UGC India for the Award of Emeritus Fellowship $No.$ $F6 - 6/2017/ - 18/ EMERITUS - 2017 - 18 - GEN - 10822/(SA - II)$. Ambily P. Mathew thanks the UGC-India for the teacher fellowship sanctioned under the Faculty Development Programme [$F.No.$ $FIP/12^{th}plan/KLMG002TF06$.

References

1. Dallery, Y., Frein, Y.: On decomposition methods for tandem queueing networks with blocking. Oper. Res. **41**, 386–399 (1993)
2. Corral, A.G.: A tandem queue with blocking and Markovian arrival process. Queueing Syst. **41**, 343–370 (2002)
3. Corral, A.G., Martos, M.E.: Performance of two-stage tandem queues with blocking: the impact of several flows of signals. Perform. Eval. **63**(9–10), 910–938 (2006)
4. Kim, C.S., Klimenok, V., Tsarenkov, G., Breuer, L., Dudin, A.: The BMAP/G/1/PH/1/m tandem queue with feedback and losses. Perform. Eval. **64**(7–8), 802–818 (2007)
5. Klimenok, V., Breuer, L., Tsarenkov, G., Dudin, A.: The BMAP/G/1/PH/1/m tandem queue with losses. Perform. Eval. **61**(1), 17–40 (2005)
6. Klimenok, V., Dudin, A., Vishnevsky, V.: On the stationary distribution of tandem queue consisting of a finite number of stations. In: Kwiecień, A., Gaj, P., Stera, P. (eds.) CN 2012. CCIS, vol. 291, pp. 383–392. Springer, Heidelberg (2012). https://doi.org/10.1007/978-3-642-31217-5_40
7. Klimenok, V., Dudin, A., Vishnevsky, V.: Tandem queueing system with correlated input and cross-traffic. In: Kwiecień, A., Gaj, P., Stera, P. (eds.) CN 2013. CCIS, vol. 370, pp. 416–425. Springer, Heidelberg (2013). https://doi.org/10.1007/978-3-642-38865-1_42
8. Latouche, G., Neuts, M.F.: Efficient algorithmic solutions to exponential tandem queues with blocking. SIAM J. Algebr. Discrete Methods **1**, 93–106 (1980)
9. Latouche, G., Ramaswami, V.: Introduction to Matrix Analytic Methods in Stochastic Modeling, vol. 5. SIAM, Philadelphia (1999)
10. Neuts, M.F., Rao, B.M.: Numerical investigation of a multiserver retrial model. Queueing syst. **7**, 169–189 (1990)

Transient Analysis of a Repairable Single Server Queue with Working Vacations and System Disasters

M. I. G. Suranga Sampath[1]($\boxtimes$) and K. Kalidass[2]

[1] Library, University of Kelaniya, Kelaniya 11600, Sri Lanka
migsuranga@kln.ac.lk
[2] Department of Mathematics, Karpagam Academy of Higher Education,
Coimbatore 641021, Tamil Nadu, India
dassmaths@gmail.com

Abstract. This study investigates the repairable single server queue with working vacations and system disasters. The server allows to take a working vacation if there is no any customers in the system. There is a possibility of breakdowns happening in a system. When the system occurs server breakdowns, the server goes to the failure state and all customers in the queue are flushed away. The repairing process starts immediately, when the server comes to the failure state. The explicit expression for system size probabilities of the queueing system is derived in terms of the modified Bessel function of first kind using the probability generating function method, Laplace transform and continued fractions. Additionally, the mean and variance for number of jobs in the system at time t are derived as the performance measures. Finally, a numerical example is presented to study the behavior of the system.

Keywords: M/M/1 queue · Repairable server · Working vacations · System disasters

1 Introduction

Applications of queueing model with vacations exist in various fields such as network service, web service, file transfer service and mail service. Working vacation is one type of the vacation policies and Servi and Finn [19] introduced this concept generalizing the classical single server vacation model. They derived the explicit expressions for the mean, variance of the number of customers in the queue. In working vacation duration, the server serves the customers with a lower rate than the normal service rate. This may be a reason to reduce the leaving of customers from the system during the vacation period. Wu and Takagi [25] have derived the expressions for the number of jobs in the queue and the response time for an arbitrary customer extending the $M/M/1/WV$ model to new $M/G/1/WV$ model. A $GI/M/1$ queue with multiple working vacations was analyzed by Baba [3] to obtain the steady state result for the system size

© Springer Nature Switzerland AG 2019
V. M. Vishnevskiy et al. (Eds.): DCCN 2019, CCIS 1141, pp. 258–272, 2019.
https://doi.org/10.1007/978-3-030-36625-4_21

in the queue both at arrival and arbitrary epochs. Banik et al. [5] discussed the finite buffer single server $GI/M/1$ queue with multiple working vacations and they presented the distribution for number of customers in the system at pre-arrival and arbitrary epoch. Do [8] obtained time independent expression for the retrial $M/M/1$ queue with working vacations. Yang et al. [26] applied the matrix-analytic method to derive the steady-state probabilities and some system characteristics of the F -policy $M/M/1/K$ queueing system with working vacation. Baba [4] studied $M^X/M/1$ queue with multiple working vacation and he obtained the probability distribution for system size and some of the performance measures for the queueing system considering that the server is in its equilibrium state. Arivudainambi et al. [2] have derived stationary result for a single server retrial queue introducing the concept of single working vacation. The $M/G/1$ queue with working vacations has been analyzed by Aissani et al. [1] to derive the expressions for joint probability distribution of the server state and system size probabilities of the queue when the server is in steady-state by using the Laplace and z- transforms. Recently, Vijaya Laxmi and Rajesh [23] analyzed single sever queue with customer impatience and variant working vacation policy. They obtained the explicit expression for system size probabilities and some performance measures at steady state.

Gelenbe [9] introduced the notion of catastrophes and it has been gaining significant scholarly attention during the last few decades since their applications are widely used in service systems, computer systems, manufacturing systems and. Catastrophes occur at random time when server is going to complete the service for all the customers at that time or the server is waiting for a new arrival. This situation can be considered as negative customer arrival in queueing system and they have a property to remove all the customers or some of them in the queueing system. It may be possible to happen either from another service station or from outside the system. A mail server with an infected virus can be considered as an example for a queueing system with catastrophes. Since this email transmit the virus during its transferring to the other processors, disasters may occur to clear the operation of all emails stored in the system. Krishna Kumar and Arivudainambi [14] analyzed the transient solution for an $M/M/1$ queue with catastrophes. Chao [6] has extended the research which has been done by Di Crescenzo et al. [7] for the $M/M/1$ queue with catastrophes to a network of queues. An $M/M/R/N$ queueing system with balking, reneging and server break-downs was analyzed by Wang and Chang [24].

Queueing systems with repairable servers often arise in the field of computer and communication switching systems and web servicing systems where the processors have to handle failing and repairing of them [16]. Therefore, the studying of queues subjected to catastrophes and breakdowns and repairable servers has got more attention of the researchers. An $M/M/1$ queue which has N servers with server breakdowns and repairs has been analyzed by Neuts and Lucantoni [17]. A single server priority queue with server failures and queue flushing has been discussed by Towsley and Tripathi [22]. The transient solution for an $M/M/1$ queue subjected to catastrophes with server failure and non-zero repairable time has been derived by Krishna Kumar and Pavavi [15]. Giorno et al. [10] has obtained jump diffusion approximation for a double ended queue with catastrophes and repairs. Kalidass and et al. [13] derived the transient solution of an N-policy single server queueing system with catastrophes and repairable server. A single server queueing system with balking, catastrophes, server failures and repairs was analyzed by Tarabia [21] extending the model of Krishna Kumar and Pavai [15] with balking feature and he has obtained transient and steady state probabilities with the use of probability generating function technique and a direct approach.

Yechiali [27] has obtained the time independent probabilities of the system size of the queue with system breakdowns and customer impatience. Expanding this model, Sudesh [20] derived the transient solutions for the probabilities of number of customers in the system with the use of generating function methods and continued fractions. Considering an $M/M/1$ queue with working vacation and multiple types of server breakdowns, the distribution for number of jobs in the system was derived by Jain and Jain [12]. An $M/M/1$ queueing system with second optional service and unreliable server has been extensively researched. Using the matrix geometric technique, Jain and Chauhan [11] have analyzed a single server queue with unreliable server and second optional service. Dealing with a feedback retrial $M/G/1$ queue with multiple working vacations and vacation interruption, Rajadurai et al. [18] has obtained the time independent probabilities for the system size and some performance measures.

In existing literature, analyzing a repairable single server queue with working vacations and system disasters in transient state is less researched. Therefore, in this research, the transient solutions of an $M/M/1$ queue with working vacations and system disasters are obtained using Laplace transform, probability generating function technique and continued fractions. As the performance measures, mean and variance of the system size are explicitly expressed. The findings of this study is applicable in manufacturing systems, computer communication systems, network systems and inventory systems etc. Therefore, the results of this research may help people who use queueing theory to deal with congestion problems in the systems.

The different sections of this paper are arranged as follows. Section 2 includes the model for a repairable single server queue with working vacations and system disasters in transient state. The transient solution of system size probabilities is derived in Sect. 3. Section 4 presents the time dependent expected values. Numerical result is presented in Sect. 5. The paper concludes in Sect. 6.

2   Model Description

A single server queueing model with system failure and working vacations is considered. The assumptions of the system are build up as follows:

Arrivals are allowed to join the system according to a Poisson process with rate λ and service takes place according to an exponential distribution with rate μ. The server takes a working vacation when there are no customers in the system. Working vacation policy has an exponential distribution with mean $1/\gamma$ and the server serves the customers with service rate $\mu_v(< \mu)$ during the working vacation. The system faces server breakdowns at a Poisson rate η. When it suffers a server breakdown, all customers in the queue are flushed away and the server goes to the failure state. The repairing process is started immediately, when server comes to the failure state and the repair time has an exponential distribution with mean $1/\nu$. It is assumed that inter-arrival times, service times, repair times and vacation times are mutually independent and the service discipline is First-In, First-Out (FIFO).

Let $\{X(t), t \geq 0\}$ denotes the total number of customers in the system at time t and let $J(t)$ denotes the state of the system at time t, which is defined as follows:

$$J(t) = \begin{cases} 0, \text{ if the server is in failure state at time } t \\ 1, \text{ if the server is in functional state at time } t \\ 2, \text{ if the server is in working vacation at time } t \end{cases}$$

Then $\{J(t), X(t), t \geq 0\}$ is a two-dimensional continuous time Markov process on the state space $S = \{(j, n); j = 0, 1, 2; n = 0, 1, 2, ...\}$. Let $P_{j,n}(t)$ be the time dependent probabilities for the system to be in the state j with n customers at time t.

Then, the set of forward Kolmogorov differential difference equations governing the process is given by

$$P'_{0,0}(t) = -(\lambda + \nu)P_{0,0}(t) + \eta \sum_{n=1}^{\infty} (P_{1,n}(t) + P_{2,n}(t)) \tag{1}$$

$$P'_{0,n}(t) = \lambda P_{0,n-1}(t) - (\lambda + \nu)P_{0,n}(t), n \geq 1 \tag{2}$$

$$P'_{1,1}(t) = -(\lambda + \mu + \eta)P_{1,1}(t) + \mu P_{1,2}(t) + \nu P_{0,1}(t) + \gamma P_{2,1}(t) \tag{3}$$

$$P'_{1,n}(t) = \lambda P_{1,n-1}(t) - (\lambda + \mu + \eta)P_{1,n}(t) + \mu P_{1,n+1}(t)$$
$$+ \nu P_{0,n}(t) + \gamma P_{2,n}(t); n \geq 2 \tag{4}$$

$$P'_{2,0}(t) = -\lambda P_{2,0}(t) + \mu_v P_{2,1}(t) + \mu P_{1,1}(t) + \nu P_{0,0}(t) \tag{5}$$

$$P'_{2,n}(t) = \lambda P_{2,n-1}(t) - (\lambda + \mu_v + \eta + \gamma)P_{2,n}(t) + \mu_v P_{2,n+1}(t); n \geq 1 \tag{6}$$

Initially, it is assumed that $P_{0,0}(0) = 0$ and $P_{2,0}(0) = 1$ and $P_{j,n}(0) = 0$ for $n \geq 1$ and $j = 0, 1, 2$.

3 Transient Probabilities

3.1 Evaluation of $P_{1,n}(t)$

Multiplying the Eqs. (3) and (4) by appropriate powers of z and summing over $n \geq 1$ and using the definition of the generating function, we can obtain

$$P(z,t) = \nu \int_0^t \left(\sum_{m=1}^{\infty} P_{0,m}(u)z^m \right) e^{-\left[\lambda(1-z)+\mu(1-z^{-1})+\eta\right](t-u)} du$$

$$+ \gamma \int_0^t \left(\sum_{m=1}^{\infty} P_{2,m}(u)z^m \right) e^{-\left[\lambda(1-z)+\mu(1-z^{-1})+\eta\right](t-u)} du$$

$$- \mu \int_0^t P_{1,1}(u)e^{-\left[\lambda(1-z)+\mu(1-z^{-1})+\eta\right](t-u)} du \qquad (7)$$

It is well known that if $\alpha = 2\sqrt{\lambda\mu}$ and $\beta = \sqrt{\frac{\lambda}{\mu}}$, then $e^{\left(\lambda z+\frac{\mu}{z}\right)t} = \sum_{n=-\infty}^{\infty} (\beta z)^n I_n(\alpha t)$ where $I_n(\cdot)$ is the modified Bessel function of the first kind.

Substituting this equation to the Eq. (7) and after some algebra, we can obtain

$$P_{1,n}(t) = \nu \int_0^t \sum_{m=1}^{\infty} P_{0,m}(u)\beta^{n-m} \left[I_{n-m}\left(\alpha(t-u)\right) - I_{n+m}\left(\alpha(t-u)\right)\right] e^{-(\lambda+\mu+\eta)(t-u)} du$$

$$+ \gamma \int_0^t \sum_{m=1}^{\infty} P_{2,m}(u)\beta^{n-m} \left[I_{n-m}\left(\alpha(t-u)\right) - I_{n+m}\left(\alpha(t-u)\right)\right]$$

$$\times e^{-(\lambda+\mu+\eta)(t-u)} du \qquad (8)$$

3.2 Evaluation of $P_{0,n}(t)$

$\hat{P}_{j,n}(s)$ represents Laplace transform of $P_{j,n}(t)$. Taking the Laplace transform of the Eq. (2) and substituting the initial value and after some algebra, we have

$$\hat{P}_{0,n}(s) = \left(\frac{\lambda}{s+\lambda+\nu} \right)^n \hat{P}_{0,0}(s) \qquad (9)$$

We can obtain the following equation after taking the Laplace transform of the Eq. (1) and applying the initial condition

$$(s+\lambda+\nu)\hat{P}_{0,0}(s) = \eta \sum_{n=1}^{\infty} \left(\hat{P}_{1,n}(s) + \hat{P}_{2,n}(s) \right) \qquad (10)$$

Clearly for $t > 0$,

$$\sum_{n=0}^{\infty} P_{0,n}(t) + \sum_{n=1}^{\infty} P_{1,n}(t) + \sum_{n=0}^{\infty} P_{2,n}(t) = 1$$

The above equation can be expressed as follows after taking Laplace transform and some algebra

$$\sum_{n=1}^{\infty} \left(\hat{P}_{1,n}(s) + \hat{P}_{2,n}(s) \right) = \frac{1}{s} - \hat{P}_{2,0}(s) - \sum_{n=0}^{\infty} \hat{P}_{0,n}(s)$$

Substituting the above equation to the Eq. (10) and after some mathematical calculations, we are able to derive

$$\hat{P}_{0,0}(s) = \hat{A}(s) \left[\frac{1}{s} - \hat{P}_{2,0}(s) \right] \tag{11}$$

where

$$\hat{A}(s) = \frac{\eta}{(s + \lambda + \nu)} \sum_{k=0}^{\infty} (-1)^k \left(\frac{\eta}{s + \nu} \right)^k$$

We will have the following equation after taking inversion of the above equation,

$$P_{0,0}(t) = A(t) * [1 - P_{2,0}(t)]$$

where

$$A(t) = e^{-(\lambda + \nu)t} \sum_{k=0}^{\infty} (-1)^k \eta^{k+1} e^{-\nu t} \frac{t^{k-1}}{(k-1)!}$$

By the Eq. (9), we have

$$\hat{P}_{0,n}(s) = \hat{A}(s) \left[\frac{1}{s} - \hat{P}_{2,0}(s) \right] \left(\frac{\lambda}{s + \lambda + \nu} \right)^n$$

After taking inverse Laplace transform transform of the above equation, we have

$$P_{0,n}(t) = \lambda^n A(t) * (1 - P_{2,0}(t)) * e^{-(\lambda + \nu)t} \frac{t^{n-1}}{(n-1)!} \tag{12}$$

where "*" denotes the convolution. The terms for $P_{0,0}(t)$ and $P_{0,n}(t)$ are expressed in terms of $P_{2,0}(t)$ which is given by the Eq. (18).

3.3   Evaluation of $P_{2,n}(t)$

Taking Laplace transform the Eq. (6) and applying the initial condition to the above equation and after some algebra, we have

$$\frac{\hat{P}_{2,n}(s)}{\hat{P}_{2,n-1}(s)} = \frac{\lambda}{(s + \lambda + \mu_v + \eta + \gamma) - \mu_v \frac{\hat{P}_{2,n+1}(s)}{\hat{P}_{2,n}(s)}}$$

It can be rewritten as follows

$$\frac{\hat{P}_{2,n}(s)}{\hat{P}_{2,n-1}(s)} = \frac{\lambda}{(s + \lambda + \mu_v + \eta + \gamma) - \Phi(s)} \tag{13}$$

where

$$\Phi(s) = \frac{\lambda\mu_v}{(s + \lambda + \mu_v + \eta + \gamma) - \frac{\lambda\mu_v}{(s+\lambda+\mu_v+\eta+\gamma)-\cdots\cdots}} \tag{14}$$

Clearly $\Phi(s)$ satisfies the quadric equation $\Phi^2(s) - (s + \lambda + \mu_v + \eta + \gamma)\Phi(s) + \lambda\mu_v = 0$. This equation has two roots $\frac{P+\sqrt{P^2-4\lambda\mu_v}}{2}$ and $\frac{P-\sqrt{P^2-4\lambda\mu_v}}{2}$. Here, since $\left|\frac{P-\sqrt{P^2-4\lambda\mu_v}}{2}\right| < 1$, it is the real root of $\Phi(s)$. Where $P = s + \lambda + \mu_v + \eta + \gamma$. Substituting $\Phi(s)$ to the Eq. (14) and after some algebra, we will have

$$\hat{P}_{2,n}(s) = \left(\frac{2\lambda}{P + \sqrt{P^2 - \theta^2}}\right)^n \hat{P}_{2,0}(s) \tag{15}$$

where $\theta = 2\sqrt{\lambda\mu_v}$.

Taking the Laplace transform of the above equation, we can obtain

$$P_{2,n}(t) = \left(\frac{2}{\theta}\right)^{n-1} \lambda^n \left[I_{n-1}(\theta t) - I_{n+1}(\theta t)\right] e^{-(\lambda+\mu_v+\eta+\gamma)t} * P_{2,0}(t) \tag{16}$$

where "*" denotes the convolution.

3.4 Evaluation of $P_{2,0}(t)$

Taking the Laplace transform of the Eq. (5) and applying the initial condition and substituting the Eq. (15) for $n = 1$, we can derive

$$\hat{P}_{2,0}(s) = \sum_{j=0}^{\infty} \frac{(2\lambda\mu_v)^j}{(s + \lambda)^{j+1} \left(P + \sqrt{P^2 - \theta^2}\right)^j} \left[1 + \mu\hat{P}_{1,1}(s) + \nu\hat{P}_{0,0}(s)\right]$$

And again, substituting the Eq. (11) to the above equation, we have

$$\hat{P}_{2,0}(s) = \left(1 + \mu\hat{P}_{1,1}(s)\right) \hat{G}_i(s)\hat{B}(s) - \frac{\hat{G}_{i+1}(s)}{s} \tag{17}$$

where

$$\hat{B}(s) = \sum_{j=0}^{\infty} \frac{(2\lambda\mu_v)^j}{(s + \lambda)^{j+1} \left(P + \sqrt{P^2 - \theta^2}\right)^j}$$

and

$$\hat{G}_n(s) = \sum_{n=0}^{\infty} (-1)^n \nu^n \left[\hat{A}(s)\right]^n \left[\hat{B}(s)\right]^n$$

Inversion of the Eq. (18) yields,

$$P_{2,0}(t) = (1 + \mu P_{1,1}(t)) * G_i(t) * B(t) - \int_0^t G_{i+1}(u)du \qquad (18)$$

where

$$B(t) = \sum_{j=0}^{\infty} \left(\frac{2}{\theta}\right)^{j-1} (\lambda \mu_v)^j \, e^{-\lambda t} \frac{t^{j-1}}{(j-1)!} * [I_{m-1}(\theta t) - I_{m+1}(\theta t)] \, e^{-(\lambda + \mu_v + \eta + \gamma)t}$$

and

$$G_n(t) = \sum_{n=0}^{\infty} (-1)^n \nu^n \, [A(s)]^{*n} * [B(s)]^{*n}$$

where "*" denotes the convolution, while "*n" represents the n-fold convolution.

3.5   Evaluation of $P_{1,1}(t)$

Substituting $n = 1$ to the Eq. (8) and using the fact that $I_{-n}(\cdot) = I_n(\cdot)$, we can obtain

$$P_{1,1}(t) = \nu \int_0^t \sum_{m=1}^{\infty} P_{0,m}(u)\beta^{1-m} [I_{m-1}(\alpha(t-u)) - I_{m+1}(\alpha(t-u))] e^{-(\lambda+\mu+\eta)(t-u)} du$$

$$+ \gamma \int_0^t \sum_{m=1}^{\infty} P_{2,m}(u)\beta^{1-m} [I_{m-1}(\alpha(t-u)). - I_{m+1}(\alpha(t-u))]$$

$$\times e^{-(\lambda+\mu+\eta)(t-u)} du$$

Using the following Bessel identity $I_{m-1}(\alpha(t-u)) - I_{m+1}(\alpha(t-u)) = 2m\frac{I_m(\alpha(t-u))}{\alpha(t-u)}$ and taking the Laplace transform of the above equation and after some algebra, we have

$$\hat{P}_{1,1}(s) = 2\nu \sum_{m=1}^{\infty} \hat{P}_{0,m}(s)\frac{\beta^{1-m}}{\alpha^{m+1}} \left(P_1 - \sqrt{P_1^2 - \alpha^2}\right)^m$$

$$+ 2\gamma \sum_{m=1}^{\infty} \hat{P}_{2,m}(s)\frac{\beta^{1-m}}{\alpha^{m+1}} \left(P_1 - \sqrt{P_1^2 - \alpha^2}\right)^m$$

where $P_1 = \lambda + \mu + \eta$.

Again, substituting the Eq. (12) to above equation, we can obtain

$$\hat{P}_{1,1}(s) = \frac{\hat{A}(s)\hat{H}(s)}{s} + \left[\hat{K}(s) - \hat{A}(s)\hat{H}(s)\right] \hat{P}_{2,0}(s)$$

Finally, we can derive the following expression for $\hat{P}_{1,1}(t)$ substituting the Eq. (17) to the above equation and doing some mathematical calculations,

$$\hat{P}_{1,1}(s) = \left\{ \frac{\hat{A}(s)\hat{H}(s)}{s} + \left[\hat{K}(s) - \hat{A}(s)\hat{H}(s)\right] \left[\hat{G}_i(s)\hat{B}(s) - \frac{\hat{G}_{i+1}(s)}{s}\right] \right\}$$
$$\times \left\{ \sum_{r=1}^{\infty} \mu^r \left[\hat{G}_i(s)\right]^r \left[\hat{B}(s)\right]^r \sum_{m=0}^{\infty} (-1)^m \binom{r}{m} \left[\hat{K}(s)\right]^m \left[\hat{A}(s)\hat{H}(s)\right]^{r-m} \right\}$$

(19)

where

$$\hat{H}(s) = 2\nu \sum_{m=1}^{\infty} \left(\frac{\lambda}{s+\lambda+\nu}\right)^m \frac{\beta^{1-m}}{\alpha^{m+1}} \left(P_1 - \sqrt{P_1^2 - \alpha^2}\right)^m$$

and

$$\hat{K}(s) = 2\gamma \sum_{m=1}^{\infty} \left(\frac{2\lambda}{P + \sqrt{P^2 - \theta^2}}\right)^m \frac{\beta^{1-m}}{\alpha^{m+1}} \left(P_1 - \sqrt{P_1^2 - \alpha^2}\right)^m$$

Inversion of the Eq. (19) provides the following results

$$P_{1,1}(t) = \left\{ \int_0^t A(u) * H(u)du + [K(t) - A(t) * H(t)] * \left[G_i(t) * B(t) - \int_0^t G_{i+1}(u)du\right] \right\}$$
$$* \left\{ \sum_{r=1}^{\infty} \mu^r [G_i(t)]^{*r} * [B(t)]^{*r} * \sum_{m=0}^{\infty} (-1)^m \binom{r}{m} [K(t)]^{*m} \right.$$
$$\left. * [A(t) * H(t)]^{*(r-m)} \right\}$$

(20)

where

$$H(t) = \nu \sum_{m=1}^{\infty} \lambda^m \beta^{1-m} e^{-(\lambda+\nu)t} \frac{t^{m-1}}{(m-1)!} * [I_{m-1}(\alpha t) - I_{m+1}(\alpha t)] e^{-(\lambda+\mu+\eta)t},$$

$$K(t) = \gamma \sum_{m=1}^{\infty} \left(\frac{2}{\theta}\right)^{m-1} \lambda^m \beta^{1-m} [I_{m-1}(\theta t) - I_{m+1}(\theta t)] e^{-(\lambda+\mu_v+\eta+\gamma)t}$$
$$* [I_{m-1}(\alpha t) - I_{m+1}(\alpha t)] e^{-(\lambda+\mu+\eta)t}$$

where '*' denotes convolution while '*r', '*m' and '*(r − m)', represent r-fold convolution, m-fold convolution and '(r − m)' convolution respectively.

4   Time Dependent Mean and Variance

In this section, time dependent expected value and variance of the system size distribution are derived.

4.1 Mean

Let $X(t)$ denotes the number of jobs in the system at time t. The average number of jobs in the system at time t is given by

$$m(t) = E(X(t)) = \sum_{n=1}^{\infty} n \left(P_{0,n}(t) + P_{1,n}(t) + P_{2,n}(t) \right)$$

$$m(0) = \sum_{n=1}^{\infty} n \left(P_{0,n}(0) + P_{1,n}(0) + P_{2,0}(t) \right) = 0$$

$$m'(t) = \sum_{n=1}^{\infty} n \left(P'_{0,n}(t) + P'_{1,n}(t) + P'_{2,n}(t) \right)$$

By the Eqs. (2), (3), (4) and (6) and after some algebra, we have the following equation

$$m(t) = \lambda t - \eta \sum_{n=1}^{\infty} n \left[\int_0^t P_{1,n}(u)du + \int_0^t P_{2,n}(u)du \right]$$

$$- \mu \sum_{n=1}^{\infty} \int_0^t P_{1,n}(u)du - \mu_v \sum_{n=1}^{\infty} \int_0^t P_{2,n}(u)du \qquad (21)$$

where $P_{1,n}(t)$ and $P_{2,n}(t)$ are given by the Eqs. (8) and (16) respectively.

4.2 Variance

Let $X(t)$ denotes the number of jobs in the system at time t. The variance of jobs in the system at time t is given by

$$Var(X(t)) = E(X^2(t)) - [E(X(t))]^2 = k(t) - [m(t)]^2 \qquad (22)$$

where

$$k(t) = E(X^2(t)) = \sum_{n=1}^{\infty} n^2 \left(P_{0,n}(t) + P_{1,n}(t) + P_{2,n}(t) \right)$$

$$k(0) = \sum_{n=1}^{\infty} n^2 \left(P_{0,n}(0) + P_{1,n}(0) + P_{2,n}(0) \right) = 0$$

$$k'(t) = \sum_{n=1}^{\infty} n^2 \left(P'_{0,n}(t) + P'_{1,n}(t) + P'_{2,n}(t) \right)$$

By Eqs. (2), (3), (4) and (6) and after some algebra, we have the following equation

$$k(t) = 2\lambda \int_0^t m(u)du + \lambda t - \eta \sum_{n=1}^{\infty} n^2 \left[\int_0^t P_{1,n}(u)du + \int_0^t P_{2,n}(u)du \right]$$

$$- 2\mu \sum_{n=1}^{\infty} n \int_0^t P_{1,n}(u)du - 2\mu_v \sum_{n=1}^{\infty} n \int_0^t P_{2,n}(u)du$$

$$+ \mu \sum_{n=1}^{\infty} \int_0^t P_{1,n}(u)du + \mu_v \sum_{n=1}^{\infty} \int_0^t P_{2,n}(u)du$$

Substituting above equation into the Eq. (22), we will have

$$Var(X(t)) = 2\lambda \int_0^t m(u)du + \lambda t - \eta \sum_{n=1}^{\infty} n^2 \left[\int_0^t P_{1,n}(u)du + \int_0^t P_{2,n}(u)du \right]$$

$$- 2\mu \sum_{n=1}^{\infty} n \int_0^t P_{1,n}(u)du - 2\mu_v \sum_{n=1}^{\infty} n \int_0^t P_{2,n}(u)du$$

$$+ \mu \sum_{n=1}^{\infty} \int_0^t P_{1,n}(u)du + \mu_v \sum_{n=1}^{\infty} \int_0^t P_{2,n}(u)du - [m(t)]^2$$

where $P_{1,n}(t)$, $P_{2,n}(t)$ and $m(t)$ are given by the Eqs. (8), (16) and (21) respectively.

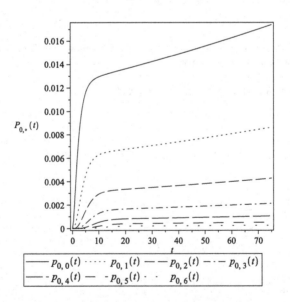

Fig. 1. Behaviour of the $P_{0,n}(t)$ against t for varying values of n

5   Numerical Illustrations

The numerical examples which illustrate the functioning of concerned model in transient state are presented in this section. Even though, this queuing system has infinite capacity, system size is limited to 25 considering the purpose of numerical solutions.

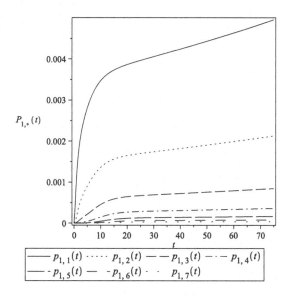

Fig. 2. Behaviour of the $P_{1,n}(t)$ against t for varying values of n

Figure 1 illustrates the behaviour of $P_{0,n}(t)$ against time t for varying values of n with parameters $\mu = 1.5$, $\gamma_1 = 0.03$, $\gamma_2 = 0.05$, $\xi = 0.01$ and $\lambda = 0.3$. It can be noticed that all the probabilities tend to settle at steady-state when time progresses.

Figure 2 is plotted to present the behaviour of $P_{1,n}(t)$ against time t for varying values of n with same parameter values. The values of $P_{1,n}(t)$ start at 0 and they reach to steady state when time progresses.

Figure 3 shows the behaviour of $P_{2,n}(t)$ against time t for varying values of n with same parameter values. Probabilities $P_{2,n}(t)$ become the steady-state when time progresses.

Figure 4 shows the variation of the mean number of the jobs in the system against time t for all values of λ (0.3, 0.5, 0.75, 1, 1.5). The expected system size increases with time t. When arrival rate is increased, expected number of customers in the queue increases. Figure 5 is plotted to explain the behaviour of the variance of the system size against time t with same parameter values for λ. When arrival rate λ is increased, the variance of the number of jobs in the queue also increases.

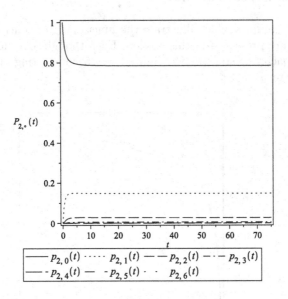

Fig. 3. Behaviour of the $P_{2,n}(t)$ against t for varying values of n

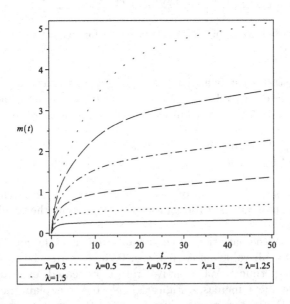

Fig. 4. Behaviour of the mean against t for varying values of λ

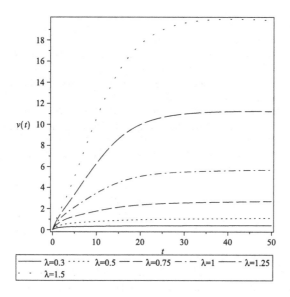

Fig. 5. Behaviour of the variance against t for varying values of λ

6   Conclusions

A repairable single server queue with working vacations and system disasters is considered in transient regime and the explicit expression for system size probabilities of the queueing system are derived in terms of the modified Bessel function of first kind. Probability generating function method, Laplace transform and continued fractions are used to derive the transient solution. Additionally, the mean and variance for number of jobs in the system at time t are derived as the performance measures. Finally, a numerical example is given to study the behavior of the system.

References

1. Aissani, A., Taleb, S., Kernane, T., Saidi, G., Hamadouche, D.: An M/G/1 retrial queue with working vacation. Adv. Intell. Syst. Comput. **240**, 443–452 (2014)
2. Arivudainambi, D., Godhandaraman, P., Rajadurai, P.: Performance analysis of a single server retrial queue with working vacation. Opsearch **51**(3), 434–462 (2014)
3. Baba, Y.: Analysis of a $GI/M/1$ queue with multiple working vacations. Oper. Res. Lett. **33**, 201–209 (2005)
4. Baba, Y.: The $M^x/M/1$ queue with multiple working vacation. Am. J. Oper. Res. **2**, 217–223 (2012)
5. Banik, A.D., Gupta, U.C., Pathak, S.S.: On the $GI/M/1/N$ queue with multiple working vacations-analytic analysis and computation. Appl. Math. Model. **31**, 1701–1710 (2007)
6. Chao, X.: A queuing network model with catastrophes and its product form solution. Oper. Res. Lett. **18**, 75–79 (1995)

7. Di Crescenzo, A., Giorno, B., Nobile, A.G.: On the M/M/1 queue with catastrophes and its continuous approximation. Queueing Syst. **43**, 329–347 (2003)
8. Do, T.V.: M/M/1 retrial queue with working vacations. Acta Informatica **47**, 47–65 (2010)
9. Gelenbe, E.: Product form queueing networks with positive and negative customers. J. Appl. Probab. **28**(3), 656–663 (1991)
10. Giorno, B., Krishna Kumar, B., Nobile, A.G.: A double-ended queue with catastrophes and repairs, and a jump-diffusion approximation. Methodol. Comput. Appl. Probab. **14**, 937–954 (2012)
11. Jain, M., Chauhan, D.: Working vacation queue with second optional service and unreliable server. IJE Trans. C **25**(3), 223–230 (2012)
12. Jain, M., Jain, A.: Working vacations queueing model with multiple types of server breakdowns. Appl. Math. Model. **34**, 1–13 (2010)
13. Kalidass, K., Gopinath, S., Gnanaraj, J., Ramanath, K.: Time dependent analysis of an M/M/1/N queue with catastrophes and a repairable server. Opsearch **49**, 39–61 (2012)
14. Krishna Kumar, B., Arivudainambi, D.: Transient solution of an M/M/1 queue with catastrophes. Comput. Math. Appl. **10**, 1233–1240 (2000)
15. Krishna Kumar, B., Pavai Madheswari, S.: Transient analysis of an M/M/1 queue subject to catastrophes and server failures. Stoch. Anal. Appl. **23**, 329–340 (2005)
16. Nagarajan, R., Kurose, J.: Grid On defining, computing and quaranteeing quality-of-service in high-speed networks. In: Proceedings of INFOCOM 1992, Florence, Italy, pp. 2016–2025 (1992)
17. Neuts, M.F., Lucantoni, D.M.: A Markovian queue with N servers subjects to breakdowns and repairs. Manag. Sci. **25**, 849–861 (1979)
18. Rajadurai, P., Saravanarajan, M.C., Chandrasekaran, V.M.: A study on M/G/1 feedback retrial queue with subject to server breakdown and repair under multiple working vacation policy. Alex. Eng. J. **2017**(00), 1–16 (2017)
19. Servi, L.D., Finn, S.G.: M/M/1 queues with working vacations. Perform. Eval. **50**, 41–52 (2002)
20. Sudhesh, R.: Transient analysis of a queue with system disasters and customer impatience. Queueing Syst. **66**, 95–105 (2010)
21. Tarabia, A.M.K.: Transient and steady-state analysis of an M/M/1 queue with balking, catastrophes, server failures and repairs. J. Ind. Manag. Optim. **7**, 811–823 (2011)
22. Towsley, D., Tripathi, S.K.: A single server priority queue with server failures and queue flushing. Oper. Res. Lett. **10**, 353–362 (1991)
23. Vijaya Laxmi, P., Rajesh, P.: Analysis of variant working vacations queue with customer impatience. Int. J. Manag. Sci. Eng. Manag. **12**(3), 186–195 (2016)
24. Wang, K.H., Chang, Y.C.: Cost analysis of a finite M/M/R queueing system with balking, reneging and server breakdowns. Math. Methods Oper. Res. **56**, 169–180 (2002)
25. Wu, D., Takagi, H.: M/G/1 queue with multiple working vacations. Perform. Eval. **63**, 654–681 (2006)
26. Yang, D.Y., Wang, K.H., Wua, C.H.: Optimization and sensitivity analysis of controlling arrivals in the queueing system with single working vacation. J. Comput. Appl. Math. **234**, 545–556 (2010)
27. Yechiali, U.: Queues with system disasters and impatient customers when system is down. Queueing Syst. **56**, 811–823 (2007)

A Queueing Inventory System with Search and Match - An Organ Transplantation Model

T. S. Sinu Lal$^{(\boxtimes)}$, A. Krishnamoorthy, and V. C. Joshua

Department of Mathematics, CMS College, Kottayam 686001, Kerala, India
{sinulal,krishnamoorthy,vcjoshua}@cmscollege.ac.in
http://www.cmscollege.ac.in

Abstract. In the current study we analyze a queueing inventory system with a special reference to the allocation of organs for transplantation. The model is composed of an inventory of organs and queue of patients who wait for transplantation. The model captures the unexpected arrival of organs and their highly perishable nature. Reneging of patients due to death is considered. We design individual search mechanisms for each of the organs to find an optimal match from the waiting line in the minimum time. The system is modelled as a level dependent quasi birth-death process and steady state distribution is obtained using matrix analytic methods. Performance characteristics are obtained and numerical illustration is provided.

Keywords: Search · Perishing · Reneging · Markovian arrival process · Matrix analytic methods

1 Introduction

Queueing models are widely applicable in health care and management for developing modelling tools that effectively meet the system requirements. Organ transplantation systems have already been widely studied in queueing literature. Organs are highly perishable items and their availability is very rare. The main concern while designing such models is on the policy for allocation of organs to suitable patients without time delay. Delay causes perishing of organs and reneging of patients. There are unexpected events of organ arrival. One of the main examples is the availability of a heart, which occurs possibly due to a brain death. Such unexpected arrivals at times complicate finding the process of finding a match. Other major constraints in finding a perfect match for an organ include physical fitness of patients, financial status, geographical factors etc.

This paper studies an organ transplantation system as a queueing inventory model with search for finding a matching patient for an inventoried organ. Search is carried out for each of the organs arrived. The queue considered is the waiting list of patients who wait for a matching organ. The inventory considered is the storage system for organs arriving at the transplantation system. The organs are subjected to perishing and the patients may renege due to death. The current

V. M. Vishnevskiy et al. (Eds.): DCCN 2019, CCIS 1141, pp. 273–287, 2019.
https://doi.org/10.1007/978-3-030-36625-4_22

study is motivated by the requirement of an efficient mechanism for finding a perfect match for an organ as soon as it has arrived. The search selects a candidate from the queue and starts only if the queue of customers is non-empty. There are possibilities for two events after the commencement of a search, which are perishing of an organ or finding a match. The search terminates only if either of these two events happens. The main feature of this model is the simultaneous running of multiple independent search mechanisms, which are meant to accelerate the process of finding a match for an organ.

Zenois [31] considers an organ allocation model with random allocation policy implemented on a system consisting of several classes of organs and several classes of patients with reneging. In this work policies for reducing the waiting times of various classes of patients are described. In [9] Boxma, David, Perry and Stadje describe an organ transplantation model as a double matching queue in which there are two connected queueing systems of the FCFS type and the system allocates a patient on his arrival epoch the oldest organ kept in storage. Also, abandonment of patients and outdating of organs are considered. Bar-Lev, Boxma, Mathijsen and Perry [4] studied a stochastic model for a blood bank. Blood is a highly perishable item and demand for blood is also impatient. For a range of perishability functions and demand impatient functions the steady state distributions of the amount of blood and its demand are obtained in this work. Bendersky and David [5] proposes a model that provides an analytical tool to aid kidney patients in deciding kidney offer admissibility based on continuous time probabilistic dynamic programming. Drekic, Stanford, Woolford and McAlister [13] analyze a self-promoting priority queueing model for patient waiting times that is dependent of the variations in health status of patients. The system is modelled as a level-dependent -quasi -birth-death process and steady state distribution is obtained using matrix analytic methods and model is calibrated with real life data.

Pearlman, Elalouf and Yechiali [30] presents a dynamic flexible resource allocation problem of random stream of resources and a random stream of arriving objects with applications to kidney cross transplantation. This paper considers two types of arriving resources waiting in separate queues and upon arrival, each resource unit is matched with a waiting object and prime objective of study is finding an optimal state dependent allocation policy. Elalouf, Pearlman and Yechiali [1] present a double ended queueing model for dynamic allocation of live organs based on a best fit criterion. This paper reveals a policy of assigning a reward at each level of fit such that higher rewards are attributed to transplants between better matched organs and candidates. In [28] Su and Zenois consider an M/M/1 queue with homogeneous patients and exponential reneging is considered. Quality of transplant organ is traced by associating each service instant a reward and the effect of interaction between patient's choice and queueing discipline is demonstrated. Berstimas, Farias and Trichakis [6] proposes a method for allocation of kidneys based on a point system that ranks patients according to some priority criteria like waiting time, medical emergency etc through a data driven analysis. David and Yechiali [11] describe a time dependent stopping problem and its applications to decision making process associated with transplantation of live organs. The main motivation behind this problem is the effect of histocompatibility between the recipient and the donor. The geographical disparities in access to deceased donor kidneys is studied by Ata, Skaro and Tayu [3] and an operational solution that offers affordable jet services to the patients, allowing them to be included them in multiple donation service areas is proposed.

The concept of search was introduced in the classical queueing context by Neuts and Ramalhoto [21]. Chakravarthy, Krishnamoorthy and Joshua [10] studied a multi server retrial queueing model with search having a negligible search time. Dudin, Deepak, Joshua, Krishnamoorthy and Vishnevsky [14] analyzed a BMAP/G/1 retrial queue with two types of searches from an orbit and the two types have differently distributed search times. Krishnamoorthy, Deepak and Joshua [17] have studied an M/G/1 retrial queue with a probabilistic search with negligible search time. Deepak, Dudin, Joshua and Krishnamoorthy [12] carried out a study on a single server retrial queueing model with two different search mechanisms for bringing customers from an orbit. In this model, one type of search selects a single customer while the other type selects a batch of customers from the orbit. Krishnamoorthy, Joshua and Mathew [19] describe a retrial queue with search for priority customers, the search is meant to minimize the customer loss due to abandonment. A tandem retrial queueing model is investigated by Mathew, Krishnamoorthy and Joshua [20], in which search is turned on when the number of customers present in a service station falls below a preassigned number. Useful descriptions of search mechanisms can also be found in [2,18].

Search mechanisms can effectively be utilized in organ transplantation models so as to reduce the delay in finding a perfect match for an arrived organ. In this paper we use search in its most generalized version to tackle the key factors that affect organ matching. The factors may be biological, financial or even geographical. The main challenge in managing organ inventory systems are the unexpected arrivals of organs and their high rate of perishability. Efficient mechanisms are called for to ensure an appropriate candidate before perishing. Even a biologically matching candidate is likely to refuse organ due to some reasons such as financial constraints or difficulties in accessing the service area on time. In this work, by search we mean an optimal mechanism that concerns all these factors in finding an optimal match for an organ before its decay. One of the main advantage in designing such a mechanism is that the rate of search can be increased with the increase in perishing rate of organs and hence number of perishing may be reduced.

Markovian arrival process (MAP) is a more generalized version of Poisson process and can effectively be used to capture the correlation and variation in arrivals. MAP was introduced by Neuts [27] and was extended by Lucantony [23] to address group arrivals. MAP is a wide class of stochastic counting process and include arrival processes like Markov modulated arrival process, phase type renewal process, super position of these processes etc. MAP can be used to approximate any point process defined on the non-negative real line as it is a dense class in the space of all point processes on $[0,\infty)$. Exponential distribution has many characteristics that help to perfectly model service times and is also well tractable. But the assumptions of exponential distribution are highly restrictive in nature. The inadequacies of the exponential distribution are overcome with the introduction of phase type distribution by Neuts [26]. A phase type distribution is defined as the distribution of the time until absorption in a Markov process with a finite state space and a single absorption state defined over nonnegative real line. A phase type distribution with transient states $\{1, 2, \ldots, n\}$ and an absorbing state $n + 1$ is represented by a two tuple of the form $(\alpha, \mathcal{T})$, where α is the probability vector of length n according to which the process selects the initial state from $\{1, 2, \ldots, n\}$ and $\mathcal{T}$ is an $n \times n$ matrix such that $\begin{pmatrix} \mathcal{T} & \mathcal{T}^0 \\ 0 & 0 \end{pmatrix}$ generates the process, given the column vector $\mathcal{T}^0$ satisfies the condition $\mathcal{T}\mathbf{e} + \mathcal{T}^0 = \mathbf{0}$. $(\alpha, \mathcal{T})$ is called the representation of the phase type distribution. The distribution F of time until the chain gets absorbed into the state $n+1$ is given by $F(x) = 1 - \alpha e^{(\mathcal{T}x)}e$, $x \geq 0$. The set of all phase type distributions is a dense subset of the set of all distributions on the non-negative real line and hence it is a best tool to approximate any arbitrary distribution in this set. For more descriptions on phase type distributions see [15]. Erlang and Coaxian distributions are special cases of phase type distributions and for more details refer [8,16]. The system studied in this paper is modelled as a level dependent quasi birth-death (LDQBD) process. For details of LDQBD see [7,24]. The system process is analyzed and steady state probability distribution is obtained using matrix analytic methods. For more about matrix analytic methods see [22,25].

The paper is organized as follows. Section 2 describes the mathematical model and its analysis. A sufficient condition for the stability of the system under consideration is also given in this section. In Sect. 3 the steady state distribution of the stochastic process governing the system is obtained. Section 4 includes key performance characteristics of the system and Sect. 5 provides a numerical illustration of the model. Section 6 concludes the study.

2   Mathematical Model

We consider a system consisting of an inventory of organs and a queue of customers. Customers in the queue are the patients who are waiting for a matching organ. The customers arrive to the system according to a MAP. The MAP is governed by an underlying stochastic process $\varphi_t, t \geq 0$, which is an irreducible continuous time Markov chain on the state space $\{1, 2 \ldots, a\}$. The transition rates of the process φ_t are described by the square matrices D_0 and D_1 of size a. The matrix D_0 corresponds to the chain transitions without generating any arrival whereas D_1 corresponds to transitions generating an arrival of a single customer. The matrix $D = D_0 + D_1$ is the infinitesimal generator of the process $\varphi_t, t \geq 0$. The stationary state distribution σ of this chain is the unique solution to the system $\sigma D = \mathbf{0}$ and $\sigma e = 1$, where $\mathbf{0}$ is a zero row vector and $\mathbf{e}$ is the column vector of 1's having appropriate dimension. The fundamental arrival rate of the MAP ν is given by $\nu = \sigma D_1 e$.

Organs arrive to the system according to a Poisson process with parameter λ. Upon arrival of an organ, a random clock starts ticking and realizes at the moment when the organ perishes. The time duration for the realization of the clock follows an Erlang distribution of order r with density function.

$$f(t) = \frac{r\mu(r\mu t)^{r-1}e^{-r\mu t}}{(r-1)!}, t \geq 0, \mu > 0$$

The perishing time is assumed to be consisting of r independent exponential stages, each stage having the mean $\frac{1}{r\mu}$. An organ is perished when the corresponding perishing clock completes all the r stages of perishing process. When the inventory is non empty and there is at least one customer in the queue, searches are started independently for each of the organs. A search will be continued until the organ perishes or a match is found. The search times are independent and identically distributed according to a Coaxian distribution of order m having phase type representation (α, C), where $\alpha = (1, 0, \ldots, 0)$ is the initial probability distribution and C is matrix giving the transition rates among the m phases. C takes the form

$$C = \begin{bmatrix} -\rho(1) & q_1\rho(1) & & & \\ & -\rho(2) & q_2\rho(2) & & \\ & & \ddots & \ddots & \\ & & & \rho(m-1) & q_{m-1}\rho(r-1) \\ & & & & -\rho(m) \end{bmatrix}.$$

where $\rho(i), 1 \leq i \leq r-1$ is the rate of transition from the phase i to $i+1$ and the process moves from the phase i to $i+1$ with probability q_i or enters the absorption state with the complimentary probability $1 - q_i$ provided the initial phase is 1. A customer in the queue may renege after a random duration. The reneging times are exponentially distributed with parameter δ.

Various notations used in the description are collectively given below for easy understanding of the coming sections.

- λ = Rate of arrival of organs.
- δ = Rate of reneging of customers.
- S = Maximum inventory level.
- r and m respectively represents the order of the Erlang and Coaxian distributions.

The descriptors of the system at time t and are as follows.

- N(t) = Number of customers in the queue.
- i(t) = Number of organs in the inventory.
- $v_{i(t)} = (l_{i(t)}, s_{i(t)})$ if both inventory and queue are non empty.
- $v_{i(t)} = l_{i(t)}$ if queue is empty.
- $s_{i(t)}$ = Phase of the perishing clock.
- $l_{i(t)}$ = Phase of search.

- $a(t)$ = Phase of MAP.
- $diag(B_1, B_2, \ldots, B_n)$ is the block diagonal matrix with blocks $B_1, B_2, \ldots, B_n$.
- 0_k is a zero square matrix of order k and 0_{pq} is the zero matrix of order $p \times q$.

The process of the system is modelled as an irreducible continuous time Markov chain $\{\mathcal{X}(t), t \in \mathbb{R}_+\}$, where

$$\mathcal{X}(t) = (N(t), i(t), v_{i(t)}, v_{i-1(t)}, \ldots, v_{1(t)}, a(t))$$

The process $\mathcal{X}(t)$ is a level LDQBD process on the state space

$$\mathcal{S} = \bigcup_{k \geq 0} \mathcal{L}(k)$$

where $L(k)$ stands for the kth level and are describes as follows.

$$\mathcal{L}(0) = \{(0, j, u_1, u_2, \ldots, u_j, w) / 1 \leq j \leq S, 1 \leq u_1, \ldots, u_j \leq r, 1 \leq w \leq a\} \cup \{(0, 0, j) / 1 \leq j \leq a\}$$

$$\mathcal{L}(k) = \mathcal{L}_1(k) \cup \mathcal{L}_2(k), k \geq 1 \text{ where } \mathcal{L}_1(k) = \{(k, 0, j) / 1 \leq j \leq a\} \text{ and}$$

$$\mathcal{L}_2(k) = \{(k, j, v_j, v_{j-1}, \ldots, v_1, w) / k \in \mathbb{N}, 1 \leq h \leq S, v_h = (l_h, s_h), 1 \leq l_h \leq m, 1 \leq s_h \leq r, 1 \leq w \leq a\}$$

The number of states in each level is given as $|\mathcal{L}(0)| = a\frac{r^{S+1}}{r-1}$ and $|\mathcal{L}(k)| = a\frac{(rm)^{S+1}}{rm-1}, k \geq 1$.

The infinitesimal generator $\mathcal{Q}$ of the process is

$$\mathcal{Q} = \begin{bmatrix} \mathcal{A}_{00} & \mathcal{A}_{01} \\ \mathcal{A}_{10} & \mathcal{A}_{11} & \mathcal{A}_0 \\ & \mathcal{A}_{20} & \mathcal{A}_{21} & \mathcal{A}_0 \\ & & \ddots & \ddots & \ddots \\ & & & & \ddots & \ddots & \ddots \end{bmatrix}.$$

The entries of $\mathcal{Q}$ are described in detail below.

$$\mathcal{A}_{00} = \begin{bmatrix} \Psi_0 & \Phi_1 \\ \Omega_1 & \Psi_1 & \Phi_2 \\ & \Omega_2 & \ddots & \ddots \\ & & \ddots & \ddots & \Phi_S \\ & & & \Omega_S & \Psi_S \end{bmatrix}$$

$\Psi_0 = D_0 - \lambda I_a$. For $1 \leq i \leq S$

$$\Psi_i = \begin{bmatrix} D_0 & \mu I_a \\ & D_0 & \mu I_a \\ & & \ddots & \ddots \\ & & & \ddots & \mu I_a \\ & & & & D_0 \end{bmatrix} - diag((\lambda + \mu)I_{k_1}, (\lambda + 2\mu)I_{k_2})$$

where $k_1 = r^2 - 1$ and $k_2 = (rm)^i - r^2 + 1$.
$$\Phi_1 = \left(\lambda I_a \, e_{r-1}^T \otimes 0_a \right), \Phi_{i+1} = \left(\lambda I_a \, e_{r-1}^T \otimes 0_a \right) \text{ for } 1 \le i \le S - 1.$$

$$\Omega_1 = \begin{pmatrix} e_{r-1} \otimes 0_a \\ \mu I_a \end{pmatrix}, \Omega_j = \begin{bmatrix} V_{11} & V_{12} & \dots & V_{1r^{j-1}} \\ V_{21} & V_{22} & \dots & V_{1r^{j-1}} \\ \vdots & & & \\ V_{r^J 1} & V_{r^J 2} & \dots & V_{r^J r^{j-1}} \end{bmatrix} \quad 2 \le j \le S$$

For $1 \le l \le r^j$ and $1 \le k \le r^{j-1}$, each V_{lk} is a square matrix of order a. Elements of V_{lk} are specified as follows: As the states of $\mathcal{A}_{00}$ corresponds to the empty queue, search is not initiated hence the corresponding components $v_{j(t)} = l_{j(t)}, 1 \le j \le S$. The transitions $(0, j, l_j \dots, l_1, h) \to (0, j, l_{j-1} \dots, l_1, h)$ correspond to an organ perishing with rate μ. For $1 \le m, n \le a$

$$V_{lk}(m, n) = \begin{cases} \mu, \text{ If } (0, j, l_j \dots, l_1, h) \to (0, j, l_{j-1} \dots, l_1, h) \\ 0, \text{ Otherwise} \end{cases}$$

$$\mathcal{A}_{01} = diag(\Lambda_1, \Lambda_2 \dots, \Lambda_S)$$

$$\Lambda_1 = D_1, \Lambda_i = \begin{bmatrix} U_{11} & U_{12} & \dots & U_{r^i (rm)^i} \\ U_{21} & U_{22} & \dots & U_{r^i (rm)^i} \\ \vdots & & & \\ U_{r^i 1} & U_{r^i 2} & \dots & U_{r^i (rm)^i} \end{bmatrix} \quad 1 \le i \le S.$$

For $1 \le h \le r^i$ and $1 \le k \le (rm)^i$, U_{hk} is a square matrix of order a. Elements of U_{lk} corresponds to transitions that result in an arrival of a patient when the queue is empty. These transitions are generally represented as $(0, i, l_i, l_{i-1}, \dots, l_1, a(t_1)) \to (1, i, (l_i, s_i), \dots, (l_j, s_j), \dots, (l_1, s_1), a(t_2))$. Now for $1 \le m, n \le a$,

$$U_{lk}(m, n) = \begin{cases} D_1(m, n), \text{ if } (0, i, l_i, l_{i-1}, \dots, l_1, m) \to (1, i, (l_i, 1), \dots, (l_1, 1), n) \\ 0 \qquad\qquad \text{Otherwise} \end{cases}$$

$$\mathcal{A}_{10} = \begin{bmatrix} \Gamma_0 & & & & \\ \Gamma_1^* & \Gamma_1 & & & \\ & \Gamma_2^* & \ddots & & \\ & & \ddots & \ddots & \\ & & & \Gamma_S^* & \Gamma_S \end{bmatrix}.$$

where $\Gamma_0 = \delta I_0$, $\Gamma_1 = e_S \otimes (\delta I_{ar})$. For $2 \le i \le S$, Γ_i are defined as follows:

$$\Gamma_i = \begin{bmatrix} e_m \otimes (I_r \otimes (e_m \otimes (\delta I_a))\mathbf{0} \dots \mathbf{0}) \\ e_m \otimes (\mathbf{0}I_r \otimes (e_m \otimes (\delta I_a))\mathbf{0} \dots \mathbf{0}) \\ \vdots \\ e_m \otimes (\mathbf{0} \dots \mathbf{0}(I_r \otimes (e_m \otimes (\delta I_a)))) \end{bmatrix}.$$

In Γ_i, **0** are zero matrices of appropriate dimension.

$$\Gamma_1^* = e_m \otimes \begin{bmatrix} q_1\rho(1) \\ q_2\rho(2) \\ \vdots \\ q_{r-1}\rho(r-1) \\ \rho(r) \end{bmatrix} \otimes I_a, \Gamma_i^* = e_m \otimes \begin{bmatrix} q_1\rho(1) \\ q_2\rho(2) \\ \vdots \\ q_{r-1}\rho(r-1) \\ \rho(r) \end{bmatrix} \otimes I_{ar}$$

$$\mathcal{A}_{11} = A^* - diag(0_a, I_{m^s r^{s-1}} \otimes diag((\rho_1, \rho_2, \ldots, \rho_r) \otimes I_a) - (\mu + \delta)I_T$$

where $T = \frac{(rm)^{S+1}-1}{rm-1}$.

$$A^* = \begin{bmatrix} B_0 & C_0 & & & \\ E_1 & B_1 & C_1 & & \\ & E_2 & \ddots & & \\ & & \ddots & \ddots & C_{S-1} \\ & & & E_S & B_S \end{bmatrix}.$$

where

$$B_0 = D_0, B_j = \begin{bmatrix} B_1^* & F_1 & & & \\ & B_2^* & F_2 & & \\ & & \ddots & & \\ & & & \ddots & F_{r-1} \\ & & & & B_r^* \end{bmatrix} \quad 1 \le j \le S.$$

$$B_i^* = \begin{bmatrix} D_0 & p_1\rho(1)I_a & & & \\ & D_0 & p_2\rho(2)I_a & & \\ & & \ddots & & \\ & & & \ddots & p_{r-1}\rho(r-1)I_a \\ & & & & D_0 \end{bmatrix} \quad 1 \le i \le r.$$

$$F_i = \mu I_{am}, 1 \le j \le r-1$$

$$C_0 = \begin{bmatrix} \lambda I_a & 0_a & 0_a & \ldots & 0_a \end{bmatrix}.$$

For $1 \le i \le S-1$, $C_i = \begin{bmatrix} \lambda I_{m^*} & 0_{m^*} & 0_{m^*} & \ldots & 0_{m^*} \end{bmatrix}$ where $m^* = (rm)^i a$.

$$E_1 = \begin{bmatrix} 0_a & \ldots & 0_a & e_m \end{bmatrix} \otimes I_a, E_i = \begin{bmatrix} E_{11} \\ E_{12} \\ \vdots \\ E_{1S} \end{bmatrix} \quad 1 \le i \le S$$

$$E_{1i}(u,v) = \begin{cases} \mu, \text{ if } (1, i, (l_i, s_i), \ldots, (l_1, s_1), n) \to (1, i-1, (l_{i-1}, s_{i-1}), \ldots, (l_1, 1), n) \\ 0, \text{ Otherwise} \end{cases}$$

where $1 \leq n \leq a$. In the state $(1, i, (l_i, s_i), \ldots, (l_1, s_1), n)$, there are i organs in the inventory. When one organ perishes the state $(1, i, (l_i, s_i), \ldots, (l_1, s_1), n)$ changes to the state $(1, i, (l_i, s_i), \ldots, (l_1, s_1), n)$.

$$A_0 = I_T \otimes D_1, T = \frac{(rm)^{S+1} - 1}{rm - 1}$$

$$A_{i0} = i\delta I_{T^*} + \Theta, i \geq 1, T^* = aT$$

$$\Theta = \begin{bmatrix} 0 & \cdots & & & 0 \\ \Theta_1 & 0 & \cdots & & 0 \\ 0 & \Theta_2 & \ddots & & \\ \vdots & & \ddots & 0 \\ 0 & \cdots & 0 & \Theta_S & 0 \end{bmatrix}.$$

$\Theta_h = [\Theta_{h1} \ldots \Theta_{hS}]^T, 1 \leq h \leq S$. The elements of Θ_h are as follows.

$$\Theta_{h1} = e_r \otimes \begin{bmatrix} q_1\rho(1) \\ q_2\rho(2) \\ \vdots \\ q_{r-1}\rho(r-1) \\ \rho(r) \end{bmatrix} \otimes I_a, \Theta_{hl} = \begin{bmatrix} q_1\rho(1) \\ q_2\rho(2) \\ \vdots \\ q_{r-1}\rho(r-1) \\ \rho(r) \end{bmatrix} \otimes I_{(mr)a}, 1 \leq l \leq S$$

$$A_{i1} = A^* - diag(0_a, I_{m^s r^{S-1}} \otimes diag((\rho_1, \rho_2, \ldots, \rho_r) \otimes I_a) - (\mu + i\delta)I_T.$$

Theorem 1. *The system is always stable.*

Proof. Let ψ be the Lyaponov test function defined by $\psi(s) = k$, where s is a state in the kth level. The mean drift y_u for a state u belonging to i^{th} level is given by

$$y_u = \sum_p \mathcal{Q}_{rp}(\psi(p) - \psi(u))$$

$$\sum_p \mathcal{Q}_{rp}(\psi(p) - \psi(u)) = \sum_{u'} \mathcal{Q}_{uu'}(\psi(u') - \psi(u)) + \sum_{u''} \mathcal{Q}_{uu''} - (\psi(u'') - \psi(u)) + \sum_{u'} \mathcal{Q}_{uu'}(\psi(u') - \psi(u))$$

In the above sum, the state u' is a state in the level $i - 1$. u'' is a state it the ith level and u''' is a state in the $i + 1^{th}$ level. $\mathcal{Q}_{ij}$ is the $(i, j)^{th}$ entry of the infinitisimal generator. So $\psi(u') = i$, $\psi(u'') = i$ and $\psi(u''') = i + 1$

$$y_u = \sum_p \mathcal{Q}_{up}(\psi(p) - \psi(u)) \sum_{u'} \mathcal{Q}_{uu'} + \sum_{u'''} \mathcal{Q}_{uu'''} = -(\mathcal{A}_{i0}e)_u + (\mathcal{A}_{i2}e)_u$$

$-(\mathcal{A}_{i0}e)_u = -q_i\rho(s_j) - i\delta \rightarrow -\infty$ as $i \rightarrow \infty$, where $-(\mathcal{A}_{i0}e)_u$ is u^{th} entry of the column vector $\mathcal{A}_{i0}e$. Since number of states in the ith level is finite, $(\mathcal{A}_{i2}e)_u$ is always bounded by a fixed positive real number. Thus given any real number $\epsilon > 0$ there exist a natural number N_0 such that if $i \geq N_0$, $y_u < -\epsilon$ for any state u belonging to the ith level. Hence the proof follows from [29].

3   Steady State Distribution

We compute the steady state probability distribution in its most general form. Since the process considered is an LDQBD, the steady state distribution is obtained using the methods proposed by Bright and Taylor [7].

The steady state probability vector has the form $\mathbf{x} = (\mathbf{x}(0), \mathbf{x}(1), \dots)$ where $x(i)$ is the subvector corresponding to the ith level described as follows.

$$\mathbf{x(i)} = (\mathbf{x(i,0)}, \mathbf{x(i,1)}, \dots, \mathbf{x(i,S)}), \mathbf{i} = \mathbf{1, 2}, \dots$$

$$x(0,0) = (x(0,0,1), x(0,0,2), \dots, x(0,0,a))$$

$$x(0,j) = (x(0,j,1), x(0,j,2), \dots, x(0,j,k^*)), 1 \leq j \leq S, k^* = ar^j$$

$$x(i,j) = (x(i,j,1), x(i,j,2), \dots, x(i,j,k^*)), i \geq 1, 1 \leq j \leq S, k^* = a(rm)^j$$

For $i \geq 1$, $x(i)$ can be expressed as a matrix product $x(i) = x(0) \prod_{l=0}^{i-1} R_i$ where the sequence of matrices $\{R_k / k \geq 0\}$ are minimal non negative solutions to the system of equations.

$$\mathcal{A}_{01} + R_0 \mathcal{A}_{11} + R_0 R_1 \mathcal{A}_{20} = 0$$

$$\mathcal{A}_0 + R_k \mathcal{A}_{k+11} + R_k R_{k+1} \mathcal{A}_{k+20} = 0$$

R_k is a matrix of dimension $\frac{(rs)^{S+1}-1}{rs_1}$. $x(0)$ may be obtained by solving the system of equations

$$x(0)(\mathcal{A}_{00} + R_0 \mathcal{A}_{10}) = 0$$

subject to the normalizing condition $x(0)e + x(0) \sum_{k=1}^{\infty} \left(\prod_{l=0}^{k-1} R_l \right) e = 1$. As there are infinitely many terms in the above summation we consider a truncated sum with a suitable choice for the level of truncation. We take the following procedure to obtain the level of truncation. We construct a dominating process as follows. For the process $\mathcal{X}(t)$, we have $(\mathcal{A}_{k0})_{ij} > 0$. Hence by [7] there exist a dominating process $\mathcal{X}^*(t)$ on the same state space as that of $\mathcal{X}(t)$. The infinitismal generator of $\mathcal{X}^*(t)$ is as follows.

$$\mathcal{Q}^* = \begin{bmatrix} \mathcal{A}_{01}^* & \mathcal{A}_1^* & & \\ \mathcal{A}_{10}^* & \mathcal{A}_{11}^* & \mathcal{A}_0^* & \\ & \mathcal{A}_{20}^* & \mathcal{A}_{21}^* & \mathcal{A}_0^* \\ & & \ddots & \ddots & \ddots \end{bmatrix}.$$

The etries of $\mathcal{Q}^*$ are described below.

$(\mathcal{A}_1^*)_{ij} = (\mathcal{A}_{01})_{ij}$, $(\mathcal{A}_0^*)_{ij} = \frac{(\mathcal{A}_0 e)_{max}}{T}$

$\mathcal{A}_{10}^* = 0$, $(\mathcal{A}_{k0}^*)_{ij} = \frac{(\mathcal{A}_{k-10} e)_{min}}{T}$, $k \geq 0$, $T = \frac{(rm)^{S+1}-1}{rm-1}$

$(\mathcal{A}_{n1}^*)_{ij} = (\mathcal{A}_{n1})_{ij}$, $n \geq 0$

where $(Ae)_{max}$ is the maximum element of the column vector Ae.

Let the marginal distributions of $\mathcal{X}(t)$ and $\mathcal{X}^*(t)$ when they are in the steady state be $\{\xi_n / n \geq 0\}$ and $\{\xi_n^* / n \geq 1\}$ respectively. Let $\theta = (\theta_1, \theta_2, \dots)$ be an invariant distribution for $\mathcal{X}^*(t)$. Denote $M_n = \theta_n e$ and $Y^{-1} = \sum_{n=1}^{\infty} M_n$.

When $Y^{-1} < \infty$ the steady state distribution of $\mathcal{X}^*(t)$ exists and ξ_n^* is given by $\xi_n^* = Y M_n$. Now $\{\xi_n^*/n \geq 1\}$ can be considered as a steady state distribution of a standard QBD on the state space $\{i \geq 1\}$. The rates of transitions $q_*(i,j)$ are described as follows.

$$q_*(0,1) = 0, q_*(1,0) = 0$$
$$q_*(i, i-1) = (\mathcal{A}_{k-1} 0 e)_{min}, q_*(i, i+1) = (\mathcal{A}_0 e)_{max}, i > 0$$

Now $\{\xi_n^*/n \geq 1\}$ can be directly found as $\xi_n^* = Y \prod_{j=1}^{n-1} \frac{q_*(i,i+1)}{q_*(i+1,i)}, n \geq 1$.

Taking summation on both sides we get $Y \sum_{n=1}^{\infty} \prod_{j=1}^{n-1} \frac{q_*(i,i+1)}{q_*(i+1,i)} = 1$ Convergence of the geomertic series in the above summation sufficiently imply $Y^{-1} < \infty$ and hence steady state distribution of $\mathcal{X}^*(t)$ exist. Hence steady state distribution of $\mathcal{X}(t)$ exist since $\mathcal{X}^*(t)$ dominates $\mathcal{X}(t)$ stochastically. The truncation level N^* can be fixed in such a way that $\sum_{n=N^*}^{\infty} \xi_n^* < \epsilon$, for any pre assigned real number $\epsilon > 0$. But $\sum_{n=N^*}^{\infty} \xi_n \leq \sum_{n=N^*}^{\infty} \xi_n^*$. For the same N^* we have $\sum_{n=N^*}^{\infty} \xi_n < \epsilon$, for any given $\epsilon > 0$. Then the steady state vector $x_{N^*}(k)$ are obtained as $x_{N^*}(k) = x_{N^*}(0) \prod_{j=0}^{k-1} R_j$, $0 \leq k \leq N^*$ with the condition $x_{N^*}(0)(\mathcal{A}_{00} + R_0 \mathcal{A}_{10}) = 1$. Now for $k \geq N^*$, $x_{N^*}(k) = x(N^*) \prod_{j=1}^{k-1} R_k$. Hence the normalizing condition may be modefied as $x_{N^*}(0) + x_{N^*}(0) \sum_{j=1}^{N^*} (\prod_{h=0}^{j-1} R_h) e + x_{N^*}(N^*+1)(I - R_{N^*})^{-1} e = 1$. The term $x_{N^*}(N^*+1)(I - R_{N^*})^{-1} e$ can be made less than any given tolerence level $\epsilon > 0$.

4   Performance Characteristics of the System

- Expected number of perishings

$$E_n(p) = \sum_{i=1}^{\infty} \sum_{j=1}^{S} j x(i,j) e_{n_j}$$

where $n_j = (mr)^j a$, $1 \leq j \leq S$.
- Expected rate of loss of organs

$$R_p = x(0,1)\mu I_a e + x(1,1)\mu I_{amr} e + \sum_{i=1}^{\infty} \sum_{j=1}^{S} x(i,j) E_j e$$

Organ perishing takesplace only when inventory is non-empty. The matrices E_i are described in Sect. 2 and entries of E_j correspond to the perishings of organs at j^{th} inventory level.
- Expected rate of matching of organs

$$R_m = \sum_{i=1}^{S} x(1,i)\Gamma_i^* e + \sum_{i=2}^{\infty} \sum_{j=2}^{\infty} x(i,j)\Theta_j e$$

Matching occurs only when inventory is non-empty and there is at least one customer in the queue. The matrices Γ_i^* and Θ_i are defined as in Sect. 2 and entries of these matrices correspond to matching between organs and patients.

- Probability that the inventory is empty, $\mathcal{P}_0 = \sum_{i=0}^{\infty} x(i,0)e$
- Probability that a customer reneges when inventory level is zero,
  $\mathcal{P}_0^* = \sum_{i=0}^{\infty} \sum_{j=1}^{a} x(i,j)e$
- Expected number of patients in the waiting list $\mathcal{E}_p = \sum_{i=0}^{\infty} ix(i)e$
- Probability that a search successfully finds a match before perishing of an organ

$$\mathcal{P}_s = \sum_{i=1}^{\infty} \sum_{k=1}^{S} \sum_{j=1}^{rm^j a} x(i,k,j)$$

- Probability that search fails to find a match for an organ before perishing.

$$\mathcal{P}_f = 1 - \sum_{i=1}^{\infty} \sum_{k=1}^{S} \sum_{j=1}^{rm^j a} x(i,k,j)$$

5   Numerical Example

For the numerical illustration of the model, we consider a queueing inventory system to which customers arrive according to a MAP described by the matrices $D_0 = \begin{pmatrix} -1.7 & 1 \\ 1 & -4.1 \end{pmatrix}$, $D_1 = \begin{pmatrix} 0.5 & 0.2 \\ 2.1 & 1 \end{pmatrix}$. Organ arrivals are assumed to be according to a Poisson process with parameter λ and search times also follow an exponential distribution with parameter θ. Reneging times of customers is exponentially distributed with parameter δ and Perishing time of an organ follows exponential distribution with parameter γ. For the computations we fix $\gamma = 2$ and $\delta = 0.2$. Rate of search is increased and the corresponding variations in mean rate of matching and mean rate of loss of organs are analyzed.

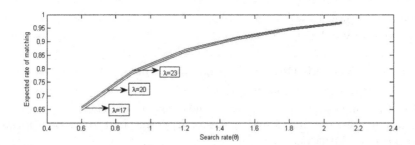

Fig. 1. Variation in expected rate of matching with respect to increase in search rate

In Fig. 1, it is obvious that the matching process gets accelerated with the increase in search rate and the phenomenon is intutively true. The experiment is carried out for different arrival rates of organs and the same behaviour is observed. Figure 2 shows that rate of loss of organs decreases with the increase in search rate. Same pattern is observed for different arrival rates of organs. These observations strongly demonstrates the impact of search in the design and control of the queueing inventory of organs.

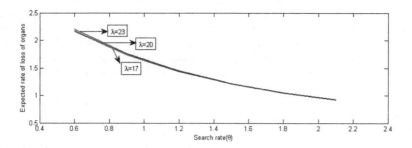

Fig. 2. Variation in expected rate of loss of organs with respect to increase in search rate

6 Conclusion

We studied an organ transplantation model as a queueing inventory system with search. The central attraction of the model is the introduction of search to find an optimal match from the waiting list of patients for each of the inventoried organ. The system is modelled as a multi dimensional Markov chain and a sufficient condition for the stability of the system is derived. Steady state probability distribution is obtained using matrix analytic methods and key performance measures are specified. From the numerical investigation of the model, it is clear that rate of loss of organs decreases and rate of matching increases with the increase in search rate. Thus the model uses search as a successful mechanism to find a matching candidate for an inventoried organ within minimum time.

Acknowledgment. Sinu Lal T S thanks University Grants Commission(UGC) of India for UGC-Junior Research Fellowship (Roll no.-432693, June 2017).

References

1. Elalouf, A., Perlman, Y., Yechiali, U.: A double-ended queueing model for dynamic allocation of live organs based on a best-fit criterion. Appl. Math. Model. **60**, 179–191 (2018)
2. Artalejo, J.R., Joshua, V.C., Krishnamoorthy A.: An M/G/1 retrial queue with orbital search by the server. In: Advances in Stochastic Modelling, pp. 41–54 (2002)
3. Ata, B., Skaro, A., Tayu, S.: OrganJet: overcoming Geographical disparities in access to deceased donor kidneys in the United States. Manag. Sci. **63**, 1–20 (2016)
4. Bar-Lev, S.K., Boxma, O., Mathijsen, B., Perry, D.: A blood bank model with perishable blood and demand impatience. Stoch. Syst. **7**(2), 237–263 (2017)
5. Bendersky, M., David, I.: Deciding kidney-offer admissibility dependent on patients' lifetime failure rate. Eur. J. Oper. Res. **251**(2), 686–693 (2016)

6. Bertsimas, D., Farias, V.F., Trichakis, N.: Fairness, efficiency, and flexibility in organ allocation for kidney transplantation. Oper. Res. **61**(1), 73–87 (2013)
7. Bright, L., Taylor, P.G.: Calculating the equilibrium distribution in level dependent quasi-birth-and-death processes. Stoch. Model. **11**(3), 497–525 (1995)
8. Fackrell, M.: Modelling healthcare systems with phase-type distributions. Health Care Manag. Sci. **12**(1), 11 (2009)
9. Boxma, O.J., David, I., Perry, D., Stadje, W.: A new look at organ transplantation models and double matching queues. Probab. Eng. Inf. Sci. **25**(2), 135–155 (2011)
10. Chakravarthy, S.R., Krishnamoorthy, A., Joshua, V.C.: Analysis of a multi-server retrial queue with search of customers from the orbit. Perform. Eval. **63**(8), 776–798 (2006)
11. David, I., Yechiali, U.: A time-dependent stopping problem with application to live organ transplants. Oper. Res. **33**(3), 491–504 (1985)
12. Deepak, T.G., Dudin, A.N., Joshua, V.C., Krishnamoorthy, A.: On an M (X)/G/1 retrial system with two types of search of customers from the orbit. Stoch. Anal. Appl. **31**(1), 92–107 (2013)
13. Drekic, S., Stanford, D.A., Woolford, D.G., McAlister, V.C.: A model for deceased-donor transplant queue waiting times. Queueing Syst. **79**(1), 87–115 (2015)
14. Dudin, A., Deepak, T.G., Joshua, V.C., Krishnamoorthy, A., Vishnevsky, V.: On a $BMAP/G/1$ retrial system with two types of search of customers from the orbit. In: Dudin, A., Nazarov, A., Kirpichnikov, A. (eds.) ITMM 2017. CCIS, vol. 800, pp. 1–12. Springer, Cham (2017). https://doi.org/10.1007/978-3-319-68069-9_1
15. He, Q.M.: Fundamentals of Matrix-Analytic Methods, vol. 365. Springer, New York (2014). https://doi.org/10.1007/978-1-4614-7330-5
16. Jain, M., Agrawal, P.K.: M/Ek/1 queueing system with working vacation. Qual. Technol. Quant. Manag. **4**(4), 455–470 (2007)
17. Krishnamoorthy, A., Deepak, T.G., Joshua, V.C.: An M— G— 1 retrial queue with nonpersistent customers and orbital search. Stoch. Anal. Appl. **23**(5), 975–997 (2005)
18. Krishnamoorthy, A., Joshua, V.C., Mathew, A.P.: A retrial queueing system with multiple hierarchial orbits and orbital search. In: Vishnevskiy, V.M., Kozyrev, D.V. (eds.) DCCN 2018. CCIS, vol. 919, pp. 224–233. Springer, Cham (2018). https://doi.org/10.1007/978-3-319-99447-5_19
19. Krishnamoorthy, A., Joshua, V.C., Mathew, A.P.: A retrial queueing system with abandonment and search for priority customers. In: Vishnevskiy, V.M., Samouylov, K.E., Kozyrev, D.V. (eds.) DCCN 2017. CCIS, vol. 700, pp. 98–107. Springer, Cham (2017). https://doi.org/10.1007/978-3-319-66836-9_9
20. Mathew, A.P., Krishnamoorthy, A., Joshua, V.C.: A retrial queueing system with orbital search of customers lost from an offer zone. In: Dudin, A., Nazarov, A., Moiseev, A. (eds.) ITMM/WRQ -2018. CCIS, vol. 912, pp. 39–54. Springer, Cham (2018). https://doi.org/10.1007/978-3-319-97595-5_4
21. Neuts, M.F., Ramalhoto, M.F.: A service model in which the server is required to search for customers. J. Appl. Probab. **21**(1), 157–166 (1984)
22. Latouche, G., Ramaswami, V.: Introduction to Matrix Analytic Methods in Stochastic Modeling, vol. 5. SIAM, Philadelphia (1999)
23. Lucantoni, D.M.: New results on the single server queue with a batch Markovian arrival process. Commun. Stat. Stoch. Model. **7**(1), 1–46 (1991)
24. Narayanan, V.C., Deepak, T.G., Krishnamoorthy, A., Krishnakumar, B.: On an (s, S) inventory policy with service time, vacation to server and correlated lead time. Qual. Technol. Quant. Manag. **5**(2), 129–143 (2008)

25. Neuts, M.F.: Matrix-Geometric Solutions in Stochastic Models: An Algorithmic Approach. Courier Corporation, North Chelmsford (1994)
26. Neuts, M.F.: Probability distributions of phase type. Liber Amicorum Prof. Emeritus H. Florin (1975)
27. Neuts, M.F.: A versatile Markovian point process. J. Appl. Probab. **16**(4), 764–779 (1979)
28. Su, X., Zenios, S.: Patient choice in kidney allocation: the role of the queueing discipline. Manuf. Serv. Oper. Manag. **6**(4), 280–301 (2004)
29. Tweedie, R.L.: Sufficient conditions for regularity, recurrence and ergodicity of Markov processes. In: Mathematical Proceedings of the Cambridge Philosophical Society, vol. 78, no. 1, pp. 125–136. Cambridge University Press (1975)
30. Perlman, Y., Elalouf, A., Yechiali, U.: Dynamic allocation of stochastically-arriving flexible resources to random streams of objects with application to kidney cross-transplantation. Eur. J. Oper. Res. **265**(1), 169–177 (2018)
31. Zenios, S.A.: Modeling the transplant waiting list: a queueing model with reneging. Queueing Syst. **31**(3–4), 239–251 (1999)

Mathematical Models of the Queuing Systems with MMPP Flow and Instantaneous Feedback

Agassi Melikov[1]($\boxtimes$) and Sevinc Aliyeva[2]

[1] Institute of Control Systems, National Academy of Science of Azerbaijan,
Baku, Azerbaijan
agassi.melikov@gmail.com
[2] Baku State University, Baku, Azerbaijan
s@aliyeva.info

Abstract. Mathematical models of the single channel queuing system with MMPP flow and instantaneous feedback are proposed. After completion of the service, calls either leave the system or, immediately return to system to receive repeated services. Service times both of primary and feedback (secondary) calls has exponential distribution but with different parameters. Primary calls have pre-emptive priority over feedback calls. Ergodicity conditions of the models are established and methods to calculation of stationary distribution of the appropriate three-dimensional Markov chains (3D MC) as well as explicit formulas to determining the performance measures of the proposed models are developed.

Keywords: Single channel queue · Instantaneous feedback · Methods for calculation

1 Introduction

The feedback is common property of communication networks in which data (packets, frames, etc.) are re-transmitted if errors occurred during their initial transmission. It also often appear in production systems where issues that are not fully machined are re-processed.

It is necessary to distinguish two types of queuing systems with feedback (Feedback Queue, FBQ): the Instantaneous Feedback Queue (IFBQ) and the Delayed Feedback Queue (DFBQ). In IFBQ, the repetition occurs immediately after the completion of the previous servicing, and in DFBQ, the repetition occurs after positive delay. The first publications devoted to the study of FBQ of both types are the works of Takach [1,2]. In them, the method of generating functions is used to study of two-dimensional Markov models of single-channel systems with an unbounded queue and an infinite orbit volume (for a model with delayed feedback). After the classical works of Takach [1,2], FBQ models did not attract the attention of researchers for a long time. In the past three decades, various authors have intensively researched these models. A review of

© Springer Nature Switzerland AG 2019
V. M. Vishnevskiy et al. (Eds.): DCCN 2019, CCIS 1141, pp. 288–301, 2019.
https://doi.org/10.1007/978-3-030-36625-4_23

some publications can be found in [3], and therefore we will not dwell on the analysis of known results. We also note the works [4–11], which were not included in the list of references in [3]. Despite the fact that FBQ models without queues and with orbits for feedback calls have been studied in sufficient detail (see, for example, [12–15] and their list of references), there are few works available in the literature that are devoted to studying models of such systems with queues. One of the first papers devoted to the study of FBQ models with queues is [16]. In this work, the matrix-geometric method [17] is used to calculate the steady-state probabilities of the corresponding four-dimensional Markov chain (4D MC) and the system characteristics (queue length distribution, call loss probability, etc.) is also found.

Simple models of single-channel IFBQ with impatient calls using different mechanisms for keeping them in the queue at the time of the completion of the allowable waiting time are studied in [18–23]; similar models of multichannel IFBQs were studied in [24]. The IFBQ model with two heterogeneous servers and a limited queue, in which incoming calls with known probabilities is assigned to servers, was studied in [25]. Here it is assumed that the probability of joining the queue depends on the total number of calls in the queue. In papers [18–25], one-dimensional birth-death processes is used as mathematical models of the investigated systems.

Models of IFBQ with single server and infinite separate queues of primary and feedback calls is been investigated extensively in recent paper [4]. Here it is assumed that calls arrive to a two-priority system according to a Poisson process and separate waiting rooms (buffers) have infinity capacity. Calls in waiting line 1 have priority over the ones in line 2. The service time is class-dependent phase type. After completion of service, high priority calls may feedback for service according to a Bernoulli process. Feedback calls are send to the low priority queue and they are served only when no calls in the line 1. Authors assumed that multiple feedback does not allowed. Both preemptive and non-preemptive priorities are analyzed. The case where there is no external entry to the line 2 is discussed as well. The matrix-geometric method [17] is used to calculate the steady-state probabilities of the models and the waiting time distribution of both type of calls are derived.

The multi-channel DFBQ model with a finite buffer size for primary calls and an infinite orbit for feedback calls, in which the intensity of feedback calls does not depend on the number of calls in orbit, is studied in [26]. Using the matrix-geometric method [17], the stationary distribution of the corresponding 2D MC is found and the system performance measures were calculated.

It is important to note that in all the papers noted above, it is assumed that the primary calls and feedback calls are identical for all parameters. At the same time, in real situations they may differ from each other in some parameters, for example, in the time of their processing, degree of importance, etc. Similar models of single-channel DFBQ with a limited queue for high-priority calls were studied in [27,28]. They assume that different types of calls differ from each other also in terms of load parameters, while low-priority calls cannot wait in a queue, and only they can require repeated servicing. The orbit for feedback calls has infinite size. In [27], it is assumed that high-priority calls do not interrupt the already started service process of a low-priority call and in [28] it is considered that such an interruption is possible. In both papers, the matrix-geometric method [17] is used to study the system.

An analysis of the available literature has shown that the vast majority of papers study FBQ models with a constant intensity of primary calls. However, in real situations it is a variable quantity. In addition, in them primary and feedback calls are considered identical for all parameters.

Based on the above facts, here we study the IFBQ models with the MMPP flow [29], in which the primary calls have preemptive priority over feedback ones. The models studied here are similar to the models studied in [4]. However, there are some differences. So, unlike [4] here it is assumed that the primary flow is MMPP, and in addition, we also consider models with a finite queue of primary calls. It is important to note that an alternative method for studying the proposed models is proposed, which is based on the principle of hierarchical merging of three-dimensional Markov chains [30].

The paper has the following structure. In Sect. 2, a description of the investigated system with instantaneous feedback and the problem statement are given. A mathematical model of the system in the form of a 3D MC and its generating matrix (GM) are developed in Sect. 3. Here, the ergodicity conditions of 3D MC are obtained and exact and approximate methods are developed for calculating the steady-state probabilities. Results of numerical experiment are shown in Sect. 4. Conclusion remarks are given in Sect. 5.

2   Description of the System with Instantaneous Feedback and the Problem Statement

The structural diagram of the investigated system is shown in Fig. 1. At the input of a single-channel system with an unlimited buffer, an MMPP flow arrives with parameters (Σ, Λ), where $\Sigma = \|\sigma_{ij}\|$ denotes a GM of the Markov chain with $N > 1$ possible states which control the intensity of the incoming flow, and the vector $\Lambda = (\lambda_1, \lambda_2, ..., \lambda_N)$ sets the values of the intensities of the incoming flow. This means that σ_{ij} determines the intensity of the transition from state i to state j, $j \neq i$; $\sigma_{ii} = -\sum_{j=1, j\neq i}^{N} \sigma_{ij}$. It is considered that when the MC is in a state n, the intensity of the incoming flow is λ_n, $n = 1, 2, ..., N$, and with a change in the state of the controlling MC, the intensity of the incoming flow changes instantaneously.

After the completion of the service, call received from outside (the primary call, the p-call) according to the Bernoulli scheme either leaves the system or instantly requires repeated service. It means that the p-call either with a probability α leaves the system or with a complementary probability $1 - \alpha$ instantly requires re-service. If at the moment of completion of service of the p-call the queue was empty, and it requires repeated service, then this call (secondary call, s-call) instantly begins to be serviced. Otherwise, the s-call joins the queue, and

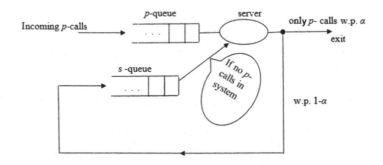

Fig. 1. Structure of the system.

it is assumed that the queues of p-calls and s-calls are separate (separate or common queues have meaning only for models with limited buffers). It is assumed that s-calls cannot require another re-service (single feedback), and servicing of s-calls starts only when there are no p-calls in the system. In addition, p-calls have preemptive priorities over s-calls, while the interrupted call returns to the queue, and it will be re-serviced.

The servicing times for both types of calls are independent and identically distributed (i.i.d.) random variables (r.v). Suppose that cumulative distribution function (CDF) of the indicated r.v. are exponential ones with the average service time for primary and secondary calls being equal $1/\mu_p$ and $1/\mu_s$, respectively.

Changes in the state of the MC that control the intensity of the incoming flow and service of calls are independent of each other random processes. The task is to find a joint distribution of the states of the MMPP flow and the number of calls of each type in the system. Solving this problem allows to find the performance measures of the system.

3   Calculation of the State Probabilities and Performance Measures

We consider both exact and approximate methods to solving indicated problem. First, consider exact method.

3.1   Exact Method

The state of the system is determined by a three-dimensional vector (n, k, r), where n is the state of the MC which control intensity of the incoming flow, k is the number of p-calls in the system (including p-call in the channel), r is the number of s-calls in the system (including s-call in the channel). Then the state space of this 3D MC is determined as follows:

$$E = \{1, 2, ..., N\} \times \{0, 1, ..., \} \times \{0, 1, ..., \} . \tag{1}$$

The intensity of the transition from state (n, k, r) to state (n', k', r') is denoted by $q((n, k, r), (n', k', r'))$. These values are calculated as:

1. the transition $(n,\, k,\, r) \to (n',k,r)$, $n' \neq n$ is carried out with intensity $\sigma_{n,n'}$ when the state of the MMPP flow changes;
2. the transition $(n,\, k,\, r) \to (n, k+1, r)$ is carried out with intensity λ_n when a p-call is arrived;
3. the transition $(n,\, k,\, r) \to (n, k-1, r)$, $k > 0$, is carried out with intensity $\mu_p \alpha$ at the completion of the service of the p-call and its departure from the system;
4. transition $(n,\, k,\, r) \to (n, k-1, r+1)$, $k > 0$, is carried out with intensity $\mu_p (1 - \alpha)$ at the completion of the service of the p-call and returning it to the system for re-servicing;
5. the transition $(n, 0, r) \to (n, 0, r-1)$, $r > 0$, is carried out with intensity μ_s at the completion of the service of the s-call;

Thus, the positive elements of the GM of the 3D MC are determined from the following relations:

$$
q\left((n,k,r),\left(n',k',r'\right)\right) = \begin{cases} \sigma_{nn'}, & \text{if } n' \neq n,\ k' = k,\ r' = r, \\ \lambda_n, & \text{if } n' = n,\ k' = k+1,\ r' = r, \\ \mu_p \alpha, & \text{if } k > 0,\ n' = n,\ k' = k-1,\ r' = r, \\ \mu_p (1 - \alpha), & \text{if } k > 0,\ n' = n,\ k' = k-1,\ r' = r+1, \\ \mu_s, & \text{if } k = 0,\ r > 0,\ n' = n,\ k' = 0,\ r' = r - 1. \end{cases}
\tag{2}
$$

Let $p(n,k,r)$ denotes the stationary probability of the state $(n,k,r) \in E$. The conditions for the existence of the stationary mode are obtained below.

The steady-state probabilities satisfy the system of equilibrium equations (SEE) of infinite dimension. This SEE is based on relations (2) and its explicit form is not given here because of the obviousness of its compilation.

Finding the steady-state probabilities is sufficient to calculate the performance measures(characteristics) of the system. Thus, the average number of p-calls (L_p) and s-calls (L_s) in the system are defined as the mathematical expectation of the corresponding r.v.:

$$
L_p = \sum_{n=1}^{N} \sum_{k=1}^{\infty} k \sum_{r=0}^{\infty} p(n,k,r) ;
\tag{3}
$$

$$
L_s = \sum_{n=1}^{N} \sum_{r=1}^{\infty} r \sum_{k=0}^{\infty} p(n,k,r) .
\tag{4}
$$

Due to the complex structure of the GM (see formulas (2)), it is not possible to find an analytical solution for the corresponding SEE for steady-state probabilities. Therefore, to solve it, one has to use the method of multidimensional generating functions, which has a number of methodological and computational difficulties for use in such systems and is rather cumbersome. So, an alternative approach for solving this problem, based on a hierarchical space merging method of multidimensional MC is developed [30].

3.2 Approximate Method. Case Unlimited Buffers for both Types of Calls

For the correct application of the developed method, suppose that the MMPP flow remains in each state for a long time before leaving it, i.e. transitions between its states occur at low intensities.

Consider the following splitting of the state space (1):

$$E = \bigcup_{n=1}^{N} E_n \ , \ E_n \bigcap E_{n'} = \emptyset \,, \text{if } n \neq n', \tag{5}$$

where $E_n = \{(n, k, r) \in E : \ k = 0, 1, ...; \ r = 0, 1, ...\} \,, \ n = 1, 2, ..., N.$

On the basis of the splitting (5), the merge function $U_1 (n, k, r) \ =< n >, (n, k, r) \in E_n$, is determined, where $< n >$ is the merged state, which includes all states from the class E_n, $n = 1, 2, ..., N$. Denote $\Omega_1 = \{< n >: \ n = 1, 2, ..., N\}$.

The steady-state probabilities of the initial model are defined as follows (see [30]):

$$p(n, k, r) \approx \rho_n (k, r) \pi_1 (< n >) \,, \tag{6}$$

where $\rho_n (k, r)$ – probability of state (k, r) inside a split model with a state space, E_n, $\pi_1 (< n >)$ –probability of an merged state $< n >\in \Omega_1$.

Since the transitions between the states of the MC which controlled the intensity of the incoming flow do not depend on the status of the channel (busy or idle) the probability of the states $\pi_1 (< n >)$, $< n >\in \Omega_1$, is determined by its GM Σ. Consequently, it will only be necessary to find the stationary distribution of 2D MC with state spaces E_n, $n = 1, 2, ..., N.$

Now to the 2D MC with state space E_n apply the procedure of merging (the second level of the hierarchy). Since all split models are identical, then we fix the value of the parameter n, and consider the split model with the state space E_n.

In a class E_n, consider the following splitting:

$$E_n = \bigcup_{r=0}^{\infty} E_n^r \ , \ E_n^r \bigcap E_n^{r'} = \emptyset \,, \text{if } r \neq r', \tag{7}$$

where $E_n^r = \{(k, r) \in E_n : \ k = 0, 1, ...\} \,, \ r = 0, 1, ...$

Further, on the basis of the splitting (7), the merge function $U_2 ((k, r)) =< r >$, if $(k, r) \in E_n^r$, is determined, where $< r >$ is the merged state, which includes all the states from the class E_n^r. Denote $\Omega_2 = \{< r >: \ r = 0, 1, ...\}$.

According to [30] we have:

$$\rho_n (k, r) \approx \rho_n^r (k) \pi_2^n (< r >) \,, \tag{8}$$

where $\rho_n^r (k)$ is the probability of the state (k, r) inside the split model with the state space E_n^r, $\pi_2^n (< r >)$ is the probability of the merged state $< r >\in \Omega_2$.

In the state vector within the class E_n^r, $r = 0, 1, ...,$ the second component is constant and equal r. Therefore, when studying split models with a space E_n^r, each state (k, r) can be defined only by the first component, i.e. hereinafter the state $(k, r) \in E_n^r$ is denoted as $k, k = 0, 1, ...$

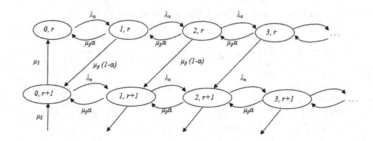

Fig. 2. Fragment of the state graph for the split model with state space E_n

Then from relations (2) we conclude that the intensities of transitions between the states of the split model with the space of states E_n^r, $r = 0, 1, ...$ do not depend on the parameter r, and are determined as follows (see. Fig. 2):

$$q_n(k, k') = \begin{cases} \lambda_n, & \text{if } k' = k + 1, \\ \mu_p \alpha, & \text{if } k' = k - 1. \end{cases} \tag{9}$$

From (9), we find that, if the condition $\lambda_n < \mu_p \alpha$ is satisfied, the state probabilities of the split models with the state space E_n^r do not depend on the index r, $r = 0, 1, ...,$ and are defined as follows (see Fig. 2):

$$\rho_n^r(k) = (1 - \nu_n) \nu_n^k, \quad k = 0, 1, ..., \tag{10}$$

where $\nu_n = \lambda_n / \mu_p \alpha$.

Note 1. Since the condition $\lambda_n < \mu_p \alpha$ must be fulfilled for each n, $n = 1, 2, ..., N$, we get the first condition for ergodicity of the model:

$$\max_{n=\overline{1,N}} \{\nu_n\} < 1 \tag{11}$$

Taking into account (2) and (10), we find that the intensity of transitions between Ω_2 states are determined as follows (see Fig. 2):

$$q_n(<r>, <r'>) = \begin{cases} \mu_p (1 - \alpha) \nu_n, & \text{if } r' = r + 1, \\ \mu_s (1 - \nu_n), & \text{if } r' = r - 1. \end{cases} \tag{12}$$

Then from relations (12) we conclude that, if the condition $\theta_n < 1$ is fulfilled then probability of merged states $\pi_2^n(<r>)$, $<r> \in \Omega_2$, are calculated as follows:

$$\pi_2^n(<r>) = \theta_n^r (1 - \theta_n), \quad r = 0, 1, ..., \tag{13}$$

where $\theta_n = (1 - \alpha) \frac{\mu_p}{\mu_s} \frac{\nu_n}{1-\nu_n}$.

Note 2. Since the condition $\theta_n < 1$ must be satisfied for each n, $n = 1, 2, ..., N$, we obtain the second condition for the ergodicity of the model:

$$\max_{n=\overline{1,N}} \left\{ \frac{\nu_n}{1 - \nu_n} \right\} < \frac{\mu_s}{\mu_p (1 - \alpha)} \tag{14}$$

Thus, relations (11) and (14) determines the ergodicity conditions of the investigated system. Finally, taking into account relations (6), (8), (10) and (13), the approximate values of the steady-state probabilities of the original 3D MC are found in a multiplicative form.

After certain mathematical calculations, we obtain the following simple formulas for an approximate calculation of the performance measures of the system:

$$L_p \approx \sum_{n=1}^{N} \frac{\nu_n}{1 - \nu_n} \pi_1 (<n>) ; \tag{15}$$

$$L_s \approx \sum_{n=1}^{N} \frac{\theta_n}{1 - \theta_n} \pi_1 (<n>) . \tag{16}$$

It can be seen from formula (15) that the average number of p-calls in the system does not depend on the service intensity of s-calls. This is an expected fact, since p-calls have preemptive priority over s-calls. It is important to note that this formula has a physical meaning: its right side is a weighted sum of the average number of calls in the $M/M/1/\infty$ systems with loads ν_n, $n = 1, ..., N$, where the weighting is done according to the stationary distribution of the MMPP flow. In other words, formula (15) is exact one, and it could be proposed without following the developed approximate procedures. This fact confirms the high accuracy of the developed approximate formulas. Note that the right side of the formula (16) is the weighted sum of the average number of calls in the $M/M/1/\infty$ systems with loads θ_n, $n = 1, ..., N,$; here the weighting is also performed by the stationary distribution of the MMPP flow. From formula (16) it can be seen that the average number of s-calls in the system depends on all system parameters, including the intensity of their service.

3.3 Approximate Method. Case Limited Buffer for Primary Calls

The developed approach can be used for models in which there are limited buffers for p-calls and/or s-calls. Due to the limited size of the paper, only a model with limited separate buffer for p-calls is considered here, and it is considered that they are impatient in the queue.

Let the maximum number of p-calls in the system be equal $R_p < \infty,$; Allowable waiting time in the queue of p-calls are r.v. that are independent of each other and have exponential distributions with parameters τ_p. For simplicity, it is assumed that reneging rate of p-calls from the queue does not depend on number of calls in queue, i.e. constant reneging rate is considered.

The state of this system is also determined by 3D vector(n, k, r), but the state space of the corresponding 3D MC is given as:

$$E_1 = \{1, 2, ..., N\} \times \{0, 1, ..., R_p\} \times \{0, 1, ...\} \ .$$

The positive elements of GM of this chain is defined similarly to (2), i.e.

$$q\left((n, k, r), (n', k', r')\right) = \begin{cases} \sigma_{nn'}, & \text{if } n' \neq n, \ k' = k, \ r' = r, \\ \lambda_n, & \text{if } n' = n, \ k' = k+1, \ r' = r, \\ \mu_p \alpha, & \text{if } k = 1, \ n' = n, \ k' = 0, \ r' = r, \\ \mu_p \alpha + \tau_p, & \text{if } k > 1, \ n' = n, \ k' = k-1, \ r' = r, \\ \mu_p (1 - \alpha), & \text{if } k > 0, \ n' = n, \ k' = k-1, \ r' = r+1, \\ \mu_s, & \text{if } k = 0, \ r > 0, \ n' = n, \ k' = 0, \ r' = r-1. \end{cases}$$
$$(17)$$

The steady-state probabilities can be determined from the SEE, which is compiled on the basis of relations (17).

In this model, a new performance measure appears – the blocking probability of p-calls (PB_p). It is calculated as follows:

$$PB_p = \sum_{(n, R_p, r) \in E_1} p(n, R_p, r) + \sum_{(n, k, r) \in E_1} \frac{\tau_p}{\lambda_n + \mu_p + \tau_p} p(n, k, r) I(k > 1) \ ,$$
$$(18)$$

where $I(A)$ is indicator function of the event A.

Calculating the steady-state probabilities and the performance measures for models of moderate dimensionality of the state space E_1 presents no difficulties. At the same time, for models of large dimension for this purpose the approximate method described above can be used.

Omitting a detailed description of the stages of applying this method, the final forms of the corresponding formulas are given below.

From (17), we find that in this model, at the second level of the hierarchy, the state probabilities of split models with the state space E_n^r do not depend on the index r, $r = 0, 1, ...$, and are defined as:

$$\rho_n(k) = \nu_n \xi_n^{k-1} \rho_n(0) \ , \quad k = 1, \ ..., \ R_p \ , \tag{19}$$

where $\xi_n = \frac{\lambda_n}{\mu_p \alpha + \tau_p}$. The probability $\rho_n(0)$ is calculated from the normalization condition $\sum_{k=0}^{R_p} \rho_n(k) = 1$. In other words,

$$\rho_n(0) = \left(1 + \nu_n \frac{1 - \xi_n^{R_p}}{1 - \xi_n}\right)^{-1} . \tag{20}$$

For this model, the probabilities of merged states at the second level of the hierarchy are calculated as follows:

$$\pi_2^n(<r>) = \zeta_n^r (1 - \zeta_n), \ r = 0, 1, ..., \tag{21}$$

where $\zeta_n = \frac{\mu_p(1-\alpha)(1-\rho_n(0))}{\mu_s \rho_n(0)}$.

Note 3. Since the condition $\zeta_n < 1$ must be satisfied for each n, $n = 1, 2, ..., N$, we obtain the ergodicity condition of the model (compare with *Note* 2):

$$\max_{n=\overline{1,N}} \left\{ \frac{1 - \rho_n (0)}{\rho_n (0)} \right\} < \frac{\mu_s}{\mu_p (1 - \alpha)} \qquad (22)$$

Further, taking into account relations (6), (8), (19)–(21), the approximate values of the steady-state probabilities of this model are found.

After certain mathematical calculations, we obtain the following formulas for an approximate calculation of the performance measures of the system with limited buffer for p-calls:

$$L_p \approx \sum_{n=1}^{N} \pi_1 (<n>) \sum_{k=1}^{R_p} k \rho_n (k) \; ; \qquad (23)$$

$$L_s \approx \sum_{n=1}^{N} \pi_1 (<n>) \frac{\zeta_n}{1 - \zeta_n} \; ; \qquad (24)$$

$$PB_p = \sum_{n=1}^{N} \pi_1 (<n>) \left(\rho_n (R_p) + \frac{\tau_p}{\lambda_n + \mu_p + \tau_p} \sum_{k=2}^{R_p} \rho_n (k) \right) . \qquad (25)$$

As in the model with unlimited queues for both types of calls, it is also clear from formulas (23) and (25) that the average number of p-calls in the system and the probability of their blocking do not depend on the service intensity of s-calls. At the same time, it is clear from formula (24) that the average number of s-calls in the system depends on all structural and load parameters of the system.

4 Numerical Results

The purpose of the numerical experiments conducted here is to study the behavior of the characteristics of the systems under consideration versus its parameters. For a system with infinite queues, for both types of calls, as a variable parameter is selected α, and for a system with a finite queue for p-calls, as a variable parameter is selected R_p.

The initial data of a hypothetical model of a system with infinite queues for both types of calls are determined as follows. The GM of the Markov chain, which controls the intensity of the incoming flow with states $N = 3$, is defined as follows:

$$\Sigma = \left\| \begin{matrix} -34 & 20 & 14 \\ 18 & -32 & 14 \\ 4 & 16 & -20 \end{matrix} \right\| .$$

Intensity values of the incoming flow $\Lambda = (15, 10, 5)$. The behavior of the characteristics of the system versus the parameter α change is shown in Fig. 3.

Both characteristics L_p and L_s are decreasing. This was to be expected, since with an increase in the probability of p-calls leaving the system without repeating service, the number of s-calls in the system decreases. A decrease in the number of s-calls in the system leads to a decrease in server load, and therefore the number of p-calls in the system decreases. Note that from Fig. 3 shows that with an increase in the service intensity of both types of calls, both characteristics L_p and L_s decrease and, at the same time, there is a relation $L_p > L_s$.

The initial data of a hypothetical model of a system with a finite queue for p-calls is selected as above for a model with infinite queues. The corresponding results of numerical experiments for this model are shown in Fig. 4, where the case $(\mu_p, \mu_s) = (55, 75)$ is considered. Figure 4 shows that the characteristics L_p and L_s are increasing, the characteristic PB_p decreases with respect to the increase in the parameter R_p. These results were expected, since with an increase in the probabilities of p-calls leaving the system without repeated service, the chances of such calls being queued increase. Note that with an increase in the parameter α, all characteristics of the system improve.

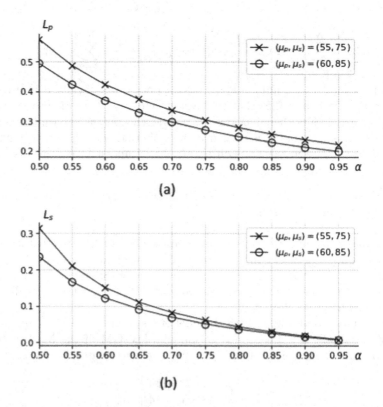

Fig. 3. Average number of p-calls (a) and s-calls (b) versus α for the model with unlimited buffers for both types of calls

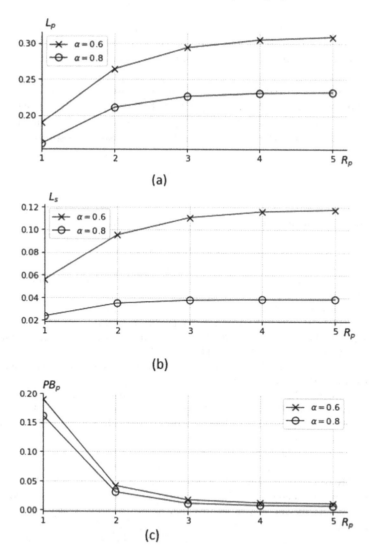

Fig. 4. Average number of p-calls (a), s-calls (b) and blocking probability of p-calls (c) versus R_p for the model with limited buffer for p-calls

5 Conclusion

The paper studies single-channel systems with unlimited and limited queues, MMPP-flow and instantaneous feedback. It is assumed that primary and feedback calls are served in the common channel, while primary calls has preemptive priority over feedback calls. It is shown that the mathematical models of the systems under study are some infinite or finite 3D MCs. For a model with unlimited queues, the conditions for ergodicity of the model are obtained. For both types of models, a unified approximate approach is proposed and simple formulas for calculating the steady-state probabilities of the corresponding 3D MCs are developed. In addition, by using these probabilities the performance measures of the investigated systems are calculated.

As directions for further research should indicate the study of such models in the presence of MAP-flow and PH-distribution for the service time of heterogeneous calls. Of interest are also problems of optimization of systems with feedback regarding the selected criteria. These issues are subject to special studies.

References

1. Takacs, L.: A single-server queue with feedback. Bell Syst. Tech. J. **42**, 505–519 (1963)
2. Takacs, L.: A queuing model with feedback. Oper. Res. **11**, 345–354 (1977)
3. Melikov, A.Z., Ponomarenko, L.A., Rustamov, A.M.: Methods for analysis of queuing models with instantaneous and delayed feedbacks. Commun. Comput. Inf. Sci. **564**, 185–199 (2015)
4. Krishnamoorthy, A., Manjunath, A.S.: On queues with priority determined by feedback. Calcutta Stat. Assoc. Bull. **70**(1), 33–56 (2018)
5. Boxma, O.J., Yechiali, U.: An M/G/1 queue with multiple types of feedback and gated vacations. J. Appl. Probab. **34**(3), 773–784 (1997)
6. Choi, B.D., Kim, B.: M/G/1 queuing system with fixed feedback policy. The ANZIAM J. **44**(2), 283–297 (2002)
7. Choi, B.D., Kim, B., Choi, S.H.: An M/G/1 queue with multiple types of feedback, gated vacations and FCFS policy. Comput. Oper. Res. **30**(9), 1289–1309 (2003)
8. Jewkes, E.M., Buzacott, J.A.: Flow time distributions in a K class M/G/1 priority feedback queue. Queuing Syst. **8**(2), 183–202 (1991)
9. Krishna, K.B., Madheswari, S.P., Vijayakumar, A.: The M/G/1 retrial queue with feedback and starting failures. Appl. Math. Model. **26**(11), 1057–1075 (2002)
10. Krishna, K.B., Vijayalakshmi, G., Krishnamoorthy, A., Sadiq, B.S.: A single server feedback retrial queue with collision. Comput. Oper. Res. **37**(7), 1247–1255 (2010)
11. Simon, B.: Priority queue with feedback. J. Assoc. Comput. Mach. **11**(1), 134–149 (1984)
12. Choi, B.D., Kulkarni, V.G.: Feedback retrial queuing systems. Stochast. Models Relat. Fields 93–105 (1992)
13. Choi, B.D., Chang, Y.: Single server retrial queues with priority calls. Math. Comput. Model. **30**, 7–32 (1999)
14. Choi, B.D., Kim, Y.C., Lee, Y.W.: The M/M/c retrial queue with geometric loss and feedback. Comput. Math. Appl. **36**(6), 41–52 (1998)
15. Lee, Y.W.: The M/G/1 feedback retrial queue with two types of customers. Bull. Korean Math. Soc. **42**(4), 875–887 (2005)
16. Dudin, A.N., Kazimirsky, A.V., Klimenok, V.I., Breuer, L., Krieger, U.: The queuing model MAP/PH/1/N with feedback operating in a Markovian random environment. Austrian J. Stat. **34**(2), 101–110 (2005)
17. Neuts, M.F.: Matrix-Geometric Solutions in Stochastic Models: An Algorithmic Approach, p. 332. John Hopkins University Press, Baltimore (1981)
18. Sharma, S.K., Kumar, R.: A Markovian feedback queue with retention of reneged customers. Adv. Model. Optim. **14**(3), 673–680 (2012)

19. Sharma, S.K., Kumar, R.: A Markovian feedback queue with retention of reneged customers and balking. Adv. Model. Optim. **14**(3), 681–688 (2012)
20. Sharma, S.K., Kumar, R.: M/M/1 feedback queueing model with retention of reneged customers and balking. Am. J. Oper. Res. **3**(2A), 1–6 (2013)
21. Sharma, S.K., Kumar, R.: A single-server Markovian feedback queuing system with discouraged arrivals and retention of reneged customers. Am. J. Oper. Res. **4**(3), 35–39 (2013)
22. Kumar, R., Jain, N.K., Som, B.K.: Optimization of an M/M/1/N feedback queue with retention of reneged customers. Oper. Res. Dec. **24**(3), 45–58 (2014)
23. Santkumaran, A., Thangaraj, V.: A single server queue with impatient and feedback customers. Inf. Manag. Sci. **11**(3), 71–79 (2000)
24. Som, B.K., Seth, S.: M/M/c/N queuing system with encouraged arrivals, reneging, retention and feedback customers. Yugoslav J. Oper. Res. **28**(3), 333–344 (2018)
25. Bouchentouf, A.A., Kadi, M., Rabhi, A.: Analysis of two heterogeneous server queuing model with balking, reneging and feedback. Math. Sci. Appl. E-notes **2**(2), 10–21 (2013)
26. Krishna-Kumar, B., Rukmani, R., Thangaraj, V.: On multi-server feedback retrial queue with finite buffer. Appl. Math. Modell. **33**, 2062–2083 (2009)
27. Ayyapan, G., Subramanian, A., Sekar, G.: M/M/1 retrial queuing system with loss and feedback under non-pre-emptive priority service by matrix geometric method. Appl. Math. Sci. **4**(48), 2379–2389 (2010)
28. Ayyapan, G., Subramanian, A., Sekar, G.: M/M/1 retrial queuing system with loss and feedback under pre-emptive priority service. Int. J. Comput. Appl. **2**(6), 27–34 (2010)
29. Fisher, W., Meier-Hellstern, K.: The Markov-modulated Poisson process (MMPP) cookbook. Perform. Eval. **18**, 149–171 (1992)
30. Melikov, A.Z., Ponomarenko, L.A., Rustamov, A.M.: Hierarchical space merging algorithm to analysis of open tandem queuing networks. Cybern. Syst. Anal. **52**(6), 867–877 (2016)

On Estimation of Weibull-Tail and Log-Weibull-tail Distributions for Modeling End-to-end Delay

Igor V. Rodionov[1,2(✉)] [ORCID]

[1] Trapeznikov Institute of Control Sciences of Russian Academy of Sciences, Profsoyuznaya ulitsa 65, 117997 Moscow, Russia
`vecsell@gmail.com`
[2] Steklov Mathematical Institute of Russian Academy of Sciences, Gubkina ulitsa 8, 117485 Moscow, Russia

Abstract. We model the distribution tails of the end-to-end delay of services with the help of two-parameter Weibull-tail and log-Weibull-tail distributions. First we discuss the role of the Weibull-tail and log-Weibull-tail distributions in statistics of extremes. Then we propose the statistical procedure based on the largest observations of a sample to decide between these two tail models and to estimate the parameters of the selected model. The proposed procedure allows us to estimate the extreme quantiles from the Weibull-tail and log-Weibull-tail distributions in an unified way. The efficiency of the introduced methods is illustrated on a small simulation study.

Keywords: Distribution tail · Extreme value analysis · Extreme quantile · Weibull-tail index · End-to-end delay distribution

1 Introduction.

The background for a systematic end-to-end delay analysis is given by video streaming services in Internet and their QoE/QoS assessment, [1]. However, in teletraffic engineering users often need the whole estimate of the cumulative distribution function (cdf) or the corresponding probability density function (pdf) of an underlying random variable (rv). In last decades, a lot of different models describing the cdf of the monitored delay have been investigated, [2–5], but regarding real data distributions of the "body" and the "tail", respectively, quite often differ and could not be described in the framework of one parametric model. Therefore, it becomes necessary to estimate the "body" and the tail of distribution separately. According to statistics of extremes [6,7], only the largest order statistics can be used for the estimation of the tail, whereas the "body", i.e. moderate values of a rv, can be estimated using standard statistical tools.

The work is supported by the Russian Science Foundation under grant No.19-11-00290.

V. M. Vishnevskiy et al. (Eds.): DCCN 2019, CCIS 1141, pp. 302–314, 2019.
https://doi.org/10.1007/978-3-030-36625-4_24

For instance, the methods of cdf and pdf estimation that combined the classical non-parametric kernel method of estimating the "body" and special tail estimation techniques were proposed in [8–10].

In this work, we model the distribution tails of the end-to-end delay of services in the Internet by means of two-parameter Weibull-tail and log-Weibull-tail distributions. We propose an effective testing procedure based on the largest order statistics of a sample to distinguish between these two classes and develop a method to estimate the distribution tail in both cases. This method also allows us to estimate the high level quantiles from both the considered models. The brief algorithm of the method is provided at the end of the Sect. 2. In Sect. 3, the finite sample behavior of the proposed estimators are evaluated on the basis of a simulation study.

1.1   Classical Approach to Tail Estimation

The classical extreme value theory deals with the rare event probabilities and studies the behavior of the largest observations in independent data. The central result in this theory is given by the extreme value theorem (cf. [11,12]). It states that if there exist sequences of constants $a_n > 0$, $b_n, n \in \mathbb{N}$, such that the cdf of the normalized maximum $M_n = \max(X_1, \ldots, X_n)$ of a sample $\{X_i\}_{i=1}^n$ tends to some non-degenerate cdf G, i.e.,

$$\lim_{n \to \infty} P(M_n \le a_n x + b_n) = G(x), \tag{1}$$

then there exist constants $a > 0$, b such that $G(ax + b) = G_\gamma(x)$, where

$$G_\gamma(x) = \exp\left(-(1 + \gamma x)^{-1/\gamma}\right), \quad 1 + \gamma x > 0, \tag{2}$$

$\gamma \in \mathbb{R}$, and for $\gamma = 0$ the right-hand side should be understood as $\exp(-e^{-x})$. The parameter γ is called the extreme value index [6] (EVI), it allows to classify the extreme-value distributions (2) into three classes. The cdf of a sample $(X_1, \ldots, X_n)$ is said to belong to the Fréchet (inverse-Weibull, Gumbel, respectively) maximum domain of attraction (MDA) if (1) holds for $\gamma > 0$ ($\gamma < 0$, $\gamma = 0$, respectively). It is known, that the Fréchet MDA contains the distributions with Pareto-like tails, whereas the distributions from the inverse-Weibull MDA have a finite right endpoint. The Gumbel MDA is in our interest. It mainly consists the distributions with the exponentially decreasing tails, including such distributions as normal, exponential, gamma, log-normal among others.

The problem of tail estimation is central in statistics of extremes. Now the most popular method of tail estimation is based on Pickands-Balkema-de Haan theorem (cf. [13,14]). It states that if the cdf F of the random variable X belongs to the MDA of G_γ (2) for some $\gamma \in \mathbb{R}$, then for x, such that $0 < x < (\min(0, -\gamma))^{-1}$, it holds

$$\lim_{t \uparrow x^*} P\left(\frac{X - t}{f(t)} > x \,|\, X > t\right) = (1 + \gamma x)^{-1/\gamma},$$

where $x_F^* = \sup\{x : F(x) < 1\}$, $f(t)$ is some positive nondecreasing function (for details, see [6], p. 65), and for $\gamma = 0$ the expression $(1 + \gamma x)^{-1/\gamma}$ should be understood as e^{-x}. Therefore, it is enough to estimate the EVI to obtain the estimate of a distribution tail. The most often used estimators of the EVI are the Hill estimator (that is consistent for $\gamma > 0$), the maximum likelihood estimator ($\gamma > -1$) and the moment estimator ($\gamma \in \mathbb{R}$), see for details [6], Chap. 3. For instance, in [8] the Hill estimator was applied to estimate the distribution tails of one-way delays and round-trip times in ATM- and IP-networks.

The latter approach yields good results for distributions belonging to the Fréchet or inverse-Weibull MDA, since these domains are fairly accurate described by the EVI. Contrariwise, $\gamma = 0$ holds for all distributions belonging to the Gumbel MDA, which is the widest of all MDAs. For instance, we cannot distinguish between the tails of the normal and log-normal distributions using the above approach and estimate the rate of decrease of distribution tails from the Gumbel MDA. Therefore, it is necessary to propose other methods and techniques to solve such problems.

1.2   Weibull-Type and Log-Weibull-type Distributions

Weibull-type and log-Weibull-type distributions constitute two rich subclasses in the Gumbel MDA. These two classes can be completely described only in semi-parametric way. We say, that cdf F is of Weibull-type, if for some $\theta > 0$

$$1 - F(x) = \exp(-x^\theta \ell(x)), \quad x > 0, \tag{3}$$

where $\ell(x)$ is the so-called slowly varying function, i.e. $\ell(\lambda x)/\ell(x) \to 1$ as $x \to \infty$ for all positive λ. The parameter θ in (3) is called the Weibull-tail index. Note, that its lower values indicate slower tail decay. Clearly, distributions such as normal, exponential, gamma and Weibull distributions are of Weibull-type. In recent years, several estimators of the Weibull-tail index have been proposed, we refer to [15–17]. In addition, a general estimation method of a distribution tail parameter proposed in [18] can be adapted for Weibull-tail index estimation. The estimator introduced by Beirlant et al. [15] was applied to evaluate the distribution tail of the end-to-end delay of advanced services in the Internet in [5].

One can define the log-Weibull-type cdf $F(x)$ as a cdf such that $F(e^x)$ is of Weibull-type. In other words, F is of log-Weibull-type, if for some $\theta > 0$

$$1 - F(x) = \exp(-(\ln x)^\theta \ell^*(x)), \quad x > 1,$$

where $\ell^*(x)$ is such that for all $\varepsilon > 0$ there exists $x_0 = x_0(\varepsilon)$ such that

$$(\ln x)^{-\varepsilon} \le \ell^*(x) \le (\ln x)^\varepsilon$$

holds for all $x > x_0$. The last relation follows from Drees' inequality, see [6], p. 369. If the value of the parameter θ, called the log-Weibull-tail index, is more than 1, then the corresponding cdf belongs to the Gumbel MDA. A cdf of log-Weibull-type lies in the Fréchet MDA in case $\theta = 1$ and does not belong to any

MDA, if $\theta \in (0,1)$. An example of a distribution of log-Weibull-type belonging to the Gumbel MDA is the log-normal distribution. The problem of log-Weibull-tail index estimation has not been considered yet in the literature separately. However, the methods, proposed in [18] and [19], allows to derive the estimate of this tail index. Moreover, estimators of the Weibull-tail index can also be applied for this problem.

The problem of distinguishing between two corresponding classes of distribution tails is of special interest. In [20], the general test to distinguish between two separable classes of distribution tails was proposed, see also Sect. 2.1 of this work. One can apply this test to the latter problem. Goegebeur and Guillou [21] introduced the goodness-of-fit test to check whether the distribution is of Weibull-type, see also [22].

2 Parameter Estimation of Weibull-Tail and Log-Weibull-tail Models

In this section we propose a statistical procedure to decide between two-parametric Weibull-tail and log-Weibull-tail models and, further, to estimate the parameters of the selected one. These two models are the particular cases of two semi-parametric classes of distribution tails introduced in the previous section. We say, that a cdf F has a Weibull tail with the Weibull-tail index $\theta > 0$ and the scale parameter $c > 0$ (hereafter $F \sim WT(\theta, c)$), if

$$1 - F(x) = \exp(-(x/c)^{\theta}), \ x > x_0$$

holds with some sufficiently large $x_0 > 0$. If x_0 is equal to 0, then we get a classical Weibull distribution. If a cdf F has the following tail

$$1 - F(x) = \exp(-(\ln(x/c))^{\theta}), \ x > x_0$$

with some sufficiently large $x_0 > c$, $\theta > 0$, $c > 0$, then we say, that F has a log-Weibull tail, $F \sim LWT(\theta, c)$. This model with $x_0 = c$ is a special case of the Weibull-Pareto model introduced in [23], called the log-Weibull distribution. Krieger and Markovich [5] modeled the response times in terms of the Weibull-Pareto model to evaluate the QoE performance of the end-to-end delay dependent services using the MOS metric. Notice, that if rv Y obeys $WT(\theta, c)$, then its transformation $X = \exp(cY)/d$ has the log-Weibull tail with the log-Weibull-tail index θ and the scale parameter d.

2.1 Model Selection

We consider the problem of distinguishing between Weibull-tail and log-Weibull-tail models for modeling the distribution tails of the observed data. We know by previous studies (cf. [8,24]), that in most practical cases the traffic characteristics, i.e. call or session durations or inter-arrival times, can be regarded as independent identically distributed (iid) observations. Let $\mathbf{X} = (X_1, \ldots, X_n)$ be an iid sample with a cdf F, and let $X_{(1)} \leq \ldots \leq X_{(n)}$ denote the nth order statistics of a sample $\mathbf{X}$. Denote for $k < n$

$$S_{k,n} = \ln \frac{k}{n} - \frac{1}{k} \sum_{i=n-k+1}^{n} \ln \left(1 - F_0 \left(\frac{X_{(i)}}{X_{(n-k)}} u_{k/n} \right) \right), \tag{4}$$

with some continuous cdf F_0 and $u_{k/n} = F_0^{-1}(1 - k/n)$, the $(1 - k/n)$-quantile of F_0. The introduced statistic $S_{k,n}$ is a slight modification of the statistic $R_{k,n}$ proposed in [20],

$$R_{k,n} = \ln(1 - F_0(X_{(n-k)})) - \frac{1}{k} \sum_{i=n-k+1}^{n} \ln \left(1 - F_0(X_{(i)}) \right),$$

see also [18]. But in contrast to $R_{k,n}$, the statistic $S_{k,n}$ is not sensitive to scale parameter changes, what makes it more preferable in our approach.

Consider the null hypothesis $H_0 : F$ has a log-Weibull tail against the alternative $H_1 : F$ has a Weibull tail. Select $F_0(x) = \{1 - \exp\{-\exp(-\sqrt{\ln x^2})\}\}I(x > 1)$, as it was recommended in [20]. Then the test to distinguish between H_0 and H_1

$$\text{if } S_{k,n} < 1 + \frac{\gamma_\alpha}{\sqrt{k}}, \text{ then reject } H_0, \tag{5}$$

has an asymptotic significance level α and is consistent against the alternative H_1, as $k \to \infty$, $k/n \to 0$, $n \to \infty$, here γ_α is the α-quantile of the standard normal distribution, [20].

The selection of the parameter k remains the problem. There is no general method to select the threshold, or equivalently the number k of the largest order statistics to be used for testing procedures related to extreme events, [25]. However, in the framework of the proposed testing procedure to distinguish between log-Weibull-tail and Weibull-tail models the values of the statistic $\sqrt{k}(S_{k,n} - 1)$ typically decrease with respect to increasing k, if $\mathbf{X} \sim WT(\theta, c)$, but there is usually no tendency in the behavior of the test statistic, if $\mathbf{X} \sim LWT(\theta, c)$, see also Fig. 1. Our recommendation is to select $k \approx \sqrt{n}$. With such values of k, the proposed testing procedure yields the best performance in the numerical simulations, [20].

2.2   Parameter Estimation of the Selected Model

Let us turn to the problem of estimating the parameters of the selected model. Let $\mathcal{F} = \{F_\theta, \theta \in \Theta\}$, $\Theta \subseteq \mathbb{R}$ be the parametric family of distribution tails. Here we assume that θ is a shape parameter. Consider the generalization of the statistic (4)

$$S_{k,n}(\theta) = \ln \frac{k}{n} - \frac{1}{k} \sum_{i=n-k+1}^{n} \ln \left(1 - F_\theta \left(\frac{X_{(i)}}{X_{(n-k)}} u_{k/n}(\theta) \right) \right), \tag{6}$$

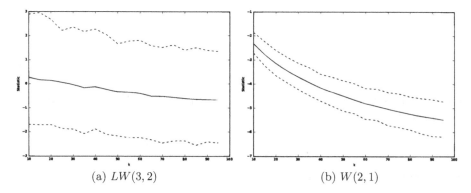

(a) $LW(3,2)$ (b) $W(2,1)$

Fig. 1. Empirical median (solid line) and empirical 95% confidence interval (dashed lines) of $\sqrt{k}(R_{k,n} - 1)$ as a functions of k, $k = 5,\ldots,100$ obtained from 1000 samples of size $n = 1000$ from (a) Log-Weibull distribution with parameters $\theta = 3$ and $c = 2$ and (b) Weibull distribution with parameters $\theta = 2$ and $c = 1$.

where $u_{k/n}(\theta) = F_\theta^{-1}(1-k/n)$, the $(1-k/n)$-quantile of F_θ. Assume, that the tail of a cdf F of a sample $\mathbf{X}$ is of $\mathcal{F}$-type, i.e. $F(x/c) = F_{\theta_0}(x)$ for some $c > 0$, $\theta_0 \in \Theta$ and all sufficiently large x. Then the following estimator of the parameter θ

$$\widehat{\theta} = \arg\{\theta : S_{k,n}(\theta) = 1\} \tag{7}$$

is consistent under some weak conditions imposed on $\mathcal{F}$ as $k \to \infty$, $\frac{k}{n} \to 0$, $n \to \infty$, [18].

This approach can be adapted to estimate the parameter θ in the framework of the Weibull-tail and log-Weibull-tail models. Indeed, assume that we check the hypothesis $H_0 : F \sim LWT(\theta, c)$ using the test (5) and accept it. To obtain the estimate of θ using the proposed estimation technique, we must choose the log-Weibull parametric family of cdfs $\mathcal{F}^{LW} = \{F_\theta^{LW}, \theta > 1\}$ with $F_\theta^{LW}(x) = 1 - \exp(-(\ln x)^\theta)$, $x > 1$, in (6) and find the solution $\widehat{\theta}$ of the equation $S_{k,n}(\theta) = 1$. This solution is unique with the probability tending to 1 as $n \to \infty$, [18]. On the contrary, if we reject the hypothesis $H_0 : F \sim LWT(\theta, c)$ and thus accept the hypothesis $H_1 : F \sim WT(\theta, c)$, we must choose the Weibull parametric family $\mathcal{F}^W = \{F_\theta^W, \theta > 0\}$ with $F_\theta^W(x) = 1 - \exp(-x^\theta)$, $x > 0$, in (6). The estimators of the Weibull-tail index mentioned above can also be applied to estimate the parameter θ in the framework of the Weibull-tail model. However, the estimator (7) shows the better simulation results in most cases, than other estimators of the Weibull-tail index, [18].

The problem to select k for Weibull-tail index estimation has not been investigated in the literature. We can only propose a heuristic method, that is similar to one accepted for EVI estimation. We suggest to plot the values of the selected θ estimator as a function of k and to choose a "stable" point of this plot, [25].

It remains to propose the method to estimate the scale parameter c. Using the estimate $\widehat{\theta}$ of the parameter θ, the estimate $\widehat{c}$ of the parameter c is derived as follows (cf. [26])

$$\widehat{c} = \left(\prod_{i=0}^{k-1} X_{(n-i)} \Big/ \prod_{i=1}^{k} u_{i/(n+1)}(\widehat{\theta}) \right)^{1/k}, \tag{8}$$

where $u_\alpha(\widehat{\theta})$ is the α-quantile of $F_{\widehat{\theta}}$ and F_θ is F_θ^W or F_θ^{LW} depending on the model selected.

2.3   Extreme Quantile Estimation

An extreme quantile u_{p_n} of order p_n is defined by $u_{p_n} = F^\leftarrow(1-p_n)$ with $p_n \to 0$ as $n \to \infty$, [27]. Here $F^\leftarrow(t) = \inf\{x : F(x) \geq t\}$, the so-called generalized reverse function, that coincides with F^{-1} if F is strictly monotone and continuous. If $np_n \to \infty$ as $n \to \infty$, then the extreme quantile u_{p_n} is larger almost surely than the maximum observation of the sample. This requires extrapolating the sample values to areas where no data are observed. We refer to applications of extreme quantiles in reliability [28], finance [29], climatology [30], among other, see also [6]. In [5], the extreme quantiles of MOS metric are estimated under the assumption that the response time obeys the Weibull-Pareto distribution.

From the definition of the families $WT(\theta, c)$ and $LWT(\theta, c)$, we easily obtain

$$u_{p_n}^W = u_{p_n}^W(\theta, c) := F^\leftarrow(1 - p_n) = c(-\ln p_n)^{1/\theta}, \tag{9}$$

with $F \sim W(\theta, c)$ and

$$u_{p_n}^{LW} = u_{p_n}^{LW}(\theta, c) = c \exp\left((-\ln p_n)^{1/\theta} \right) \tag{10}$$

with $F \sim LW(\theta, c)$. Thus plugging the estimates of the parameters θ and c in (9) and (10) one derives the estimates $\widehat{u}_{p_n}^W$ and $\widehat{u}_{p_n}^{LW}$ of the extreme quantiles $u_{p_n}^W$ and $u_{p_n}^{LW}$ respectively.

Another approach to extreme quantile estimation, similar to those proposed in [17,27], allows us to suggest the estimator depending only on the estimate of the shape parameter θ. Let $F \sim WT(\theta, c)$. From (9) for quantiles u_s^W and u_t^W of F with $0 < s < t < 1$ we have

$$\ln u_s^W - \ln u_t^W = \frac{1}{\theta}(\ln(-\ln s) - \ln(-\ln t)). \tag{11}$$

Substituting in (11) $t = k/n$, $s = p_n$, the empirical quantile $X_{(n-k)}$ and some estimate $\widehat{\theta}$ instead of u_t^W and θ respectively, an estimator of the extreme quantile $u_{p_n}^W$ can be derived by

$$\widetilde{u}_{p_n}^W = X_{(n-k)} \left(\frac{\ln p_n}{\ln(k/n)} \right)^{1/\widehat{\theta}}.$$

Similar reasoning leads us to the following estimator of the extreme quantile $u_{p_n}^{LW}$

$$\widetilde{u}_{p_n}^{LW} = X_{(n-k)} \exp\left\{ (-\ln p_n)^{1/\widehat{\theta}} - (\ln(n/k))^{1/\widehat{\theta}} \right\}.$$

A comparison of the proposed estimators by numerical simulation is given in Sect. 3.

2.4 Algorithm

Finally, we provide a brief computational algorithm to implement our estimation method efficiently:

1. Select $k : \sqrt{n} \leq k \leq 2\sqrt{n}$, e.g. $k = \sqrt{n}$. Select the significance level α, e.g. $\alpha = 0.05$.
2. Use the largest values $x_{(n-k)} \leq \ldots \leq x_{(n)}$ of the observations $(x_1, \ldots, x_n)$ to compute the statistics $S_{k,n}$

$$S_{k,n} = -\ln \frac{n}{k} + \frac{1}{k} \sum_{i=n-k+1}^{n} \exp \left(\sqrt{2 \ln \left(\frac{x_{(i)}}{x_{(n-k)}} u_{k/n} \right)} \right),$$

where $u_{k/n} = \exp \left(\frac{1}{2} (\ln \ln \frac{n}{k})^2 \right)$.
3. Let γ_α be the α-quantile of $N(0,1)$ distribution. If $S_{k,n} < 1 + \frac{\gamma_\alpha}{\sqrt{k}}$, select the Weibull-tail model, if not, the log-Weibull-tail model.
4. If the Weibull-tail model is selected, let $F_\theta(x) = 1 - \exp(-x^\theta)$, $x > 0$. If not, $F_\theta(x) = 1 - \exp(-(\ln x)^\theta)$, $x > 1$.
5. For each $k = 5m$, $2 \leq m \leq \sqrt{n}/5$, compute the statistic $S_{k,n}(\theta)$ (6) as a function of θ and derive $\hat{\theta}_k$ as a solution of the equation $S_{k,n}(\theta) = 1$. If the Weibull-tail model is selected, then $u_{k/n}(\theta) = (\ln(n/k))^{1/\theta}$ should be substituted in (6); if the log-Weibull-model is selected, then $u_{k/n}(\theta) = \exp((\ln(n/k))^{1/\theta})$.
6. Plot the values of $\hat{\theta}_k$ as a function of k and derive the estimate of θ from the stable point $(\hat{k}, \hat{\theta})$ of this plot.
7. Using the estimate $\hat{\theta}$ and the optimal number of the largest observations $\hat{k}$, derive the estimate of the scale parameter c, (8).

3   Simulation Study

In this section, we compare numerical properties of the proposed estimator (7) adapted to the Weibull-tail index estimation with three other estimators of the Weibull-tail index and study the behavior of the introduced extreme quantile estimators $\hat{u}_{p_n}^{W}$, $\hat{u}_{p_n}^{LW}$, $\tilde{u}_{p_n}^{W}$ and $\tilde{u}_{p_n}^{LW}$.

First, we consider the estimator of Girard [31], denoted by $\hat{\theta}_1$, the estimator of Beirlant et al. [15], denoted by $\hat{\theta}_2$, and the estimator of Gardes et al. [17], denoted by $\hat{\theta}_3$, of the Weibull-tail index θ. The comparison of these estimators with our estimator $\hat{\theta}_0 := \hat{\theta}$ (7) is provided for 4 different distributions:

1. an exponential distribution with 5 as scale parameter ($\theta = 1$);
2. a Weibull distribution with 4 as shape parameter and 3 as scale parameter ($\theta = 4$);

3. a normal distribution $N(0, 100)$ $(\theta = 2)$;
4. a Gamma distribution with 2 as shape and scale parameters $(\theta = 1)$.

For each considered distribution, $M = 300$ samples of size $n = 1000$ were simulated. For each sample, the estimates $\hat{\theta}_{l,j}$, $l = 0, 1, 2, 3$, $1 \leq j \leq M$ are computed for $k = 10, 15, \ldots, 150$, the empirical means and empirical mean squared errors of each estimator are built by plotting the pairs $\left(k, \frac{1}{M} \sum_j \hat{\theta}_{l,j}\right)$ and $\left(k, \frac{1}{M} \sum_j (\hat{\theta}_{l,j} - \theta)^2\right)$. The true value of θ is indicated by a solid line (Figs. 2 and 3).

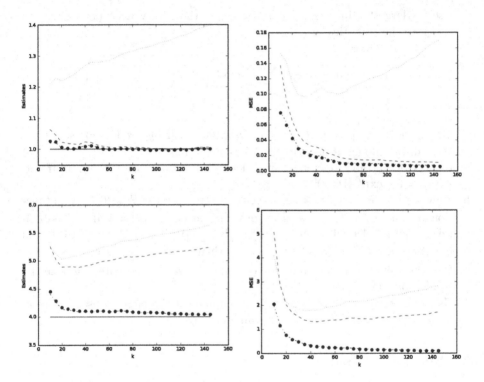

Fig. 2. Empirical mean and empirical MSE of $\hat{\theta}_0$ (circles), $\hat{\theta}_1$ (dotted line), $\hat{\theta}_2$ (dashed line) and $\hat{\theta}_3$ (dotted-dashed line) as a functions of k, $k = 10, \ldots, 150$ obtained from $Exp(5)$ (upper) and Weibull distribution with parameters $\theta = 4$ and $c = 3$ (bottom).

It appears that our estimator outperforms the estimators $\hat{\theta}_1$ and $\hat{\theta}_2$ and seems a bit better than the estimator $\hat{\theta}_3$ of Gardes et. al. However, it should be recalled that the proposed method can be applied not only for estimation of the Weibull-tail index, but for estimation of the shape parameter of an arbitrary family of distributions belonging to the Gumbel MDA.

The numerical comparison of the proposed empirical quantile estimators is provided on Figs. 4 and 5 for 4 different distributions:

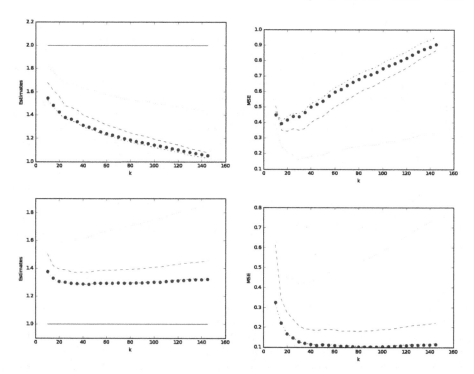

Fig. 3. Empirical mean and empirical MSE of $\hat{\theta}_0$ (circles), $\hat{\theta}_1$ (dotted line), $\hat{\theta}_2$ (dashed line) and $\hat{\theta}_3$ (dotted-dashed line) as a functions of k, $k = 10, \ldots, 150$ obtained from $N(0, 100)$ (upper) and $\Gamma(2, 2)$ (bottom).

1. a log-normal distribution with 0 as location parameter and $1/3$ as scale parameter;
2. a log-Weibull distribution with 3 as shape parameter and 2 as scale parameter;
3. a normal distribution $N(0, 9)$;
4. a Weibull distribution with 0.5 as shape parameter and 5 as scale parameter.

For each considered distribution, $M = 300$ samples of size $n = 500$ were simulated. For each sample, the estimates $\hat{u}^{LW}_{p_n,j}$, $\tilde{u}^{LW}_{p_n,j}$, $1 \leq j \leq M$ for the first two distributions and $\hat{u}^{W}_{p_n,j}$, $\tilde{u}^{W}_{p_n,j}$, $1 \leq j \leq M$ for the last two distributions are computed for $p_n = 0.95, 0.951, \ldots, 0.999$ with $k = 50$. The empirical means of each estimator are built by plotting the pairs $(p_n, \frac{1}{M} \sum_j u_j^*)$. The true quantile values are indicated by a solid line.

It appears that the extreme quantile estimates obtained with and without use of the scale parameter estimator $\hat{c}$ have almost the same behavior and both show a good numerical performance.

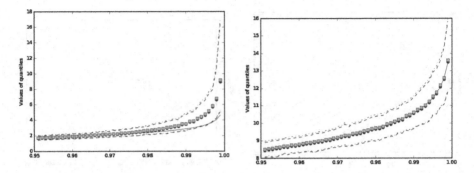

Fig. 4. Empirical means and empirical 95% confidence intervals of $\hat{u}_{p_n}^{LW}$ (squares and dotted-dashed lines) and $\tilde{u}_{p_n}^{LW}$ (circles and dashed lines) as a functions of p_n, $p_n = 0.95,\ldots,0.999$ obtained from $LN(0,1/9)$ (left) and $LW(3,2)$ (right).

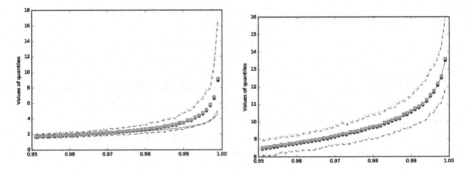

Fig. 5. Empirical means and empirical 95% confidence intervals of $\hat{u}_{p_n}^{W}$ (squares and dotted-dashed lines) and $\tilde{u}_{p_n}^{W}$ (circles and dashed lines) as a functions of p_n, $p_n = 0.95,\ldots,0.999$ obtained from $N(0,9)$ (left) and $W(0.5,5)$ (right).

4 Conclusion

The real data distribution cannot always be well described within the framework of one parametric model, since distributions of the moderate observations and the "tail", respectively, quite often differ. Whereas moderate sample values can be estimated using standard statistical tools like non-parametric kernel method, estimation of the tail often remains a parametric problem.

In order to model the distribution tails of the end-to-end delay of user services in the Internet the Weibull-tail and log-Weibull-tail distributions can be considered. Firstly, we discuss the role of these two tail models in statistics of extremes. Next, we propose the statistical procedure based on the largest order statistics of a sample to distinguish between these two models and to estimate the parameters of the selected model. The proposed procedure also allows us to estimate the extreme quantiles of the Weibull-tail and log-Weibull-tail distributions in a similar way. Finally, we provide a compact computational algorithm to implement our estimation method efficiently. The numerical performance of the proposed methods is evaluated by a simulation study.

References

1. Hoßfeld, T., Schatz, R., Krieger, U.R.: QoE of YouTube video streaming for current internet transport protocols. In: Fischbach, K., Krieger, U.R. (eds.) MMB&DFT 2014. LNCS, vol. 8376, pp. 136–150. Springer, Cham (2014). https://doi.org/10.1007/978-3-319-05359-2_10
2. Zhang, W., He, J.: Modeling end-to-end delay using Pareto distribution. In: Second International Conference on Internet Monitoring and Protection, ICIMP 2007, 4271767 (2007)
3. Wang, Y., Vuran, M.C., Goddard, S.: Cross-layer analysis of the end-to-end delay distribution in wireless sensor networks. IEEE/ACM Trans. Networking 20(1), 305–318 (2012)
4. Hernandez, J.-A., Phillips, I.W.: Weibull mixture model to characterise end-to-end Internet delay at coarse time-scales. IEEE Proc. - Commun. 153(2), 295–304 (2006)
5. Krieger, U.R., Markovich, N.M.: Estimation of a heavy-tailed weibull-pareto distribution and its application to QoE modeling. In: Vishnevskiy, V.M., Kozyrev, D.V. (eds.) DCCN 2018. CCIS, vol. 919, pp. 21–30. Springer, Cham (2018). https://doi.org/10.1007/978-3-319-99447-5_3
6. de Haan, L., Ferreira, A.: Extreme Value Theory: An Introduction. Springer Series in Operations Research and Financial Engineering. Springer, New York (2006). https://doi.org/10.1007/0-387-34471-3
7. Beirlant, T., Goegebeur, Y., Teugels, J., Segers, J.: Statistics of Extremes: Theory and Applications. Wiley, New York (2004)
8. Markovitch, N.M., Krieger, U.R.: The estimation of heavy-tailed distributions, their mixtures and quantiles. Comput. Netw. 40(3), 459–474 (2002)
9. Barron, A.R., Györfi, L., van der Meulen, E.: Distribution estimation consistent in total variation and in two types of information divergence. IEEE Trans. Inf. Theory 38, 1437–1454 (1992)
10. Markovich, L.A.: Light- and heavy-tailed density estimation by Gamma-Weibull Kernel. In: 3rd ISNPS, Avignon, France, pp. 145–158 (2016)
11. Fisher, R.A., Tippett, L.H.C.: Limiting forms of the frequency distribution in the largest particle size and smallest number of a sample. Proc. Cambridge Philos. Soc. 24, 180–190 (1928)
12. Gnedenko, B.V.: Sur la distribution limite du terme maximum d'une serie aleatoire. Ann. Math. 44, 423–453 (1943)
13. Pickands III, J.: Statistical inference using extreme order statistics. Ann. Statist. 3, 119–131 (1975)
14. Balkema, A.A., de Haan, L.: Residual life time at great age. Ann. Probab. 2, 792–804 (1974)
15. Beirlant, J., Broniatowski, M., Teugels, J.L., Vynckier, P.: The mean residual life function at great age: applications to tail estimation. J. Statist. Plan. Inference 45, 21–48 (1995)
16. Balakrishnan, N., Kateri, M.: On the maximum likelihood estimation of parameters of Weibull distribution based on complete and censored data. Stat. Probab. Lett. 78, 2971–2975 (2008)

17. Gardes, L., Girard, S., Guillou, A.: Weibull tail-distributions revisited: a new look at some tail estimators. J. Statist. Plan. Infer. **141**(4), 429–444 (2011)
18. Rodionov, I.: On parametric estimation of distribution tails. In: 4th ISNPS. Salerno, Italy (2018, in press)
19. El Methny, J., Gardes, L., Girard, S., Guillou, A.: Estimation of extreme quantiles from heavy and light tailed distributions. J. Statist. Plann. Inf. **142**(10), 2735–2747 (2012)
20. Rodionov, I.V.: On discrimination between classes of distribution tails. Probl. Inf. Transm. **54**(2), 124–138 (2018)
21. Goegebeur, J., Guillou, A.: Goodness-of-fit testing for Weibull-type behavior. J. Statist. Plan. Infer. **140**(6), 1417–1436 (2010)
22. Rodionov, I.V.: A discrimination test for tails of Weibull-type distributions. Theory Probab. Appl. **63**(2), 327–335 (2018)
23. Alzaatreh, A., Famoye, F., Lee, C.: Weibull-Pareto distribution and its applications. Commun. Stat. Theor. Methods **42**(9), 1673–1691 (2013)
24. Roppel, C.: Estimating cell transfer delay and cell delay variation in ATM networks: measurement techniques and results. Eur. Trans. Telecomm. **10**(1), 13–21 (1999)
25. Gomes, M.I., Guillou, A.: Extreme value theory and statistics of univariate extremes: a review. Int. Statist. Rev. **83**(2), 263–292 (2015)
26. Akhtyamov, P.I., Rodionov, I.V.: On estimation of the scale and location parameters of distribution tails. Fundam. Appl. Math. 23(1) (2019, in press)
27. Weissman, I.: Estimation of parameters and larger quantiles based on the k largest observations. J. Am. Stat. Assoc. **73**(364), 812–815 (1978)
28. Ditlevsen, O.: Distribution arbitrariness in structural reliability. In: Structural Safety and Reliability, Balkema, Rotterdam, pp. 1241–1247 (1994)
29. Embrechts, P., Klüppelberg, C., Mikosch, T.: Modelling Extremal Events: for Insurance and Finance. AM, vol. 33. Springer, Heidelberg (1997). https://doi.org/10.1007/978-3-642-33483-2
30. Hassanzadeh, E., Nazemi, A., Adamowski, J., Nguyen, T.-H., Van-Nguyen, V.-T.: Quantile-based downscaling of rainfall extremes: notes on methodological functionality, associated uncertainty and application in practice. Adv. Water Resour. **131**, 103371 (2019)
31. Girard, S.: The Hill-type of the Weibull tail-coefficient. Comm. Stat. Theor. Meth. **33**(2), 205–234 (2004)

Asymptotic Analysis of the Output Process in Retrial Queue with Markov-Modulated Poisson Input Under Low Rate of Retrials Condition

Ivan Lapatin$^{(\boxtimes)}$(iD) and Anatoly Nazarov(iD)

Institute of Applied Mathematics and Computer Science,
National Research Tomsk State University, 36 Lenina Avenue, Tomsk 634050, Russia
ilapatin@mail.ru, nazarov.tsu@gmail.com

Abstract. We consider retrial queue with Markov-modulated Poisson input process and exponential probability distribution of service durations. The object of our research is output flow of the system. We use asymptotic analysis method under low rate of retrials limit condition to obtain probability distribution of the number of served customers at the moment t. The obtained formulae has explicit expression and contains matrix exponential. Furthermore, we show that the output belongs to the class of Markovian arrival processes.

Keywords: Output process · Retrial queue · Markov-modulated poisson process · Markovian arrival process · Queueing system · Asymptotic analysis method

1 Introduction

Classic queueing models [8] arose, first of all, as an analytic tool for making decisions in production management. Moreover, the nature of the simulated processes can be very diverse (for example, students waiting for a consultation with their professor or servicing insurance company contracts). At the same time, two main classes of models were distinguished - systems with expectation and systems with losses.

The rapid development of information technologies has led to the need to simulate the operation of telecommunication systems, mobile networks, call centers, etc., the functioning of which uses various modifications of random multiple access protocols. To simulate such processes, another class of queuing models with repeated calls or the RQ-system (Retrial Queueing System) is used.

A feature of this class of models is that the application received in the system while another application is being serviced is not lost, but joins an orbit

The reported study was funded by RFBR according to the research project No 18-01-00277.

V. M. Vishnevskiy et al. (Eds.): DCCN 2019, CCIS 1141, pp. 315–324, 2019.
https://doi.org/10.1007/978-3-030-36625-4_25

and after a while repeats the attempt to occupy the server. Various modifications of such phenomena arise in cellular communication systems, random access communication systems and others. Such situations can be caused not only by the lack of free servers at the receipt time of application, but by some technical reasons.

The first works on similar models appeared in the middle of the twentieth century [5,16], but most of the work falls on recent decades. A detailed description of retrial queues and a review of the classical results of their research are presented in monographs [1,6]. Currently, there are various modifications of service disciplines in retrial queues, which allow more adequately simulate various processes in applied problems. For example, in [2,13], retrial queues with two-way communication, in which the server processes both incoming and outgoing calls, are considered.

Most of the studies on the aforementioned models are devoted to the numerical analysis, asymptotic analysis and simulation of the number of customers in the system or in the orbit. Although one of the main characteristics that determines the quality of the communication system is the number of customers served by the system per unit of time. Information on the characteristics of the output process is of great practical interest, since often the output process of one system is input to another. The research results of outgoing flows in queuing networks are widely used in the modeling of computer systems, in the design of data transmission networks and in the analysis of complex multi-stage production processes.

Note that to date, insufficient attention has been paid to the study of output processes. This is due to the fact that the output process depends on the functioning of the entire system, which greatly complicates the analytical study. However, there are no general approaches to their study, although information on their characteristics is of great interest - in particular, in the study of tandem models and multiphase systems.

The main results of an analytical study of the output processes in the framework of the classical theory were made in the middle of the twentieth century [3,7,15]. An analysis of the output flow in systems with cyclic servicing can be found in [14]; in [9], the output of a single server system with correlated input process is considered. A more detailed review of the outgoing flows can be found in [10].

In [11], an asymptotic probability distribution of the number of events of the outgoing flow in a classical retrial queue with the Poisson input process was obtained. In this paper, we propose a study of the output process of Markovian retrial queue with Markov modulated Poisson input, which has a more complex structure and can more adequately describe real telecommunication processes.

Analysis of the proposed model is carried out by the method of asymptotic analysis. The used approach makes it possible to find the form of the limiting distribution of the number of events in the output flow in the $MMPP/M/1$ retrial queue under low rate of retrials asymptotic condition.

2   Mathematical Model and Problem Definition

We consider single server retrial queue with Markov-modulated Poisson input [4,12]. Let $k(t)$ is a continuous time Markov chain with finite set of states $k = 1, 2, \ldots, K$ and denotes an underlying process of MMPP. The infinitesimal generator of $k(t)$ is defined by matrix $\mathbf{Q}$ of q_{ij} elements. Diagonal matrix $\mathbf{\Lambda}$ contains conditional arrival rates λ_k.

Upon arrival a customer occupies the server if it is idle. Service duration is an exponentially distributed random variable with rate μ. The customer that finds the server busy joins the orbit and makes a random delay for an exponentially distributed time with rate σ and repeats his request for service.

Our goal is to obtain probability distribution of the number of served customers in the system.

3   Kolmogorov System of Equations

We introduce following denotes to clarify the analysis:

$k(t)$ is the state of MMPP underlying process;

$i(t)$ is the number of customers in the system at the moment t;

$n(t)$ is the state of server at the moment t: 0 if the server is idle, 1 if the server is busy;

$m(t)$ is the number of served customers at the moment t.

We denote the probability distribution of the three-dimensional process

$$P\left\{n(t) = n, k(t) = k, i(t) = i, m(t) = m\right\} = P_n(k, i, m, t). \tag{1}$$

As the process $\{n(t), k(t), i(t), m(t)\}$ is Markovian then the probability distribution (1) is the unique solution of Kolmogorov system of differential equations

$$\frac{\partial P_0(k,i,m,t)}{\partial t} = -(\lambda_k + i\sigma)P_0(k,i,m,t) + \mu P_1(k,i+1,m-1,t)$$

$$+ \sum_{v=1}^{K} P_0(v,i,m,t)q_{vn},$$

$$\frac{\partial P_1(k,i,m,t)}{\partial t} = -(\lambda_k + \mu)P_1(k,i,m,t) + i\sigma P_0(k,i,m,t)$$

$$+ \lambda_k P_0(k,i-1,m,t) + \lambda_k P_1(k,i-1,m,t) + \sum_{v=1}^{K} P_1(v,i,m,t)q_{vn}. \tag{2}$$

Let $H_n(k, u_1, u, t)$ denotes the partial characteristic functions

$$H_n(k, u_1, u, t) = \sum_{i=0}^{\infty} \sum_{m=0}^{\infty} e^{ju_1 i} e^{jum} P_n(k, i, m, t), \tag{3}$$

where $j = \sqrt{-1}$. Then using (2) we obtain

$$\frac{\partial H_0(k,u_1,u,t)}{\partial t} = -\lambda_k H_0(k,u_1,u,t) + j\sigma \frac{\partial H_0(k,u_1,u,t)}{\partial u_1}$$

$$+ \mu e^{-ju_1} e^{ju} H_1(k,u_1,u,t) + \sum_{v=1}^{K} H_0(k,u_1,u,t) q_{vn},$$

$$\frac{\partial H_1(k,u_1,u,t)}{\partial t} = \lambda_k e^{ju_1} H_0(k,u_1,u,t) - j\sigma \frac{\partial H_0(k,u_1,u,t)}{\partial u_1}$$

$$- (\lambda_k + \mu) H_1(k,u_1,u,t) + \lambda e^{ju_1} H_1(k,u_1,u,t) + \sum_{v=1}^{K} H_1(k,u_1,u,t) q_{vn}.$$

Denoting

$$\mathbf{H}_n(u_1,u,t) = \{H_n(1,u_1,u,t), H_n(2,u_1,u,t),..,H_n(K,u_1,u,t)\},$$

we rewrite the system in following form

$$\frac{\partial \mathbf{H}_0(u_1,u,t)}{\partial t} =$$

$$\mathbf{H}_0(u_1,u,t)(\mathbf{Q} - \mathbf{\Lambda}) + \mu e^{-ju_1} e^{ju} \mathbf{H}_1(u_1,u,t) + j\sigma \frac{\partial \mathbf{H}_0(u_1,u,t)}{\partial u_1}, \tag{4}$$

$$\frac{\partial \mathbf{H}_1(u_1,u,t)}{\partial t} = e^{ju_1} \mathbf{H}_0(u_1,u,t)\mathbf{\Lambda}$$

$$+ \mathbf{H}_1(u_1,u,t)(\mathbf{Q} + (e^{ju_1} - 1)\mathbf{\Lambda} - \mu\mathbf{I}) - j\sigma \frac{\partial \mathbf{H}_0(u_1,u,t)}{\partial u_1},$$

where $\mathbf{I}$ is unit matrix of K dimensions.

The obtained system (4) completely describes the process of states changes in $MMPP/M/1$ retrial queue which includes the output process. Some characteristics of the random processes $k(t)$, $n(t)$, $i(t)$ and $m(t)$ can be derived using the obtained system (4) with marginal distribution consistency condition.

4   Asymptotic Analysis Method

We will investigate the obtained system of equations (4) using asymptotic analysis method under low rate of retrials condition ($\sigma \to 0$).

Denoting $\sigma = \varepsilon$ we introduce the following notations in the system (4)

$$u_1 = \varepsilon w, \quad \mathbf{H}_n(u_1,u,t) = \mathbf{F}_n(w,u,t,\varepsilon),$$

to obtain the system of equations (5)

$$\frac{\partial \mathbf{F}_0(w,u,t,\varepsilon)}{\partial t} =$$

$$\mathbf{F}_0(w,u,t,\varepsilon)(\mathbf{Q} - \mathbf{\Lambda}) + \mu e^{-j\varepsilon w} e^{ju} \mathbf{F}_1(w,u,t,\varepsilon) + j \frac{\partial \mathbf{F}_0(w,u,t,\varepsilon)}{\partial w}, \tag{5}$$

$$\frac{\partial \mathbf{F}_1(w,u,t,\varepsilon)}{\partial t} = e^{j\varepsilon w} \mathbf{F}_0(w,u,t,\varepsilon)\mathbf{\Lambda}$$

$$+ \mathbf{F}_1(w,u,t,\varepsilon)(\mathbf{Q} + (e^{j\varepsilon w} - 1)\mathbf{\Lambda} - \mu\mathbf{I}) - j \frac{\partial \mathbf{F}_0(w,u,t,\varepsilon)}{\partial w}.$$

The solution of the system of equations (5) is formulated in Lemma and Theorem.

Lemma 1. *Let $i(t)$ is the number of customers in $MMPP/M/1$ retrial queue, then in the stationary regime we obtain*

$$\lim_{\varepsilon \to 0}\{\mathbf{F}_0(w,0,t,\varepsilon) + \mathbf{F}_1(w,0,t,\varepsilon)\} = \lim_{\sigma \to 0} Me^{jw\sigma i(t)} = e^{jw\kappa}, \tag{6}$$

where κ is positive root of equation

$$\mathbf{r}\Lambda\mathbf{e} - \mu + \mu^2\mathbf{r}(\Lambda + (\kappa + \mu)\mathbf{I} - \mathbf{Q})^{-1}\mathbf{e} = 0, \tag{7}$$

Vector $\mathbf{R}$ has K dimensions and occurs to be the stationary probability distribution of the states of underlying process $k(t)$. Vector $\mathbf{R}$ is the unique solution of the system $\mathbf{RQ} = 0$, $\mathbf{Re} = 1$, where $\mathbf{e}$ is unit vector.

Proof. In the system (5) we set $u = 0$, which removes the process $m(t)$ from consideration. The resulting system of equations for the process $\{k(t), n(t), i(t)\}$ we consider in stationary regime, which allows us to get away from the time derivatives.

Denoting

$$\mathbf{F}_n(w,\varepsilon) = \lim_{t \to \infty} \mathbf{F}_n(w,0,t,\varepsilon),$$

we rewrite the system in following form

$$\mathbf{F}_0(w,\varepsilon)(\mathbf{Q} - \Lambda) + \mu e^{-j\varepsilon w}\mathbf{F}_1(w,\varepsilon) + j\frac{\partial \mathbf{F}_0(w,\varepsilon)}{\partial w} = 0,$$

$$e^{j\varepsilon w}\mathbf{F}_0(w,\varepsilon)\Lambda + \mathbf{F}_1(w,\varepsilon)(\mathbf{Q} + (e^{j\varepsilon w} - 1)\Lambda - \mu\mathbf{I}) - j\frac{\partial \mathbf{F}_0(w,\varepsilon)}{\partial w} = 0. \tag{8}$$

To derive an additional equation we sum up the equations of the system (8) and multiply the result by a unit column vector $\mathbf{e}$ of K dimensions on the right. Denoting $\mathbf{F}(w,\varepsilon) = \mathbf{F}_0(w,\varepsilon) + \mathbf{F}_1(w,\varepsilon)$, we obtain

$$(e^{j\varepsilon w} - 1)\mathbf{F}(w,\varepsilon)\Lambda\mathbf{e} + \mu(e^{-j\varepsilon w} - 1)\mathbf{F}_1(w,\varepsilon)\mathbf{e} = 0. \tag{9}$$

Taking the limit as $\varepsilon \to 0$ in the system (8), denoting the following functions $\mathbf{F}_n(w) = \lim_{\varepsilon \to 0} \mathbf{F}_n(w,\varepsilon)$, we write

$$\mathbf{F}_0(w)(\mathbf{Q} - \Lambda) + \mu\mathbf{F}_1(w) + j\frac{\partial \mathbf{F}_0(w)}{\partial w} = 0,$$

$$\mathbf{F}_0(w)\Lambda + \mathbf{F}_1(w)(\mathbf{Q} - \mu\mathbf{I}) - j\frac{\partial \mathbf{F}_0(w)}{\partial w} = 0. \tag{10}$$

We assume the solution of the system (10) can be obtained in the following form

$\mathbf{F}_n(w) = \mathbf{R}_n\Phi(w)$, then we rewrite this system as follows

$$\mathbf{R}_0(\mathbf{Q} - \Lambda) + \mu\mathbf{R}_1 + j\frac{\partial\Phi'(w)}{\partial\Phi(w)} = 0,$$

$$\mathbf{R}_0\Lambda + \mathbf{R}_1(\mathbf{Q} - \mu\mathbf{I}) - j\frac{\partial\Phi'(w)}{\partial\Phi(w)} = 0. \tag{11}$$

Denote, that vectors $\mathbf{R}_0$ and $\mathbf{R}_1$ are two-dimensional probability distribution of the random process $\{k(t), n(t)\}$. Since the relation $\Phi'(w)/\Phi(w)$ doesn't depend on w the function $\Phi(w)$ itself is an exponential

$$\Phi(w) = e^{j\kappa w}, \frac{\Phi'(w)}{\Phi(w)} = j\kappa.$$

Then the system (11) has following form

$$\mathbf{R}_0(\mathbf{Q} - \mathbf{\Lambda}) + \mu\mathbf{R}_1 - \kappa\mathbf{R}_1 = 0,$$

$$\mathbf{R}_0\mathbf{\Lambda} + \mathbf{R}_1(\mathbf{Q} - \mu\mathbf{I}) + \kappa\mathbf{R}_1 = 0. \tag{12}$$

From (12) we obtain the expression for $\mathbf{R}_0$

$$\mathbf{R}_0 = \mu\mathbf{R}(\mathbf{\Lambda} + (\kappa + \mu)\mathbf{I} - \mathbf{Q})^{-1}. \tag{13}$$

Returning back to the Eq. (9) we lay out the exhibitors in Taylor series, divide both sides of the equation by $j\varepsilon w$ and take the limit as $\varepsilon \to 0$. Given the form of the solution for $\mathbf{F}_n(w) = \mathbf{R}_n\Phi(w)$, we obtain

$$\mathbf{R}\mathbf{\Lambda}\mathbf{e} = \mu - \mu\mathbf{R}_0\mathbf{e}. \tag{14}$$

Vector $\mathbf{R} = \mathbf{R}_0 + \mathbf{R}_1$ occurs to be the stationary probability distribution of the states of underlying process $k(t)$. Vector $\mathbf{R}$ is the unique solution of the system $\mathbf{R}\mathbf{Q} = 0$, $\mathbf{R}\mathbf{e} = 1$, where $\mathbf{e}$. Substituting the expression (11) for $\mathbf{R}_0$ in (14), we obtain the equations for κ

$$\mathbf{r}\mathbf{\Lambda}\mathbf{e} - \mu + \mu^2\mathbf{r}(\mathbf{\Lambda} + (\kappa + \mu)\mathbf{I} - \mathbf{Q})^{-1}\mathbf{e} = 0$$

which coincides with (7). The lemma is proved.

Theorem 1. *Let $m(t)$ is the number of served customers in $MMPP/M/1$ retrial queue during t, then*

$$\lim_{\varepsilon \to 0}\{\mathbf{F}_0(0, u, t, \varepsilon) + \mathbf{F}_1(0, u, t, \varepsilon)\} = \lim_{\sigma \to 0} Me^{jum(t)} = \mathbf{R}e^{\mathbf{G}(u)t}\mathbf{e}, \tag{15}$$

where the matrix $\mathbf{G}(u)$ contains four blocks of $K \times K$ dimensions given as follows

$$\mathbf{G}(u) = \begin{bmatrix} (\mathbf{Q} - \mathbf{\Lambda} - \kappa\mathbf{I}) & (\mathbf{\Lambda} + \kappa\mathbf{I}) \\ \mu e^{ju}\mathbf{I} & (\mathbf{Q} - \mu\mathbf{I}) \end{bmatrix},$$

vector $\mathbf{R} = \{\mathbf{R}_0, \mathbf{R}_1\}$ has $2K$ dimensions and its blocks $\mathbf{r}_0$ and $\mathbf{r}_1$ are two-dimensional probability distribution of the random process $\{k(t), n(t)\}$ and appear as the solution of the system

$$\mathbf{R}_0 = \mu\mathbf{r}(\mathbf{\Lambda} + (\kappa + \mu)\mathbf{I} - \mathbf{Q})^{-1},$$

$$\mathbf{R}_1 = \mathbf{r} - \mathbf{r}_0. \tag{16}$$

Proof. We consider the system (5). Taking the limit as $\varepsilon \to 0$ in this system, denoting the following functions

$$\mathbf{F}_n(w, u, t) = \lim_{\varepsilon \to 0} \mathbf{F}_n(w, u, t, \varepsilon)$$

we rewrite the system in following form

$$\frac{\partial \mathbf{F}_0(w,u,t)}{\partial t} = \mathbf{F}_0(w, u, t)(\mathbf{Q} - \mathbf{\Lambda}) + \mu e^{ju}\mathbf{F}_1(w, u, t) + j\frac{\partial \mathbf{F}_0(w,u,t)}{\partial w},$$

$$\frac{\partial \mathbf{F}_1(w,u,t)}{\partial t} = \mathbf{F}_0(w, u, t)\mathbf{\Lambda} + \mathbf{F}_1(w, u, t)(\mathbf{Q} - \mu\mathbf{I}) - j\frac{\partial \mathbf{F}_0(w,u,t)}{\partial w}. \tag{17}$$

We assume the solution of the system (17) can be obtained in the following form

$$\mathbf{F}_n(w, u, t) = \Phi(w)\mathbf{F}_n(u, t).$$

Then we rewrite this system as follows

$$\frac{\partial \mathbf{F}_0(u,t)}{\partial t} = \mathbf{F}_0(u, t)(\mathbf{Q} - \mathbf{\Lambda}) + \mu e^{ju}\mathbf{F}_1(u, t) + j\frac{\partial \Phi'(w)}{\partial \Phi(w)},$$

$$\frac{\partial \mathbf{F}_1(u,t)}{\partial t} = \mathbf{F}_0(u, t)\mathbf{\Lambda} + \mathbf{F}_1(u, t)(\mathbf{Q} - \mu\mathbf{I}) - j\frac{\partial \Phi'(w)}{\partial \Phi(w)}. \tag{18}$$

Here the function $\Phi(w)$ the function makes sense of the asymptotic approximation of the characteristic function of the process $i(t)$. Its form has been found in the proof of Lemma.

$$\Phi(w) = e^{j\kappa w}, \frac{\Phi'(w)}{\Phi(w)} = j\kappa.$$

Taking it into account, the system (18) has following form

$$\frac{\partial \mathbf{F}_0(u,t)}{\partial t} = \mathbf{F}_0(u, t)(\mathbf{Q} - \mathbf{\Lambda} - \kappa\mathbf{I}) + \mu e^{ju}\mathbf{F}_1(u, t),$$

$$\frac{\partial \mathbf{F}_1(u,t)}{\partial t} = \mathbf{F}_0(u, t)(\mathbf{\Lambda} + \kappa\mathbf{I}) + \mathbf{F}_1(u, t)(\mathbf{Q} - \mu\mathbf{I}). \tag{19}$$

After that we move on to the vector notation for the component $n(t)$. Note that the system under consideration (19) is written in vector form with respect to the component $k(t)$ and the vector rows $\mathbf{F}_n(u, t)$ of K dimension. We denote vector $\mathbf{FF}(u, t) = \{\mathbf{F}_0(u, t), \mathbf{F}_1(u, t)\}$, which has $2K$ dimension, and matrix $\mathbf{G}(u)$

$$\mathbf{G}(u) = \begin{bmatrix} (\mathbf{Q} - \mathbf{\Lambda} - \kappa\mathbf{I}) & (\mathbf{\Lambda} + \kappa\mathbf{I}) \\ \mu e^{ju}\mathbf{I} & (\mathbf{Q} - \mu\mathbf{I}) \end{bmatrix},$$

blocks of which correspond to the matrix coefficients for unknowns in the system (19). Then this system can be presented in form of ordinary differential equation in the matrix form

$$\frac{\partial \mathbf{FF}(u, t)}{\partial t} = \mathbf{FF}(u, t)\mathbf{G}(u). \tag{20}$$

To solve this equation we write an initial condition. At the moment of output process observation start ($t = 0$), the system was functioning in a stationary mode and the process $m(t) = 0$. Therefore, the initial condition can be written as follows

$$\mathbf{FF}(u, 0) = \mathbf{R}. \tag{21}$$

vector $\mathbf{R} = \{\mathbf{R}_0, \mathbf{R}_1\}$ has $2K$ dimensions and its blocks $\mathbf{R}_0$ and $\mathbf{R}_1$ are two-dimensional probability distribution of the random process $\{k(t), n(t)\}$. We write the solution of the obtained Cauchy problem (20)–(21).

$$\mathbf{FF}(u, t) = \mathbf{R}e^{\mathbf{G}(u)}.$$

Now we can find the asymptotic approximation of the characteristic function of the process $m(t)$

$$\lim_{\sigma \to 0} Me^{jum(t)} = \lim_{\varepsilon \to 0}\{\mathbf{F}_0(0, u, t, \varepsilon) + \mathbf{F}_1(0, u, t, \varepsilon)\}\mathbf{e}$$

$$= \mathbf{R}e^{\mathbf{G}(u)t}\mathbf{ee}.$$

here $\mathbf{ee}$ is a unit column vector of $2K$ dimensions. The obtained expression is equal to (15).

As the right part of (8) has the same form as the formulae for characteristic function of the number of events in Markovian arrival process [4, 12], then we can take the corresponding laying out and formulate the following Corollary.

Corollary 1. *Output process of considered retrial queue under low rate of retrials condition ($\sigma \to 0$) is synchronous Markovian arrival process [12] defined by infinitesimal generator* $\mathbf{Q1}$ *of underlying Markov chain of $2K$ dimensions*

$$\mathbf{Q1} = \begin{bmatrix} (\mathbf{Q} - \boldsymbol{\Lambda} - \kappa\mathbf{I}) & (\boldsymbol{\Lambda} + \kappa\mathbf{I}) \\ \mu\mathbf{I} & (\mathbf{Q} - \mu\mathbf{I}) \end{bmatrix},$$

and probability matrix $\mathbf{D}$ *of event occurrence in MAP at the moment of state changes of underlying Markov chain*

$$\mathbf{D} = \begin{bmatrix} \boldsymbol{0} & \boldsymbol{0} \\ \mathbf{I} & \boldsymbol{0} \end{bmatrix}, B = \mathbf{Q1} * \mathbf{D} = \begin{bmatrix} \boldsymbol{0} & \boldsymbol{0} \\ \mu\mathbf{I} & \boldsymbol{0} \end{bmatrix},$$

where sign $*$ is Hadamard product.

Proof. It was shown in [12] that the equation defining the characteristic function of the probability distribution of the number of events that occurred in the MAP for some time t has the form (15). Moreover, the matrix $\mathbf{G}(u)$ is determined using the matrices defining the MAP as follows

$$\mathbf{G}(u) = \mathbf{Q1} + (e^{ju} - 1)[\boldsymbol{\Lambda}\mathbf{1} + \mathbf{Q1} * \mathbf{D}].$$

The matrix $\mathbf{Q1}$ is an infinitesimal generator of the underlying Markov process; $\boldsymbol{\Lambda}\mathbf{1}$ is a matrix of events conditional intensities on the intervals of the control circuit states constancy; $\mathbf{D}$ is a probability matrix of events occurring during state switch of underlying Markov process. Now, the matrix $\mathbf{G}(u)$ obtained in the Theorem can be expressed in terms of the corresponding matrices defining the MAP.

$$\mathbf{G}(u) = \begin{bmatrix} (\mathbf{Q} - \boldsymbol{\Lambda} - \kappa\mathbf{I}) & (\boldsymbol{\Lambda} + \kappa\mathbf{I}) \\ \mu e^{ju}\mathbf{I} & (\mathbf{Q} - \mu\mathbf{I}) \end{bmatrix} = \begin{bmatrix} (\mathbf{Q} - \boldsymbol{\Lambda} - \kappa\mathbf{I}) & (\boldsymbol{\Lambda} + \kappa\mathbf{I}) \\ ((e^{ju} - 1)\mu\mathbf{I} + \mu\mathbf{I}) & (\mathbf{Q} - \mu\mathbf{I}) \end{bmatrix}$$

$$= \begin{bmatrix} (\mathbf{Q} - \boldsymbol{\Lambda} - \kappa\mathbf{I}) & (\boldsymbol{\Lambda} + \kappa\mathbf{I}) \\ \mu\mathbf{I} & (\mathbf{Q} - \mu\mathbf{I}) \end{bmatrix} + (e^{ju} - 1)\begin{bmatrix} 0 & 0 \\ \mu\mathbf{I} & 0 \end{bmatrix}.$$

In our case we obtain

$$\mathbf{Q1} = \begin{bmatrix} (\mathbf{Q} - \boldsymbol{\Lambda} - \kappa\mathbf{I}) & (\boldsymbol{\Lambda} + \kappa\mathbf{I}) \\ \mu\mathbf{I} & (\mathbf{Q} - \mu\mathbf{I}) \end{bmatrix},$$

and probability matrix $\mathbf{D}$ of event occurrence in MAP at the moment of state changes of underlying Markov chain

$$\mathbf{D} = \begin{bmatrix} 0 & 0 \\ \mathbf{I} & 0 \end{bmatrix}, \mathbf{Q1} * \mathbf{D} = \begin{bmatrix} 0 & 0 \\ \mu\mathbf{I} & 0 \end{bmatrix}, \boldsymbol{\Lambda}1 = \begin{bmatrix} 0 & 0 \\ 0 & 0 \end{bmatrix}$$

where sign $*$ is Hadamard product.

Since we expressed the matrix $\mathbf{G}(u)$ in terms of the matrices that determine the MAP, the output of the considered system when the asymptotic condition of a low rate of retrials is fulfilled belongs to the class of MAP. As the matrix of conditional intensities $\boldsymbol{\Lambda}1$ in this case turned out to be zero, this point process belongs to the class of synchronous Markov arrival processes.

5   Conclusion

In this paper, we have considered retrial queue with MMPP input and exponentially distributed service times. We have been provided the output process research using asymptotic analysis method. The results of the research has been formulated in Lemma, Theorem and Corollary.

In Lemma we have been derived an equation to obtain the normalized asymptotic mean number of customers in the system. This result is an auxiliary to the research of an output process and can be used to obtain the characteristics of retrial queue.

The main contribution of this paper we have been formulated in Theorem and its Corollary. During the proof of the Theorem we have been obtained an explicit expression (15) for an asymptotic approximation of characteristic function of the number of occurred events in $MMPP/M/1$ retrial queue output process. This formula requires calculating the matrix exponent. During conducting numerical experiments, we have been used the method of spectral decomposition of the matrix, in which the matrix was reduced to a diagonal form with eigenvalues on the diagonal using the matrix of eigenvectors. In this form, the matrix function can be calculated directly.

A very important fact is proved in the Corollary. We have been shown that it is possible not only to find the asymptotic distribution of the number of events in the output process, but also to tell which class of flows it belongs to. This is a very important result, which can greatly simplify the research of tandem systems.

References

1. Artalejo, J.R., Gómez-Corral, A.: Retrial Queueing Systems: A Computational Approach. Springer, Heidelberg (2008). https://doi.org/10.1007/978-3-540-78725-9
2. Artalejo, J., Phung-Duc, T.: Single server retrial queues with two way communication. Appl. Math. Model. **37**(4), 1811–1822 (2013)
3. Burke, P.J.: The output of a queuing system. Oper. Res. **4**(6), 699–704 (1956)
4. Dudin, A., Klimenok, V.: Corellated Flow Queueing Systems, p. 175. BSU Publications, Minsk (2000)
5. Elldin, A.: Approach to the theoretical description of call retries. Ericssion Tech. **23**(3), 345–407 (1967)
6. Falin, G., Templeton, J.G.: Retrial Queues, vol. 75. CRC Press, Boca Raton (1997)
7. Finch, P.: The output process of the queueing system m/g/1. J. R. Stat. Soc. Ser. B (Methodol.) **21**(2), 375–380 (1959)
8. Gnedenko, B.V., Kovalenko, I.N.: Introduction to Queuing Theory. Science (1987)
9. Green, D.A.: Departure processes from MAP/PH/1 queues. Ph.D. thesis (1999)
10. Lapatin, I.L.: Research of mathematical models of the output flows in infinite-server queueing systems. Ph.D. thesis, Tomsk State University (2012)
11. Lapatin, I.L., Nazarov, A.A.: Research of the output process in retrial queue m/m/1 under low rate of retrials limit condition. In: Vishnevskiy, V.M., Kozyrev, D.V. (eds.) Distributed Computer and Communication Networks: Control, Computation, Communications (DCCN-2018), pp. 246–253. Springer, Cham (2018). https://doi.org/10.1007/978-3-319-99447-5
12. Lopuhova, S.V.: Asymptotic and numerical methods of research of special flows of homogeneous events. Ph.D. thesis, Tomsk State University (2008)
13. Nazarov, A.A., Paul, S., Gudkova, I.: Asymptotic analysis of markovian retrial queue with two-way communication under low rate of retrials condition. In: Proceedings 31st European Conference on Modelling and Simulation, pp. 678–693 (2017)
14. Projdakova, E., Fedotkin, M.: Definition of the existance conditions for the stationary distribution of output processes in queueing system with cyclic control. Bulletin of the Nizhny Novgorod University named after N. I. Lobachevsky Series Mathematics (1), 92–102 (2006)
15. Reich, E.: Waiting times when queues are in tandem. Ann. Math. Stat. **28**(3), 768–773 (1957)
16. Wilkinson, R.I.: Theories for toll traffic engineering in the USA. Bell Syst. Tech. J. **35**(2), 421–514 (1956)

An Upper Bound of the Large Deviation Probability in Multi-server Constant Retrial Rate System

Evsey Morozov[1,2] and Ksenia Zhukova[1,2(✉)]

[1] Karelian Research Centre, Institute of Applied Mathematical Research,
Pushkinskaya 11, Petrozavodsk, Russia
emorozov@karelia.ru, kalininaksenia90@gmail.com
[2] Petrozavodsk State University, Lenina 33, Petrozavodsk, Russia

Abstract. In this paper, a multiserver retrial system with constant retrial policy is considered. The input is assumed to be a general renewal process, service times are iid with a general distribution and the retrial attempts follow an exponential distribution. The system is described with a regenerative process, and we focus on the logarithmic asymptotics of the large deviation (overflow) probability that the orbit size reaches a high level N within regeneration cycle. To analyze the upper bound of this probability, we interpret the retrial system as a buffered system with service times of a special type. Simulation results show the accuracy of the obtained bound for $Weibull/M/2$ retrial system.

Keywords: Retrial system · Large deviations · Overflow probability · Upper bound · Constant retrial rate · Logarithmic asymptotics

1 Introduction

In this paper, we study the asymptotic of the large deviation probability in a multiserver retrial queueing system with constant retrial rate. In such systems, retrial rate remains constant and does not depend on the orbit size as long as the orbit is non idle. The retrial models are highly motivated by numerous applications in the modern telecommunications systems, see for instance [7–10]. In this work, we consider the *large deviation (overflow)* probability that the orbit size of the retrial system reaches a (high) level N within regeneration cycle. The large deviation asymptotics of the overflow probability, being an important QoS indicator, plays a critical role in the analysis of modern computer and telecommunication systems, see [2,3]. The problem of estimating the overflow probability in retrial systems has been previously considered, in particular, in [4,5] for the

The research was carried out under state order to the Karelian Research Centre of the Russian Academy of Sciences (Institute of Applied Mathematical Research KarRC RAS) and supported by the Russian Foundation for Basic Research, projects 18-07-00147, 18-07-00156 and 19-07-00303.

© Springer Nature Switzerland AG 2019
V. M. Vishnevskiy et al. (Eds.): DCCN 2019, CCIS 1141, pp. 325–337, 2019.
https://doi.org/10.1007/978-3-030-36625-4_26

queue size process in $MAP/G/1$ systems. In a recent work [14], an exponential large deviation asymptotics of the overflow probability *during regeneration cycle*, in the retrial systems with general renewal input and both classic and constant retrial rates, is established. More exactly, the lower and upper bounds for the decay rate of this probability are obtained based on the analysis developed in the recent paper [6] and on the well-known large deviation asymptotics for the queue size in classic buffered systems proved in [1]. In [14], the authors establish upper and lower bounds for the logarithmic asymptotic of the stationary overflow probabilities and the overflow probability during regeneration cycle, in a single-server retrial queue. Moreover, the asymptotical analysis in [14] covers also the multi-server case, however only for the case of overflow probability during the so-called *full busy cycle*, but not for a regeneration cycle. It is because the fundamental result obtained in [1] holds namely for the full busy cycle. On the other hand, the key idea of the approach in [14] was, based on the results of the work [1], to modify appropriately the original retrial system and interpret it as a classic buffered system.

We touch upon the approximation of the retrial system by a classical buffered system. A heuristic approach allowing to construct a classic buffered system which is being very close to a single-server system with constant retrial rate has been suggested in the work [13]. In the paper [14], the interpretation of a single-server retrial queue as a classic system is used to develop a large deviation analysis of the overflow probability during a regenerative cycle. While the estimation the overflow probability within full busy cycle is motivated by applications to the models of the modern telecommunication systems and computer networks, it is also quite natural to develop large deviation analysis of the overflow probability during classic regeneration cycle of the system. However the main difficulty to analyze the latter probability in the multiserver retrial system is that there are no results for this quantity which would be similar to that have been obtained in [1] for the overflow probability within full busy cycle.

This paper addresses this problem and develops large deviation asymptotics for the overflow probability during regeneration cycle using an alternative approach. The main contribution of this research is a large deviation analysis of the multiserver system based on the *moment properties of the regeneration cycle* of classic buffered systems. More exactly, we consider the approach based on the moment properties of the regeneration cycle of a classic multiserver system, obtained by Thorisson [11] which is combined with the asymptotic properties of the renewal processes (see, for instance, Asmussen [12]). Using this approach we obtain the *upper bound* of logarithmic asymptotic of the overflow probability during a regenerative cycle in multi-server retrial queue with general service times and with a special class of inter-arrival time distributions.

The paper is organized as follows. Section 2 contains description of the retrial system and the basic regenerative process. Section 3 contains the main analysis of the paper concerning the upper bound of the overflow probability during a regeneration cycle. In Sect. 4, simulation results are presented. Section 5 contains summary of the main results.

2 Multiserver Retrial System

We consider a $GI/GI/m$-type retrial system, denoted Σ, with a renewal input of customers arriving at the instants $\{t_n\}$, with the iid interarrival times $\tau_n := t_{n+1} - t_n$, $t_1 = 0$, and iid service times S_n, $n \geqslant 1$. We omit serial index to denote a generic element of an iid sequence. For instance, τ and S denote generic interarrival time and generic service time, respectively. We denote the input rate $\lambda = 1/\mathsf{E}\tau$ and service rate $\mu = 1/\mathsf{E}S$. Also introduce the traffic intensity $\rho = \lambda/\mu$. It is assumed that the system has $m \geqslant 1$ identical servers working in parallel. The system follows a retrial policy, namely, if a new customer finds the server busy, he joins an infinite-capacity virtual orbit and attempts to occupy server after an exponentially distributed time with rate γ. We consider *constant retrial policy*. It means that the total intensity of retrials from the orbit remains constant and equals γ regardless of the orbit size. In such a system, we may assume that customers in the orbit form a First-Come-First-Served queue, where only the *oldest customer* makes the attempts to enter server.

Define the basic process $\mathcal{Q} = \{Q(t), t \geqslant 0\}$, where $Q(t)$ is the number of customers in the system (in orbit and in server) at instant t^-. Denote $Q(t_k) = Q_k$, the number of customers which customer k finds upon arrival, $k \geqslant 1$. Now we define the classic *regenerations* of the process $\mathcal{Q}$. These regenerations occur when an arrival finds fully empty system. More exactly, let $Z_0 = 0$, and define the instances

$$Z_{n+1} = \min(t_k > Z_n : Q_k = 0), \quad n \geqslant 0, \tag{1}$$

which are (classic) the regenerations of the process $\mathcal{Q}$ [12]. Denote T the generic length of regeneration cycle, that is, $T =_{st} Z_{n+1} - Z_n$ for each n, where $=_{st}$ means stochastic equality. By construction, the regeneration cycle lengths are iid. In what follows we need the following condition

$$\mathsf{P}(\tau > S) > 0, \tag{2}$$

which is used to construct classic regenerations in the system Σ [9]. We assume that the retrial system Σ is *positive recurrent*, meaning that the length of regeneration cycle has finite mean, $\mathsf{E}T < \infty$. It is well-known that, under mild conditions, positive recurrence implies the existence of the stationary distribution of the regenerative process $\mathcal{Q}$ (for more detail see [12]). Thus, positive recurrence, implying *stability* of the system, is the main assumption required for our analysis.

In what follows, we study the logarithmic asymptotics of the (overflow) probability, that the number of customers in the stationary retrial system Σ reaches a level N during a regeneration cycle, as $N \to \infty$.

3 Asymptotic of the Overflow Probability During Regeneration Cycle

The logarithmic asymptotics of the overflow probability in the multi-server retrial system has been studied previously, in particular, the stationary probability P_N that the number of customers reaches a level N during a *full busy cycle* has been considered [14]. Note that a full busy cycle starts and ends by an arrival which meets exactly $m - 1$ busy servers. In the keystone paper [1]), J. Sadowsky obtained the logarithmic asymptotic of the overflow probability during full busy cycle in *classic multi-server system*. In turn, in the paper [14], the authors have extended this result to the large deviation analysis of the multi-server *retrial systems*. However, it seems difficult or impossible to extend this analysis further to capture the probability of overflow during classic regeneration cycle in the retrial system. On the other hand, such a generalization would be valuable because regenerations divide the paths of queueing process on iid segments (random elements), and the performance analysis of such a system is reduced to analysis of the system over regeneration cycle [12]. By this reason, the main purpose of the subsequent analysis is to obtain the upper bound of the logarithmic asymptotics of the probability that stationary queue Q in a multi-server retrial system with constant retrial rate exceeds a threshold N, that is

$$\mathcal{P}_N := \mathsf{P}(Q \geqslant N \text{ during regeneration cycle}). \tag{3}$$

It is useful for the subsequent analysis to recall the asymptotics of the stationary overflow probability during a full busy cycle found in [14].

For a random variable X, we denote the logarithmic (log) moment generating function, $\Lambda_X(\theta) = \log \mathsf{E} e^{\theta X}$, $\theta > 0$. Assume that function $\Lambda_S(\theta) = \log \mathsf{E} e^{\theta S}$ exists for some $\theta > 0$ and denote

$$\hat{\theta} = \sup(\theta > 0 : \mathsf{E} e^{\theta S} < \infty) > 0. \tag{4}$$

The following result has been proved in [14].

Theorem 1. *Assume that conditions (2) and*

$$\mathsf{P}(S > m\tau) > 0, \tag{5}$$

$$\rho + \frac{\lambda}{\gamma} < m \tag{6}$$

hold true. Then the system Σ is positive recurrent and the decay rate of the overflow probability satisfies the following lower and upper bounds

$$\Lambda_\tau(-m\theta_*) \leqslant \limsup_{N \to \infty} \frac{1}{N} \log \mathsf{P}_N \leqslant \Lambda_\tau(-m\theta^*), \tag{7}$$

where

$$\theta_* = \sup \left(\theta \in (0, \hat{\theta}) : \Lambda_\tau(-m\theta) + \Lambda_S(\theta) \leqslant 0 \right), \tag{8}$$

$$\theta^* = \sup(\theta > 0 : \Lambda_\tau(-m\theta) + \Lambda_S(\theta) + \log \frac{\gamma}{\gamma - \theta} \leqslant 0). \tag{9}$$

This result is based on comparison the original retrial system Σ with a classic *minorant and mojorant* multi-server systems, which in turn are studied in [1]. In the analysis developed in [14], the original retrial system, is replaced by a FCFS queueing system which is stochastically equivalent to original retrial system.

In the new (buffered) system, each new customer joins the "end" of the orbit-queue regardless of the state of server. This is a key point of analysis, and me focus on it in more detail.

If the server is busy upon an arrival, then the behaviour of this customer is identical in both systems. Otherwise, if a customer meets idle server upon arrival, then the oldest customer in the orbit-queue jumps to server to be served immediately, instead of the new arrival. The idle time of server after departure is then interpreted as a part of the service time of the departed customer. This replacement does not change distributional properties of the queue size, and both systems are equivalent from the point of view of stability. Such an interpretation of the retrial system allows to apply monotonicity properties to compare the number of customers in retrial system and classic FCFS queues and, as a result, to apply the analysis developed in [1].

Now we apply this idea for the analysis of the probability $\mathcal{P}_N$ defined by (3). To guarantee the existence of classic regenerations in the system, we need condition (2) to be held, while to accumulate an arbitrary large queue in the system, to study *large deviation*, we also need condition

$$\mathsf{P}(S > m\tau) > 0. \tag{10}$$

(For a more detailed discussion these conditions, see [14].)

For the subsequent analysis, we restrict the class of the renewal input processes. In this regard we recall that a distribution F is called *New-Better-than-Used (NBU)* if, for any $x, y \geqslant 0$, its tail $\bar{F} = 1 - F$ satisfies inequality (see [17])

$$\bar{F}(x + y) \leqslant \bar{F}(y)\bar{F}(x). \tag{11}$$

Note that Weibull distribution with density,

$$f(x) = \frac{a}{b}\left(\frac{x}{b}\right)^{a-1} e^{-(x/b)^a}, \; x \geqslant 0, \, b > 0, \, a \geqslant 1. \tag{12}$$

and the uniform distribution are examples of NBU distributions.

Introduce the total time which the process $\mathcal{Q}$ spends at or above level N during a regeneration cycle,

$$\mathcal{T}_N = \int_0^T \mathbf{1}(Q(t) \geqslant N)dt, \tag{13}$$

where $\mathbf{1}$ is an indicator function. It is easy to show that $\mathcal{T}_N$ is bounded from below by a positive constant independently of N. Denote the event

$$A = \{Q \geqslant N \text{ within a regeneration cycle}\}.$$

The proof of the following statement is similar to the proof of Lemma 2 in [14].

Lemma 1. *If (10) holds, then there exists a constant $c \in (0, \infty)$ such that, for all $N \geqslant 1$,*

$$\mathsf{E}[\mathcal{T}_N | A] \geqslant c. \tag{14}$$

For the positive recurrent process Q, there exists the weak limit (in distribution) $Q_n \Rightarrow Q_\infty$ that represents the stationary number of customers in the system *just before arrival*. Moreover, if the interarrival time τ is non-lattice, then there exists the weak limit $Q(t) \Rightarrow Q(\infty)$ representing the stationary number of customers in the system at *arbitrary time* [12].

Theorem 2. *Assume that, in the system Σ, interarrival time τ is non-lattice and has NBU distribution. Moreover assume that conditions (2) and (6) hold true. Then the system Σ is positive recurrent and the decay rate of the overflow probability (3) satisfies the following upper bound:*

$$\limsup_{N\to\infty} \frac{1}{N} \log \mathcal{P}_N \leqslant \limsup_{N\to\infty} \frac{1}{N} \log \mathsf{P}(Q(\infty) \geqslant N) \tag{15}$$

$$\leqslant \limsup_{N\to\infty} \frac{1}{N} \log \mathsf{P}(Q_\infty \geqslant N) \tag{16}$$

$$\leqslant \Lambda_\tau(-\tilde{\theta}), \tag{17}$$

where parameter $\tilde{\theta}$ is defined as

$$\tilde{\theta} = \min(\hat{\theta}, \gamma). \tag{18}$$

Proof. Following [14], first we show positive recurrence. Instead of the original retrial system we consider its modification described in previous Section. We do not distinguish these two systems since they are equivalent from the point of view of our analysis. Thus we keep the notation Σ for the modified system. Consider a classic FCFS m-server system $GI/GI/m$, denoted $\hat{\Sigma}$, with the same input as in the original system Σ, and with the iid service times $\{\hat{S}_n\}$ that are distributed as $\hat{S} = S + \xi$, where S_n is the service time of the nth customer in the original system and ξ is exponential with parameter γ. We remind that, since we consider a constant retrial rate, the delay of the top orbital customer, before to jump in server, is always distributed as exponential variable ξ. Hence, the following inequality holds

$$\hat{S}_n \geqslant S_n, \ n \geqslant 1. \tag{19}$$

Then it is easy to see that the well-known (classic) stability condition of the buffered system $\hat{\Sigma}$

$$\mathsf{E}\hat{S} < m\mathsf{E}\tau,$$

can be rewritten as condition (6). By (19), the queue size process in the system $\hat{\Sigma}$ dominates the corresponding process in the original system Σ [12,20]. Then it follows that both systems are positive recurrent regenerative and all required weak limits exist.

To prove inequalities (15) and (16), we can repeat the steps of the the proof of Lemma 3 in [14] (developed for the single-server retrial system). First, by the positive recurrence, we apply a regenerative argument, to obtain, as $t \to \infty$,

$$\frac{1}{t} \int_0^t \mathbf{1}(Q(u) \geqslant N) du \to \frac{\mathsf{E}\mathcal{T}_N}{\mathsf{E}\mathcal{T}} = \mathsf{P}(Q(\infty) \geqslant N). \tag{20}$$

Since $\mathsf{E}\mathcal{T}_N = \mathsf{E}[\mathcal{T}_N|A]\mathcal{P}_N$, then, combining (20), and (14), we obtain the same upper bound as in [14]:

$$\mathcal{P}_N \leqslant \frac{1}{c}\mathsf{E}\mathcal{T}\,\mathsf{P}(Q(\infty) \geqslant N). \tag{21}$$

To prove inequality (16), we show, as in [14], that

$$\mathsf{P}(Q(\infty) \geqslant N) \leqslant \mathsf{P}(Q_\infty + 1 \geqslant N). \tag{22}$$

Now, as in [14], denote by V_k the sojourn time of customer k in stationary regime, in which case the queue size Q_k is distributed as Q_∞. Note that $Q_k \geqslant N$ if the sojourn time V_{k-N+1} of customer $k-N+1$ exceeds the sum of $N-1$ interarrival times. Then it is a key observation that, unlike single-server case, now we obtain the following *inequality*:

$$\mathsf{P}(Q_k \geqslant N) \leqslant \mathsf{P}\left(V_{k-N+1} \geqslant \sum_{i=k-N+1}^{k-1} \tau_i\right), \tag{23}$$

where the sojourn time V_{k-N+1} and the $\sum_{i=k-N+1}^{k-1}\tau_i$ are *independent*. Then, as in [14], we obtain, by Chernoff bound [18], that for any $\theta > 0$,

$$\mathsf{P}\left(V_{k-N+1} \geqslant \sum_{i=k-N+1}^{k-1} \tau_i\right) \leqslant \mathsf{E}\left[e^{\theta V_{k-N+1}}\right]\mathsf{E}\left[e^{-\theta\sum_{i=k-N+1}^{k-1}\tau_i}\right]. \tag{24}$$

Recall that the sojourn time $V_{k-N+1} =_{st} V_\infty$. To prove inequality (17), we need to show the finiteness of the expectation $\mathsf{E}\left[e^{\theta V_{k-N+1}}\right]$. Namely in this point we apply another approach than that is developed in the single-server case in [14]. In the following discussion we assume that $\theta \in (0, \theta^*)$. In particular, then

$$\mathsf{E}e^{\theta\hat{S}} = \mathsf{E}e^{\theta S}\frac{\gamma}{\gamma-\theta} < \infty. \tag{25}$$

Denote $\hat{T}$ the length of regeneration period in the system $\hat{\Sigma}$, and let $F_{\hat{T}}$ be its distribution function. By the positive recurrence, $\mathsf{E}\hat{T} < \infty$. Moreover, the length $\hat{T}$ has the *same finite moments as service time* $\hat{S}$ [19]. Then, by (25), we obtain

$$\mathsf{E}e^{\theta\hat{T}} < \infty. \tag{26}$$

Define the remaining regeneration time at instant t,

$$\hat{T}(t) = \inf_k(t - \hat{T}_k : t - \hat{T}_k > 0),\ t \geqslant 0.$$

Since the interarrival time τ is non-lattice, then the weak limit $\hat{T}(t) \Rightarrow \hat{T}(\infty)$ exists and is the *stationary remaining regeneration time* with the distribution function [12]:

$$\mathsf{P}(\hat{T}(\infty) \leqslant x) = \frac{1}{\mathsf{E}\hat{T}}\int_0^x \mathsf{P}(\hat{T} > u)du$$

$$= \frac{1}{\mathsf{E}\hat{T}}\int_0^x (1 - F_{\hat{T}}(u))du =: G(x),\ x \geqslant 0.$$

Then, by (26),

$$
\mathsf{E}e^{\theta \hat{T}(\infty)} = \int_0^\infty e^{\theta x} dG(x) = \frac{1}{\mathsf{E}\hat{T}} \int_0^\infty e^{\theta x}(1 - F_{\hat{T}}(x))dx
$$

$$
= \frac{1}{\mathsf{E}\hat{T}\,\theta} \int_0^\infty e^{\theta x} dF_{\hat{T}}(x) = \frac{\mathsf{E}e^{\theta \hat{T}}}{\mathsf{E}\hat{T}\,\theta} < \infty, \tag{27}
$$

where we apply integration by part. In the system $\hat{\Sigma}$ we denote: the stationary waiting time $\hat{W}(\infty)$ and the stationary sojourn time $\hat{V}(\infty)$ at an arbitrary instant; the stationary waiting time $\hat{W}_\infty$ and the stationary sojourn time $\hat{V}_\infty$ at the arrival instants. Also let W_∞ and V_∞ be, respectively, the stationary waiting time and stationary sojourn time at the arrival instants in the original system Σ. (V_∞ in fact is used in (23)). The following stochastic relations are then evident:

$$
\hat{W}(\infty) \leqslant_{st} \hat{T}(\infty), \quad \hat{V}(\infty) =_{st} \hat{W}(\infty) + \hat{S}.
$$

Then it follows by (4), (18) and (27) that

$$
\mathsf{E}e^{\theta \hat{W}(\infty)} < \infty, \ \mathsf{E}e^{\theta \hat{V}(\infty)} = \mathsf{E}e^{\theta \hat{W}(\infty)}\,\mathsf{E}e^{\theta \hat{S}} < \infty, \tag{28}
$$

where the independence between $\hat{W}(\infty)$ and $\hat{S}$ is used. Since the interarrival time τ has NBU distribution, then $\hat{W}_\infty \leqslant_{st} \hat{W}(\infty)$, see [15–17], and thus, by (28),

$$
\mathsf{E}e^{\theta \hat{W}_\infty} \leqslant \mathsf{E}e^{\theta \hat{W}(\infty)} < \infty. \tag{29}
$$

Because $\hat{V}_\infty =_{st} \hat{W}_\infty + \hat{S}$, then (29) implies

$$
\mathsf{E}e^{\theta \hat{V}_\infty} = \mathsf{E}e^{\theta \hat{W}_\infty}\,\mathsf{E}e^{\theta \hat{S}} \leqslant C(\theta), \tag{30}
$$

where a constant $C(\theta) < \infty$ depends on θ. On the other hand, $\hat{S} \geqslant_{st} S$, and then $W_\infty \leqslant_{st} \hat{W}_\infty$ by the monotonicity of the waiting time and its stationary version, see [12,20,21]. This implies

$$
V_\infty =_{st} W_\infty + S \leqslant_{st} \hat{W}_\infty + \hat{S} =_{st} \hat{V}_\infty. \tag{31}
$$

Now, by (30), (31), we obtain that

$$
\mathsf{E}e^{\theta V_\infty} \leqslant \mathsf{E}e^{\theta \hat{V}_\infty} \leqslant C(\theta) < \infty \tag{32}
$$

for each $\theta \in (0, \tilde{\theta})$. Because

$$
V_k =_{st} V_\infty, \ Q_k =_{st} Q_\infty,
$$

then we finally obtain

$$
\limsup_{N \to \infty} \frac{1}{N} \log \mathsf{P}(Q_\infty \geqslant N) \leqslant \limsup_{N \to \infty} \frac{1}{N}\left(\log C(\theta) + \log \mathsf{E}e^{-\theta \sum_{i=k-N+1}^{k-1} \tau_i}\right)
$$

$$
= \Lambda_\tau(-\theta).
$$

Then the tightest bound (17) follows with $\theta = \tilde{\theta}$ [1].

4   Simulation Results

In this section we present simulation results to demonstrate the accuracy of the bound of the overflow probability $\mathcal{P}_N$ in a two-server $Weibull/M/2$ constant retrial rate system. We note that in this system conditions (2) and (10) always hold true.

Experiment 1. We simulate the retrial system with Weibull inter-arrival time (12) with parameters $a = 2$, $b = 1$, service rate $\mu = 2$ and two values of retrial rate, $\gamma = 10, 100$, which satisfy condition (6), and thus the statements of Theorem 2 hold true. Although it is a hard problem in general to derive function $\Lambda_\tau(\theta)$ in an explicit form, in this setting the upper bound is easily available (see [14] for details), implying $\tilde{\theta} = \hat{\theta} = \mu = 2$, see (18).

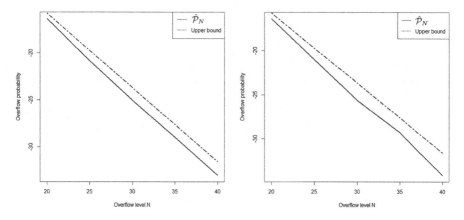

Fig. 1. The estimate $\hat{\mathcal{P}}_N$ of the overflow probability $\mathcal{P}_N$ vs. upper bound, for $Weibull/M/2$ retrial system with $\gamma = 10$ (left) and $\gamma = 100$ (right), logarithmic scale.

Figure 1 shows that the estimate is indeed located under the upper bound. Note that the upper bound does not depend on γ since in both cases $\gamma > \mu = \hat{\theta} = \min(\hat{\theta}, \gamma) = \tilde{\theta}$.

Experiment 2. Now we take $a = 2$, $b = 1$, $\mu = 10$ and $\gamma = 7$, $\gamma = 20$, so condition (6) is satisfied in both cases and the statements of Theorem 2 hold true as well. Figure 2 shows that the estimate of the overflow probability is located under the bound. If $\gamma = 7$ (and thus $\gamma < \mu = 10$) then $\tilde{\theta} = \min(\hat{\theta}, \gamma) = \gamma = 7$. In this case the upper bound depends on γ, and both bounds are quite tight, see Fig. 2 (left). When $\gamma = 20$, we obtain $\tilde{\theta} = \mu = 10$.

It is worth mentioning that the upper bound does not depend on γ, when $\gamma > \mu$, and does not depend on μ when $\gamma < \mu$.

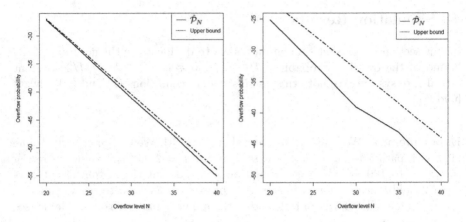

Fig. 2. The estimate $\hat{\mathcal{P}}_N$ of the overflow probability vs. upper bound, for $Weibull/M/2$ retrial system with $\gamma = 7$ (left) and $\gamma = 20$ (right), logarithmic scale.

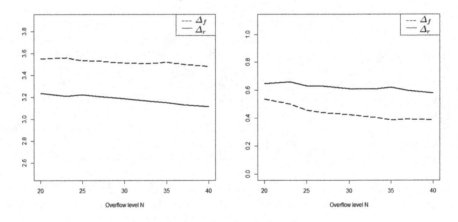

Fig. 3. The accuracy of the upper bounds, Δ_f vs. Δ_r, in $Weibull/M/2$ retrial system with $\gamma = 10$ (left) and $\gamma = 100$ (right).

Experiment 3. Again, we simulate the retrial system with parameters $a = 2$, $b = 1$, $\mu = 2$ and $\gamma = 10, 100$ to compare the accuracy of the upper bound $L := \Lambda_\tau(-m\theta^*)$ in (7) for the stationary overflow probability P_N *during full busy cycle* (studied in [14]), and the upper bound $\mathcal{L} := \Lambda_\tau(-\tilde{\theta})$ in (17) for the overflow probability $\mathcal{P}_N$ (during regeneration cycle). In this setting, the conditions of Theorems 1 and 2 hold true, and the upper bounds (7) and (17) can be easily calculated (see Experiment 1 and [14]). Then we calculate the estimates $\hat{\mathsf{P}}_N$ and $\hat{\mathcal{P}}_N$ of the probabilities P_N and $\mathcal{P}_N$, respectively, and consider the *accuracy* of the bound:

$$\Delta_f = L - \hat{\mathsf{P}}_N, \tag{33}$$

$$\Delta_r = \mathcal{L} - \hat{\mathcal{P}}_N. \tag{34}$$

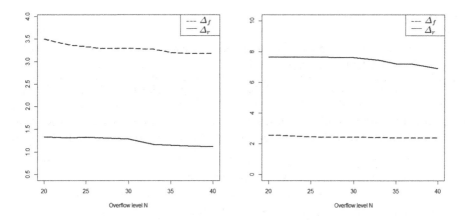

Fig. 4. The accuracy of the upper bounds, Δ_f vs. Δ_r, in $Weibull/M/2$ retrial system with $\gamma = 7$ (left) and $\gamma = 20$ (right).

Simulation results on Fig. 3 show that $\Delta_f > \Delta_r$ for $\gamma = 10$, so the upper bound of $\mathcal{P}_N$ is tighter than that for P_N, and situation is opposite for $\gamma = 100$.

Experiment 4. Now we repeat the steps of the Experiment 3 in the settings of the Experiment 2, that is we simulate the retrial system with parameters $a = 2$, $b = 1$, $\mu = 6$ and $\gamma = 7, 20$ and again we compare the accuracy of the upper bounds L and $\mathcal{L}$. Then we calculate values Δ_f (33) and Δ_r (34) and compare the accuracy of the bounds, see Fig. 4.

Thus, both experiments show that the upper bound of $\mathcal{P}_N$ is tighter than that for P_N, when rate γ is rather small, so it indicates that the upper bound $\mathcal{L}$ has a promising potential in large deviation analysis of the retrial systems.

5   Conclusion

A large deviation analysis of a multi-server retrial system with constant retrial policy is considered. Using the regenerative approach and renewal theory we construct an upper bound of the logarithmic asymptotics of the stationary overflow probability that the orbit size reaches a high level N within regeneration cycle. The key element of the analysis is the interpretation of the retrial system as a classic (buffered) system. We also present simulation results showing the accuracy of the bound for $Weibull/M/2$ retrial system with constant retrial rate.

References

1. Sadowsky, J.S.: Large deviations theory and efficient simulation of excessive backlogs in a GI/GI/m queue. IEEE Trans. Autom. Control **36**(12), 1383–1394 (1991)
2. Elwalid, A.I., Mitra, D.: Effective bandwidth of general Markovian traffic sources and admission control of high speen networks. IEEE/ACM Trans. Netw. **1**, 329–343 (1993)
3. Kelly, F.: Notes on effective bandwidths. In: Stochastic Networks: Theory and Applications. Royal Statistical Society Lecture Notes Series, vol. 4, pp. 141–168 (1996)
4. Kim, J., Kim, B.: Tail asymptotics for the queue size distribution in the MAP/G/1 retrial queue. Queueing Syst. **66**, 79–94 (2010)
5. Kim, J., Kim, B., Ko, S.-S.: Tail asymptotics for the queue size distribution in an M/G/1 retrial queue. J. Appl. Prob. **44**, 1111–1118 (2007)
6. Buijsrogge, A., de Boer, P.-T., Rosen, K., Scheinhardt, W.: Large deviations for the total queue size in non-Markovian tandem queues. Queueing Syst. **85**, 305–312 (2017)
7. Phung-Duc, T., Rogiest, W., Takahashi, Y., Bruneel, H.: Retrial queues with balanced call blending: analysis of single-server and multiserver model. Ann. Oper. Res. **239**, 429–449 (2016). https://doi.org/10.1007/s10479-014-1598-2
8. Aissani, A., Phung-Duc, T.: Profiting the idleness in single server system with orbit-queue. In: Proceedings of VALUETOOLS, Italy, pp. 5–7 (2007). https://doi.org/10.1145/3150928.3150929
9. Morozov, E.: A multiserver retrial queue: regenerative stability analysis. Queueing Syst. **56**, 157–168 (2007)
10. Morozov, E., Phung-Duc, T.: Stability analysis of a multiclass retrial system with classical retrial policy. Perform. Eval. **112**, 15–26 (2017)
11. Thorisson, H.: Coupling, Stationarity, and Regeneration. Springer, New York (2000)
12. Asmussen, S.: Applied Probability and Queues, 2nd edn. Springer, New York (2003). https://doi.org/10.1007/b97236
13. Zhukova, K.: Large deviations in retrial queues with constant retrial rates. In: Proceedings of the First International Workshop on Stochastic Modeling and Applied Research of Technology, Petrozavodsk, Russia, pp. 54–61 (2018)
14. Morozov, E., Zhukova, K.: A large deviation analysis of retrial models with constant and classic retrial rates. Perform. Eval. **135** (2019). https://doi.org/10.1016/j.peva.2019.102021, ISSN: 0166-5316
15. Morozov, E., Rumyantsev, A., Kalinina, K.: Inequalities for workload process in queues with NBU/NWU input. In: Gruca, A., Czachórski, T., Harezlak, K., Kozielski, S., Piotrowska, A. (eds.) ICMMI 2017. AISC, vol. 659, pp. 535–544. Springer, Cham (2018). https://doi.org/10.1007/978-3-319-67792-7_52
16. Miyazawa, M.: A Formal approach to queueing processes in the steady state and their applications. J. Appl. Probab. **16**, 332–346 (1979)
17. König, D., Schmidt, V.: Stochastic Inequalities between customer-stationary and time-stationary characteristics of queueing systems with point processes. J. Appl. Prob. **17**, 768–777 (1980). https://doi.org/10.2307/3212970
18. Chang, C.-S.: Performance Guarantees in Communication Networks. Springer, London (2000). https://doi.org/10.1007/978-1-4471-0459-9
19. Thorisson, H.: The queue GI/G/k: finite moments of the cycle variables and uniform rates of convergence. Stoch. Models Appl. **1**, 221–238 (1985)

20. Muller, A., Stoyan, D.: Comparison Methods for Stochastic Models and Risks. Wiley, Hoboken (2002)
21. Whitt, W.: Comparing counting processes and queues. Adv. Appl. Probab. **13**, 207–220 (1981)

Method of Moments for the Estimation of the Probability Density Parameters in Correlated Semi-synchronous Event Flow of the Second Order

Lyudmila Nezhelskaya and Diana Tumashkina[✉]

National Research Tomsk State University, 36 Lenin Ave., 634050 Tomsk, Russia
ludne@mail.ru, diana1323@mail.ru

Abstract. We consider a correlated semi-synchronous event flow of the second order with two states; it is one of the mathematical models for an incoming stream of claims (events) in modern digital integral servicing networks, telecommunication systems and satellite communication networks. We obtain an explicit form for a probability density of the values of the interval duration between the moments of the events occurrence and an explicit form of the joint probability density of the values of the adjacent intervals durations. We solve the problem of estimating the probability density parameters by the method of moments for general and special cases of setting the flow parameters. The results of statistical experiments performed on a flow simulation model are given.

Keywords: Doubly stochastic event flow · Correlated semi-synchronous event flow · Probability density function · Joint probability density function · Estimation of the parameters · Method of moments

1 Introduction

Nowadays when describing and analyzing real economic, technical, physical, and other processes and systems, it is quite often necessary to use mathematical models of queueing theory. The main task of queueing theory is to establish the relationship between the probabilistic characteristics that determine the functional capabilities of the queueing system and the effectiveness of its functioning. Thanks to the fast development of information technologies, another important fields of queueing theory applications are the design and creation of digital integrated service networks (DISN). Since in practice, the parameters defining the event flow change randomly in time, the doubly stochastic event flows are adequate mathematical models of information flows of messages operating in the DISN [1–8]. These flows are characterized by double randomness: the moments when events occur are random and the intensity of the flow is a random process.

However, the operating conditions of real processes and systems are such that the parameters of the incoming flows are unobservable, only the moments of

© Springer Nature Switzerland AG 2019
V. M. Vishnevskiy et al. (Eds.): DCCN 2019, CCIS 1141, pp. 338–351, 2019.
https://doi.org/10.1007/978-3-030-36625-4_27

the flow events occurrence are observed. Therefore, it is of considerable interest to solve the problems of estimating the states [9,10] and parameters [11–14] of doubly stochastic event flows from observations of the moments of events occurrence.

In this paper, we consider a doubly stochastic event flow, the accompanying random process of which is a piecewise constant with a finite number of states equal to two. Depending on how the transition from state to state occurs, these event flows can be divided into three types: (1) synchronous flows, the transition from state to state in which depends directly on the occurrence of the event; (2) asynchronous flows, the transition from state to state in which does not depend on whether an event has occurred or not; (3) semi-synchronous flows, for which the definition of the first type is true for one state and the second type for the second state. A semi-synchronous event flow of the second order is the object of studying in this work.

The problem of optimal states estimation for the considered event flow under its complete observability was solved in [10] and with partial observability in [15].

In this paper, we obtain the explicit form of the probability density of the values of the interval duration between the moments of the events occurrence for general and special cases of setting the parameters of a correlated semi-synchronous event flow of the second order, and we also obtain an explicit form of the joint probability density of the values of the durations of adjacent intervals. The problem of estimating density parameters is solved by the method of moments for each considered case. The quality of estimation is established by conducting a series of experiments on a simulation model of the flow under consideration.

2   Problem Setting

We consider the stationary operation mode of a semi-synchronous doubly stochastic event flow of the second order (hereinafter flow), the accompanying random process of which is a piecewise constant process $\lambda(t)$ with two states S_1 and S_2. Hereinafter, the ith state of the process is understood as the state S_i, $i = 1, 2$.

The duration of the interval between the flow events at the first state is determined by the random variable $\eta = min(\xi^{(1)}, \xi^{(2)})$, where random variable $\xi^{(1)}$ has distribution function $F_1^{(1)}(t) = 1 - e^{-\lambda_1 t}$, random variable $\xi^{(2)}$ has distribution function $F_1^{(2)}(t) = 1 - e^{-\alpha_1 t}$; $\xi^{(1)}$ and $\xi^{(2)}$ are independent random variables.

At the moment of the flow event occurrence, the process $\lambda(t)$ transits from the state S_1 to S_j either with probability $P_1^{(1)}(\lambda_j|\lambda_1)$, or with probability $P_1^{(2)}(\lambda_j|\lambda_1)$, depending on what value the random variable η has taken, $j = 1, 2$. Here $\sum_{j=1}^{2} P_1^{(k)}(\lambda_j|\lambda_1) = 1$, $k = 1, 2$. The duration of the interval between the flow events at the first state is random variable with distribution function $F(t) = 1 - e^{-(\lambda_1 + \alpha_1)t}$.

The time during which the process $\lambda(t)$ remains at the second state is random variable with distribution function $F_2(t) = 1 - e^{-\alpha_2 t}$. During the time when the process $\lambda(t)$ is in the second state, there is a Poisson event flow with parameter λ_2.

Hereinafter, it is assumed that the state S_i (ith state) of the process $\lambda(t)$ takes place if $\lambda(t) = \lambda_i$, $i = 1, 2$; $\lambda_1 > \lambda_2 \geq 0$.

We note that the flow is called semi-synchronous, because the change of state from S_1 to S_2 takes place only at the moment of the flow event occurrence (synchrony), while the change of state from S_2 to S_1 occurs at an arbitrary time moment that is not associated with the moment of the event occurrence (asynchrony).

Since the process $\lambda(t)$ is unobservable in principle, and we can only observe time moments $t_1, t_2, \ldots$ when events occur in the flow, then $\lambda(t)$ is a hidden Markov process or an unobservable accompanying Markov process. The sequence $\{\lambda(t_k)\}$ at the time moments $t_1, t_2, \ldots, t_k, \ldots$ of events occurrence is an embedded Markov chain.

We denote by $\tau_k = t_{k+1} - t_k$, $k = 1, 2, \ldots$, the value of interval duration between neighboring events, and by $p(\tau)$ the probability density of the value of interval duration between neighboring events in the observed flow. Since we consider the stationary operation mode of the observed flow then $p(\tau_k) = p(\tau)$ for all $k = 1, 2, \ldots$, $\tau \geq 0$. Then we can let the moment of event occurrence t_k equal to zero without loss of generality, i.e. the moment of the event occurrence is $\tau = 0$.

3   Derivation of Probability Density $p(\tau)$

We introduce the conditional probability $p_{ij}(\tau)$ that there are no events on the interval $(0, \tau)$ and that the process value $\lambda(\tau) = \lambda_j$ at the time moment τ, provided that the process value $\lambda(0) = \lambda_i$, $i, j = 1, 2$ at the time moment $\tau = 0$ [16].

3.1   Probability Density for the General Case of Setting Flow Parameters

Lemma 1. *The conditional probabilities $p_{ij}(\tau)$, $i, j = 1, 2$, in a correlated semi-synchronous event flow of the second order are given by the following*

$$p_{11}(\tau) = e^{-(\lambda_1 + \alpha_1)\tau}, \; p_{12}(\tau) = 0, \; p_{22}(\tau) = e^{-(\lambda_2 + \alpha_2)\tau},$$

$$p_{21}(\tau) = \frac{\alpha_2}{(\lambda_1 + \alpha_1) - (\lambda_2 + \alpha_2)}[e^{-(\lambda_2 + \alpha_2)\tau} - e^{-(\lambda_1 + \alpha_1)\tau}], \; \tau \geq 0, \qquad (1)$$

where $(\lambda_1 + \alpha_1) - (\lambda_2 + \alpha_2) \neq 0$.

Proof. We note that, in accordance with the definition of flow, the transition of a process $\lambda(\tau)$ from the state S_1 to the state S_2 is accompanied by the event occurrence. In this regard, the probability $p_{12}(\tau) = 0$. The system of differential equations to find conditional probabilities $p_{11}(\tau)$, $p_{22}(\tau)$, $p_{21}(\tau)$ is the following:

$$p'_{ii}(\tau) = -(\lambda_i + \alpha_i)p_{ii}(\tau), \ i = 1, 2, \ p'_{21}(\tau) = -(\lambda_1 + \alpha_1)p_{21}(\tau) + \alpha_2 p_{22}(\tau),$$

$$p_{11}(0) = 1, \ p_{22}(0) = 1, \ p_{21}(0) = 0.$$

Solving this system of differential equations taking into account the initial conditions, we get (1).

The proof is carried out by solving differential equations for $p_{ij}(\tau)$, $i, j = 1, 2$.

Lemma 2. *The probability densities $\tilde{p}_{ij}(\tau)$, $i, j = 1, 2$, in a correlated semi-synchronous event flow of the second order are given by the following formulas*

$$\tilde{p}_{1j}(\tau) = [\lambda_1 P_1^{(1)}(\lambda_j|\lambda_1) + \alpha_1 P_1^{(2)}(\lambda_j|\lambda_1)]e^{-(\lambda_1+\alpha_1)\tau}, \ j = 1, 2,$$

$$\tilde{p}_{21}(\tau) = \frac{\alpha_2[\lambda_1 P_1^{(1)}(\lambda_1|\lambda_1) + \alpha_1 P_1^{(2)}(\lambda_1|\lambda_1)]}{(\lambda_1 + \alpha_1) - (\lambda_2 + \alpha_2)}[e^{-(\lambda_2+\alpha_2)\tau} - e^{-(\lambda_1+\alpha_1)\tau}], \quad (2)$$

$$\tilde{p}_{22}(\tau) = \frac{\alpha_2[\lambda_1 P_1^{(1)}(\lambda_2|\lambda_1) + \alpha_1 P_1^{(2)}(\lambda_2|\lambda_1)]}{(\lambda_1 + \alpha_1) - (\lambda_2 + \alpha_2)}[e^{-(\lambda_2+\alpha_2)\tau} - e^{-(\lambda_1+\alpha_1)\tau}]+$$

$$+ \lambda_2 e^{-(\lambda_2+\alpha_2)\tau},$$

where $\tau \geq 0$, $(\lambda_1 + \alpha_1) - (\lambda_2 + \alpha_2) \neq 0$.

Proof. We introduce the joint probability that without flow events occurrence at the interval $(0, \tau)$ the process $\lambda(\tau)$ transited from the first state to the first at this interval, then the event occurred at the half-interval $[\tau, \tau + \Delta\tau)$ with probability $1 - e^{-\lambda_1 \Delta\tau}$ and, at the moment of the flow event occurrence, the process $\lambda(\tau)$ transited from the first state to the first $(S_1 \rightarrow S_1)$ with probability $P_1^{(1)}(\lambda_1|\lambda_1)$ or, on the half-interval $[\tau, \tau + \Delta\tau)$ a flow event occurred with probability $1 - e^{-\alpha_1 \Delta\tau}$ and at the moment of the flow event occurrence the process $\lambda(\tau)$ transited from the first state to the first $(S_1 \rightarrow S_1)$ with probability $P_1^{(2)}(\lambda_1|\lambda_1)$. This joint probability is as follows $p_{11}(\tau)[\lambda_1 P_1^{(1)}(\lambda_1|\lambda_1) + \alpha_1 P_1^{(2)}(\lambda_1|\lambda_1)]\Delta\tau + o(\Delta\tau)$. We note that the joint probability under consideration can be represented as $p_{11}(\tau)[\lambda_1 P_1^{(1)}(\lambda_1|\lambda_1) + \alpha_1 P_1^{(2)}(\lambda_1|\lambda_1)]\Delta\tau + o(\Delta\tau) = \tilde{p}_{11}(\tau)\Delta\tau + o(\Delta\tau)$, where $\tilde{p}_{11}(\tau)$ is the probability density corresponding to the joint probability. Writing the last equality in the form $\tilde{p}_{11}(\tau) + o(\Delta\tau)/\Delta\tau = [\lambda_1 P_1^{(1)}(\lambda_1|\lambda_1) + \alpha_1 P_1^{(2)}(\lambda_1|\lambda_1)]p_{11}(\tau) + o(\Delta\tau)/\Delta\tau$ and tending $\Delta\tau$ to zero, we find $\tilde{p}_{11}(\tau) = [\lambda_1 P_1^{(1)}(\lambda_1|\lambda_1) + \alpha_1 P_1^{(2)}(\lambda_1|\lambda_1)]p_{11}(\tau)$. We define the remaining joint probabilities in the same way. Thus, the probability density $p_{ij}(\tau)$ that the process $\lambda(\tau)$ transits from the state S_i to S_j, $i, j = 1, 2$, at the interval $(0, \tau)$, without the flow event occurrence at this interval and with the event occurrence at the moment τ, is written in the form

$$\tilde{p}_{1j}(\tau) = [\lambda_1 P_1^{(1)}(\lambda_j|\lambda_1) + \alpha_1 P_1^{(2)}(\lambda_j|\lambda_1)]p_{11}(\tau), \ j = 1, 2,$$

$$\tilde{p}_{21}(\tau) = [\lambda_1 P_1^{(1)}(\lambda_1|\lambda_1) + \alpha_1 P_1^{(2)}(\lambda_1|\lambda_1)]p_{21}(\tau),$$

$$\tilde{p}_{22}(\tau) = [\lambda_1 P_1^{(1)}(\lambda_2|\lambda_1) + \alpha_1 P_1^{(2)}(\lambda_2|\lambda_1)]p_{21}(\tau) + \lambda_2 p_{22}(\tau). \quad (3)$$

Substituting (1) into (3), we obtain (2).

We define p_{ij}, $i, j = 1, 2$, the probability of the process $\lambda(\tau)$ transition from the state S_i to S_j for a time which will pass from the moment $\tau = 0$ until the next flow event occurrence. Since τ is an arbitrary time moment, then these transition probabilities are defined as

$$p_{ij} = \int_0^\infty \tilde{p}_{ij}(\tau)d\tau, \ i, j = 1, 2. \tag{4}$$

Lemma 3. *The transition probabilities p_{ij}, $i, j = 1, 2$, in a correlated semi-synchronous event flow of the second order are given by the following*

$$p_{1j} = (\lambda_1 P_1^{(1)}(\lambda_j|\lambda_1) + \alpha_1 P_1^{(2)}(\lambda_j|\lambda_1))/(\lambda_1 + \alpha_1), \ j = 1, 2,$$

$$p_{21} = (\alpha_2[\lambda_1 P_1^{(1)}(\lambda_1|\lambda_1) + \alpha_1 P_1^{(2)}(\lambda_1|\lambda_1)])/[(\lambda_1 + \alpha_1)(\lambda_2 + \alpha_2)],$$

$$p_{22} = (\alpha_2[\lambda_1 P_1^{(1)}(\lambda_2|\lambda_1) + \alpha_1 P_1^{(2)}(\lambda_2|\lambda_1)])/[(\lambda_1 + \alpha_1)(\lambda_2 + \alpha_2)] + \lambda_2/(\lambda_2 + \alpha_2). \tag{5}$$

Proof. Substituting (2) into (4), we obtain (5).

Let us consider the conditional stationary probability $\pi_i(0)$ that the process $\lambda(\tau)$ is in the state S_i at the time $\tau = 0$, provided that a flow event has occurred at the time $\tau = 0$, $i, j = 1, 2$, $\pi_1(0) + \pi_2(0) = 1$.

Lemma 4. *The conditional stationary probabilities $\pi_i(0)$, $i = 1, 2$, in a correlated semi-synchronous event flow of the second order are given by the following formulas*

$$\pi_1(0) = \frac{\alpha_2[\lambda_1 P_1^{(1)}(\lambda_1|\lambda_1) + \alpha_1 P_1^{(2)}(\lambda_1|\lambda_1)]}{\lambda_2[\lambda_1 P_1^{(1)}(\lambda_2|\lambda_1) + \alpha_1 P_1^{(2)}(\lambda_2|\lambda_1)] + (\lambda_1 + \alpha_1)\alpha_2},$$

$$\pi_2(0) = \frac{(\lambda_2 + \alpha_2)[\lambda_1 P_1^{(1)}(\lambda_2|\lambda_1) + \alpha_1 P_1^{(2)}(\lambda_2|\lambda_1)]}{\lambda_2[\lambda_1 P_1^{(1)}(\lambda_2|\lambda_1) + \alpha_1 P_1^{(2)}(\lambda_2|\lambda_1)] + (\lambda_1 + \alpha_1)\alpha_2}, \ \pi_1(0) + \pi_2(0) = 1. \tag{6}$$

Proof. Since the sequence of the moments $t_1, t_2, ..., t_k, ...$ forms an embedded Markov chain $\{\lambda(t_k)\}$ then the following equations are valid for probabilities $\pi_i(0)$

$$\pi_1(0) = p_{11}\pi_1(0) + p_{21}\pi_2(0), \ \pi_2(0) = p_{12}\pi_1(0) + p_{22}\pi_2(0), \tag{7}$$

where the probabilities p_{ij}, $i, j = 1, 2$, are defined by (5).
From (7), taking into account that $\pi_1(0) + \pi_2(0) = 1$, we find

$$\pi_1(0) = p_{21}/(p_{12} + p_{21}), \ \pi_2(0) = p_{12}/(p_{12} + p_{21}). \tag{8}$$

Substituting (5) into (8), we obtain (6).

Lemmas 2 and 4 yield the following theorem.

Theorem 1. *The probability density of the value of interval duration between neighboring events in a correlated semi-synchronous event flow of the second order is given by the following*

$$p(\tau) = \gamma z_1 e^{-z_1 \tau} + (1 - \gamma) z_2 e^{-z_2 \tau}, \ \tau \geq 0, \tag{9}$$

$$\gamma = \frac{\pi_1(0)(\lambda_1 + \alpha_1 - \lambda_2) - \alpha_2}{(\lambda_1 + \alpha_1) - (\lambda_2 + \alpha_2)}, \ z_1 = \lambda_1 + \alpha_1, \ z_2 = \lambda_2 + \alpha_2,$$

where $(\lambda_1 + \alpha_1) - (\lambda_2 + \alpha_2) \neq 0$, *and the probability* $\pi_1(0)$ *is defined in (6).*

Proof. Due to the fact that the process $\lambda(t)$ has a Markov property, if its evolution is considered starting from the time moment t_k, $k = 1, 2, ...$, of the flow event occurrence, the probability density $p(\tau)$ of the value of interval duration between neighboring events in the flow under consideration is determined as

$$p(\tau) = \sum_{i=1}^{2} \pi_i(0) \sum_{j=1}^{2} \tilde{p}_{ij}(\tau), \ \tau \geq 0. \tag{10}$$

Substituting the expressions (2) and (6) into (10), after the necessary transformations, we obtain (9).

3.2 Probability Density for the Special Case of Setting Flow Parameters

Let us consider the case when the coefficient $(\lambda_1 + \alpha_1) - (\lambda_2 + \alpha_2) = 0$ in (9).

Lemma 5. *The conditional probabilities* $p_{ij}(\tau)$, $i, j = 1, 2$, *in a correlated semi-synchronous event flow of the second order in the case when* $(\lambda_1 + \alpha_1) - (\lambda_2 + \alpha_2) = 0$ *are given by the following*

$$p_{11}(\tau) = e^{-(\lambda_1 + \alpha_1)\tau}, \ p_{12}(\tau) = 0, \ p_{22}(\tau) = e^{-(\lambda_1 + \alpha_1)\tau},$$

$$p_{21}(\tau) = (\lambda_1 + \alpha_1 - \lambda_2)\tau e^{-(\lambda_1 + \alpha_1)\tau}, \ \tau \geq 0. \tag{11}$$

Proof. Similarly as in the general case of setting parameters, the system of differential equations for $p_{ij}(\tau)$, $i, j = 1, 2$, in the case when $(\lambda_1 + \alpha_1) - (\lambda_2 + \alpha_2) = 0$ is following:

$$p'_{11}(\tau) = -(\lambda_1 + \alpha_1)p_{11}(\tau), \ p'_{22}(\tau) = -(\lambda_1 + \alpha_1)p_{22}(\tau),$$

$$p'_{21}(\tau) = -(\lambda_1 + \alpha_1)p_{21}(\tau) + (\lambda_1 + \alpha_1 - \lambda_2)p_{22}(\tau),$$

$$p_{11}(0) = 1, \ p_{22}(0) = 1, \ p_{21}(0) = 0,$$

note that $p_{12}(\tau) = 0$. Solving this system of differential equations we get (11).

Then, based on Lemmas 4 and 5, we formulate the following theorem.

Theorem 2. *The probability density of the value of interval duration between neighboring events in a correlated semi-synchronous event flow of the second order in the case when* $(\lambda_1 + \alpha_1) - (\lambda_2 + \alpha_2) = 0$ *is given by the following*

$$p(\tau) = [(\lambda_1 + \alpha_1) - \pi_2(0)(\lambda_1 + \alpha_1 - \lambda_2)(1 - (\lambda_1 + \alpha_1)\tau)]e^{-(\lambda_1 + \alpha_1)\tau}, \ \tau \geq 0, \ (12)$$

where the probability $\pi_2(0)$ *is defined in (6).*

Proof. Substituting (11) into (3), we find the densities $\tilde{p}_{ij}(\tau)$; substituting (3) into (4) and (4) into (8), we obtain the probabilities $\pi_i(0)$. Substituting $\tilde{p}_{ij}(\tau)$ and $\pi_i(0)$, $i, j = 1, 2$, into (10), as a result of the necessary transformations we obtain (12).

4   Derivation of Joint Probability Density $p(\tau_1, \tau_2)$

Let us consider two adjacent time intervals (t_1, t_2), (t_2, t_3) with the values of intervals durations $\tau_1 = t_2 - t_1$ and $\tau_2 = t_3 - t_2$, respectively. We denote the joint probability density by $p(\tau_1, \tau_2)$, $\tau_1 \geq 0$, $\tau_2 \geq 0$ [16].

Theorem 3. *A semi-synchronous event flow of second-order is correlated in general and the joint probability density of the values of the adjacent intervals durations is the following*

$$p(\tau_1, \tau_2) = p(\tau_1)p(\tau_2) + \gamma(1 - \gamma)\left(1 - \frac{\lambda_1 P_1^{(1)}(\lambda_2|\lambda_1) + \alpha_1 P_1^{(2)}(\lambda_2|\lambda_1)}{\lambda_1 + \alpha_1}\right) \times$$

$$\times \frac{\lambda_2}{\lambda_2 + \alpha_2}[z_1 e^{-z_1\tau_1} - z_2 e^{-z_2\tau_1}][z_1 e^{-z_1\tau_2} - z_2 e^{-z_2\tau_2}], \quad (13)$$

where $\tau_1 \geq 0$, $\tau_2 \geq 0$, γ, z_1, z_2, $p(\tau_k)$ *are defined for* $\tau = \tau_k$, $k = 1, 2$, *in (9).*

Proof. Since the sequence of the time moments of the flow events occurrence $t_1, t_2, ..., t_k, ...$ forms an embedded Markov chain $\{\lambda(t_k)\}$, then the following holds for density $p(\tau_1, \tau_2)$

$$p(\tau_1, \tau_2) = \sum_{i=1}^{2} \pi_i(0) \sum_{j=1}^{2} \tilde{p}_{ij}(\tau_1) \sum_{k=1}^{2} \tilde{p}_{jk}(\tau_2), \quad (14)$$

where $\tilde{p}_{ij}(\tau_1)$, $\tilde{p}_{ij}(\tau_2)$ are the probability densities that correspond to the transition probabilities $p_{ij}(\tau_1)$, $p_{ij}(\tau_2)$ and are calculated by the formulas (2) for $\tau = \tau_1$ and $\tau = \tau_2$.

Substituting the expressions for $\tilde{p}_{ij}(\tau_1)$, $\tilde{p}_{ij}(\tau_2)$, then for $\pi_i(0)$, $i = 1, 2$, defined by formulas (6), into (14), after the necessary transformations, we obtain (13).

Remark 1. It can be shown that if either $1 - \frac{\lambda_1 P_1^{(1)}(\lambda_2|\lambda_1) + \alpha_1 P_1^{(2)}(\lambda_2|\lambda_1)}{\lambda_1 + \alpha_1} = 0$, or $\lambda_2 = 0$, or $\gamma(1 - \gamma) = 0$ then the probability density $p(\tau_1, \tau_2, ..., \tau_n) = \prod_{k=1}^{n} p(\tau_k)$, i.e. a semi-synchronous event flow of the second order is recurrent.

5   Estimation of the Distribution Parameters by the Method of Moments

Let us consider the statistics $C_l = \frac{1}{n}\sum_{k=1}^{n} \tau_k^l$, where $\tau_k = t_{k+1} - t_k$.

5.1   General Case of Setting Flow Parameters

Let us have a sample $\tau_1, \tau_2, ..., \tau_n$ from the distribution $p(\tau|z_1, z_2, \gamma)$ depending on three unknown parameters z_1, z_2, γ. Let $M(\tau^l) = \int_0^\infty \tau^l p(\tau|z_1, z_2, \gamma)d\tau$ be the initial theoretical moment of the lth order which is a function of the unknown parameters. Then it is close to the corresponding selective moment $\bar{\tau}^l$ which is the $C_l = \frac{1}{n}\sum_{k=1}^{n}\tau_k^l$ statistics. For the first three initial moments we write the moment equations

$$M(\tau^l) = C_l, \ l = 1, 2, 3. \tag{15}$$

Given the kind of density (9), we get $M(\tau^l) = l!\gamma/z_1^l + l!(1-\gamma)/z_2^l, l = 1, 2, 3$. Then the system (15) takes the following form

$$z_1 z_2 C_1 - z_2\gamma - z_1(1-\gamma) = 0, \ (z_1 + z_2)C_1 - z_1 z_2 C_2/2 = 1, \tag{16}$$
$$(z_1 + z_2)C_2 - z_1 z_2 C_3/3 = 2C_1.$$

Solving the system of equations (16), we find parameter estimates of $p(\tau)$

$$\hat{z}_{1,2} = \frac{1}{2}\left(-\frac{2(C_3 - 3C_1 C_2)}{3C_2^2 - 2C_1 C_3} \pm \sqrt{\left(\frac{2(C_3 - 3C_1 C_2)}{3C_2^2 - 2C_1 C_3}\right)^2 + 4\frac{6(C_2 - 2C_1^2)}{3C_2^2 - 2C_1 C_3}}\right). \tag{17}$$

Since $z_1 = \lambda_1 + \alpha_1$, $z_2 = \lambda_2 + \alpha_2$, where $\lambda_1 > \lambda_2 \geq 0$, and the relationship between parameters α_1 and α_2 is unknown, then the relationship between parameters z_1 and z_2 is unknown too. Thus, in order to determine which root of equation (17) can be chosen as the parameter estimate $\hat{z}_1$ and which parameter as the estimate $\hat{z}_2$, additional information about the flow is needed.

From the first equation of system (16), we obtain the estimate of the parameter γ

$$\hat{\gamma} = \hat{z}_1(1 - C_1\hat{z}_2)/(\hat{z}_1 - \hat{z}_2), \ \hat{z}_2 \neq \hat{z}_1. \tag{18}$$

Let us introduce the density $p(\tau)$ in the following form

$$p(\tau) = \gamma(\lambda_1 + \alpha_1)e^{-(\lambda_1 + \alpha_1)\tau} + (1 - \gamma)(\lambda_2 + \alpha_2)e^{-(\lambda_2 + \alpha_2)\tau}, \ \tau \geq 0.$$

And let us solve the problem of estimating parameters $\lambda_1, \lambda_2, \alpha_1, \alpha_2$ of the density function by the method of moments with a known value of the parameter γ. We note that the parameters under estimation coincide with the flow parameters.

Let us have a sample $\tau_1, \tau_2, ..., \tau_n$ from the distribution $p(\tau|\lambda_1, \lambda_2, \alpha_1, \alpha_2)$ depending on four unknown parameters. So for estimating $\lambda_1, \lambda_2, \alpha_1, \alpha_2$ it is necessary to have four moment equations, i.e. $M(\tau^l) = C_l, \ l = 1, 2, 3, 4$. As a result of the necessary transformations, the system of moments equations takes the form

$$(\lambda_1 + \alpha_1)(\lambda_2 + \alpha_2)C_1 - (\lambda_2 + \alpha_2)\gamma - (\lambda_1 + \alpha_1)(1 - \gamma) = 0,$$
$$((\lambda_1 + \alpha_1) + (\lambda_2 + \alpha_2))C_1 - 1/2(\lambda_1 + \alpha_1)(\lambda_2 + \alpha_2)C_2 = 1,$$
$$((\lambda_1 + \alpha_1) + (\lambda_2 + \alpha_2))C_2 - 1/3(\lambda_1 + \alpha_1)(\lambda_2 + \alpha_2)C_3 = 2C_1, \quad (19)$$
$$((\lambda_1 + \alpha_1) + (\lambda_2 + \alpha_2))C_3 - 1/4(\lambda_1 + \alpha_1)(\lambda_2 + \alpha_2)C_4 = 3C_2.$$

Theorem 4. *The system (19) for the unknown parameters λ_1, λ_2, α_1, α_2 of a semi-synchronous event flow of the second order is incompatible.*

Proof. We reduce the system (19) to a linearly inhomogeneous system by letting

$$x_1 = \lambda_2 + \alpha_2 - \lambda_1 - \alpha_1, \ x_2 = \lambda_1 + \alpha_1, \ x_3 = \lambda_2 + \alpha_2 + \lambda_1 + \alpha_1, \ x_4 = (\lambda_1 + \alpha_1)(\lambda_2 + \alpha_2).$$

The resulting system of linear inhomogeneous equations for four unknowns x_i, $i = 1, 2, 3, 4$, is the following

$$\gamma x_1 + x_2 - C_1 x_4 = 0, \ C_1 x_3 - C_2 x_4/2 = 1, \ C_2 x_3 - C_3 x_4/3 = 2C_1, \quad (20)$$
$$C_3 x_3 - C_4 x_4/4 = 3C_2.$$

The system (20) has no solutions, i.e. the system is incompatible and it is impossible to estimate the flow parameters λ_1, λ_2, α_1, α_2 having only information about the density $p(\tau)$.

5.2   Special Case of Setting Flow Parameters

Let us introduce the density (12) in the following form

$$p(\tau) = [z - a(1 - z\tau)]e^{-z\tau}, \ z = \lambda_1 + \alpha_1, \ a = \pi_2(0)\alpha_2, \ \tau \geq 0. \quad (21)$$

Let us have a sample $\tau_1, \tau_2, ..., \tau_n$ from the distribution $p(\tau|z, a)$ depending on two unknown parameters z, a. Let $M(\tau^l) = \int_0^\infty \tau^l p(\tau|z, a)d\tau$ be the initial theoretical moment of the lth order which is a function of the unknown parameters. Considering the form of the density (21), we obtain $M(\tau^l) = l!/z^i [1 + a/z]$, $l = 1, 2$.

Then the system of moments equations takes the following form

$$[z + a]/z^2 = C_1, \ [z + 2a]/z^3 = C_2. \quad (22)$$

Solving the system (22), we find

$$\hat{z}^{(1)} = \frac{1}{C_2}\left(2C_1 - \sqrt{4C_1^2 - 2C_2}\right)/C_2, \ \hat{z}^{(2)} = \frac{1}{C_2}\left(2C_1 + \sqrt{4C_1^2 - 2C_2}\right). \quad (23)$$

In this case, the following conditions must be satisfied

$$\hat{z}^{(1)}\hat{z}^{(2)} = 2/C_2 > 0, \ \hat{z}^{(1)} + \hat{z}^{(2)} = 4C_1/C_2 > 0, \ 4C_1^2 - 2C_2 \geq 0.$$

Substituting the estimates $\hat{z}^{(1)}$ and $\hat{z}^{(1)}$ into the first equation of (22), we obtain

$$\hat{a}^{(1)} = C_1(\hat{z}^{(1)})^2 - \hat{z}^{(1)}, \ \hat{a}^{(2)} = C_1(\hat{z}^{(2)})^2 - \hat{z}^{(2)}. \tag{24}$$

The question naturally arises of which pair $\{\hat{z}^{(1)}, \hat{a}^{(1)}\}$ and $\{\hat{z}^{(2)}, \hat{a}^{(2)}\}$ to choose as a solution of the problem. Taking into account the explicit form of the parameters $z = \lambda_1 + \alpha_1$, $a = \pi_2(0)\alpha_2$ and the condition $(\lambda_1 + \alpha_1) - (\lambda_2 + \alpha_2) = 0$, it is easy to show that $\hat{z}$ and $\hat{a}$ must satisfy the conditions $\hat{a} > 0$ and $\hat{z} - \hat{a} > 0$. During the analytical test of the obtained pairs for the fulfillment of these conditions, the following conclusions were made. If $C_2/2 \leq C_1^2 < 2C_2/3$ then both pairs are equal and any of them can be chosen as an estimate; if $C_1^2 \geq 2C_2/3$ then only $\{\hat{z}^{(1)}, \hat{a}^{(1)}\}$ can be chosen as the solution of the estimation problem.

Remark 2. It can be shown that for the special case of setting flow parameters $(\lambda_1 + \alpha_1) - (\lambda_2 + \alpha_2) = 0$, a system of three or more moment equations is incompatible.

6   Results of Numerical Calculations

In order to establish the quality of estimation, we developed the algorithm for calculating parameter estimates of the density. The algorithm consists of two stages. The simulation of the semi-synchronous event flow of the second order is performed directly at the first stage of the implementation algorithm. Parameter estimates are calculated at the second stage of the algorithm using the formulas obtained.

Let us consider general case of setting the parameters of the flow. For this case we compute the estimates $\hat{z}_1$, $\hat{z}_2$, γ by the formulas (17), (18) and we obtain sample averages $\hat{M}(\hat{\theta}) = \frac{1}{N}\sum_{k=1}^{N}\hat{\theta}^{(k)}$ and offset estimates $\delta(\theta) = |\hat{M}(\hat{\theta}) - \theta|$, where $\theta \in \{z_1, z_2, \gamma\}$, $\hat{\theta} \in \{\hat{z}_1, \hat{z}_2, \hat{\gamma}\}$.

In the first statistical experiment, we consider the dependence of $\hat{M}(\hat{\theta})$, $\delta(\theta)$ from the values $\lambda_1 = 2, 3, 4, 5, 6$ for simulation time $T = 100$ units of time, $N = 100$, probabilities $P_1^{(1)}(\lambda_1|\lambda_1) = P_1^{(2)}(\lambda_2|\lambda_1) = 0,4$, $P_1^{(1)}(\lambda_2|\lambda_1) = P_1^{(2)}(\lambda_1|\lambda_1) = 0,6$ and parameters $\lambda_2 = 0,8$, $\alpha_1 = 2$, $\alpha_2 = 0,8$. The experiment results are given in Table 1, where the last three rows are the real values of the parameters under evaluation.

In the second statistical experiment, we consider the dependence of $\hat{M}(\hat{\theta})$, $\delta(\theta)$ from the values $\alpha_1 = 2, 3, 4, 5, 6$ for $T = 100$ units of time, $N = 100$, probabilities $P_1^{(1)}(\lambda_1|\lambda_1) = P_1^{(2)}(\lambda_2|\lambda_1) = 0,4$, $P_1^{(1)}(\lambda_2|\lambda_1) = P_1^{(2)}(\lambda_1|\lambda_1) = 0,6$ and parameters $\lambda_1 = 2$, $\lambda_2 = 0,8$, $\alpha_1 = 0,8$. The results of this experiment are given in Table 2, where the last three rows are the real values of the parameters under evaluation.

The results of these statistical experiments are illustrated in Fig. 1, where the behavior of $\delta(z_1)$ is indicated by a gray marker, $\delta(z_2)$ by black, $\delta(\gamma)$ by white.

Analyzing the numerical results given in Tables 1 and 2 and in Fig. 1, we can make the following conclusions. There is a displacement in the obtained estimates by an amount $\delta(\theta)$, $\theta \in \{z_1, z_2, \gamma\}$, relative to the initial values of the

Table 1. The results of the first statistical experiment for z_1, z_2, γ

λ_1	2	3	4	5	6
$\hat{M}(\hat{z}_1)$	4,3298	5,2167	6,1552	7,0959	8,0483
$\delta(z_1)$	0,3298	0,2167	0,1552	0,0959	0,0483
$\hat{M}(\hat{z}_2)$	3,1902	3,0743	3,0377	2,9545	2,9151
$\delta(z_2)$	0,3902	0,2743	0,2377	0,1545	0,1151
$\hat{M}(\hat{\gamma})$	0,3621	0,3393	0,3291	0,2909	0,2717
$\delta(\gamma)$	0,2510	0,1846	0,1513	0,0989	0,0701
z_1	4	5	6	7	8
z_2	2,8	2,8	2,8	2,8	2,8
γ	0,1111	0,1547	0,1778	0,1920	0,2016

Table 2. The results of the second statistical experiment for z_1, z_2, γ

α_1	2	3	4	5	6
$\hat{M}(\hat{z}_1)$	4,3502	5,2681	6,1798	7,1102	8,0611
$\delta(z_1)$	0,3502	0,2681	0,1798	0,1102	0,0611
$\hat{M}(\hat{z}_2)$	1,9802	1,9213	1,8103	1,7699	1,7145
$\delta(z_2)$	0,3802	0,3213	0,2103	0,1699	0,1145
$\hat{M}(\hat{\gamma})$	0,3578	0,3788	0,3968	0,3978	0,3851
$\delta(\gamma)$	0,2467	0,1801	0,1489	0,1183	0,0834
z_1	4	5	6	7	8
z_2	1,6	1,6	1,6	1,6	1,6
γ	0,1111	0,1987	0,2479	0,2795	0,3017

estimated parameters. The offset $\delta(\theta)$ decreases with the increase of the parameter λ_1 (Table 1). This is due to the fact that the frequency of transitions from the first state to the second of the process $\lambda(t)$ increases with the increase of the parameter λ_1, which has a positive effect on the conditions of states distinguishability. In other words, the convergence of the values of the parameters λ_1 and λ_2 affects negatively on the conditions of states distinguishability therefore $\delta(\theta)$ has large values with $\lambda_1 - \lambda_2$ decreasing.

The offset $\delta(\theta)$ decreases with the increase of the parameter α_1 (Table 2), which is also explained by the frequency of state changes.

Let us consider special case of setting the parameters of the flow. For this case we compute the estimates $\hat{z}$, $\hat{a}$, γ by the formulas (23), (24) and we obtain sample averages $\hat{M}(\hat{\theta})$ and offset estimates $\delta(\theta)$, where $\theta \in \{z, a\}$, $\hat{\theta} \in \{\hat{z}, \hat{a}\}$.

In the third statistical experiment, we consider the dependence of $\hat{M}(\hat{\theta})$, $\delta(\theta)$ from the simulation time values T_m for fixed $N = 100$, probabilities $P_1^{(1)}(\lambda_1|\lambda_1) = P_1^{(2)}(\lambda_1|\lambda_1) = 0,65$, $P_1^{(1)}(\lambda_2|\lambda_1) = P_1^{(2)}(\lambda_2|\lambda_1) = 0,35$ and flow

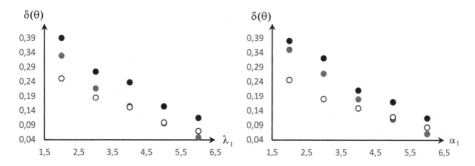

Fig. 1. The dependence of $\delta(\theta)$, $\theta \in \{z_1, z_2, \gamma\}$, on the parameter λ_1 values (on the left) and on the α_1 values (on the right)

parameters $\lambda_1 = 4$, $\lambda_2 = 1,5$, $\alpha_1 = 0,5$, $\alpha_2 = 3$. The results of the experiment are given in Table 3, where the last two rows are the real values of the parameters under evaluation, and are also illustrated on Fig. 2, where the evolution of the $\hat{M}(\hat{\theta})$ is shown by a dotted line, and the behavior of the real value of the parameter by continuous line.

Table 3. The results of the third statistical experiment for z, a

T_m	50	100	150	200	...	900	950	1000
$\hat{M}(\hat{z})$	4,8894	4,7864	4,7664	4,7463	...	4,5398	4,5397	4,5388
$\delta(z)$	0,3894	0,2864	0,2664	0,2463	...	0,0398	0,0397	0,0388
$\hat{M}(\hat{a})$	1,6402	1,5211	1,4984	1,4698	...	1,3777	1,3775	1,3769
$\delta(a)$	0,2998	0,1807	0,1580	0,1294	...	0,0373	0,0371	0,0365
z	4,5	4,5	4,5	4,5	...	4,5	4,5	4,5
a	1,3404	1,3404	1,3404	1,3404	...	1,3404	1,3404	1,3404

Analysis of the numerical results given in Table 3 and in Fig. 2 shows that the offset $\delta(\theta)$, $\theta \in \{z, a\}$, decreases with the increase of the T_m, which is quite normal. In other words, the quality of estimating the density parameters is the better (in the sense of reducing the offset estimates), the larger the simulation time T_m.

In the fourth statistical experiment, we consider the dependence of $\hat{M}(\hat{\theta})$, $\delta(\theta)$ from the parameter $\lambda_1 = 2$, 3, 4, 5, 6 for simulation time $T = 100$ units of time, $N = 100$, probabilities $P_1^{(1)}(\lambda_1|\lambda_1) = P_1^{(2)}(\lambda_2|\lambda_1) = 0,4$, $P_1^{(1)}(\lambda_2|\lambda_1) = P_1^{(2)}(\lambda_1|\lambda_1) = 0,6$ and flow parameters $\lambda_2 = 1$, $\alpha_1 = 0,8$. And the parameter α_2 is calculated by the formula $\alpha_2 = \lambda_1 + \alpha_1 - \lambda_2$. The results of this experiment are given in Table 4, where the last two rows are the real values of the parameters under evaluation.

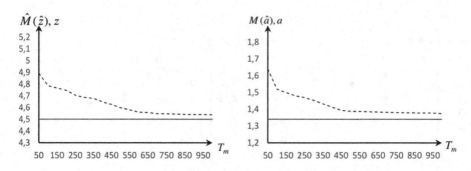

Fig. 2. The dependence of $\hat{M}(\hat{z})$, z (on the left) and $\hat{M}(\hat{a})$, a (on the right) on the T_m values

Table 4. The results of the fourth statistical experiment for z, a

λ_1	2	3	4	5	6
α_2	1,8	2,8	3,8	4,8	5,8
$\hat{M}(\hat{z})$	3,1823	3,9902	4,9281	5,8682	6,8279
$\delta(z)$	0,3823	0,1902	0,1281	0,0682	0,0279
$\hat{M}(\hat{a})$	1,5620	1,9810	2,4835	3,0254	3,5955
$\delta(a)$	0,3942	0,2133	0,1165	0,0592	0,0299
z	2,8	3,8	4,8	5,8	6,8
a	1,1678	1,7677	2,3670	2,9662	3,5656

Analyzing the numerical results obtained in Table 4, we can make the following conclusions. With the increase of the parameter value λ_1 and, respectively, with the increase of the parameter α_2, the offset value $\delta(\theta)$ decreases, since the states of the process $\lambda(t)$ become more distinguishable ($\lambda_1 > \lambda_2$).

7   Conclusion

In this paper, the semi-synchronous event flow of the second order was considered under its complete observability; the explicit form of the probability density of the values of the interval duration between the moments of the events occurrence was obtained for general and special cases of setting the flow parameters, the explicit form of the joint probability density of the values of the adjacent intervals durations was obtained as well. The estimates of probability density parameters were found by the method of moments. The expressions for parameter estimates are obtained explicitly, which allows for calculations without the use of numerical methods. The algorithm for calculating estimates of density parameters is implemented in C# with Visual Studio 2013. In order to establish the quality of estimation, statistical experiments were performed, the numerical results of which do not contradict the physical interpretation.

References

1. Basharin, G.P., Kokotushkin, V.A., Naumov, V.A.: On the equivalent substitutions method for computing fragments of communication networks. Izv. Akad. Nauk USSR. Tekhn. Kibern. **6**, 92–99 (1979)
2. Neuts, M.F.: A versatile Markov point process. J. Appl. Probab. **16**, 764–779 (1979)
3. Cox, D.R.: The analysis of non-Markovian stochastic processes by the inclusion of variables. Proc. Camb. Philos. Soc. **51**(3), 433–441 (1955)
4. Lucantoni, D.M.: New results on the single server queue with a bath Markovian arrival process. Commun. Stat. Stoch. Models **7**, 1–46 (1991)
5. Dudin, A.N., Klimenok, V.I.: Queueing Systems with Correlated Flows. Belarus Gos. Univ., Minsk (2000)
6. Basharin, G.P., Gaidamaka, Y.V., Samouylov, K.E.: Mathematical theory of tele-traffic and its application to the analysis of multiservice communication of the next generation networks. Autom. Control Comput. Sci. **47**(2), 62–69 (2013)
7. Klimenok, V., Dudin, A., Vishnevsky, V.: Tandem queueing system with correlated input and cross-traffic. In: Kwiecień, A., Gaj, P., Stera, P. (eds.) CN 2013. CCIS, vol. 370, pp. 416–425. Springer, Heidelberg (2013). https://doi.org/10.1007/978-3-642-38865-1_42
8. Vishnevsky, V.M., Semenova, O.V.: Polling Systems: Theory and Applications for Broadband Wireless Networks. Academic Publishing, London (2012)
9. Nezhelskaya, L.: Optimal state estimation in modulated MAP event flows with unextendable dead time. In: Dudin, A., Nazarov, A., Yakupov, R., Gortsev, A. (eds.) ITMM 2014. CCIS, vol. 487, pp. 342–350. Springer, Cham (2014). https://doi.org/10.1007/978-3-319-13671-4_39
10. Nezhelskaya, L., Tumashkina, D.: Optimal state estimation of semi-synchronous event flow of the second order under its complete observability. In: Dudin, A., Nazarov, A., Moiseev, A. (eds.) ITMM/WRQ -2018. CCIS, vol. 912, pp. 93–105. Springer, Cham (2018). https://doi.org/10.1007/978-3-319-97595-5_8
11. Kalyagin, A.A., Nezhelskaya, L.A.: Comparison of MP- and MM-estimates of the duration of the dead time in generalized semi-synchronous event flow. Tomsk. State Univ. J. Control. Comput. Sci. **3**(32), 23–32 (2015)
12. Nezhel'skaya, L.: Probability density function for modulated MAP event flows with unextendable dead time. In: Dudin, A., Nazarov, A., Yakupov, R. (eds.) ITMM 2015. CCIS, vol. 564, pp. 141–151. Springer, Cham (2015). https://doi.org/10.1007/978-3-319-25861-4_12
13. Gortsev, A.M., Nezhel'skaya, L.A.: Estimate of parameters of synchronously alternating Poisson stream of events by the moment method. Telecommun. Radio Eng. **50**(1), 56–63 (1996)
14. Gortsev, A.M., Klimov, I.S.: Estimation of the parameters of an alternating Poisson stream of events. Telecommun. Radio Eng. **48**(10), 40–45 (1993)
15. Nezhelskaya, L.A., Tumashkina, D.A.: Optimal state estimation of semi-synchronous event flow of the second order with non-extending dead time. Tomsk. State Univ. J. Control. Comput. Sci. **46**, 73–82 (2019)
16. Nezhelskaya, L.A.: The joint probability density of the duration of the intervals of modulated MAP-flow of events and the conditions of the flow recurrence. Tomsk. State Univ. J. Control. Comput. Sci. **1**(30), 57–67 (2015)

Resource Queueing System with the Requirements Copying at the Second Phase

Anastasia Galileyskaya[1], Ekaterina Lisovskaya[1,2(✉)],
and Ekaterina Fedorova[1]

[1] National Research Tomsk State University,
36 Lenina Avenue, Tomsk 634050, Russian Federation
n.galileyskaya@bk.ru, moiskate@mail.ru
[2] Peoples' Friendship University of Russia (RUDN University),
6 Miklukho-Maklaya Street, Moscow 117198, Russian Federation
lisovskaya-eyu@rudn.ru

Abstract. In this paper, we propose the modification of the dynamic screening method for the resource queueing systems with customers copying at the second phase. We obtain the characteristic function of the studied three-dimensional Markov chain and the main numerical characteristics. The obtained analytical results are compared with the simulation ones, the high accuracy of probability is demonstrated, the recommended limit value of the system and the loss probability are found.

Keywords: Queueing system · Arbitrary service time · Poisson arriving · Customers copying · Characteristic function

1 Introduction

The interest to the study of resource queueing systems (RQS) is determined by the possibility of their application in the modelling of modern technical devices, next-generation data networks and computer systems, including cloud computing systems. Obviously, the physical objects resources, including radio resources in data transmission networks are limited, that is, the incoming demand is lost if the system does not have enough resources to service it. However, rarely it is possible to obtain adequate analytical results. In this regard, the group of authors of this paper proposes methods for studying RQS mathematical models without failures, that is, with an unlimited resource amount and an unlimited devices number. Using the result, it is easy to estimate the sufficient resource amount in the systems, to assess the loss probability and minimize it.

The publication has been prepared with the support of the "RUDN University Program 5–100" (recipient E. Lisovskaya, mathematical model development). The reported study was funded by RFBR according to the research project No. 19-41-703002 (recipient E. Fedorova, initial moments method and numerical analysis).

© Springer Nature Switzerland AG 2019
V. M. Vishnevskiy et al. (Eds.): DCCN 2019, CCIS 1141, pp. 352–363, 2019.
https://doi.org/10.1007/978-3-030-36625-4_28

The growing wireless network popularity necessitates the creation of new approaches to assessing the telecommunication operators services quality [4,5]. In these networks, each active connection requires a certain amount of each of radio resources, which are deliberately limited, provided to the request at the time of its receipt (call, message, video recording) and are released at the connection end [14]. The required resources amount is determined by a pre-defined probability distribution, which can take into account the various radio resource distribution schemes features in analyzing the wireless networks performance [6,9]. Wireless communication networks modeling using RQS, starting with [7] and other, is represented by a large publications number. However, most of the results were obtained under simplifying assumptions: deterministic resources requests, exponentially distributed service time, Poisson arrival, simplest QS configuration, which is associated with the complexity of constructing the corresponding random processes (see [2,8,11,13] and the review therein).

In addition, RQS, where clients require a server and a certain resources amount during their service, allow to simulate any the resources distribution features in modern wireless networks. The signals inclusion that trigger the resources redistribution makes it possible to take into account the subscribers mobility [1,15,16]. Multicast technology offers a possible solution to the transferring the same data to different devices problem, which leads to a significant improvement in the spectral efficiency and throughput of a wireless network. Since the same frequency band is used for several devices in the multicast mode, the data transfer rate can reach higher values compared to the unicast mode, where only a small separate frequency band is allocated for each device [3]. Simultaneous servicing at different stations is modeled using RQS with parallel servicing, sequential servicing at different stations is modeled using multiphase RQS [10]. As well as the possibility to take into account the requests heterogeneity [8] and their requirements for a resources variety [12].

The rest of the current paper is organized as follows. Section 2 is devoted to mathematical model of the two phase resource queueing system with three service units and Poisson input flow. The duration of the service time in each unit is arbitrary distributed. In Sect. 3, we obtain the Kolmogorov integro-differential equation describing the system behaviour and solve this equation to obtain the explicit expression of the characteristic function of the total volume of requirements in the system. Furthermore, we derive the expression for the average service time in the system. Section 4 is devoted to estimate optimal resource size for models with a limited resource. In Sect. 5 we obtain this values for one example and find the loss probability. Section 6 is devoted to some concluding remarks.

2 Mathematical Model

Let us consider a queueing system with unlimited servers number and arbitrary service time. Customers arrive according to a Poisson process with parameter λ. We will assume that each requirement is characterized by some random volume. Each arriving customer immediately occupies any free server on the first phase and requires resource. The service time distribution is $B_1(\tau)$ and the volume distribution given by probability function $G_1(y)$. After servicing at the first phase, the application proceeds to the next two phases by copying. The application takes a random amount of a certain resource with the distribution function $B_2(\tau)$ at the second phase and $B_3(\tau)$ at the third one, respectively. The service times on the second and the third phases don't depend on each other and have corresponding arbitrary distribution functions $G_2(y)$ and $G_3(y)$. When the service is complete in the second and third phases, the customer leaves the system. Resource amount and service times are mutually independent and do not depend on the epochs of customer arrivals. Figure 1 shows the structure of the system.

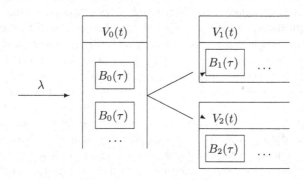

Fig. 1. Queueing system with unlimited servers number and the customers copying at the second phase

Denote by $V_k(t)$ the total resource amount at the k-th phase at time $t, (k = 1, 2, 3)$. Our goal is to derive the probabilistic characterization of the 3-dimensional process $\mathbf{V}(t) = \{V_1(t), V_2(t), V_3(t)\}$. This process, in general, is not Markovian. Therefore, we use the dynamic screening method for its investigation.

Consider four time axes that are numbered from 0 to 3 (Fig. 2). Let axis 0 shows the epochs of customers' arrivals, axes 1, 2 and 3 correspond to the first, second and third screened processes respectively.

We introduce functions (dynamic probability) $S_k(t)$, that satisfy the condition $0 \le S_k(t) \le 1$. Let the system be empty at moment t_0, and let us fix some arbitrary moment T in the future. $S_1(t)$ represents the probability that a customer arriving at the time $t > t_0$ will be serviced at the system by moment T. It is easy to show that

$$S_1(t) = 1 - B_1(T - t)$$

for $t_0 \le t \le T$.

The probability that the customer that came to the system at time $t > t_0$ by time T will finish service at the first and the third phases, but will not be finish

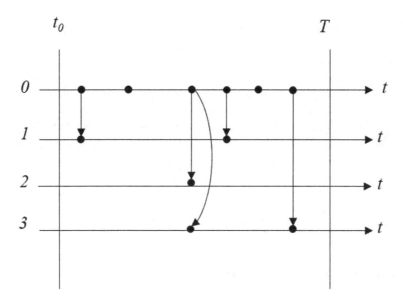

Fig. 2. Screening of the customers arrivals

it at the second phase (it's screened at the second axis) is equal to

$$S_2(t) = (B_3 * B_1)(T - t) - \int_0^{T-t} B_2(T - t - x)B_3(T - t - x)dB_1(x).$$

The probability that the customer that came to the system at time $t > t_0$ by time T will finish service at the first and the second phases, but will not be finish it at the third phase (it's screened at the third axis) is equal to

$$S_3(t) = (B_2 * B_1)(T - t) - \int_0^{T-t} B_2(T - t - x)B_3(T - t - x)dB_1(x).$$

The probability that the customer that came to the system at time $t > t_0$ by time T will finish service at the first and will not be finish it at the second and the third phases (it's screened at the second and the third axes) is equal to

$$S_{23}(t) = B_1(T - t) - (B_2 * B_1)(T - t) - (B_3 * B_1)(T - t)+$$

$$\int_0^{T-t} B_2(T - t - x)B_3(T - t - x)dB_1(x).$$

According to the initial condition

$$S_1(t) + S_2(t) + S_3(t) + S_{23}(t) \leq 1.$$

Denote by

$$S_0 = 1 - S_1(t) - S_2(t) - S_3(t) - S_{23}(t).$$

Denote by $W_i(t)$ the total amount of resources screened on axis k, $(k = 1, 2, 3)$. It is easy to prove that

$$
\begin{aligned}
P\left\{V_1(t) < z_1, V_2(t) < z_2, V_3(t) < z_3\right\} = \\
P\left\{W_1(t) < z_1, W_2(t) < z_2, W_3(t) < z_3\right\},
\end{aligned}
\tag{1}
$$

for $z_1, z_2, z_3 > 0$. We use equality (1) to investigate the process $\{V_1(t), V_2(t), V_3(t)\}$ via the analysis of the process $\{W_1(t), W_2(t), W_3(t)\}$.

3   Kolmogorov Integro-Differential Equation

We introduce the notation for the probability distribution of the Markov process

$$P\left\{W_1(t) < z_1, W_2(t) < z_2, W_3(t) < z_3\right\} = P(z_1, z_2, z_3, t).$$

For this distribution, we make up a direct system of Kolmogorov integro-differential equations using the Δt-method. By the formula of total probability we get

$$P(z_1, z_2, z_3, t + \Delta t) = \lambda \Delta t S_1(t) \int_0^{z_1} P(z_1 - y, z_2, z_3, t) dG_1(y) +$$

$$\lambda \Delta t S_2(t) \int_0^{z_2} P(z_1, z_2 - y, z_3, t) dG_2(y) +$$

$$\lambda \Delta t S_3(t) \int_0^{z_3} P(z_1, z_2, z_3 - y, t) dG_3(y) +$$

$$\lambda \Delta t S_{23}(t) \int_0^{z_2} \int_0^{z_3} P(z_1, z_2 - y_2, z_3 - y_3, t) dG_2(y_2) dG_3(y_3) +$$

$$P(z_1, z_2, z_3, t)(1 - \lambda \Delta t) + P(z_1, z_2, z_3, t)\lambda \Delta t S_0(t) + o(\Delta t),$$

and we can write the following system of Kolmogorov integro-differential equations

$$\frac{\partial P(z_1, z_2, z_3, t)}{\partial t} = \lambda S_1(t) \left[\int_0^{z_1} P(z_1 - y, z_2, z_3, t) dG_1(y) - P(z_1, z_2, z_3, t) \right] +$$

$$\lambda S_2(t) \left[\int_0^{z_2} P(z_1, z_2 - y, z_3, t) dG_2(y) - P(z_1, z_2, z_3, t) \right] +$$

$$\lambda S_3(t) \left[\int_0^{z_3} P(z_1, z_2, z_3 - y, t) dG_3(y) - P(z_1, z_2, z_3, t) \right] +$$

$$\lambda S_{23}(t) \left[\int_0^{z_2} \int_0^{z_3} P(z_1, z_2 - y_2, z_3 - y_3, t) dG_2(y_2) dG_3(y_3) - P(z_1, z_2, z_3, t) \right],$$

with the initial condition

$$P(z_1, z_2, z_3, t_0) = \begin{cases} 1, z_1 = z_2 = z_3 = 0, \\ 0, \text{ otherwise.} \end{cases}$$

We introduce the partial characteristic function

$$h(v_1, v_2, v_3, t) = \int_0^\infty e^{j v_1 z_1} \int_0^\infty e^{j v_2 z_2} \int_0^\infty e^{j v_3 z_3} P(dz_1, dz_2, dz_3, t),$$

where $j = \sqrt{-1}$ is the imaginary unit.

The following theorem has been formulated.

Theorem 1. *Characteristic function of the total volume of the occupied resource at each phase of the system is as follows*

$$h(v_1, v_2, v_3, t) = exp \left\{ \lambda \left[(G_1^*(v_1) - 1) \int_{t_0}^t S_1(\tau) d\tau + (G_2^*(v_2) - 1) \int_{t_0}^t S_2(\tau) d\tau + \right. \right.$$

$$\left. \left. (G_3^*(v_3) - 1) \int_{t_0}^t S_3(\tau) d\tau + (G_2^*(v_2) G_3^*(v_3) - 1) \int_{t_0}^t S_{23}(\tau) d\tau \right] \right\}.$$

Proof. We can write the following differential equation

$$\frac{\partial h(v_1, v_2, v_3, t)}{\partial t} = \lambda h(v_1, v_2, v_3, t) \left[S_1(t) (G_1^*(v_1) - 1) + S_2(t) (G_2^*(v_2) - 1) + \right.$$

$$\left. S_3(t) (G_3^*(v_3) - 1) + S_{23}(t) (G_2^*(v_2) G_3^*(v_3) - 1) \right],$$

where

$$G_k^*(\alpha) = \int_0^\infty e^{j \alpha y} dG_k(y), \quad k = 1, 2, 3.$$

The solution of this equation has the form

$$h(v_1, v_2, v_3, t) = exp \left\{ \lambda \left[(G_1^*(v_1) - 1) \int_{t_0}^{t} S_1(\tau)d\tau + (G_2^*(v_2) - 1) \int_{t_0}^{t} S_2(\tau)d\tau + \right. \right.$$

$$\left. \left. (G_3^*(v_3) - 1) \int_{t_0}^{t} S_3(\tau)d\tau + (G_2^*(v_2)G_3^*(v_3) - 1) \int_{t_0}^{t} S_{23}(\tau)d\tau \right] \right\}.$$

Corollary 1. *The characteristic function of the process* $\{V_1(t), V_2(t), V_3(t)\}$ *in the steady-state regime*

$$h(v_1, v_2, v_3) = \exp \left\{ \lambda \left[(G_1^*(v_1) - 1) b_1 + (G_2^*(v_2) - 1) b_2 + \right. \right.$$

$$\left. \left. (G_3^*(v_3) - 1) b_3 + (G_2^*(v_2)G_3^*(v_3) - 1) b_{23} \right] \right\}, \tag{2}$$

where

$$b_1 = \int_0^\infty (1 - B_1(\tau)) \, d\tau,$$

$$b_2 = \int_0^\infty \left[(B_3 * B_1)(\tau) - \int_0^T B_2(\tau - y)B_3(\tau - y)dB_1(y) \right] d\tau,$$

$$b_3 = \int_0^\infty \left[(B_2 * B_1)(\tau) - \int_0^T B_2(\tau - y)B_3(\tau - y)dB_1(y) \right] d\tau,$$

$$b_{23} = \int_0^\infty \left[B_1(\tau) - (B_2 * B_1)(\tau) - (B_3 * B_1)(\tau) + \int_0^T B_2(\tau - y)B_3(\tau - y)dB_1(y) \right] d\tau.$$

Proof. It is easy to show, using (1), when $t = T$ and $t_0 \to -\infty$.

Using the result, it is easy to estimate the amount of a sufficient resource amount in the systems, to assess the probability of loss and minimize it.

4   Initial Moments Method

We use the initial moments method for deriving the mean and the variance of the total volume of the occupied resource on each systems block. These values will be necessary to obtain the recommended value for the resource limit.

Given that

$$\left. \frac{\partial h(v_1, v_2, v_3)}{\partial v_i} \right|_{v_1=v_2=v_3=0} = jm_1^{(i)},$$

$$\left. \frac{\partial^2 h(v_1, v_2, v_3)}{\partial^2 v_i} \right|_{v_1=v_2=v_3=0} = j^2 m_2^{(i)},$$

where $m_1^{(i)}$ is the stationary expectation of $V_i(t)$, $m_2^{(i)}$ is the stationary second initial moment of $V_i(t)$. So, from (2), we get:

$$
\frac{\partial h(v_1, v_2, v_3)}{\partial v_1}\Bigg\|_{v_1=v_2=v_3=0} = \Bigg[\exp\Big\{\lambda\Big[\left(G_1^*(v_1) - 1\right)b_1 + \left(G_2^*(v_2) - 1\right)b_2 +
$$

$$
\left(G_3^*(v_3) - 1\right)b_3 + \left(G_2^*(v_2)G_3^*(v_3) - 1\right)b_{23}\Big]\Big\} \cdot \lambda\left(G_1^*(v_1)\right)' b_1\Bigg]\Bigg\|_{v_1=v_2=v_3=0},
$$

$$
\frac{\partial h(v_1, v_2, v_3)}{\partial v_2}\Bigg\|_{v_1=v_2=v_3=0} =
$$

$$
\Bigg[\exp\Big\{\lambda\Big[\left(G_1^*(v_1) - 1\right)b_1 + \left(G_2^*(v_2) - 1\right)b_2 +
$$

$$
\left(G_3^*(v_3) - 1\right)b_3 + \left(G_2^*(v_2)G_3^*(v_3) - 1\right)b_{23}\Big]\Big\} \cdot
$$

$$
\Big[\lambda\left(G_2^*(v_2)\right)' b_2 + G_3^*(v_3) * \left(G_2^*(v_2)\right)' b_{23}\Big]\Bigg]\Bigg\|_{v_1=v_2=v_3=0},
$$

$$
\frac{\partial h(v_1, v_2, v_3)}{\partial v_3}\Bigg\|_{v_1=v_2=v_3=0} =
$$

$$
\Bigg[\exp\Big\{\lambda\Big[\left(G_1^*(v_1) - 1\right)b_1 + \left(G_2^*(v_2) - 1\right)b_2 +
$$

$$
\left(G_3^*(v_3) - 1\right)b_3 + \left(G_2^*(v_2)G_3^*(v_3) - 1\right)b_{23}\Big]\Big\} \cdot
$$

$$
\Big[\lambda\left(G_3^*(v_3)\right)' b_3 + G_2^*(v_2) * \left(G_3^*(v_3)\right)' b_{23}\Big]\Bigg]\Bigg\|_{v_1=v_2=v_3=0}.
$$

Note that for an infinitesimal quantity ε, it is hold:

$$
G_i^*(\varepsilon) = \int_0^\infty e^{j\varepsilon y} dG_1(y) \approx \int_0^\infty \left(1 + j\varepsilon y + \frac{(j\varepsilon y)^2}{2}\right) dG_i(y) =
$$

$$
1 + j\varepsilon a_1^{(i)} + \frac{(j\varepsilon)^2}{2} a_2^{(i)}, \tag{3}
$$

then

$$
\left(G_i^*(0)\right)' = ja_1^{(i)}, \quad G_i^*(0) = 1.
$$

We obtain

$$
jm_1^{(1)} = j\lambda a_1^{(1)} b_1,
$$

$$
jm_1^{(2)} = j\lambda a_1^{(2)}(b_2 + b_{23}),
$$

$$
jm_1^{(3)} = j\lambda a_1^{(3)}(b_3 + b_{23}).
$$

From the second derivative (2), we obtain:

$$\frac{\partial^2 h(v_1, v_2, v_3)}{\partial^2 v_1}\bigg\|_{v_1=v_2=v_3=0} =$$

$$\bigg[\exp\bigg\{\lambda\bigg[(G_1^*(v_1) - 1)\,b_1 + (G_2^*(v_2) - 1)\,b_2+$$

$$(G_3^*(v_3) - 1)\,b_3 + (G_2^*(v_2)G_3^*(v_3) - 1)\,b_{23}\bigg]\bigg\}\cdot$$

$$\left(\lambda\,(G_1^*(v_1))'\,b_1\right)^2 + \lambda\,(G_1^*(v_1))''\,b_1\bigg]\bigg\|_{v_1=v_2=v_3=0},$$

$$\frac{\partial^2 h(v_1, v_2, v_3)}{\partial^2 v_2}\bigg\|_{v_1=v_2=v_3=0} =$$

$$\bigg[\exp\bigg\{\lambda\bigg[(G_1^*(v_1) - 1)\,b_1 + (G_2^*(v_2) - 1)\,b_2 +$$

$$(G_3^*(v_3) - 1)\,b_3 + (G_2^*(v_2)G_3^*(v_3) - 1)\,b_{23}\bigg]\bigg\}\cdot$$

$$\bigg[\big[\lambda\,(G_2^*(v_2))'\,b_2 + G_3^*(v_3) * (G_2^*(v_2))'\,b_{23}\big]^2 +$$

$$\lambda\,(G_2^*(v_2))''\,b_2 + G_3^*(v_3)\,(G_2^*(v_2))''\,b_{23}\bigg]\bigg]\bigg\|_{v_1=v_2=v_3=0},$$

$$\frac{\partial^2 h(v_1, v_2, v_3)}{\partial^2 v_3}\bigg\|_{v_1=v_2=v_3=0} =$$

$$\bigg[\exp\bigg\{\lambda\bigg[(G_1^*(v_1) - 1)\,b_1 + (G_2^*(v_2) - 1)\,b_2 +$$

$$(G_3^*(v_3) - 1)\,b_3 + (G_2^*(v_2)G_3^*(v_3) - 1)\,b_{23}\bigg]\bigg\}\cdot$$

$$\bigg[\big[\lambda\,(G_3^*(v_3))'\,b_3 + G_2^*(v_2) * (G_3^*(v_3))'\,b_{23}\big]^2 +$$

$$\lambda\,(G_3^*(v_3))''\,b_3 + G_2^*(v_2) * (G_3^*(v_3))''\,b_{23}\bigg]\bigg]\bigg\|_{v_1=v_2=v_3=0}.$$

Here, from (3), we get:

$$(G_i^*(0))'' = \frac{j^2 a_2^{(i)}}{2},$$

then

$$j^2 m_2^{(1)} = \left(j\lambda a_1^{(1)} b_1\right)^2 + \frac{j^2 \lambda a_2^{(1)} b_1}{2},$$

$$j^2 m_2^{(2)} = \left(j\lambda a_1^{(2)}(b_2 + b_{23})\right)^2 + \frac{j^2 \lambda a_2^{(2)}(b_2 + b_{23})}{2},$$

$$j^2 m_2^{(3)} = \left(j\lambda a_1^{(3)}(b_3 + b_{23})\right)^2 + \frac{j^2 \lambda a_2^{(3)}(b_3 + b_{23})}{2}.$$

Therefore, the characteristics are presented in Table 1.

Table 1. Numerical characteristics

Unit i	$E\{V_i(t)\}$	$Var\{V_i(t)\}$
1	$\lambda a_1^{(1)} b_1$	$\dfrac{\lambda a_2^{(1)} b_1}{2}$
2	$\lambda a_1^{(2)}(b_2 + b_{23})$	$\dfrac{\lambda a_2^{(2)}(b_2 + b_{23})}{2}$
3	$\lambda a_1^{(3)}(b_3 + b_{23})$	$\dfrac{\lambda a_2^{(3)}(b_3 + b_{23})}{2}$

Then, according to the "three sigma rule", we conclude: in order for the probability of a denial of the service to be negligible, it is necessary to set the limit per resource on each block in the amount of

$$R_i = E\{V_i(t)\} + 3 * \sqrt{Var\{V_i(t)\}}$$

or, in case, when the resource has total buffer

$$R = R_1 + R_2 + R_3.$$

5   Numerical Example

Let Poisson arrival has the parameter $\lambda = 1$, the customers service time characterized by Gamma distribution with parameters $\alpha_1 = \beta_1 = 1.5$, $\alpha_2 = \beta_2 = 2.5$ and $\alpha_3 = \beta_3 = 3.5$. So, let resource distributions have uniform on $[0; 3]$, $[0; 2]$ and $[0; 1]$, respectively. You can see the results at Table 2.

Table 2. Numerical characteristics and resource limits

Unit i	$E\{V_i(t)\}$	$Var\{V_i(t)\}$	R_i
1	1.5	1.500	5.174
2	1.0	0.667	3.449
3	0.5	0.167	1.725
Total R	–	–	10.348

In the simulation model, we limit the resource capabilities of each block as R_i. In this case, the request will be refused if the free resource is not enough for maintenance. We estimate the loss probability as "the customers number that were refused"/"total number of all customers". Then, the loss probability will be equal 0.005. We can control this probability by changing the limit on the total resource amount. We can reduce it to ensure the best quality of connections, we can increase it to minimize the cost of buying and installing the equipment.

6 Conclusion

In the present work, the resource QS was studied with the customers copying at the second system phase and with the Poisson arrival. The joint characteristic function of the occupied resource on the blocks total volume was found, from which the probability distribution can be obtained using the inverse Fourier transform. In addition, from the obtained characteristic function, the main total volume numerical characteristics of the occupied resource were found in the paper, which made it possible to calculate the required resource amount in RQS with rejects and to find the loss probability.

References

1. Ageev, K., Sopin, E., Konstantin, S.: Simulation of the limited resources queuing system with signals. In: 10th International Congress on Ultra Modern Telecommunications and Control Systems and Workshops (ICUMT), pp. 1–5 (2018). https://doi.org/10.1109/ICUMT.2018.8631246
2. Basharin, G.P., Samouylov, K.E., Yarkina, N.V., Gudkova, I.A.: A new stage in mathematical teletraffic theory. Autom. Remote Control 70(12), 1954–1964 (2009). https://doi.org/10.1134/S0005117909120030
3. Beschastnyi, V., Savich, V., Ostrikova, D., Gudkova, I., Araniti, G., Shorgin, V.: Analysis of machine-type communication data transmission by multicasting technology in 5G wireless networks, vol. 2116, p. 090005 (2019). https://doi.org/10.1063/1.5114070
4. Buturlin, I.A., Gaidamaka, Y.V., Samuylov, A.K.: Utility function maximization problems for two cross-layer optimization algorithms in ofdm wireless networks. In: 2012 IV International Congress on Ultra Modern Telecommunications and Control Systems, pp. 63–65 (2012). https://doi.org/10.1109/ICUMT.2012.6459745
5. Galinina, O., et al.: Capturing spatial randomness of heterogeneous Cellular/WLAN deployments with dynamic traffic. IEEE J. Sel. Areas Commun. 32(6), 1083–1099 (2014). https://doi.org/10.1109/JSAC.2014.2328172
6. Galinina, O., Andreev, S., Turlikov, A., Koucheryavy, Y.: Optimizing energy efficiency of a multi-radio mobile device in heterogeneous beyond-4G networks. Perform. Eval. 78, 18–41 (2014). https://doi.org/10.1016/j.peva.2014.06.002
7. Gimpelson, L.: Analysis of mixtures of wide- and narrow-band traffic. IEEE Trans. Commun. Technol. 13(3), 258–266 (1965). https://doi.org/10.1109/TCOM.1965.1089121
8. Gorbunova, A.V., Naumov, V.A., Gaidamaka, Y.V., Samouylov, K.E.: Resource queuing systems as models of wireless communication systems. Informatika i ee Primeneniya 12(3), 48–55 (2018). https://doi.org/10.14357/19922264180307
9. Gudkova, I., et al.: Analyzing impacts of coexistence between M2M and H2H communication on 3GPP LTE system. In: Mellouk, A., Fowler, S., Hoceini, S., Daachi, B. (eds.) WWIC 2014. LNCS, vol. 8458, pp. 162–174. Springer, Cham (2014). https://doi.org/10.1007/978-3-319-13174-0_13
10. Lisovskaya, E., Moiseeva, S., Pagano, M.: Infinite-server tandem queue with renewal arrivals and random capacity of customers. In: Vishnevskiy, V.M., Samouylov, K.E., Kozyrev, D.V. (eds.) DCCN 2017. CCIS, vol. 700, pp. 201–216. Springer, Cham (2017). https://doi.org/10.1007/978-3-319-66836-9_17

11. Lisovskaya, E., Moiseeva, S., Pagano, M.: Multiclass GI/GI/∞ queueing systems with random resource requirements. In: Dudin, A., Nazarov, A., Moiseev, A. (eds.) ITMM/WRQ -2018. CCIS, vol. 912, pp. 129–142. Springer, Cham (2018). https://doi.org/10.1007/978-3-319-97595-5_11

12. Lisovskaya, E.Y., Moiseev, A.N., Moiseeva, S.P., Pagano, M.: Modeling of mathematical processing of physics experimental data in the form of a Non-Markovian multi-resource queuing system. Russ. Phys. J. **61**(12), 2188–2196 (2019). https://doi.org/10.1007/s11182-019-01655-6

13. Naumov, V., Samouylov, K.: Analysis of multi-resource loss system with state-dependent arrival and service rates. Probab. Eng. Inf. Sci. **31**(4), 413–419 (2017). https://doi.org/10.1017/S0269964817000079

14. Naumov, V.A., Samuilov, K.E., Samuilov, A.K.: On the total amount of resources-occupied by serviced customers. Autom. Remote Control **77**(8), 1419–1427 (2016). https://doi.org/10.1134/S0005117916080087

15. Samuvlov, A., Moltchanov, D., Krupko, A., Kovalchukov, R., Moskaleva, F., Gaidamaka, Y.: Performance analysis of mixture of unicast and multicast sessions in 5G NR systems. In: 10th International Congress on Ultra Modern Telecommunications and Control Systems and Workshops (ICUMT), pp. 1–7 (11 2018). https://doi.org/10.1109/ICUMT.2018.8631230

16. Sopin, E., Vikhrova, O., Samouylov, K.: LTE network model with signals and random resource requirements. In: International Congress on Ultra Modern Telecommunications and Control Systems and Workshops, pp. 101–106 (2017). https://doi.org/10.1109/ICUMT.2017.8255155

Resource Queueing System with Dual Requests and Their Parallel Service

Tatyana Bushkova[iD], Elena Danilyuk$^{(\boxtimes)}$[iD], Svetlana Moiseeva$^{(\boxtimes)}$[iD], and Ekaterina Pavlova[iD]

National Research Tomsk State University, Lenina Avenue, 36, 634050 Tomsk, Russia
bushkova70@mail.ru, daniluc_elena@sibmail.com, smoiseeva@mail.ru,
pavlovakatya_2010@mail.ru

Abstract. In this paper we consider a mathematical model of parallel servicing of dual customers in the form of a queuing system with two service units each of them contains of an unlimited number of devices. Input flow is the Poisson flow of dual customers. Each customer is characterized by a random total capacity which is independent of the service time. Based on the method of moments it is possible to deduce the expressions for characteristic function of the process of the total amount of resource in two-service unit system. The mathematical models of this type could be of great interest in terms of application in telecommunication, for example, for modeling wireless network, enhancing the existing and designing new ones.

Keywords: Queueing system · Server · Resource · Dual customer · Random capacity

1 Introduction

Nowadays resources and a problem of their optimal allocation (or sharing) are of interest and call for the exact solution with mathematical instruments. There are many examples of such resources in computer networks and telecommunications: bit-rate, packet size of information, energy demand in mobile devices, size of storage space of cloud or other memory for media content transmission. In common case, every claim (or customer) have different random capacity requirement on the same resource (that is limited in the most of cases) at the same time. The goal is, on the one hand, to service all customers in parallel way and to give them demanded amount of the resource, on the other hand. The queuing system is the most convenient mathematical tool for studying behaviour of such real systems because the models take account of that input claims are stochastic. There are many papers devoted to resource queueing system research and applications [1–21]. The detailed overview of the resource queuing systems is given by authors Gorbunova, Naumov, Gaidamaka, Samuylov in the article "Resource Queuing

Supported by RFBR according to the research project No. 19-41-703002.

V. M. Vishnevskiy et al. (Eds.): DCCN 2019, CCIS 1141, pp. 364–374, 2019.
https://doi.org/10.1007/978-3-030-36625-4_29

Systems as Models of Wireless Communication Systems" [14]. The authors show the state of studying of such systems and mark that "there have been very few works devoted to their analysis until recently, which was due to the complexity of constructing a random process to describe their functioning and, accordingly, of obtaining the numerical results. However, in recent years, there has been a significant shift in the study of the resource systems - new methods for their analysis have been proposed, which made it possible to construct recursive algorithms suitable for the numerical calculations". In this paper they also consider the models of wireless communication systems based on resource queuing systems without waiting space with exponentially distributed service time.

Our research is devoted to queueing systems with splitting (or copying) of arriving customers and their parallel service. Such models is used to describe real processes in multi-service networks and telecommunication systems. Queueing systems with service units which work in parallel is presented in papers by Basharin, Samouylov, Movaghar, Knessl, Morrison [15–17] and others. The different systems with parallel service such as queueing systems with one server and finite or infinite buffer, priority-service discipline, impatient customers and general arrivals; queueing systems with two and more service units with finite numbers of servers and finite queue are studied in the research. Among the works on analysis of data network resource sharing with two types of traffic it can be mentioned the paper by Boussetta, Beylot [18].

There are many works where resource queueing systems with Poison incoming flow and exponential distribution of the service time are investigated with analytical methods but the most of real systems are described by more complex models and, therefore, require different ways of studying (for example, asymptotic analysis method) [3,5,9,19]. In the present paper we use Poisson arrival process and not exponential distribution of service time but any one with distribution function $B(x)$. To study the system with such complicated structure we preliminary use the dynamic screening method and then apply method of moments [19]. Analogical problems of analysis of queueing systems with splitting of arrivals and infinite numbers of servers for different types of input flows are considered in [20,21].

All our results were obtained in assumption that service time and customers capacity volume are independent quantities (this assumption satisfies the real computer and communication systems [3]). For instance, in [6] performance analysis of LTE networks is carried out in terms of flow level dynamics and the amount of required radio resources does not depend on the duration of the flow.

One more point to vary mathematical models of described systems is the number of service devices. Single server queueing systems for customers with random bandwidth with the simplest input flow, where service time is distributed exponentially arbitrarily under the assumption that customer capacity and service time are independent were considered in [2]. Similar results were established in [3] for infinite-server system with exponential and arbitrary distributed service times. We study queueing system with two type of infinite-server units.

The rest of the paper is organized as follows. The general information about mathematical model of the studied resource queueing system with dual requests and their parallel service and the problem statement are presented in Sect. 2. In Sect. 2, the system of Kolmogorov differential equations is also performed and obtained expressions of characteristic functions that are used in the method of moments are given. Section 3 consists of the application of the method of moments for solving the problem under consideration. Section 4 concludes the paper.

2   Statement of the Problem

2.1   Mathematical Model

Consider the queueing system with two service units each of which contains of unlimited number of servers. There are two types of resources and each customer requires a random quantity of the both of them. We use arrival process as Poisson process with rate λ. At the time of occurrence of the event in the arriving flow each customer splits (flow of original customers and flow of customers copies), and we have two flow of claims in the system (or we can say that one customer induces two types of claims or customers). Each customer goes to a free server in the first and the second service units, where its service is performed during a random time distributed with any function $B_i(x)$, $i = 1, 2$, corresponding to the type of unit.

Figure 1 shows the model of the resource queueing system with dual requests and their parallel service.

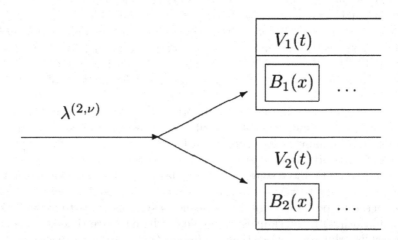

Fig. 1. Heterogeneous queue $M^{(2,v)}|G^{(2,v)}|\infty$ with random customers capacities

Let each customer with number j requires some random capacity $\nu_i^{(j)} > 0$, $i = 1, 2$, with distribution function $G_i(y)$, $i = 1, 2$.

Denote by $\nu = \left\{ \nu_1^{(j)}(t), \nu_2^{(j)}(t) \right\}$ capacity of each customer on each service unit in the system at time t. Set the problem of exploring of two-dimensional stochastic process $\{V_1(t), V_2(t)\}$ of total capacity on each service unit in the system at time t, where

$$V_i(t) = \sum_{j=1}^{\infty} \nu_i^{(j)}, \quad i = 1, 2.$$

This process is not Markovian, therefore, we use the dynamic screening method for its investigation (Fig. 2).

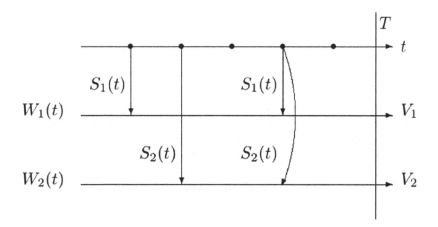

Fig. 2. Screening of the customers arrivals

We assume the system is empty at the moment t_0, and let us fix some arbitrary moment T in the future. $S_i(t)$ represents the probability that a customer arriving at time t will be serviced in the system by the moment T on i-type of service unit

$$S_i(t) = 1 - B_i (T - t),$$

$i = 1, 2, t_0 \leq t \leq T$.

Denote by $\{W_1(t), W_2(t)\}$ the total capacity of arrivals screened before the moment t.

As it is shown in [5], the probability distribution of the capacity of the customer in the system at the moment T coincides with the probability distribution of the capacity of screened arrivals on the axis

$$P\{V_1(T) < w_1, V_2(T) < w_2\} = P\{W_1(T) < w_1, W_2(T) < w_2\}.$$

2.2 Kolmogorov Differential Equations

Consider the process $\{W_1(t), W_2(t)\}$ and denote by

$$P(w_1, w_2, t) = P\{W_1(t) < w_1, W_2(t) < w_2\},$$

We can write for two-dimensional Markovian process the following system of Kolmogorov differential equations

$$\frac{\partial P(w_1, w_2, t)}{\partial t} = -\lambda(S_1(t) + S_2(t) - S_1(t)S_2(t))P(w_1, w_2, t)$$

$$+ \lambda S_1(t)(1 - S_2(t)) \int\limits_0^{w_1} P(w_1 - y_1, w_2, t)dG_1(y_1)$$

$$+ \lambda(1 - S_1(t))S_2(t) \int\limits_0^{w_2} P(w_1, w_2 - y_2, t)dG_2(y_2) \tag{1}$$

$$+ \lambda S_1(t)S_2(t) \int\limits_0^{w_1}\int\limits_0^{w_2} P(w_1 - y_1, w_2 - y_2, t)dG_2(y_1)dG_2(y_2)$$

with initial condition

$$P(k, w_1, w_2, t_0) = \begin{cases} 1, & if \quad w_1 = w_2; \\ 0, & if \quad w_1 \neq w_2; \end{cases}$$

where $w_i > 0, i = 1, 2$.

We introduce the characteristic function

$$h(u_1, u_2, t) = \int\limits_0^\infty e^{ju_1 w_1} \int\limits_0^\infty e^{ju_2 w_2} P(dw_1, dw_2, t),$$

where $j = \sqrt{-1}$ hereinafter is the imaginary unit.

Then we can rewrite the system (1)

$$\frac{\partial h(u_1, u_2, t)}{\partial t} = -\lambda(S_1(t) + S_2(t) - S_1(t)S_2(t))h(u_1, u_2, t)$$

$$+ \lambda S_1(t)(1 - S_2(t))G_1^*(u_1)h(u_1, u_2, t) \tag{2}$$

$$+ \lambda(1 - S_1(t))S_2(t)G_2^*(u_2)h(u_1, u_2, t)$$

$$+ \lambda S_1(t)S_2(t)G_1^*(u_1)G_2^*(u_2)h(u_1, u_2, t),$$

where

$$G_i^*(u_i) = \int\limits_0^\infty e^{ju_i y_i}dG_i(y_i), \quad i = 1, 2.$$

Let us transform and simplify the system (2)

$$
\frac{\partial h(u_1, u_2, t)}{\partial t} = \lambda h(u_1, u_2, t)[S_1(t)(G_1^*(u_1) - 1) + S_2(t)(G_2^*(u_2) - 1)
$$
$$
+ S_1(t)S_2(t)(G_1^*(u_1) - 1)(G_2^*(u_2) - 1)]. \tag{3}
$$

In the Eq. (3) we separate the variables

$$
\frac{dh(u_1, u_2, t)}{h(u_1, u_2, t)} = \lambda[S_1(t)(G_1^*(u_1) - 1) + S_2(t)(G_2^*(u_2) - 1)
$$
$$
+ S_1(t)S_2(t)(G_1^*(u_1) - 1)(G_2^*(u_2) - 1)]dt. \tag{4}
$$

Then we obtain the solution of the differential equation (4)

$$
h(u_1, u_2, t) = C \cdot \exp\left\{ \lambda \int_{t_0}^{t} \left[S_1(\tau)(G_1^*(u_1) - 1) + S_2(\tau)(G_2^*(u_2) - 1) \right.\right.
$$
$$
\left.\left. + S_1(\tau)S_2(\tau)(G_1^*(u_1) - 1)(G_2^*(u_2) - 1) \right] d\tau \right\}. \tag{5}
$$

In the resulting solution (5) we set $t = T$, perform the limit $t_0 \to -\infty$ and define a constant from the initial condition $C = 1$, then we can rewrite

$$
h(u_1, u_2, T) = \exp\left\{ \lambda(G_1^*(u_1) - 1) \int_{-\infty}^{T} (1 - B_1(T - \tau))d\tau \right.
$$
$$
+ \lambda(G_2^*(u_2) - 1) \int_{-\infty}^{T} (1 - B_2(T - \tau))d\tau + \lambda(G_1^*(u_1) - 1)(G_2^*(u_2) - 1) \tag{6}
$$
$$
\left. \times \int_{-\infty}^{T} (1 - B_1(T - \tau))(1 - B_2(T - \tau))d\tau \right\}.
$$

In the expression (6) perform the exchange of variables $w = T - t$ and let us denote

$$
\int_{-\infty}^{T} (1 - B_i(T - \tau))d\tau = \int_{0}^{\infty} (1 - B_i(w))dw = b_i, \quad i = 1, 2,
$$

$$
\int_{-\infty}^{T} (1 - B_1(T - \tau))(1 - B_2(T - \tau))d\tau = \int_{0}^{\infty} (1 - B_1(w))(1 - B_2(w))dw = b_{12},
$$

where b_i, $i = 1, 2$, is average service time on first and second service units, b_{12} is average service time of the dual customer in both service units.

Taking into account the introduced notation we finally obtain the form of the characteristic function of the customers capacity of the required resource in two service units

$$h(u_1, u_2) = \exp \left\{ \lambda \left[(G_1^*(u_1) - 1) \, b_1 + (G_2^*(u_2) - 1) \, b_2 \right. \right.$$
$$\left. \left. + (G_1^*(u_1) - 1) \, (G_2^*(u_2) - 1) \, b_{12} \right] \right\}. \tag{7}$$

3 Method of Moments

To find probabilistic characteristics, we use the following properties of the characteristic function

$$\left. \frac{\partial h(u_1, u_2)}{\partial u_1} \right|_{u_1=0} = jE\{V_1\}, \quad \left. \frac{\partial h(u_1, u_2)}{\partial u_2} \right|_{u_2=0} = jE\{V_2\},$$

$$\left. \frac{\partial^2 h(u_1, u_2)}{\partial u_1{}^2} \right|_{u_1=0} = -\,E\{V_1{}^2\}, \quad \left. \frac{\partial^2 h(u_1, u_2)}{\partial u_2{}^2} \right|_{u_2=0} = -E\{V_2{}^2\}.$$

Let us differentiate last equation for $h(u_1, u_2)$ (7) by variable u_1

$$\frac{\partial h(u_1, u_2)}{\partial u_1} = \lambda \left[G_1'^*(u_1) b_1 + G_1'^*(u_1)(G_2^*(u_2) - 1) b_{12} \right]$$
$$\times \exp \left\{ \lambda \left[(G_1^*(u_1) - 1) \, b_1 + (G_2^*(u_2) - 1) \, b_2 \right. \right. \tag{8}$$
$$\left. \left. + (G_1^*(u_1) - 1) \, (G_2^*(u_2) - 1) \, b_{12} \right] \right\}.$$

and by variable u_2

$$\frac{\partial h(u_1, u_2)}{\partial u_2} = \lambda \left[G_2'^*(u_2) b_2 + (G_1^*(u_1) - 1) G_2'^*(u_2) b_{12} \right]$$
$$\times \exp \left\{ \lambda \left[(G_1^*(u_1) - 1) b_1 + (G_2^*(u_2) - 1) b_2 \right. \right. \tag{9}$$
$$\left. \left. + (G_1^*(u_1) - 1)(G_2^*(u_2) - 1) b_{12} \right] \right\}.$$

and consider at the point $u_1 = u_2 = 0$

$$\left. \frac{\partial h(u_1, u_2)}{\partial u_1} \right|_{u_1=u_2=0} = j\lambda a_1^{(1)} b_1, \quad \left. \frac{\partial h(u_1, u_2)}{\partial u_2} \right|_{u_1=u_2=0} = j\lambda a_2^{(1)} b_2,$$

where

$$a_i^{(1)} = \int_0^\infty y_i dG_i(y_i), \quad i = 1, 2.$$

We obtain the expressions for the first moments

$$E\{V_i\} = \lambda a_i^{(1)} b_i, \quad i = 1, 2.$$

To obtain the second moments, we differentiate the Eqs. (8) and (9) one more time by variables u_1

$$
\begin{aligned}
\frac{\partial^2 h(u_1, u_2)}{\partial u_1^2} &= \exp\left\{\lambda\Big[(G_1^*(u_1) - 1)\, b_1 + (G_2^*(u_2) - 1)\, b_2 \right. \\
&\quad + (G_1^*(u_1) - 1)\,(G_2^*(u_2) - 1)\, b_{12}\Big]\Big\} \\
&\quad \times \Big[\lambda j (G_1'^*(u_1) b_1 + G_1'^*(u_1)(G_2^*(u_2) - 1) b_{12})\Big]^2 \\
&\quad + \exp\left\{\lambda\Big[(G_1^*(u_1) - 1) b_1 + (G_2^*(u_2) - 1) b_2 \right. \\
&\quad + (G_1^*(u_1) - 1)(G_2^*(u_2) - 1) b_{12}\Big]\Big\} \\
&\quad \times \Big[\lambda (G_1''^*(u_1) b_1 + G_1''^*(u_1)(G_2^*(u_2) - 1) b_{12})\Big].
\end{aligned}
\tag{10}
$$

and by variable u_2

$$
\begin{aligned}
\frac{\partial^2 h(u_1, u_2)}{\partial u_2^2} &= \exp\left\{\lambda\Big[(G_1^*(u_1) - 1) b_1 + (G_2^*(u_2) - 1) b_2 \right. \\
&\quad + (G_1^*(u_1) - 1)(G_2^*(u_2) - 1) b_{12}\Big]\Big\} \\
&\quad \times \Big[\lambda j (G_2'^*(u_2) b_2 + (G_1^*(u_1) - 1) G_2'^*(u_2) b_{12})\Big]^2 \\
&\quad + \exp\left\{\lambda\Big[(G_1^*(u_1) - 1) b_1 + (G_2^*(u_2) - 1) b_2 \right. \\
&\quad + (G_1^*(u_1) - 1)(G_2^*(u_2) - 1) b_{12}\Big]\Big\} \\
&\quad \times \Big[\lambda j (G_2''^*(u_2) b_2 + (G_1^*(u_1) - 1) G_2''^*(u_2) b_{12})\Big].
\end{aligned}
\tag{11}
$$

and consider (10), (11) at the point $u_1 = u_2 = 0$

$$
\left.\frac{\partial^2 h(u_1, u_2)}{\partial u_1^2}\right|_{u_1 = u_2 = 0} = (j\lambda a_1^{(1)} b_1)^2 + \lambda j^2 a_1^{(2)} b_1,
$$

$$
\left.\frac{\partial^2 h(u_1, u_2)}{\partial u_2^2}\right|_{u_1 = u_2 = 0} = (j\lambda a_2^{(1)} b_2)^2 + \lambda j^2 a_2^{(2)} b_2,
$$

where

$$
a_i^{(2)} = \int\limits_0^\infty y_i^2 \, dG_i(y_i), \quad i = 1, 2.
$$

We obtain the expressions for the second moments

$$
E\{V_i^2\} = \lambda a_i^{(2)} b_i + (\lambda a_i^{(1)} b_i)^2, \quad i = 1, 2.
$$

Then we can write expressions for variance

$$
Var\{V_i\} = E\{V_i^2\} - E\{V_i\}^2 = \lambda a_i^{(2)} b_i, \quad i = 1, 2.
$$

To obtain correlation let us calculate the mixed derivative of the equation for $h(u_1, u_2)$

$$
\frac{\partial^2 h(u_1, u_2)}{\partial u_1 \partial u_2} = \exp\left\{\lambda\left[(G_1^*(u_1) - 1)b_1 + (G_2^*(u_2) - 1)b_2\right.\right.
$$
$$
\left.\left. + (G_1^*(u_1) - 1)(G_2^*(u_2) - 1)b_{12}\right]\right\} \lambda G_1'^*(u_1)G_2'^*(u_2)b_{12}
$$
$$
+ \exp\left\{\lambda\left[(G_1^*(u_1) - 1)b_1 + (G_2^*(u_2) - 1)b_2 + (G_1^*(u_1) - 1)(G_2^*(u_2) - 1)b_{12}\right]\right\}
$$
$$
\times \lambda^2 \left[G_1'^*(u_1)b_1 + G_1'^*(u_1)(G_2^*(u_2) - 1)b_{12}\right]\left[G_2'^*(u_2)b_2 + (G_1^*(u_1) - 1)G_2'^*(u_2)b_{12}\right].
$$

and consider at the point $u_1 = u_2 = 0$

$$
E\{V_1 V_2\} = \left.\frac{\partial^2 h(u_1, u_2)}{\partial u_1 \partial u_2}\right|_{u_1 = u_2 = 0} = \lambda a_1^{(1)} a_2^{(1)} b_{12} + \lambda^2 a_1^{(1)} a_2^{(1)} b_1 b_2.
$$

We obtain the expression of correlation

$$
r = \frac{E\{V_1 V_2\} - E\{V_1\}E\{V_2\}}{\sqrt{Var\{V_1\}}\sqrt{Var\{V_2\}}} = \frac{a_1^{(1)} a_2^{(1)} b_{12}}{\sqrt{a_1^{(2)} a_2^{(2)} b_1 b_2}}.
$$

4   Conclusion

Thus, a mathematical model of $M^{(2,v)}|GI^{(2,v)}|\infty$ type queuing system was constructed, expressions were found for the characteristic function, first and second moments, variance and correlation.

Explicit form of generating function lets to find the law of probability for given functions $G_i(y)$ and $B_i(x)$, $i = 1, 2$. Obtained numerical results show that coefficient of correlation between $V_1(t)$ and $V_2(t)$ increases when parameters of servicing are close and decrease if otherwise (for example, means are differ by a factor of 10). However, for the systems with non-Poison incoming flow it is possible to write the Kolmogorov differential equations that will not be resolved with analytical methods. In this case we should use the method of asymptotic analysis, and it is our further study.

References

1. Tikhonenko, O., Kempa, W.M.: The generalization of AQM algorithms for queueing systems with bounded capacity. In: Wyrzykowski, R., Dongarra, J., Karczewski, K., Waśniewski, J. (eds.) PPAM 2011. LNCS, vol. 7204, pp. 242–251. Springer, Heidelberg (2012). https://doi.org/10.1007/978-3-642-31500-8_25
2. Tikhonenko, O., Kempa, W.M.: Queue-size distribution in M/G/1-type system with bounded capacity and packet dropping. In: Dudin, A., Klimenok, V., Tsarenkov, G., Dudin, S. (eds.) BWWQT 2013. CCIS, vol. 356, pp. 177–186. Springer, Heidelberg (2013). https://doi.org/10.1007/978-3-642-35980-4_20

3. Lisovskaya, E., Moiseeva, S., Pagano, M., Potatueva, V.: Study of the MMPP/GI/∞ queueing system with random customers' capacities. Inform. Appl. **11**(4), 111–119 (2017)
4. Naumov, V., Samouylov, K., Samouylov, A.: On the total amount of resources occupied by serviced customers. Autom. Remote Control **77**(8), 1419–1427 (2016). https://doi.org/10.1134/S0005117916080087
5. Moiseev, A., Nazarov, A.: Asymptotic analysis of a multistage queuing system with a high-rate renewal arrival process. Optoelectron. Instrum. Data Process. **50**(2), 163–171 (2014)
6. Naumov, V., Samouylov, K., Sopin, E., Andreev, S.: Two approaches to analysis of queueing systems with limited resources. In: Ultra-Modern Telecommunications and Control Systems and Workshops Proceedings, Piscataway, NJ, USA, pp. 485–488. IEEE (2014)
7. Sopin, E.S., Ageev, K.A., Markova, E.V., Vikhrova, O.G., Gaidamaka, Y.V.: Performance analysis of M2M traffic in LTE network using queuing system with random resource requirements. Autom. Control Comput. Sci. **52**(5), 345–353 (2018)
8. Sopin, E., Samouylov, K., Vikhrova, O., Kovalchukov, R., Moltchanov, D., Samuylov, A.: Evaluating a case of downlink uplink decoupling using queuing system with random requirements. In: Galinina, O., Balandin, S., Koucheryavy, Y. (eds.) NEW2AN/ruSMART -2016. LNCS, vol. 9870, pp. 440–450. Springer, Cham (2016). https://doi.org/10.1007/978-3-319-46301-8_37
9. Lisovskaya, E., Moiseeva, S., Pagano, M.: Multiclass GI/GI/∞ queueing systems with random resource requirements. In: Dudin, A., Nazarov, A., Moiseev, A. (eds.) ITMM/WRQ -2018. CCIS, vol. 912, pp. 129–142. Springer, Cham (2018). https://doi.org/10.1007/978-3-319-97595-5_11
10. Basket, F., Chandy, K.M., Muntz, R.R., Palasios, F.G.: Open, closed, and mixed networks of queues with different classes of customers. J. ACM **22**(2), 248–260 (1975)
11. Galinina, O., Andreev, S., Turlikov, A., Koucheryavy, Y.: Optimizing energy efficiency of a multi-radio mobile device in heterogeneous beyond-4G networks. Perfomance Eval. **78**, 18–41 (2014)
12. Naumov, V., Samouylov, K.: Analysis of multi-resource loss system with state dependent arrival and service rates. Probab. Eng. Inf. Sci. **31**(4), 413–419 (2017)
13. Sopin, E.S., Ageev, K.A., Markova, E.V., Vikhrova, O.G., Gaidamaka, Y.V.: Performance analysis of M2M traffic in LTE network using queuing systems with random resource requirements. Autom. Control Comput. Sci. **52**(5), 345–353 (2018). https://doi.org/10.3103/S0146411618050127
14. Gorbunova, A.V., Naumov, V.A., Gaidamaka, Y.V., Samuylov, K.E.: Resource queuing systems as models of wireless communication systems. Inform. Appl. **12**(3), 48–55 (2018). https://doi.org/10.14357/19922264180307
15. Basharin, G.P., Gaidamaka, Y.V., Samouylov, K.E.: Mathematical theory of teletraffic and its application to the analysis of multiservice communication of next generation networks. Autom. Control Comput. Sci. **47**(2), 62–69 (2013)
16. Movaghar, A.: Analysis of a dynamic assignment of impatient customers to parallel queues. Queueing Syst. **67**(3), 251–273 (2011)
17. Knessl, C., Morrison, J.A.: Heavy traffic analysis of two coupled processors. Queueing Syst. **43**(3), 173–220 (2003)
18. Boussetta, K., Beylot, A.-L.: Multirate resource sharing for unicast and multicast connections. In: Tsang, D.H.K., Kuhn, P.J. (eds.) Broadband Communications 1999, Hong Kong, pp. 561–570 (1999)

19. Moiseev, A., Nazarov, A.: Queueing network MAP-(GI/∞)K with high-rate arrivals. Eur. J. Oper. Res. **254**(1), 161–168 (2016). https://doi.org/10.1016/j.ejor.2016.04.011
20. Ivanovskaya (Sinyakova), I., Moiseeva, S.: Investigation of the queuing system $MMP^{(2)}|M_2|\infty$ by method of the moments. In: The Third International Conference on Problems of Cybernetics and Informatics Proceedings, Elm, Baku, pp. 196–199 (2010)
21. Sinyakova, I., Moiseeva, S.: Investigation of queuing system $GI^{(2)}|M_2|\infty$. In: Queues: Flows, Systems, Networks: Proceedings of the International Conference on Modern Probabilistic Methods for Analysis and Optimization of Information and Telecommunication Networks, Minsk, pp. 219–225 (2011)

Organization of a Reserved Inter-machine Exchange with General and Separate Queues Distributed over Computer Nodes for Access to Aggregated Channels

Vladimir Bogatyrev, Ivan Slastikhin, and Aleksey Derkach$^{(\boxtimes)}$

Saint-Petersburg National Research University of Information Technologies,
Mechanics and Optics, Saint Petersburg, Russia
alexitmo1@gmail.com

Abstract. The possibilities of improving the reliability and timeliness of the computer systems' interaction as a result of redundant transmissions through aggregated channels are investigated. The implementation options for computer nodes of distributed separate queues for access to each channel, or a general queue for access to all channels, are taken into account. Simulation models of the considered variants of the reserved exchange through aggregated channels are built and the effectiveness of the use of the analyzed variants is determined. For the first exchange option, there are separate queues for access to each channel in each computer node. The incoming request is copied k times depending on the criticality to the waiting time, and each copy is placed in one of the queues. For the second variant, there is one common queue to all channels in each computer node, into which each incoming packet is entered (request for its transmission). K copies of the transmitted packet are generated when a request is issued from the common queue. Each copy is transmitted through one of the n channels when the node gets access to channel.

Keywords: Transfer redundancy · Aggregated channels · Queues · Multiple access · Reliability · Accuracy · Transfer timeliness

1 Introduction

The fundamental task of designing infocommunication systems is to ensure their reliability, security [1–5] and timeliness of processing, data transfer and storage [6–10]. It is necessary to ensure not only the structural reliability and fault tolerance of the infocommunication subsystem for a stable (reliable) inter-machine exchange of a computer system, but also its functional reliability. Functional reliability is determined not only by the operability of nodes and communication channels involved in data exchange, but also by the ability of the network, as a queuing system, to maintain the timeliness of unmistakable delivery of data

© Springer Nature Switzerland AG 2019
V. M. Vishnevskiy et al. (Eds.): DCCN 2019, CCIS 1141, pp. 375–388, 2019.
https://doi.org/10.1007/978-3-030-36625-4_30

(packets) in conditions of failures, halting and external destructive impacts. Increasing the probability of unmistakable and timely execution of requests (especially critical for real-time systems) is possible based on reserve maintenance requests [11], under certain conditions, despite the increase in the overall system load. Copies are made that are sent to different resources for execution with reserve maintenance, for each request. The effect is achieved due to the fact that at least one copy will be made earlier than the others, which increases the probability of the request being executed within the maximum permissible time. It also reduces the average time it waits for at least one of the copies to be executed. The principle of reserve maintenance can be applied both in cluster data processing systems [11] and in data transfer systems for the backup transfer of packets' copies along different paths (routes, channels) [12–15]. Reservation of transfers for ensuring timeliness of inter-machine exchange can be combined with traffic prioritization and balancing, with packet segmentation, etc. When reserving transfers, there are copies of the packets that are transmitted through different channels (paths), which makes it possible to increase the probability of unmistakable and timely delivery of at least one of the generated copies of the packets during the inter-machine exchange in systems that exclude the possibility of retransfers. In multipath reserve transfer, a path is understood as the minimum set of elements whose operability ensures the required interaction of computers in the system, while the failure of any of the elements included in the minimum path leads to loss of connectivity through it.

In the studied systems, the multiplicity of reserve (copying) is set depending on the criticality of requests (packets) to allowable delays [11].

The effectiveness of inter-machine exchange through aggregated channels is largely determined by the influence of the maintenance's organization of queuing distributed over computer nodes on reserve exchanges and depends, among other things, on the implementation of multiple access used to provide authority to service requests from queues of different nodes.

The purpose of the work is to study the possibilities of increasing the reliability and timeliness of inter-machine exchange as a result of reserved transfers through aggregated channels, taking into account the implementation options in computer nodes of distributed separate queues for access to each channel, or a common distributed queue for access to all channels.

The research method is based on the construction of simulation models [16] of the reserved exchange's variants through aggregated channels and the determination of the effectiveness and the area of expedient use of each of the analyzed variants.

2   Options for Reserved Inter-machine Exchange

Consider two options for the organization of the reserved exchange (interaction of nodes) differing in the organization of the queues distributed among the nodes for access to the aggregated channels:

Option A: with separate distributed queues for access to channels: Each computer node has n separate queues, each for access to a separate channel.

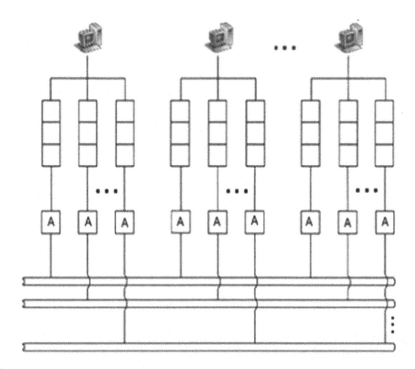

Fig. 1. A variant of a reserved inter-machine exchange with the organization of queues distributed over computer nodes to each channel

The incoming request is copied k times (depending on time-criticality), and each copy is placed in one of the n queues (Fig. 1). We present this variant in the form of n single-channel queuing systems (QS) [17,18] with queues distributed between nodes.

Option B: with a shared distributed queue for access to channels: One common queue is organized in each computer node into which each incoming packet is entered (request for its transfer). When issuing a request from a queue, k copies of the packet are formed, each of which is transmitted through one of the channels as the node receives access rights to the channels (Fig. 2). Thus, an asynchronous multichannel transfer of a given number of copies is realized as far as the unused channels are provided to the node (the channels can be provided based on multiple access, carried out independently and asynchronously for each channel). This option can be represented by multi-channel QS with a common distributed queue and n maintaining devices (channels).

A feature of reserved maintenance is the formation for each request k of its copies issued for service in different channels. A maintenance is considered successfully completed if at least one of the k created copies of the request is executed without error (at least one copy of the packet is transmitted without error).

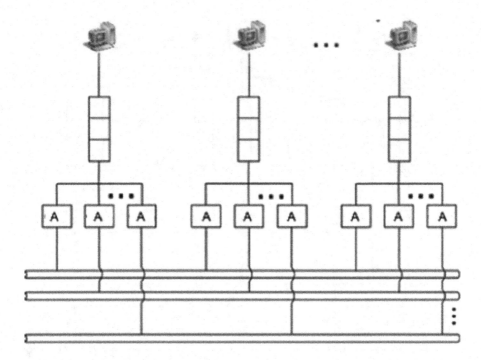

Fig. 2. Options for reserved inter-machine exchange with the organization of the queue, distributed among computer nodes for access to all channels

For variants of inter-machine exchange with reserved transfer of packet copies, modification is possible when channels are combined into groups with the implementation of a single multiple access procedure for all grouped channels (options A1 and B1), and thus the simultaneous authorization of reserved transfers of all copies through channels of the group is realized. As a result of such a combination, the reserved transfer of copies by the packet is realized simultaneously for all channels and the effect is achieved only as a result of an increase in the probability of unmistakable delivery of at least one copy of the k transfers.

We represent variant A1 in the form of N single-channel QS, with each group of channels considered as one single-channel QS with queues distributed between nodes.

The number of channels grouped into different groups may not coincide, while:

$$\sum_{i=1}^{N} k_i = n$$

Option B1 can be represented by a multi-channel QS with a common distributed queue and N maintaining devices (each service channel represents k_i physical channels).

3   Reliability, Timeliness and Accuracy of Reserved Transfers

The studied real-time intermachine exchange processes are sensitive to packet transmission delays, which in some cases excludes the possibility of packet retransmissions if they are lost or distorted. Such cases are characteristic of real-time control systems, including cyberphysical systems critical to the continuity of computational processes. The effectiveness of the systems under consideration is evaluated by the readiness of the structure to perform the required tasks, the probability of faultlessness and timeliness of maintenance requests including the transmission of packets through the network.

For the systems under consideration, without taking into account the requirements of stationarity and timeliness of exchange through the network, the probability of operability (readiness) of at least one of the n aggregated channels for data transmission is determined as:

$$P = 1 - (1 - p)^n,$$

in this case, if a subscriber needs to be connected to a channel to transmit a packet through a channel, then $p = p_a p_c$ and if the exchange between a pair of subscribers, then $p = p_a^2 p_c$, where p_a and p_c are the probability of operability of the adapter and the communication line (channel).

When reserving communication channels and functionally similar servers, the probability of the subscriber's computer node being connected to at least one of the n_s servers, at least one through the n channels is calculated as:

$$P = 1 - (1 - p_a p_c [1 - (1 - p_a)^{n_s}])^n,$$

If there are n_s servers with different types of functionality and the requirement that the subscriber's computer is connected with each type of server, at least through one channel will find how:

$$P = \sum_{i=1}^{n} C_n^i (p_a p_c)^i (1 - p_a p_c)^{n-i} [1 - (1 - p_a)^i]^{n_s},$$

The probability of the system being ready for exchange between a pair of computers in the stationary mode with the multiplicity of packet reservation k_i, which has a value not less than the number of healthy channels i, is calculated as

$$P = \sum_{i=1}^{n} \delta_i C_n^i p^i (1 - p)^{n-i},$$

Wherein stationary condition

$$\delta_i = \begin{cases} 1, if(\Lambda k_i \nu / i) < 1), \\ 0, if(\Lambda k_i \nu / i) \geq 1), \end{cases} \tag{1}$$

$$k_i = \begin{cases} i, & if \ (i < k), \\ k, & if \ (i \geq k), \end{cases} \qquad (2)$$

where k is the redundancy ratio in the initial state without accumulation of failures, the average transfer time of a packet of length L (bit) and speed of the channel s *(bit/s)* is $\nu = L/s$.

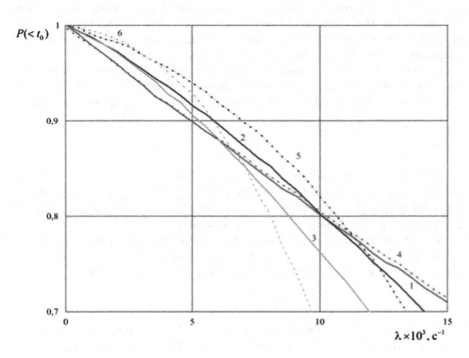

Fig. 3. The probability of timely delivery (a) and average delivery time (b) of at least one copy of packets to the addressee

The probability of the system being ready for exchange in a stationary mode with the multiplicity of packet's reserve k and the condition of unmistakable delivery of at least one copy of the packet is calculated as:

$$P = \sum_{i=1}^{n} \delta_i \{1 - [1 - (1-b)^L]^{k_i}\} C_n^i p^i (1-p)^{n-i}, \qquad (3)$$

where b is the probability of bit errors, and the multiplicity of packet reserve k_i is determined by the formula (2).

With the adaptive assignment of the reserve transfers' ratio, depending on the number of healthy channels and the intensity of the request stream, the probability of unmistakable backup transfern is determined by the formula (3).

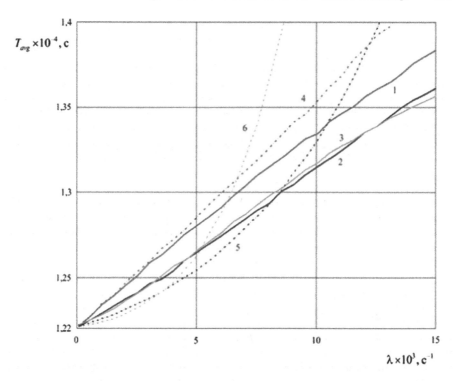

Fig. 4. The probability of timely delivery (a) and average delivery time (b) of at least one copy of packets to the addressee

In this case, the multiplicity factor of the reservation k_i in the case of the health of the i channels is defined as the value of j $(j = 1, 2, ..., i)$, at which the maximum probability of unmistakable transfer is reached in the stationary mode:

$$Max_j(\delta_{ij}\{1 - [1 - (1 - b)^L]^j\}), \tag{4}$$

$$\delta_{ij} = \begin{cases} 1, if(\Lambda j\nu/i) < 1), \\ 0, if(\Lambda j\nu/i) \geq 1), \end{cases}$$

If the requirement of waiting for requests is no more than t_0, the probability of the system being ready for exchange with the packet reservation ratio k and the condition of timely delivery of at least one copy of the packet in the initial state (if all n communication channels and network adapters are operational) assuming the simplest input stream and exponential service time (packet transfer) is calculated as [11]:

$$P(t_0) = 1 - [\frac{\Lambda k\nu}{n}e^{(\frac{\Lambda k}{n} - \frac{1}{\nu})t_0}]^k,$$

For the option without redundant transmissions, the probability of waiting for packets less than the maximum permissible time is calculated as [6]:

$$P(t_0) = 1 - \frac{\Lambda\nu}{n}e^{(\frac{\Lambda}{n}-\frac{1}{\nu})t_0},$$

Considering possible failures of network adapters and communication channels:

$$P(t_0) = \sum_{i=1}^{n}[1 - (\frac{\Lambda k_i \nu}{n}e^{(\frac{\Lambda k_i}{n}-\frac{1}{\nu})t_0})^{k_i}]C_n^i p^i (1-p)^{n-i},$$

where k_i is determined by the formula (2).

For real-time systems, the optimal reserve ratio k_i in case of i channels health can be assigned based on the achievement of the maximum probability of a timely, unmistakable transfer in stationary mode, which in the case of nonexponential distribution of the packet transmission time can be determined based on simulation.

As a generalized indicator of the effectiveness of the organization of reserved transfers for the efficiency status of n channels, we use an indicator reflecting the delay in unmistakable packet delivery.

$$M(n, k, \Lambda) = B(k)(t_0 - T(n, k, \Lambda)), \tag{5}$$

at the same time, $B(k)$ is the probability of unmistakable delivery of at least one of k transmitted copies of packets, $T(n, k, \Lambda)$ is the average residence time of a packet's copy in the system, with the intensity of requests for transfer packets Λ, t_0 is the maximum allowable residence time package in the system.

Taking into account the possible working states of the system with a different number of failed channels, the indicator $M(n, k, \Lambda)$ can be modified as:

$$M(n, k, \Lambda) = \sum_{i=1}^{n} \delta_i B(k_i)(t_0 - T(i, k_i, \Lambda)), \tag{6}$$

at the same time, the condition of stationarity of the exchange mode δ_i is determined by the formula (1) and the reserve ratio k_i can be found by (2) or when optimizing it based on finding the value k_i at which the maximum probability of unmistakable transferis reached with i channels that have preserved operability.

4   Simulation Experiments

Simulation models were built in the AnyLogic 7 modeling environment. Existing simulation models with multi-path reserved transfer [19] do not take into account the options for organizing distributed queues for access to aggregated channels, therefore, in this paper, models are developed for conducting simulation experiments taking into account the specified features of the study. For variants of the organization of reserved maintenances, simulation experiments were performed with varying the intensity of the input stream with the values of O and the multiplicity of reserved transfers.

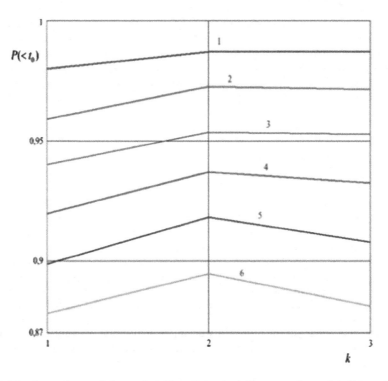

Fig. 5. The dependence of the probability of timely delivery on the multiplicity of reservation of transfers in the implementation of the general distributed (a) and individual queues to channels (b).

For variants A and B of the organization of reserved transmissions, a number of simulation experiments were carried out to determine the dependence of the probability of timely delivery $P(< t_0)$ on the total intensity of the input stream; at the maximum allowable delay $t_0 = 2.28 \cdot 10^{-4}$ s; bandwidth of communication channels $L = 1\,\mathrm{Mbit/s}$; average packet length $N = 1024$ bits. The probability of timely delivery $P(< t_0)$ is defined as the ratio of the number of packets delivered over time to the total number of packets transmitted during this time. A package is considered to be transmitted successfully if at least one of its copies is delivered on time. Modeling is carried out for random and marker multiple access methods.

The dependence of the probability of timely delivery for error-free transmissions on the total intensity of the incoming packet stream when implementing individual queues to channels with transmission redundancy ratios $k = 1, 2, 3$ is represented by the curves 1–3, and when implementing a common queue to all channels, the curves 4–6 in Fig. 3 with a random multiple access method.

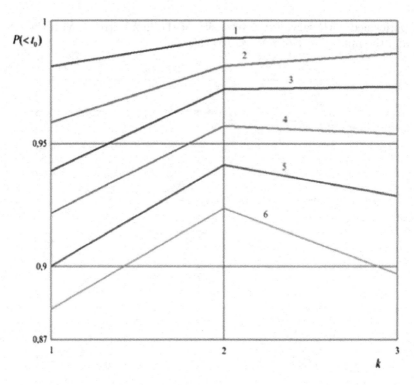

Fig. 6. The dependence of the probability of timely delivery on the multiplicity of reservation of transfers in the implementation of the general distributed (a) and individual queues to channels (b).

As can be seen from the graph, reserved packet transmission can increase the likelihood of timely delivery even in the absence of bit errors in the communication channels. With reserved transmission, taking into account the possibility of bit errors, the probability of timely delivery of at least one copy of the packet to the recipient increases. It has been established that redundancy of transmissions allows to reduce the average delay time with sufficient delivery of at least one copy of the transmitted packet to the addressee. So the results of simulation to determine the average delivery delays of the first copy for the above source data are shown in Fig. 4. The figures show the presence of a reserved transmission efficiency area.

In Figs. 5 and 6 shows the dependences of the probability of timely delivery on the multiplicity of reserved transmissions when implementing the general distributed (a) and individual queues to channels (b). Simulation results at query flow intensities $O = 1000, 2000.3000.4000, 5000s^{-1}$ are represented by the curves 1–6, respectively. The results were obtained with the above source data and the probability of bit errors in the communication channels $b = 10^{-5}$.

From the presented dependencies, the existence of an optimal multiplicity of reserved transfers is seen. For the given initial data, the optimal redundancy ratio is $k = 2$.

The simulation results for determining the efficiency of reserved transfers using the complex criterion (5) of the input stream intensity O are presented in Fig. 7. The results were obtained with the maximum allowable network delay $t_0 = 2.28 \cdot 10^{-4}$ s; communication channel bandwidth $L = 1$ Mbit/s; - average packet length $N = 1024$; - probability of bit errors in communication channels $b = 10^{-5}$. The dependencies of the efficiency of reserved transfers when implementing individual queues for access to channels with transfer reserve ratios $k = 1, 2, 3$ are represented by the curves 1–3, and when implementing a common queue, for all channels of transmission reserve ratios $k = 1, 2, 3$ by the curves 4, 5, 6.

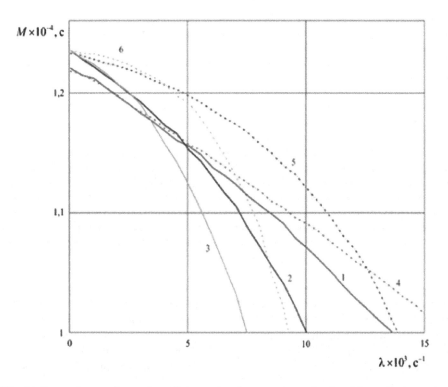

Fig. 7. Dependence of transfer efficiency by the complex criterion M on the intensity of the input packet stream

In Fig. 8 there are dependency graphs of the efficiency of the system from the redundancy ratio k (with option B), in which curves 1–3 correspond to the intensity of the input stream $\Lambda = 300, 500, 800$ 1/s. Experiments were carried out with the number of communication channels $n = 4$; bandwidth of each channel $s = 1$ Mbit/s; the probability of bit errors $b = 10^{-4}$; the average length of packets arriving at the system is $L = 2048$ bits; The maximum allowable delivery time is $t_0 = 0.0005$ s.

From the presented graph shows that there is an optimal multiplicity of reserved transfers, the value of which depends on the intensity of the stream.

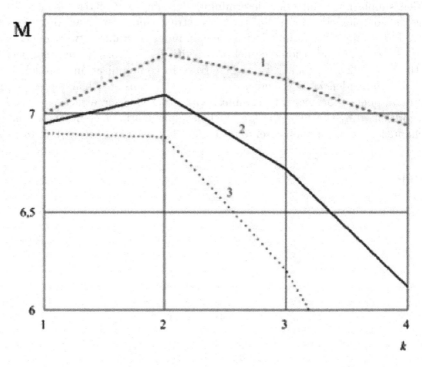

Fig. 8. The dependence of the multiplicative criterion M under option B ($\lambda = 300\,1/s$ - curve 1, $\lambda = 500\,1/s$ - curve 2, $\lambda = 800\,1/s$ - curve 3) on the redundancy ratio k

Let us compare the efficiency of the exchange options A and B by the average residence time, depending on the intensity of the stream of requests. In Fig. 9 for options A and B, the curves 1 and 2 represent the dependencies of the average times T of packets in the system on the intensity of the input packet stream, and the curve 3 their differences. From the presented graph it can be seen that, depending on the intensity of the steam of requests, there is an area of effective use of options A and B.

From the presented graph shows that there is an optimal multiplicity of reserved transfers, the value of which depends on the intensity of the stream.

Let us compare the efficiency of the exchange options A and B by the average residence time, depending on the intensity of the stream of requests. In Fig. 4 for options A and B, the curves 1 and 2 represent the dependencies of the average times T of packets in the system on the intensity of the input packet stream, and the curve 3 their differences. From the presented graph it can be seen that, depending on the intensity of the steam of requests, there is an area of effective use of options A and B.

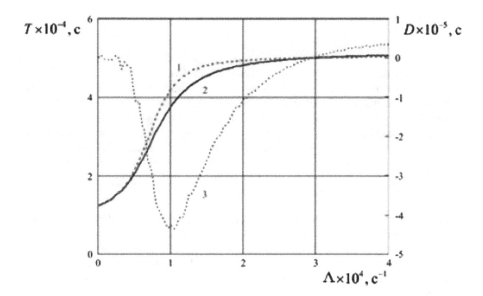

Fig. 9. Dependence of the average residence time T on the total intensity of the input streams Λ

5   Conclusion

The possibilities of improving the reliability and timeliness of inter-machine exchange based on reserved transfers through aggregated channels were studied.

The effectiveness of reserved exchanges implemented in computer nodes was analyzed. The efficiency with distributed separate queue for access to each channel and the general distributed queue for access to all channels was investigated.

The area of considered variants' effectiveness of the reserved inter-machine exchange is defined.

References

1. Kopetz, H.: Real-Time Systems: Design Principles for Distributed Embedded Applications. Springer, Heidelberg (2011). https://doi.org/10.1007/978-1-4419-8237-7
2. Sorin, D.: Fault Tolerant Computer Architecture. Morgan & Claypool, Madison (2009)
3. Shooman, M.: Reliability of Computer Systems and Networks: Fault Tolerance, Analysis, and Design. Wiley, New York (2002)
4. Coolen, F.P.A., Utkin, L.V.: Robust weighted SVR-based software reliability growth model. Reliab. Eng. Syst. Saf. **176**, 93–101 (2018)
5. Bogatyrev, V.A., Vinokurova, M.S.: Control and safety of operation of duplicated computer systems. Commun. Comput. Inf. Sci. **700**, 331–342 (2017)

6. Vishnevskiy, V.M.: Theoretical Fundamentals of Computer Network Design. Tekhnosfera, Moscow (2003)
7. Aliev, T.: The synthesis of service discipline in systems with limits. In: Vishnevsky, V., Kozyrev, D. (eds.) DCCN 2015. CCIS, vol. 601, pp. 151–156. Springer, Cham (2016). https://doi.org/10.1007/978-3-319-30843-2_16
8. Tatarnikova, T., Kolbanev, M.: Statement of a task corporate information networks interface centers. In: Structural Synthesis IEEE EUROCON 2009, pp. 1883–1887 (2009)
9. Kabatiansky, G., Krouk, E., Semenov, S.: Error Correcting Coding and Security for Data Networks. Analysis of the Super channel Concept. Wiley, New York (2005)
10. Krouk, E., Semenov, S.: Application of coding at the network transport level to decrease the message delay. In: Proceedings of 3rd International Symposium on Communication Systems Networks and Digital Signal Processing, pp. 109–112 (2002)
11. Bogatyrev, V.A., Bogatyrev, A.V.: Functional reliability of a real-time redundant computational process in cluster architecture systems. Autom. Control Comput. Sci. **49**(1), 46–56 (2015). https://doi.org/10.3103/S0146411615010022
12. Arustamov, S.A., Bogatyrev, V.A., Polyakov, V.I.: Back up data transmission in real-time duplicated computer systems. In: Abraham, A., Kovalev, S., Tarassov, V., Snášel, V. (eds.) IITI 2016. AISC, vol. 451, pp. 103–109. Springer, Cham (2016). https://doi.org/10.1007/978-3-319-33816-3_11
13. Bogatyrev, V.A.: Protocols for dynamic distribution of requests through a bus with variable logic ring for reception authority transfer. Autom. Control Comput. Sci. **33**(1), 57–63 (1999)
14. Bogatyrev, V.A.: On interconnection control in redundancy of local network buses with limited availability. Eng. Simul. **16**(4), 463–469 (1999)
15. Bogatyrev, A.V., Bogatyrev, S.V., Bogatyrev, V.A.: Analysis of the timeliness of redundant service in the system of the parallel-series connection of nodes with unlimited queues. In: 2018 Wave Electronics and its Application in Information and Telecommunication Systems (WECONF) (2018)
16. Kutuzov, O.I., Tatarnikova, T.M.: On the acceleration of simulation modeling. In: XXI International Conference on Soft Computing and Measurements, SCM 2018 (2018)
17. Kleinrock, L.: Queueing Systems: Volume 1: Theory. Wiley, New York (1975). https://doi.org/10.1002/net.3230060210
18. Kleinrock, L.: Queueing Systems: Volume II: Computer Applications. Wiley, New York (1976)
19. Bogatyrev, V.A., Parshutina, S.A., Poptcova, N.A., Bogatyrev, A.V.: Efficiency of redundant service with destruction of expired and irrelevant request copies in real-time clusters. In: Vishnevskiy, V.M., Samouylov, K.E., Kozyrev, D.V. (eds.) DCCN 2016. CCIS, vol. 678, pp. 337–348. Springer, Cham (2016). https://doi.org/10.1007/978-3-319-51917-3_30

Solution of Lindley Integral Equation for Correlated Traffic

Igor Kartashevskiy$^{(\boxtimes)}$ (iD)

Povolzhskiy State University of Telecommunications and Informatics,
23 Lev Tolstoy str., Samara, Russia
ivk@psuti.ru

Abstract. The paper considers estimating the average waiting time of a request in a queue when processing correlated traffic in a hypothetical queuing system $M/M/1$. We assume that inter-arrival and service time are correlated with each other. The solution based on the representation of the joint cumulative distribution function through the copula function according to the Sklar's theorem. A variant has been proposed for establishing the analytical relationship between the mutual correlation coefficient of the considered time sequences and the Farlie-Gumbel-Morgenstern copula parameter. To achieve a quantitative result, the spectral method for solving the Lindley integral equation is used. It is shown that the presence of mutual correlation decreases the average time of a request in a queue compared to the case when inter-arrival and service time are independent.

Keywords: Queuing system · Correlated traffic · Copula function

1 Introduction

It is known [1,2] that traffic in computer and telecommunication networks has fractal properties. From a theoretical point of view, this means that traffic as a discrete random process has a correlation function with a long-term dependence and hyperbolic attenuation. As a result, the series of the consecutive values of the correlation function is nonsummable and diverges. So, we cannot get satisfactory results characterizing the operation of network devices strictly following the assumption that traffic is really a self-similar random process.

On the other hand, there are many examples that the correlation function has a certain number of significant decreasing counts, which allows us to consider the correlation function to be finite in time. For example [3], IPTV traffic at the access level has pronounced correlation properties with a finite correlation function. Paper [4] describes an experiment based on the simultaneous processing different types of traffic (Internet of things and video) in a network device. This experiment showed that the presence of insignificant HTTP traffic of Internet of things greatly increases the correlation of the arrival flow but the correlation function also remains finite. Thus, a self-similar random process is not always

© Springer Nature Switzerland AG 2019
V. M. Vishnevskiy et al. (Eds.): DCCN 2019, CCIS 1141, pp. 389–400, 2019.
https://doi.org/10.1007/978-3-030-36625-4_31

suitable for describing traffic with correlation properties due to the nonsummable correlation function. Also, using self-similar process as an arrival flow (traffic) for a queuing system (QS) describing the operation of any network device assumes asymptotic approximations. Therefore, QS with a self-similar arrival flow is not suitable for analytical research [5]. However, we have to take into account the correlation of the arrival flow in the QS because of presence of positive counts of the correlation function leads to the bursty traffic, significantly distorts the queue statistics and worsens the main characteristics of QS [6]. There are a large number of papers devoted to the analysis of QS with a correlated arrival flow. A detailed analysis of these works is given in [5].

It is also possible to use renewal correlated flows to simulate correlated traffic. These flows allow reducing the task of processing the correlated traffic to processing traffic with independent time intervals with modified marginal distributions using index of dispersion of the arrival flow [7,8].

The variety of physical causes indicated in [9] generates traffic where the inter-arrival time and service time in QS also turn out to be correlated. This situation is discussed below.

2   Lindley Integral Equation Solution for Estimating the Waiting Time

The classical queuing theory is based on the assumption that the inter-arrival time and service time are stationary sequences of random variables with independent elements [10,11]. Moreover, independence postulated for elements both within each sequence and between elements of different sequences. At the same time, the most important parameter characterizing the quality of service in the general type QS $(G/G/1)$, namely, the waiting time of a request in a queue can be determined with the solution of the Lindley integral equation:

$$F(y) = \int_{0_-}^{\infty} K(y - x)dF(x), \quad y > 0,$$

where $F(y)$ is the cumulative distribution function (CDF) of waiting time for service requests in the queue, $K(y)$ is the kernel of the integral equation, written as:

$$K(y) = \int_{0}^{\infty} B(y + x)dA(x), \tag{1}$$

where $B(\cdot)$ is the CDF of service time and $A(\cdot)$ is the CDF of inter-arrival time.

The value of $F(0)$ for a single server QS determines the probability that the system is empty and has the form $F(0) = 1 - \rho$ [10], where ρ is the utilization coefficient, $\rho < 1$. Essential attention is given to a random variable $u = \eta - \nu$ [10] representing the time difference between the service of the n-th request and the time interval between the receipt of the n-th and the $(n + 1)$-th requests.

As shown in [10] the distribution of the random variable u assumed to be the same for all n.

The CDF of random variable u is the kernel $K(y)$ of an integral equation assuming independence of the sequences defining the random variable u.

If the random variables η and ν are correlated (with the correlation coefficient R), then the probability density function (PDF) associated with the kernel (1) is

$$\frac{d}{dy} K(y) = w_u(y),$$

and can only be determined through the joint PDF $w_{\eta, \nu}(\cdot, \cdot)$ of random variables η and ν:

$$w_u(y) = \int_0^\infty w_{\eta, \nu}(x, y + x) dx. \tag{2}$$

3 Synthesis of a Joint Distribution Based on the Copula Function

If the traffic monitoring system allows to estimate the one-dimensional distributions $w_\eta(\cdot)$ and $w_\nu(\cdot)$, as well as the correlation coefficient R for sequences η and ν then we can use the Sklar's theorem to synthesize the joint PDF [12,13]. This theorem states that using the concept of a copula, the joint PDF of correlated random variables X and Y is:

$$w_{XY}(x, y) = c(W_X(x), W_Y(y)) w_X(x) w_Y(y), \tag{3}$$

where $c(u, v) = \frac{\partial^2 C(u, v)}{\partial u \partial v}$ is the derivative of the copula (density).

$C(u, v)$ is the copula function defined on a set $[0, 1] \times [0, 1]$, and is essentially a joint CDF on a given set with variables $u = W_X(x)$ and $v = W_Y(y)$. $W_X(x)$, $W_Y(y)$ are the distribution functions for X and Y.

Copulas are of interest for the synthesis of multidimensional distributions with the certain dependence of traffic counts. We can describe this dependence not only by the Pearson correlation coefficient R, but also by others, for example, the Kendall's rank correlation coefficient and the Spearman's coefficient R_s. These factors operate with the finer structure of the dependence of the traffic counts.

It is important to choose the correct copula for expression (3), which allow to obtain an adequate and simple description for the joint distribution. There are an overview in [13] of some types of copulas and the possibilities of their use for modeling joint distributions of random variables. In particular, the Farlie-Gumbel-Morgenstern family with polynomial extension [12] in the form:

$$C(u, v) = uv \left[1 + \theta(1 - u^p)(1 - v^p) \right], \tag{4}$$

where p is an integer, $p = 2, 3, \ldots$ The parameter θ of this copula characterizes the dependence of traffic counts and satisfies the condition:

$$-\left(\max\{1, \ p^2\} \right)^{-2} \leq \theta < p^{-1}.$$

At the same time, the Spearman's correlation coefficient $R_s(u, v)$ related to the θ parameter by the ratio:

$$R_s(u, v) = 3\theta \left(\frac{p}{p+2} \right)^2.$$

For example, when $p = 3$ the parameter θ can be within $-0.012 \leq \theta < 0,333$, and the maximum value of $R_s(u, v)$ for a given copula is $R_{s\,max} = 0,36$. The PDF $w_{XY}(x, y)$ with $p = 3$ and copula (4) is:

$$w_{XY}(x, y) = \left[1 + \theta \left(1 - 3u^2 \right) \left(1 - 3v^2 \right) \right] w_X(x) w_Y(y), \tag{5}$$

where u and v are chosen CDFs.

4 Relationship Between the Correlation Coefficient and the Copula Parameter

Since the interdependence of variables in joint density (5) is determined by the parameter θ, it is necessary to establish a relationship between the correlation coefficient R and this parameter. The following version of this relationship proposed in [13].

The correlation coefficient of random variables X and Y is:

$$R(X, Y) = \frac{E(XY) - E(X)E(Y)}{\sqrt{\sigma_X^2 \sigma_Y^2}}. \tag{6}$$

where the mutual average $E(XY)$ with the selected PDF (5) is:

$$E(XY) = \int_0^\infty \int_0^\infty xy dW_{XY}(x, y) = \int_0^\infty \int_0^\infty xy W'_X(x) W'_Y(y) dx dy$$

$$+ \theta \int_0^\infty \int_0^\infty xy W'_X(x) W'_Y(y) \left[1 - 3W_X^2(x) \right] \left[1 - 3W_Y^2(y) \right] dx dy. \tag{7}$$

The average values and variances from (6) can be easily found for known one-dimensional distributions $w_X(x)$ and $w_Y(y)$. After substituting (7) into (6) we can determine the value of the parameter θ.

We should also note that expression (6) loses its meaning with one-dimensional distributions with heavy tails when variance may tend to infinity, so it is more expedient to characterize the relationship between X and Y with the Kendall's rank correlation coefficient τ.

5 $M/M/1$ Queuing System with Correlated Traffic

For the sake of simplicity, consider a hypothetical QS $M/M/1$ where the inter-arrival time and service time are correlated with the correlation coefficient R (according to Pearson):

$$W_X(x) = (1 - e^{-\lambda x}), W_Y(y) = (1 - e^{-\mu y}),$$

The expression (5) will take the form:

$$w_{XY}(x, y) = \lambda\mu \left[1 + \theta \left(1 - 3(1 - e^{-\lambda x})^2\right) \left(1 - 3(1 - e^{-\mu y})^2\right)\right] e^{-\lambda x} e^{-\mu y}.$$

After calculating the integrals in expression (7), the relationship between R and θ will be $\theta = \frac{R}{2,592}$. For example, when $R = 0,7$ will be $\theta = 0,27$, less than the allowed value of 0.333 for selected $p = 3$ for this copula. Figures 1 and 2 show the joint PDF $w_{XY}(x, y)$ for $\rho = 0,5$ with different $R = 0$ ($\theta = 0$) and $R = 0,7$ ($\theta = 0,27$).

In accordance with (2), the PDF $w_u(y)$ satisfying the normalization condition can now be written as:

$$w_u(y) = \frac{1}{\Delta} \left\{ \frac{\lambda}{1+\rho} e^{-\mu y} + \lambda\theta \left[\alpha e^{-\mu y} + \beta e^{-2\mu y} + \gamma e^{-3\mu y}\right] \right\}, \qquad (8)$$

where $\rho = \lambda/\mu$, $\alpha = \frac{2(8\rho^2+5\rho+1)}{(1+\rho)(1+2\rho)(1+3\rho)}$, $\beta = \frac{-2(8\rho^2+5\rho+2)}{(1+\rho)(2+\rho)(2+3\rho)}$, $\gamma = \frac{(8\rho^2+5\rho+1)}{3(1+\rho)(3+\rho)(3+2\rho)}$, Δ is the normalizing factor equal to the value of a definite integral within the limits $(0, \infty)$ of the function inside the curly brackets.

Figure 3 shows the $w_u(y)$ with $\lambda = 0,2$; $\mu = 0,4$ and $R = 0,7$ ($\theta = 0,27$). The first term in (8), depicted by a black solid line, characterizes the density $w_u(y)$ for noncorrelated η and ν.

We have to note that it is difficult to distinguish between graphs in Figs. 1, 2 and 3, related respectively to correlated and noncorrelated traffic, although the difference in the expressions is significant. We can explain this by saying that in either case we use exponential distributions. The difference will be more noticeable for quantitative results for non-exponential types of traffic.

If we also use the spectral method for solving the Lindley integral equation as in [10], then the main problem for obtaining $F(y)$ is the factoring problem of the right side of the expression:

$$K'(s) - 1 = \frac{\psi_+}{\psi_-},$$

where the functions ψ_+ and ψ_- should be analytical without right half plane zeros for ψ_+ and in the $\text{Re}(s) < D$ half plane for ψ_- ($D > 0$ was selected during decomposition). Therefore, the choice of reasonable approximations for $K'(s)$ is always justified, leading to the expression $K'(s) - 1$ being a rational function of s. For example, we can use hyperexponential distributions that can approximate densities $w_\eta(\cdot)$ and $w_t(\cdot)$.

The kernel derivative (8) has been already represented as the sum of exponential functions. The Laplace transform from (8) is:

$$K'(s) = \frac{\lambda}{\Delta} \left\{ \frac{[1 + \alpha\theta(1+\rho)]}{1+\rho} \cdot \frac{1}{s+\mu} + \frac{\beta\theta}{s+2\mu} + \frac{\gamma\theta}{s+3\mu} \right\}$$

$$= \frac{k_1}{s+\mu} + \frac{k_2}{s+2\mu} + \frac{k_3}{s+3\mu}, \qquad (9)$$

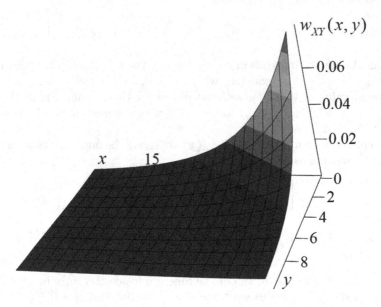

Fig. 1. Joint exponential distribution with $R = 0$ ($\theta = 0$).

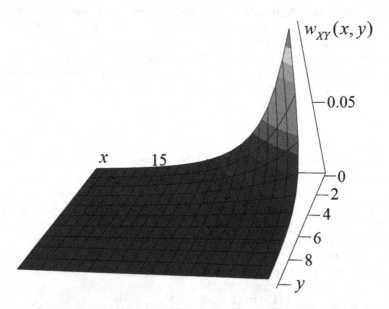

Fig. 2. Joint exponential distribution with $R = 0,7$ ($\theta = 0,27$).

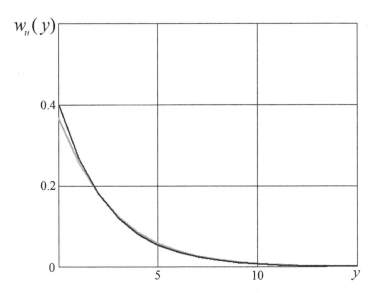

Fig. 3. Probability density function $w_u(y)$.

where the k_1, k_2 and k_3 are obvious from the first line of the expression.

As a result, for decomposition $K'(s) - 1 = \frac{\psi_+}{\psi_-}$ we can write:

$$\frac{\psi_+}{\psi_-} = \frac{(s - s_1)(s - s_2)(s - s_3)}{(s + \mu)(s + 2\mu)(s + 3\mu)},$$

where the roots s_i, $i = 1, \ 2, \ 3$ can be found from the solution of the equation:

$$s^3 - s^2(k_1 + k_2 + k_3 - 6\mu) - s\mu(5k_1 + 4k_2 + 3k_3 - 11\mu) - \mu^2(6k_1 + 3k_2 + 2k_3 - 6\mu) = 0.$$

The numerical solution gives $s_1 = -1, 1$; $s_2 = -0, 87$; $s_3 = 0, 03$. This allows us to select functions ψ_+ and ψ_- as

$$\psi_+ = \frac{(s - s_1)(s - s_2)(s - s_3)}{(s + \mu)(s + 2\mu)(s + 3\mu)}, \psi_- = \frac{1}{(s - s_3)}.$$

Following [10], we introduce the function Φ_+ - the Laplace transform of the CDF of the waiting time of a request in a queue. Also Φ_+ can be defined as $\Phi_+ = \frac{K_0}{\psi_+}$, where $K_0 = \frac{(1-\rho)}{\lambda} \psi_-(0)$.

Finally, we can find the average waiting time of a request in a queue using a derivative of the characteristic function $\varphi(s) = s\Phi_+(s)$ of PDF of the waiting time in a queue:

$$E(W_R) = -\left.\frac{d\phi(s)}{ds}\right|_{s=0}. \tag{10}$$

Now we can compare $E(W_R)$ with the $E(W)$ - the average waiting time of a request in a queue for $M/M/1$ system with noncorrelated traffic using expression $E(W) = \rho/\mu(1 - \rho)$. So, $E(W) = 2, 5$ with $\rho = 0, 5$; $\mu = 0, 4$ and using expressions (10) and (9) $E(W_R) \approx 2, 03$ with $R = 0, 7$ and $\rho = 0, 5$; $\mu = 0, 4$.

6 $M/G/1$ Queuing System with Correlated Traffic

This section focuses on the effect of correlated inter-arrival time and service time for $M/G/1$ queuing system. Here for the $M/G/1$ QS the Weibull distribution is selected as a distribution of service time intervals:

$$w(x) = \frac{\alpha}{\beta}\left(\frac{x}{\beta}\right)^{\alpha-1} e^{-\left(\frac{x}{\beta}\right)^{\alpha}}, \quad x \geq 0,$$

and it follows from expression (5) that:

$$w_2(x, y) = \left[1 + \theta\left(1 - 3\left(1 - e^{-\left(\frac{x}{\beta}\right)^2}\right)^2\right)\left(1 - 3(1 - e^{-\lambda y})^2\right)\right]$$

$$\times \frac{2}{\beta}\left(\frac{x}{\beta}\right) e^{-\left(\frac{x}{\beta}\right)^2} \lambda e^{-\lambda y}.$$

Here λ is the arrival rate, and $\alpha = 2$.

The following integrals from [14]:

$$\int_0^\infty \exp\left(-\frac{x^2}{4\eta} - \gamma x\right) dx = \sqrt{\pi\eta}\, e^{\eta\gamma^2}\left[1 - \Phi(\gamma\sqrt{\eta}\,\right], \quad \operatorname{Re}\eta > 0,$$

$$\int_0^\infty x \exp(-\mu x^2 - 2\nu x) dx = \frac{1}{2\mu} - \frac{\nu}{2\mu}\sqrt{\frac{\pi}{\mu}} e^{\frac{\nu^2}{\mu}}\left[1 - \Phi\left(\frac{\nu}{\sqrt{\mu}}\right)\right],$$

$$\operatorname{Re}\mu > 0, \ |\arg\nu| < \frac{\pi}{2},$$

where $\Phi(x) = \frac{2}{\sqrt{\pi}}\int_0^x e^{-t^2} dt$, helps to obtain the derivative of the kernel of the Lindley integral equation:

$$K'(u) = \lambda e^{-\frac{u^2}{\beta^2}}\left\{1 - \frac{\lambda\beta\sqrt{\pi}}{2}\exp\left(\frac{(2u + \lambda\beta^2)^2}{4\beta^2}\right)\left[1 - \Phi\left(\frac{2u + \lambda\beta^2}{2\beta}\right)\right]\right\}. \quad (11)$$

The graph of the derivative of the kernel according to (11) is shown in Fig. 4 for $\beta = 1$ and $\lambda = 1$. It coincides with the plot of PDF of the random value $u = \eta - \nu$ for independent η and ν.

Now, using (2) it can be written for $w_2(y, x + y)$:

$$w_2(y, x + y) = S + \theta \cdot S \times$$

$$\left[1 - 3\left(1 - e^{-\left(\frac{x+y}{\beta}\right)^2}\right)^2 - 3(1 - e^{-\lambda y})^2 + 9(1 - e^{-\lambda y})^2\left(1 - e^{-\left(\frac{x+y}{\beta}\right)^2}\right)^2\right],$$

$$(12)$$

where $S = 2\frac{\lambda}{\beta}\left(\frac{x+y}{\beta}\right) e^{-\lambda y} e^{-\left(\frac{x+y}{\beta}\right)^2}$.

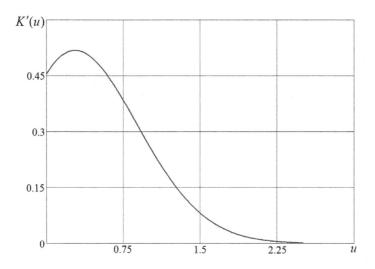

Fig. 4. The derivative $K'_R(u)$ of the kernel of the integral equation

We made the following approximation for $\Phi(\cdot)$ [15] for calculating the two-dimensional PDF (12):

$$\Phi(x) = 1 - (a_1 t + a_2 t^2 + a_3 t^3)e^{-x^2}, t = \frac{1}{1+qx},$$

with parameters $q = 0,47047$; , $a_1 = 0,34802$, $a_2 = -0,09587$, $a_3 = 0,74785$. For $K'_R(u)$, the obtained result is presented in Fig. 5 for different values of the θ parameter ($\theta = 0,1; 0,2; 0,45$). The graph for $K'(u)$ (11), corresponding to the case $\theta = 0$ value, is also given. The difference between these graphs characterizes the effect of correlation on the form of the kernel of the Lindley integral equation and on the distribution of the delay time of a request in the $M/G/1$ QS with correlated traffic.

The correlation coefficient R can be determined according to the expressions (6) and (7). For the Weibull distribution the mean and variance are:

$$E(X) = \beta\Gamma\left(\frac{3}{2}\right) = \frac{\beta\sqrt{\pi}}{2}, \quad \sigma_X^2 = \beta^2\left[\Gamma(2) - \Gamma^2\left(\frac{3}{2}\right)\right] = \beta^2\left[1 - \left(\frac{\sqrt{\pi}}{2}\right)^2\right],$$

and for exponential distribution:

$$E(Y) = \frac{1}{\lambda} \quad \sigma_Y^2 = \frac{1}{\lambda^2}.$$

Now, the relation between R and θ can be written in the form:

$$R = \frac{\frac{\beta}{\lambda}\left(\frac{\sqrt{\pi}}{2} + \theta \cdot 0,337\right) - \frac{\beta\sqrt{\pi}}{2\lambda}}{\sqrt{\frac{\beta^2}{\lambda^2}\left[1 - \left(\frac{\sqrt{\pi}}{2}\right)^2\right]}} = \theta \cdot 0,728. \tag{13}$$

with the values $\beta = 1$ and $\lambda = 1$.

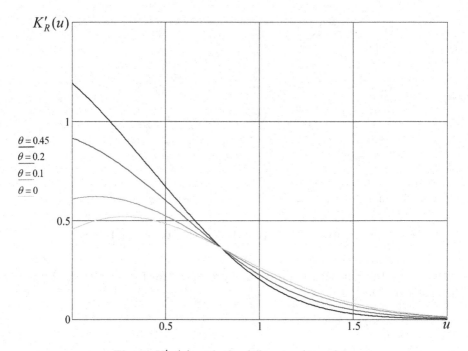

Fig. 5. $K'_R(u)$ with the different values of θ.

It follows from (13) that for the chosen copula and parameter $\alpha = 2$ for the Weibull distribution, expression (13) is valid for any choice of the remaining parameters of distributions. Inequality $-1 \leq \theta < 0,5$ defines the boundaries for the change of θ parameter for a given copula and it follows from (13) that:

$$-0,728 \leq R < 0,364.$$

Now it is possible to determine the θ parameter and solve the Lindley equation with any kernel defined (Fig. 5) for a given value of the correlation coefficient R. As it was shown before, the Laplace transform from the derivative of the kernel is desired to be a rational function for the solution of Lindley integral equation. So, $K'_R(u)$ is approximated here with the sum of decaying exponentials. The form of the distributions presented in Fig. 5 allows to use such an approximation:

$$\hat{K}'_R(u) \approx \sum_{i=1}^{N} q_i e^{-\gamma_i u},$$

where q_i, γ_i, $i = \overline{1, N}$ are determined provided that $\hat{K}'_R(u)$ and $K'_R(u)$ coincide with a minimum approximation error.

The following approximation procedure can be suggested here. N points on the curve for $K'_R(u)$ are selected where the curves need to be matched, and a system of $2N$ nonlinear equations relative to unknowns q_i, γ_i is created. The system is solved by the Newton-Kantorovich method, for example. The form of the curve $K'_R(u)$ with $\theta = 0,45$ allows us to consider the simplest case for $N = 2$. The solution gives an approximation in the form:

$$\hat{K}'_R(u) \approx q_1 e^{-\gamma_1 u} + q_2 e^{-\gamma_2 u}, \tag{14}$$

with parameters $q_1 = 0,05; q_2 = 1,09; \gamma_1 = 2,47; \gamma_2 = 1,8$.

The idea of using approximation in the form of the sum of decaying exponents is in the extreme simplicity of the spectral method of solving the Lindley integral equation at approximation of this type.

The determination of the average waiting time of a request in the queue is reduced now to the analysis of some equivalent QS $M/\hat{G}/1$, where the correlation properties are determined by the parameters of distribution $\hat{G}$, i.e. parameters $\hat{K}'_R(u)$.

For approximation (14) using the spectral method of solving the Lindley equation, it can be shown that:

$$\hat{K}'_R(s) = \frac{q_1}{\gamma_1 + s} + \frac{q_2}{\gamma_2 + s}, \psi_+(s) = \frac{(s - s_1)(s - s_2)s}{(s + \mu_1)(s + s_2)},$$

where $s_1 = -2,445$ and $s_2 = -0,685$ are the roots of the numerator of the expression $\frac{q_1}{\gamma_1+s} + \frac{q_2}{\gamma_2+s} - 1$.

In this case, the characteristic function of the waiting time of the request in the queue is determined as $\phi(s) = \frac{ks}{\psi_+(s)}$, and the constant k is $k = \lim\limits_{s\to 0} \frac{\psi_+(s)}{s}$. After calculating $\phi(s)$, the average waiting time in the queue is calculated using the expression:

$$E(W_R) = -\left.\frac{d\phi(s)}{ds}\right|_{s=0}$$

For our case, it is easy to obtain: $E(W_R) = 0,914$ units of time.

In the absence of correlation, the value of $E(W)$ for the $M/G/1$ QS is determined with the Pollaczek-Khinchine formula:

$$E(W) = \frac{\lambda\langle\tau^2\rangle}{2(1 - \rho)},$$

where $\langle\tau^2\rangle$ is the second moment of the distribution of the service time.

For the Weibull distribution used with the parameters $\alpha = 2$, $\beta = 1$:

$$\langle\tau^2\rangle = \beta^2 \Gamma\left(1 + \frac{2}{\alpha}\right) = \Gamma(2) = 1$$

Considering $\rho = \lambda\langle\tau\rangle = \lambda\beta\Gamma\left(1 + \frac{1}{\alpha}\right) = \Gamma(1,5) = \frac{\sqrt{\pi}}{2}$ (since $\lambda = 1$ was accepted above), $E(W) = 4,386$ units of time. The analysis shows that for the $M/G/1$ QS, the correlation between the inter-arrival time and the service time leads to a more noticeable decrease in the waiting time of the request in the queue than in the $M/M/1$ QS, compared to the case when the time intervals considered are independent.

7   Conclusion

This result shows that the presence of a correlation between the inter-arrival time and service time decreases waiting time of the request in the queue. It is not unexpected because of certain "consistency" of the mentioned intervals can occur due to presence of a correlation. Therefore, this may lead to decrease in the waiting time for a request in the queue.

Finally, use of the copula function allows us to investigate the behavior of any network node simulated by the general type queuing system processing correlated traffic, which certainly expands the possibilities of the classical queuing theory.

References

1. Chakraborty, D., Ashir, A., Suganuma, T., Mansfield, K.G., Roy, T.K., Shiratori, N.: Self-similar and fractal nature of Internet traffic. Int. J. Netw. Manag. **14**(2), 119–129 (2004)
2. Sheluhin, O., Teniakshev, A., Osin, A.: Fractal processes in telecommunications. Radiotekhnika, Moscow (2003). (in Russian)
3. Buzov, A.L., Bukashkin, S.A.: Special radio. Development and modernization of equipment and facilities, Radiotekhnika, Moscow (2017). (in Russian)
4. Kartashevskiy, I.V., Volkov, A.N., Kirichek, R.V.: Analysis of the average delay time in the queuing system when processing correlated traffic. Elektrosvyaz **3**, 41–50 (2019)
5. Vishnevskii, V.M., Dudin, A.N.: Queueing systems with correlated arrival flows and their applications to modeling telecommunication networks. Autom. Remote Control **78**(8), 1361–1403 (2017)
6. Jagerman, D.L., Balcioglu, B., Altiok, T., Melamed, B.: Mean waiting time approximations in the G/G/1 queue. Queueing Syst. **46**, 481–506 (2004)
7. Balcioglu, B., Jagerman, D.L., Altiok, T.: Merging and splitting autocorrelated arrival processes and impact on queueing performance. Perform. Eval. **65**(8), 653–669 (2008)
8. Kartashevskiy, I.V., Saprykin, A.V.: Waiting time analysis for the request in general queuing system. T-Comm. Telecommun. Transp. **12**(2), 4–10 (2018)
9. Smith, R.D.: The dynamics of internet traffic: self-similarity, self-organization, and complex phenomena. Adv. Complex Syst. **14**(6), 905–949 (2011)
10. Kleinrock L.: Queueing Systems, Volume I: Theory. Wiley, New York (1975)
11. Saaty, T.L.: Elements of Queueing Theory: with Applications. McGraw-Hill, New York (1961)
12. Balakrishnan, N., Chin-Diew, L.: Continuous Bivariate Distributions, 2nd edn. Springer, New York (2009). https://doi.org/10.1007/b101765
13. Kartashevskiy, I.: Telecommunication traffic analysis by using Copulas. Infokommunikacionnye tehnologii **14**(4), 405–412 (2016)
14. Gradstein, I., Ryjik, I.: Table of Integrals, Series, and Products. Academic Press, San Diego (2000)
15. Abramowitz, M., Stegun, I.A. (eds.): Handbook of Mathematical Functions. Applied Mathematics Series 55. NBS, Washington, D.C. (1964)

On Verification of Stability of Multi-orbit System with General Retrials: Simulation Approach

Ruslana Nekrasova[1,2](✉)

[1] Institute of Applied Mathematical Research Karelian Research Centre RAS,
Petrozavodsk, Russia
ruslana.nekrasova@mail.ru
[2] Petrozavodsk State University, Petrozavodsk, Russia

Abstract. We explore multi-orbit retrial system with classical retrial policy. Arrivals enter the system according to renewal input. A random customer belongs to class j with a given probability p_j, $j = 1, \cdots, K$. If new arrival, meets the server busy, it joins to corresponding infinite capacity orbit and then retries to attack server after a random time interval distributed as $\xi^{(j)}$. The main feature of considered system is general distribution of retrial times. Our goal in this paper is to verify the stability by exploring the behavior of orbit dynamics for different distributions of $\xi^{(j)}$ under the sufficient stability condition, obtained for the systems with exponential and NBU retrials.

Keywords: Retrial system · Classical retrial policy · Stability · General retrials

1 Introduction

The paper deals with a single-server K-class retrial model with a classical policy. Arriving customers enter the system according to renewal input with a given rate λ. A random arrival belongs to a class $j = 1, \cdots, K$ with a probability $p_j > 0$ and has class dependent service time distributed as $S^{(j)}$. In case the server is busy, a customer joins the corresponding j-th virtual buffer (so called orbit) and retry to access server after a random time interval $\xi^{(j)}$ until the server is found idle. Such intervals are considered independent and identically distributed within the class. As retrial discipline is classical, all orbital costumers make independent attempts and total class j retrial (or orbit) rate is proportional to corresponding orbit size.

Retrial queuing systems are successfully used in simulation of a wide class of models like modern wireless telecommunication systems, cellular mobile networks or multi-access protocols, where rejected packages are sent again after some waiting period. Other important field of application are call centers. In such a model customer, who finds a busy operator, decides to call again after

The research is partly supported by Russian Foundation for Basic Research, projects 18-07-00156, 19-07-00303.

V. M. Vishnevskiy et al. (Eds.): DCCN 2019, CCIS 1141, pp. 401–412, 2019.
https://doi.org/10.1007/978-3-030-36625-4_32

some time and independently of other callers. Thus, with a growth of failures, the number of "secondary" calls (analogue of orbit customers) also increases, and that leads to a bigger attack to the server. The significant feature of considered model is a general distribution of retrial times $\xi^{(j)}$. Unlike usually-mentioned models with exponential retrials, the presented system could be useful for a wider class of applications.

Regarding to the overview of retrial theory it is worth mention such a significant works as [1–4]. Note, that most of research in this field is dedicated to more widespread single-class models with exponential retrials like in [5–8]. In that case authors may rely on Marcovian theory and establish explicit statements for stability conditions or stationary characteristics.

One of the first results, obtained for single-server single-class retrial model with non-exponential retrials is presented in [9]. Authors analyzed stability region for a general retrial queue. Stability analysis of multi-server single-class system with classical retrial discipline was extended in [11]. The research was based on regenerative approach. Later, stability of a single-class multi-server retrial model with generally distributed retrial times has been analyses in [10], where method was based on fluid model equations. And the original model was reduced to its deterministic analogue.

The stability of general *multi-class* retrial queueing systems with the renewal input, general class-dependent service time and with *exponential* class-dependent retrial times were studied in [12]. Authors established that considered system is stable if and only if the number of servers is greater, than load coefficient. The necessary condition for such a system was obtained basing on balance equation for the workload process, while the proof of sufficiency relied on regenerative method. Results, presented in [12], were extended in recent work [13] for more general case of multi-class system, where distributions of retrial times $\xi^{(j)}$ has so-called New Better than Used (NBU) property.

Our goal in this paper is to explore multi-class retrial system and verify by simulation the stability of general-retrials case, basing on known results for corresponding exponential and NBU retrials system. Namely, we investigate system behaviour under sufficient stability conditions, obtained in [12] for multi-server system with exponential distribution of $\xi^{(j)}$ and extended in [13] for single-server system with NBU retrials.

The paper is organized as follows. Section 2 contains the detailed description of presented model. Section 3 includes some preliminary results for the stability of related multi-class model with exponential and NBU retrials. Section 4 presents simulation results for two-class system in which retrial times $\xi^{(j)}$ have class-dependent distribution. We demonstrate, that under the sufficient condition, obtained for exponential and NBU cases, orbits in considered system are also stable. Moreover, we illustrate that, varying orbit rates in stable regime, we can manage the behavior of the system. Section 5 concludes the paper.

2   Description of the Model

We construct a single-server bufferless retrial system with classical retrial policy. Consider a sequence of inter-arrival instants $\{t_n, n \geq 1\}$, assume $t_0 = 0$. As input stream is renewal, inter-arrival times $\tau_n = t_n - t_{n-1}$ are independent and identically distributed (i.i.d.). We define the generic arrival interval by τ and obtain the input rate as $\lambda = 1/\mathsf{E}\tau$.

The system accepts $K \geq 1$ classes of customers. Note, that a random arrival belongs to class $j = 1, \ldots, K$ with a probability $p_j > 0$ and assume $\{S_n^{(j)}, n \geq 1\}$ are i.i.d. service times for the corresponding class (with a generic element $S^{(j)}$). Thus, the j-th class input and service rates are defined as

$$\lambda_j = p_j \lambda, \ \mu_j = 1/\mathsf{E}S^{(j)}, \tag{1}$$

respectively.

If the j-th class arrival meets the server busy, it is not lost, but joins the corresponding virtual infinite capacity orbit and then tries to call the server again after a random time interval distributed as $\xi^{(j)}$ (so called inter-retrial time). Define by $N^{(j)}(t)$ the size of orbit-j at time instant t. Thus, the total orbit is obtained as follows:

$$N(t) = N^{(1)}(t) + \cdots + N^{(K)}(t). \tag{2}$$

The considered model is organized according to classical retrial policy: all orbit customers makes independent attempts. Thus, total retrial rate is proportional to orbit size (number of orbit customers). Namely, every orbit customer attacks the server with a rate $g_j = 1/\mathsf{E}\xi^{(j)}$ until successful attempt. Although the value g_j is often called retrial (or orbit) rate, it represents the behavior of individual calls. Actual (or total) rate from class-j orbit at instant t consists of summary attempts of all $N^{(j)}(t)$ customers.

When $N(t) > 0$ after a departure, the server would stay idle up to the first orbital attempt. Namely, the operator stays empty, while customers are still waiting in the system, such an effect reduces the capacity. The classical retrial policy implies, that with a growth of $N(t)$, orbit streams become more intensive, and gaps between attempts decrease. It means, that more loaded system becomes more effective.

Note, that the only reason and indicator of stability is infinite growth of orbit size $N(t)$. The stability region (appropriate values of input parameters λ_j, μ_j, g_j, which provide stable orbit) could be establish analytically for Markov case. For more general models we can rely on regenerative approach. Consider a queue process $\{\nu(t), t \geq 0\} \in \{0, 1\}$. Namely, it indicates the server state (idle/busy) at instant t. Then construct the process $X(t) = \nu(t) + N(t)$ – total number of customers at instant t. For its discrete analogue $X_n = X(t_n^-)$ consider the sequence as follows:

$$\beta_{n+1} = \inf_k \left(k > \beta_n : X_k = 0\right), \qquad \beta_0 = 0, \ n \geq 0. \tag{3}$$

Thus, β_n indicates the actual number of arrival, which joins totally empty system. At time instants $\{t_{\beta_n}\}$ the system starts over in stochastic sense or *regenerates*. The sequence $\{\beta_n, n \geq 0\}$ denotes regeneration points. The lengths of regeneration cycles $\alpha_{n+1} = \beta_{n+1} - \beta_n$ are i.i.d. Denote generic length by α. We consider zero initial state $X(0) = 0$. If the mean cycle length is finite:

$$\mathsf{E}\alpha < \infty, \tag{4}$$

then basic regenerative process X is called *positive recurrent*. The demand of positive recurrence is a necessary condition for application of regenerative method, which could be effectively used for analysis of a wide class of stochastic models and estimation of QoS parameters. Namely, the fulfillness of (4) is equivalent, that considered retrial system is stable.

3   Preliminary Results

In this section we present some stability results, established for related systems. We rely on these results in further analysis. First, denote the class-j load coefficient by $\rho_j := \lambda_j/\mu_j$, which implies the total load coefficient

$$\rho := \rho_1 + \cdots + \rho_K. \tag{5}$$

In [12] was obtained that m-server multi-class retrial system with *exponential* retrials is stable if and only if

$$\max_{1 \leq j \leq K} \mathsf{P}(\tau > S^{(j)}) > 0, \tag{6}$$

$$\rho < m. \tag{7}$$

The proof of necessity is based on the balance equation for the workload process and does not depend on retrials-times distribution. Assume, that $A_j(t)$ is the number of class-j arrivals, $D(t)$ and $R(t)$ are total departed and remaining work, respectively in the interval $[0, t)$. It is easy to evaluate, that

$$\sum_{j=1}^{K} \sum_{n=1}^{A_j(t)} S_n^{(j)} = R(t) + D(t). \tag{8}$$

Dividing both parts of (8) by t and taking $t \to \infty$, we obtain (7). The condition (6) is automatically implied for unbounded τ (holds for the most considered models). Necessary stability condition from [12] could be easily extended to the case of general retrials. Thus, if considered single server system is stable, then $\rho < 1$.

The recent paper [13] explored the stability of multi-class system with a specific distribution of inter-retrial times. The sufficient stability condition was obtained for the case, when distributions $\xi^{(j)}$ have so called *New Better than*

Used (NBU) property. Namely, for $x, y \geq 0$ and $j = 1, \ldots, K$ the following inequality holds

$$P(\xi^{(j)} > x + y | \xi^{(j)} > y) \leq P(\xi^{(j)} > x). \tag{9}$$

Authors in [13] also assumed the unbounded inter-arrival time τ:

$$P(\tau > x) > 0, \qquad x \geq 0, \tag{10}$$

zero initial state $X(0) = 0$ and established the following result.

Consider $\xi^{(j)}$ belongs to NBU distributions and

$$\inf_{x} P(\xi^{(j)} \leq x) > 0, \qquad j = 1, \cdots, K. \tag{11}$$

If the condition

$$\rho < 1 \tag{12}$$

holds, then single-server K-class retrial system is stable.

The proof is based on regenerative approach. As the stability is equivalent to the positive recurrence of a basic regenerative process X, its enough to show, that mean cycle length is finite $E\alpha < \infty$. In this framework, under the condition (12) authors established the negative drift of the orbit size process.

Let $N_n = N(t_n^-)$ define a discrete analogue of total orbit process, considered in arrival instants $\{t_n, n \geq 0\}$. The NBU retrials and condition (11) are technical requirements, which are used to establish, that mean interval between retrial attempts from all K orbits converges to zero with a growth of total orbit size. That makes possible to show, that under the assumption $N_n \Rightarrow \infty$, the mean idle period $E\Delta_n$ among the arrival interval τ_n convergence to zero, as $n \to \infty$. Meanwhile, the condition $\rho < 1$ implies $E\Delta_n \not\to 0$. Thus, the assumption leads to contradiction and

$$N_n \not\Rightarrow \infty, \qquad n \to \infty, \tag{13}$$

The process N_n visits a bounded set infinitely often. This makes possible to show, that remaining regeneration time

$$\beta(n) = \min_{k}(\beta_k - n : \beta_k - n > 0) \tag{14}$$

does not go to infinity. By the regenerative theory, the result

$$\beta(n) \not\to \infty, \qquad n \to \infty \tag{15}$$

is equivalent to condition $E\alpha < \infty$ and the system is stable.

Basing on stability analysis from [12,13], we can assume, that $\rho < 1$ is a sufficient stability condition of considered multi-class system with general retrials. In this way we had to rely on simulation and explore the orbit dynamics $N_n^{(j)}$ in case the load coefficient is less than number of servers.

4   Simulation

In this section we present some numerical results related to single-server retrial system with two classes (two orbits). With no loss of generality, we consider Poisson input and exponential service times and investigate orbit dynamics for different distributions of inter-retrial times $\xi^{(1)}$, $\xi^{(2)}$. In all configurations we consider the fulfillness of condition

$$\rho_1 + \rho_2 < 1 \tag{16}$$

and expect, that considered system is stable. As the orbit behavior illustrates the stability, we explore the dynamics of mean orbit sizes $N_n^{(1)}$, $N_n^{(2)}$ for both classes of costumers.

In all experiments we consider $10\,000$ arrivals ($n = 1, \cdots, 10\,000$) and construct average orbit sizes over $k = 300$ independent paths for each arrival instant as follows:

$$MN[j]_n = \frac{1}{k} \sum_{i=1}^{k} N_n^{(j)}(i), \tag{17}$$

where i indicates the number of replication, $j = 1, 2$ is associated with the class and n shows the number of arrival.

We explore a few distributions of inter-retrial times to verify the stability of considered system. It is expected, that light load (small values of ρ) will provide bounded mean orbit sizes, thus we are interested in configurations which shows the system near to the stability border and consider ρ close to 1.

First, assume Weibull distribution of $\xi^{(j)}$. We define scale parameter fixed and equal to 1 and shape parameter denoted by w_j:

$$\mathsf{P}(\xi^{(j)} \le x) = 1 - \exp(-x^{w_j}), \qquad j = 1, 2. \tag{18}$$

The corresponding orbit rates are defined via Γ (Gamma-function) as follows

$$g_j = \frac{1}{E\xi^{(j)}} = \left[\Gamma\left(\frac{1}{w_j} + 1\right)\right]^{-1}. \tag{19}$$

Weibull distribution changes its stochastic properties for different values of w_j. Namely, for $w_j > 1$, the property (9) holds and inter-retrial times belong to the NBU class. For $w_j < 1$ the distribution of $\xi^{(j)}$ belongs to opposite *New Worse Than Used* (NWU) class and

$$\mathsf{P}(\xi^{(j)} > x + y | \xi^{(j)} > y) > \mathsf{P}(\xi^{(j)} > x). \tag{20}$$

Note, that for $w_j = 1$ retrials are exponential and we obtain equality in (9). As the sufficient condition for NBU retrials was established in [13], we explore NWU case.

Consider retrial rate as a function of corresponding shape parameter $g_j = g_j(w_j)$ and let $w_j \in (0, 1]$. By the properties of Gamma-function, g_j monotonically increases with respect to w_j and

$$\lim_{w_j \to 0} g_j(w_j) = 0, \qquad g_j(1) = 1. \tag{21}$$

We assumed $\rho_1 < \rho_2$ (the 2-nd class customers provides bigger load) and explore different relations of retrial rates g_1, g_2. We illustrate just a few numerical examples, as tests for other configurations had shown rather similar results.

The set of parameters, which corresponds to the system with Weibull retrials is presented in Table 1.

Table 1. Configuration for Weibull retrials system.

Test	j	λ_j	μ_j	ρ_j	ρ	w_j	g_j
1.	1	0.50	2.00	0.25	0.91	0.90	0.92
	2	1.00	1.50	0.66		0.40	0.30
2.	1	0.50	2.00	0.25	0.91	0.40	0.30
	2	1.00	1.50	0.66		0.90	0.92

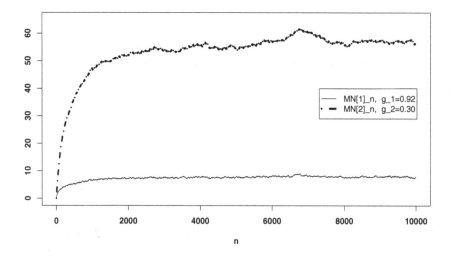

Fig. 1. Weibull retrials, test 1.: $\rho_1 = 0.25$, $\rho_2 = 0.66$.

Figure 1 illustrates the case of more intensive 1-st class retrials $g_1 > g_2$. The 2-nd class customers arrives to the system and load the server more actively $\rho_2 > \rho_1$. Thus, the second mean orbit systematically dominates the first one. The simulation results had shown, that both orbits are stable.

Figure 2 presents the system with the same set of load parameters, as on Fig. 1, but consider opposite relation of retrial rates: $g_1 < g_2$. However $\rho_1 < \rho_2$, the 1-st class orbit is more loaded. This phenomenon is explained by rate values: the 2-nd class retrial attempts are more intensive and unload the orbit faster. Thus, we can conclude that the appropriate choice of orbit rates can redistribute server attacks between two classes.

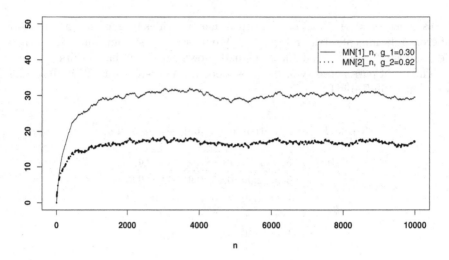

Fig. 2. Weibull retrials, test 2.: $\rho_1 = 0.25$, $\rho_2 = 0.66$.

Next we explore the system with Pareto retrials. Define the basic parameter by $\alpha_j > 1$ and the scale parameter fixed equal to 1. Thus,

$$\mathsf{P}(\xi^{(j)} \leq x) = 1 - x^{-\alpha_j}, \, x \geq 1, \tag{22}$$

the orbit rate is obtained as

$$g_j = \frac{\alpha_j - 1}{\alpha_j} \in [0, 1) \tag{23}$$

and monotonically increases with respect to α_j. The set of parameters for the system with Pareto distribution of inter-retrial times is presented in Table 2.

Choosing the configuration, we relied on the same approach as for the Weibull retrials system. First test describes the case of more loaded 2-nd class orbit: more intensive arrivals $\lambda_2 > \lambda_1$, slower service $\mu_2 < \mu_1$ and less aggressive orbit attempts $g_2 < g_1$. In second test we consider more intensive orbit 2.

Figure 3 illustrates joint results for both experiments. Black lines present the case, when the 1-st class attempts is more intensive ($\alpha_1 = 20$, $\alpha_2 = 1.2$), and grey lines are related to opposite case ($\alpha_1 = 1.2$, $\alpha_2 = 20$). Note, that we

Table 2. Configuration for Pareto retrials system.

Test	j	λ_j	μ_j	ρ_j	ρ	α_j	g_j
1.	1	0.40	2.00	0.20	0.90	1.20	0.95
	2	0.70	1.00	0.70		20.00	0.17
2.	1	0.40	2.00	0.20	0.90	20.00	0.17
	2	0.70	1.00	0.70		1.20	0.95

simulate the system for $n = 10\,0000$ arrivals, but present $MN[j]_n$ dynamics only for $n = 5\,000$, as further behavior does not significantly influence for the results.

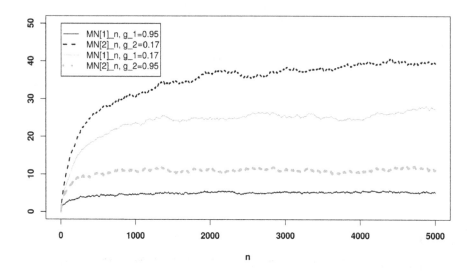

Fig. 3. Pareto retrials: two systems comparison $\rho_1 = 0.2$, $\rho_2 = 0.7$.

As in Weibull case, experiments for Pareto retrials illustrate that both orbits are stable, even the 2nd orbit in first test (black dash line). And rates g_j significantly influences to orbit behavior: in 2-nd test $MN[1]_n$ systematically dominates $MN[2]_n$, while 1-st class load coefficient is less then the 2-nd one.

Last, we present simulation results for a model with Uniform distribution ($\mathcal{U}$) of inter-retrial times. Note, that considered system is quite similar to the multi-class queueing system with infinite buffer, where the demand $\rho < 1$ is also a stability criterion. The main difference between systems is that in retrial model rejected customers does not enter the server immediately, as it became idle. We additionally construct 2-class infinite-buffer and single server queueing system with the same inter-arrival times and same distribution of service times, as in original retrial system. Note, that total queue in such auxiliary model contains of two components $Q^{(1)}(t)$, $Q^{(2)}(t)$, which correspond to different classes. We can expect, that total orbit $N(t)$ will dominate the total queue in corresponding queuing system at time instant t. Our goal is to simulate both systems and present for $n = 5\,000$ arrivals the dynamics of mean orbits $MN[j]_n$ and mean queue components (among $k = 300$ independent paths):

$$MQ[j]_n = \frac{1}{k} \sum_{i=1}^{k} Q_n^{(j)}(i), \qquad (24)$$

Table 3. Configuration for Uniform retrials.

j	λ_j	μ_j	ρ_j	ρ	$\xi^{(j)}$	g_j
1	0.30	1.00	0.30	0.90	$\mathcal{U}[0, 0.5]$	4.00
2	1.20	2.00	0.60		$\mathcal{U}[4,\ 5]$	0.22

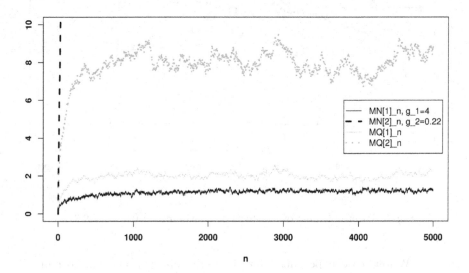

Fig. 4. Retrial system vs. Queuing system, $\rho_1 = 0.3$, $\rho_2 = 0.6$. (Uniform retrials.)

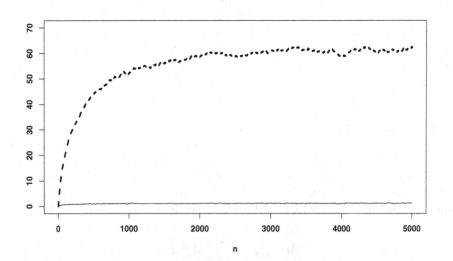

Fig. 5. Uniform retrials, $MN[2]_n$ dynamics.

where $Q_n^{(j)}(i)$ denotes the j-class queue component just before time instant t_n in i-th independent replication. The set of common parameters for considered systems is presented in Table 3.

Figure 4 demonstrates results related to both systems. Note, that in this configuration, the 2-nd class orbit is significantly grater than the 1-st class orbit and queue size components. The dynamics of $MN[2]_n$ is presented in Fig. 5 in more appropriate scale, the second orbit is stable. The main interest relates to the 1-st class. The situation is rather surprising: the queue size dominates the orbit. As in previous cases, the obtained results illustrates, that varying the orbit rate, we can unload (at least) one of the orbits and manage the system behavior.

5   Conclusion

We explored the behavior of multi-orbit (multi-class) retrial system with general distributions of inter-retrial times under the sufficient stability condition obtained in [12,13] for the system with exponential and NBU retrials, respectively. All experiments had shown that orbits are stable. Thus, we can expect, that criterion $\rho < 1$ could be considered the sufficient stability condition of presented system. We also demonstrated, that appropriate choice of orbit parameters could redistribute server attacks among different classes and unload the required orbit, independently of marginal load coefficients. Such an approach could be useful in modeling of systems with priority-like policy.

References

1. Artalejo, J.R.: Accessible bibliography on retrial queues. Math. Comput. Model. **30**, 1–6 (1999)
2. Artalejo, J.R.: A classified bibliography of research on retrial queues: progress in 1990–1999. Top **7**, 187–211 (1999)
3. Artalejo, J.R., Gomez-Corral, A.: Retrial Queueing Systems: A Computational Approach. Springer, Cham (2008). https://doi.org/10.1007/978-3-540-78725-9
4. Falin, G.I., Templeton, J.G.D : Retrial Queues. Chapman and Hall, London (1997)
5. Artalejo, J.R., Phung-Duc, T.: Markovian retrial queues with two way communication. J. Ind. Manag. Optim. **8**, 781–806 (2012)
6. Artalejo, J.R., Phung-Duc, T.: Single server retrial queues with two way communication. Appl. Math. Model. **37**, 1811–1822 (2013)
7. Sakurai, H., Phung-Duc, T.: Two-way communication retrial queues with types of outgoing calls. TOP **23**, 466–492 (2015)
8. Kim, J., Kim, B.: A survey of retrial queueing systems. Ann. Oper. Res. **247**, 1–34 (2015)
9. Altman, E., Borovkov, A.A.: On the stability of retrial queues. Queueing Syst. **26**, 343–363 (1997)
10. Kang, W.: Fluid limits of many-server retrial queues with non-persistent customers. Queueing Syst. **79**(2), 183–219 (2015)
11. Morozov, E.: A multiserver retrial queue: regenerative stability analysis. Queueing Syst. **56**, 157–168 (2007)

12. Morozov, E., Phung-Duc, T.: Stability analysis of a multiclass retrial system with classical retrial policy. Perform. Eval. **112**, 15–26 (2017)
13. Morozov, E., Nekrasova, R.: Stability conditions of a multiclass system with NBU retrials. In: Phung-Duc, T., Kasahara, S., Wittevrongel, S. (eds.) QTNA 2019. LNCS, vol. 11688, pp. 51–63. Springer, Cham (2019). https://doi.org/10.1007/978-3-030-27181-7_4

Reliability Analysis of Systems with Hybrid Recovery and Imperfect Built-in-Test

Armen Stepanyants and Valentina Viktorova$^{(\boxtimes)}$ (iD)

V. A. Trapeznikov Institute of Control Sciences of Russian Academy of Sciences,
65 Profsoyuznaya Street, 117997 Moscow, Russia
ray@ipu.ru, v.s.viktorova@ieee.org
http://www.ipu.ru

Abstract. The paper describes approach to reliability analysis of systems with cyclic mode of operation, in which recovery actions after failures become possible only in offline mode. Theoretical basis of this investigation is continuous-time discrete state Markov processes. Complex reliability measure which is both cumulative and instantaneous is created for investigation of operative availability of such systems. Step by step schema for calculating operative availability index is suggested. Results of analysis of main redundant configurations of these systems with imperfect fault coverage are presented.

Keywords: Availability · Built-in-test · Fault coverage · Hybrid recovery · Latent faults · Markov process · Reliability · Testability

1 Introduction

We consider reusable cyclic systems with built-in-test (BIT) whose duty cycle consists of a target operation phases (TOP), repair phases (RP), inspection and maintenance phases (IMP). Mission profile of such systems is shown in Fig. 1. During the TOP phase, the system is on-line and operates as intended. When the system is online (target operation phase), recovery is not possible, and information about failures detected by the BIT is recorded in the onboard computer memory. Recovery becomes possible only in offline mode after performing operation phase. During repair phase, only those components whose failures were detected by the built-in-test are recovered. The imperfect fault coverage of the BIT results in imperfect system recovery during RP phase (system renewal percentage < 100%). Full system recovery occurs during IMP after every m cycles each of which includes operation phase (TOP) and repair phase (RP). At the phase of inspection and maintenance all faults are eliminated, including latent (hidden) faults that were not detected by the BIT. Detailed inspection, functional checks, servicing, repairing or replacing of faulty components support

Supported by organization LLC UAC - Aggregation Center.

100% system renewal during IMP. Systems with such a maintenance program
are called hybrid recovery systems (HRS). Classical logical-probabilistic reliability analysis methods including fault-trees and reliability block diagrams has been
intensively applied to engineering problems [1–3]. Logical-probabilistic methods
are based on the construction of a Boolean function relating the state of the system with the states of its elements. The resulting Boolean function is transformed
to a form that allows to replace logical variables by corresponding probabilities.
Static models of these methods for the case of repairable systems allow us to
calculate only the differential (instantaneous) reliability indicators determined
at the time instant t and not describe system reliability behavior as the process
developing in time. So, they are not suitable for investigation of the systems
with hybrid recovery.

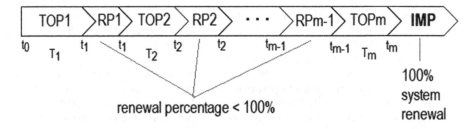

Fig. 1. Mission profile of hybrid recovery system.

Markov modeling allows taking into account dynamic features of reliability
behavior, imperfect built-in test equipment leading to the presence of undetected
failures, smart recovery strategies. In addition, dynamic Markov models allow
you to calculate all the main dependability measures both for repairable and non-
repairable systems. These measures are instantaneous indices (e.g. availability
at the time instant); interval indices (e.g. reliability during the time interval),
time-independent stationary indicators (e.g. mean time between failures). Consequently, Markov processes can be directly applied to reliability analysis of HRS.
However, the state space explosion; stiffness of transient solution of Kolmogorov
equations (or ill-conditionality for stationary case); limitation to exponential
distribution of random time variables of the model are well known drawbacks
inherent in Markov models. The problem of large size of the Markov model can
be solved by decomposition and aggregation of the model parts and by automation of the model construction. Successful implementations of this approach are
dynamic failure trees [4] and Boolean logicdriven Markov processes [5]. The technique of merging states of a Markov model is described in [3, pp. 141–144].

Different approaches to numerical evaluation of Markov model transient
behavior were presented in [6,7]. They are uniformization, an explicit solution
method (Runge-Kutta) and special stable implicit method (TR-BDF2). Here we
apply efficient method based on evaluation of matrix exponential at a small step

[8, pp. 81–112]. This method ensures a fast and accurate transient solution of stiff systems of differential equations.

In our opinion, the restriction of the Markov model to the exponential distribution is not so important. At the design stage, we dont have, as a rule, objective information about the distribution functions of random time variables of the model. In addition, all generally accepted standards of reliability prediction [9] and Data Bases [10,11] of failure parameters of electronic and nonelectronic parts are oriented specifically to the exponential distribution. Therefore, we believe that the assumption of exponentiality is quite acceptable and appropriate, especially in terms of practical calculations of system reliability.

2 Computational Schemes for Calculating Reliability Indices of HRS

The main criterion of dependability of a system with hybrid recovery is the probability that at the beginning of the $(j + 1)^{th}$ TOP it is available and will be operational during the time interval (T_{j+1})

$$A(t_j, T_{j+1}) = \sum_{i \in \Omega_g} A_i(t_j) R_i(T_{j+1}) \tag{1}$$

where Ω_g is the set of operational system states; T_{j+1} is duration of target operation phase $(j + 1)$; $j = \overline{0, m}$; $R_i(T_{j+1})$ is system reliability for the time interval (t_j, t_{j+1}) provided that $(j + 1)^{th}$ TOP starts from state i; $A_i(t_j)$ is the probability that the system state is i at the time instant t_j.

Let the number of states of Markov reliability model of the HRS be n. Let Ω_f be the set of failed system states. For reliability calculation the states from Ω_f are absorbing states, so they can be merged into one state. Let it be the state with the number n. Then the system of Kolmogorov differential equations is

$$P'(t) = P(t)\Lambda \tag{2}$$

where Λ is $n \times n$ infinitesimal matrix of transition rates; $P(t) = (P_1(t), P_2(t), \ldots, P_n(t))$ is row-vector of state probability at time instant t; initial condition vector is $P(0) = (P_1(0), P_2(0), \ldots, P_n(0))$.

Our task is to assess the trend of operative availability based on the results of calculating the index (1) at intervals $(T_1, T_2, \ldots, T_m)$. To do this, we need to solve system (2) m times, each time forming a new vector of initial conditions. The vector of initial conditions is formed on the basis of the solution at the previous step. It takes into account that all failures detected by the BIT are eliminated during the recovery phase. Therefore, the probability of finding the system in the original good state $(i = 1)$ increases by the total probability of states with detectable failures. Thereby $P^{(0)}(0) = (1, 0, \ldots, 0)$ for the first interval T_1. For subsequent intervals $(T_2, T_3, \ldots, T_m)$, the components of the vector $P^{(j)}(0)$ are determined by the rule

$$P_i^{(j)}(0) = \begin{cases} P_1(t_j) + \sum_{i \in \Omega_{bit}} P_i(t_j) & \text{if } i = 1 \\ P_i(t_j) & \text{if } i \in \Omega_{nbit} \\ 0, & \text{if } i \in \Omega_{bit} \end{cases} \tag{3}$$

Hear Ω_{bit} is the set of system states with at least one detected by BIT failure; Ω_{nbit} is the set of system states with only latent failures; $i = \overline{1, n}$; $j = \overline{1, m-1}$.

The method of calculating initial vector will be demonstrated on specific examples of the next section.

System reliability over the time interval T_j under initial conditions calculated by (3) is operative availability

$$R(T_{j+1}) = \sum_{i=1}^{n-1} P_i(t_{j+1}) = A(t_j, T_{j+1}) \qquad (4)$$

As a result of step-by-step calculation of the complex indicator (1), a curve of operative availability of a system with a hybrid recovery is formed.

3 Markov Reliability Models of Main Redundant Structures

Hybrid recovery cannot be implemented without the use of structural and/or functional redundancy. The redundancy of the components of technical or functional structure allows the system to perform the task on the time interval of operational phase in the presence of both hidden and BIT detected failures. The failure rate of channel (element) of redundant structure is denoted by λ.

Let us construct Markov reliability models of the main redundant structures duplicate (1 out of 2), triple (1 out of 3), majority (2 out of 3). When constructing the model, a channel (element) is considered as consisting of two parts - covered and uncovered by BIT. In general, the BIT coverage (fault detection percentage) can be defined as the conditional probability of detecting a failure, provided that the failure has occurred: $\eta = \frac{1 - e^{-\int_0^t \lambda_{bit}(\tau)d\tau}}{1 - e^{-\int_0^t \lambda(\tau)d\tau}}$, where λ_{bit} failure rate of covered by BIT part; λ failure rate of channel (element). In the case of the exponential distribution of random time to failure and $\lambda t \ll 1$ $\eta = \frac{\lambda_{bit}}{\lambda}$. Thereby failure rate of the covered part is $\eta\lambda$, the failure rate of uncovered part is $(1 - \eta)\lambda$.

3.1 Duplicate and Majority Structures

Reliability Block Diagram (RBD) of 1 out of 2 and 2 out of 3 structures and Markov graph of reliability behavior of duplicate and majority systems at TOP time interval are shown in Fig. 2.

Descriptions of the model states:

1 - all elements are operational;

2 - latent failure in one of reserved channels (elements) occurs, the other channels are operational, the system is operable in a whole, but there is a hidden failure of one of the channels (elements), this failure will not be eliminated at RP stage;

3 - detected failure in one of reserved channels (elements) occurs, the other channels are operational, the system is operable in a whole, but there is a failure

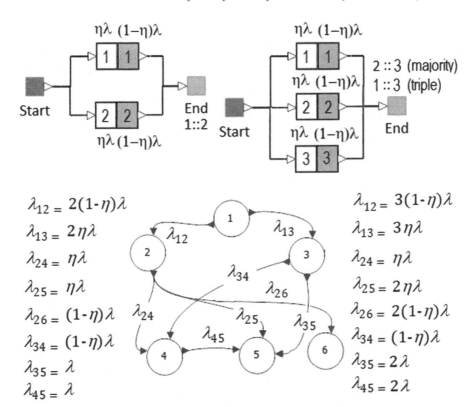

Fig. 2. Markov reliability model of duplicate and majority hybrid recovery system at TOP time interval.

of one of the channels (elements), which is covered by BIT, this failure will be eliminated at RP stage;

4 - there are two failures in one of the channels (elements), one is detected and another is undetected, the other channels are operational, the system is operable in a whole, the faulty channel (element) will be recovered at the RP stage;

5 - in two channels (elements) there are detected failures, the system is inoperable in a whole, but is recoverable at the RP stage;

6 - in two channels (elements) there are undetected failures, the system in a whole is inoperable and is not recoverable at the RP stage.

In the presence of only latent failures (states 2, 6) the system 100% restoration will be possible only at heavy maintenance stage (IMP).

Vector of initial condition for every j^{th} calculation step are defined by (3) as follows:

$$P_1^{(j)}(0) = P_1(t_j) + P_3(t_j) + P_4(t_j) + P_5(t_j);$$
$$P_2^{(j)}(0) = P_2(t_j);$$
$$P_3^{(j)}(0) = P_4^{(j)}(0) = P_5^{(j)}(0) = 0;$$
$$P_6^{(j)}(0) = P_6(t_j).$$

3.2 Triple Module Structure

Markov graph of reliability behavior of triple module redundancy system at TOP
time interval is shown in Fig. 3. RBD of 1 out of 3 structure is shown in Fig. 2.

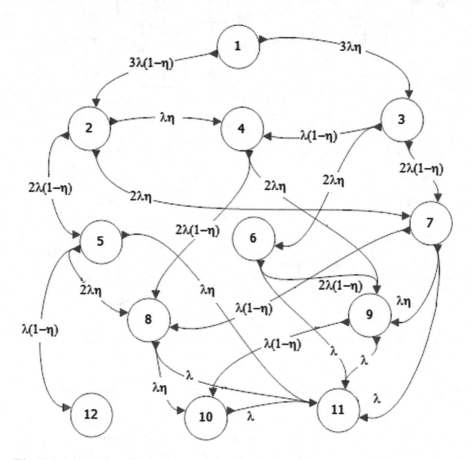

Fig. 3. Markov reliability model of triple hybrid recovery system at TOP time interval.

Descriptions of the model states:

States (1–4) related to the failure of one channel (element) are similar to the
previous model (Fig. 2);

5 - in two out of three channels (elements) there are undetected failures, the
system in a whole is operable, these failures will not be eliminated at the RP
stage;

6 - in two out of three channels (elements) there are detected by BIT failures,
the system in a whole is operable, these failures will be eliminated at the RP
stage;

7 - in two out of three channels (elements) there are failures, one failure is detected by BIT, another failure is latent, system in a whole is operable, the system will be recovered at the RP stage;

8 - state with three failures in two channels (elements), one channel has two failures, one of which is detected by BIT, the other is hidden, the second channel has an undetected failure, the third channel and the system in a whole are operable;

9 - a state with three failures in two channels (elements), one channel has two failures, one of which is detected by BIT, the other is hidden, the second channel has detected failure, the third channel and the system in a whole are operable;

10 - a state with four failures in two channels (elements), each channel has two failures, one of which is detected by BIT, the other is hidden, the third channel and the system in a whole are operable;

11 - in this state, there are failures in all three channels, therefore the structure is inoperable; but among these failures there are detected ones, therefore, restoration to the original fully operational state is possible at the RP stage;

12 - in this state, there are latent failures in all three channels, therefore the structure is inoperable, elimination of these failures is not possible at the RP stage.

In the presence of only latent failures (states 2, 5, 12) the system 100% restoration will be possible only at heavy maintenance stage (IMP).

Vector of initial condition for every j^{th} calculation step are defined as follows:

$$P_1^{(j)}(0) = P_1(t_j) + P_3(t_j) + P_4(t_j) + P_6(t_j) + P_7(t_j) + P_8(t_j) + P_9(t_j) + P_{10}(t_j) + P_{11}(t_j);$$

$$P_2^{(j)}(0) = P_2(t_j);$$
$$P_3^{(j)}(0) = P_4^{(j)}(0) = 0;$$
$$P_5^{(j)}(0) = P_5(t_j).$$
$$P_6^{(j)}(0) = P_7^{(j)}(0) = P_8^{(j)}(0) = P_9^{(j)}(0) = P_{10}^{(j)}(0) = P_{11}^{(j)}(0) = 0;$$
$$P_{12}^{(j)}(0) = P_{12}(t_j).$$

4   Operative Availability Analysis

We analyze the operative availability as a function of time. Two configurations are considered and compared - a system without redundancy and a duplicate system. Calculations of $A(t_j, T_{j+1})$ for duplicate system was performed in accordance with the procedure described in the previous section. Model of non-redundant system is presented in Fig. 4. This hybrid recovery mode can be implemented in multi-functional systems. As a result of analytical solution of the system of Kolmogorov equations of non-redundant HRS, we obtain

$$P_1(t_{j+1}) = C_1(t_j) \exp\left(-\lambda T_{j+1}\right) = A(t_j, T_{j+1});$$
$$P_3(t_{j+1}) = C_3 \exp\left(-\eta \lambda T_{j+1}\right) - C_1(t_j) \exp\left(-\lambda T_{j+1}\right);$$
$$P_2(t_{j+1}) = 1 - (P_1(t_{j+1}) + P_3(t_{j+1})),$$

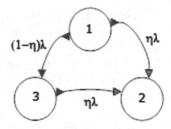

Fig. 4. Markov reliability model of non-redundant hybrid recovery system at TOP time interval.

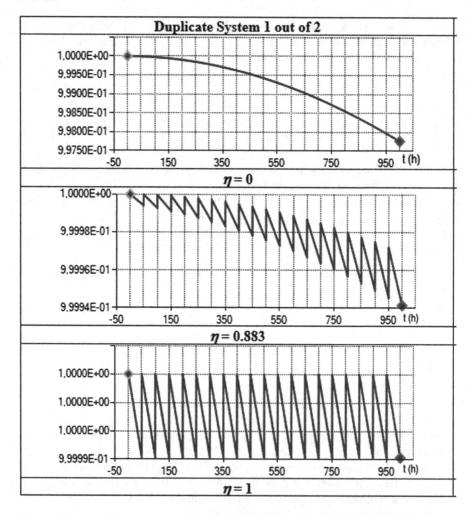

Fig. 5. Operative availability trend between heavy maintenance of duplicate system.

where arbitrary constants are defined as
$$C_1(t_j) = P_1(t_j) + P_2(t_j);$$
$$C_3 = 1.$$
Initial conditions are
$$P_1^{(j)}(0) = P_1(t_j) + P_2(t_j);$$
$$P_2^{(j)}(0) = 0;$$
$$P_3^{(j)}(0) = P_3(t_j).$$
Cureves of operative availability of duplicate and non-redundant systems
are shown in Figs. 5 and 6 respectively. The curves $A(t_j, T_{j+1})$ versus time are
plotted for each TOP stages between heavy maintenance. Interval between heavy
maintenance $T_\Sigma = \sum_{j=1}^{m} T_j$.

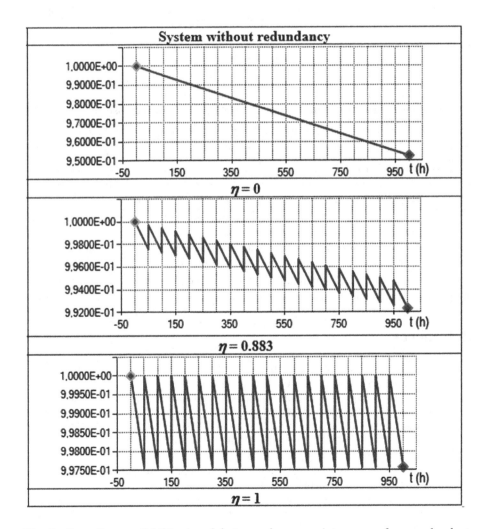

Fig. 6. Operative availability trend between heavy maintenance of non-redundant
system.

$T_\Sigma = 1000\,\text{h}$; TOP duration $T_j = 20\,\text{h}$; channel (element) failure rate = $4.85 \cdot 10^{-5}$ /hour.

The curves are plotted for three values of BIT coverage:

- $\eta = 0$: completely uncovered system;
- $\eta = 0.883$: real data for avionics channels;
- $\eta = 1$: fully BIT covered system.

Results of analysis show:

- when $\eta = 0$, the system with a hybrid recovery turns into a non-recoverable system, and operative availability is monotonically decreasing function of time;
- when $\eta = 1$, operative availability has a sawtooth form, the availability of the system at the beginning of each TOP interval is equal to 1, since all failures are detected, and therefore are eliminated at RP stages;
- when $0 < \eta < 1$, operative availability has a sawtooth form, $P_1^{(0)}(0) = 1$; $P_1^{(j)}(0) < 1$ for all $j > 0$, and $P_1^{(j)}(0) > P_1^{(j+1)}(0)$, which is due to the accumulation of latent, unrecoverable failures;
- "aging" inherent in redundant structures leads to an increase in the rate of decline of $A(t_j, T_{j+1})$ on each subsequent TOP interval when $0 < \eta < 1$.

5   Conclusions

The hybrid recovery mode is typical for objects of aircraft industry, shipbuilding, hazardous technological production. The proposed approach to the analysis of operative availability can be used to set the requirements for fault coverage of the built-in test and determine the maintenance intervals that guarantee the safe operation of facilities. The described computational scheme for the calculation of operative availability is implemented in the software for analyzing the testability of aviation equipment [12, 13].

References

1. Henley, E.J., Kumamoto, H.: Probabilistic Risk Assessment and Management for Engineers and Scientists, 2nd edn. IEEE Press (1996)
2. Ryabinin, I.A.: Logical probabilistic analysis and its history. Int. J. Risk Assess. Manag. 18(3/4), 256–265 (2015)
3. Viktorova, V.S., Stepanyants, A.S.: Models and Methods of Reliability Analysis of Technical Systems, 2nd edn. LENAND Publ, Moscow (2016)
4. Meshkat, L., Dugan, J.B., Andrews, J.D.: Dependability analysis of systems with on-demand and active failure modes, using dynamic fault trees. IEEE Trans. Reliab. 51(2), 240–251 (2002)
5. Bouissou, M., Bon, J.-L.: A new formalism that combines advantages of fault-trees and Markov models: boolean logic driven Markov processes. Reliab. Eng. Syst. Saf. 82(2), 149–163 (2003)

6. Reibman, A.L., Smith, R., Trivedi, K.S.: Markov and Markov reward model transient analysis: an overview of numerical approaches. Eur. J. Oper. Res. **40**(2), 257–267 (1989)
7. Lindemann, C., Malhotra, M., Trivedi, K.S.: Numerical methods for reliability evaluation of Markov closed fault-tolerant systems. IEEE Trans. Reliab. **44**(4), 694–704 (1995)
8. Rakitskiy, Y.V., Ustinov, S.M., Chernorutskiy, I.G.: Numerical Methods for Solving Stiff Systems. Nauka Publ, Moscow (1979)
9. HDBK-217Plus^{TM}: 2015, Notice 1
10. Electronic Parts Reliability Data (EPRD-97). Reliability Analysis Center (RAC) (1997)
11. Nonelectronic Parts Reliability Data (NPRD-95). Reliability Analysis Center (RAC) (1995)
12. Viktorova, V.S., Lubkov, N.V., Stepanyants, A.S.: Software for testability analysis of aircraft functional systems: certificate of state registration of computer programs 2017660269 RF; dated 20 September 2017
13. Viktorova, V.S., Stepanyants, A.S.: Analysis of the trend of operational availability of aircraft systems: certificate of state registration of computer programs 2019618035 RF; dated 26 June 2019

Random Graph Node Classification by Extremal Index of PageRank

Natalia M. Markovich$^{(\boxtimes)}$ and Maxim S. Ryzhov

V. A. Trapeznikov Institute of Control Sciences Russian Academy of Sciences,
Profsoyuznaya Street 65, 117997 Moscow, Russia
markovic@ipu.rssi.ru

Abstract. Taking account for the graph randomness, our purpose is a node classification by their extremal indexes (EI) as the local dependence measure of node influence characteristics. The EI was calculated by node PageRanks of the local tree related to the node, which is a kind of Thorny Branching Tree (TBT). The blocks estimator was used for the EI estimation by sliding and disjoint block definitions. The classification by the node EI value and the average block size for the local node TBT was introduced for simulated graphs by the Forest Fire and Erdős-Rényi Models and the Berkeley-Stanford dataset as a real example. The new classification methodology is proposed irrespective on the graph structure.

Keywords: PageRank · Extremal index · Bootstrap · Random graph · Sliding block · Disjoint block

1 Introduction

A node classification in the random directed graph, graph with the random number of links between nodes, is a well-known problem with a lot of applications. The node influence characteristic is an important constrain of the statistical analysis. One of the popular node influence is the in-degree, i.e. a number of the ingoing links to a node [1]. The alternative way is PageRank [1]. By Google's definition [2] PageRank is determined as the rank $R(p_i) = R_i$ of node (Web page) p_i by

$$R(p_i) = \sum_{p_j \in N(p_i)} \frac{c}{D_j} R(p_j) + (1-c)q_i, \quad i = \overline{1, v}, \tag{1}$$

where $N(p_i)$ is a set of nodes, in-degree connected with current one p_i, D_j is an out-degree of node p_j (a number of the outgoing links from a node), $c \in (0, 1)$ is a damping factor, $q_i \geq 0$ is a personalization probability of node p_i, $v = |V|$ is a number of nodes in the complete directed graph $G = (V, E)$.

The reported study was partly funded by RFBR, project number 19-01-00090 (recipient N.M. Markovich, conceptualization, mathematical model development, methodology development; recipient M. S. Ryzhov, numerical analysis, validation).

V. M. Vishnevskiy et al. (Eds.): DCCN 2019, CCIS 1141, pp. 424–435, 2019.
https://doi.org/10.1007/978-3-030-36625-4_34

The damping factor is often assumed to be equal to $c = 0.85$ as an average probability to browse a web-page connected with current one [2, 7]. PageRank is a numeric measure of inter-relations between nodes. Thus, PageRank reflects the local network structure.

In previous work [13], one of approaches was statistical clustering according to node extremal indexes (EI). The stationary sequence $\{X_n\}_{n\geq 1}$ with the distribution function $F(x)$ is said to have EI $\theta \in [0, 1]$ if for each $0 < \tau < \infty$ there is a sequence of real numbers $u_n = u_n(\tau)$ such that it holds mixing condition $D(u)$ and

$$\lim_{n\to\infty} n(1 - F(u_n)) = \tau, \tag{2}$$

$$\lim_{n\to\infty} P\{M_n \leq u_n\} = e^{-\tau\theta}, \tag{3}$$

where $M_{k,l} = max\{X_{k+1}, \ldots, X_l\}$ is a maximum of subsequence, $0 \leq k < l$, $M_n = M_{0,n}$ [8]. The EI shows relations between the distributions of extremes and a single random variable [8]

$$P(M_n \leq u_n) = (F(u_n))^{n\theta} + o(1), n \to \infty. \tag{4}$$

For independent random variables the EI is equal to one, other cases indicate about dependence. In our work conclusions were made that the EI describes the dependence of extremes in the graph and shows the ability to attract highly ranked nodes in the orbit of a randomly selected node.

Our main task is to classify nodes of a random directed graph which corresponds to a structure of complex graphs. Common classification methods, as for example Within-Network Classification [5] and PageRank-based classification [3], need a partial initial classification, and they are generally applicable to the determined graphs. On the first side, an assumption can be made that the node EI is a measure of the current node dependence on neighbours in the graph G. On the other hand, the consideration were given in [4] that PageRank in a random graph is an autoregressive process with random coefficients and a random depth of dependence on it. Thus, a local graph structure can be defined as a branching tree with the investigated node as a root, a kind of Thorny Branching Tree (TBT) [6], based on PageRank relations between nodes. That means the node EI value should be found with an attribute sequence $\{X_n\}_{n\geq 1}$ for nodes, belonged to the associated node TBT. According to mentioned above the EI is chosen as an appropriate dependence measure of nodes for the classification.

The paper is organized as follows. In Sect. 2 a theoretical basis of our study and an explanation of chosen assumptions are given. There we propose an adaptation of the blocks estimator of the node EI. In Sects. 3.1 and 3.2 a classification is introduced for a graph from Stanford datasets [15] or modelled by the Forest Fire [9] and Erdős-Rényi [14] models respectively. In Sect. 4 the exposition is finalized by some conclusions.

2   Related Work

In practice, the rank estimation is provided by the iterative formula, which could be used for dangling nodes [6]:

$$R_i^{(0)} = 1, R_i^{(k)} = \sum_{j \to i} \frac{c}{D_j} R_j^{(k-1)} + (1-c)q_i, \quad i = \overline{1, v}, \quad k > 1, \qquad (5)$$

where (k) is an iteration number. After a limited number of the iterations the result sequence will have the same distribution as PageRanks received by (1) [6].

The EI value is empirically calculating by an appropriate estimator. One of the common used estimators is the blocks estimator [11]

$$\widehat{\theta}_{Bl,n}(u) = \frac{n \sum_{i=1}^{l} 1\left(M_{(i-1)r,ir} > u\right)}{rl \sum_{i=0}^{n-1} 1\left(X_i > u\right)}, \qquad (6)$$

where $r = \left[\frac{n}{l}\right]$ is a block size, l is a number of blocks. The calculated EI may be interpreted as a number of the clusters divided with a number of the exceedances, exceeding over a chosen level u [11]:

$$\frac{1}{\theta} \approx the \;\; mean \;\; cluster \;\; size = \frac{A \;\; number \;\; of \;\; the \;\; exceedances}{A \;\; number \;\; of \;\; the \;\; clusters}. \qquad (7)$$

Here a cluster of the exceedances, or a cluster, is a block contained at least one observation over a chosen level u.

Obviously, the blocks estimator can be applied for the graph after the redefinition of the blocks and clusters sequences. Then a block is a group of the graph nodes having in-degree edges with the same parent node. Their PageRanks structured in the blocks are suitable for the further estimations. However, there can be interconnections between nodes within the same or/and other blocks, that is why additional assumptions are needed:

1. *Sliding blocks.* If a node attends in different blocks, its copy should be added in each one.
2. *Disjoint blocks.* Block intersections should be excluded from the blocks. Their unions are interpreted as new blocks.

Examples of the blocks are shown in Fig. 1. In the same way, a cluster is the block of nodes which PageRanks may exceed a predefined threshold. For sliding model block sizes are equal to in-degrees. For disjoint model it becomes more complex, but block sizes stay be random variables. This means that the blocks estimator (6) should be rewritten as

$$\widehat{\theta}_{rBL,n}(u) = \frac{n^* \sum_{j=1}^{l} 1(\widehat{M}_j > u)}{r_{mn}l \sum_{i=0}^{n-1} d_i 1(X_i > u)}, \qquad (8)$$

where $n^* = \sum_{i=1}^{n} d_i$, $d_i \leq D_i$ is a count of blocks which i node belongs to, what is not greater then node out-degree count, $r_{mn} = \frac{1}{l}\sum_{i=1}^{l} r_i$, $\widehat{M}_j$ is a maximum

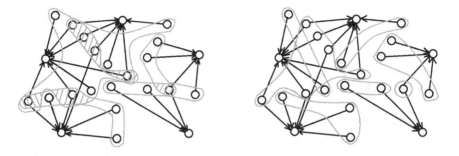

Fig. 1. Examples of the block definition: sliding block (left), disjoint block (right).

from r_j block. Since any EI estimator including the blocks one depends on the threshold parameter u, one can estimate it by bootstrap method as it is proposed in [13].

As result the following algorithm of the node EI estimation is presented:

Algorithm 1

---

1. *Estimate the PR (1) values of each directed graph $G = (V, E)$ node by the method (5).*
2. *For the chosen node p_i receive a node sequence $\{p_{j_k} : k \geq 1, p_{j_1} = p_i\}$ associated with the sequence of the k largest values $\{\epsilon_{j_k}^i\}_{k \geq 1}$, where [12]*

$$
\epsilon_j^i = \begin{cases} 1, & p_i = p_j, \\ \dfrac{c^{k-1} R_j}{\prod_{m=1}^{k-1} D_{j_m} R_i}, & \exists (p_{j_1}, \ldots, p_{j_k}) = min\{(p_{j_1}, \ldots, p_{j_f}): \\ & \forall m = \overline{2, f} \rightarrow (p_{j_{m-1}}, p_{j_m}) \in E, p_{j_1} = p_j, p_{j_k} = p_i\}, \\ 0, & otherwise. \end{cases} \tag{9}
$$

3. *The EI value is empirically calculating by the (8) estimator according to the sliding or disjoint blocks model (Fig. 1). Level u is chosen with the bootstrap method [13].*
4. *Estimating the EI for the enlarging length $k, k+1, \ldots$ node sequence until a stable value is reached.*

Here it should be noticed that the presented definitions of the blocks type 2 are not the only possible ones. The idea naturally arises to break up the nodes into dependent or independent communities. The division into independent community-like blocks were applied like presented in Leskovec work [9]. In order to take account into the intersections, highly cohesive substructures (HCSs) described in the work [10] were applied. The latter were chosen because of the similarity with the concept of the node local graph constructing, a kind of TBT, for the EI determining. The study showed three possible cases: the identical equality of node EI to one, the presence of only one community or communities consisted of one node, which makes such definition unsuitable in our case. However, their study have an interest in the case of an undirected graph, which is the goal for the future work.

3   Node Classification

3.1   Simulation Study

First, a set of graphs was modeled by the Forest Fire Model [9], in which edges are added via a recursive "burning" mechanism: adding and removing of edges are determined by geometrical distributions with the parameters p_α and p_β. Next, for each node a subgraph related to the whole network that is a kind of Thorny Branching Tree (TBT) [6] was found [12]. The EI was estimated by (8) for each node with the sequence of PageRanks (5) provided by the local TBT. The block configuration is defined by the sliding and disjoint models. All of these steps are described in the Algorithm 1.

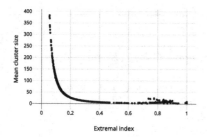 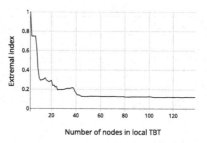

Fig. 2. Example of the mean cluster size (7) against the node EI calculated by (8) (left). Example of the node EI calculated by (8) against the number of nodes in its TBT (right).

The condition (7) is fulfilled for each modeled graph as in Fig. 2 (left), which allowed our results to be theory-based. The EI value is selected corresponding to the stability interval of the plot in Fig. 2 (right).

Next part of our investigation was an EI comparison with other node and graph properties. The EI value relates to the graph structure. In Fig. 3 one can see similar trends reflecting an inverse proportionality between considered properties. This approves that the EI describes a local node structure and a dependence between nodes. Received results are similar for both disjoint and sliding blocks models in Fig. 3 (the first raw). However, the dependence of the EI and the number of nodes in the local TBT is different for mentioned models, see Fig. 3 (the second raw). The latter is caused by the intersections counting. In case of intersected blocks, a node from intersections is related to the new block to apply the disjoint block estimator. The different value of nodes in the local TBT is arisen due to different stability intervals, as in Fig. 2 (right) for

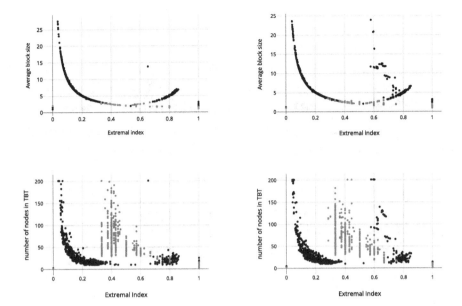

Fig. 3. The average block size (the first raw) and the number of nodes (the second raw) in the local TBT versus the node EI calculated by (8) for two Forest Fire Model graphs with 1000 nodes: $p_\alpha = 0.3$ and $p_\beta = 0.8$ (left column), $p_\alpha = 0.8$ and $p_\beta = 0.3$ (right column). The black and grey dotes show the sliding and disjoint blocks models, respectively.

different block definitions. According to the sliding blocks model, the small EI value corresponds to a large number of nodes in the TBT. Nodes with the same EI value may have different PageRanks and in-degrees (Fig. 4). Important is that sliding models allow separating into classes better than the disjoint model.

Same results were provided for the Erdős–Rényi random graph model [14]. On this step we decided to exclude the disjoint block model from the investigation due to higher probability of node partition on the one node-sized blocks (see, Fig. 5 for sliding block model).

According to the better definition on large graph scales the following nodes classes can be introduced, see Fig. 6, where $E_{TBT}(N_i) = \frac{n^*}{n}$ is the average block size in the local TBT:

1. $\theta \leq 0.5, \theta \approx \frac{1}{E_{TBT}(N_i)} = \frac{n}{n^*}$ corresponds to the strong local dependence and a large number of nodes in a block.
2. $\theta > 0.5, \theta \approx \frac{1}{E_{TBT}(N_i)} = \frac{n}{n^*}$ corresponds to the weak local dependence and a small number of nodes in a block.
3. $\theta > 0.5, \theta \approx 1 - \frac{1}{E_{TBT}(N_i)} = 1 - \frac{n}{n^*}$ corresponds to the weak local dependence and a large number of nodes in a block.
4. $\theta \leq 0.5, \theta \approx 1 - \frac{1}{E_{TBT}(N_i)} = 1 - \frac{n}{n^*}$ corresponds to the strong local dependence and a small number of nodes in a block (the case was not found in the modelled graphs).

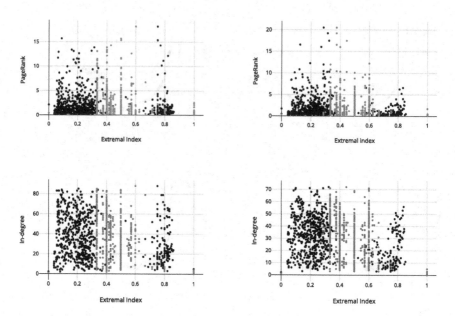

Fig. 4. The node PageRank (the first raw) and the node in-degree (the second raw) versus the node EI calculated by (8) for two Forest Fire Model graphs with 1000 nodes: $p_\alpha = 0.3$ and $p_\beta = 0.8$ (left column), $p_\alpha = 0.8$ and $p_\beta = 0.3$ (right column). The black and grey dotes show the sliding and disjoint blocks models, respectively.

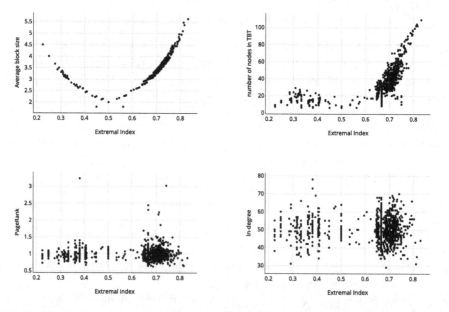

Fig. 5. The average block size (left top), the number of nodes (right top), the node PageRank (left bottom), the node in-degree (right bottom) versus the node EI calculated by Algorithm 1 and the sliding block model for $G(n, p)$ model graph, $p = 0.1$ with 1000 nodes.

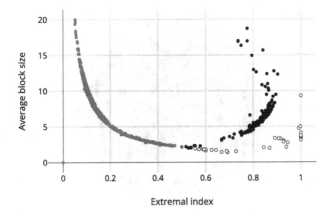

Fig. 6. Example of the node classification for the sliding blocks model for the Fire Forests Model graph with 10000 nodes, $p_\alpha = 0.3$ and $p_\beta = 0.8$. The grey, black and white nodes belong to the first, the second and the third classes, respectively.

On Figs. 7, 8 and 9 examples of the node classification are given according to above mentioned classes. It can be seen that there is a dependence of the distribution of the nodes classes and the graph sparsity: the first and second classes of nodes prevail in the more sparse graphs, otherwise the third class prevails. This means that knowledge of the nodes EI can describe the graph structure.

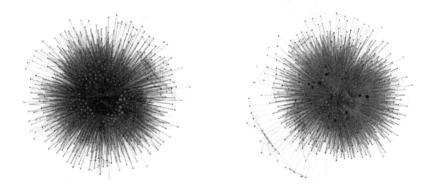

Fig. 7. Example of the node classification for two Forest Fire Model graphs with 1000 nodes: $p_\alpha = 0.3$ and $p_\beta = 0.8$ (left), $p_\alpha = 0.8$ and $p_\beta = 0.3$ (right). The grey, white and black nodes belong to the first, the second and the third class, respectively. The node sizes are proportional to their PageRanks.

Fig. 8. Example of the node classification for $G(n,p)$ model graph with 1000 nodes, $p = 0.1$. The grey, white and black nodes belong to the first, the second and the third class, respectively.

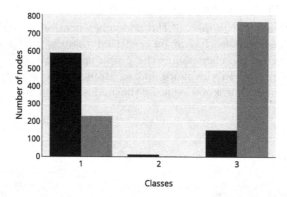

Fig. 9. The node classification for $G(n,p)$ model, $p = 0.1$ (grey), and Forest Fire Model, $p_\alpha = 0.3$ and $p_\beta = 0.8$ (black), graphs with 1000 nodes.

3.2   Web Graph Dataset

The proposed classification method is verified on a real network data. For this purpose we chose Berkeley-Stanford web graph dataset (685,230 nodes and 7,600,595 edges) [15]. Since the nodes EIs calculation becomes more time-spending due to the graph size, the decision was made to obtain sets of nodes independently for the Berkeley and Stanford subgraphs by random walk with a random transition [16]. The results are visible on Figs. 10 and 11.

The classification shows that both subgraphs obtained by random walk are sparse: the nodes from the first two classes are most often found. It is also seen that the neighbour nodes, if they belong to a strong-connected community, have close EI values and belong to the same class. This affects the fact that EI can be used for random graph node clustering.

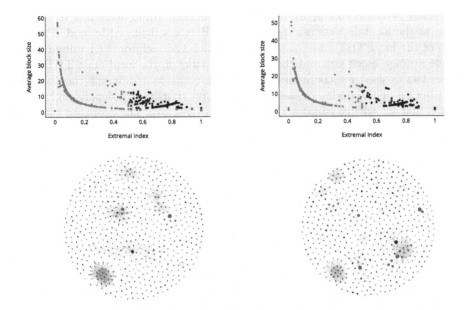

Fig. 10. The average block size versus the node EI (the first raw) calculated by Algorithm 1 and the sliding block model, the node classification (the second raw) for Berkeley (left column) and Stanford (right column) subgraphs. The grey, white and black nodes belong to the first, the second and the third class, respectively. The nodes were received by the random walk process.

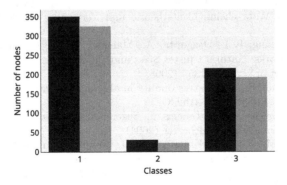

Fig. 11. The node classification for Berkeley (grey) and Stanford (black) subgraphs, $n = 1000$.

4   Conclusion

Our main task was to classify nodes of the random graph according to the EI value as the local dependence measure. The EI was calculated by node PageRanks of the local TBT tree related to the node. The reciprocal EI value relates to the average block size and the number of nodes in the local node TBT. The latter fact is used for the nodes partition into the four different classes. The distribution of the nodes classes and the graph sparsity are dependent, which means that knowledge can describe the graph structure. The proposed tool is a novel classification method.

References

1. Fortunato, S., Boguñá, M., Flammini, A., Menczer, F.: Approximating pagerank from in-degree. In: Aiello, W., Broder, A., Janssen, J., Milios, E. (eds.) WAW 2006. LNCS, vol. 4936, pp. 59–71. Springer, Heidelberg (2008). https://doi.org/10.1007/978-3-540-78808-9_6

2. Brin, S., Page, L.: The anatomy of a large-scale hypertextual web search engine. Computer Networks and ISDN Systems, pp. 107–117 (1998)

3. Avrachenkov, K., Gonçalves, P., Legout, A., Sokol, M.: Graph based classification of content and users in BitTorrent. In: Proceeding of NIPS Big Learning Workshop, December 2011

4. Markovich, N.M.: Extremes in Random Graphs Models of Complex Networks, 5 April 2017. arXiv:1704.01302v1 [math.ST]

5. Desrosiers, C., Karypis, G.: Within-network classification using local structure similarity. In: Buntine, W., Grobelnik, M., Mladenić, D., Shawe-Taylor, J. (eds.) ECML PKDD 2009. LNCS (LNAI), vol. 5781, pp. 260–275. Springer, Heidelberg (2009). https://doi.org/10.1007/978-3-642-04180-8_34

6. Chen, N., Litvak, N., Olvera-Cravioto, M.: Ranking algorithms on directed configuration networks, 12 October 2014. arXiv:1409.7443v2 [math.PR]

7. Volkovich, Y., Litvak, N.: On the exceedance point process for a stationary sequence. Adv. Appl. Probab. **42**, 577–604 (2010)

8. Leadbetter, M.R.: Extremes and local dependence in stationary sequences. Zeitschrift für Wahrscheinlichkeitstheorie und Verwandte Gebiete, pp. 291–306 (1983)

9. Leskovec, J., Lang, K.J., Dasgupta, A., Mahoney, M.W.: Community Structure in Large Networks: Natural Cluster Sizes and the Absence of Large Well-Defined Clusters, eprint arXiv:0810.1355 (2008)

10. Diestel, R., Oum, S.: Tangle-tree duality in abstract separation systems, 26 April 2018. arXiv:1701.02509v3 [math.CO]

11. Beirlant, J., Goegebeur, Y., Teugels, J., Segers, J.: Statistics of Extremes: Theory and Applications. Wiley, Chichester (2004)

12. Markovich, N.M., Ryzhov, M., Krieger, U.R.: Nonparametric analysis of extremes on web graphs: pagerank versus max-linear model. In: Vishnevskiy, V.M., Samouylov, K.E., Kozyrev, D.V. (eds.) DCCN 2017. CCIS, vol. 700, pp. 13–26. Springer, Cham (2017). https://doi.org/10.1007/978-3-319-66836-9_2

13. Markovich, N.M., Ryzhov, M.S., Krieger, U.R.: Statistical clustering of a random network by extremal properties. In: Vishnevskiy, V.M., Kozyrev, D.V. (eds.) DCCN 2018. CCIS, vol. 919, pp. 71–82. Springer, Cham (2018). https://doi.org/10.1007/978-3-319-99447-5_7

14. Erdős, P., Rényi, A.: On random graphs. Publicationes Mathematicae **6**, 290–297 (1959)
15. Leskovec, J., Krevl, A.: SNAP Datasets: Stanford Large Network Dataset Collection (2014). http://snap.stanford.edu/data
16. Leskovec, J., Faloutsos, C.: Sampling from large graphs (2006). https://doi.org/10.1145/1150402.1150479

Group Polling Method Upon the Independent Activity of Sensors in Unsynchronized Wireless Monitoring Networks

Ivan Tsitovich[(✉)](iD)

Institute for Information Transmission Problems (Kharkevich Institute) RAS,
Bolshoy Karetny per. 19, Build. 1, 127051 Moscow, Russia
cito@iitp.ru

Abstract. We propose a new method of a sensor signal coding for an alarm signalization in large monitoring networks with working independently sensors when sensors do not be synchronized in time. This method bases on the method of group polling for alarming sensors identification for a synchronized network and has similar characteristics of complexity. Our method has more complicated structure than previous ones and includes a special part of the signal named as the starting code. Based on numerical simulations, it is found that the proposed group polling method may be effective for unsynchronized networks with thousands or more sensors and the decoding algorithm may be realized on-time using parallel executions. Recommended length of the starting code is proposed.

Keywords: Wireless sensor network · Sensor for an alarm
signalization · Group polling · Unsynchronized time

1 Introduction

The exponential increase in the number of devices on the Internet of Things (IoT) is leading to an increase in the M2M connections traffic and, consequently, to an exponential increase of the traffic in the wireless data networks that provide these connections. Most IoT devices are devices that keep monitoring and telemetry systems functioning, therefore, it is actual the problem to limit the traffic of such devices.

Such systems are increasingly infiltrating all areas of human activity, also due to the success of microelectronics, which have led to the creation of low-cost sensors that transmit information about the state of objects. Modern wireless sensor networks (WSN) should be able to monitor the state of various systems and objects and allow detecting violations in their functioning in the shortest

Supported by grant AAAA-A19-119022590088-5.

possible time. Here, the natural unit of time is the time of response to an emergency: the time of an active sensor detecting should be the same order.

At the same time, it is necessary to strive to ensure that the cost of operating this kind of WSN was low, while the volume of the WSN can contain thousands or more sensors. The principles of the possible organization and operation of WSN have been actively researched recently (see, for example, [1] and [2]). Methods of information exchange between sensors and the control center (CC) have a significant impact on the physical device of sensors, protocols of information exchange, and on the WSN traffic.

This paper examines the interaction of sensors with the CC via the WSN, where three sensor states are possible: the sensor transmits information about its current state according to a given schedule sharing information with the CC; the sensor reports the occurrence of an emergency on the communication channel which is common to all sensors; the sensor transmits information about the emergency at the request of the CC through the dedicated communication channel. This interaction organization minimizes the WSN no predicted traffic, as only a narrowband channel is reserved for reports of an emergency that is common to all sensors.

If a sensor detects an emergency then it sends the alarm signal. We call such sensor as an active one. It is evident that WSN has a few such sensors. Remaining sensors are no active and so not send a signal. Let us t be a number of sensors in the network and s be a number of active sensors. It is natural to assume that s is small compared to t, because the probability of an emergency in any point of the domain by the WSN is low and a time of an active sensor detection is small. By this reason we investigate asymptotic properties of our method of active sensor detecting when $t \to \infty$ and s is fixed.

This paper examines monitoring and telemetry systems consisting of thousands or more sensors that transmit information about the occurrence of emergencies at different parts of the network. Since sensors in such networks transmit information only in the event of an emergency that is low likely, the methods developed for high-activity sensor systems are not effective, as the time of survey in such systems are proportional to the number of sensors and linearly grows as their number increases. Therefore, the time of an active sensor detecting is very large in WSN servicing millions sensors and it is necessary to develop survey methods based on group polling, where multiple sensors are surveyed simultaneously on the same bandwidth, thereby significantly reducing the time required to detect active sensors.

The mathematical basis for this task is the planning of the screening experiments [3, 4]. In this case a number of necessary experiments for detecting s_0 active elements between t elements is $O(s_0 \log t)$. This is contrast with the traditional methods of polling when a time of active elements detecting is $O(t)$ [5].

In the paper [6], it is proposed the method of group polling for detecting of active sensors in the monitoring network where properties of this method are investigated under the assumption of independent activity of the alarming sensors. It is supposed that the monitoring network is very large and contains thousands of sensors but all sensors synchronize in time their alarm signals. The last demand is difficult for its practical realization. But proposed in [6] method ensures the fulfillment of a short time of alarming sensors detection, i.e. the detection time is $O(\log t)$.

In the papers [7] and [8], it is proposed a generalization of this method onto a case of unsynchronized in time alarming signal sending. It is showed that the group polling method for alarming sensors identification is applicable in this case but its computation complexity is such that it is difficult to use this method for an online detection.

The goal of our paper is to propose a more complex profiles of an alarming sensor signal which give us possibility to detect alarming sensors in time similar to WSN with synchronized by time alarming signal sending and to investigate some properties of this method by numerical modeling.

The formulation of the problem is presented in Sect. 2. In Sect. 3 it is described the algorithm of WSN output signal modeling when alarming sensors begin their signals in random time moments and take into account digitization in time of the output signal.

In Sect. 4 we examine our new element of an alarming sensor profile that is named as the starting point of the alarming signal. In the next sections we present the calculated results with variations in such model parameters as the threshold level of accepting a sensor as an active one L_0, the number of sensors in network t, the parameter of noise intensity in communication channel α, the real value of active sensors s_r in contrast with supposed in the algorithm value of active sensors s_0. We examine the characteristics P—the detection of redundant alarming sensors is not an error and $\bar{s}$—the mean number of identified passive sensors as active ones.

Finally, we present conclusions and recommendations on the group polling of sensors.

2   Setting of the Problem

We will follow, in general, the notations of [7]. We develop a polling strategy aimed at the fastest identification of s sensors that are ready for data transmission and named as alarming sensors. We assume that $s \ll t$ (a relatively small number of the active sensors in the network). Such scenario is typical for the sensor network that is located at a relatively large area where the probability of local emergency is very low.

The i-th alarming sensor begins to send the signal in a random time u_i, $i = 1, \ldots, s$, and is sending it during a time U. After the time $u_i + U$ it may repeat the signal by a special scenario while an operator of CC does not identify its and switches its working state. After this moment the sensor does not alarm. Therefore, the number of simultaneously alarming sensors is no less then the number of sensors with u_i such that $|u_i - u_j| \leq O(U)$ and may be small.

The alarming signal is dropped onto short time intervals with the length Δ, in every of this intervals the sensor may send or not send the special signal.

This interprets as sending 1 or 0, respectively. The generated sequence of symbols 0 or 1 is named as the profile of this sensor alarming signal.

For the alarming signal creating every sensor has its unique code $\mathbf{a} = (a^1, \ldots, a^N)$, $a^i = 0$ or 1, and for the i-th sensor its code is denoted by $\mathbf{a}_i$, N is the code length. The first question is how to construct the vectors $\mathbf{a}_i, i = 1, \ldots, t$. According with [3], it is proposed the procedure for the Boolean matrix $\mathbf{A} = (\mathbf{a}_i, i = 1, \ldots, t)$, $\mathbf{a}_i \neq \mathbf{a}_j$ if $i \neq j$, with near to optimal properties, where a_i^j are independent random numbers 0 or 1 with a proper probability p^0 (see (5)) for 1 in the matrix for maximizing the channel capacity.

Using a small time of alarming signal we must identify the active sensors in such a way that the mean probability of the false identification of one sensor does not exceed a predetermined level (the averaging is performed with respect to a priori uniform distribution of alarming sensors on the set $T = \{1, \ldots, t\}$ denoted by $\mathbf{P}$). We use two criteria for an admissibility of alarming sensors identification denoted by values P_1 and $\bar{s}$: P_1 is the probability to loss an active sensor and $\bar{s}$ is the mean number of identified passive sensors as active ones.

The mathematical model of the sensors activity may be described as follow. The state of t sensors to be alarming or not is described using variables $x_1, \ldots, x_t$ that can be 0 or 1 (a passive sensor, which is not alarming, and an active sensor, which is alarming, respectively). The variables with numbers $i_1, \ldots, i_s$ are unities, and the remaining variables are zeros. Let S be the ordered set of alarming sensors, i.e. $S = \{i_1, \ldots, i_s\}$. Let $\hat{S}$ be the set of identified sensors then $P_1 = 1 - \mathbf{P}(\hat{S} \supseteq S)$.

In the group polling, we simultaneously receive signals from several sensors. The j-th column of $\mathbf{A}$ gives us the group of sensors involving in the j-th polling (if $a_i^j = 1(0)$ then the i-th sensor is (is not) involved in the polling). When the group contains at least one alarming sensor, we receive the signal that is interpreted as 1. If the group does not contain alarming sensors, we do not receive signals and the result is 0. Thus, response of the sensors of the j-th group is represented as

$$f_j = (a_1^j \wedge x_1) \vee \cdots \vee (a_t^j \wedge x_t),$$

where $\wedge$ is the Boolean product and $\vee$ is the Boolean sum.

We assume that data transmission errors are possible in the network. This means that the value f_j is known with a certain error. In each polling session, the result is distorted regardless of the remaining polling sessions in accordance with the stochastic transition matrix

$$\mathbf{W} = \begin{pmatrix} 1 - \alpha_0, & \alpha_0 \\ \alpha_1, & 1 - \alpha_1 \end{pmatrix} \tag{1}$$

where α_0 is the probability of false zero (i.e., detection of 1 instead of 0 as the output signal) and α_1 is the probability of false unity (i.e., detection of 0 instead of 1). Therefore, the result of the j-th polling is $\hat{f}_j$, which is 0 or 1 in accordance with matrix $\mathbf{W}$ regardless of the results in the remaining sessions provided that the values of f_j are fixed.

3   Profiles for Signals of Active Sensors

The code $\mathbf{a}_i$ generates a profile for a signal of the i-th sensor when this sensor is active by the following way. The profile consists of two part; every part has three portions.

Let us begin with the first part. Let us suppose that the i-th sensors begins to be active at time u_i. Therefore, on the first portion the sensor is passive, i.e.

$$a_i(u) = 0, \text{ if } u < u_i.$$

On the second potion the sensor is sending a special signal

$$a_i(u) = 1, \text{ if } u_i \leq u < u_i + L\Delta, \tag{2}$$

where L is one of parameters for the profile. This is a new element in the profile in contrast with [7].

On the third portion we followe in general to the profile from [7] by the following

$$a_i(u) = \begin{cases} 1, & \text{if } a_i^0 = 0 \text{ for } u_i + \Delta L \leq u < u_i + \Delta(L+1) \text{ and} \\ 0 & \text{for } u_i + (L+1)\Delta \leq u < u_i + (L+k+1)\Delta, \\ a_i^j, & \text{if } a_i^j = a_i^{j-1} \text{ for the next time interval } k\Delta, \\ a_i^{j-1}, & \text{if } a_i^j \neq a_i^{j-1} \text{ for the next time interval } \Delta \text{ and} \\ a_i^j & \text{for the next time interval } (k+1)\Delta, \\ & \text{and so on,} \end{cases} \tag{3}$$

where $k, k > 1$, is a number of repeating of code symbols and is a parameter of the profile. Therefore, the length of the third portion depends on the vector $\mathbf{a}_i$ and is denoted by $n_i\Delta$ and n_i is a random value with the mean $k + p^0(1 - p^0)$.

On the second part the first portion of the profile is as for the first part but it has a random length v with the exponential distribution with a parameter $\lambda = O(n_i\Delta)$. This pause gives us a possibility to limit an ifluence of another active sensors on detecting this sensor if it is active.

After them the sensor repeats the portios 2 and 3 of the first part and does not send a signal late.

Thus, we get response of the sensors at an arbitrary time is represented as

$$f(u) = (a_1(u) \wedge x_1)\vee, \dots, \vee(a_t(u) \wedge x_t). \tag{4}$$

The result of output continuous signal is dropped onto short time intervals with the length Δ. Therefore, the continuous function $f(u)$ drops onto the group of observations $(f_1, f_2 \dots)$, that can be 0, 1 or nil.

The value nil we introduce as in [7] by the following reason. We observe 0 value of the function $f(u)$ for a time interval of digitization in the situation of absence of signals during the time from the interval, that means $f(u) \equiv 0$ on the interval. We observe 1 value of the function $f(u)$ at the interval if $f(u) \equiv 1$ on the interval. When the function $f(u)$ at the interval changes its value (from

0 to 1, or 1 to 0), i.e. $\exists u_1, u_2 : f(u_1) \neq f(u_2)$, we cannot interpret the received signal correctly. This situation is a conflict and we should mark such signals as nil. Thus, for the j-th Δ-interval $f(u)$ from (4) transforms to

$$
f_j = \begin{cases} 1, & \text{if } f(u) \equiv 1, \\ 0, & \text{if } f(u) \equiv 0, \\ nil, & \text{if } \exists u_1, u_2 : f(u_1) \neq f(u_2). \end{cases}
$$

This conflicts is determined by the functions $a_i(u)$ for alarming sensors also. By the same way we construct discrete values $\hat{a}_i(j)$ that can be also 0, 1 or nil (i is the number of the sensor and j is the number of the corresponding time interval). The value

$$
N_i = 2(L + n_i) + [v/\Delta]
$$

is the length of the i-th sensor profile after its digitization.

Then the result of an output signal for the j-th interval is

$$
f_j = (\hat{a}_1(j) \wedge x_1) \vee, \ldots, \vee (\hat{a}_t(j) \wedge x_t),
$$

providing that $1 \vee nil \equiv 1$ and $0 \vee nil \equiv nil$.

Finally, results 0 or 1 of f_j we transform in accordance with the matrix $\mathbf{W}$ and we get the vector of observation $\hat{\mathbf{f}}$. If $f_j = nil$ then $\hat{f}_j$ equals 0 or 1 with probability 0.5 and the observations $f_j = nil$ cannot help for an alarming sensor detection. Therefore, we lose a part of information in contrast with the case from [6] and need to have longer sensors codes for an alarming sensor identification with the same quality. By this reason, we use p^0 such as

$$
p^0 = \frac{2 + k - 4p_0 - \sqrt{(2+k)^2 - 16(1+k)p_0(1-p_0)}}{4 - 8p_0} \tag{5}
$$

for $p_0 < 0.5$ and $p^0 = 0.5$ for $p_0 = 0.5$, where

$$
p_0 = 1 - \sqrt[s_0]{\frac{\frac{1}{2} - \alpha_0}{1 - \alpha_0 - \alpha_1}} \tag{6}
$$

and s_0 is a supposing number of active sensors.

In [8] it was proposed to use $k = 3$ and we use this value of k in all subsequence numerical simulations for the algorithm properties investigation.

4   The Algorithm of Active Sensors Detecting

4.1   Start Points Detecting

As followed from a sensor profile (see (2)), it begins with the sequence of L symbols 1. This sequence in the profile of an active sensor generates outputs $f_j = 1, j = 1, \ldots, L$. Such as we suppose that $\alpha_i, i = 0, 1$, are small then it does not exceed $\alpha_1^2 O(L^2)$ the probability that the corresponding sequence

$\hat{f}_j, j = 1, \ldots, L$, has more then one 0. Therefore, if we observe in the sequence $\hat{\mathbf{f}}$ a subsequence of L consecutive elements which contains no more then one 0 then we suppose that this subsequence is generated by starting symbols of an active sensor profile. The first index of this subsequence in $\hat{\mathbf{f}}$ is named as a start point of the time when an active sensor begins to send its signal. By this way we have the sequence start points $j_1, \ldots, j_{m'}$ based on the vector of observation $\hat{\mathbf{f}}$.

As it is followed from the next subsection, a number of such start points needs be a small as possible but real start points must not be pass. Therefore we need to choose a suitable value of L which is not so small that we find many wrong start points and is not so big that we loss real start points. Based on numerical simulations we propose L when start points are founded as described above.

In Table 1 we present results of numerical simulation the model of WSN with parameters as in [8] with the new signal profiles. In the table s_0 is a number of active sensors, t is the number of sensors in the network, L is a code length of the second portion of a profile with symbols 1, M is the mean value of the number of detected starting points m', σ is its standard deviation, and P is the probability to detect not all alarming sensors.

Table 1. Mean and standard deviation of number of starting points

s_0	t	L	M	σ	P	s_0	t	L	M	σ	P
2	1000	10	67.8	22.7	0.015	3	1000	18	63.2	19.2	0.136
2	1000	12	48.2	18.0	0.019	3	1000	20	58.0	17.9	0.146
2	1000	14	37.9	13.8	0.017	3	1000	25	52.3	17.3	0.174
2	1000	16	32.8	12.5	0.019	3	2000	18	64.8	20.7	0.059
2	1000	18	29.6	10.6	0.018	2	10000	18	32.6	11.8	0.034
2	2000	18	30.1	11.0	0.005	2	100000	18	35.3	12.8	0.256

It is followed from the results in Table 1 that for $L = 18$ we have stable values of P and M for all t if $s_0 = 2$. Therefore we use this value of L for simulations in the next subsection where we investigate properties of our algorithm for active sensors detection. If L is less than 18 then M is significantly reduced with increasing L. If $s_0 = 3$ then the growth of the magnitude of L does not increase the accuracy of start points detection. To address this circumstance in the proposed method of coding sensor signal transmission retry is provided with pause a random duration on the assumption that part of active sensors to the re-transmission time is detected, thus reducing the magnitude of s_0.

The computational complexity of the sensor detecting method from [7,8] for the i-th sensor is $O(N_i^3)$ per one sensor in contrast with [6], where it is $O(N)$. For our method (see the next section) the computational complexity is $O(m' N_i)$. The main result followed from Table 1 consists in that the computational complexity of our method grows approximately in 30 times in comparison with synchronized WSN in contrast with the method from [7,8] where the complexity is approximately in 10^6 times greater than for the synchronized case such as $N_i \approx 10^3$.

For more stable detection of real start points we add to the set

$$J = \{j_1, \ldots, j_{m'}\} \; \bigcup_{i=1,\ldots,m'} \{j_i - 1\} \; \bigcup_{i=1,\ldots,m'} \{j_i + 1\}, \tag{7}$$

and form by this way the final set of start points $j_1, \ldots, j_m$. Therefore the probability to miss a start point does not exceed $\alpha_1^2 O(L^2)$.

Such as generated by the start symbols of an active sensor profile observations are not informative for active sensors detection it is necessary to exclude this elements in $\hat{\mathbf{f}}$ in an algorithm of active sensors detection in contrast with the algorithms in [7,8], and [6]. Therefore, we form the set of indexes

$$J_1 = \bigcup_{j_1,\ldots,j_m} \{j_i, j_i + 1, \ldots, j_i + L - 1\}$$

as the set of no informative observations and its complement J_0.

4.2   Active Sensors Detecting

The algorithm of active sensors detecting has the following steps.

(i) The algorithm is described by the input parameters: t is the number of sensors in WSN, s_0 is the supposed number of active sensors in WSN, α_0 is the parameters of the noise in the WSN channel (1) ($\alpha_1 = \alpha_0$), N_i is the length of the i-th sensor alarm signal profile (3), $i = 1, \ldots, t$, L is the length the starting code of the alarm signal profile (2), k is the number of repeating of code symbols (3), m is the number of elements in the set J (7), L_0 is the threshold for accepting a sensor as the active one and is determined later (8).

To simplify the identification procedure, we make a decision on the activity of the given sensor using the factor analysis with the aid of the maximum likelihood method of [3].

(ii) The decision on the activity of the sensor is based on the following data. For any start point j_k and any sensor i for the window of observations $[j_k, j_k + l], l \leq n_i$, we calculate: $x_{00}(i, l)$ is the number of observations in which the i-th sensor is not interrogated, i.e. $a_i^j = 0$, the polling result is $\hat{f}_j = 0$, and $j \in J_0$; $x_{10}(i, l)$ is the number of observations in which $a_i^j = 0$, $\hat{f}_j = 1$, and $j \in J_0$; $x_{01}(i, l)$ is the number of observations in which $a_i^j = 0$, the polling result is $\hat{f}_j = 0$, $j \in J_0$; and finally $x_{11}(i, l)$ is the number of observations in which $a_i^j = 1$, $\hat{f}_j = 1$, $j \in J_0$.

We calculate

$$L_i(l) = a_{00}x_{00}(i, l) + a_{10}x_{10}(i, l) + a_{01}x_{01}(i, l) + a_{11}x_{11}(i, l),$$

where

$$a_{00} = \log \frac{1 - \alpha_0 - \hat{p}(1 - \alpha_0 - \alpha_1)}{1 - \alpha_0 - p^*(1 - \alpha_0 - \alpha_1)},$$

$$a_{01} = \log \frac{\hat{p}(1 - \alpha_0 - \alpha_1) + \alpha_0}{p^*(1 - \alpha_0 - \alpha_1) + \alpha_0},$$

$$a_{10} = \log \frac{\alpha_1}{1 - \alpha_0 - p^*(1 - \alpha_0 - \alpha_1)},$$

$$a_{11} = \log \frac{1 - \alpha_1}{\alpha_0 + p^*(1 - \alpha_0 - \alpha_1)},$$

$$\hat{p} = 1 - (1 - p_0)^{s_0 - 1}, \quad p^* = 1 - (1 - p_0)^{s_0}.$$

For justification of the values a_{ij} see [6].

(iii) We compare logarithm of the likelihood ratio $L_i(l)$ with the threshold L_0 for each l, $l \leq n_i$. If it exists l such that $L_i(l)$ is greater than the threshold level, we conclude that the i-th sensor is active and must be ready for the transmission of data of emergency.

(iv) Based on all start points we get quantities $\hat{i}_1, \ldots, \hat{i}_{\hat{s}}$ as the output parameters of the algorithm, where $\hat{s}$ is the number of identified active sensors and $\hat{i}_1, \ldots, \hat{i}_{\hat{s}}$ are the numbers of identified sensors.

(v) The quality of algorithm is characterized by the probability of correct identification of active sensors. Let $\hat{S} = \{\hat{i}_1, \ldots, \hat{i}_{\hat{s}}\}$ then P_1 is the probability of missing of active sensor, so that $S \subseteq \hat{S}$ (the detection of redundant active sensors is not an error. The probabilities are calculated on the assumption that a uniform distribution is determined on the set of the possible values of S.

5   Numerical Results of the Method Effectiveness Simulating

In this section we outline the results of computation modeling for different input parameters of the network model.

5.1   Determining of the Parameter L_0

We begin with determining the main parameter

$$L_0 = cCN, \tag{8}$$

where

$$C = 1 - p^0 h(\alpha_0) - (1 - p^0) h(\alpha_1)$$

is the capacity of the communication channal,

$$h(\alpha) = -(\alpha \log(\alpha) + (1 - \alpha) \log(1 - \alpha)), 0 < \alpha < 1.$$

Based on the result of [6] we need to find the value c which does not depend on t and C.

It is obvious that with the increase in the threshold of L_0 it is increased the probability of losing active sensors, and at its small values it is increased the number of mistakenly identified sensors as active. Therefore, it is necessary to

Table 2. Accuracy of the algorithm for active sensors detecting in depending of L_0 when $\alpha_0 = \alpha_1 = 0.01$, $t = 1000$, $s_0 = s_r = 2$, $N = 126$, $L = 18$

c	P_1	P_2	L_0
6.3	0.0001	323.9	121.6
6.6	0.0003	282.3	127.4
6.9	0.0014	188.5	133.2
7.2	0.0037	47.3	139.0
7.5	0.0083	3.0	144.8
7.8	0.0177	0.1	150.6

determine the value of L_0, at which errors of the first and second types will be acceptable. The Table 2 provides numerical studies of the impact of the L_0 on the quality of the algorithm. The acceptable value of c is 7.5.

We get a very short interval of acceptable values for c. It is necessary to increase the value N for a more wide interval of acceptable values for c.

This value c will be used in subsequent studies of the impact of other parameters of the algorithm on the quality of its work.

5.2   Numerical Simulations of Active Sensors Detecting in Depending of the Number of Sensors t

In Table 3 we outline results of the numerical simulations of active sensors detecting in depending of the number of sensors t. It is followed from the Table that the accuracy of the algorithm increases with the number of sensors. This effect was outlined and analysed in [6]. If for $t = 1000$ the profile has the length 2520 or 2.5 per one sensor but $P_1 \approx 0.008$, for $t = 100000$ the profile has the length 4320 or 0.0432 per one sensor but $P_1 \approx 0.0004$. Therefore, the method effectiveness grows with increasing of t and proposed method is effective for networks with thousands or million sensors since the length of sensor profile grows logarithmical of t.

Table 3. Accuracy of the algorithm for active sensors detecting in depending of t when $\alpha_0 = \alpha_1 = 0.01$, $c = 7.5$, $s_0 = s_r = 2$, $L = 18$

t	P_1	$\overline{s}$	L_0	N
1000	0.0083	3	144.8	126
5000	0.0026	9	179.3	156
10000	0.0023	1	193.0	168
25000	0.0012	2	213.7	186
100000	0.0004	0	248.2	216

5.3   Numerical Simulations of Active Sensors Detecting in Depending of the Noise Intensity α

In this section it is explored the stability of the algorithm of active sensors search to the amount of noise in the CC communication channel. A WSN with 100000 sensors has been selected for certainty, of which only 2 are active. The increase in noise values has little impact on the growth of the sensor profile length, and the characteristics of the algorithm only improve. This paradox is because the required number of observations for finding active sensors is redundant and as noise intensity increases the redundancy increases.

Table 4. Accuracy of the algorithm for active sensors detecting in depending of α_0 when $\alpha_1 = \alpha_0$, $t = 100000$, $c = 7.5$, $s_0 = s_r = 2$, $L = 18$

α_0	P_1	$\bar{s}$	N	m	σ
0.01	0.0002	0	216	38	14.9
0.02	0.0003	0	228	111	38.3
0.05	0.0000	0	276	105	36.3
0.10	0.0038	0	372	103	34.7

At the same time, for $\alpha_0 = 0.1$ there is a deterioration in the properties of the algorithm: the probability of missing the active sensor increases. This effect is because it is increased the probability of missing a start point of the signal transmission by an active sensor. The latter is because it is permissible to distort only one signal when transmitting a starting combination of signals; at $\alpha_0 = 0.1$ the probability of such an event is approximately 0.03. The fact that the P_1 is an order of magnitude less is provided by duplication of signal transmission.

The results of Table 4 show the sustainability of the proposed method of detecting active sensors to interference in the communication channel.

5.4   Numerical Simulations of Active Sensors Detecting in Depending of Active Sensors Real Value s_r

This section looks at a situation where assumptions about the state of the WSN network do not match the actual ones, meaning the number of suspected active sensors s_0 does not match the actual ones s_r. If $s_r < s_0$, as the first line of the Table 5 shows, the algorithm for detecting of active sensors gives a good result. This is because the number of observations significantly exceeds the number of observations required for this, although observations are not as informative as for the case $s_r = s_0$ because the choice of p_0 is incorrect: in (5), (6) we use s_0 instead of s_r.

In the case $s_r > s_0$ the situation changes: the probability of missing active sensors increases dramatically. This is because the required number of observations is greater such as $N = s_0 \log(t)$ instead of requirement $N = s_r \log(t)$.

Table 5. Accuracy of the algorithm for active sensors detecting in depending of s_r when $\alpha_0 = \alpha_1 = 0.01$, $t = 100000$, $c = 7.5$, $s_0 = 2$, $L = 18$

s_r	P_1	$\bar{s}$	m	σ
1	0.0000	0	12	5.5
2	0.0004	0	38	14.9
3	0.4325	0	344	76.8
4	0.7783	0	507	63.3

In addition, the observations themselves are not informative due to the incorrect choice of the magnitude of p_0. However, a certain number of active sensors are detected: for the case $s_r = 3$ the average number of detected active sensors is at least 1.7, and for $s_r = 4$ it is at least 0.8. Since the signal transmission algorithm involves the re-independent transmission of the signal after a random time, during this time the detected sensors will be switched to the transmission of information about the emergency and will not transmit the alarm signal, then the new value s_r may be valid for detecting the remaining active sensors.

However, this situation requires more researches to ensure greater stability of the proposed algorithm.

6   Conclusion

The main result consists in that for unsynchronized in time WSN may be constructed a group polling with $O(\log t)$ time for alarming sensors detection. The constant before $\log t$ is more that the analogous constant for the synchronized WSN because we loss same information per one observation.

Computational complexity is growing no so much as in the case of [8]. Therefore, our group polling method has near to [6] parameters of effectiveness and complexity.

Recommended in [8] number of the code signal repetitions k is valid for our method.

The decoding procedure is more complicated for no synchronized WSN and requires additional investigations for WSN with dependent alarming sensors as, for example, it is outlined in [9] for more accurate choose of algorithm parameters.

References

1. Dargie, W., Poellabauer, C.: Fundamentals of Wireless Sensor Networks: Theory and Practice. Wiley, Hoboken (2010)
2. Sohraby, K., Minoli, D., Znati, T.: Wireless Sensor Networks: Technology, Protocols, and Applications. Wiley, Hoboken (2007)
3. Malyutov, M.B.: Mathematical models and results in the theory of screening experiments. In: Malyutov, M.B. (ed.) Theoretical Problems of Experimental Design, pp. 5–69. Sov. Radio, Moscow (1983). (In Russian)

4. Malyutov, M.: Search for sparse active inputs: a review. In: Aydinian, H., Cicalese, F., Deppe, C. (eds.) Information Theory, Combinatorics, and Search Theory. LNCS, vol. 7777, pp. 609–647. Springer, Heidelberg (2013). https://doi.org/10.1007/978-3-642-36899-8_31

5. Vishnevsky, V., Semenova, O.: Polling Systems: Theory and Applications for Broadband Wireless Networks. LAMBERT Academic Publishing, London (2012)

6. Malikova, E.E., Tsitovich, I.I.: Group polling upon the independent activity of sensors in the monitoring networks. J. Commun. Technol. Electron. **56**, 1556–1563 (2011)

7. Tsitovich, I.I., Shtokhov, A.N.: Method of group polling upon the independent activity of sensors in nonsynchronized monitoring networks. Inf. Process. **16**, 237–245 (2016)

8. Shtokhov, A., Tsitovich, I., Poryazov, S.: On the method of group polling upon the independent activity of sensors in unsynchronized wireless monitoring networks. In: Vishnevskiy, V.M., Samouylov, K.E., Kozyrev, D.V. (eds.) DCCN 2016. CCIS, vol. 678, pp. 266–278. Springer, Cham (2016). https://doi.org/10.1007/978-3-319-51917-3_24

9. Malikova, E.E., Tsitovich, I.I.: Analysis of the efficiency of group polling upon the dependent activity of sensors in the monitoring networks. J. Commun. Technol. Electron. **56**, 1552–1555 (2011)

On Implied Volatility Surface Construction for Stochastic Investment Models

S. G. Shorokhov[(✉)] and M. B. Fomin

RUDN University, 6 Miklukho-Maklaya Street, Moscow 117198, Russian Federation
{shorokhov-sg,fomin-mb}@rudn.ru
http://www.rudn.ru

Abstract. We study the problem of implied volatility surface construction when asset prices are determined by a stochastic model, different from Black-Scholes constant volatility model. Implied volatility of a European call option is determined using Nesterov-Nemirovsky version of damped Newton's method or Levenberg-Marquardt method. Initial approximation for implied volatility is given by Brenner-Subrahmanyam formula. The suggested algorithm for construction of implied volatility surface is implemented in Python using NumPy, SciPy and Matplotlib packages. The implementation is used for construction of implied volatility surfaces for option prices in shifted-lognormal, Cox-Ross and hyperbolic-sine local volatility models.

Keywords: Option pricing · Black-Scholes model · Implied volatility surface · Local volatility model

1 Introduction

Stochastic investment models forecast dynamics of financial asset returns and prices. In continuous time stochastic investment models are usually represented by stochastic differential equations (SDE).

The most significant stochastic model by Black and Scholes (BS model) [2, 20] is given by SDE

$$\frac{dS}{S} = r\,dt + \sigma\,dW, \tag{1}$$

where S is the price at time t of a non-dividend financial asset (stock) in risk-neutral framework, $r > 0$ is the annualized continuously compounded risk-free interest rate, $\sigma > 0$ is the volatility (standard deviation) of the asset returns, W is a standard Wiener process.

In BS model [2] the fair value (price) of a European call option with strike price K and maturity T is equal to

$$c_{BS}\left(S, t, K, T, r, \sigma\right) = S\,\Phi\left(d_{+}\right) - K\,e^{-r\,(T-t)}\,\Phi\left(d_{-}\right), \tag{2}$$

The paper has been prepared with the support of the "RUDN University Program 5-100" (IVS construction methodology). The paper was funded by RFBR, grant No. 16-08-00558 (IVS algorithm implementation).

$$d_\pm = \frac{\ln\left(\frac{S}{K}\right) + \left(r \pm \frac{\sigma^2}{2}\right)(T-t)}{\sigma\sqrt{T-t}}, \; \Phi(x) = \int\limits_{+\infty}^{x} \frac{1}{\sqrt{2\pi}} e^{-\frac{u^2}{2}} \, du.$$

BS model is widely used for derivative pricing, market risk management, credit risk modeling and in many other areas, including telecommunications [18]. But unfortunately BS model fails to explain some important financial market phenomena, including the so-called volatility smiles [11] and fat tails of financial data distributions [23]. Therefore BS model was extended [4,8,9,12] to local volatility models with a more general SDE for the underlying asset price

$$\frac{dS}{S} = r\,dt + \sigma(S,t)\,dW \tag{3}$$

with volatility σ being a function of asset price S and time t.

When European call option price c_E is determined by financial market or some theoretical option pricing model, the implied volatility σ_{imp} of the option [13,15] is the volatility of the underlying asset, for which BS model produces the fair value of the option c_{BS}, equal to the given option price c_E, i.e. implied volatility σ_{imp} of the option is a solution to the following equation

$$c_{BS}(S,t,K,T,r,\sigma_{imp}) = c_E \tag{4}$$

with left-hand side given by (2).

Implied volatility surface (IVS) is the 3d plot of dependence of implied volatility σ_{imp} on strike price K and maturity T with other components of the option fair value in (2) fixed.

Usually IVS is determined from some restricted data set of European call option market prices. We consider the problem of IVS construction for the case when option prices are determined by a local volatility model with extended range of strike prices and maturities.

2   Implied Volatility

2.1   Methods of Implied Volatility Construction

Put-call parity of European call and put options [14] on a non-dividend asset implies that the price of a call option c_E satisfies the following double inequality

$$\max(0, \, S - Ke^{-r(T-t)}) \le c_E \le S \tag{5}$$

and hence belongs to the half-strip area in Fig. 1.

The first derivative of call option price c_{BS} by volatility σ is equal to

$$\Lambda = \frac{\partial c_{BS}}{\partial \sigma} = \frac{S\sqrt{T-t}}{\sqrt{2\pi}} e^{-\frac{1}{2}(d_+)^2} \tag{6}$$

and is called the vega of call option [14]. Since $\Lambda > 0$ for $S > 0$, $T > t$, the price of European call option c_{BS} is an increasing function of volatility σ.

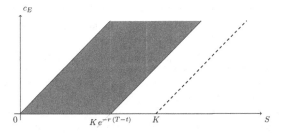

Fig. 1. European call option prices

If volatility σ approaches 0 from the right, then $d_\pm$ tends to $\pm\infty$ depending on the sign of expression $\ln\left(\frac{S}{K}\right) + r\,(T-t)$ and c_{BS} tends to 0 when $\ln\left(\frac{S}{K}\right) + r\,(T-t) < 0$ and tends to $S - K\,e^{-r\,(T-t)}$ when $\ln\left(\frac{S}{K}\right) + r\,(T-t) \geq 0$.

If volatility σ tends to $+\infty$, then d_+ tends to $+\infty$ and d_- tends to $-\infty$, thus c_{BS} tends to S (from the left).

So for fixed strike price K and maturity T each point inside the area, determined by inequalities (5) and shown in gray in Fig. 1, corresponds to some combination of underlying asset price S and volatility σ.

The Eq. (4) for the determination of implied volatility σ_{imp} is a nonlinear equation without analytically tractable closed-form solutions, therefore we have to use approximations and/or numerical methods [22] to obtain a solution.

Implied volatility σ_{imp} from Eq. (4) can be approximated in a number of ways, including approaches by Brenner, Subrahmanyam [3], Bharadia, Christofides, Salkin [1], Chance [6], Corrado, Miller [7], Li [17]. When using Brenner-Subrahmanyam formula [3], implied volatility is approximated by expression

$$\sigma_{imp} \approx \sqrt{\frac{2\pi}{T}}\,\frac{c_E - \delta}{S}, \quad \delta = \frac{S - K\,e^{-r\,(T-t)}}{2}. \tag{7}$$

2.2 Algorithm of Implied Volatility Construction

The most basic version of the Newton's root-cfunction $f(x)$ is the iterative procedure

$$x_{n+1} = x_n - \frac{f(x_n)}{f'(x_n)}, \tag{8}$$

which ends, when desired level of accuracy ε is reached, i.e. $|x_{n+1} - x_n| < \varepsilon$.

If the derivative $f'(x)$ takes values close to zero, then various modifications of the Newton's method are used to solve the equation $f(x) = 0$.

In damped Newton's method [10] a modified iterative procedure

$$x_{n+1} = x_n - \alpha_n \frac{f(x_n)}{f'(x_n)}, \tag{9}$$

with adaptive step α_n is used. Here $0 < \alpha_n \le 1$ and the step α_n is chosen to achieve decrease of $|f(x_n)|$, in other words, the step α_n is reduced to ensure the condition $|f(x_{n+1})| < |f(x_n)|$.

In Nesterov-Nemirovsky version of damped Newton's method [21] the step α_n is equal to

$$\alpha_n = 1, |\delta_n| \le \frac{1}{4}, \alpha_n = \frac{1}{1+\delta_n}, |\delta_n| > \frac{1}{4}, \tag{10}$$

where $\delta_n = \frac{f(x_n)}{\sqrt{f'(x_n)}}$.

If $f'(x_n)$ is close to zero, instead of procedure (9) we can use the iterative procedure of Levenberg–Marquardt method [16,19]:

$$x_{n+1} = x_n - \frac{f(x_n)}{\alpha_n + f'(x_n)}. \tag{11}$$

The values of α_n in (11) are selected from inequality $|f(x_{n+1})| < |f(x_n)|$.

Hereby, for the determination of implied volatility σ_{imp} we build a sequence $\{\sigma_n\}$, which converges to σ_{imp}. The initial term of the sequence σ_0 is calculated by Brenner-Subrahmanyam formula (7). The iteration process stops if $|\sigma_{n+1} - \sigma_n| < \varepsilon$, where ε is the given accuracy.

The vega value (6) measures the call option's sensitivity to volatility. Vega of a call option is always positive, but may be quite close to zero in some cases.

Assume that asset price $S = 1$, current time $t = 0$, strike-price K varies from 0,5 to 1,5, maturity T varies from 0 to 0,25, risk-free rate $r = 0,05$, volatility $\sigma = 0,4$.

Table 1. European call option vega values ($S = 1, t = 0, r = 0,05, \sigma = 0,4$)

Vega values	$T = 0,0$	$T = 0,05$	$T = 0,1$	$T = 0,15$	$T = 0,2$	$T = 0,25$
$K = 0,5$	0,000	4,6e−15	2,2e−08	3,9e−06	5,5e−05	0,0003
$K = 0,6$	0,000	4,9e−09	2,4e−05	0,0004	0,0020	0,0050
$K = 0,7$	0,000	2,3e−05	0,0018	0,0081	0,0181	0,0300
$K = 0,8$	0,000	0,0033	0,0221	0,0453	0,0676	0,0881
$K = 0,9$	0,000	0,0408	0,0814	0,1117	0,1362	0,1573
$K = 1,0$	0,000	0,0889	0,1255	0,1533	0,1765	0,1969
$K = 1,1$	0,000	0,0545	0,1021	0,1371	0,1655	0,1899
$K = 1,2$	0,000	0,0129	0,0515	0,0889	0,1218	0,1507
$K = 1,3$	0,000	0,0015	0,0181	0,0452	0,0745	0,1030
$K = 1,4$	0,000	9,9e-05	0,0048	0,0191	0,0396	0,0628
$K = 1,5$	0,000	4,3e−06	0,0010	0,0069	0,0188	0,0350

From Table 1 it follows that vega values are zero (or are close to zero) when the call option is near maturity or strike-price K is far from current asset price S.

When the vega of call option (6) is far from zero, we perform the iteration using the Nesterov-Nemirovsky version of damped Newton's method (9), (10):

$$\sigma_{n+1} = \sigma_n - \alpha_n \frac{S\,\Phi\left(d_+^{(n)}\right) - K\,e^{-r\,(T-t)}\,\Phi\left(d_-^{(n)}\right) - c_E}{\frac{S\sqrt{T-t}}{\sqrt{2\pi}}e^{-\frac{1}{2}\left(d_+^{(n)}\right)^2}}, \tag{12}$$

$$d_\pm^{(n)} = \frac{\ln\left(\frac{S}{K}\right) + \left(r \pm \frac{\sigma_n^2}{2}\right)(T-t)}{\sigma_n\sqrt{T-t}}, \quad \alpha_n = 1, |\delta_n| \le \frac{1}{4}, \alpha_n = \frac{1}{1+\delta_n}, |\delta_n| > \frac{1}{4},$$

$$\delta_n = \frac{S\,\Phi\left(d_+^{(n)}\right) - K\,e^{-r\,(T-t)}\,\Phi\left(d_-^{(n)}\right) - c_E}{\sqrt{\frac{S\sqrt{T-t}}{\sqrt{2\pi}}e^{-\frac{1}{2}\left(d_+^{(n)}\right)^2}}}.$$

When the vega of call option (6) is close to zero, i.e. $\Lambda < \epsilon$, where ϵ is some vega threshold, we perform the iteration using Levenberg-Marquardt method (11):

$$\sigma_{n+1} = \sigma_n - \frac{S\,\Phi\left(d_+^{(n)}\right) - K\,e^{-r\,(T-t)}\,\Phi\left(d_-^{(n)}\right) - c_E}{\alpha_n + \frac{S\sqrt{T-t}}{\sqrt{2\pi}}e^{-\frac{1}{2}\left(d_+^{(n)}\right)^2}},$$

where $\alpha_n^2 > \frac{1}{4}\frac{S\sqrt{T-t}}{\sqrt{2\pi}}e^{-\frac{1}{2}\left(d_+^{(n)}\right)^2}\frac{d_+d_-}{\sigma_n}\left(-S\Phi(d_+^{(n)}) + Ke^{-r(T-t)}\Phi(d_-^{(n)}) + c_E\right) > 0.$

3   Implied Volatility Surfaces for Local Volatility Models

3.1   IVS for Shifted-Lognormal Model

The volatility function of shifted-lognormal (SL) model [4]

$$\sigma\left(S, t\right) = \sigma_{SL}\left(1 - \frac{\alpha\,e^{rt}}{S}\right), \sigma_{SL} > 0, \tag{13}$$

is a constant, when parameter α vanishes, thus SL model is a direct generalization of BS model. SL model behaves differently depending on the sign of parameter α.

The European call option price in SL model [4] is equal to

$$c_{SL}\left(S, t, K, T, r, \sigma_{SL}, \alpha\right) = \left(S - \alpha\,e^{rt}\right)\Phi\left(d_+^{(SL)}\right)$$

$$- \left(K - \alpha\,e^{rT}\right)e^{-r\,(T-t)}\,\Phi\left(d_-^{(SL)}\right), \tag{14}$$

where

$$d_\pm^{(SL)} = \frac{\ln\left(\frac{S-\alpha\,e^{rt}}{K-\alpha\,e^{rT}}\right) + \left(r \pm \frac{\sigma_{SL}^2}{2}\right)(T-t)}{\sigma_{SL}\sqrt{T-t}}.$$

Calculation of European call option prices using (14) with asset price $S = 1$, current time $t = 0$, strike prices K from 0,5 to 1,5, maturity T from 0 to 0,25,

Table 2. European call option prices in shifted-lognormal model with $\alpha < 0$

Option prices	$T = 0,0$	$T = 0,05$	$T = 0,1$	$T = 0,15$	$T = 0,2$	$T = 0,25$
$K = 0,5$	0,500	0,5012	0,5025	0,5040	0,5058	0,5080
$K = 0,6$	0,400	0,4015	0,4033	0,4060	0,4095	0,4136
$K = 0,7$	0,300	0,3020	0,3060	0,3117	0,3183	0,3252
$K = 0,8$	0,200	0,2048	0,2148	0,2256	0,2361	0,2461
$K = 0,9$	0,100	0,1181	0,1368	0,1526	0,1665	0,1790
$K = 1,0$	0,000	0,0547	0,0780	0,0962	0,1116	0,1253
$K = 1,1$	0,000	0,0197	0,0397	0,0565	0,0712	0,0845
$K = 1,2$	0,000	0,0055	0,0180	0,0310	0,0434	0,0551
$K = 1,3$	0,000	0,0012	0,0074	0,0160	0,0253	0,0348
$K = 1,4$	0,000	0,0002	0,0027	0,0078	0,0142	0,0214
$K = 1,5$	0,000	0,0000	0,0009	0,0036	0,0077	0,0128

risk-free rate $r = 0,05$, volatility multiplier $\sigma_{SL} = 0,4$ and negative parameter $\alpha = -0,5$ gives us the following table of option prices (Table 2).

All the values in Table 2 satisfy inequalities (5), therefore we are able to determine implied volatilities, corresponding to option prices in Table 2.

For the construction of implied volatility surface we have to determine implied volatilities for option prices and visualize the obtained implied volatilities as a 3d plot. Table 1 shows that vega values for some combinations of strike prices

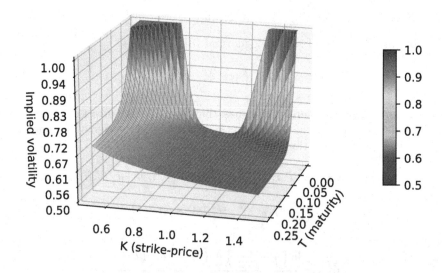

Fig. 2. IVS for shifted-lognormal model ($\alpha < 0$)

and maturities from Table 2 are close to zero, so we have to use the algorithm from Sect. 3 for the determination of implied volatility.

The IVS for SL model with negative α is visualized in Fig. 2 for strike prices K from 0,5 to 1,5 and maturity T from 0 to 0,25.

The shape of IVS in Fig. 2 indicates that for short time to maturity SL model with $\alpha < 0$ captures the smile effect of volatility and for longer time to maturity the surface is flattened into a plane with some skew effect for small strike prices.

Calculation of European call option prices using (14) with positive parameter $\alpha = 0,5$ and all other parameters, taken from the previous calculation, gives us the following table of option prices (Table 3).

When $\alpha > 0$, the volatility function (13) takes lower values than when $\alpha < 0$. This leads to lower option prices, because option price is an increasing function of volatility. Some values in the first row of Table 3 are not computed due to a negative argument to logarithmic function.

Table 3. European call option prices in shifted-lognormal model with $\alpha > 0$

Option prices	$T = 0,0$	$T = 0,05$	$T = 0,1$	$T = 0,15$	$T = 0,2$	$T = 0,25$
$K = 0,5$	0,500	–	–	–	–	–
$K = 0,6$	0,400	0,4015	0,4030	0,4045	0,4060	0,4075
$K = 0,7$	0,300	0,3017	0,3035	0,3052	0,3070	0,3087
$K = 0,8$	0,200	0,2020	0,2040	0,2060	0,2080	0,2100
$K = 0,9$	0,100	0,1023	0,1052	0,1085	0,1120	0,1156
$K = 1,0$	0,000	0,0191	0,0277	0,0345	0,0405	0,0459
$K = 1,1$	0,000	0,0004	0,0027	0,0059	0,0094	0,0131
$K = 1,2$	0,000	0,0000	0,0001	0,0006	0,0016	0,0029
$K = 1,3$	0,000	0,0000	0,0000	0,0000	0,0002	0,0005
$K = 1,4$	0,000	0,0000	0,0000	0,0000	0,0000	0,0001
$K = 1,5$	0,000	0,0000	0,0000	0,0000	0,0000	0,0000

The shape of IVS for SL model with $\alpha > 0$ in Fig. 3 demonstrates strong smile effect for short maturities which becomes smoother for longer maturities.

3.2 IVS for Cox-Ross Model

The volatility function of Cox–Ross (CR) local volatility model [9] is equal to

$$\sigma (S, t) = \frac{\sigma_{CR}}{S}, \ \sigma_{CR} > 0. \tag{15}$$

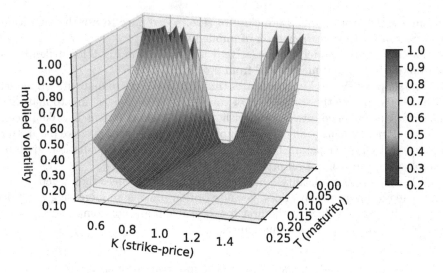

Fig. 3. IVS for shifted-lognormal model ($\alpha > 0$)

The European call option price in CR model is equal to

$$c_{CR}\left(S,t,K,T,r,\sigma_{CR}\right) = \left(S - K\,e^{-r\,(T-t)}\right)\Phi\left(d_{+}^{(CR)}\right) +$$

$$+\,\frac{\sigma_{CR}\sqrt{1 - e^{-2\,r\,(T-t)}}}{2\sqrt{\pi\,r}}\,e^{-\frac{1}{2}\left(d_{+}^{(CR)}\right)^2}, \tag{16}$$

where

$$d_{+}^{(CR)} = \sqrt{2\,r}\,\frac{S\,e^{r\,(T-t)} - K}{\sigma_{CR}\sqrt{e^{2\,r\,(T-t)} - 1}}.$$

Calculation of European call option prices using (16) with asset price $S = 1$, current time $t = 0$, strike-prices K from 0,5 to 1,5, maturity T from 0 to 0,25, risk-free rate $r = 0,05$, volatility multiplier $\sigma_{CR} = 0,4$ gives us the Table 4.

Basically, option prices in Table 4 are between option prices in Tables 2 and 3 for SL model with negative and positive parameter α.

Following the algorithm from Sect. 3 for the determination of implied volatility we receive the IVS for CR model in Fig. 4 for strike-prices K from 0,5 to 1,5 and maturity T from 0 to 0,25.

The shape of IVS for CR model in Fig. 4 is similar to the IVS for SL model with negative parameter α with lower level of volatility and greater surface slope for longer time to maturity.

Table 4. European call option prices in Cox-Ross model

Option prices	$T = 0,0$	$T = 0,05$	$T = 0,1$	$T = 0,15$	$T = 0,2$	$T = 0,25$
$K = 0,5$	0,500	0,5012	0,5025	0,5038	0,5051	0,5066
$K = 0,6$	0,400	0,4015	0,4030	0,4047	0,4067	0,4089
$K = 0,7$	0,300	0,3018	0,3038	0,3066	0,3100	0,3139
$K = 0,8$	0,200	0,2024	0,2068	0,2125	0,2186	0,2248
$K = 0,9$	0,100	0,1079	0,1189	0,1291	0,1384	0,1470
$K = 1,0$	0,000	0,0369	0,0529	0,0654	0,0761	0,0857
$K = 1,1$	0,000	0,0063	0,0166	0,0263	0,0352	0,0435
$K = 1,2$	0,000	0,0004	0,0034	0,0080	0,0133	0,0188
$K = 1,3$	0,000	0,0000	0,0004	0,0018	0,0040	0,0068
$K = 1,4$	0,000	0,0000	0,0000	0,0003	0,0009	0,0021
$K = 1,5$	0,000	0,0000	0,0000	0,0000	0,0002	0,0005

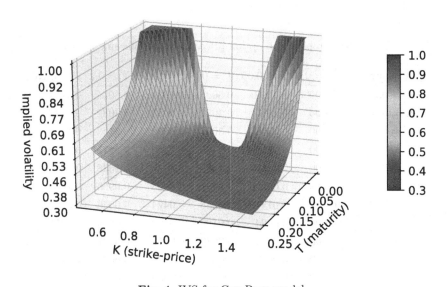

Fig. 4. IVS for Cox-Ross model

3.3 IVS for Hyperbolic-Sine Model

The volatility function of hyperbolic-sine (HS) local volatility model [24]

$$\sigma\left(S, t\right) = \sqrt{\frac{\lambda^2}{S^2} + 2\,r} \tag{17}$$

is an exact solution of PDE, derived by Carr, Tari, Zariphopoulou [5].

The European call option price in HS model is equal to

$$c_{HS}(S, t, K, T, r, \lambda) = \frac{1}{2}S\left(\Phi\left(\sqrt{2r(T-t)} - K^*\right) + \Phi\left(-\sqrt{2r(T-t)} - K^*\right)\right) +$$

$$+ \frac{1}{2}\sqrt{\frac{\lambda^2}{2r} + S^2}\left(\Phi\left(\sqrt{2r(T-t)} - K^*\right) - \Phi\left(-\sqrt{2r(T-t)} - K^*\right)\right) -$$

$$- e^{-r(T-t)}K\Phi(-K^*), \tag{18}$$

where

$$K^* = \frac{1}{\sqrt{2r(T-t)}}\left(\operatorname{arsinh}\left(\frac{\sqrt{2r}}{\lambda}K\right) - \operatorname{arsinh}\left(\frac{\sqrt{2r}}{\lambda}S\right)\right).$$

Calculation of European call option prices using (18) with asset price $S = 1$, current time $t = 0$, strike-prices K from 0,5 to 1,5, maturity T from 0 to 0,25, risk-free rate $r = 0,05$, hyperbolic-sine model parameter $\lambda = 0,4$ gives us the following table of option prices (Table 5).

Table 5. European call option prices in hyperbolic-sine model

Option prices	$T = 0,0$	$T = 0,05$	$T = 0,1$	$T = 0,15$	$T = 0,2$	$T = 0,25$
$K = 0,5$	0,500	0,5012	0,5025	0,5039	0,5055	0,5074
$K = 0,6$	0,400	0,4015	0,4032	0,4053	0,4082	0,4115
$K = 0,7$	0,300	0,3018	0,3048	0,3092	0,3143	0,3199
$K = 0,8$	0,200	0,2034	0,2108	0,2193	0,2279	0,2363
$K = 0,9$	0,100	0,1132	0,1285	0,1419	0,1538	0,1646
$K = 1,0$	0,000	0,0467	0,0667	0,0823	0,0956	0,1075
$K = 1,1$	0,000	0,0131	0,0288	0,0424	0,0546	0,0657
$K = 1,2$	0,000	0,0024	0,0102	0,0193	0,0285	0,0375
$K = 1,3$	0,000	0,0003	0,0030	0,0078	0,0137	0,0201
$K = 1,4$	0,000	0,0000	0,0007	0,0028	0,0061	0,0101
$K = 1,5$	0,000	0,0000	0,0001	0,0009	0,0025	0,0048

Option prices in Table 5 are above option prices in Table 3 for SL model with positive parameter α and Table 4 for CR model, but are below option prices in Table 2 for SL model with negative parameter α.

The IVS for HS model is visualized in Fig. 5 for strike-prices K from 0,5 to 1,5 and maturity T from 0 to 0,25.

The shape of IVS for HS model in Fig. 5 is similar to the IVS for SL model with negative parameter α and IVS for CR model and captures smile effect for short maturities and skew effect for longer maturities.

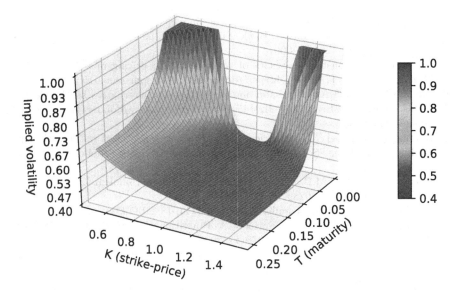

Fig. 5. IVS for hyperbolic-sine model

4   Conclusion

We obtained the IVS for option prices in several local volatility models, including shifted-lognormal model with positive and negative model parameter, Cox-Ross model and hyperbolic-sine model. IVS for these models demonstrate that all the models capture both smile and skew effects of implied volatility with some peculiarities. This suggests that alternative volatility models can be used instead of BS model in various practical applications of stochastic modelling, including derivative pricing and risk management.

References

1. Bharadia, M., Christofides, N., Salkin, G.: Computing the Black-Scholes implied volatility: generalization of a simple formula. Adv. Futures Options Res. **8**, 15–30 (1995)
2. Black, F., Scholes, M.: The pricing of options and corporate liabilities. J. Polit. Econ. **81**(3), 637–654 (1973). https://doi.org/10.1086/260062
3. Brenner, M., Subrahmanyan, M.G.: A simple formula to compute the implied standard deviation. Financ. Anal. J. **44**(5), 80–83 (1988). https://doi.org/10.2469/faj.v44.n5.80
4. Brigo, D., Mercurio, F.: Fitting volatility skews and smiles with analytical stock-price models. In: Seminar paper, Institute of Finance, University of Lugano (2000)
5. Carr, P., Tari, M., Zariphopoulou, T.: Closed form option valuation with smiles. Preprint. NationsBanc Montgomery Securities (1999)
6. Chance, D.M.: A generalized simple formula to compute the implied volatility. Financ. Rev. **31**(4), 859–867 (1996). https://doi.org/10.1111/j.1540-6288.1996.tb00900.x

7. Corrado, C.J., Miller, T.W.: A note on a simple, accurate formula to compute implied standard deviations. J. Bank. Financ. **20**(3), 595–603 (1996). https://doi.org/10.1016/0378-4266(95)00014-3

8. Cox, J.C.: The constant elasticity of variance option pricing model. J. Portfolio Manag. **23**(5), 15–17 (1996). https://doi.org/10.3905/jpm.1996.015

9. Cox, J.C., Ross, S.A.: The valuation of options for alternative stochastic processes. J. Financ. Econ. **3**(1–2), 145–166 (1976). https://doi.org/10.1016/0304-405x(76)90023-4

10. Dennis, J., Schnabel, R.: Numerical methods for unconstrained optimization and nonlinear equations. Soc. Ind. Appl. Math. (1996). https://doi.org/10.1137/1.9781611971200

11. Derman, E., Miller, M.B.: The Volatility Smile. Wiley, Hoboken (2016). https://doi.org/10.1002/9781119289258

12. Dupire, B.: Pricing with a smile. Risk Mag. **7**(1), 18–20 (1994)

13. Gatheral, J.: The Volatility Surface: A Practitioner's Guide. Wiley, Hoboken (2012). https://doi.org/10.1002/9781119202073

14. Hull, J.C.: Options, Futures, and Other Derivatives, 10th edn. Pearson, London (2018)

15. Javaheri, A.: Inside Volatility Filtering. Wiley, Hoboken (2015). https://doi.org/10.1002/9781118949092

16. Levenberg, K.: A method for the solution of certain nonlinear problems in least squares. Q. Appl. Math. **2**, 164–168 (1944)

17. Li, S.: A new formula for computing implied volatility. Appl. Math. Comput. **170**(1), 611–625 (2005). https://doi.org/10.1016/j.amc.2004.12.034

18. Mak, W.H.J., Cassidy, S., Clarkson, P.J.: Towards an assessment of resilience in telecom infrastructure projects using real options. In: Proceedings of the 21st International Conference on Engineering Design (ICED17), vol. 2, pp. 487–496. The University of British Columbia, Vancouver (August 2017)

19. Marquardt, D.: An algorithm for least squares estimation of nonlinear parameters. SIAM J. Appl. Math. **11**, 431–441 (1963)

20. Merton, R.C.: Theory of rational option pricing. Bell J. Econ. Manag. Sci. **4**(1), 141–183 (1973). https://doi.org/10.2307/3003143

21. Nesterov, Y., Nemirovskii, A.: Interior-point polynomial algorithms in convex programming. Soc. Ind. Appl. Math. (1994). https://doi.org/10.1137/1.9781611970791

22. Orlando, G., Taglialatela, G.: A review on implied volatility calculation. J. Comput. Appl. Math. **320**, 202–220 (2017). https://doi.org/10.1016/j.cam.2017.02.002

23. Rachev, S.T., Menn, C., Fabozzi, F.J.: Fat-Tailed and Skewed Asset Return Distributions: Implications for Risk Management, Portfolio Selection, and Option Pricing. Wiley, Hoboken (2005)

24. Shorokhov, S., Buuruldai, A.: On hyperbolic-sine local volatility model. In: Proceedings of Conference Information and Telecommunication Technologies and Mathematical Modeling of High-Tech Systems, pp. 404–406. RUDN University (May 2018)

Negative Binomial Approximation in Retrial Queue M/M/1 with Collisions and Impatient Calls

Elena Danilyuk$^{(\boxtimes)}$ and Ekaterina Fedorova

National Research Tomsk State University,
36 Lenina Avenue, Tomsk, Russian Federation
daniluc_elena@sibmail.com, moiskate@mail.ru

Abstract. In the paper, the retrial queueing system of $M/M/1$ type with collisions and impatient calls is considered. The impatience of calls in the orbit is exponential distributed. The process of the number of calls in the orbit is analyzed. We propose the method of the negative binomial approximation using the first and the second moments of the distribution which were obtained asymptotically. The numerical analysis of comparison exact (obtained by simulation) and approximate distributions for different values of the system parameters are presented.

Keywords: Retrial queueing system · Negative binomial distribution · Collisions · Impatient calls

1 Introduction

Nowadays, an analysis and an optimization of various telecommunication systems are an important problem. One of the most used mathematical model for description of communication systems controlled via random multiple access protocol, mobile networks, call centers and other technical systems is retrial queueing model [1–8]. The main feature of such models is the presence of repeated calls to a device in some random time after failure to get service. Such cases can be produced not only by lack of free servers at arrival time, but also by technical reasons. The books of Artalejo and Gómez-Corral [9], Falin and Templeton [10] give detailed description of retrial queues.

More complicated models are retrial queues with collisions examined in [11–13]. Usually collisions happen in networks when a message is transmitted during the transmission of another message. As the result, this messages "collide". It is supposed that both messages are deformed and they go into the orbit, where they try to get service again after a random delay.

One else research direction of retrial queueing theory is investigating of the retrial queues with losses, where unserviced calls leave the system. Such retrial queues were studied by Cohen [3], Falin [10], Yang, Templeton [14],

Supported by RFBR according to the research project No. 19-41-703002.

V. M. Vishnevskiy et al. (Eds.): DCCN 2019, CCIS 1141, pp. 461–471, 2019.
https://doi.org/10.1007/978-3-030-36625-4_37

Krishnamoorthy [15], Kim [16], Aissani [17], Kumar [18], etc. In mentioned works, after an unsuccessful attempt to receive the service, a call with probability p goes into orbit and with probability $1 - p$ leaves the system (so called p–persistence). In this paper, we consider the retrial queue with impatient calls that leave the system after a random time of waiting in the orbit (without necessarily calling on the device). The same model was studied in papers [13, 19–21].

The majority of studies devoted to RQ-systems investigation are performed by matrix methods [9, 16, 22] and further numerical analysis or computer simulation [23, 24] etc. Analytical results are obtained only for the simplest models, e.g. retrial queues with a stationary Poisson arrival process and an exponential distribution of the service time [10].

Various asymptotic methods and approaches in queuing theory are described in [10, 25]. The Tomsk research school develops asymptotic method for investigating queueing systems of various configurations and under different limit conditions, including retrial queues [11–13, 19–21]. Such method allows obtaining asymptotic expressions of necessary system's characteristics in cases when pre-limit research of considered RQ-system is impossible. And obtained results are acceptable for practice.

In the previous paper [13], we studied the retrial queue with collision by the asymptotic analysis method with assumption of a long delay and high patience of calls in the orbit. We have proved the theorem about the Gaussian form of the probability distribution of the number of calls in the orbit and the asymptotic mean and variance of the distribution of the number of calls in the orbit was been obtained. There we have compared asymptotic and simulated distributions and have made the conclusion about the asymptotic method applicability area. In this paper, we will use obtained formulas and perform the comparison of the asymptotic and the exact moments.

In addition, it is known that the probability distribution of the number of calls in the simplest retrial queues has the negative binomial distribution (it is also called as the discrete gamma distribution). Also this distribution is close enough to the exact distribution even for retrial queues with MMPP arrivals [26]. Thus in this paper, we propose to apply the negative binomial approximation for more complicated models – a retrial queue with collision.

The rest of the paper is organized as follows. The general information about mathematical model of the studied retrial queueing system and the problem statement are presented in Sect. 2. In Sect. 3, the main theorem about asymptotic formulas for the first and the second moments of the studied process is formulated and the numerical analysis of this formulas accuracy is performed. Section 4 consists of the definition of the negative binomial distribution and the description of the approximation method. Some numerical results of analysis of approximation method applicability are presented in Sect. 5. Section 6 concludes the paper.

2   Model Description

We consider an retrial queue with one server and Poisson arrival process with rate λ. If an arriving call (or customer) founds the service device free, call takes it for the service for random time distributed exponentially with parameter μ. If the device is busy, arriving and services calls enter into a "collision" and go into the orbit. On the orbit each call waits during exponential distributed random time with parameter σ, and then again tries to get access the device. If the device is free, the call occupies it for random servicing time. If the device is busy, we have a "collision" again, so both calls immediately go into the orbit and wait once more random time. Moreover, a call from the orbit leaves the system after exponential distributed time with parameter α, demonstrating the "impatience" property.

Figure 1 shows the model of the retrial queueing system of the $M|M|1$ type with collisions and impatient calls.

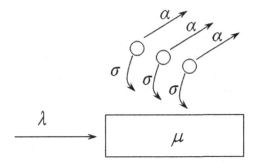

Fig. 1. Retrial queue $M|M|1$ with collisions and impatient calls

Let us denote $i(t)$ the process of the number of calls in the orbit at the moment t. The problem is to study the stationary distribution $P(i)$ of the number of calls in the orbit for the described system. It is worth noting that the Kolmogorov's system of differential equations for $P(i)$ can not be directly solved analytically and we use some limit (asymptotic) condition for finding a decision.

3   Asymptotic Results

For the approximation constructing by negative binomial distribution, we need to know values of the first and the second moments of the studied process. But in the considered model, the exact formulas for the moments can not be obtained. Thus we use the results from paper [13], where the following theorem about the Gaussian form of the distribution of the number of call in the orbit has been proved by means of applying the asymptotic analysis method.

Theorem 1. *The stationary distribution of the number of calls in orbit in the RQ-system M/M/1 with collisions and impatient calls (with the Poisson arrival process of intensity λ, exponential servicing distribution with parameter μ, exponential distribution law of the random delay parameter σ, exponential distribution of a call's impatience with parameter $\alpha = q\sigma$, and constant $q > 0$) is an asymptotically normal distribution under the long delay and the high patience of calls condition with mean $E\{i(t)\}$ and variance $Var\{i(t)\}$*

$$E\{i(t)\} = \frac{\lambda - \mu R_1}{q\sigma}, \tag{1}$$

$$Var\{i(t)\} = \frac{-\mu f_1 + \lambda R_1 + q\sigma E\{i(t)\}\,(1 - R_1)}{q\sigma}, \tag{2}$$

where

$$f_1 = b/\left(1 + a\right),$$

$$a = \frac{\lambda\left(1 + q\right) + 2R_1\left(q\sigma E\{i(t)\} - \lambda\right)}{q\left(\lambda + \mu + \sigma E\{i(t)\}\right) + \left(1 - 2R_1\right)\left(\mu + q\sigma E\{i(t)\} - \lambda\right)},$$

$$b = \frac{\left[\left(1 - 2R_1\right)\left(\lambda - q\sigma E\{i(t)\}\right) - q\sigma E\{i(t)\}\left(1 + q\right)\right] R_1}{q\left(\lambda + \mu + \sigma E\{i(t)\}\right) + \left(1 - 2R_1\right)\left(\mu + q\sigma E\{i(t)\} - \lambda\right)},$$

and R_1 is the probability that the device is occupied in the stationary mode of system operation, which is determined by equation

$$2\mu R_1^2 - \left(2\lambda + \mu\right)\left(1 + q\right) R_1 + \lambda\left(1 + q\right) = 0, \quad R_1 \in [0;1].$$

The numerical analysis has shown that the value of the Kolmogorov distance between the asymptotic and simulated distributions decreases with the growth of the system load and with the increase in delay time of orders in orbit ($\sigma \to 0$). Some examples showing the quality of approximation are presented on Figs. 2 and 3. In [13] we made the conclusion that the asymptotic results can be applied for $\sigma \leq 0.01$ if $\rho = 0.8$.

Table 1. Relative precision for the asymptotic mean for various values of ρ and σ

	$\sigma = 0.005$	$\sigma = 0.01$	$\sigma = 0.05$	$\sigma = 0.1$	$\sigma = 1$
$\rho = 0.4$	0.00054	0.00107	0.00481	0.00853	0.02800
$\rho = 0.5$	0.00056	0.00110	0.00506	0.00918	0.03400
$\rho = 0.6$	0.00053	0.00105	0.00492	0.00907	0.03700
$\rho = 0.7$	0.00049	0.00097	0.00460	0.00859	0.03800
$\rho = 0.8$	0.00044	0.00088	0.00422	0.00796	0.03700
$\rho = 0.9$	0.00040	0.00080	0.00383	0.00727	0.03600
$\rho = 1.0$	0.00036	0.00072	0.00347	0.00665	0.03500
$\rho = 2.0$	0.00704	0.00028	0.00140	0.00276	0.02000

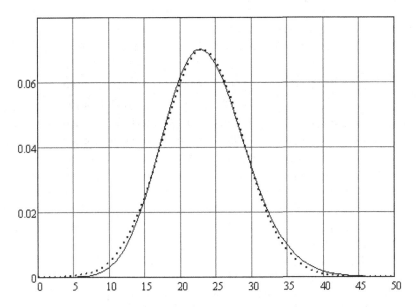

Fig. 2. Comparison of the asymptotic (dashed line) and the simulated (solid line) distributions for $\sigma = 0.01$ and $\rho = 0.8$

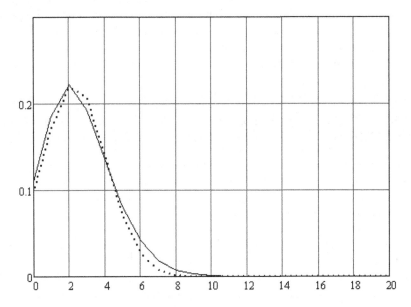

Fig. 3. Comparison of the asymptotic (dashed line) and the simulated (solid line) distributions for $\sigma = 0.1$ and $\rho = 0.8$

Table 2. Relative precision for the second moment for various values of ρ and σ

	$\sigma = 0.005$	$\sigma = 0.01$	$\sigma = 0.05$	$\sigma = 0.1$	$\sigma = 1$
$\rho = 0.4$	0.00055	0.00109	0.00503	0.00912	0.03100
$\rho = 0.5$	0.00030	0.00591	0.00289	0.00550	0.02200
$\rho = 0.6$	0.00006	0.00013	0.00786	0.00177	0.02200
$\rho = 0.7$	0.00014	0.00026	0.00103	0.00155	0.00108
$\rho = 0.8$	0.00029	0.00057	0.00249	0.00430	0.01200
$\rho = 0.9$	0.00040	0.00080	0.00363	0.00650	0.02300
$\rho = 1.0$	0.00049	0.00097	0.00449	0.00821	0.03200
$\rho = 2.0$	0.11400	0.00131	0.00638	0.01200	0.07500

For formulas (1)–(2) accuracy estimating, we present values of relative precisions for the mean and the variance in Tables 1, 2 respectively.

We see that the asymptotic moments are close enough to exact ones for various values of the system parameters.

Distributions on Fig. 2 have not Gaussian form, so to improve research results, we will use formulas (1) and (2) for negative binomial approximation constructing.

4   Negative Binomial Approximation

It is known that the probability distribution of the number of calls in the system in the simplest retrial queues has the negative binomial distribution (it is also calls as the discrete gamma distribution [26]). Thus we offer to construct the negative binomial approximation for more complicated retrial queues studying such as retrial queus with collision and impatient calls.

First of all, we present the definition of negative binomial distribution.

Definition 1. *The negative binomial distribution is a discrete probability distribution $Pg(i)$ for $i \geq 0$, which characteristic function has the following form*

$$G(u) = \left(\frac{1 - \gamma}{1 - \gamma e^{ju}} \right)^{\eta},$$

with parameters $\eta > 0$ and $0 < \gamma < 1$, $j = \sqrt{-1}$.

It is easy to show that the parameters η and γ are expressed in terms of the mean $E\{i(t)\}$ and the variance $\text{Var}\{i(t)\}$ of the distribution $Pg(i)$ as follows

$$\gamma = 1 - \frac{E\{i(t)\}}{\text{Var}\{i(t)\}}, \quad \eta = E\{i(t)\} \cdot \frac{1 - \gamma}{\gamma}. \tag{3}$$

The method of the approximation consists in approximating the probability distribution of the number of calls in the orbit $P(i)$ by the negative binomial distribution $Pg(i)$ which parameters are calculated via known mean and variance of $P(i)$ by formulas (3).

For demonstrating the applicability area of the proposed approximation, we present some numerical examples in the following section.

5   Numerical Analysis

We perform the solving of the system evolution by simulation [27] and compare statistical results with analytical ones derived in the paper. For the demonstrating the applicability area of the negative binomial approximation, let us compare the probability distribution of the number of calls in the retrial queueing system $P(i)$ calculated via simulation and its approximation $Pg(i)$ constructed by using the asymptotic moments (1)–(2) for different values of the system parameters.

In the example, let the service rate be $\mu = 1$, $\alpha = 2\sigma$. We variate parameters $\rho = \lambda/\mu$ and σ for the results demonstrating. To evaluate the result we use Kolmogorov distance between respective distribution functions:

$$d = \max_{i \geq 0} \left| \sum_{l=0}^{i} [P(l) - Pg(l)] \right|.$$

The comparison of the distributions is shown in Figs. 4, 5 and 6. Values of Kolmogorov distance are presented in Table 3.

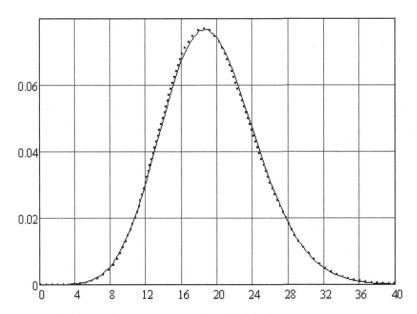

Fig. 4. Comparison of the approximate (dashed line) and the simulated (solid line) distributions for $\sigma = 0.01$ and $\rho = 0.8$

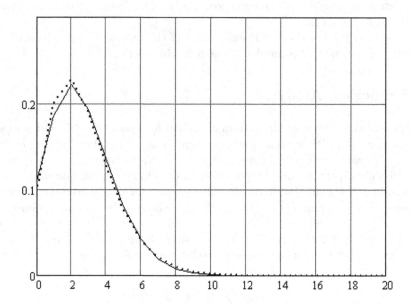

Fig. 5. Comparison of the approximate (dashed line) and the simulated (solid line) distributions for $\sigma = 0.1$ and $\rho = 0.8$

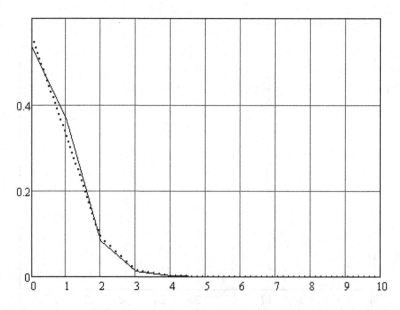

Fig. 6. Comparison of the approximate (dashed line) and the simulated (solid line) distributions for $\sigma = 1$ and $\rho = 0.8$

Table 3. Kolmogorov distances d for various values of the parameter ρ and σ

	$\sigma = 0.005$	$\sigma = 0.01$	$\sigma = 0.05$	$\sigma = 0.1$	$\sigma = 1$
$\rho = 0.4$	0.0050	0.0069	0.0200	0.0160	0.0077
$\rho = 0.5$	0.0040	0.0056	0.0180	0.0150	0.0120
$\rho = 0.6$	0.0034	0.0047	0.0130	0.0150	0.0170
$\rho = 0.7$	0.0029	0.0040	0.0087	0.0150	0.0210
$\rho = 0.8$	0.0025	0.0035	0.0072	0.0130	0.0250
$\rho = 0.9$	0.0022	0.0031	0.0060	0.0110	0.0280
$\rho = 1.0$	0.0019	0.0027	0.0055	0.0085	0.0300
$\rho = 2.0$	0.0180	0.0012	0.0026	0.0035	0.0203

Thus, Kolmogorov distances between the distributions have values $d \leq 0.03$ for different σ and ρ. We obtain the same results for other values of the system parameters.

6 Conclusions

In this regard, the method of the negative binomial approximation for the probability distribution of the number of calls in the orbit is offered for the retrial queue of $M/M/1$ type with collisions and impatient calls. The numerical comparison of the distributions obtained via simulation and approximate one for different values of the system parameters shows the wide range of the method application.

References

1. Aguir, S., Karaesmen, F., Askin, O.Z., Chauvet, F.: The impact of retrials on call center performance. OR Spektrum **26**, 353–376 (2004)
2. Almási, B., Bérczes, T., Kuki, A., Sztrik, J., Wang, J.: Performance modeling of finite-source cognitive radio networks. Acta Cybernetica **22**(3), 617–631 (2016)
3. Cohen, J.W.: Basic problems of telephone traffic and the influence of repeated calls. Philips Telecommun. Rev. **18**(2), 49–100 (1957)
4. Gosztony, G.: Repeated call attempts and their effect on traffic engineering. Budavox Telecommun. Rev. **2**, 16–26 (1976)
5. Elldin, A., Lind, G.: Elementary Telephone Traffic Theory. Ericsson Public Telecommunications, Stockholm (1971)
6. Phung-Duc, T., Kawanishi, K.: Multiserver retrial queue with setup time and its application to data centers. J. Ind. Manag. Optim. **15**(1), 15–35 (2017). https://doi.org/10.1142/S0217595914400089
7. Roszik, J., Sztrik, J., Kim, C.: Retrial queues in the performance modelling of cellular mobile networks using MOSEL. Int. J. Simul. **6**, 38–47 (2005)
8. Wilkinson, R.I.: Theories for toll traffic engineering in the USA. Bell Syst. Tech. J. **35**(2), 421–507 (1956). https://doi.org/10.1002/j.1538-7305.1956.tb02388.x

9. Artalejo, J.R., Gómez-Corral, A.: Retrial Queueing Systems. A Computational Approach. Springer, Heidelberg (2008). https://doi.org/10.1007/978-3-540-78725-9
10. Falin, G.I., Templeton, J.G.C.: Retrial Queues. Chapman & Hall, London (1997)
11. Sudyko, E.A., Nazarov, A.A.: A study of a Markov RQ-system with call conflicts and elementary incoming stream. Vestn. Tomsk. Gos. Univ., Upravlen., Vychisl. Tekh. Inf. 3(12), 97–106 (2010)
12. Nazarov, A., Sztrik, J., Kvach, A.: Comparative analysis of methods of residual and elapsed service time in the study of the closed retrial queuing system M/GI/1//N with collision of the customers and unreliable server. In: Dudin, A., Nazarov, A., Kirpichnikov, A. (eds.) ITMM 2017. CCIS, vol. 800, pp. 97–110. Springer, Cham (2017). https://doi.org/10.1007/978-3-319-68069-9_8
13. Danilyuk, E.Y., Fedorova, E.A., Moiseeva, S.P.: Asymptotic analysis of an retrial queueing system M/M/1 with collisions and impatient calls. Autom. Remote Control 79(12), 2136–2146 (2018). https://doi.org/10.1134/S0005117918120044
14. Yang, T., Posner, M., Templeton, J.: The M/G/1 retrial queue with non-persistent customers. Queueing Syst. 7(2), 209–218 (1990)
15. Krishnamoorthy, A., Deepak, T., Joshua, V.: An M/G/1 retrial queue with non-persistent customers and orbital search. Stoch. Anal. Appl. 23, 975–997 (2005). https://doi.org/10.1080/07362990500186753
16. Kim, J.: Retrial queueing system with collision and impatience. Commun. Korean Math. Soc. 4, 647–653 (2010)
17. Aissani, A., Taleb, S., Hamadouche, D.: An unreliable retrial queue with impatience and preventive maintenance. In: Proceedings of the 15th Applied Stochastic Models and Data Analysis (ASMDA 2013), Mataró (Barcelona), Spain, pp. 1–9 (2013)
18. Kumar, M.S., Arumuganathan, R.O.: Performance analysis of single server retrial queue with general retrial time, impatient subscribers, two phases of service and bernoulli schedule. Tamkang J. Sci. Eng. 13(2), 135–143 (2010)
19. Fedorova, E., Voytikov, K.: Retrial queue M/G/1 with impatient calls under heavy load condition. In: Dudin, A., Nazarov, A., Kirpichnikov, A. (eds.) ITMM 2017. CCIS, vol. 800, pp. 347–357. Springer, Cham (2017). https://doi.org/10.1007/978-3-319-68069-9_28
20. Danilyuk, E., Vygoskaya, O., Moiseeva, S.: Retrial queue M/M/N with impatient customer in the orbit. In: Vishnevskiy, V.M., Kozyrev, D.V. (eds.) DCCN 2018. CCIS, vol. 919, pp. 493–504. Springer, Cham (2018). https://doi.org/10.1007/978-3-319-99447-5_42
21. Vygovskaya, O., Danilyuk, E., Moiseeva, S.: Retrial queueing system of MMPP/M/2 type with impatient calls in the orbit. In: Dudin, A., Nazarov, A., Moiseev, A. (eds.) ITMM/WRQ -2018. CCIS, vol. 912, pp. 387–399. Springer, Cham (2018). https://doi.org/10.1007/978-3-319-97595-5_30
22. Dudin, A.N., Klimenok, V.I.: Queueing system BMAP/G/1 with repeated calls. Math. Comput. Modell. 30(3–4), 115–128 (1999). https://doi.org/10.1016/S0895-7177(99)00136-3
23. Artalejo, J.R., Pozo, M.: Numerical calculation of the stationary distribution of the main multiserver retrial queue. Ann. Oper. Res. 116, 41–56 (2002). https://doi.org/10.1023/A:1021359709489
24. Neuts, M.F., Rao, B.M.: Numerical investigation of a multiserver retrial model. Queueing Syst. 7(2), 169–189 (1990). https://doi.org/10.1007/BF01158473
25. Borovkov, A.A.: Asymptotic Methods in Queueing Theory. Wiley, New York (1984)

26. Fedorova, E., Nazarov, A., Paul, S.: Discrete gamma approximation in retrial queue MMPP/M/1 based on moments calculation. In: Rykov, V.V., Singpurwalla, N.D., Zubkov, A.M. (eds.) ACMPT 2017. LNCS, vol. 10684, pp. 121–131. Springer, Cham (2017). https://doi.org/10.1007/978-3-319-71504-9_12
27. Moiseev, A., Demin, A., Dorofeev, V., Sorokin, V.: Discrete-event approach to simulation of queueing networks. Key Eng. Mater. **685**, 939–942 (2016). https://doi.org/10.4028/www.scientific.net/KEM.685.939

Distributed Systems Applications

On Physical Web for Social Networks

Dmitry Namiot[1(✉)] and Manfred Sneps-Sneppe[2]

[1] Faculty of Computational Mathematics and Cybernetics,
Lomonosov Moscow State University,
GSP-1, 1-52, Leninskiye Gory, Moscow 119991, Russia
`dnamiot@gmail.com`
[2] Ventspils International Radio Astronomy Centre, Ventspils University of Applied
Sciences, Inzenieru 101a, Ventspils LV-3601, Latvia
`manfreds.sneps@gmail.com`

Abstract. The article discusses the use of Physical Web approaches for the expansion of social networks. This implies the presentation of data from social networks in a real (physical) context, as well as the inverse task of using information about a real physical context in querying and analyzing data from social networks. First of all, mobile phones of social network users are considered as real objects that will be used both in data dissemination and in gathering information about the context. In this case, the purpose of consideration is to build a "natural" extension, when the implementation does not require the creation of a special type of social network entries. The general scheme or model of implementation is based on the minimization (or even complete absence) of requesting additional rights to access the social network, the absence of marks in the social network, and the use of basic functionality and standard protocols for mobile devices.

Keywords: Physical Web · Network proximity · Social networks

1 Introduction

The term Physical Web refers to the ability to interact with physical objects. The word "interaction" here means the possibility of obtaining any information related to or associated with real objects [1]. In general, we are talking about projects related to proximity services. And proximity, in this case, will be determined in relation to other objects (close to other objects). Any interaction with physical objects is understood as the possibility of obtaining any information from closely located (located somewhere nearby) real physical objects. To determine this proximity, it must somehow be measured. In other words, we need to introduce some metrics [2].

The measurement (determination) of proximity is a separate problem. The first thing that comes to mind in this situation is the use of proximity sensors on mobile devices (mobile phones). The proximity sensor is a contactless sensor that allows you to determine the object in front of you and the distance to it [3]. As a rule, it is an infrared emitter and receiver [4]. If the receiver does not receive the infrared beam from the emitter reflected by the object, then this means that

© Springer Nature Switzerland AG 2019
V. M. Vishnevskiy et al. (Eds.): DCCN 2019, CCIS 1141, pp. 475–487, 2019.
https://doi.org/10.1007/978-3-030-36625-4_38

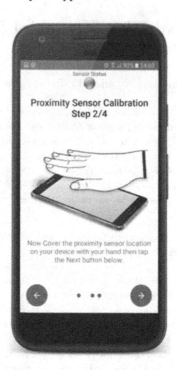

Fig. 1. Proximity sensors [6].

there are no objects in front of the sensor, and if the signal is reflected, then there is an object. In fact, the principles of action may be different [5] (Fig. 1).

However, the practical use of such sensors today is difficult. First, they may not be present on all mobile devices. Secondly, there is no standard API here. Many measurement methods can be inconvenient to use. For example, it can interrupt or even fully eliminate the work of other applications. Proximity sensors cannot determine what is nearby. For example, they cannot determine the presence of another phone. The picture is quite real when sensors based on different physical signs are better suited for different types of objects.

All this leads to the fact that in practice, the proximity of objects is understood in terms of network proximity. This is the ability to receive signals of any wireless networks with limited range. This is where the terms such as Bluetooth distance, Wi-Fi distance, etc. came from. In other words, if the device registers ("sees") the Bluetooth network, then this means, for example, that the device (most often it is a mobile phone) is located in the interval of 1–10 m from the source. Thus, signal reception is an indication of proximity, and various measured characteristics of such a signal (for example, the RSSI - signal strength) can be used in calculating metrics. The network proximity and its applications are described in many of our papers [7,8].

The data transfer for nearby devices is also connected with the physical proximity [9]. Initially, it all started with RFID-based projects, then large companies began to offer solutions that became de-facto standards. First, here we need to mention Apple's iBeacon [10]. These are Bluetooth tags that send a couple of integers. Applications can take on these values and, depending on this, take some action (Fig. 2).

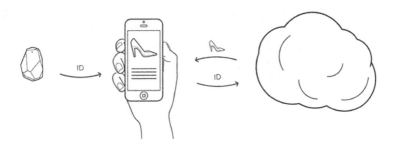

Fig. 2. iBeacon: how it works [11].

The interpretation of these values depends on the application. For example, for an information system in a store, the first value received from a tag is the number of the sales area, the second value is the shelf in this room. It is possible to build some global directories with the data of the installed tags (for example, for navigation in airports).

An alternative solution from another large company (Google) is also a Bluetooth tag (Eddystone), which can send a URL as content. Actually, Google and coined the term physical web. A web link (URL) is associated with a certain object, which is represented by the Bluetooth tag (Fig. 3).

Everywhere further in the work under the physical web, we will understand just such a model. A physical object (a tag representing it or a tag that is associated with it) will distribute some URL. The physical web describes exactly how that URL turns out (how it is formed) and how it is distributed. And its processing upon receipt is, of course, already the specificity of a particular application service. The above-mentioned EddyStone distributes URLs associated with commercial services (Fig. 4). As it will be explained below, our model will distribute links from social networks.

Also, we should note a very important practical point. The tags mentioned here can be created (can be emulated) programmatically. iBeacon is not only a hardware tag but also, for example, an iPhone mobile phone. Bluetooth point can be opened on Android phone programmatically and so on.

The next point on which it is necessary to stop in the introduction is the nature of data our approach is dealing with and the distribution of which is discussed in this paper. The article discusses the possibilities of expanding (in terms of information dissemination) social networks. Needless to say about the great penetration of social networks. All Internet users, one way or another are present

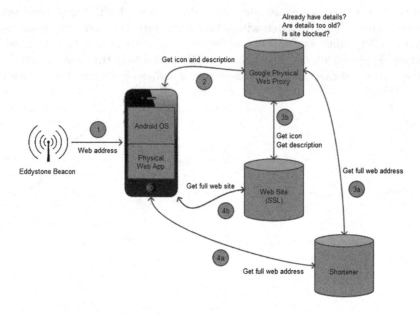

Fig. 3. The Physical Web: how it works [12].

Fig. 4. The Physical Web: business models [13].

in social networks. In reality, the younger generation "lives" on social media. These can be public networks (Facebook), professional networks (Linkedin), networks that specialize in some type of content (Instagram), etc. In any case, social networks today are a huge distribution and consumption of content. At the same time (according to the concept and current statistics of mobile first) content is created (and consumed), first of all, on mobile devices. Accordingly, this article discusses approaches that allow sharing content from social networks among mobile subscribers (mobile users) located in a certain geographically-limited area (near the author). But we should note here an important addition. The traditional approach of social networks in this issue is to use the so-called check-in - marks (records) in the social network, which contain information about the location. Then simply select the records that contain close coordinates. This paper discusses an approach that is not associated with any special publications (entries) on the social network. We are based on the direct dissemination of data by mobile devices. This can be called a natural extension of the social network.

The next important note is that almost all mobile applications (services) are context-sensitive (context-aware) or should be, as a result, such. Context is any measurable data that can be added to a location. Information about the physical environment of a mobile device that transmits data is, of course, part of the context (in fact, most of it) [14]. Accordingly, the expansion of the social network should also be context sensitive.

The remainder of the article is structured as follows. In Sect. 2, we consider passive extension methods. Section 3 deals with active extension methods.

2   Context-Aware QR-Codes

In this section, we want to focus on passive distribution data (links) from social networks. The passive methods just links (associate) data with physical objects and make them available by requests.

A QR code is currently a popular and well-known way of presenting data for mobile users. Classically, a QR code is a two-dimensional bar code that encodes information for quick recognition using the camera on a mobile phone. It is currently a widely-used approach in practical applications [15].

QR codes are usually associated with some physical objects. A picture with a QR code is applied (pasted) on some object (object). This is the most famous use. Essentially - some static information is presented. Just like regular bar codes that are applied, for example, on a packaging. By virtue of its architecture, QR codes provide more information than a regular bar code. Accordingly, it allows you to enter some formats (interpretation) for the provided text. In addition to the actual text, it can be a link (URL), SMS dialogue, vCard information, etc.

In the context of this work, we are interested in the representation of links (URL). The idea of including additional information in such a presentation was considered in several papers. In addition to the above-mentioned work [15], the paper [16] can be highlighted here. In the paper [16], QR code is a location source too. In this case, some symbolic location code is encoded into the URL.

The context server is then used to map locations to symbolic locations. There are several ways to describe symbolic locations. For example, Geohash codes were designed to be used in short URLs to identify locations. Usually, symbolic codes should represent an area, not a point, where the size of the area is variable.

In the paper [17] authors create context-aware QR codes from two parts: "traditional" QR code info and XML-based description for messages that should be provided to the reader.

In our approach, we propose to transfer contextual information through the transmission of data about the wireless environment. It is about creating a customized version of the QR-code reader, which can work in the modes of adding and checking the availability of contextual information. Information about available wireless networks can be represented in the form of a so-called fingerprint. This approach is widely used, for example, in context-sensitive services that are based on network proximity approaches. The fingerprint includes information about the available nodes of wireless networks (for example, their MAC addresses) and, optionally, signal characteristics (for example, RSSI). For example, a list of MAC addresses, a list of MAC addresses and two limits for RSSI (minimal and maximal value), etc. All these examples are valid fingerprints.

A custom QR code reader can automatically add information about available wireless networks to the recognized link (for example, as GET parameters). Accordingly, when accessing the specified URL, the site (CGI script) in the request will receive information about the wireless environment in the place where the QR code was scanned. This can be used, for example, for positioning a scanning application (if you know where the wireless node/network is located) or for any additional checks (if other networks are present).

Using the information about wireless networks has its advantages over using global positioning systems (GPS):

- GPS request is very energy intensive.
- When positioning in the premises, the request for GPS coordinates may not be possible. Using the information on wireless networks, we can provide greater accuracy than using commercially available GPS systems.
- GPS signal may be muffled/changed.
- Access to a service based only on geographic coordinates is more difficult to limit (a location request is available to everyone).
- One of the most important advantages: geographical coordinates - do not change. And if we base the service on proximity to the object (object), which itself moves, then we get a completely new class of services, which was impossible with geographic coordinates. The service (available information, possible actions, etc.) is tied to the current location of the reference object and moves with it. By analogy with the geo-grid, we can talk about proximity-fence.

In another use case, a customized QR-code reader can compare information about a wireless environment with information that is actually available at a given time and location. Model service that allows using an account (profile) in a social network for physical authorization:

- Mobile user is authorized in the social network right from the own phone.
- As a result of authorization, our application will get a link to a user profile in a social network.
- For the received profile (URL), a QR code is built and a wireless fingerprint is used as a contextual information.
- This QR code is presented for verification right from the mobile phone screen. The code is simply displayed on the screen. And the code will also be scanned directly from the screen. The code screen is displayed in the same way as a paper ID document.
- When scanning a QR code (on the checking phone), the profile of the user being checked in the social network (for example, you can compare the profile picture with the original) and, most importantly, the data on the wireless environment presented in the QR code are compared with the live environment as it is seen from the checking phone. The coincidence of these data (within one or another metric) indicates that the user has logged in to the social network directly at the inspection site, and does not show a previously prepared picture.

Note that this approach does not require any marks on the social network. Here the application does not write (and also does not read anything) on the social network. The network is used only for authorization. It just confirms of authorship. This is an extension of the social network where identification information is used in the real (physical) space.

3 Wireless Node Advertising as a Data Sharing Tool

Unlike the previous passive approach, where the user identification in the social network (link to his profile) in some way was presented on demand, this approach can be called active. Here we intend to distribute through the network links to profiles of social network users located nearby. As a tool for distribution, it is proposed to use the mobile phone of a user who is authorized in a social network. In essence, the proposed approach turns a mobile phone into a tag similar to Google Eddystone, which will distribute some URL. In our case, it is a link to a user profile in a social network. In other words, it will be a classic Physical Web in its very original definition.

The idea is to override the function (procedure) of announcing wireless nodes. Nodes in wireless networks will learn about the existence (presence) of other nodes due to the presence of the presentation procedure. The node informs other nodes about its existence. Thus, mobile devices will recognize, for example, the presence of a Wi-Fi access point, a Bluetooth scan on a mobile phone will indicate the presence of other nodes, and so on. This all represents the process of advertising of wireless nodes and, accordingly, the mechanism of disclosure (search, definition) of wireless nodes.

So, our idea is to use the advertising process to distribute information defined by the end user. The advantages of this approach are obvious - we use regular mechanisms of wireless networks, and there is no need to somehow modify the system software on a mobile device to receive any information from neighboring nodes. Details of this approach are described in our works [18].

The key point here is just a simple detection (discovery) of nearby devices (mobile users), without the need for users to perform any special actions. That is why the process of searching (discovering) for devices is an ideal candidate. We also note that new technologies, such as, for example, Wi-Fi Direct, support fairly advanced presentation technologies (Wi-Fi Direct thus represents available services). This opens up great opportunities for distributing user-generated content in these service messages.

As per technical details, a Bluetooth 4.0 device, for example, may operate in three different modes depending on required functionality: advertising, scanning and initiating. It is illustrated in Fig. 5. A device in advertising mode, named advertiser, periodically transmits advertising information in three channels [19].

For example, the SSID (node's name) is a 32-byte string. For this service, the first three bytes are reserved for the standard prefix. Accordingly, when scanning, only Bluetooth nodes with such prefixes are selected. If we consider this as the transmission of data on the network, the following figures can be cited. Changing the name of the Bluetooth node and recognizing the new name by surrounding devices takes 500–1000 ms. Accordingly, from the point of view of information

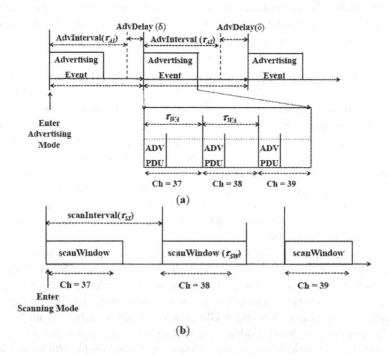

Fig. 5. On advertising (a) and scanning (b) in Bluetooth 4.0 [20]

transfer, this corresponds to a transmission rate of 32 bytes per second [18]. However, the amount of data transmitted in such a "network" is very small, the approach itself is 100% compatible with all models of model phones. In addition, this approach will work even in the absence of a telecom operator.

In other words, the representation (advertising) of a wireless node is used to transmit user information. This creates a new quality for services. Receiving such an advertisement indicates closeness to the source of the message and at the same time receives some data from it. If you compare this approach with geo-location services, for example, there, after receiving the coordinates on the mobile device with them, as a rule, you need to contact the server to obtain data related to these coordinates. The proposed model combines the definition of proximity (this replaces the acquisition of geo-coordinates) and request information. This direction has been developed by the authors of the article for a long time. This bibliography illustrates our various works in this field.

What the service looks like:

1. In the mobile application, the user is authorized in the social network.
2. After authorization, the application receives a user ID. This ID allows you to uniquely create a link to a user profile in a social network. For example, this is something like https://network.server.com/ID.php. The details of the URL can vary, but the general approach remains the same for all networks (Facebook, Vkontakte, Linkedin, etc.). Actually, it is the standard template for web applications - the so-called well-known URL.
3. Using the received ID, we can programmatically create (open) a Bluetooth point directly on the author's phone, using the received ID from the social network as a name for our node. We can make this node discoverable. It could be done via Android API too. This is the representation of the wireless node. Accordingly, other mobile devices that are nearby (Bluetooth distance) can "see" this ID.
4. Io is the same mobile application, which is used for authorization in a social network in parallel scans the available ("visible") Bluetooth nodes. This operation, in fact, is available in some local neighborhood - the so-called Bluetooth distance [23]. In most cases - up to 10 m. Among the available Bluetooth nodes, you can select those that broadcast ID from the social network. This is achieved by adding a prefix to the point name during translation.
5. Having collected the available ("visible") identifiers, the application can obtain publicly available information for each of them. This also works uniformly for all networks – another well-known URL will allow you to get a public image (avatar) of a profile.
6. As a result, you can create an illustrated list of profiles of social network users who are nearby. The list is made up of pictures of profiles and clickable links to profiles. Accordingly, if necessary, you can go to the profile of the selected neighbor and there already interact with it as allowed by a specific social network. Note that in this way the application will work in all social networks, where you can form a "well known URL" for the user profile. The figure below from the diploma project carried out in Lomonosov Moscow

State University under the direction of the author illustrates the work in the network VK.com - the most popular social network in Russia.

As a simple business model of an application, you can call an analog badge for a conference or other social event, where instead of a physical card with a name and a photo, there is a link to a social profile with a name and a photo.

Also, we should note that this approach does not require to make any entries in the social network or even getting any permissions to read data from the social network. Other users of the network can not get information about such a test. In fact, an ID from a social network is shown only to those who physically see its owner.

For social networks, the proposed model is a typical example of customized check-ins [21]. Classically, a check-in record in a social network is some message (post, status, etc.) linked to the particular location (to the particular geographical place). In other words, it is some geo-located message, presented on a social network [24]. These marks are made in order to introduce the social network user to other users nearby (according to their own check-ins). In this model, we can generally avoid any marks on social networks, and directly present the user to his neighbors.

Traditionally, "places" in social networks are described via classical geo-location info (latitude and longitude pair). Of course, it is already problematic for indoor applications. With this model, we can remove "places" from social networks and make them completely dynamic. Any new "place" is just a network fingerprint [25]. In our case, it is a set of "visible" Bluetooth nodes. And the "places" in this definition are completely dynamic. Turning a Bluetooth node on or off defines a new "location" (Fig. 6).

This design can also be called serverless check-in. A social network user does not actually report their location to the social network. He shares his representation in the social network with those who at the moment can see it physically and only with them. In reality, the social network here can be replaced by any other service that confirms authorship.

Speaking of security and privacy issues, two points should be noted. Firstly, we are not talking about monitoring problems. This is a conscious and benevolent representation of a user's link to his or her profile on a social network. The closest analogue is to wear a badge with your name on it at the conference. The user fully controls when he provides such information (takes off and puts on a badge). Accordingly, we are not talking about privacy violations here. The information thus can receive only those who are in physical proximity (literally - can see the author). When we talk about security, it is necessary to note the following. The proposed approach does not imply any connection between the devices. This is exactly what provides security.

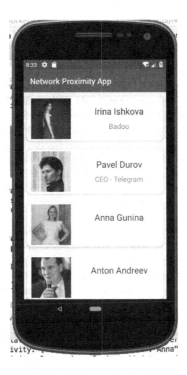

Fig. 6. Main screen [22]

4   Conclusion

In this article, we examined the mechanisms for expanding social networks in the physical space. This term was understood as the direct use of data (links) from social networks without the need to make any marks in the profiles of social network users. The basic model of the services presented here can be described as the use of social network identification in real (physical) space. In the given model examples, confirmed user identification in a social network, performed in close proximity to other mobile users, became available to these users. This accessibility was provided by the dissemination of the URL (web link) to the user profile on the social network. At the same time, the proposed approach does not require any entries (marks) in the social network.

The proposed model is an illustration of the general approach to the development of services based on proximity, when customized advertising of wireless nodes is used to redistribute user-defined information.

References

1. Sneps-Sneppe, M., Namiot, D.: On physical web models. In: 2016 International Siberian Conference on Control and Communications (SIBCON). IEEE (2016)
2. Namiot, D., Sneps-Sneppe, M.: On Bluetooth proximity models. In: Advances in Wireless and Optical Communications (RTUWO). IEEE (2016)
3. Kim, W., et al.: On target tracking with binary proximity sensors. In: Proceedings of the 4th International Symposium on Information Processing in Sensor Networks. IEEE Press (2005)
4. Sabatini, A.M., et al.: A low-cost, composite sensor array combining ultrasonic and infrared proximity sensors. In: Proceedings 1995 IEEE/RSJ International Conference on Intelligent Robots and Systems. Human Robot Interaction and Cooperative Robots, vol. 3. IEEE (1995)
5. Han, M., Lee, Y.-K., Lee, S.: Comprehensive context recognizer based on multi-modal sensors in a smartphone. Sensors 12(9), 12588–12605 (2012)
6. Proximity Sensor Repair. https://proximity-sensor-repair-reset.ru.aptoide.com/. Accessed Apr 2019
7. Namiot, D., Sneps-Sneppe, M.: Context-aware data discovery. In: 2012 16th International Conference on Intelligence in Next Generation Networks (ICIN). IEEE (2012)
8. Namiot, D., Sneps-Sneppe, M.: Geofence and network proximity. In: Balandin, S., Andreev, S., Koucheryavy, Y. (eds.) NEW2AN/ruSMART -2013. LNCS, vol. 8121, pp. 117–127. Springer, Heidelberg (2013). https://doi.org/10.1007/978-3-642-40316-3_11
9. Schneps-Schneppe, M., et al.: Wired smart home: energy metering, security, and emergency issues. In: 2012 IV International Congress on Ultra Modern Telecommunications and Control Systems. IEEE (2012)
10. Namiot, D., Sneps-Sneppe, M.: The physical web in smart cities. In: 2015 Advances in Wireless and Optical Communications (RTUWO). IEEE (2015)
11. Ibeacon Insidder. http://www.ibeacon.com/what-is-ibeacon-a-guide-to-beacons/. Accessed Apr 2019
12. How EddyStone Works. https://www.beaconzone.co.uk/HowEddystoneWorks/. Accessed Apr 2019
13. EddyStone vs iBeacon. https://blog.beaconstac.com/2016/08/eddystone-vs-physical-web-vs-ibeacon-why-eddystone-will-rule-the-beacon-space/. Accessed Apr 2019
14. Namiot, D.: Context-aware browsing–a practical approach. In: 2012 Sixth International Conference on Next Generation Mobile Applications, Services and Technologies. IEEE (2012)
15. Namiot, D., Sneps-Sneppe, M., Skokov, O.: Context-aware QR codes. World Appl. Sci. J. 25(4), 554–560 (2013)
16. Lyardet, F., Szeto, D.W., Aitenbichler, E.: Context-aware indoor navigation. In: Aarts, E., et al. (eds.) AmI 2008. LNCS, vol. 5355, pp. 290–307. Springer, Heidelberg (2008). https://doi.org/10.1007/978-3-540-89617-3_19
17. Rouillard, J.: Contextual QR codes. In: 2008 The Third International Multi-Conference on Computing in Global Information Technology (ICCGI 2008). IEEE (2008)
18. Namiot, D., Sneps-Sneppe, M.: On proximity-based information delivery. In: Vishnevskiy, V.M., Kozyrev, D.V. (eds.) DCCN 2018. CCIS, vol. 919, pp. 83–94. Springer, Cham (2018). https://doi.org/10.1007/978-3-319-99447-5_8

19. Liu, J., Chen, C., Ma, Y.: Modeling and performance analysis of device discovery in Bluetooth low energy networks. In: Proceedings of the IEEE on Global Communications Conference (GLOBECOM), Anaheim, CA, USA, 3–7 December 2012, pp. 1538–1543 (2012)
20. Liu, J., Chen, C., Ma, Y.: Modeling neighbor discovery in Bluetooth low energy networks. IEEE Commun. Lett. **16**, 1439–1441 (2012)
21. Namiot, D., Sneps-Sneppe, M.: Customized check-in procedures. In: Balandin, S., Koucheryavy, Y., Hu, H. (eds.) NEW2AN/ruSMART -2011. LNCS, vol. 6869, pp. 160–164. Springer, Heidelberg (2011). https://doi.org/10.1007/978-3-642-22875-9_14
22. Makarychev, I.: Using physical web as an extension for social networks. Diploma thesis, Lomonosov Moscow State University (2019)
23. Jung, J., Kang, D., Bae, C.: Distance estimation of smart device using bluetooth. In: ICSNC 2013 - The Eighth International Conference on Systems and Networks Communications, pp. 13–18 (2013)
24. Stefanidis, A., Crooks, A., Radzikowski, J.: Harvesting ambient geospatial information from social media feeds. GeoJournal **78**(2), 319–338 (2013)
25. Faragher, R., Harle, R.: Location fingerprinting with Bluetooth low energy beacons. IEEE J. Sel. Areas Commun. **33**(11), 2418–2428 (2015)

The General Renovation as the Active Queue Management Mechanism. Some Aspects and Results

Viana C. C. Hilquias[1]([✉])[iD], I. S. Zaryadov[1,2][iD], V. V. Tsurlukov[1][iD],
T. A. Milovanova[1][iD], E. V. Bogdanova[1][iD], A. V. Korolkova[1][iD],
and D. S. Kulyabov[1,3][iD]

[1] Department of Applied Probability and Informatics, Peoples' Friendship University
of Russia (RUDN University), Miklukho-Maklaya street 6, Moscow 117198, Russia
{1042195028,zaryadov-is,1032181900,milovanova-ta,
korolkova-av,kulyabov-ds}@rudn.ru
[2] Institute of Informatics Problems, FRC CSC RAS, IPI FRC CSC RAS,
44-2 Vavilova Street, Moscow 119333, Russia
[3] Laboratory of Information Technologies, Joint Institute for Nuclear Research,
Joliot-Curie 6, Dubna, Moscow region 141980, Russia

Abstract. This work is devoted to some aspects of using the general renovation (the definition and brief overview are given) as the active queue management mechanism (like RED (Random Early Detection) algorithms).

The attention is paid to the queuing system in which a threshold mechanism and the general renovation mechanism are implemented. This allows to adjust the number of packets in the system by dropping (resetting) the packets from the queue depending on the ratio of a certain control parameter with specified thresholds. But in contrast to standard RED-like systems, a possible reset occurs not at the time of arriving of the next packets in the system, but at the time of the end of service on the device (server). Numerical results for the main probability characteristic (stationary loss probability) are presented.

Keywords: Random early detection · Active queue management · Queuing system · General renovation · Threshold mechanism

The publication has been prepared with the support of the "RUDN University Program 5–100" (A. V. Korolkova—the analysis of RED algorithms, Viana C. C. Hilquias—mathematical model development, V. V. Tsurlukov — the scientific review preparation, E. V. Bogdanova—simulation modelling and numerical analysis). Also the publication has been funded by Russian Foundation for Basic Research (RFBR) according to the research project No. 19-01-00645 (D. S. Kulyabov—simulation model of RED algorithms development), No. 18-07-00692 (I. S. Zaryadov—statement of the problem, the mathematical model development, T. A. Milovanova—the mathematical model development) and No. 19-07-00739 (I.S. Zaryadov, T.A. Milovanova—numerical analysis based on the obtained analytical results).

© Springer Nature Switzerland AG 2019
V. M. Vishnevskiy et al. (Eds.): DCCN 2019, CCIS 1141, pp. 488–502, 2019.
https://doi.org/10.1007/978-3-030-36625-4_39

1   Introduction

The problem of mitigation of congestion and congestion avoidance in modern communication networks is the actual task for researches and practitioners, and, as may be seen [1], this problem does not have a satisfying solution.

According to RFC 7567 [1] active queue management (AQM) is considered as a best practice of network congestion avoidance (reducing) in Internet routers. The active queue management is a based on some rules (algorithms such as random early detection (RED) [2–20], Random Exponential Marking (REM), Blue [21] and stochastic fair Blue (SFB), Adaptive virtual queue algorithm [18–20,22], Explicit Congestion Notification (ECN) [3,23], or controlled delay (CoDel) [24–26]) technique of intelligent drop of network packets inside a buffer associated with a network interface controller (NIC), when that buffer becomes full or gets close to becoming full.

But, as was mentioned before, the problem of congestion avoidance is still actual [31–36], so there exists the IETF working group on "Active Queue Management and Packet Scheduling" [27], where some more novel AQM algorithms are investigated and standardised. A numerous number of AQM schemes have been proposed [19,28–37]. The performance analysis of the most of them is performed by simulation (for example, [34,38]) and the bridges between the available use-case results and analytic results, as well as between the available analytic results are very few (see, for example, [39–43]).

We will consider only the case of RED algorithm and some its modifications. RED has the ability to absorb bursts and also is simple, robust and quite effective at reducing persistent queues [2–5].

In this paper the mathematical model of RED-like algorithm with thresholds will be presented. The word "RED-like" is used because in contrast to standard RED algorithm, when a possible reset occurs at the time of arriving of the next packet in the system [2] and the control parameter is an exponentially weighted average queue length as for classic RED [2], our model is based on completely different idea: the decision about a possible packet drop is synchronised with the service completions. Thus we will use the so called renovation mechanism [44,57–60,63,64].

Some experimental results [61,62] in this direction show that the use of the renovation mechanism in the finite-capacity single server queues under heavy overload conditions allows one to achieve at least the same performance level, as the one guaranteed by the classical random early detection scheme.

The structure of the article is following: the Sect. 2 gives the brief description of the classic RED algorithm, the Sect. 3 is devoted to the general renovation mechanism, some results concerning RED and renovation, and our mathematical model based on general renovation mechanism with thresholds. The last section concludes the paper with the short discussion.

2   The Brief Description of RED Algorithm Module

The classic RED [2] is a queueing discipline with two thresholds (Q_{min} and Q_{max}) and a low-pass filter to calculate the average queue size $\hat{Q}$:

$$\hat{Q}_{k+1} = (1 - w_q)\hat{Q}_k + w_q\hat{Q}_k, \quad k = 0, 1, 2, \ldots, \tag{1}$$

where w_q, $0 < w_q < 1$ is a weight coefficient of the exponentially weighted moving-average and determines the time constant of the low-pass filter. In the optimal bounds for w_q are presented.

RED monitors the average queue size and drops (or marks when used in conjunction with ECN) packets based on statistical probabilities $p(\hat{Q})$ [2].

$$p(\hat{Q}) = \begin{cases} 0, & 0 \le \hat{Q} \le Q_{min}, \\ \dfrac{\hat{Q} - Q_{min}}{Q_{max} - Q_{min}} p_{max}, & Q_{min} < \hat{Q} \le Q_{max}, \\ 1, & \hat{Q} > Q_{max}, \end{cases} \tag{2}$$

p_{max} is the maximum level of packages to be dropped (marked or reset).

RED is more fair than tail drop when the incoming packet is dropped only if the buffer is full. Also RED does not possess a bias against bursty traffic that uses only a small portion of the bandwidth. But, as shown in [7], RED has a number of problems, one of which is that it need tuning and has a little guidance on how to set configuration parameters.

The RED modifications are well presented in [37], but we will specify some of them.

Weighted RED (WRED) [8]—in this algorithm different probabilities for different types of traffic with different priorities (IP precedence, DSCP) and/or queues may be defined. The modification of WRED is Distributed Weighted RED (DWRED)[9].

Adaptive RED or active RED (ARED) [10]—was designed in order to make RED algorithm (based on the observation of the average queue length) more or less aggressive. If the average queue length $\hat{Q}$ oscillates around Q_{min} minimum threshold then early detection considers to be aggressive. If the average queue length $\hat{Q}$ oscillates around Q_{max} threshold then early detection is being too conservative. The drop probability is changed by the algorithm according to how aggressively it senses it has been discarding traffic.

Robust RED (RRED) [11]—is proposed for TCP throughput improvement against DoS (Denial-of-Service) attacks, especially LDoS (Low-rate Denial-of-Service) attacks. The basic idea behind the RRED is to detect and filter out LDoS attack packets from incoming flows before they feed to the RED algorithm. When loss of a sent packet is detected by the source then there will be a transmit delay, so a packet which was sent within a short-range after a loss detection will be suspected to be an attacking packet. This is the basic idea of the detection algorithm of Robust RED (RRED).

EASY RED [12]—is a simpler variant of RED. the drop probability is defined not by average queue length but by instantaneous queue length. The reason is to inform the sender about congestion as soon as possible. The EASY RED parameters are the minimum threshold Q_{min} and the drop probability p_{drop}, which is a constant and used only when the instantaneous queue length is greater or equal to Q_{min}.

Stabilized RED (SRED) [13]—aims at stabilizing buffer occupation by estimating the number of active connections in order to set the drop probability as a function of the number of the active flows and of the instantaneous queue length.

Flow RED (FRED) [14]—uses per-active-flow accounting (based on minimum and maximum limits on the packets that a each flow may have in the queue) to impose on each incoming flow a loss rate (depends on the degree of buffer usage by a flow), it also uses a more aggressive drop against the flows that violates the maximum bound. The state information about active connections also needs to be maintained in the routers.

Balanced RED (BRED) [15]—is proposed to regulate the bandwidth of a flow also by doing per-active-flow accounting for the buffer, similar to FRED but with a different approach: in BRED, two variables (the measures of the packet number for one flow in the buffer and the packet number accepted from this flow since the previous packet dropping) for each flow having packets in the buffer are maintained, which are. As a result the decision of packet drop or acceptance is based on before mentioned two flow state variables.

Dynamic RED (DRED) [16]—is proposed to discard packets with a load dependent probability. The drop probability is updated by employing an integral controller (the input of the controller is the difference between the average queue length and the target buffer level, the output is the drop probability).

The more information about RED and its modifications (Gentle RED (GRED), RED with In and Out (RIO), WRED with thresholds (WRT), Exponential RED (EXPRED), Double Slope RED (DSRED), Random Early Dynamic Detection (REDD)), as other AQM algorithms (Random Exponential Marking (REM), Blue [21] and stochastic fair Blue (SFB), Adaptive virtual queue algorithm) is available at Sally Floyd webpage [17], or in [18–20], new approaches to AQM [22].

3 General Renovation as an Active Queue Management Scheme

3.1 The Definition of General Renovation Mechanism

The renovation mechanism was first defined in the paper [44] and the idea of this mechanism was following: at the moment of the end of its service the packet on the server may either just leave the system with some non-zero probability p, or may empty the buffer with the renovation probability $q = 1 - p$. In [44] the steady-state probability distributions for several types of queueing systems were presented.

Some applications of the renovation mechanism in finance other fields were shown in [45]. So, the queues with renovation are similar to queues with disasters (or negative customers), when the incoming flow of signals cause the buffer to drop some or all the packets, or to queues with unreliable servers, which cause the packet dropping ([46–56]).

In [57] the mathematical model of renovation mechanism with repeated service (or feedback) was proposed by P. P. Bocharov. It means that the served packet after emptying the buffer with probability q enters the server for another round of service. The main characteristics in matrix-analytical form were obtained.

Later on the renovation mechanism was further generalised by Pechinkin [58], who proposed the mathematical model of general renovation: at the end of service the packet discards from the buffer of capacity $0 < r < \infty$ exactly i, $i \geq 1$, other packets with probability $q(i)$ and leaves the system, or just leaves the system without any effect on it with the complementary probability $p = 1 - \sum_{i=1}^{r} q(i)$. In [58–60] the multiserver $GI|M|n|\infty$ and $GI|M|n|r$ queueing systems with different renovation and service disciplines were studied. The general renovation with feedback (the retrial queueing system with general renovation and recurrent input flow) was investigated in [63,64].

3.2   The Mathematical Model of RED-like Algorithm by Queueing System with Renovation and Thresholds

In this part of the article we will discuss another model of RED-like algorithm based on queueing systems with renovation and thresholds.

In [65] the queueing system with two thresholds ($q_{min} < q_{max}$), recurrent input flow of packets and exponentially identically distributed service times on C servers with renovation mechanism was considered.

The idea of mechanism of renovation with thresholds is following. At the moment of the end of a packet service the current queue length $\tilde{q}$ is compared with thresholds q_{min} and q_{max}, and if $\tilde{q} \leq q_{min}$ then no one of the packets from the buffer is dropped. If $q_{min} + 1 \geq \tilde{q} \leq q_{max}$ then the last packet in the buffer is dropped with probability $p(\tilde{q})$, $0 < p(\tilde{q}) < p_{max}$. If $\tilde{q} \geq q_{max} + 1$ also the last packet in the buffer is dropped but with maximal probability p_{max}.

The steady-state probability distribution of packets in the system(for the imbedded upon arrival moments Markov chain) were obtained in geometric form:

$$p_0 = \sum_{j=1}^{C-1} p_{j-1} A_{j,0} + p_{C-1} A_0^*, \tag{3}$$

$$p_i = \sum_{j=i-1}^{C-1} p_j A_{j+1,i} + p_{C-1} A_i^*, \quad i = \overline{1, C-1}, \tag{4}$$

$$p_{C+j} = p_{C-1} \prod_{m=0}^{j} g_{q_{min}-m}, \quad j = \overline{0, q_{min}-1}, \tag{5}$$

$$p_{C+q_{\min}+j} = p_{C-1} \prod_{m=1}^{q_{\min}} g_m \prod_{l=0}^{j} \tilde{g}_{q_{\max}-q_{\min}-l}, \quad j = \overline{0, q_{\max} - q_{\min} - 1}, \quad (6)$$

$$p_{C+q_{\max}+j} = p_{C-1} \hat{g}^{j+1} \prod_{m=1}^{q_{\min}} g_m \prod_{l=0}^{q_{\max}-q_{\min}} \tilde{g}_l, \quad j \geq 0, \quad (7)$$

where $A_{(i,j)}$, $i, j \geq 0$ are elements of the transition probability matrix of the embedded Markov chain [65], A_i^*, $i = \overline{0, C-1}$ are auxiliary values [65], $\hat{g}$ is the unique solution of the equation $\hat{g} = \alpha(C\mu(1 - \hat{g}))$ which belongs to the interval $(0; 1)$, $\alpha(\cdot)$ is the Laplace-Stieltjes transformation, the values g_i and $\tilde{g}_i$ are defined by

$$g_i = \frac{\alpha(C\mu)}{L_i}, \quad i = \overline{1, q_{\min}}, \quad \tilde{g}_i = \frac{C\mu}{K_i}, \quad i = \overline{1, q_{\max} - q_{\min} - 1}, \quad (8)$$

and L_i, $i = \overline{1, q_{\min}}$, K_i, $i = \overline{1, q_{\max} - q_{\min} - 1}$ are also fully defined in [65].

The probability p^{loss} that the incoming packet will be dropped from the system by one of the served packets is:

$$p^{loss} = p_{C-1}(1 - \alpha(C\mu)) \left(\sum_{i=q_{\min}}^{q_{\max}-1} p(\hat{q}) \prod_{m=1}^{q_{\min}} g_m \prod_{l=0}^{i-q_{\min}} \tilde{g}_{q_{\max}-q_{\min}-l} + \right.$$
$$\left. + \prod_{m=1}^{q_{\min}} g_m \prod_{l=0}^{q_{\max}-q_{\min}} \tilde{g}_l \frac{\hat{g}}{1 - \hat{g}} \right). \quad (9)$$

The probability p^{serv} that the incoming packet will be served is:

$$p^{serv} = 1 - p^{loss}. \quad (10)$$

The mean waiting time of a served packet w^{serv} is

$$w^{serv} = \frac{p_{C-1}}{p^{serv}} \left(\sum_{i=0}^{q_{\min}-1} \frac{i+1}{C\mu} \prod_{m=0}^{i} g_{q_{\min}-m} + + \prod_{m=1}^{q_{\min}} g_m \sum_{i=q_{\min}}^{q_{\max}-1} \prod_{l=0}^{i-q_{\min}} \tilde{g}_{q_{\max}-q_{\min}-l} \right.$$
$$\left(\frac{(i+1)(1 - p(\tilde{q}))}{C\mu} - \alpha^{(1)}(C\mu) + \frac{(i+1)\alpha(C\mu)}{C\mu} \right) +$$
$$\left. + \prod_{m=1}^{q_{\min}} g_m \prod_{l=0}^{q_{\max}-q_{\min}} \tilde{g}_l \left(\frac{\alpha(C\mu)\hat{g}q_{\max}}{C\mu(1 - \hat{g})} + \frac{\alpha(C\mu)\hat{g}}{C\mu(1 - \hat{g})^2} - \frac{\alpha^{(1)}(C\mu)\hat{g}}{(1 - \hat{g})} \right) \right). \quad (11)$$

The recommendations on the optimal values of $(q_{\min} < q_{\max})$ (based on the numerical analysis of the obtained characteristics) were similar as in [5].

3.3   The Mathematical Model of RED-like Algorithm by Queueing System with General Renovation and One or Two Thresholds

But we want to consider the model with general renovation — when a group of packets may be dropped from the buffer. For example, the authors work on the $G|M|1|\infty$ system with only one threshold q_{min}. If the current queue size $\tilde{q} \leq q_{min}$, then served packet just leave the system. But if $q_{min} + 1 \geq \tilde{q}$ then three different types of dropping mechanism may be applied:

1. with probability $q(i)$ $(i \geq 1)$ the group of i packets from the buffer is dropped (if there are less than i packets in buffer, the buffer will be emptied);
2. with probability $q(i)$ $(i \geq 1)$ the group of i packets from the buffer is dropped if there were $q_{min} + i$ packets or only q_{min} packets will remain in the buffer;
3. the virtual threshold q^* is introduced and with probability $q(i)$ $(i \geq 1)$ the group of i packets from the buffer is dropped if $\tilde{q} - i \leq q^*$ or only q^* packets will remain in the buffer.

The minus of the first drop mechanism is that too many packets may be dropped. The minus of the second drop mechanism is that the buffer may remain overflowed. The third mechanism is the combination of the previous ones without minuses of the previous ones, but it may be difficult for analytical investigation.

Our goal is to construct mathematical models for all three cases and to compare such characteristics as the probability p^{loss} of an arbitrary arrived packet being dropped from the system, and mean sojourn times w^{loss} and w^{serv} for lost (dropped) packets and served packets.

For the first model we obtained the steady-state probability distribution of packets p_i $(i \geq 0)$ in the system(for imbedded Markov chain), some probabilities are represented by geometric form (when the threshold q is overcomed):

$$p_i = \sum_{j=i-1}^{q} p_j(-\mu)^{j+1-i} \alpha^{(j+1-i)}(\mu) +$$

$$+ p_{q+1} g^{i-q-2} \left(g - \alpha(\mu) - \int_0^\infty A(g,x)e^{-\mu x} dA(x) \right),$$

$$i = \overline{1, q+1}, \quad (12)$$

$$p_i = p_{q+1} g^{i-(q+1)}, i > q+1, \quad (13)$$

$$p_0 = \sum_{i=0}^{\infty} p_i p_{i,0}, \quad (14)$$

$$p_{i,0} = 1 - \sum_{j=1}^{i} (-\mu)^j \alpha^{(j)}(\mu), 0 < i \leq q, p_{i,0} = 1 - \sum_{j=1}^{i} \pi(j, i-j) \alpha^{(j)}(\mu), i > q. \quad (15)$$

where g is the unique solution of the equation

$$g = \alpha \left(\mu(1 - gQ(g)) \right), \tag{16}$$

and belongs to interval $(0; 1)$, $\alpha(s)$ is the Laplase-Stieltjes transformation of interrarrival time distribution function $A(x)$, $Q(g)$ is the probability generating function

$$Q(l, g) = \sum_{k=0}^{\infty} \pi(l, k)g^k = Q^l(g), \quad Q(g) = p + \sum_{k=1}^{\infty} \pi(1, k)g^k. \tag{17}$$

for probabilities $\pi(l, k)$, $l \geq 1, k \geq 0$, that l packets will be served and k packets will be dropped from the buffer.

The probabilities $\pi(l, k)$, $l \geq 1$, $k \geq 0$, may be defined by following. If k packets are served and none of the packets are dropped from the buffer, then

$$\pi(l, 0) = p^l, \quad l \geq 1, \quad \pi(0, k) \equiv 0, \quad k \geq 1, \tag{18}$$

because the packets may be dropped from the queue only at the moment of the end of the service, but

$$\pi(0, 0) \equiv 1, \tag{19}$$

if the service on the server has not ended, then no packet can be dropped from the queue.

$$\pi(1, k) = q(k), \quad k \geq 1, \tag{20}$$

is the probability that a packet at the moment of the end of the service will drop from the queue k other packets. And the general formula is

$$\pi(l, k) = \sum_{n=0}^{k} \pi(1, n)\pi(l - 1, k - n), \quad l \geq 1, \quad k \geq 0. \tag{21}$$

$$A(g, x) = \sum_{l=1}^{q+1-i} \frac{(\mu x g)^l}{l!} \sum_{j=0}^{q+1-i-l} \pi(l, j)g^j. \tag{22}$$

Also the probability that the arriving packet will be dropped and sojourn time characteristics for dropped packets are obtained in form of integral equations.

$$p^{loss} = \sum_{i=1}^{\infty} p_{i,0}^{loss} p_i, \tag{23}$$

where $p_{i,j}^{loss}$ is the the probability that the "selected" packet will be dropped if there are i, $i \geq 1$, packets before it and j, $j \geq 0$ packets behind it; $p_{i,0}^{loss}$ is the probability that the arriving packet will be dropped if there are i, $i \geq 1$, packets in the system.

$$p_{i,j}^{loss} = \int_0^{\infty} \int_0^x \sum_{k=0}^{i} \frac{(\mu y)^k}{k!} e^{-\mu y} dA(y) p_{i-k,j+1}^{loss}(x - y)dx, \quad 0 \leq i + j \leq q, \tag{24}$$

$$p_{0,j}^{loss} \equiv 0, j \geq 0, \tag{25}$$

because the packet could not be dropped being on service.

$$p_{i,j}^{loss} = \int\limits_0^\infty \bar{A}(x) \sum_{m=1}^i \frac{\mu^m x^{m-1}}{(m-1)!} e^{-\mu x} \pi^*(m, i+j) dx +$$

$$+ \int\limits_0^\infty \int\limits_0^x \sum_{m=0}^i p^m \frac{(\mu y)^m}{m!} e^{-\mu y} dA(y) p_{i-m,j+1}^{loss}(x-y) dx, \quad i+j > q, \tag{26}$$

where $\pi^*(m, i)$ is the auxiliary probability that m, $m \geq 1$, served packets will completely empty the buffer of the system if there were i, $i \geq 0$, packets in it:

$$\pi^*(1, i) = \sum_{k=i}^\infty q(k), \quad \pi^*(m, i) = \sum_{k=0}^i \pi(1, k) \pi^*(m-1, i-k), \quad m > 1, i \geq 0. \tag{27}$$

The mean value of packets in the system is defined by the following formula:

$$N = \sum_{i=0}^\infty i p_i = \sum_{i=1}^q i p_i + p_{q+1} \frac{q(1-g)+1}{(1-g)^2}. \tag{28}$$

The probability distribution of the time in the system for dropped packet $W^{loss}(x)$ may de derived by the same way as p^{loss}.

$$W^{loss}(x) = \sum_{i=0}^\infty W_{i,0}^{loss}(x) p_i, \tag{29}$$

where $W_{i,0}^{loss}(x)$ is the probability that the packet will be dropped from the buffer for time less than x if there were i, $i \geq 0$, other packets in the system at the arrival moment. In terms of Laplace-Stieltjes (29) takes form:

$$\omega^{loss}(s) = \sum_{i=0}^\infty \omega_{i,0}^{loss}(s) p_i, \tag{30}$$

where

$$\omega_{i,j}^{loss}(s) = \sum_{m=1}^i \frac{(-\mu)^m \alpha^{(m)}(\mu s)}{m!} \omega_{i-m,j+1}^{loss}(s), \quad i+j \leq q, \tag{31}$$

$$\omega_{i,j}^{loss}(s) = \sum_{m=1}^i \frac{(-1)^{m-1} \mu^m \bar{\alpha}^{(m-1)}(\mu s)}{(m-1)!} \pi^*(m, i+j) +$$

$$+ \sum_{m=0}^i \frac{(-\mu p)^m \alpha^{(m)}(\mu s)}{m!} \omega_{i-m,j+1}^{loss}(s), \quad i+j > q. \tag{32}$$

3.4   The Comparison of RED Algorithm with General Renovation and General Renovation with Feedback

In this section we will compare the loss probability p^{loss} (the probability of packet being dropped from the system) for RED and TailDrop algorithms (the values of p^{loss} are presented in [40,43]) and values of the probability p^{loss}, obtained by formulas derived for queueing system with general renovation [61,62], queueing system with general renovation and feedback [63,64], renovation with two thresholds (Sect. 3.2) and general renovation with one threshold (Sect. 3.3). Also we will use results obtined by the members of our authors group in their Master's thesises.

Table 1. The values of p^{loss} for Taildrop, RED, general renovation and general renovation with feedback

Loss probability					
Taildrop	RED	renov	ren-fd	ren-2-th	ren-1-th.
$\rho = 0.5$					
0	0.002	0.002	0.0002	0.00021	0.000205
$\rho = 1$					
0.051	0.091	0.104	0.11	0.067	0.074
$\rho = 2$					
0.500	0.500	0.502	0.54	0.503	0.500
$\rho = 3$					
0.667	0.667	0.667	0.71	0.668	0.666

As can be see from the Table 1, according to the values of the p^{loss}, all types of renovation mechanism can perform as good as RED in the wide range of the offered load ρ, but for the case of general renovation fine and accurate tuning of general renovation probabilities $q(i)$ is required,

4   Conclusion

Even though the idea behind the renovation-type AQM is completely different from the idea behind RED-type AQM, renovation-type AQM may allow one to achieve in some cases at least the same system performance level as guaranteed by RED-type AQM.

The presented numerical experiments show that the results remain qualitatively the same for RED-type AQM with other dropping functions. Being defined by N parameters, the renovation mechanism is very flexible and this constitutes its strength and weakness. By varying the values of the renovation probabilities $q(i)$, it is possible to carry out conditional optimisation, but good searching procedures are required here.

Implementation of the renovation as a packet dropping mechanism requires a priori tuning and/or operational configuration of its parameters. Thus, whether it is appropriate to use renovation as a packet dropping mechanism or not in practice heavily depends on the use case. Although the tuning of the renovation parameters qi can be made on the fly during operation, with respect to the recommendations of the RFC 7567 [1], renovation mechanism is not the proper choice for the network congestion control unless simple recommendations on how to set up the renovation parameters are given. We believe this can be done based on more deep and insightful numerical experiments.

At the end, it is worthy of mention that there is another approach to the analysis of behaviour of networks with burst traffic — the method based on hysteretic thresholds load control [66–69] and it will interesting to investigate systems with hysteretic control of renovation probabilities, especially for the case of Markov arrival process [69]. The authors will try to combine the method of hysteretic thresholds load control with renovation mechanism.

References

1. Baker, F., Fairhurst, G.: IETF recommendations regarding active queue management. RFC 7567. Internet Engineering Task Force. https://tools.ietf.org/html/rfc7567. Accessed 29 Apr 2019
2. Floyd, S., Jacobson, V.: Random early detection gateways for congestion avoidance. IEEE/ACM Trans. Netw. 4(1), 397–413 (1993). https://doi.org/10.1109/90.251892
3. Ramakrishnan, K., Floyd, S., Black, D.: The addition of explicit congestion notification (ECN) to IP. RFC 3168. Internet Engineering Task Force. https://tools.ietf.org/html/rfc3168. Accessed 29 May 2019
4. Floyd, S., Gummadi, R., Shenker, S.: Adaptive RED: An Algorithm for Increasing the Robustness of RED's Active Queue Management (2001). http://www.icir.org/floyd/papers/adaptiveRed.pdf
5. Floyd, S.: RED: Discussions of Setting Parameters (1997). http://www.aciri.org/floyd/REDparameters.txt
6. Korolkova, A.V., Zaryadov, I.S.: The mathematical model of the traffic transfer process with a rate adjustable by RED. In: International Congress on Ultra Modern Telecommunications and Control Systems and Workshops (ICUMT), pp. 1046–1050. IEEE. Moscow. Russia (2010) https://doi.org/10.1109/ICUMT.2010.5676505
7. Jacobson, V., Nichols, K., Poduri, K.: RED in a Different Light (2019). http://citeseerx.ist.psu.edu/viewdoc/summary?doi=10.1.1.22.9406. Accessed 1 Sept 2019
8. Class-Based Weighted Fair Queueing and Weighted Random Early Detection (2019). http://www.cisco.com/c/en/us/td/docs/ios/12_0s/feature/guide/fswfq26.html. Accessed 1 Sept 2019
9. Cisco IOS Quality of Service Solutions Configuration Guide, Release 12.2. https://www.cisco.com/c/en/us/td/docs/ios/qos/configuration/guide/12_2sr/qos_12_2sr_book.pdf. Accessed 1 Sept 2019
10. Floyd, S., Gummadi, R., Shenker, S.: Adaptive RED: An Algorithm for Increasing the Robustness of RED's Active Queue Management (2001). http://www.icir.org/floyd/papers/adaptiveRed.pdf. Accessed 1 Sept 2019

11. Changwang, Z., Jianping, Y., Zhiping, C., Weifeng, C.: RRED: robust RED algorithm to counter low-rate Denial-of-Service attacks. IEEE Commun. Lett. **14**(5), 489–491 (2010). https://doi.org/10.1109/LCOMM.2010.05.091407
12. Grieco, L.A., Mascolo, S.: TCP westwood and easy RED to improve fairness in high-speed networks. In: Carle, G., Zitterbart, M. (eds.) PfHSN 2002. LNCS, vol. 2334, pp. 130–146. Springer, Heidelberg (2002). https://doi.org/10.1007/3-540-47828-0_9
13. Ott, T.J., Lakshman, T.V., Wong, L.H.: SRED: stabilized RED. In: Proceedings IEEE INFOCOM 1999, vol. 3, pp. 1346–1355. IEEE, New York, NY, USA (1999). https://doi.org/10.1109/INFCOM.1999.752153
14. Lin, D., Morris, R.: Dynamics of random early detection. Comput. Commun. Rev. **27**(4), 127–137 (1997)
15. Anjum, F.M., Tassiulas, L.: Balanced RED: An Algorithm to Achieve Fairness in the Internet. Technical Research Report (1999). http://www.dtic.mil/dtic/tr/fulltext/u2/a439654.pdf
16. Aweya, J., Ouellette, M., Montuno, D.Y.: A control theoretic approach to active queue management. Comput. Netw. **36**, 203–235 (2001)
17. Sally Floyd Website. http://www.icir.org/floyd/. Accessed 1 Sept 2019
18. Chrysostomoua, C., Pitsillidesa, A., Rossidesa, L., Polycarpoub, M., Sekercioglu, A.: Congestion control in differentiated services networks using fuzzy-RED. Control Eng. Pract. **11**, 1153–1170 (2003)
19. Feng W.-C.: Improving Internet Congestion Control and Queue Management Algorithms. http://thefengs.com/wuchang/umich_diss.html
20. Al-Raddady, F., Woodward, M.: A new adaptive congestion control mechanism for the internet based on RED. In: 21st International Conference on Advanced Information Networking and Applications. AINAW 2007 Workshops (2007)
21. Feng, W., Kandlur, D.D., Saha, D., Shin, K.G.: BLUE: A New Class of Active Queue Management Algorithms. UM CSE-TR-387-99 (1999). https://www.cse.umich.edu/techreports/cse/99/CSE-TR-387-99.pdf
22. Baldi, S., Kosmatopoulos, E.B., Pitsillides, A., Lestas, M., Ioannou, P.A., Wan, Y.: Adaptive optimization for active queue management supporting TCP flows. In: 2016 American Control Conference (ACC), pp. 751–756 (2016)
23. Floyd, S.: TCP and Explicit Congestion Notification. https://www.icir.org/floyd/papers/tcp_ecn.4.pdf. Accessed 29 May 2019
24. Nichols, K., Jacobson, V., McGregor, A., Iyengar, J.: Controlled Delay Active Queue Management. RFC 8289. Internet Engineering Task Force. https://tools.ietf.org/html/rfc8289. Accessed 29 Aug 2019
25. Nichols, K., Jacobson, V.: Controlling Queue Delay ACM Queue. ACM Publishing **10**(5), 2020–2034 (2012). queue.acm.org/detail.cfm?id=2209336
26. Hoeiland-Joergensen, T., McKenney, P., Taht, D., Gettys, J., Dumazet, E.: The flow queue codel packet scheduler and active queue management algorithm. Internet Engineering Task Force (2018). https://tools.ietf.org/html/rfc8290
27. IETF Working Group on Active Queue Management and Packet Scheduling (AQM): Description of the Working Group. http://tools.ietf.org/wg/aqm/charters. Accessed 29 Aug 2019
28. McKenney, P.E.: Stochastic fairness queueing. In: Proceedings of IEEE International Conference on Computer Communications, vol. 2, pp. 733–740. IEEE, San Francisco, CA, USA (1990). https://doi.org/10.1109/INFCOM.1990.91316
29. Adams, R.: Active queue management: a survey. IEEE Commun. Surv. Tutorials **15**(3), 1425–1476 (2013)

30. Paul, A.K., Kawakami, H., Tachibana, A., Hasegawa, T.: An AQM based congestion control for eNB RLC in 4G/LTE network. In: 2016 IEEE Canadian Conference on Electrical and Computer Engineering (CCECE), pp. 1–5. IEEE. Vancouver, BC, Canada (2016). https://doi.org/10.1109/CCECE.2016.7726792

31. Dai, Yu., Wijeratne, V., Chen, Y., Schormans, J.: Channel quality aware active queue management in cellular networks. In: Computer Science and Electronic Engineering (CEEC) 2017, pp. 183–188. IEEE. Colchester, UK (2017). https://doi.org/10.1109/CEEC.2017.8101622

32. Beshay, J.D., Nasrabadi, A.T., Prakash, R., Francini, A.: On active queue management in cellular networks. In: 2017 IEEE Conference on Computer Communications Workshops (INFOCOM WKSHPS), pp. 384–389. IEEE, Atlanta, GA, USA (2017) https://doi.org/10.1109/INFOCOMW.2017.8116407

33. Ali Rezaee, A., Pasandideh, F.: A fuzzy congestion control protocol based on active queue management in wireless sensor networks with medical applications. Wirel. Personal Commun. **98**(1), 815–842 (2018). https://doi.org/10.1007/s11277-017-4896-6

34. Menth, M., Veith, S.: Active queue management based on congestion policing (CP-AQM). In: German, R., Hielscher, K.-S., Krieger, U.R. (eds.) MMB 2018. LNCS, vol. 10740, pp. 173–187. Springer, Cham (2018). https://doi.org/10.1007/978-3-319-74947-1_12

35. Adesh, N.D., Renuka, A.: Adaptive receiver-window adjustment for delay reduction in LTE networks. J. Comput. Netw. Commun. **2019**, 17 (2019). https://doi.org/10.1155/2019/3645717. Article ID 3645717

36. Irazabal, M., Lopez-Aguilera, E., Demirkol, I.: Active queue management as quality of service enabler for 5G networks. In: European Conference on Networks and Communications (EuCNC), pp. 421–426. IEEE, Valencia, Spain (2019). https://doi.org/10.1109/EuCNC.2019.8802027

37. Korolkova, A.V., Kulyabov, D.S., Chernoivanov, A.I.: On the classification of RED algorithms. Bulletin of Peoples' Friendship University of Russia. Series "Mathematics. Information Sciences. Physics". No. 3, 34–46 (2009)

38. Korolkova, A.V., Velieva, T.R., Abaev, P.O., Sevastianov, L.A., Kulyabov, D.S.: Hybrid simulation of active traffic management. In: Proceedings of 30th European Conference on Modelling and Simulation (ECMS), pp. 692–697. ECMS. Regensburg, Germany (2016). https://doi.org/10.7148/2016-0685

39. Bonald, T., May, M., Bolot, J.: Analytic evaluation of RED performance. In: Proceedings IEEE INFOCOM 2000 Conference on Computer Communications, vol. 3, pp. 1415–1424. IEEE, Tel Aviv, Israel (2000). https://doi.org/10.1109/INFCOM.2000.832539

40. Chydzinski, A., Chrost, L.: Analysis of AQM queues with queue size based packet dropping. Int. J. Appl. Math. Comput. Sci. **21**(3), 567–577 (2011). https://doi.org/10.2478/v10006-011-0045-7

41. Zhernovyi, Y., Kopytko, B., Zhernovyi, K.: On characteristics of the $M/G/1/m$ and $M/G/1$ queues with queue-size based packet dropping. J. Appl. Math. Comput. Mech. **13**(4), 163–175 (2014)

42. Chydzinski, A., Mrozowski, P.: Queues with dropping functions and general arrival processes. PLoS One, 11(3) (2016). https://doi.org/10.1371/journal.pone.0150702

43. Konovalov, M.G., Razumchik, R.V.: Numerical analysis of improved access restriction algorithms in a $GI|G|1|N$ system. J. Commun. Technol. Electron. **63**(6), 616–625 (2018). https://doi.org/10.1134/S1064226918060141

44. Kreinin, A.: Queueing systems with renovation. J. Appl. Math. Stochast. Anal. **10**(4), 431–443 (1997). https://doi.org/10.1155/S1048953397000464

45. Kreinin, A.: Inhomogeneous random walks: applications in queueing and finance. In: CanQueue 2003, Fields Institute, Toronto (2003)

46. Gelenbe, E.: Product-form queueing networks with negative and positive customers. J. Appl. Prob. **28**(3), 656–663 (1991). https://doi.org/10.2307/3214499

47. Pechinkin, A.V., Razumchik, R.V.: The stationary distribution of the waiting time in a queueing system with negative customers and a bunker for superseded customers in the case of the LAST-LIFO-LIFO discipline. J. Commun. Technol. Electron. **57**(12), 1331–1339 (2012). https://doi.org/10.1134/S1064226912120054

48. Razumchik, R.V.: Analysis of finite capacity queue with negative customers and bunker for ousted customers using chebyshev and gegenbauer polynomials. Asia-Pacific J. Oper. Res. **31**(04), 1450029 (2014). https://doi.org/10.1142/S0217595914500298

49. Semenova, O.V.: Multithreshold control of the $BMAP/G/1$ queuing system with MAP flow of Markovian disasters. Autom. Remote Control **68**(1), 95–108 (2007). https://doi.org/10.1134/S0005117907010092

50. Li, J., Zhang, L.: $M^X|M|c$ queue with catastrophes and state-dependent control at idle time. Front. Math. China **12**(6), 1427–1439 (2017). https://doi.org/10.1007/s11464-017-0674-8

51. Kim, C., Klimenok, V.I., Dudin, A.N.: Analysis of unreliable $BMAP|PH|N$ type queue with Markovian flow of breakdowns. Appl. Math. Comput. **314**, 154–172 (2017). https://doi.org/10.1016/j.amc.2017.06.035

52. Gudkova, I., et al.: Modeling and analyzing licensed shared access operation for 5G network as an inhomogeneous queue with catastrophes. In: International Congress on Ultra Modern Telecommunications and Control Systems and Workshops, pp. 282–287. IEEE, Lisbon, Portugal (2016). https://doi.org/10.1109/ICUMT.2016.7765372

53. Dudin, A., Klimenok, V., Vishnevsky, V.: Analysis of unreliable single server queueing system with hot back-up server. In: Plakhov, A., Tchemisova, T., Freitas, A. (eds.) EmC-ONS 2014. CCIS, vol. 499, pp. 149–161. Springer, Cham (2015). https://doi.org/10.1007/978-3-319-20352-2_10

54. Krishnamoorthy, A., Pramod, P.K., Chakravarthy, S.R.: Queues with Interruptions: a survey. TOP **22**(1), 290–320 (2014). https://doi.org/10.1007/s11750-012-0256-6

55. Vishnevsky, V.M., Kozyrev, D.V., Semenova, O.V.: Redundant queuing system with unreliable servers. In: International Congress on Ultra Modern Telecommunications and Control Systems and Workshops, pp. 283–286. IEEE, St. Petersburg, Russia (2014) https://doi.org/10.1109/ICUMT.2014.7002116

56. Rykov, V.V., Kozyrev, D.V.: Analysis of renewable reliability systems by markovization method. In: Rykov, V.V., Singpurwalla, N.D., Zubkov, A.M. (eds.) ACMPT 2017. LNCS, vol. 10684, pp. 210–220. Springer, Cham (2017). https://doi.org/10.1007/978-3-319-71504-9_19

57. Bocharov, P.P., Zaryadov, I.S.: Probability distribution in queueing systems with renovation. Bulletin of Peoples' Friendship University of Russia Series "Mathematics Information Sciences Physics". No. 1–2, 15–25 (In Russian) (2007)

58. Zaryadov, I.S., Pechinkin, A.V.: Stationary time characteristics of the $GI/M/n/\infty$ system with some variants of the generalized renovation discipline. Autom. Remote Control **70**(12), 2085–2097 (2009). https://doi.org/10.1134/S0005117909120157

59. Zaryadov, I.S.: Queueing systems with general renovation. In: ICUMT 2009 - International Conference on Ultra Modern Telecommunications, pp. 1–6. IEEE, St. Petersburg, Russia (2009) https://doi.org/10.1109/ICUMT.2009.5345382

60. Zaryadov, I., Razumchik, R., Milovanova, T.: Stationary waiting time distribution in G—M—n—r with random renovation policy. In: Vishnevskiy, V.M., Samouylov, K.E., Kozyrev, D.V. (eds.) DCCN 2016. CCIS, vol. 678, pp. 349–360. Springer, Cham (2016). https://doi.org/10.1007/978-3-319-51917-3_31

61. Konovalov, M., Razumchik, R.: Queueing systems with renovation vs. queues with RED. Supplementary Material (2017). https://arxiv.org/abs/1709.01477

62. Konovalov, M., Razumchik, R.: Comparison of two active queue management schemes through the M/D/1/N queue. Informatika i ee Primeneniya (Informatics and Applications), 12(4), 9–15 (2018). https://doi.org/10.14357/19922264180402

63. Bogdanova, E.V., Zaryadov, I.S., Milovanova, T.A., Korolkova, A.V., Kulyabov, D.S.: Characteristics of lost and served packets for retrial queueing system with general renovation and recurrent input flow. In: Vishnevskiy, V.M., Kozyrev, D.V. (eds.) DCCN 2018. CCIS, vol. 919, pp. 327–340. Springer, Cham (2018). https://doi.org/10.1007/978-3-319-99447-5_28

64. Zaryadov, I., Bogdanova, E., Milovanova, T., Matushenko, S., Pyatkina, D.: Stationary characteristics of the $GI|M|1$ queue with general renovation and feedback. In: 10th International Congress on Ultra Modern Telecommunications and Control Systems and Workshops (ICUMT), article No. 8631244. IEEE, Moscow, Russia (2019) https://doi.org/10.1109/ICUMT.2018.8631244

65. Zaryadov, I.S., Korolkova, A.V.: The application of model with general renovation to the analysis of characteristics of active queue management with random early detection (RED). T-Comm. No. 7, 84–88 (2011). (In Russian)

66. Abaev, P., Gaidamaka, Y., Samouylov, K., Pechinkin, A., Razumchik, R., Shorgin, S.: Hysteretic control technique for overload problem solution in network of SIP servers. Comput. Inform. 33(1), 218–236 (2014)

67. Abaev, P., Khachko, A., Beschastny, V.: Queuing model for SIP server hysteretic overload control with K-state MMPP bursty traffic. In: International Congress on Ultra Modern Telecommunications and Control Systems and Workshops, pp. 495–500. IEEE, St. Petersburg, Russia (2015). https://doi.org/10.1109/ICUMT.2014.7002151

68. Pechinkin, A.V., Razumchik, R.R., Zaryadov, I.S.: First passage times in $M_2^{[X]}|G|1|R$ queue with hysteretic overload control policy. In: AIP Conference Proceedings, vol. 1738, article No. 220007. American Institute of Physics Inc., Rhodes (2016). https://doi.org/10.1063/1.4952006

69. Razumchik, R.: Analysis of finite $MAP|PH|1$ queue with hysteretic control of arrivals. In: International Congress on Ultra Modern Telecommunications and Control Systems and Workshops, article No. 7765373, pp. 288–293. IEEE, Lisbon. Portugal (2016) https://doi.org/10.1109/ICUMT.2016.7765373

Human-Computer Mobile Distributed Computing for Time Series Forecasting

Rumen Ketipov, Georgi Kostadinov, Plamen Petrov, Iliyan Zankinski, and Todor Balabanov$^{(\boxtimes)}$

Institute of Information and Communication Technologies,
Bulgarian Academy of Sciences,
acad. Georgi Bonchev Str., block 2, 1113 Sofia, Bulgaria
iict@bas.bg
http://iict.bas.bg/

Abstract. Distributed computing became very popular in the last two decades and in many cases such projects are based on a donated calculation power. With the expansion of the mobile devices in the last decade it become relevant some donated distributed computing solutions to be developed as mobile applications. Such a solution was developed at IICT-BAS [2], which is based on an Android Live Wallpaper technology. This research proposes an extension of the work done at IICT-BAS in the direction of human-computer based distributed computing by providing software capabilities of the users to vote for future financial changes. At the beginning, a brief overview of the topic with special emphasis on ANNs/GAs and their strengths/weaknesses when applied for the problem's solution is given. After that the study treats the aspect of extension of a distributed computing system based on mobile devices to human-computer distributed computing. Experiments and results are presented in Sect. 3 and the final Sect. 4 concludes and provides some further work suggestions.

Keywords: Distributed computing · Time series forecasting · Artificial neural networks · Evolutionary algorithms

1 Introduction

One of the most famous donated distributed computing project is SETI@home, which is related to deep space signals processing in attempt to find an alien life [17]. But donated distributed computing has its influence also in other areas like the time series forecasting. The MoneyBee project was a desktop screensaver application, which was calculating financial time series forecasting by training of artificial neural networks with evolutionary algorithms [1]. Financial time series forecasting is an attractive and intensively researched area [3], whereat financial forecasting has two common components - direction of the change and magnitude of the change. In many situations it is enough direction of the change

This work was supported with private funding by Velbazhd Software LLC.

© Springer Nature Switzerland AG 2019
V. M. Vishnevskiy et al. (Eds.): DCCN 2019, CCIS 1141, pp. 503–509, 2019.
https://doi.org/10.1007/978-3-030-36625-4_40

to be predicted. After that decision makers are able to do their decisions. If magnitude of the change is also predicted decisions taken would be much more accurate and to be able to predict the magnitude of the change is crucial to predict the direction of the next serious change in the opposite direction.

One of the most interesting research directions in this field is related to the usage of artificial neural networks combined with evolutionary algorithms [4]. In some studies evolutionary algorithms are used for artificial neural network topology optimization [5], but in other studies evolutionary algorithms are used for artificial neural network weights optimization [6]. When evolutionary algorithms are used for weights optimization it is very common the training to be combined with back-propagation algorithm [7]. Even when such a hybrid training algorithm is used, it is in the class of supervised training. Training examples are fed to the artificial neural network and its weights are modified according the selected training algorithm. Generally, evolutionary algorithms have a common advantage compared to the exact numerical methods and that is the possibility for parallel calculations [8]. When the parallel calculations are done on separate mobile devices it is the case of mobile distributed computing [9]. It is very common the financial forecasting to be done by the usage of the past values in the time series, but it can be also combined with human evaluation of the future values.

In this research a human-computer based mobile distributed computing solution is proposed. The users are using an Android application which trains artificial neural network with evolutionary algorithms and back-propagation, but users are capable to vote for increase or decrease of the future values. The most successful forecast votes of the users are used in artificial neural network training process.

2   Model Proposition

As already mentioned above, a famous example in the area of artificial neural network is the project MoneyBee [1]. The MoneyBee screensaver (Fig. 1) was organized as a desktop application, which runs in background when the computer is in idle mode and the artificial neural networks were trained with the usage of evolutionary algorithms. The main advantage of this approach is the usage of donated computation power of many participants. In the same time the biggest disadvantage is that the screensaver is used relatively rare, only when the user is not using his/her computer. The other disadvantage is that personal computers are not in operation 24/7.

These disadvantages are addressed in the VitoshaTrade project (Fig. 2), which is developed in IICT-BAS (Institute of Information and Communication Technologies - Bulgarian Academy of Sciences) [13]. The implementation is done under Android operating system as an active wallpaper. Inside the forecasting module a multilayer perceptron is used. The training algorithm is a combination between back-propagation and differential evolution. An active wallpaper Android application executes all calculations and it works 24/7 in a background

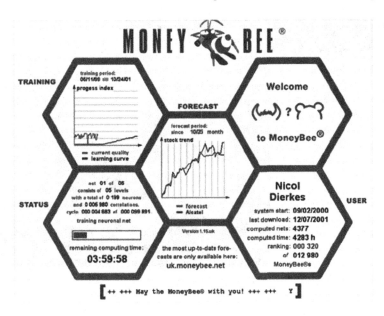

Fig. 1. MoneyBee screensaver

mode. When better artificial neural network weights are found a communication with a remote server is established. Active wallpapers are used to visualize information on a full screen behind all other visual components.

In both projects only a pure mathematical information is used for forecasts calculations. Such limitation was overcome in ElectricSheep [12] project. In ElectricSheep is implemented a rating system whereat the users vote for different fractal generated animations. Voting is done in two directions - thumb up and thumb down. Each vote increases or decreases individual rating, which is directly related to its survival chances. The rating is not constant and it goes down with the time passing. In the proposed model a direction voting is implemented similar to the ElectricSheep project. Users vote only up if they think that the price will go up, and down if they think that the price will drop (Fig. 3), whereat there is no limit how many times the user will press up or down button. The intensity of the single user voting in one time interval (daily intervals for example) does not count. Only the first push of up/down button is taken in account. Users' voting is irregular, because in real-time operation the operator of the system has no control over donated human-computer participation.

This study proposes a combination of the projects described above. The VitoshaTrade [13] project is extended in the direction of additional voting system. The Android operating system offers software modules called widgets, which have graphical user interface and they occupy small part of the visible working area. Most commonly, widgets are used for representation of small pieces of information, such astronomical time, weather forecast, calendar, and others. Widgets are much more interactive than the active wallpaper. The user is capable to

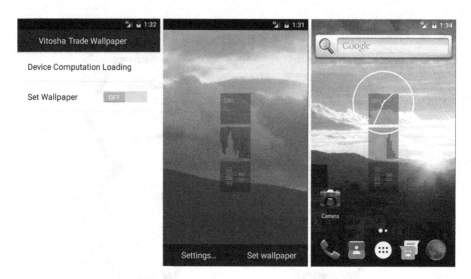

Fig. 2. VitoshaTrade active wallpaper

input small pieces of information as voting, for example. Another advantage of the widgets is that users have much better control over visualization and placement of the widgets. In this study a composite widget is proposed, which shows the thicker of the currency pair (EUR/USD in the example provided), the value of the rate between the two currencies, and arrows for vote up or vote down.

Fig. 3. Voting widget positioned over active wallpaper

Voting of the users is collected on PHP/MySQL based web server [2] as part of the server side implementation of the VitoshaTrade project. On the server side additional analysis are done. Votes which were more accurate are better taken for next generations in the global distributed differential evolution population.

3   Experiments

Experiments are done with ForEx financial time series of EUR/USD currency pair (Fig. 4). Time series is disassembled in training examples (7 values for lag and 3 values for lead) and a back-propagation combined with differential evo-lution training of multilayer perceptron is additionally done. As topology for the multilayer perceptron 7-5-3 is used an implementation with Encog Machine Learning Framework [10] for Java. The input has value 7, the output has value 3 and the hidden layer is selected to be $5 = (7 + 3)/2$. The training process is continuous, which means that the forecasting is done parallel with the training. The voting of the user directly influences the fitness value of the individuals in the distributed differential evolution population. Some users are better in their vision and are more successful during the voting process, therefore the tracking of the users is of great importance. More reliable votes are taken with higher coefficient for the further predictions. According to some regulations, like GDPR (General Data Protection Regulation) in EU (European Union) [11], a special care should be taken for the personal information collected on the servers side.

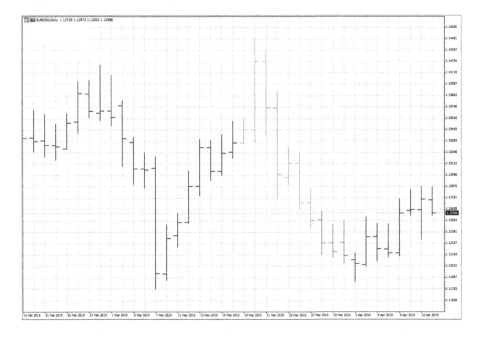

Fig. 4. ForEx EUR/USD time series.

There is no limit how many people can participate in the proposed human-computer distributed computing system. The system itself is highly scalable, but experiments in this study were done in a closed group of the team by IICT-BAS (8 users), which is working on the research.

4   Conclusion

Pure technical analysis in time series field has its limits. Values measured in time are just one projection of the process which goes under it. Any computer based calculation nowadays is not capable to act like human intuition. Involving human intuition in the forecasting process is the main advantage of the proposed model. The mechanisms behind human intuition are not investigated enough and they are still very difficult to be recreated in the modern computers. Because of these reasons the proposed financial time series forecasting solution advances the pure technical analysis with minimal influence of human impact. The proposed simple voting system consolidates the power of machine learning algorithms and the human sixth sense feeling in an efficient and cost effective financial forecasting infrastructure.

As a future research it will be interesting to investigate different communication capabilities as described in [14] and working processes optimization as described in [15]. Also it will be interesting generalized artificial neural networks [16] to be used instead of multilayer perceptron. More attention should be paid to the security, how to increase the dependability of distributed systems by providing better protection against deliberate attacks. Higher reliability, ubiquity, and better security are only some aspects, which in further research are desired.

References

1. Bohn, A., Guting, T., Mansmann, T., et al.: MoneyBee: a new product to predict stock market developments using artificial intelligence and increased calculation capacity (in German). Wirtschaftsinf 45(3), 325–333 (2003)
2. Tomov, P., Zankinski, I., Barova, M.: Mobile alternative of the MoneyBee project for financial forecasting. In: Proceedings of the Annual University Scientific Conference of the National Military University Vasil Levski, Veliko Tarnovo, pp. 1085–1089 (2018)
3. Nava, N., Di Matteo, T., Aste, T.: Financial time series forecasting using empirical mode decomposition and support vector regression. Risks 6(1), 7 (2018)
4. Zhang, R., Tao, J.: A nonlinear fuzzy neural network modeling approach using an improved genetic algorithm. IEEE Trans. Ind. Electron. 65(7), 5882–5892 (2018)
5. Kapanova, K., Dimov, I., Sellier, J.M.: A genetic approach to automatic neural network architecture optimization. Neural Comput. Appl. 29(5), 1481–1492 (2016)
6. Aljarah, I., Faris, H., Mirjalili, S.: Optimizing connection weights in neural networks using the whale optimization algorithm. Soft Comput. 22(1), 1–15 (2016)
7. Balabanov, T., Genova, K.: AJAX distributed system for evolutionary algorithms based artificial neural networks training. In: XXIV International Symposium on Management of Thermal Energy Facilities and Systems, Management of Energy, Industrial and Environmental Systems, John Atanasoff Union of Automation and Informatics, pp. 49–52 (2016)

8. Altinoz, O.T., Deb, K.: Late parallelization and feedback approaches for distributed computation of evolutionary multi-objective optimization algorithms. Neural Comput. Appl. **30**(3), 723–733 (2016)

9. Sharma, R., Rodriguez, T.: Distributed computing for portable computing devices, US10282470B2, United States (2019)

10. Heaton, J.: Encog machine learning framework. http://heatonresearch.com/encog/

11. Hristov, P., Dimitrov, W.: The blockchain as a backbone of GDPR compliant frameworks. In: Proceedings of 8th International Multidisciplinary Symposium - Challenges and Opportunities For Sustainable Development Through Quality and Innovation in Engineering and Research Management, vol. 20, no. 1, pp. 305–310 (2018)

12. Draves, S.: The electric sheep screen-saver: a case study in aesthetic evolution. In: Rothlauf, F., et al. (eds.) EvoWorkshops 2005. LNCS, vol. 3449, pp. 458–467. Springer, Heidelberg (2005). https://doi.org/10.1007/978-3-540-32003-6_46

13. Balabanov, T.: VitoshaTrade Project - vitoshatrade.veldsoft.eu. http://github.com/VelbazhdSoftwareLLC/VitoshaTrade

14. Alexandrov, A.: Comparative analysis of IEEE 802.15.4 based communication protocols used in wireless intelligent sensor systems. In: Proceedings of the International conference RAM, pp. 51–54 (2014)

15. Atanasov, J., Atanasova, T.: Optimizing the management and control of apparel enterprise by information technologies. In: 19th Telecommunications Forum, TELFOR 2011 - Proceedings of Papers, pp. 1245–1248 Belgrade (2011). Article no. 6143777

16. Tashev, T., Hristov, H.: MoneyBee: aktienkursprognose mit kuenstlicher intelligenz bei hoher rechenleistung. Wirtschaftsinformatik **3**(2), 92–104 (2003)

17. University of California, SETI@home. https://setiathome.berkeley.edu/

Investigation of a Hybrid Sensor- and Computational Network for Numerical Weather Prediction Calculations

Ádám Vas[✉] and László Tóth

Faculty of Informatics, University of Debrecen,
Kassai Street 26, Debrecen 4028, Hungary
{vas.adam,toth.laszlo}@inf.unideb.hu

Abstract. In this paper we investigate a hybrid approach to integrate our Distributed Sensor Network for Prediction Calculations (DSN-PC) system's measurements into a hybrid computational network which assimilates them with publicly available atmospheric analysis dataset. The forecast calculations are based on a simple and relatively easily implementable algorithm. The parallelization is provided on the mathematical layer which is the highest possible level of parallelizing an algorithm for the solution of a system of differential equations. Our future goal is to move to solely DSN-PC measurement-based computations to make weather forecasts, but we do not have sufficient number of stations installed yet. However, a hybrid solution where we included NOAA Global Forecast System (GFS) data in the production of the initial values made it possible to test our sensor network's capability of performing such calculations. To evaluate the viability of our approach we analyzed the error of the forecasts. Below the results of the distributed calculations are shown along with the method of the assimilation of the two datasets.

Keywords: Sensor network · Distributed computing · Numerical calculations · Weather prediction · Data assimilation

1 Introduction

Weather forecast models require input measurements that have time- and spatial resolution as high as possible. Atmospheric measurement techniques is an intensely researched field and the quality of these measurements continue to grow. Measurement of upper-air parameters are performed by radiosondes, radars, satellites, LIDARs etc. which are quite expensive to operate and to maintain. However, for the surface layer the usage of cheaper, widely available devices such as mobile phones or microcontroller-based sensors can make a difference, mainly because of their low price and simplicity. Nowadays, the network availability in outdoor areas is very high due to technologies like Wi-Fi or 3G/4G/5G mobile networks which makes these small stations able to communicate with any

This work was supported by the construction EFOP-3.6.3-VEKOP-16-2017-00002. The project was co-financed by the Hungarian Government and the European Social Fund.

other node on the Internet. While these small stations are not as accurate as more advanced systems and usually do not have that amount of sensors, their size makes them much easier to deploy to nearly any location. This way a much higher spatial resolution can be achieved on the surface-level for certain atmospheric parameters.

Following our previous work [1,2] where we used only publicly available measurements to test our network's capability of performing weather forecast calculations in a distributed, co-operative way, we improved our system by including real-time measurements by our DSN-PC weather stations installed over Hungary in the calculations. The focus of this paper is to test the possibility of assimilating our measurements and publicly available datasets and to evaluate the results of numerical weather prediction calculations based on these initial data.

2 System and Model Description

2.1 The Investigated Geographical Area

We implemented a virtual sensor network which covers a European area and consists of 20 × 20 nodes forming a regular grid on a map using polar stereographic projection (see Fig. 1). The properties of the grid are:

- Coordinates of the lower-left grid point: 39°N, 2.6°W
- Coordinates of the upper-right grid point: 54.1371°N, 38.6715°E
- Grid step at North Pole: 150 km
- Central angle of the map: 0°

2.2 The Two Sources of Input Data

The initial values for the computations are assimilated from 2 sources. The first source is our DSN-PC weather stations distributed over Hungary (see Table 1). They are equipped with temperature, atmospheric pressure and relative humidity sensors. The detailed description of the hardware can be found in [3].

Table 1. The geographic locations of our currently operational DSN-PC weather stations

Station ID	Latitude (°N)	Longitude (°E)	Altitude (m)
1	48.17	20.42	254
2	46.92	19.67	119
3	46.65	21.29	87
4	47.31	18.01	268
5	46	18.68	91

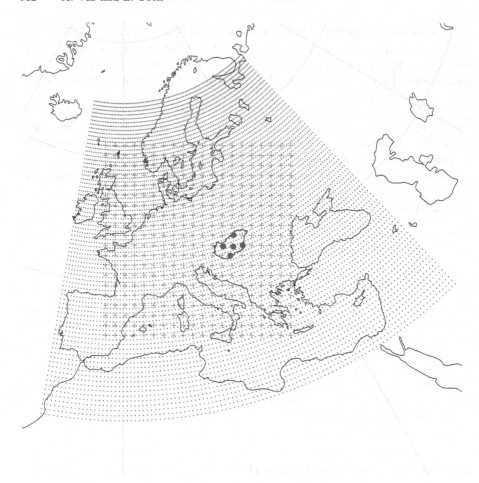

Fig. 1. The regular grid of our simulated computational network (marked with +), the grid points of the NOAA GFS dataset (marked with ·) and the locations of our DSN-PC weather stations in Hungary (marked with o).

Based on these measured parameters the stations estimate an upper-air atmospheric parameter, the 500 hPa geopotential height based on the hypsometric equation [4]:

$$z_{500} - z_{sl} = \frac{R_d \overline{T_v}}{g} \cdot \ln\left(\frac{p_{sl}}{500hPa}\right), \tag{1}$$

where

- z_{500} and z_{sl} (m) are the geometric heights at pressure levels 500 hPa and station-level pressure,
- $R_d = 287.058 \frac{J}{kg \cdot K}$ is the specific gas constant for dry air,
- $\overline{T_v}$ (K) is the mean virtual temperature,
- $g = 9.80665 \frac{m}{s^2}$ is the gravitational acceleration.

To cover the computational grid we also integrated the data from the publicly available GFS-ANL database [5]. The Global Forecast System (GFS) is a weather forecast model produced by the National Centers for Environmental Prediction (NCEP). In GFS gridded data are available for download through the NOAA National Operational Model Archive and Distribution System (NOMADS). In our case the 0.5° resolution data was downloaded and narrowed down to the grid points whose coordinates are between (30°N; 67°N) latitude and (10°W; 44°E) longitude. For our simple model we extracted the 500 hPa geopotential height analysis data from the dataset.

2.3 Data Assimilation

The combination of these two data sources were done by performing natural neighbor interpolation method [6]. The z_{500} values of the GFS grid points and those from our DSN-PC weather stations were used to calculate the z_{500} values for the 20 × 20 computational grid. Before the interpolation method the geographic latitude(°) and longitude(°) coordinates of the GFS and DSN-PC grid points were converted to (x,y) coordinate pairs using polar stereographic projection [7]:

$$r = \frac{\cos(latitude)}{1 + \sin(latitude)} \cdot 2a, \tag{2}$$

where $a = \frac{4 \cdot 10^7}{2\pi}$ is the radius of the Earth (m). Then

$$x = r \cdot \sin(longitude) \tag{3}$$

$$y = -r \cdot \cos(longitude) \tag{4}$$

The first step of the interpolation is to create the Voronoi-diagram on the map where the map plane is partitioned into cells for each GFS and DSN-PC grid point. Each grid point has a corresponding cell consisting of all points closer to it than to any other. After that the following steps are executed for each point in the 20 × 20 computational grid (see Fig. 2):

1. Create the Voronoi-diagram again including the current grid point.
2. Take the intersection of this point's cell and the original cells. For those of the original cells which have an intersection with this point's cell (neighbor cells) calculate the intersection's area, then divide this value with the sum of the intersection areas. These values are considered the weights of the neighbor cells.
3. The z_{500} value of this point is then calculated by

$$z_{500} = \sum_i w_i \cdot z_{500,i} \tag{5}$$

where w_i is the weight and $z_{500,i}$ is the 500 hPa geopotential height of the neighbor cell i.

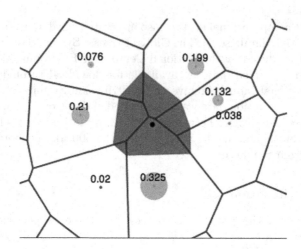

Fig. 2. Natural neighbor interpolation. The new point is assigned a value based on the values of neighbor grid points [8].

2.4   The Forecast Algorithm

After the preparations the 24 h forecast results were calculated by applying the algorithm based on the barotropic vorticity equation developed by Charney, Fjørtoft, and von Neumann (CFvN) [7]. After connecting to their 4 neighbors the nodes exchange measured and calculated values between each other several times. By the end of the calculations each node has a dz_{500}/dt value which represents the change of the 500 hPa geopotential height. The detailed description of the distributed algorithm may be found in [1].

3   Numerical Forecast Results

We investigated the period between 2019.03.21. and 2019.03.27. Measurements were available for 00:00 UTC each day. The time step of the numerical algorithm

Table 2. Mean Absolute Error values of the forecast calculations performed by the CFvN algorithm and the persistence method

Date	CFvN MAE (m)	Persistence MAE (m)
2019.03.21. 00:00 UTC	59.52	31.29
2019.03.22. 00:00 UTC	50.63	44.33
2019.03.23. 00:00 UTC	73.76	49.23
2019.03.24. 00:00 UTC	37.88	94.54
2019.03.25. 00:00 UTC	85.71	101.07
2019.03.26. 00:00 UTC	46.33	60.31
2019.03.27. 00:00 UTC	35.87	61.54

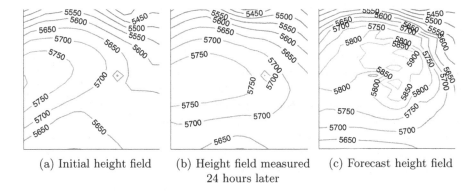

(a) Initial height field   (b) Height field measured
24 hours later   (c) Forecast height field

Fig. 3. The result of the forecast performed on 2019.03.21. 00:00 UTC data.

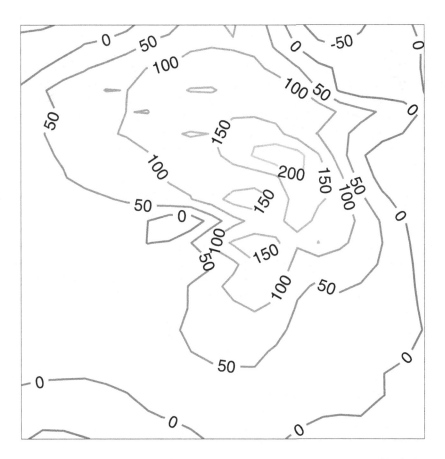

Fig. 4. The error of the CFvN forecast performed on 2019.03.21. 00:00 UTC data.

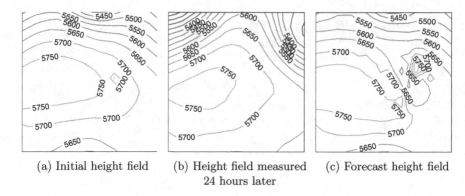

(a) Initial height field

(b) Height field measured 24 hours later

(c) Forecast height field

Fig. 5. The result of the forecast performed on 2019.03.22. 00:00 UTC data.

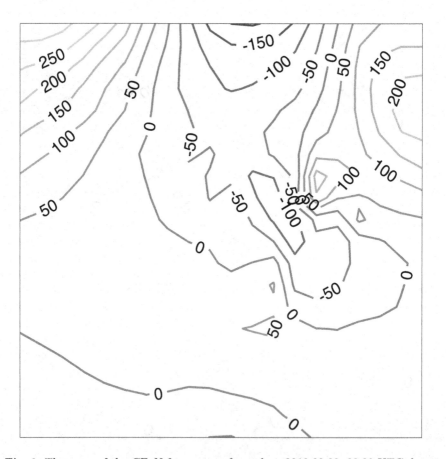

Fig. 6. The error of the CFvN forecast performed on 2019.03.22. 00:00 UTC data.

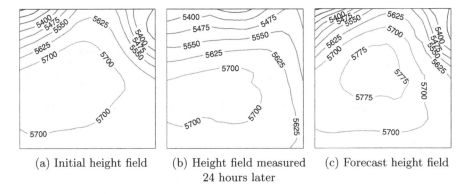

(a) Initial height field   (b) Height field measured   (c) Forecast height field
                                24 hours later

Fig. 7. The result of the forecast performed on 2019.03.23. 00:00 UTC data.

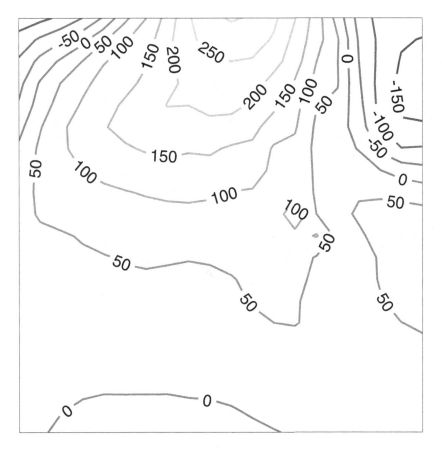

Fig. 8. The error of the CFvN forecast performed on 2019.03.23. 00:00 UTC data.

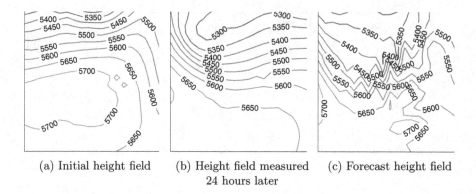

(a) Initial height field   (b) Height field measured   (c) Forecast height field
                                    24 hours later

Fig. 9. The result of the forecast performed on 2019.03.24. 00:00 UTC data.

Fig. 10. The error of the CFvN forecast performed on 2019.03.24. 00:00 UTC data.

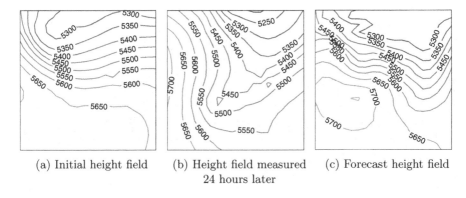

(a) Initial height field   (b) Height field measured   (c) Forecast height field
                                 24 hours later

Fig. 11. The result of the forecast performed on 2019.03.25. 00:00 UTC data.

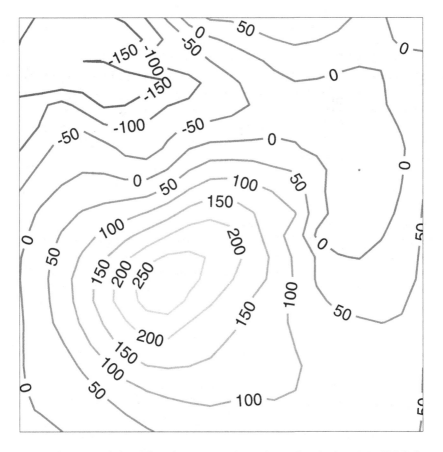

Fig. 12. The error of the CFvN forecast performed on 2019.03.25. 00:00 UTC data.

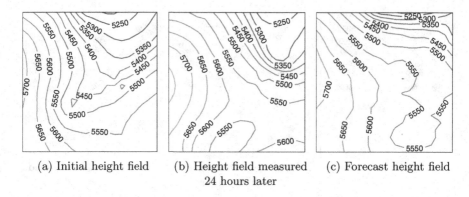

(a) Initial height field   (b) Height field measured   (c) Forecast height field
                                24 hours later

Fig. 13. The result of the forecast performed on 2019.03.26. 00:00 UTC data.

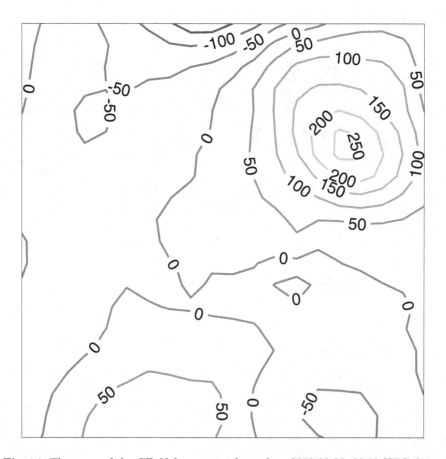

Fig. 14. The error of the CFvN forecast performed on 2019.03.26. 00:00 UTC data.

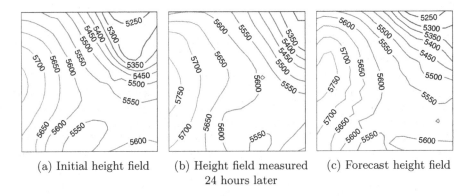

(a) Initial height field   (b) Height field measured   (c) Forecast height field
                                 24 hours later

Fig. 15. The result of the forecast performed on 2019.03.27. 00:00 UTC data.

Fig. 16. The error of the CFvN forecast performed on 2019.03.27. 00:00 UTC data.

were chosen on the basis of the details in [7] and our empirical results. We ultimately set it to 0.01 h which provided adequate numerical stability. The forecast length was 24 h. The Mean Absolute Error (MAE) was calculated for each forecast by

$$MAE = \frac{1}{18 \cdot 18} \sum_{i=1}^{18} \sum_{j=1}^{18} |z_{500,i,j} - z'_{500,i,j}|, \tag{6}$$

where $z'_{500,i,j}$ is the calculated and $z_{500,i,j}$ is the measured 500 hPa geopotential height measured after 24 h. The boundary points were not taken into account here because they are handled in a special way during the calculations [7].

The MAE values of the forecast calculations are summarized in Table 2. The CFvN algorithm remained numerically stable in each case and provided similar MAE values to those produced by the persistence method. On Figs. 3, 5, 7, 9, 11, 13, 15 the initial, the analysis and the forecast height fields are shown for the calculations related to the different days of the investigated time period. The error maps of the forecast are shown on Figs. 4, 6, 8, 10, 12, 14, 16.

4   Conclusion

Following our previous results when we succeeded to implement a distributed numerical weather prediction algorithm on our virtual DSN-PC system, we could step forward and include real-time measurements by our weather stations in the calculations. After the assimilation of these measurements with the NOAA GFS dataset distributed calculations were run on a computational grid over a European area. The results show the possibility of getting numerically stable and reasonable results from this hybrid sensor network. The next steps include the investigation of more advanced data assimilation techniques regarding our DSN-PC measurements and GFS analysis data. After that it may be worth testing a method which only involves boundary conditions from GFS dataset for the different time steps of our algorithm, for which different data smoothing algorithms have yet to be tested.

Ackowledgements. We wish to thank Ficsor Endre, Perlaki Csaba, Szabó Sándor, Vas Ferenc and the Baptist Church of Kecskemét for providing place for our DSN-PC weather stations and thus supporting our research. We also thank László Elemér from ATOMKI for assistance with accessing public analysis data and for comments that improved the manuscript.

References

1. Vas, Á., Fazekas, Á., Nagy, G., Tóth, L.: Distributed sensor network for meteorological observations and numerical weather prediction calculations. Carpathian J. Electron. Comput. Eng. **6**(1), 56–63 (2013)

2. Vas, Á., Tóth, L.: Evaluation of a simulated distributed sensor- and computational network for numerical prediction calculations. In: Vishnevskiy, V.M., Kozyrev, D.V. (eds.) DCCN 2018. CCIS, vol. 919, pp. 9–20. Springer, Cham (2018). https://doi. org/10.1007/978-3-319-99447-5_2
3. Vas, Á., Nagy, G., Tóth, L.: Networkable sensor station for DSN-PC system. Carpathian J. Electron. Comput. Eng. **8**(2), 37–40 (2015)
4. Wallace, J.M., Hobbs, P.V.: 3 - Atmospheric Thermodynamics. In: Atmospheric Science: An Introductory Survey. 2nd edn., pp. 63–111. Academic Press, Cambridge (2006). https://doi.org/10.1016/B978-0-12-732951-2.50008-9
5. NOAA Global Forecast System (GFS) Analysis. https://www.ncdc.noaa.gov/data-access/model-data/model-datasets/global-forcast-system-gfs. Accessed 10 Sep 2019
6. Sibson, R.: Interpolating Multivariate Data: Chapter 2: A Brief Description of Natural Neighbor Interpolation, pp. 21–36. Wiley, New York (1981)
7. Charney, J.G., Fjørtoft, R., Von Neumann, J.: Numerical integration of the barotropic vorticity equation. Tellus **2**, 237–254 (1950)
8. File: Natural-Neighbors-Coefficients-Example.png - Wikimedia Commons. http://commons.wikimedia.org/wiki/File:Natural-neighbors-coefficients-example.png. Accessed 10 Sep 2019

Methods for Processing of Heterogeneous Data in IoT Based Systems

Tatiana Atanasova$^{(\boxtimes)}$ (iD)

Institute of Information and Communication Technologies – BAS, Sofia, Bulgaria
atanasova@iit.bas.bg
http://iinf.bas.bg/en/atanasova.htm

Abstract. The concept of the Internet of Things (IoT) is based on the idea of a permanent connection between the physical and digital world, which is now technologically feasible. The IoT can describe a scenario in which a large number of objects have built-in uniquely identifiable computing devices connected to the Internet that allow them to collect, store, share and analyze data and to be managed remotely via other devices with an Internet connection. It is important to provide adequate processing of the data to see what is behind it and to assess the situation. The lack of Reference Modelling IoT Architecture prevents a common approach to processing the generated data. The data from various sources has different nature, range, rate and volume. The need to retrieve and analyze this data from the IoT complex systems in real-time requires the application of wide scope of methods and tools. This paper discusses different approaches to process the data from various sources according to different goals: sensing, data analytics, and machine learning with hope that using of IoT will improve all aspects of our life.

Keywords: IoT · Heterogeneous data · Data management · Machine learning

1 Introduction

The intensive information transformation of various industries is underway. It creates connected environment of data, people, processes, services, systems and industries through the Internet. The generation and use of information as a way and means of realizing the intelligent ecosystems of industrial innovation and collaboration goes ahead to Industry 4.0. Industry 4.0 takes a holistic approach consisting of evaluating the transformation process, application management, data management, asset management and organizational alignment.

Cyber-physical systems and the Internet of Things (IoT) underpin this transformation and provide new opportunities in areas such as product and service design, prototyping and development, remote control and diagnostics, status monitoring, proactive and predictable support, tracking and planning, innovation capability, flexibility, real-time applications, and more.

© Springer Nature Switzerland AG 2019
V. M. Vishnevskiy et al. (Eds.): DCCN 2019, CCIS 1141, pp. 524–535, 2019.
https://doi.org/10.1007/978-3-030-36625-4_42

IoT is expanding every day. IoT is the network of physical objects and network connectivity that enables these objects to collect and exchange data. Different views of the IoT concept exist and different applications are developing as avalanche (see Fig. 1).

Research in IoT focuses to a large extent on real-time data collection, effective data detection, on-site data manipulation, and real-time visualization. The availability of data generated by various sources opens up new opportunities for innovative applications in various fields such as intelligent transport systems, intelligent buildings, healthcare, energy, logistics and overall decision-making systems.

Fig. 1. IoT tree, adapted from IoT-A, European FP7 Research Project.

The Internet of Things as a growing research area faces many challenges. One challenge is to ensure timely and reliable processing of data sequences coming from the networked devices.

In the paper different approaches to process the data from various sources according to different goals: sensing, data analytics, and machine learning are discussed.

2   IoT System Architecture

Different definitions of IoT exist, as well as many different IoT system implementations. However, three main parts of an IoT system can be distinguished - things, connectivity, and processing, or using the received data.

The "things" (Fig. 2) in the Internet of Things consist of a variety of hardware specifications, communication capabilities and service quality, making IoT diverse in its nature. "Things" can be anything from any non-living object to any living object, whether human or animal, the devices in this technology may differ as physical things and virtual things. The outputs of the things can be analog or digital. Power consumption, bandwidth capacity, signal penetration and module cost are of great importance in their computational capabilities, the allocation and management of energy needed, memory specifications, and reliability management.

Fig. 2. The "things" in the Internet of Things.

Things with embedded computing devices become recognizable and intelligent. They can take action or offer context-sensitive solutions. They also have the ability to provide information about themselves and access summary information from other "smart devices". These smart devices can interact with each other through wireless (Fig. 3) or cable connections and unique addressing schemes. Researches on these things concern the development of novel sensors, low energy consumption computing devices, with small sizes, and wide embedded capabilities. The purpose of the Internet of Things is to enable objects to be connected at anytime, anywhere. Different aspects of IoT data processing, or using the received data are shown on Fig. 4. They cover enabling technologies, applications, business models together with social and environments impacts.

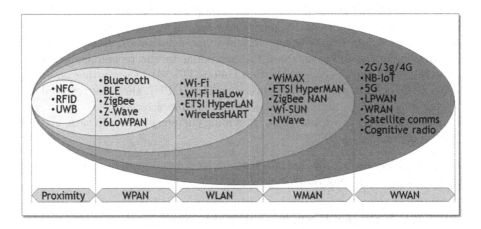

Fig. 3. The wireless connectivity in the Internet of Things.

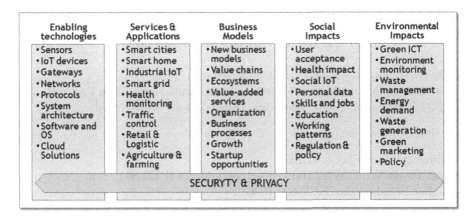

Fig. 4. All aspects of using data from IoT.

The architecture of the IoT system consists of several layers:

- The Sensing Layer - The edge technology layer includes IoT devices, all hardware components such as sensors, readers, RFID tags, drives and more. Sensors are mainly responsible to collect and transduce the required data.

- Access gateway layer - This layer is used to provide communication between devices and the cloud. Transceivers facilitate the collection of the message sent by the local and remote computing nodes or other associated devices.
- The Network layer. It can be considered as mist, fog and cloud computing. On the mist level some decisions are taken at the source which allows discarding useless information and to speed up data processing at destination as processing is done on the level of the sensors. Fog and cloud computing provide different levels of data analysis and processing.
- The Middleware layer is a two-way communication channel that includes an access portal, internet and backend server or/and storage. The main tasks of this layer are data management, data analysis, aggregation, filtering devices, device discovery and access control, according to the recommendations of the International Telecommunication Union (ITU) [1]. Computing nodes on this layer contain the central processing unit (CPU), they are required for processing the data and information received from a sensor.
- Application layer is responsible for providing customer service in various areas and locations.

IoT protocols are mainly divided into short-range communication, medium-range communication, and long-range communication.

IoT Network Protocols Stack consists of:

- On the Physical and MAC Layer it is IEEE 802.15.4.
- On the Adaptation Layer the 6LoWPAN, an acronym for IPv6 over low power wireless personal area networks, enables communication using IPv6 over the IEEE 802.15.4.
- On the Network Layer the RPL for Low Power and Lossy Networks (LLNs) is used.
- On the Transport Layer UDP is preferred over TCP.
- On the Application Layer the CoAP (Constrained Application Protocol) and MQTT (Message Queue Telemetry Transport) are widely used.

And non-IP networks:

- Bluetooth Low Energy (BLE).
- Low Power WiFi (WiFi HaLow) is based on the IEEE 802.11ah.
- Zigbee based on the IEEE 802.15.4.
- RFID.
- LPWAN.

Researchers in the communication field often focus mainly on the those aspects of the Internet of Things that relate to the development of new communication protocols and networking capabilities.

Operating Systems for IoT Devices. Ubuntu Core is widespread in drones, gateways, robots and more. MbedOS (ARM) is commonly used in wearable devices. Zephyr is mainly used in low resource and low power devices. Apache Mynewt works primarily on devices that use the BLE protocol. RIOT is free, open source and very friendly. RealSense can be found in many different sensors and cameras. Nucleus RTOS (Real Time Operating System) is completed with database storage, management, USB, multimedia, and basic GUI support. It can be often seen in industrial and aerospace applications. Brillo is Android based and is used on embedded devices. Contiki is used in low-power wireless devices such as street lighting, sound systems, and surveillance systems. Integrity RTOS is designed for a multi-core computer. It is widely used in the military applications. TinyOS aimed at low-power wireless devices. Fuchsia OS created for the Internet of Things devices by Google.

3   Heterogeneous Data

IoT data is inherently heterogeneous and noisy by nature because of different hardware, operating systems, used software and gateway requirements.

The different information sources in IoT provide diverse types of data. These information sources come from the physical nature of the source of the data, which is collected and processed. These may be scalar measurements, time series, samples, signals, images, videos, etc. Heterogeneous data sources offer heterogeneous, imprecise and potentially incomplete data. Examples are temperature, pressure, voltage, current, velocity, acceleration, gas analysis, sound measures, heart rate, blood pressure, illumination, humidity, noise levels, weight, coordinates, etc.

A heterogeneous domain is defined as a Cartesian product of a collection of source sets: $H^n = \Psi_1 \times ... \times \Psi_n ...$, where $n > 0$ is the number of information sources to consider. Heterogeneous variables are composed of mixtures of nominal, ordinal, interval, ratio, fuzzy variables, and entire empirical probability distributions [2]. Uncertainties and incompleteness of different degrees are additive to the heterogeneity.

The Internet itself is an example of a heterogeneous network as it conforms diverse hardware and software interfaces used in common by different products.

The existence of heterogeneous IoT data sources makes it difficult for the existing tools to analyze them.

4   Diverse Applications

The main domain areas [4–8] and examples of IoT corresponding applications are shown on Fig. 5.

Health	Monitoring, Prognosis, Ambient assisted living, Neo-natal care, Elderly care
Transport	Traffic Control and Routing, Smart Parking, Pedestrian Detection, Intelligent vehicles, Autonomous vehicles
Living	Smart buildings, Smart cities, Cultural behavior, Public safety, Memory augmentation, Lifestyle monitoring, Drone surveillance systems, Wearable personal digital assistant, Humans 2.0
Environment	Smart energy, Intelligent sensor agents, Complex event processing, Disaster detection and response, Wind forecasting, Fire hazard prevention
Industry	Logistic and supply chain management, Smart agriculture, Remote control, Chemical process, Manufacturing systems, M2M, Intelligent sensors
Science & Education	Space and Earth exploration, AR&VR, Social IoT

Fig. 5. Domain areas and IoT corresponding applications.

5   Adding Intelligence to IoT

With a lot of diverse areas of applications it is more essential to develop IoT-based projects with less 'wow' and a more 'what and how?'. So the serious trend towards introducing methods from Artificial Intelligence (AI) and Machine Learning (ML) into the various IoT developments is observed (Fig. 6). It is important to find new methods and intelligent computing techniques to increase the efficiency of the use of data provided by devices with Internet connectivity. The IoT generates large amount of heterogeneous data and a lot of complex problems that can be solved by using AI with a data which is streamed from IoT devices. As the demand for these solutions increases, the methods to handle these problems are becoming more widely available and convenient to use.

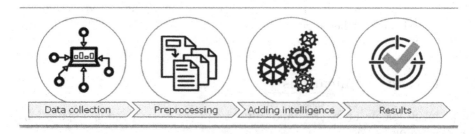

Fig. 6. Adding intelligence to IoT.

The whole range of AI and ML methods can be applied [9]:

- Data Mining
- Association Rule Mining
- Pattern Recognition and Extraction
- Hierarchical Clusters
- Knowledge Discovery
- Prediction
- Principle Component Analysis
- Classification.

Knowledge of data and well-formulated goals are an important element in the overall process of determining the type of ML algorithm required.

General methodology of applying ML methods against IoT dataset includes collection, transmission, intelligent processing and visualization of gathered data:

- Collecting data from physical devices;
- Organizing and grouping data according to predefined rules and user needs;
- Cloud data processing and analysis;
- The results obtained are shaped as ready-to-visual logical blocks of data in the various user interfaces.

A standardized data processing route - OSEMN (Fig. 7) - gives a clear order of activities - Obtain, Scrub, Explore, Model data and iNterpret the data [3].

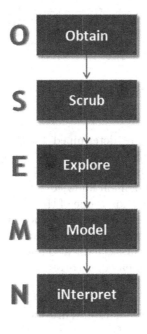

Fig. 7. OSEMN process.

Machine Learning is an iterative process (Fig. 8) that consists of:

– Creating experiments;
– Split data for training and verification;
– Train model;
– Confirm model results against dataset;
– Determine quality factors;
– Evaluate model.

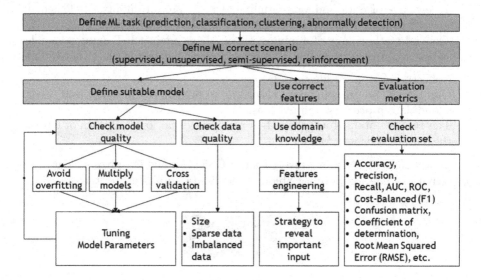

Fig. 8. ML iterative process.

Appropriate algorithms are determined by the data in use [10,11]. The most widely used Supervised, Unsupervised and several specific types of Semi-Supervised learning algorithms are shown on Fig. 9.

Feature engineering uses domain knowledge to increase the predictive power of ML algorithm. Feature selection is one of the main concepts in machine learning, which has a significant impact on the model's performance. Depending on the model type the evaluation metrics are exploited, for example: classification – AUC; regression - R2; multi-class classification model - confusion matrix, binary classification model - accuracy based on correct answers (Area under ROC curve - AUC).

As Performance Metrics the following can be used: Root Mean Squared Error (RMSE); Coefficient of determination; Threshold and Precision = $TP/(TP + FP)$; Recall = $TP/(TP + FN)$; Cost-Balanced (F1).

Handling of imbalanced data often is crucial. Real data is mostly imbalanced. This is the case where there are significantly more examples of one class than of

Supervised learning		Unsupervised learning		Semi-Supervised learning		Reinforcement learning
Classification	Regression	Clustering	Association	Classification	Clustering	Markov Decision Process
• Support Vector Machine (SVM) • Logistic Regression • Linear Discriminant Analysis • Naïve Bayes • k-Nearest Neighbor	• Linear Regression • Non- linear Regression	• K-Medoids, Fuzzy C-Means • Hierarchical • Gaussian Mixture • Hidden Markov Model • Density-Based Spatial clustering of Applications with Noise • Neural Networks • Recommender system • PCA	Apriori algorithms for association rule: • Apriori • FP Growth.	Text classification	Adversarial training	• Q-Learning • Temporal Difference (TD) • Deep Adversarial Networks
• Decision Trees • Ensemble Methods • Neural Networks • Random Forest • AdaBoost • Gradient-boosting trees		• k-means clustering, • Association Rules		• Graph-based methods • Heuristic approaches • Low-density separation • Generative models		

Fig. 9. The most widely used Supervised, Unsupervised and several specific types of Semi-Supervised learning algorithms.

any others. Prediction of the minority class (for example, failures or abnormalities) is of vital importance. However, less than 10% of the data may belong to this minority class. Thus the performance of standard ML algorithm may be compromised. Accuracy of 99% does not mean useful model for 1% minority class. Two methods which allow better learning can be considered: Over-sampling (generate examples combining features of target with features of neighbors) [12] and Under-sampling.

6 Tendencies and Barriers to IoT Systems Development

6.1 Tendencies

- Analytics plays a role in many IoT applications across many domains. Machine learning methods like Neural Networks and Deep learning are targeting in image recognition, natural language processing (NLP), feature extraction and overall decision making.
- In an IoT ecosystem, most of the communication is in the form of Ma-chine-to-Machine (M2M) interactions. Hence Blockchain enabled enhanced IoT security can be proposed.
- In the context of the IoT, self-adaptation is a salient property of smart objects. It allows them to be self-configured and adapted to extreme conditions while ensuring the target system objectives such as comfort, automation, security and safety goals [13].

– Semantic enrichment of the IoT data is an ongoing topic of research [14]. Thing description relies on defining state properties, actions and events exposed by a Web of Things. Semantic Web of Things should support various application domains with diverse characteristics for better discoverability.
– Using of high spectrum frequency can increase user benefits by widening of potential applications.

6.2   And Barriers

– Various communication challenges exist [15]. IoT is a collection of many parts and systems they are fundamentally different.
– Many data mining and machine learning methods do not handle IoT data heterogeneity well. Data and algorithms alone don't actually add very much on their own. There is a need for appropriate interpretation of ML models results.
– Lack of standards for authentication and authorization of IoT edge devices.
– Security and privacy risks, exceedingly complex to manage, with all sorts of unpredictable vulnerabilities, tech products have a short reliable life span. Privacy/Security/Safety (PSS) is always an issue with every new technology or concept.
– Ethical and legal issues also have to be taken under consideration with IoT expanding.

7   Conclusion

The tasks of information processing in IoT systems from various fields of applications are still challenging due to the high complexity of these systems. The development of mathematical methods and the means of artificial intelligence to retrieve data show the potential opportunity to provide information to help understand how certain processes for better management in today's complex environments are being implemented. Detecting regularities and abnormalities, forecasting of the system states and constructing classification are becoming increasingly important in wide range of IoT applications.

The unresolved problems that exist on the Internet of Things systems largely predetermine the major threats posed by the IoT systems. Solution of real world IoT cases is possible by using ML with AI approaches, developed in the last few years. In the paper some trends and methods are pointed out which may bring the physical and information world closer together.

References

1. Internet of Things (IoT). https://www.itu.int/en/ITU-T/ssc/resources/Pages/topic-001.aspx. Accessed 8 August 2019
2. Valdés, J.J.: Extreme learning machines with heterogeneous data type. Neurocomputing **277**, 38–52 (2018)

3. Dineva, K., Atanasova, T.: OSEMN process for working over data acquired by IoT devices mounted in beehives. Curr. Trends Nat. Sci. **7**(13), 47–53 (2018)

4. Siow, E., Tiropanis, T., Hall, W.: Analytics for the Internet of Things: a survey. ACM Comput. Surv. **51**(4), 36 pages (2018). https://doi.org/10.1145/3204947

5. Mezghani, E., Expósito, E., Drira, K.: A model-driven methodology for the design of autonomous and cognitive IoT-based systems: application to healthcare. IEEE Trans. Emerg. Top. Comput. Intell. **1**(3), 224–234 (2017)

6. Dineva, K., Atanasova, T.: ICT-based beekeeping using IoT and machine learning. In: Vishnevskiy, V.M., Kozyrev, D.V. (eds.) DCCN 2018. CCIS, vol. 919, pp. 132–143. Springer, Cham (2018). https://doi.org/10.1007/978-3-319-99447-5_12

7. Alexandrov, A.: Ad-hoc Kalman filter based fusion algorithm for real-time wireless sensor data integration. Flexible Query Answering Systems 2015. AISC, vol. 400, pp. 151–159. Springer, Cham (2016). https://doi.org/10.1007/978-3-319-26154-6_12

8. Perwej, Y., Haq, K., Parwej, F., Mumdouh, M., Hassan, M.: The internet of things (IoT) and its application domains. Int. J. Comput. Appl. **182**(49), 36–49 (2019). https://doi.org/10.5120/ijca2019918763

9. Abu-Elkheir, M., Hayajneh, M., Ali, N.A.: Data management for the Internet of Things: design primitives and solution. Sensors (Basel) **13**(11), 15582–15612 (2013)

10. Tashev, T., Monov, V.: Large-scale simulation of uniform load traffic for modeling of throughput on a crossbar switch node. In: Lirkov, I., Margenov, S., Waśniewski, J. (eds.) LSSC 2011. LNCS, vol. 7116, pp. 638–645. Springer, Heidelberg (2012). https://doi.org/10.1007/978-3-642-29843-1_73

11. Balabanov, T., Zankinski, I., Shumanov, B.: Slot machines RTP optimization with genetic algorithms. In: Dimov, I., Fidanova, S., Lirkov, I. (eds.) NMA 2014. LNCS, vol. 8962, pp. 55–61. Springer, Cham (2015). https://doi.org/10.1007/978-3-319-15585-2_6

12. Chawla, N.V., Bowyer, K.W., Hall, L.O., Kegelmeyer, W.P.: SMOTE: synthetic minority over-sampling technique. J. Artif. Intell. Res. **16**, 321–357 (2002)

13. Gatouillat, A., Badr, Y., Massot, B.: QoS-driven self-adaptation for critical IoT-based systems. In: Braubach, L., et al. (eds.) ICSOC 2017. LNCS, vol. 10797, pp. 93–105. Springer, Cham (2018). https://doi.org/10.1007/978-3-319-91764-1_8

14. Bali, A., Al-Osta, M., Abdelouahed, G.: An ontology-based approach for IoT data processing using semantic rules. In: Csöndes, T., Kovács, G., Réthy, G. (eds.) SDL 2017. LNCS, vol. 10567, pp. 61–79. Springer, Cham (2017). https://doi.org/10.1007/978-3-319-68015-6_5

15. Zikria, Y.B., Kim, S.W., Hahm, O., Afzal, M.K., Aalsalem, M.Y.: Internet of Things (IoT) operating systems management: opportunities, challenges, and solution. Sensors **19**, 1793 (2019)

Calculation of Probability Density Distribution of Ultracold Atoms and Molecules in Waveguide-Like Traps

Vladimir S. Melezhik[1,2](✉)📷 and Leonid A. Sevastianov[1,2]📷

[1] Bogoliubov Laboratory of Theoretical Physics, Joint Institute for Nuclear Research, Joliot-Curie 6, Moscow Region, Dubna 141980, Russia
melezhik@theor.jinr.ru
[2] Peoples' Friendship University of Russia (RUDN University), Miklukho-Maklaya str. 6, Moscow 117198, Russia
sevastianov_la@rudn.ru
http://theor.jinr.ru/~melezhik

Abstract. The quantitative description of the confined quantum scattering in waveguide-like atomic traps is an actual problem of modern physics of ultracold atoms and molecules. It is a challenging problem of computational mathematics to integrate the few-dimensional Schrödinger equation describing such scattering. In the present work we show how the split-operator method developed by V.S. Melezhik in discrete-variable representation can be parallelized and extended for calculation of probability density distribution in such quantum systems. By using as an example the confined collision of Li atoms with Yb ions in a hybrid atom-ion trap we demonstrate calculation of the time-evolution of the atom-ion probability density distribution. The calculated function is an important parameter for analysis of this reaction. Due to resent development of unique experimental technique in this field it becomes actual experimental analysis of cold low-dimensional few-body systems. However, interpretation and planning of the experiments demand quantitative description of the systems. The present work opens promising perspective in the development of this direction.

Keywords: Schrödinger equation · Splitting methods · Discrete-variable representation · Waveguides · Probability density

1 Introduction

The quantitative description of the confined quantum scattering in waveguide-like atomic traps is an actual problem of modern physics of ultracold atoms and molecules (see, for example [1,2]). It is a challenging problem of computational mathematics too. Actually, the conventional few-body theory for a free space is

Supported by the "RUDN University Program 5-100".

© Springer Nature Switzerland AG 2019
V. M. Vishnevskiy et al. (Eds.): DCCN 2019, CCIS 1141, pp. 536–546, 2019.
https://doi.org/10.1007/978-3-030-36625-4_43

no longer valid here and the development of the low-dimensional theory, including the influence of the confinement, is needed. In works of V.S. Melezhik with co-authors an efficient computational method for the time-dependent Schrödinger equation describing pair collisions in tight atomic waveguides was developed and several novel effects were found [3–6].

The key element of the computational scheme is a component-by-component split-operator method based on a nondirect product discrete variable representation (npDVR) [7–11]. With the developed method the three-dimensional (one radial and two angular variables) [9–13] and four-dimensional (two radial and two angular variables) [3–5] equations were integrated.

In the present work we discuss an extension of the method to the confined collision of Li atoms with Yb ions in a hybrid atom-ion trap. We calculate the time-evolution of the atom-ion probability density distribution. The calculated function is an important parameter for analysis of this reaction. These calculations become actual for interpretation and planning experiments in this fast growing field of research. We also discuss how the computational scheme can be parallelized and extended to a higher dimension of the Schrödinger equation (up to two radial and four angular variables) with the use of distributed computations. Such extension opens new possibilities for modeling new class of low-dimensional few-body processes in confined geometry of atomic and classical optical waveguides.

2   Splitting-Up Method in DVR for Few-Dimensional Time-Dependent Schrödinger Equation

2.1   Pair Collision of Identical Atoms in Waveguide-Like Trap

The description of pair collisions of ultracold identical atoms ($m_1 = m_2 = m$) in optical waveguides, when trap frequencies are identical for both atoms (ω_x and ω_y) and the center-of-mass of the pair can be separated [4], was reduced in [3,4] to the system of N Schrödinger-like equations ($\hbar = 1$)

$$i\frac{\partial}{\partial t}\psi_j(r,t) = \sum_{j'=1}^{N} H_{jj'}(r)\psi_{j'}(r,t). \tag{1}$$

The system (1) approximates the initial 3D Schrödinger equation at the 2D npDVR on the angular grid $\Omega_j = \{\theta_{j_\theta}, \phi_{j_\phi}\}$. The construction of the npDVR basis functions $f_j(\Omega)$

$$f_j(\Omega) = \sum_{\nu=1}^{N} Y_\nu(\Omega)(Y^{-1})_{\nu j} \tag{2}$$

is described in [6,9–11], where also $Y_\nu(\Omega)$ is defined as a spherical harmonic slightly corrected by a small additive of linear combination of higher harmonics. In this representation the number of basis functions N ($j = (j_\theta, j_\phi) = 1, 2, ..., N; \nu = (l, m) = 1, 2, ..., N$) is equal to the number of angular grid points

$N = N_\theta \times N_\phi$ $(j_\theta = 1, 2, ..., N_\theta, j_\phi = 1, 2, ..., N_\phi)$, what permits to define the matrix $(Y^{-1})_{j\nu}$ inverse to the quadratic $N \times N$ matrix $Y_\nu(\Omega_j)$.

For the propagation $\psi_j(r, t_n) \to \psi_j(r, t_{n+1})$ in time $t_n \to t_{n+1} = t_n + \Delta t$ V.S. Melezhik has suggested [9,10] a computational scheme [9–11] based on the component-by-component split-operator method of Marchuk [14]. In this approach the effective Hamiltonian $\hat{H}(r)$ splits into the following two parts

$$H_{jj'}(r) = h_{jj'}(r) + U_j(r)\delta_{jj'} \tag{3}$$

where

$$h_{jj'}(r) = -\frac{\delta_{jj'}}{2\mu}\frac{d^2}{dr^2} + \delta_{jj'}V(r) - \frac{1}{2\mu r^2 \sqrt{\lambda_j \lambda_{j'}}} \sum_\nu^N (Y)_{j\nu}^{-1} l(l+1)(Y^{-1})_{\nu j'},$$

$$U_j(r) = \frac{1}{2}\mu(\omega_x^2 x_j^2 + \omega_y^2 y_j^2).$$

Here the diagonal in the npDVR potentials $V(r)$ and $U(r)$ describe the spherically symmetric interatomic interaction and the interaction of the atoms with the waveguide-like trap, $x_j = r \sin\theta_{j_\theta} \cos\phi_{j_\phi}$, $y_j = r \sin\theta_{j_\theta} \sin\phi_{j_\phi}$, where λ_j are weights of Gaussian quadratures over θ and ϕ and $\mu = m/2$.

The time-step $\psi_j(r, t_n) \to \psi_j(r, t_{n+1} = t_n + \Delta t)$ was approximated according to

$$\psi(t_n + \Delta t) = \exp(-\frac{i}{2}\Delta t \hat{U}) \exp(-i\Delta t \hat{h}) \exp(-\frac{i}{2}\Delta t \hat{U})\psi(t_n) + O(\Delta t^3). \tag{4}$$

The time evolution is described in detail in [6,10,11] where the fact that the npDVR on the basis-functions $f_j(\Omega)$ (2) gives the diagonal representation for $\hat{U}$ and the $Y_\nu(\Omega)$-representation gives the diagonal representations for angular part of the kinetic-energy operator in $\hat{h}$ was used for constructing an efficient computational algorithm. The time evolution was proceeded as follows. For the first and the last steps according to the relation (4) we write the function ψ and the operators $\exp(-i\Delta t \hat{U}/2)$ in our 2D npDVR (2) on the grid $\Omega_j = (\theta_{j_\theta}, \phi_{j_\phi})$. Since the potential $\hat{U}(r) = U_j(r)\delta_{jj'}$ is diagonal in this representation the first and last steps represent simple multiplications of the diagonal matrices $\exp(-\frac{i}{2}\Delta t U_j(r))\delta_{jj'}$. The intermediate step in (4) depending on $\hat{h}$ is treated in the basis $Y_\nu(\Omega)$ where the matrix operator $\hat{h}(r)$ is diagonal with respect to the indices m and l $(\nu = (l, m))$. For that we approximate the exponential operators according to

$$\exp(-i\Delta t \hat{h}) \approx (1 + \frac{i}{2}\Delta t \hat{h})^{-1}(1 - \frac{i}{2}\Delta t \hat{h}) + O(\Delta t^2), \tag{5}$$

which ensures the desired accuracy of the numerical algorithm (4). Thus, after the discretization of r with the help of finite-differences the matrix $\hat{h}$ possesses a band structure and we arrive at the following boundary-value problems

$$(1 + \frac{i}{2}\Delta t \hat{h})\psi(t_n + \frac{3}{4}\Delta t) = (1 - \frac{i}{2}\Delta t \hat{h})\psi(t_n + \frac{1}{4}\Delta t), \tag{6}$$

which can be solved rapidly due to the band structure of the matrix $\hat{h}$.

This computational scheme is unconditionally stable [14], preserves unitarity and is very efficient, i.e. the computational time is proportional to the total number N of grid points and basis functions in (1) [6,10,15].

The efficiency of the computational procedure is based on the fast transformation with help of the unitary matrix $S_{j\nu} = \lambda_j^{1/2} Y_\nu(\Omega_j)$ between the two relevant representations: 2D npDVR (2) and $Y_\nu(\Omega)$-representation.

2.2 Atom-Ion Collision in Hybrid Atom-Ion Trap

The dynamics of colliding atom-ion pair with coordinates $\mathbf{r_A}$ and $\mathbf{r_I}$ and masses m_A and m_I in the harmonic waveguide with the transverse potential

$$U(\rho_A, \rho_I) = \frac{1}{2}(m_A \omega_A^2 \rho_A^2 + m_I \omega_I^2 \rho_I^2)$$

(where $\rho_i = r_i \sin\theta_i$) is described by the 4D time-dependent Schrödinger equation ($\hbar = 1$)

$$i\frac{\partial}{\partial t}\psi(\rho_R, \mathbf{r}, t) = H(\rho_R, \mathbf{r})\psi(\rho_R, \mathbf{r}, t)$$

with the Hamiltonian

$$H(\rho_R, \mathbf{r}) = H_{CM}(\rho_R) + H_{rel}(\mathbf{r}) + W(\rho_R, \mathbf{r}). \tag{7}$$

Here

$$H_{CM} = -\frac{1}{2M}\left(\frac{\partial^2}{\partial \rho_R^2} + \frac{1}{\rho_R^2}\frac{\partial^2}{\partial \phi^2} + \frac{1}{4\rho_R^2}\right) + \frac{1}{2}(m_A\omega_A^2 + m_I\omega_I^2)\rho_R^2 \tag{8}$$

and

$$H_{rel} = -\frac{1}{2\mu}\frac{\partial^2}{\partial r^2} + \frac{L^2(\theta, \phi)}{2\mu r^2} + \frac{\mu^2}{2}\left(\frac{\omega_A^2}{m_A} + \frac{\omega_I^2}{m_I}\right)\rho^2 + V_{AI}(r) \tag{9}$$

describe the center-of-mass (CM) and relative (rel) atom-ion motions. The potential $V_{AI}(r)$ describes the atom-ion interaction, ρ_R and $\mathbf{r} = \mathbf{r_A} - \mathbf{r_I} \to (r, \theta, \phi) \to (\rho, \phi, z)$ are the polar radial CM and the relative coordinates and $M = m_A + m_I$, $\mu = m_A m_I / M$. The term

$$\frac{L^2(\theta, \phi)}{2\mu r^2} = -\frac{1}{2\mu r^2 \sin\theta}\left(\frac{\partial}{\partial\theta}\sin\theta\frac{\partial}{\partial\theta} + \frac{1}{\theta}\frac{\partial^2}{\partial\phi^2}\right)$$

represents the angular part of the kinetic energy operator of the relative atom-ion motion. The term

$$W(\rho_R, \mathbf{r}) = \mu(\omega_A^2 - \omega_I^2)r\rho_R \sin\theta \cos\phi \tag{10}$$

in the Hamiltonian leads for the atom (A) and ion (I), that feel different confining frequencies $\omega_A \neq \omega_I$, to a coupling of the CM and relative atom-ion motion, i.e. to the nonseparability of the quantum two-body problem in confined geometry of the harmonic trap.

The problem is to integrate the Schrödinger equation from time $t = 0$ to the asymptotic region $t \to +\infty$ with the initial wave-packet

$$\psi(\rho_R, \mathbf{r}, t = 0) = Nr\sqrt{\rho_R} \exp\{-\frac{\rho_A^2}{2a_A^2} - \frac{\rho_I^2}{2a_I^2} - \frac{(z - z_0)^2}{2a_z^2} + ik_0 z\} \qquad (11)$$

representing noninteracting atom and ion in the transversal ground state of the waveguide with $a_A = (1/m_A\omega_A)^{1/2}$ and $a_I = (1/m_I\omega_I)^{1/2}$ respectively and the overall normalization constant N defined by $< \psi(0)|\psi(0) >= 1$. We choose $z_0 \to -\infty$ to be far from the origin $z = 0$ and $a_z \to \infty$ to obtain a narrow width in momentum and energy space for the initial wave-packet. The wave-packet moves with a positive interatomic velocity $v_0 = k_0/\mu = \sqrt{2\epsilon_{\parallel}/\mu}$ (defined by the relative longitudinal colliding energy $\epsilon_{\parallel}$) to the scattering region at $z = 0$ and splits up after the scattering into two parts moving in opposite directions $z \to \pm\infty$. If the initial conditions permit opening inelastic channels it can lead to the collisional excitation of the transverse vibrations [6] or formation of molecular bound states [5]. Multichannel character of the scattering is also developing at the region of the so-called confinement-induced resonances (CIRs) where resonance in the closed channel leads to the resonant behaviour of the elastic scattering amplitude if the initial colliding energy coincides with the resonant energy of molecular state in the closed channel [2–5].

The modeling of the long-range atom-ion interaction $V_{AI}(r)$ is discussed in the following Sect. 3.

The splitting-up method in the DVR representation described above demands some modification for the problem (7). Thus, to apply this procedure we split the Hamiltonian (7) into the following three parts

$$H_{jj'}(\rho_R, r) = h_{jj'}^{(0)}(\rho_R) + h_{jj'}^{(1)}(r) + U_j(\rho_R, \rho)\delta_{jj'}, \qquad (12)$$

where

$$h_{jj'}^{(0)}(\rho_R) = -\frac{\delta_{jj'}}{2M}(\frac{\partial^2}{\partial\rho_R^2} + \frac{1}{4\rho_R^2}) + \frac{1}{2M\rho_R^2\sqrt{\lambda_j\lambda_{j'}}}\sum_{\nu}^{N}(Y)_{j\nu}^{-1}m^2(Y^{-1})_{\nu j'},$$

$$h_{jj'}^{(1)}(r) = -\frac{\delta_{jj'}}{2\mu}\frac{\partial^2}{\partial r^2} + \delta_{jj'}V_{AI}(r) - \frac{1}{2\mu r^2\sqrt{\lambda_j\lambda_{j'}}}\sum_{\nu}^{N}(Y)_{j\nu}^{-1}l(l+1)(Y^{-1})_{\nu j'},$$

$$U_j(\rho_R, \rho) = \frac{1}{2}(m_1\omega_1^2 + m_2\omega_2^2)\rho_R^2 + \frac{\mu^2}{2}(\frac{\omega_1^2}{m_1} + \frac{\omega_2^2}{m_2})\rho^2 + \mu(\omega_1^2 - \omega_2^2)\rho\rho_R\cos\phi_j.$$

Subsequently, we can approximate the time-step $\psi_j(\rho_R, r, t_n) \to \psi_j(\rho_R, r, t_{n+1})$ where $t_n \to t_{n+1} = t_n + \Delta t$ according to

$$\psi(t_n + \Delta t) = \exp(-\frac{i}{2}\Delta t \hat{U}) \exp(-i\Delta t \hat{h}^{(1)}) \exp(-i\Delta t \hat{h}^{(0)}) \qquad (13)$$

$$\times \exp(-\frac{i}{2}\Delta t \hat{U})\psi(t_n) + O(\Delta t^3).$$

The time evolution proceeds as follows. For the first and the last steps according to the above relation (13) we write the function ψ and the operator $\exp(-i\Delta t \hat{U}/2)$ in our 2D npDVR (2) on the 2D grid $\{\Omega_j\} = \{\theta_{j_\theta}, \phi_{j_\phi}\}$. Since the potential $U_j(\rho_R, \rho)$ is diagonal in this representation the first and last steps represent simple multiplications of the diagonal matrices $\exp(-\frac{i}{2}\Delta t U_j(\rho_R, \rho))$. Two intermediate steps in (13) depending on $\hat{h}^{(0)}$ and $\hat{h}^{(1)}$ are treated in the basis Y_ν where the matrix operators $\hat{h}^{(0)}(\rho_R)$ and $\hat{h}^{(1)}(r)$ are diagonal with respect to the indices m and l. For that we approximate the exponential operators according to

$$\exp(-i\Delta t \hat{A}) \approx (1 + \frac{i}{2}\Delta t \hat{A})^{-1}(1 - \frac{i}{2}\Delta t \hat{A}) + O(\Delta t^2), \qquad (14)$$

which ensures the desired accuracy of the numerical algorithm. Thus, after the discretization of r (or ρ_R) with the help of finite-differences the matrix $\hat{A}$ possesses a band structure and we arrive at the following boundary-value problems

$$(1 + \frac{i}{2}\Delta t \hat{A})\psi(t_n + \frac{\Delta t}{4}) = (1 - \frac{i}{2}\Delta t \hat{A})\psi(t_n),$$

which can be solved rapidly due to the band structure of the matrix $\hat{A}$. This computational scheme is unconditionally stable [14], preserve unitarity and is very efficient, i.e. the computational time is approximately proportional to the total number N of grid points over the radial and angular variables [6,10,15] (see Fig. 1). The efficiency of the computational procedure is based (as well as in the previous subsection) on the fast transformation with help of the unitary matrix $S_{j\nu} = \lambda_j^{1/2} Y_{j\nu}$ between the two relevant representations: the 2D npDVR (2) and the Y_ν-representation.

3   Numerical Example: Time-Evolution of Atom-Ion Probability Density in Pair Confined Collisions

The computational scheme, described above, was successfully applied in a number of actual problems of physics of ultracold atoms. More specifically, it has permitted us to construct theoretical models for quantitative description of different resonant atomic processes in confined geometry of optical traps [4–6]. The computational difficulty of constructing these models is in including into consideration of the effect of atomic interaction with the "walls" of the optical trap. It leads to increasing the dimensionality of the solving problem (i.e. the dimensionality of the Schrödinger equation describing a process). Actually, if the interaction between two colliding quantum particles is central, then in a free space one can separate the center-of-mass of the system as well as the angular variables. Such a way, the problem of collision of two quantum particles in

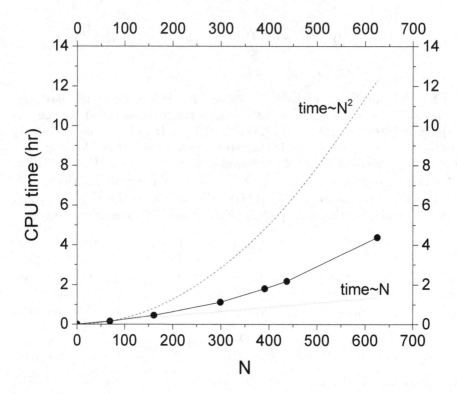

Fig. 1. The dependence of the CPU-time on the number of basis functions N of the angular DVR. The solid curve connects the calculated points.

a free space becomes reducible to a one-dimensional radial Schrödinger equation which is the simplest problem of quantum mechanics (one of milestones of classical quantum mechanics). However, if we consider the collision of two quantum particles in a waveguide-like trap, the problem becomes substantially more complicated. Interaction of the colliding particles with the trap does not permit separation of the center-of-mass and angular variables. In general case it needs to integrate the six-dimensional Schrödinger equations. In some special cases, it is possible to reduce the problem to an equation of smaller dimension, if colliding particles are identical and the interaction of particles with the trap is harmonic. A number of such actual problems of atomic collisions in confined geometry of waveguide-like traps was analysed with the developed computational methods by Melezhik with colleagues [3–5,7].

In recent years, there has been interest to hybrid confined atom-ion systems [16]. An actual problem is a collision of a rather hot ion confined in electromagnetic Paul trap with atoms from the "cloud" surrounding the ion. This problem attracts great interest due to possible important applications: investigation of the dynamics of chemical processes in new and extremal conditions, creating exotic systems such as polaron, modeling some effects of solid state physics and even modeling elements of a quantum computer [16]. Quantitative description of such hybrid systems demands further improvement of our approach. First, the different geometrical configuration of Paul and atomic traps does not permits reduction of dimensionality of the problem anymore. Second difficulty is in long-range character of the atom-ion polarization interaction

$$V_{AI}(r) \sim -1/r^4$$

with respect to the short-range Wan-der-Waals interaction $\sim -1/r^6$.

We made a first step in extension of our method to such hybrid systems by choosing the traps of a special form for colliding atom and ion. We suppose that an atom is confined in a transverse direction by an optical harmonic waveguide-like trap and can move freely in longitudinal direction. Inside the atomic trap is situated an ion harmonic waveguide-like trap which confines transversal ion motion and does not confine ion motion in longitudinal direction. We have considered such atom-ion confined collisions. The chose of the special geometry of such hybrid system permitted us to separate two angular variables in the problem and reduce it to the 4D time-dependent Schrödinger equation. The idea of this separation of variables is described in our papers [4,5] for the confined scattering of two distinguishable atoms in waveguide traps. Here we extend the scheme to atom-ion collision in confined geometry of waveguide-like atomic and ion traps.

As a numerical example we present here our calculation of the time-evolution of the probability density distribution

$$W(r, \rho_R, t) = \int |\psi(\mathbf{r}, \rho_R, t)|^2 \sin\theta d\theta d\phi \qquad (15)$$

of the colliding ^{6}Li atom and ^{171}Yb$^+$ ion confined in a hybrid atom-ion trap (see Fig. 2). The function $W(r, \rho_R, t)$ was calculated with the computer code based on the method discussed above. Here, r, θ and ϕ are the relative radial and angular variables between the atom and ion, ρ_R is the transversal radial variable of the center-of-mass of the colliding two-body systems [5,16]. We assume that the collision occurs in waveguide-like atom-ion harmonic trap. Interaction between confining trap and colliding quantum particles is described by the harmonic potential

$$U(\rho_A, \rho_I) = \frac{1}{2}(m_A \omega_A^2 \rho_A^2 + m_I \omega_I^2 \rho_I^2) \qquad (16)$$

and the atom-ion interaction - by standard C12-C4 long-range interaction

$$V_{AI}(r) = \frac{C_{12}}{r^{12}} - \frac{C_4}{r^4}. \qquad (17)$$

Here, the indexes A and I correspond to atomic and ion variables.

The calculated structure of the function $W(r, \rho_R, t)$ at large times $t \to \infty$ gives population of the analysing two-body quantum system (atom-ion) in the final state after the pair collision. The considered here example clearly demonstrates formation of molecular ion at a final state of the collision. It is clear that the formed molecule is in a highly excited state: the function $W(r, \rho_R, t \to \infty)$ has several minima with respect to the radial variable r. Moreover, as a result of the collision, the released energy went to the excitation of oscillations of the center-of-mass of the molecule (function $W(r, \rho_R, t \to \infty)$ also has minima with respect to the center-of-mass variable ρ_R). This analysis of the calculated function $W(r, \rho_R, t \to \infty)$ shows that the molecular ion was formed according to the mechanism predicted in work [5].

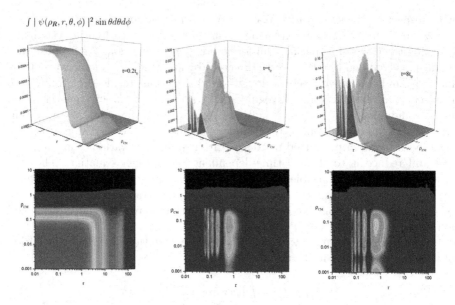

$$\int |\psi(\rho_R, r, \theta, \phi)|^2 \sin\theta d\theta d\phi$$

Fig. 2. Evolution in time of the atom-ion probability density distribution $W(r, \rho_R, t) = \int |\psi(\mathbf{r}, \rho_R, t)|^2 \sin\theta d\theta d\phi$ in the process of the confined collision of the ^{6}Li atoms with ^{171}Yb$^+$ ions. Top row of graphs demonstrates the probability density $W(r, \rho_R, t)$ calculated at three time points during the collision and the bottom row is the corresponding contour plots of these probabilities $W(r, \rho_R, t)$. The time unit $t_0 = 2\pi/\omega_A$ is defined by the lowest frequency ω_A of this dynamical quantum system.

We have to underline here the following peculiarity of the computational scheme which is important element for acceleration of the computations. The remaining most time-consuming part of the computational scheme (4–6) for integration the system of N differential equations (1) is the integration of the boundary-value problem for the system of equations (6). However, the system (6) is diagonal in $Y_\nu(\Omega)$-representation. This fact permits transparent procedure of parallelization. Actually, because the Eq. (6) are uncoupled, every equation of the system can be integrated independently on a separate processor.

The suggested modification of the computational scheme (4–6) considerably increases its efficiency. With parallelization and use of distributed computations the method can be extended for integration of the Schrödinger equations of a higher dimension. This fact makes the scheme very promising for its extension to confined atom-ion collisions with more realistic form of the atomic and ion traps. We suppose to extend the method for integration of the six-dimensional time-dependent Schrödinger equation and to analyse important effects of anharmonicity of the traps and ion "micromotion" [16]. Such an extension opens also perspectives for application of the method to other actual problems of low-dimensional few-body processes in atomic and classical optical waveguides.

4 Conclusion

The efficiency of the splitting-up method based on the npDVR for the time-dependent Schrödinger equation with the suggested simple procedure for its parallelization makes the method very promising in application to actual problems of low-dimensional few-body physics in atomic waveguide-like traps. In this respect one can mention the problem of ultracold atomic collisions in anharmonic and asymmetric wavegudes and in quasi-2D confining traps, collisional three-body problem (Efimov resonances) in tight traps, and non-linear time-dependent Schrödinger equation with a few spatial variables arising in physics of Bose-Einstein condensates. All these questions are hot problems of modern quantum physics.

References

1. Wenz, A.N., Zürn, G., Murmann, S., Brouzos, I., Lompe, T., Jochim, S.: From few to many: observing the formation of a Fermi sea one atom at a time. Science **342**(6157), 457–460 (2013)
2. Haller, E., et al.: Confinement-induced resonances in low-dimensional quantum systems. Phys. Rev. Lett. **104**, 153203-1–153203-4 (2010)
3. Kim, J.I., Melezhik, V.S., Schmelcher, P.: Suppression of quantum scattering in strongly confined systems. Phys. Rev. Lett. **97**, 193203-1–193203-4 (2006)
4. Melezhik, V.S., Kim, J.I., Schmelcher, P.: Wave-packet dynamical analysis of ultracold scattering in cylindrical waveguide. Phys. Rev. A **76**, 053611-1–053611-15 (2007)
5. Melezhik, V.S., Schmelcher, P.: Quantum dynamics of resonant molecule formation in waveguides. New J. Phys. **11**, 073031-1–073031-10 (2009)
6. Melezhik, V.S.: Multi-channel computations in low-dimensional few-body physics. In: Adam, G., Buša, J., Hnatič, M. (eds.) MMCP 2011. LNCS, vol. 7125, pp. 94–107. Springer, Heidelberg (2012). https://doi.org/10.1007/978-3-642-28212-6_8
7. Melezhik, V.S.: New method for solving multidimensional scattering problem. J. Comput. Phys. **92**, 67–81 (1991)
8. Melezhik, V.S.: Three-dimensional hydrogen atom in crossed magnetic and electric fields. Phys. Rev. A **48**, 4528–4538 (1993)
9. Melezhik, V.S.: Polarization of harmonics generated from a hydrogen atom in a strong laser field. Phys. Lett. A **230**, 203–208 (1997)
10. Melezhik, V.S.: A computational method for quantum dynamics of a three-dimensional atom in strong fields. In: Atoms and Molecules in Strong External Fields, pp. 89–94. Plenum, New-York & London (1998)
11. Melezhik, V.S., Baye, D.: Nonperturbative time-dependent approach to breakup of halo nuclei. Phys. Rev. C **59**, 3232–3239 (1999)

12. Melezhik, V.S., Schmelcher, P.: Quantum energy flow in atomic ions moving in magnetic fields. Phys. Rev. Lett. **84**, 1870–1873 (2000)
13. Melezhik, V.S., Cohen, S.J., Hu, C.Y.: Stripping and excitation in collisions between p and He^+ ($n \leq 3$) calculated by a quantum time-dependent approach with semiclassical trajectories. Phys. Rev. A **69**, 032709-1–032709-13 (2004)
14. Marchuk, G.I.: Methods of Numerical Mathematics. Springer-Verlag, New York (1975)
15. Melezhik, V.S.: Mathematical modeling of ultracold few-body processes in atomic traps. EPJ Web of Conf. **108**, 01008-1–01008-10 (2016)
16. Melezhik, V.S., Negretti, A.: Confinement-induced resonances in ultracold atom-ion systems. Phys. Rev. A **94**, 022704-1–022704-9 (2016)

Mathematical Aspects of Stable State Estimation of the Radio Equipment in Terms of Communication Channel Functioning

Vladimir Fedorenko[1] , Vladimir Samoylenko[1(✉)] , Alexey Vinogradenko[2] ,
Irina Samoylenko[3] , Ildar Sharipov[3] , and Sergey Anikuev[3]

[1] North-Caucasus Federal University, Stavropol, Russian Federation
vvsamoylenko@ncfu.ru
[2] Marshal Semyon Budyonny Military Signals and Communications Corps Academy,
St. Petersburg, Russian Federation
[3] Stavropol State Agrarian University, Stavropol, Russian Federation

Abstract. The state of the object in terms of its functioning in the system can not be estimated without the construction of multi-level modeling. A cross-correlation coefficient of the reference and distorted signals has been suggested as the indicator for radio equipment. The dependence of the indicator of reliable message transmission in the communication channel on the value of the correlation coefficient of signals (CCS) has been described. A mathematical model has been developed to consider CCS changes with parametric failures in the elements of the signal generator in the radio equipment. The application of solution methods of conditionally correct problems is established for the stable evaluation class of the technical condition of radio equipment. We have analyzed the influence of correlation function samples on the solution stability of the redefined system of linear algebraic equations obtained by sampling the distorted and reference signals.

Keywords: Radio facilities · Signal shaper · Parametric failures · Signal correlation · Inverse problem of mathematical physics

1 Introduction

For each combination of communication type (modulation/detection), channel fading model and diversity type we obtain an average bit error rate (BER) and/or symbol error rate (SER). We get an expression in a form that can be easily evaluated [1].

Traditionally, the bit error probability p_{ber} in the communication channel is determined by the type of signal, exceeding of the signal level above the noise level at the receiving point, the method of signal processing in the receiving device and the characteristics of the signal propagation medium [2].

© Springer Nature Switzerland AG 2019
V. M. Vishnevskiy et al. (Eds.): DCCN 2019, CCIS 1141, pp. 547–559, 2019.
https://doi.org/10.1007/978-3-030-36625-4_44

But the wrong decision about the type of information bit can be the result of signal distortion in the hardware part of the communication channel.

Until recently, no one has demonstrated a unified approach to assess the joint impact of fluctuation noise in the communication channel and hardware signal distortion on the probability of bit errors. This approach may simplify the previously obtained complex links for telecommunication systems indicators both analytically and computationally. In addition, it allows to obtain new multi-level models for special cases (telecommunication network – communication channel – modem – signal conditioner), which still have been solved in a complicated form [3].

In this paper we develop methodological foundations of multi-level modeling of radio facilities, which operate in discrete communication channels as the objects of control. The purpose of the article is to determine the conditions for a stable assessment of the signal generator state in a modem according to its functioning in the communication channel.

2   The Channel Model of Incoherent Messages Receiving with Hardware Signal Distortions

Construction of the model begins with definition of the input and output variables, relationship between which should be described by the model of the control object. Signal-noise mixture is formed from the output of the communication channel to the input of the demodulator receiver:

$$y(t) = \mu S_{rD}(t) + \xi(t), \tag{1}$$

where S_{rD} is a time function, which defines r-th distorted signal; μ is an amplitude transmission coefficient of the signal in the channel; $\xi(t)$ is a time function of the Gaussian noise interference.

The main method of signal reception in the channel with variable parameters is incoherent reception without extracting information about the initial phase of the high-frequency signal filling. Decisive schemes of incoherent signal processing based on matched filters or correlators have become the most widespread application, which implement the operator of the form [2]:

$$V_r = \frac{2\mu}{T}\{[\int_0^T y(t)s_{rE}(t)dt]^2 + [\int_0^T y(t)\tilde{s}_{rE}(t)dt]^2\}, t \in [0,T]; r = \overline{1,m} \tag{2}$$

where s_{rE} is the time function for r-th variant of reference signal; $\tilde{s}_{rE}$ is a function, coupled by Hilbert with s_{rE}; T is a duration of the reference signal. For binary signals ($r = \overline{1,m}$) the probability of bit error of optimal incoherent reception in Gaussian noise is determined by the expression:

$$p_{ber} = \frac{1}{2}\int_{O(\mu)} W(\mu)\{P[V_1 < V_2|s_1(t)] + [V_1 > V_2|s_2(t)]d\mu\} \tag{3}$$

where $O(\mu)$ and $W(\mu)$ are integration domain and joint probability density of the parameter μ. We assume that the signals $s_{1E}(t)$ and $s_{2E}(t)$ satisfy the orthogonality condition in the enhanced sense:

$$\int_0^T s_{1D}s_{2E}(t) = 0; \int_0^T s_{1D}\tilde{s}_{2E}(t) = 0; \qquad (4)$$

the transmission coefficient $\mu = const$, and we consider that in general case $s_{rD}(t) \neq s_{rE}(t)$, it is easy to show that the error probability of non-coherent element-by-element reception of the distorted signal is [4].

$$p_{ber} = \frac{1}{2}\sum_{r=1}^2 \{\frac{1}{2}\exp[-\frac{h^2}{2}(g_r^{(r)} + g_r^{(l)})]I_0(h^2\sqrt{g_r^{(r)}g_r^{(l)}})$$
$$+ [1 - Q(h\sqrt{g_r^{(r)}}; h\sqrt{g_r^{(l)}})]\}, r, l = \overline{1,2}; r \neq l,$$

where h^2 is the ratio of the signal energy to the spectral density of the fluctuation noise; $I_0(\alpha)$ - is modified zero-order Bessel function; $Q(\alpha; \beta)$ is Markum function;

$$g_r^{(r)} = \{[\int_0^T s_{rD}s_{rE}(t)]^2 + [\int_0^T s_{rD}\tilde{s}_{rE}(t)]^2\} \qquad (5)$$

$$g_r^{(l)} = \{[\int_0^T s_{rD}s_{rE}(t)]^2 + [\int_0^T s_{rD}\tilde{s}_{rE}(t)]^2\} \qquad (6)$$

are correlation coefficients of signals (CCS) of the r-th variant of the distorted signal and, respectively, the r-th and l-th variants of the reference signals; $K = \mu^4/P_E T^2$; P_E is the power of the reference signal. Since s_{rE} is the reference signal with nominal values of the parameters, the coefficient (6) characterizes the degree of the signal distortion s_{rE}. When $g_r^{(l)} = 0$ (which in practice is provided by the choice of orthogonal frequencies of the transmitter), the expression (5) takes the form

$$p_{ber} = \frac{1}{2}\exp(-\frac{h^2}{2}g_r) \qquad (7)$$

where $g_r^{(r)}$ is a coefficient of instrumental distortions $0 < g_r^{(r)} < 1$.

In absence of hardware distortions of the output signal, and, consequently, equality of coefficients, a known formula for the probability of error at optimal Gaussian noise incoherent element-by-element reception of messages follows from the expression (8) [5]:

$$p_{ber} = \frac{1}{2}\exp(-\frac{h^2}{2}) \qquad (8)$$

Let us present the signals in a complex form:

$$S_{rD}(t) = s_{rD}(t) + js_{rD}(t); S_{rD}^*(t) = s_{rD}(t) - js_{rD}(t). \qquad (9)$$

where $j = \sqrt{-1}$

3   Mathematical Model of Multifrequency Signal Shaper

The evaluation of signal shaper (SS) parameters determining its technical condition is one of the problems solved during their status identification. Moreover, identification of failures using a complex quality index is seen as a promising trend of technical system diagnostics [6].

The problem of evaluating the technical object condition using a complex quality index can be considered as an inverse problem of mathematical physics, where the cause (element failure) is necessary to define provided that the result (SS output signal distortion) of observation consequence is known [7]. The stability of the inverse problem solution is connected with the construction of approximate determination methods for θ solution that are close to the desired value based on the existing approximately defined initial information. The elimination of the instability of inverse problem solution is based on the use of a priori information about the problem solution of θ parameter vector location. Such a priori information makes it possible to reduce the class of θ parametric space elements with the exact solution to certain set where the inverse problem solution will be stable.

Numerous works about the diagnostics of the radio-technical systems have been published where the quality indices of their functional utilizing are used as diagnostic characters. Considering the signal shaper (SS) of different structure, the coefficient of cross-correlation of distorted at the SS output signal and reference signal with nominal values of the parameters can be used as such an index [4]:

$$g = \frac{\left| \int\limits_0^T S_D(t) S_D^*(t) dt \right|^2}{4 P_D P_E T^2}, t \in [0, T] \tag{10}$$

where $S_E^*(t)$ is the function complexly conjugated with $S_E(t)$; P_E and P_D are the powers of reference and distorted signals, correspondingly; T is the duration of reference signal element.

The dependency of this index on one of the reference signal parameters (for example, time delay of $S_E(t)$ relative to $S_D(t)$) is the correlation function (CF) which dependency graph $g(\tau)$ is determined by the nature of SS failure. Thus, for example, for the multifrequency signal shaper, described by the expressions [4]:

$$S_E(t, \tau) = \exp\left[j(\omega_c - \frac{k_1 + k_N}{2}\omega_o) \right] \sum_{k=k_1}^{k_N} U_k \exp\left[j[k\omega_0(t + \tau) + \psi_k] \right], t \in [0, T]; \tag{11}$$

$$S_E(t, \tau) = \exp\left[j(\omega_c - \frac{l_1 + l_N}{2}\omega_o) \right] \sum_{k=k_1}^{k_N} U_k \exp\left[j[l\omega_0 t + \psi_l] \right], \tag{12}$$

where ω_c is the carrier frequency, at which all kinds of signals are formed in typical generators for information transmission; $\omega_c = 2\pi/T$; U_k and ψ_k are the amplitude and initial phase of k-th component of the reference signal, correspondingly; U_l and ψ_l are the amplitude and phase of l-th component of the distorted signal $\psi_k = \pi d_k$, the correlation function will be described by the expression obtained when substituting (12) and (13) into the formula (11):

$$g(\tau) = \frac{1}{4 P_E P_D} \left| \sum_{k=k_1}^{k_N} \sum_{l=l_1}^{l_N} U_k U_l \sin[\frac{(\Omega_k l - \Omega) T}{2}] \right| \cdot$$
$$\exp\left[j[\pi(\frac{k - l - 2l\tau}{T}) + \Delta\psi_k l] \right], \tau \in [0, T];$$

where $\Omega = [(k_l + k_N)/2 - (l_l + l_N)/2]\omega_0$ is the frequency detuning between the medium spectral frequencies of the distorted and reference signals; is the phase displacement of the l-th component of the distorted signal relatively to the k-th component of reference signal; $\Delta\psi_{kl} = \psi_k - \psi_l$.

During the shaping of parallel complex signal, according to the expression (12), it is necessary to synthesize the oscillations with $k\omega_0$ frequencies, on which $s_k(t)$ elementary signals with the defined U_k amplitudes and the initial ψ_k phases are formed, and summarize the elementary signals in the total load. Figure 1 indicates a diagram of the multifrequency SS construction using the band-pass filters [8]. Any defect appearing in any element of this device causes the corresponding parameter alteration, such as frequency, amplitude or the phase of the $s_D(t)$ signal component, and, consequently, it causes the distortion of the output signal form. Figure 2 represents the $g(\tau/T)$ curves of parallel complex signals with coded by the Legendre sequences [8]: $[\psi_k] = 0, 0, \pi, 0, \pi, \pi, \pi$. The signals correspond to the following distortions of the $s_D(t)$ signal parameters with different defects in the SS: 1 - undistorted initial phases; 2 - distorted phases in the first phase inverter $\Delta\Psi_{11} = \pi/2$; 3 - distorted phases in the second phase inverter $\Delta\Psi_{22} = \pi$; solid lines imply that all the amplifiers work properly ($[U_l] = U_E$); dashed lines imply the two-fold exceeding of second amplifier transmission coefficient ($U_2 = 2U_E$); dotted-dashed line implies the absence of the signal component at the second amplifier output ($U_2 = 0$). The analysis of

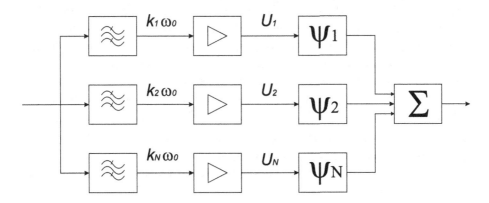

Fig. 1. Block diagram of the multifrequency signal shaper

these graphs indicates that the form of $g(\tau)$ dependency is determined by the parameters of radio equipment values, and, therefore, by its technical condition.

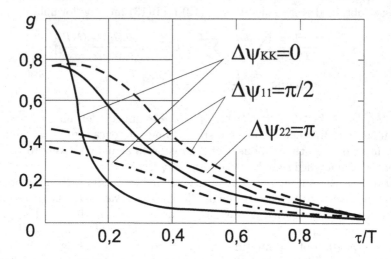

Fig. 2. Graphs of correlation functions $g(\tau)$ for different distortions of the multifrequency signals

4   Aspects of the Correct Solution of the Signal Shaper Diagnostic Problem

Solution of inverse problem is conducted within the mathematical model of the diagnostic object and consists of the Θ model parameters determination:

$$\left| \int_0^T S_E^*(\tau, t) S_D(t|\Theta) dt \right|^2 = g(\tau)/K \tag{13}$$

according to the existing measured data about $g(\tau)$ CF.

Here $K = (4P_D P_E T^2)^{-1}$. Thus, the inverse problem, comprising the determination of the $S_D(t|\Theta)$ distorted signal parameters using the recorded $g(\tau)$ function, presents the task of solving the integral equation (15), in which K coefficient and $S_E^*(\tau, t)$ and $g(\tau)$ functions are defined, and $S_D(t|\Theta)$ is unknown. Three following questions arise when solving inverse problems:

(1) the existence of the problem solution;
(2) the uniqueness of the solution, if it exists;
(3) the stability of the solution, i.e., the permanent dependency of problem solution on the initial data.

A question about the existence of the Eq. (15) solution is tightly connected with the conditions, defined for the $S_E^*(\tau, t)$ function and the right part of the equation $g(\tau)/K$. We can assume that $S_E^*(\tau, t)$ function is measurable and belongs to L_2 class across the square $0 \leq \tau, t \leq T$:

$$\int_0^T \int_0^T |S_E^{*2}(\tau, t)|^2 d\tau dt < \infty \tag{14}$$

i.e. is the Hilbert-Schmidt nucleus (operator) [9].

$\int_0^T |S_E^{*2}(\tau, t)|^2$ integral exists in view of Fubini's theorem [10]. And condition (16) almost for all τ. In other words, $S_E^*(\tau, t)$ as function of t with almost all τ belongs to $L_2[0, T]$.

Since the product of functions with the summarized square is summarizable, the integral in the right part of (15) exists for almost all τ, i.e. function $g(\tau)$ is determined almost everywhere. Due to Cauchy-Bunyakovsky inequality [11] for almost all τ we have:

$$\left| \int_0^T S_E^*(\tau, t) \cdot S_D(t|\Theta) dt \right|^2 = g(\tau)/K$$

$$\leq \int_0^T S_E^*|(\tau, t)|^2 dt \cdot \int_0^T S_D|(t|\Theta)|^2 dt = \|S_D\|^2 \cdot \int_0^T S_E^*|(\tau, t)|^2 dt.$$

Integrating by τ, and substituting iterated integral of $S_E^*|(\tau, t)|^2$ by double, we obtain $\int_0^T g(\tau) d\tau \leq K\|S_D\|^2 \int_0^T \int_0^T |S_E^*(\tau, t)|^2 d\tau dt$ inequality, which gives the integrability of $g(\tau)$.

If nucleus has continuous derivative by τ, than the right side of the Eq. (15) must have continuous derivative by τ. But if the right side contains points, at which the function $g(\tau)$ does not have derivative (when graph $g(\tau)$ occurs as a broken line), then with the presence of nucleus with the continuous derivative the Eq. (15) does not have a solution in the classical sense. Therefore, the existence of the solution depends on what classes (spaces) of functions S ($s_D(t) \in S$) and G ($g(\tau) \in G$) the signal $s_D(t)$ and the response $g(\tau)$ belong to.

As it follows from [12], the Fredholm operator $AS_D = \int_0^T S_E^*(t, \tau) \cdot S_D(t) dt$ is compact and self-adjoined. Consequently, Hilbert-Schmidt theorem is valid for it. In view of this theorem, known also as "the theorem of decomposability" [13], for existence of the Eq. (15) bsolution the $g(\tau)$ function must be decomposed into eigenfunctions of the nucleus $S_E^*|(\tau, t)$:

$$g(\tau) = \left| \sum_i q_i \varphi_i(\tau) \right|^2, \tag{15}$$

where q_i is the coefficients of the $g(\tau)$ function expansion relative to the eigenfunctions of nucleus $|q_i|^2 = \int_0^T g(\tau)\varphi_i(\tau)dt \equiv (g, \varphi_i)$, and eigenfunctions must match the integral equations:

$$\int_0^T S_E^*(\tau, t)\varphi_i(t)dt = \lambda_i^{-1}\varphi_i(\tau), \qquad (16)$$

where λ_i is the eigenvalue of the nucleus $S_E^*(\tau, t)$.

Equation (15) has a solution, which belongs to $L_2[0, T]$, and is single if and only if the system of the eigenfunctions of symmetrical nucleus $S_E^*(\tau, t)$ is complete, and series (17) is convergent, and, moreover, $g(\tau) \in L_2[0, T]$.

Let us examine how these conditions are fulfilled in the practice of SS diagnosis according to the index of the form (11). Regarding the practical tasks of determining the distorted signals at the output of radio-technical circuit, there is always a confidence in existence of the function $S_D(t|\Theta)$ under the integral on the left side of the Eq. (15). The absence of the solution in such tasks can be explained only by the inadequacy of mathematical model to the real functioning of the object.

Uniqueness of the Eq. (15) solution, i.e. the realization of the second condition of the correct solution of this problem depends on the certain form of the nucleus $S_E^*(\tau, t)$. According to [13], the nucleus $S_E^*(\tau, t)$ will be symmetrical, if for all $0 \leq \tau, t \leq T$ the identity of $S_E^*(\tau, t) \equiv S_E^*(t, \tau)$ is fulfilled.

In practical realization of the SS diagnostic procedure in terms of the values of the quality index g, the measurement of this index is accomplished by the fixed value of the argument $\tau = (\tau_1, \tau_2, ..., \tau_n)$ of the nucleus $S_E^*(\tau, t)$. For each value τ_i, $i = \overline{1, n}$, the right side of the expression (15) will be represented by the fixed values of the integrals $g(\tau_i)/K$, $i = \overline{1, n}$. In this case the solution of the inverse problem having the form (15) can be reduced to the determination of the function of one variable by the values of its integrals:

$$\left| \int_0^T S_D(t)S_{Ei}^*(t)dt \right|^2 = |q_i|^2, \qquad (17)$$

where $\{S_{Ei}^*(t)\} = S_E^*(\tau_i, t)$ is the defined set of functions, complexly combined with the functions of reference signals; $|q_i|^2 = g(\tau_i)/K$, $i = \overline{1, n}$. Taking into account (18), it is possible to express $S_{Ei}(t)$ through the eigenfunctions: $S_{Ei}^*(t) = \varphi_i(t)/\lambda_i$.

If proposed problem is presented as the task of solving the operator equation, then $AS_D = \{q_i\}$, where the operator A, determined by equalities (19), acts from the space $L_2[0, T]$, in which we assume the set of functions $\{S_{Ei}^*\}$ as complete orthonormalized, into the l_2 space of existing $\{q_i\}$ values, than the task of solving this equation is correct [15]. Actually, the equation solution exists and is unique for any right side $\{q_i\} \in l_2$, and the continuous dependency of $S_D(t)$ on $\{q_i\}$ follows from the Parseval's equality [8]. However, the main problem is to explain how to obtain the complete system of functions $\{S_{Ei}^*(t)\}$ from the function of reference signal $S_E(t)$.

The task of investigating the uniqueness of the solution of the linear integral Fredholm equation of the 1st kind can be reduced to a study of the completeness of $\{S_{Ei}^*(t)\}$ function system. Actually, if the nucleus $S_E^*(\tau_i, t)$ of the equation:

$$\int_0^T S_E^*(\tau, t) S_D(t) dt = q(\tau), 0 \leqq \tau \leqq t_k, \tag{18}$$

is such that in the section $[0, t_K]$ there is a such sequence of points τ_i so the set of functions $S_{Ei}^*(t) = S_E^*(\tau_i, t)$ is complete in the space $L_2[0, T]$, the solution of Eq. (10) is singular in $L_2[0, T]$. In practice this problem is solved by the selection of such a parameter and its values $\{\tau_i\}$, which provides the orthogonality of functions $S_{Ei}^*(t) = S_E^*(\tau_i, t)$, i.e. $\int_0^T S_{Ei}^*(t) S_{Ek}^*(t) dt = 0, i \neq k$.

Let us examine the possibility of applying the methods of solution of incorrect problems for determining the parametric region, which corresponds to Θ_{za} set and represents a certain Z_a class of technical condition of diagnosed SS in the physical interpretation. The idea of the correctness concept by Tikhonov was expressed for the purpose of substantiation of the widely used trial and error method with the solution of the inverse problems of [14]. The essence of the trial and error method during the solution of diagnostic problems provides for the stages of classification and identification of failures and consists of the following.

The researcher on the basis of the SS models as the object of diagnostics and intentionally introduced into the equipment failures constructs a physical model for different failures. Further, the direct problem of classification, which corresponds to the diagnostic model, is solved: the values $g_a(\tau)$ of the right side of the Eq. (5) for different failures $a = \overline{1, \Lambda}$ are calculated.

The operator in the process of the malfunction search compares the diagnostic models and the measurement data of the right side. Based on this comparison, a new model is selected so the data of solution of direct problem according to the new model would be closer to the experimental. At the basis of trial and error method underlies the assumption about the fact that investigated parametric region, which corresponds to the specific failures of equipment, is not too complex, the number of these failures is limited, the statistical characteristics of the parameters are described by known laws and are located within certain limits. The noted assumptions are the hypotheses about the desired solutions belonging to specific compact correctness sets.

5   Evaluation of the Sampling Correlation Function

Inserting the discretization into the arguments t and τ, we obtain a system of n algebraic equations in L unknown values:

$$\sum_{l=1}^L S_E^{(il)} S_D^{(l)} = q_i, i = \overline{1, n}, \tag{19}$$

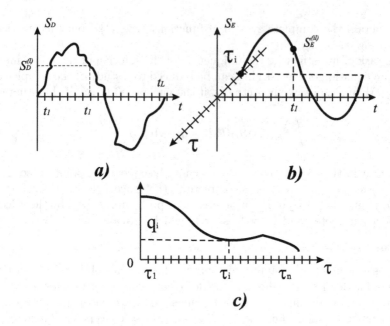

Fig. 3. Graphic illustration of the sampling process of the distorted signal (a), the reference $S_E(\tau, t)$ signal delayed by time τ (b), the correlation function (c)

where $S_E^{(il)} = p_l S_E(\tau_i, t_l)$ is elements of some matrix S_E of $n \times L$; $S_D^{(l)} = S_D(t_l)$; $q_i = q(t_i)$; factor $p_l = \Delta t/2$ - at $l = 1$ or $l = L$, $p_l = \Delta t$ is otherwise. A graphic illustration of the discretization process is shown in Fig. 3. The expression (21) can be represented in the operator form:

$$S_E \cdot s_D = q, s_D \in S, q \in Q, \tag{20}$$

where S_E is a given continuous operator; s_D is a desired solution; q is the specified right side; S and Q are some Hilbert spaces. Depending on the size ratio of the matrix S_E, the system of linear algebraic equations (SLAE) of the form (21) can be overridden ($n > L$), defined ($n = L$) or underdetermined ($n < L$).

The presence of fluctuating noise in the measurement channel causes a random nature of the results. The main method of random noise control is the statistical processing of the observation results. Besides, the reliability of the assessment increases with the sample size n of the single results, i.e. with a decrease of the sampling step $\Delta \tau$. However, if $\Delta \tau$ has a small value, the differences across adjacent values q_i (as well as $S_E^{(il)}$) are so small that these values are highly duplicative, linearly related, and provide poor new information about the desired function . In addition, the differences between the "true" values $q_i(S_D^{(il)})$ can be almost undetected against the errors in the sample measurement of correlation function (CF) σ_q and formation of the etalon signal σ_E, so that the SLAE can be strongly "noisy". Moreover, let us demonstrate the effect of instability (ill-conditioning) of the linear system (21), approximating the integral equation (20) of the FS identification using the linear algebra.

If, SLAE (21) is overridden and a pseudosolution should be applied to the problem of FS evaluation. Generally, SLAE pseudo-solution is called the solution s_D with the smallest Euclidean norm:

$$|S_E \cdot s_D - q| = \min_{s_D} \qquad (21)$$

i.e. is a normal SLAE solution. Here $|X|$ is a norm of a matrix X, and $|X| = \sqrt{\sum_{i,l=1}^{L} |x_{iL}|^2}$. However, since the matrix $S_E \cdot s_D - q$ is not square ($i = \overline{1, n}; n \neq L$), it is necessary to derive a new SLAE from the condition (17). Therefore, we solve a minimization task:

$$|S_E \cdot S_D - q|^2 = \min_{s_D} \qquad (22)$$

the left part of the expression (24) is derivated to zero by s_D. As a result, we get [16]:
$2S_E^*(S_E \cdot s_D - q) = 0$, where:

$$s_D = (S_E^* \cdot S_E)^{-1} S_E^* q, \qquad (23)$$

where S_E^* is a matrix, complex conjugated and transposed with S_E. Using the rule of matrix and vector multiplication, the redefined SLAE (22) transforms to the normal one with a solution [16]:

$$A \cdot s_D = g, s_D \in S, g \in G, \qquad (24)$$

where $A = S_E^* \cdot S_E$; $g = S_E^* q$, the elements of the square matrix A $m \times n$ and vector g (as a result of multiplying of the measured CF ordinates by discrete levels of the function $S_E^*(\tau, t)$, complexly coupled with the function of the reference signal $S_E(\tau, t)$) in case of realness S_E have the form:

$$A_{il} = \sum_{k=1}^{n} S_E^{(ik)T} \cdot S_E^{(kl)} = \sum_{k=1}^{n} S_E^{(ki)} \cdot S_E^{(kl)}; \qquad (25)$$

$$q_i = \sum_{k=1}^{n} S_E^{(ik)T} \cdot q_k = \sum_{k=1}^{n} S_E^{(ki)} \cdot q_k. \qquad (26)$$

The square matrix A in a certain SLAE (26) can be solved by Cramer rule, Gauss method, Kraut-Cholesky method, etc. Besides, if instead of g and A we have the results of their measurement $\widetilde{g}$ and $\widetilde{A}$ at $|\widetilde{g} - g| \leq \delta$, $\left|\widetilde{A} - A\right| \leq \xi$, where δ and ξ are the errors of the right part of the equation (20) (CF measurement error) and the matrix A (the error of the etalon signal formation), then the estimation of the solution error is usually used as [17]:

$$|\delta \cdot s_D| / |s_D| \leq cond(A) \cdot (\delta / |g| + xi/ |A|) \qquad (27)$$

where $cond(A) = |A| \cdot |A^{-1}| = \frac{\mu(A)_{max}}{\mu(A)_{min}} \geqq 1$ is matrix A condition; norms of vectors s_D, g and square matrix A have a form:

$$|S_D| = \sqrt{\sum_{l=1}^{n} |s_D^{(l)}|^2}, \quad |g| = \sqrt{\sum_{l=1}^{n} |g_i|^2}; \quad |A| = \sqrt{\sum_{l=1}^{n} |A_{il}|^2} \quad \mu_i(A) = \sqrt{\lambda_i(A^* \cdot A)}$$

are singular numbers in descending order $\mu_1 \geqq \mu_2 \geqq ... \geqq \mu_r \geqq ... \geqq \mu_n = 0$ (r is a rank of the matrix); λ_i is eigenvalues of the characteristic equation of the square matrix; A^* is a matrix, coupled with the A.

6   Conclusions

The complex SS index is the correlation coefficient of reference and SS output signals. The dependency of studied reference signal index on one of varied indexes is a diagnostic character, which allows the SS element failure to be identified. However, such method should be considered as inverse problem of mathematical physics. In order to provide the correct realization of diagnostic method according to complex index it is necessary to fulfill the conditions of practical existence, uniqueness and stability of problem solution. The existence of the problem solution is confirmed by the experimentally obtained graphs of dependencies of the correlation coefficients on the SS variable parameters (similar to Fig. 2). The uniqueness of inverse problem solution is provided due to the selection of the control parameter values, with which the functions of the reference signal description will be mutually orthogonal, and the system of these functions will be complete in the L_2 function space. To provide the conventional stability of the problem, its solution must be adopted from a certain set of SS technical condition classes having a priori information about the accuracy of measuring equipment and required authenticity of estimation.

References

1. Mahender, K., Kumar, T.A., Ramesh, K.S.: SER and BER performance analysis of digital modulation schemes over multipath fading channels. J. Adv. Res. Dyn. Control Syst. **9**(2), 287–291 (2017)
2. Giordano, A.A., Levesque, A.H.: Digital communications BER performance in AWGN (FSK and MSK). In: Modeling of Digital Communication Systems Using SIMULINK, pp. 101–117. Wiley Telecom (2015). https://doi.org/10.1002/9781119009511.ch5.
3. Fedorenko, V., Samoylenko, I., Kononov, Yu., Samoylenko, V.: Modeling of discretecommunication channel. In: Proceedings of the International Conference of Young Researchers in Electrical and Electronic Engineering (EIConRus). St. Peterburg, pp. 132–134 (2017). https://doi.org/10.1109/EIConRus.2017.7910511
4. Listova, N.V., Fedorenko, V.V., Samoylenko, I.V., Emelyanenko, I.V., Samoylenko, V.V.: The communications channels models in wireless sensor networks, based on the structural-energetic interaction between signals and interferences. In: Moscow Workshop on Electronic and Networking Technologies (MWENT), pp. 1–4 (2018). https://doi.org/10.1109/MWENT.2018.8337298
5. Ha, T.T.: Theory and Design of Digital Communication Systems, p. 668. Cambridge University Press, New York (2011)

6. Ye, K.Z., Bain, A.M.: Methods to improve the detection of failures and troubleshooting for technical diagnostics in instrument. Manuf. Sci. Technol. **1**(2), 31–35 (2013). https://doi.org/10.13189/mst.2013. 010202

7. Beilina, L., Shestopalov, Y.V. (eds.): Inverse Problems and Large-Scale Computations. Springer Proceedings in Mathematics & Statistics, vol. 52. Springer, Cham (2013). https://doi.org/10.1007/978-3-319-00660-4. 223 p

8. Rohling, H.: OFDM Concepts for Future Communication Systems. Springer, Heidelberg (2011). https://doi.org/10.1007/978-3-642-17496-4. 268 P

9. Ruijsenaars, S.: On positive Hilbert-Schmidt operators. Integr. Equ. Oper. Theory **75**(3), 393–407 (2013)

10. Begmatov, A.H., Muminov, M.E., Ochilov, Z.H.: The problem of integral geometry of volterra type with a weight function of a special type. Math. Stat. **3**(5), 113–120 (2015). https://doi.org/10.13189/ms.2015.030501

11. Allen, R.L., Mills, D.W. (eds.): Signal Analysis: Time, Frequency, Scale, and Structure. IEEE Press, Wiley, New York (2004). 929 p

12. Kolmogorov, A.N., Fomin, S.V.: Elements of the Theory of Functions and Functional Analysis, p. 416. Dover Publications, New York (1999)

13. Vishnevski, M.P., Priimenko, V.I.: Mathematical problems of electromagnetoelastic interactions. Boletim da Sociedade Paranaense de Matematica **25**(1–2), 55–66 (2007). https://doi.org/10.5269/bspm.v25i1-2.7424

14. Arsenin, V.Y.: Methods of Mathematical Physics and Special Functions. Nauka, Moscow (1987). 430 p

15. Kirsch, A.: An Introduction to the Mathematical Theory of Inverse Problems. Springer, Heidelberg (2011). https://doi.org/10.1007/978-1-4419-8474-6. 538 p

16. Deergha Rao, K.: Channel Coding Techniques for Wireless Communications. Springer, New Delhi (2015). https://doi.org/10.1007/978-81-322-2292-7. 407 p

17. SizikovV, S.: Mathematical Methods of Measurement Results Processing. Politekhnika, St. Petersburg (2001). [in Russian]

Approach to the Intellectual Monitoring of the Technical Condition of Difficult Dynamic Objects on the Basis of the Systems of a Polling

Pavel Budko[1], Vladimir Fedorenko[2] (ID), Alexey Vinogradenko[1] (ID), Vladimir Samoylenko[2][✉] (ID), and Alexey Pedan[1] (ID)

[1] Marshal Semyon Budyonny Military Signals and Communications Corps Academy, St. Petersburg, Russian Federation
vinogradenkoao@rambler.ru
[2] North-Caucasus Federal University, Stavropol, Russian Federation
vladstv26@mail.com

Abstract. The increase in the requirements to the quality of functioning of difficult dynamic objects (robotic complexes, self-driving cars and aircraft, etc.) and also to their safety and reliability made especially relevant a problem of monitoring of their technical condition, taking into account different impact of the attacks and the destabilizing factors, aging and technological dispersion of parameters. In article new approach to intellectual monitoring of a condition of such objects is offered. This approach is based on the interval estimation of parameters, use of the knowledge base about critical and regular conditions and application of a system of a polling in the course of transfer of measuring information. The architecture and realization of an intellectual system of monitoring of a condition of difficult autonomous technical objects is considered. The carried-out experimental assessment of the offered approach showed that use of the systems of a polling at poll of applications refusals in the course of serviceability check of objects allows to make accurate fixing of their technical states that increases the accuracy and reliability of results of identification of states and also to expand possibilities of use of technical means of control and diagnostics in the systems of monitoring.

Keywords: Intellectual estimation · Polling · Difficult dynamic object · Technical condition · Knowledge base

1 Introduction

Nowadays, telemetric systems become one of the main directions of development of the distributed control systems (DCS). It is caused by a territorially distributed of objects of the distributed system in which sensors, are connected among themselves by digital information flows with limited capacity. It is possible to carry to such systems as the system of industrial scale – SCADA-systems,

© Springer Nature Switzerland AG 2019
V. M. Vishnevskiy et al. (Eds.): DCCN 2019, CCIS 1141, pp. 560–573, 2019.
https://doi.org/10.1007/978-3-030-36625-4_45

and local application there are systems of observation of difficult dynamic objects (DDO), control systems of agrotechnical systems, independent power supply systems, mobile robots (autonomous flying and submersibles, vehicles, nanosatellites) and other DDO [1].

Continuously increasing flows of the telemetric data at control of technological processes and objects result in need of reduction of volume of the messages transferred on the communication channels (CC).

Functioning of DCS, by means of realization of a number of methods [2,3], is carried out consistently: collecting and processing of the telemetric information (TMI) arriving from DDO on one or several peripheral control elements then, after delivery on CC, in the central control element (control office of the management (COM)) is made full processing of TMI with delivery of results to operators of group of the analysis and management of DDO. Collecting diagnostic information in the similar subjects to control (SC) needs to be carried out online, without allowing transition of the precritical technical condition (TC) of DDO to crash. At the same time problems of estimation of the TC and identification of the place of refusal in DDO [4,5] are solved. Features are that, first, there is a tendency to increase of the volume of the processed TMI, secondly, taking into account limited resources of standard control elements, it is often necessary to process several streams of TMI in a row and, thirdly, not always in the place of reception of TMI (COM) there is all necessary control and measuring information and there are specialists in the analysis of a condition of DDO.

Given this, need for new approaches to assessment and exchange of TMI for the distributed TMS increases.

For reduction of redundancy of MI, when maintaining completeness of control of objects, on CC is expedient to transfer not all results of the measurement of parameters but only messages about an exit of the SC parameters out of limits of the set thresholds. The systems realizing this method of collecting telemetric data in some sources are called the adaptive systems of prestart control. Achievement by the controlled parameter of threshold level in crashal timepoint is an event on formation of an emergency signal. At the same time the value of emission of parameter over threshold level is a random variable as well [6,7].

In difference from the traditional TMS divided on a way of receiving telemetric signals into the systems of the remote signaling (SRS) and telemetry (STI) the following model of the processing of alarm signals reflecting process of integration of the existing classes of systems is offered: the emergency signal is formed only in case of excess by controlled parameter x the established threshold level (as in SRS) with the subsequent measurement (as in STI) emission sizes over a threshold [6]. In the integrated system random variables are the moments of formation of emergency signals and levels of these signals.

For implementation of intellectual monitoring in the real work the approach based on transfer of the created emergency signals characterizing the TC of DDO including process of collecting TMI and, realized by the system of ordered poll (polling) of refusals is offered.

The system of the polling represents a kind of the systems of mass service (SMS) with several turns and one general or several serving devices (servers).

The special way of collecting AI is presented in the article – two-stage collecting MI, with use of methods of the theory of mass service (TMS), taking into account priority of information that allows to increase the capacity of the communication channels (CC) and to reduce redundancy of MI taking into account mistakes 1 and 2 sorts that increases also reliability of control.

In the first phase of functioning of a system of monitoring poll of turns where requests are submitted by the separate controlled parameters which went beyond thresholds is carried out. The MI assessment taking into account a priority SC, etc. is carried out.

In the second phase – poll of turns where the applications represent already complex sizes (coefficients) created in the form of TI-TS signals taking into account the priority determined by dynamics of change of parameters and to the SC type. Further complex assessment of the DDO element taking into account a priority, mistakes 1 both 2 sorts and identification of the TC SC is carried out.

Besides, the efficiency of control increases the integration of STS and STI into the uniform system of monitoring. The remote signaling signals fixing only emission are used only on the first phase. The telemetry signals characterizing the measured value of the controlled parameter are used on the second phase.

The novelty of the work consists in realization of the new and perspective approach to expeditious monitoring of the distributed autonomous DDO based on collecting and transfer of MI realized on the basis of the two-phase system of a polling (TMS methods) with priority policy of service.

The theoretical and practical contribution is as follows: the offered approach on monitoring of the TC of DDO allows to increase the efficiency of transfer of TMI, CC capacity due to reduction of redundancy of MI and, in general, increase in productivity of a system at the expense of the adaptive mechanism of poll of applications on the basis of polling models.

2   Related Works

The analysis of the works [1–5] shows that ensuring effective functioning of DDO at simultaneous depreciation of their life cycle requires at all its stages introduction of means and methods of the automated control and diagnostics of the TC, application of effective ways and means of the ensuring of safety and reliability of their functioning.

At the same time the specifics of functioning of the systems of monitoring taking into account operating modes SC rely on use of nondestructive ways of control and diagnostics applied in various industries of industrial electronics and electrical equipment [2–6], medicine [8] and in other industries. Treat shortcomings of the devices realizing these methods the high probability of refusal in the service caused by the fact that purpose of thresholds of operation of a control system is carried out without the general condition of the communication system and level of loading of buffer devices in switching knots that causes blocking of knots in the loaded network and also rather low productivity and high coefficient of idle time as for control of DDO and identification of their TC it is necessary to perform measurement, transformation and processing of a large number of parameters that is quite often connected with shutdown of a system and its idle standing.

Monitoring of the TC of DDO has to be carried out also taking into account external factors, including plurality of structure and the difficult environment of functioning [9]. The similar way is based on use of the device of indistinct sets and the analysis of hierarchies. The conducted researches [10] show also a possibility of the solution of similar tasks by development of the compound index of risk.

Among the existing ways (strategy) of monitoring by the most optimum the control of the TC focused on reliability SC, providing adjusting and the anticipating, that is preventive, predicting control methods is considered [1].

Monitoring of the autonomous objects is, as a rule, characterized by the automated wireless exchange of MI [1–3] that allows to reduce considerably a time resource and participation of the person. At the same time, the configuration of the system of expeditious monitoring allows to carry out adjustment of parameters, by response to sudden fluctuations is able SC or sharp changes of needs for resources [11].

Alternative to above-mentioned methods are the methods of collecting, processing and exchange of MI realized in the multilevel systems of monitoring of the TC of the territorial distributed DDO in which collecting and exchange of MI is based on work of systems of a polling [12–14] and integration of SRS and STI, and processing of MI – on its complex assessment. Use of systems of a polling with priority policy of the service allows to carry out a survey of turns and service of the applications which are in them the serving element by a certain rule, and integration of STS and STI – expeditious collection of data on violation of the controlled processes (emissions) that promotes increase in efficiency of transfer of TMI, CC capacity due to reduction of redundancy of MI and, in general, to increase in productivity of a system at the expense of the adaptive mechanism of poll of applications on the basis of polling models.

In general, the conducted researches in the field of monitoring of the TC of DDO, recognition of types of refusals and their forecasting are characterized by quite wide range of approaches in this subject domain.

3 Theoretical Part of Work: Formation of Results of Control

3.1 The Description of the Systems of Monitoring, as Systems of a Polling with an Absolute Priority

Exchange in the system of monitoring of the TC of DDO of TMI signals during tests and operation assumes use of models and methods of a research of the stochastic systems with cyclic poll.

In the presented approach need of the accounting SC of various degree of importance and also various modes, working conditions (operation) assume consideration (use) of the priority systems representing the systems of a polling with priority policy of service.

In the offered system of monitoring the measured values of excesses by the controlled parameters of some technological thresholds, are defined as emissions of crashal processes. The moments of emergence of these emissions are random variables and represent a flow of applications with the Poisson law of distribution.

Let's present single-server quetch (SMS) with expectation to which come N independent Poisson input of applications with parameters $\lambda_1, ..., \lambda_N$ respectively. Duration of the service of applications of k of a stream have function of distribution $B_k(x)(k = \overline{1, N})$. The server (poll element) at the same time can serve no more than one application, and if it served the application of a stream of i, then in order that it could begin service of the application of a stream of k, it is required to spend some time of, $i \neq k$, it is required to spend some time of S_k for switching (orientation) of the server. Switching duration from i stream to a stream of k $(i = \overline{1, N}, i \neq k)$ there are random variables with function of distribution $S_k(x)$.

Streams are numbered in decreasing order of priorities. Absolute priority. It means that on the released server the application of the top priority from available in a system arrives; switching to turn and service of the application are interrupted by the arrived application of higher priority to which switching at once begins. Due to "fate" of the interrupted service (the interrupted switching) various disciplines of switching $(1, 2, 3, ..., w)$ and service $(1, 2, 3, ..., v)$ [12–14]. Various combinations of disciplines of switching and service can be presented as the scheme with double numbering (for example, scheme 2.1) where the first figure specifies number of discipline, and the second – disciplines of service can be considered.

There is some way of switching of the server at the moments when the system is free from applications: as soon as the server appears in a free state (when it is not occupied with either switching, or service), switching "is instantly dumped". Thus, at receipt in the free system of the application of any priority before the server starts its service, switching of the server is necessary. At the same time it is supposed that time of switching of the server from a zero state to service of the application of a stream of j, $(0 \rightarrow j)$ is equal to switching time $(\rightarrow j)$. Search of joint distribution of number of the applications of each priority which are in a system at any moment is carried out. It is supposed that at the initial moment the system is free from applications.

3.2   Representation of the Results of Control Taking Into Account Mistakes 1 and 2 Sorts

Control of the DDO separate parameters, without their interrelations, or does not provide the required size of reliability of control, or excessively overestimates operational indicators, at the same time numerous false signals of crash are possible [4–7].

For the management of the controlled objects on the several correlated indicators in works [1–7] multidimensional methods of the statistical analysis are used that means correcting of values of the SC parameters during its operation by results of the selective control for maintenance of statistically operated and stable process of work SC, however there is no error check 1 and 2 sorts.

In the modulated system of monitoring, at a deviation of controlled parameters of the DDO elements, comparison of parameters with threshold values within area of the working capacity is made. By the results of comparison the normal state of DDO decides on probability p_1 or its abnormal state with probability p_2. And TC recognition (critical condition) of DDO is carried out taking into account errors of the first (α) also by the second (β) sorts which correspond to probability to "false crash" and "the threshold of violation".

Minimization of the probability α comes down to the creation of the entered rectangle B_B of the maximum area (Fig. 1b) also decides by means of a method of diagonals. According to this method of top of the entered parallelepiped (Fig. 1d) are in a point of intersection of diagonals of the described parallelepiped with an ellipsoid (Fig. 1c). For this purpose values of the parties of the entered parallelepiped are defined (Fig. 1d), corresponding to thresholds on parameters of the DDO elements when ensuring zero probability of a mistake β and minimum possible mistake α.

Minimization of the probability β is close to zero if the area $\Delta\Theta$ of possible values of the DDO parameters is approximated by the entered parallelepiped B_B (Fig. 1b) so that it was completely enclosed in this area, but the area at the same time is used not completely.

Providing the required ratio α/β between mistakes 1 and 2 sorts, at approximation of area $D_{threshold}$ of permissible values of parameters (the shaded described ellipsoid on Fig. 1a), defines such threshold on parameters at which a certain probability of undetected refusal of the DDO elements is caused in advance or control system cost at implementation of the set requirements to an indicator of the TC of a controlled object (quality of functioning) is minimized.

Integral use of MI, received from the diverse sources, presented in three-dimensional space, at ellipsoidal approximation, is most applicable for the solution of problems of control of the TC of the objects in the conditions of uncertainty. Such approach promotes increase in reliability of MI about a condition of observed objects in the systems of monitoring, to expansion of a scope of technical means of control and diagnostics and also decrease in redundancy of MI at stages of its transfer that in general promotes increase in efficiency of the process of control.

4   Implementation: Modeling of a System of Monitoring of the TC of the DDO

Practice of the operation of DDO demands development and creation of a control system (monitoring) of objects, including – subsystems of assessment of the TC. For determination of architecture of a system of monitoring of the TC of DDO we will define basic data and the main stages (procedures) of functioning of a system.

The analysis of various approaches to the solution of a problem of control of the TC of DDO on the basis of the system of a polling with an absolute priority assumes the following basic data: (1) the studied TMI about regular behavior of DDO (within the range of thresholds of controlled parameters); (2) list of the controlled parameters (temperature, tension of the electromagnetic field, tension, humidity, etc.); (3) frequency of poll of the DDO telemetric sensors; (4) the greatest possible dynamics (frequency, speed) of change of the telemetered parameter; (5) knowledge base (alphabet of multidimensional images of the TC of the DDO elements: statistical values of controlled parameters and speed of their change in limits of range of thresholds); (6) numbering (distribution) of controlled objects on priorities for further poll.

For definition of the TC of DDO MI, DDO received when functioning, it is necessary to transfer control on COM.

Main stages of the functioning of a system of monitoring

The solution of a problem of the monitoring of the TC of DDO is presented in the form of the sequence of the following stages:

I. Preprocessing of MI on the remote terminal:

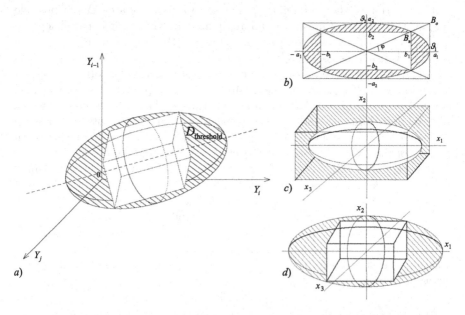

Fig. 1. Three-dimensional representation of area of working capacity: (a) working capacity ellipsoid (informational content); (b) quality ellipse; (c) the described parallelepiped; (d) the entered parallelepiped

stage 1 – group poll controllers (servers) of sensors on the DDO elements;

stage 2 – primary estimation of values of the received group of signals (remote signaling) from the DDO controllable element: evaluating degradations loud-speakers of controlled parameter;

stage 3 – correction of priority of the application according to a priority SC and dynamics of change of parameter;

stage 4 – formation of a stream of signals of TI-TS (turn), assignment of the corresponding priority (according to the most critical parameters, dynamics of their change);

stage 5 – priority transfer of TMI on COM.

II. Processing of signals of TMI (TI-TS) on COM:

stage 6 – poll of turns from various DDO according to priorities;

stage 7 – service of applications of the interviewed turn;

stage 8 – secondary complex estimation of the received group of signals of TI-TS (turn): (a) formation of a multidimensional complex image of the TC of DDO (in the most critical parameters); (b) comparison of the received image with reference values of images of signals from the knowledge base;

stage 9 – identification of the TC of DDO taking into account mistakes 1 and 2 sorts.

Block Diagram of a System of Monitoring. We will present the system of monitoring of the TC of DDO in the form of an intellectual system which basic elements will be knowledge base and a set of the modules realizing rules and procedures of performance of stages 1–9. Basic elements in architecture of a system of intellectual monitoring (Fig. 2) are: element of poll of sensors, element of assessment of change of parameter, elements of memory 1 and 2, element of correction of a priority, shaper of a stream of signals of TI-TS, radio transmitter, distribution environment, radio receiver, element of poll of turns, element of complex assessment, element of memory 2 and calculator.

Model of a System of a Polling. Researches of systems with priorities in TMS, showed that in such systems, as well as in the modulated system of monitoring, are N available flows of the priority applications proceeding from each sensor which is on the DDO element where to each stream there corresponds the priority. These systems are a kind of systems of a polling with a priority order of the poll of turns, and the turn can be served, only if more priority turns do not contain applications.

Thus, the specifics of construction and functioning of a system of monitoring allow to consider them as the two-phase system of a polling where a role of the serving elements at the first stage is played by the shaper of a stream of signals of TI-TS 6, and at the second stage – calculator 13 (Fig. 2) providing service of applications for some time in proportion to duration of transmission of messages on CC.

In case of an exit of different controlled parameters of objects out of the allowable limits, in the sensor located directly on elements of subjects to control the signal of the alarm status of such objects is formed. In the existing

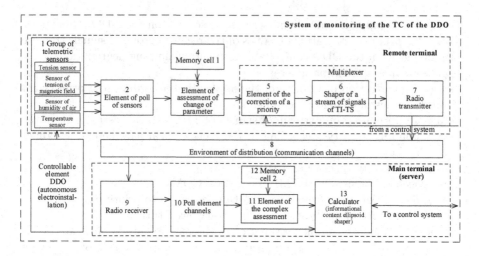

Fig. 2. Block diagram of an intellectual system of the monitoring

systems of telemetry each parameter of an object is controlled with the period T_0, irrespective of the speed of its change. However at increase of speed of the change of separate parameters they can reach permissible values in time smaller the fixed period T_0. In this case the control system will not be able to react in due time to the inadmissible changes of parameter that will lead to the refusal of a controlled object. For the efficiency of control of a condition of an object measurement and the subsequent assessment of parameter is carried out with a frequency of proportional speed of change of parameter. Depending on exit speed (time Δt_1, Δt_2 achievement of permissible value) controlled parameter U out of the allowable limits the priority of a signal which is set thanks to the multilevel system of thresholds (the more threshold level is defined, the application priority is higher).

MI arrives on an element of poll of sensors 2 and an element of poll of channels 10 (Fig. 2), not regularly, and in the form of the random-signal flow of a condition of the equipment passing through certain threshold levels which in case of an exit of values of process parameters out of limits of thresholds form packages of applications and allocate them with the status of priority.

5   Experiment Results

For high-quality development and assessment of functioning of the modulated system of monitoring of DDO in general it is necessary to develop and use its mathematical model. Within the considered approach for this purpose it is offered to use a simulation model.

The computing experiment was made by c use of the software of AnyLogic 8.2.3 which supports various approaches to creation of simulation models: process focused (discrete and event), system and dynamic, agent-based and also their combinations. The uniqueness, flexibility and power of language of the modeling provided to AnyLogic allows to consider any aspect of the modulated system with any level of specification. The graphic AnyLogic interface, tools and libraries allow creating quickly models for a wide range of tasks from modeling of production, logistics to strategic models of development of the companies and the markets [14].

As subject to control autonomous electro installations (power plants, wind generators) for which monitoring of the TC the offered approach was used were used. When carrying out an experiment accounting of internal parameters of electro installations (output voltage, temperature of heating of the generator (anchor), vibration and frequency of rotation of a rotor) was made. MI about change of controlled parameters was transferred by means of TI-TS signals on-line.

Various sensors the excesses by controlled parameters of threshold values placed on SC and fixing acted as sources of MI (as in STS). Taking into account that failure of the functioning object – a casual event was chosen random origin type of sources of MI. At random origin sources of the application are generated in a random way according to the law of distribution.

Functioning of the Modulated System of the Monitoring. Let's consider the first phase of functioning of the modulated system of monitoring. For further processing, applications from sensors, arrive on an entrance of an element of poll which are characterized by the size of a stream λ_i and volume of information ν_i, and from its exit – on an element of assessment of change of parameter where there is a comparison of comers of applications to number of a controlled object (on a priority of the object) from an element of memory of 1 (Fig. 2). Pass of the application corresponding to a signal of a parameter exit out of limits of thresholds that is emergency or a preemergency, with the subsequent formation of turn of applications is a result of comparison. Then the application arrives on an element of correction of a priority and the shaper of a stream of signals of TI-TS which is the serving device and carries out the analysis of the applications which arrived on its entrance from turn, and an element of correction of a priority – correction of their priorities. The shaper of a stream of signals (multiplexer) forms packages of messages for sending according to priority of applications, for example, in the beginning a package of applications of the highest priority, then lower, etc. for the subsequent sending. Correction of the priorities is carried out by the rule: the speed (dynamics) of change of values of parameter is higher, the frequency of its poll is higher, and, respectively, the priority and vice versa is higher. Besides, correction of a priority of the application can be carried out on the operating signal from the control unit, for example, at change of criticality of the SC parameters (the category of the consumer changed). Depending on load of the processor of the shaper, the application will be processed with some intensity of service μ. After processing of applications the shaper sends the processed information to a transmitter entrance (the most critical applications are transferred on the allocated CC, the others – in process of release of busy channels). The relevance of applications is defined by waiting time T_{wait} in turn that is caused by service of more priority applications in the beginning, then less priority.

On the second phase of functioning of the modulated system of monitoring as the systems of a polling, the transmitted TI-TS signals, arrive on an entrance of the receiver and from its exit – on an entrance of an element of poll of channels which is carrying out their poll for some time of poll t_{samp}, in proportion to signaling duration on CC which also serves as turn and are characterized by the size of a stream λ_i volume of information ν_i. From an exit of an element of poll of channels of the application arrive on an element of complex assessment where there is a complex comparison of the received signals (the come applications) with their reference values from an element of memory 2 that is necessary for accounting of distortions of the received signals. Complex estimation is necessary for the subsequent drawing up full "picture" of the TC of DDO in the calculator, for increase in informational content and accounting of the parameters characterizing this or that element and SC in general. The pass of the application in the calculator which is the serving device (it is characterized by a stream of the service of the applications μ) and carrying out the formation of an ellipsoid of informational content and its comparison with reference, calculation of dynamics of complex degradation of group of observed parameters, calculation of mistakes 1 and 2 sorts and also identification of concrete refusal from the list various and forecasting of other refusals on an ellipsoid, for the subsequent sending to a control system (the system of support of decision-making) is result of comparison.

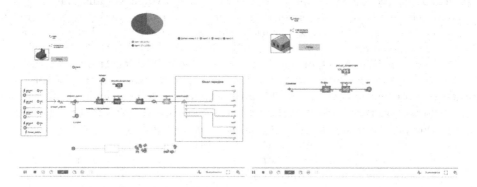

Fig. 3. Working windows of a simulation model of a system of the monitoring of the TC of DDO

In the considered model a number of restrictions was entered. First, the success of processing of applications depends on T_{wait} value. The T_{wait}, less, is possible to process less applications. Therefore it is necessary to pick up such T_{wait}, value at which the processed information will be relevant. Secondly, the control system which reduces the intensity of service has an impact on an element of correction of a priority 5 (on the remote terminal) and calculator 13 (on the main terminal), as shown in Fig. 3, loading processors of these elements various tasks (change of a priority SC, accounting only of the chosen parameters, etc.). Therefore, for the purpose of selection of the required parameters of functioning of the modulated system of monitoring, is necessary to make a number of experiments.

The option of a simulation model of a system of monitoring of the TC of the DDO developed in the environment of AnyLogic is presented in Fig. 3. In this example the turn realized in an element of correction of a priority of 5 (Fig. 2), which is formed as a result of receipt of emissions from four random origin sources of MI (sensors of tension, vibration, temperature, rotor frequency) was used. The intensity of service of applications and all entrance streams are assumed by accidental. The coefficient of the solution of the other tasks is equal in a control system 0.1 (10%).

The calculations made on the offered option of a simulation model show that for the chosen values of basic data the share of the raw applications equals 5% that is very big. It is necessary to take measures for its reduction, in particular, due to increase in T_{wait}, which is taken away under necessary operating conditions of a system of monitoring.

Calculations show that by transfer complex MI about the TC of one DDO element on the first phase is an observed increase in reliability of control at the minimum redundancy of MI, due to integration of STI and STS. When passing MI through the second phase of the system of monitoring holding time of applications taking into account extra time increases by accounting of dynamics of degradation of parameters, however at the same time the reliability of the accepted signals taking into account mistakes 1 and 2 sorts increases.

In general, two-phase functioning of a system of monitoring shows a considerable prize in reliability and efficiency of transfer of MI with use of the offered approach (Fig. 4).

Thus, results of the made experiment showed that complex idea of MI of the TC of DDO is increased by reliability and informational content of the resultant information obtained by COM and also increase efficiency of its delivery by reduction of redundancy due to integration of STS and STI.

The analysis of the results of modeling (Fig. 4) shows that the prize in reduction of volume of TMI in comparison with works [7, 9, 10] depends on informational content of signs of recognition on the second and the subsequent phases as MI volume on the first phase is fixed, and is defined by the size of free buffer space which size can strictly be controlled according to local information from knots (servers) of a system. However the increase in informational content of signs at the subsequent stages is connected with measurements in a system which volume defines quality of decision-making at incremental control. These measurements for increase in informational content are connected with need of attraction of additional measuring resources and increase in time of the analysis.

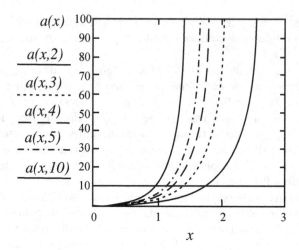

Fig. 4. Reduction of volumes of telemetric information in the system of monitoring due to increase in number of phases of control

6 Conclusions

Results of a research show that for implementation of the distributed monitoring of the SDD parameters of, especially, uninhabited autonomous objects, various instruments of control providing increase in reliability of results of identification and sensitivity to detection of emergency (preemergency) situations can be used.

The approach of the complex control of the technical condition of SDD presented in the article on the basis of integration of indications of several types of sensors can be used for creation of the universal automated complex of the control of uninhabited autonomous objects of technological systems allowing to estimate operability of the wide nomenclature of the radio-electronic equipment with high reliability. Complex submission of MI, taking into account its transfer on in a system with the integrated STI and STS promotes decrease in redundancy of MI in the system of monitoring, and the solution of a problem of priority of poll of applications application of systems of a polling to promote increase in efficiency of estimation of technical condition of SDD and further decision-making, for example, in the systems of the support of decision-making.

In general, similar approach will allow carrying out at the dispatching level decision-making support online on elimination of critical conditions.

References

1. Yang, W., Tavner, P.J., Crabtree, C.J., Feng, Y., Qiu, Y.: Wind turbine condition monitoring: technical and commercial challenges. Wind Energy **17**(5), 673–693 (2014)
2. Sokolov, B.V., Yusupov, R.M.: Conceptual foundations of quality estimation and analysis for models and multiple-model systems. J. Comput. Syst. Sci. Int. **46**(6), 831–842 (2004)
3. Takayama, K., Kariya, S.: Autonomous measuring by sensing node in telemetry system. Meas. Sci. Rev. **3**(3), 29–32 (2003)
4. Abramov, O.V.: Choosing optimal values of tuning parameters for technical devises and systems. Autom. Remote Control **77**(4), 594–603 (2016)
5. Abramov, O.V., Dimitrov, B.N.: Reliability design in gradual failures: a functional-parametric approach. Reliab. Theory Appl. **12**(4), 39–48 (2017)
6. Fedorenko, V.V., Vinogradenko, A.M., Kononov, Y.G., Samoylenko, V.V., Samoylenko, I.V.: The time-probability characteristics of a telemetrie signal with the variable number of bits. In: Proceedings of the 2017 IEEE II International Conference on Control in Technical Systems (CTS), pp. 146–149 (2017)
7. Katzel, J.: Managing alarms. Control Eng. **54**(2), 50–54 (2007)
8. Jovanov, E.: Wireless technology and system integration in body area networks for m-health applications. In: Proceedings of the 2005 IEEE Engineering in Medicine and Biology 27th Annual Conference, Conference Location: Shanghai, China (2005)
9. Fedorenko, V.V., Budko, P.A., Vershkov, N.A.: Mathematical model of discrete communication channel under the influence of destabilizing factors. Eng. Simul. **15**(1), 77–83 (1998)
10. Husmeier, D.: Neural Networks for Conditional Probability Estimation. Springer, London (1999). https://doi.org/10.1007/978-1-4471-0847-4
11. Lawson, B.G., Smirni, E., Puiu, D.: Self-adapting backfilling scheduling for parallel systems. In: Proceedings International Conference on Parallel Processing, Conference Location: Vancouver, BC, Canada, 21 August 2002. IEEE Xplore (2002)
12. Borst, S.C.: Polling sistems. Amsterdam, Stichting Mathematisc Centrum (1996)
13. Boxma, O.J.: Analysis and optimization of polling systems. In: Queueing Performance and Control of ATM, pp. 173–183. North-Holland (1991)
14. Zhang, Y., Wang, Y., Wu, L.: Research on demand-driven leagile supply chain operation model: a simulation based on anylogic in system engineering. Syst. Eng. Procedia **3**, 249–258 (2012)

Algorithm for Embedding Digital Watermarks in Wireless Sensor Networks Data with Control of Embedding Distortions

Oleg Evsutin[1,2(✉)] , Roman Meshcheryakov[2] , Vladimir Tolmachev[3],
Andrey Iskhakov[2] , and Anastasia Iskhakova[2,3]

[1] Higher School of Economics, 20 Myasnitskaya street, 101000 Moscow, Russia
oevsyutin@hse.ru
[2] Institute of Control Sciences RAS, 65 Profsoyuznaya street, 117997 Moscow, Russia
[3] Tomsk State University of Control Systems and Radioelectronics,
40 Lenina prospect, 634050 Tomsk, Russia

Abstract. The article presents a new algorithm for embedding digital watermarks in the data of wireless sensor networks. The algorithm is designed to protect against the substitution of the data source in such networks. An important distinguishing feature of the proposed algorithm is the ability to control the level of distortions introduced as a result of embedding. This allows us to recommend this algorithm for use in wireless sensor networks, including sensor nodes of various types, designed to measure physical quantities of different nature. Computing experiments were carried out by means of simulation modeling in the OMNeT++ framework. The results of the experiments showed that the obtained algorithm provides statistical indistinguishability of sensory data samples before embedding digital watermarks into them and after embedding.

Keywords: Data hiding · Digital watermarking · Wireless networks

1 Introduction

One of the promising areas of information technology, serving as the basis for creating digital economy, is the Internet of things. It is a global dynamic network infrastructure where physical and virtual "things" have identifiers and physical attributes and are integrated into the information network using various interfaces. The implementation of the Internet of things concept can lead to a qualitative change in many spheres of human life. The use of intelligent devices in various fields of activity at the state level: in medicine, for monitoring the environmental situation is very promising.

The Internet of things is based on a number of scientific and technical solutions, among which we can single out the technology of wireless sensor networks (Fig. 1).

V. M. Vishnevskiy et al. (Eds.): DCCN 2019, CCIS 1141, pp. 574–585, 2019.
https://doi.org/10.1007/978-3-030-36625-4_46

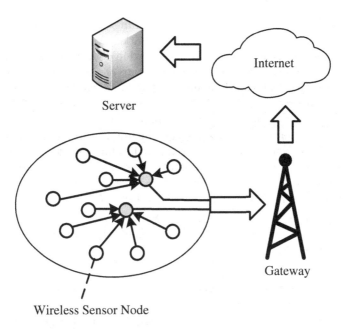

Fig. 1. Technology of wireless sensor networks.

The coverage area of the sensor network can be from several meters to several kilometers due to the ability to relay messages from one network element to another. The sensor network is capable of relaying messages along the chain from one node to another, which allows organizing the transmission of information through neighboring nodes in case of failure of one of the nodes without loss of quality. A typical network architecture includes three types of nodes (end sensor node, router, coordinator). The network itself determines the optimal route for the movement of information flows. Sensor nodes can be fixed stationary, and also have relative mobility, that is, arbitrarily move relative to each other in some space, without violating the logical connectivity of the network. In the latter case, the sensor network does not have a fixed constant topology, and its structure dynamically changes over time.

Wireless sensor networks represent a class of distributed networks that are self-organizing. A wireless sensor network consists of a set of sensor nodes, which interact with each other via radio channel [1]. The most effective application of wireless sensor networks is the monitoring of various processes, including those in remote areas. Examples include smoke monitoring and the detection of forest fires, condition monitoring and remote control of the perimeter of objects in security systems, military surveillance.

However, the use of this technology is complicated by security concerns. An intruder's interference in the network can compromise the monitoring results and create a threat to people's lives and health. One of the tasks of ensuring the security of wireless sensor networks is to protect against the data source substitution.

Protection against data source substitution in wireless sensor networks can be provided with the help of digital watermarks. A digital watermark is an information sequence that is implicitly embedded in digital data and may contain some information about the origin or initial state of this data [2]. The original purpose of the digital watermarks embedding methods was to protect multimedia data. However, at present, due to the emergence and development of the Internet of things technology, such methods have become relevant for protecting data generated in the Internet of things systems, including wireless sensor networks.

The aim of this work is to create an algorithm for embedding digital watermarks in the data of wireless sensor networks, applicable to sensor data of various nature. The rest of the article is organized as follows. Section 2 provides a literature review on the research topic. Section 3 describes the proposed algorithm for embedding digital watermarks in sensory data. Section 4 presents the results of computing experiments with the proposed algorithm and their discussion. Section 5 summarizes the present research.

2   Review of Previous Works

The original purpose of digital watermarks was to protect digital objects from such threats as falsification, data source substitution, as well as unauthorized distribution. Let us further consider exclusively digital images as digital objects since digital images are the most common type of multimedia data.

In general, a digital watermark is a fixed-length data set that contains some information about the protected image, the conditions for its creation, the identifier of the image owner, etc.

In order to embed a digital watermark in a digital image, some manipulations are made to its constituent data elements. The image obtained after these manipulations should not differ significantly from the original image. The ability to fulfill this requirement is determined by the presence of spatial redundancy in the digital image, which is expressed in the proximity of the values of neighboring pixels to each other.

However, not only multimedia data may be redundant. Sensory data also has this property.

In particular, sensory data taken from one sensor will have time redundancy, and data taken from sensors located nearby will have spatial redundancy.

It is advisable to use the noted properties when solving data protection problems in wireless sensor networks. The adequacy of this statement is confirmed by the appearance and by quite active development of a scientific field related to the design of algorithms for embedding additional information of various purposes in the data of wireless sensor networks. Let us consider some recent works in this area.

In [3], a digital watermark is used to authenticate data in sensor networks. In this paper, an original approach to sensory data aggregation is proposed, according to which a set of values coming from each group of sensors is presented in the form of a matrix. The authors of the study propose considering each of these matrices as a pseudo-image. Since adjacent sensors in general case record close values of the measured quantity, such a pseudo-image will have spatial redundancy, like a regular image. Aggregation consists in JPEG-like compression of the generated pseudo-image. Before compression, a digital watermark is embedded in the pseudo-image using the direct spread spectrum sequence (DSSS) method, according to which one bit of the digital watermark is embedded in a block of adjacent "pixels". The embedding operation consists in the fact that the values of the "pixels" in the block are changed by small quantities under the control of a pseudo-random sequence. Since the individual sensors do not interact with each other, the embedding of digital watermark bits for them is carried out independently.

The research [4] proposes a method for checking the integrity of data from various sources (sensors monitoring health or the environment), based on embedding digital watermarks into data streams. To embed digital watermarks, a spread spectrum technique using pseudo-random orthogonal codes is used. Due to the application of the proposed method, the integrity of data streams can be verified by extracting digital watermarks, even if the data goes through several stages of the aggregation process. The proposed watermark scheme preserves the natural correlations, that may exist between multiple data streams, which is an important factor in the context of the necessity for data aggregation.

The paper [5] proposed a scheme for ensuring safety of data received from sensors. The scheme is based on the combined use of digital watermarks and the compressed sensing method, which serves to restore the full signal from its sparse or compressed representation. In [5], this method is used to obtain a sparse signal on the sensor side and then restore the original signal on the side of the base station. On the sensor side, a digital watermark is generated based on the protected data using a hash function, and then it is embedded in the original signal, the elements of which are supplemented with empty symbols depending on the value of the digital watermark bit. The wireless channel transmits data sparse by the method of Compressed Sensing to the base station. The base station decoder restores the full signal, then the digital watermark is removed from the container. To verify the integrity of the data, the extracted digital watermark is compared with a digital watermark obtained from the received data.

The combination of embedding a digital watermark and an aggregation operation or ensuring the stability of a digital watermark to an aggregation operation is not considered in all works. In the article [6] it is proposed to embed one digital watermark in the source data received from sensor sensors, verify the authenticity of these data before aggregation, and after aggregation, embed a second digital watermark in the aggregated data. Digital watermark generation is performed using a pseudo random number generator.

In certain cases, digital watermarks reflect specific types of attacks. The work [7] is devoted to the identification of cyber-reproduction attacks aimed at networked control industrial systems. It is an attempt of an intruder to intervene in the control of the system by reproducing previously captured data sequences.

The main contribution of this work is not the embedding algorithm, which is taken from previous works, but the strategy of using this algorithm to protect against an intruder.

[8] presents synchronization scheme for embedding fragile digital watermarks in sensory data, which is resistant to loss of synchronization between the data sender and receiver. To solve this problem, the authors propose a dual-chaining watermark scheme. To do this, the sensory data is divided into groups of variable length, depending on the key. The development and embedding of digital watermark chains are carried out for pairs of adjacent groups. One chain of watermarks is used to authenticate the sensory data itself. Each watermark in this chain is generated as a hash value depending on two groups of data, and is embedded in the first group of the pair. The second chain of watermarks encodes the delimiters between the groups and provides synchronization between the sender and receiver of the data. A distinctive feature of the proposed scheme is the reversibility of the embedding method. Due to this, it is possible to restore the original data without changes and repeat the hashing process at the stage of extracting watermarks.

The idea to create a digital watermark depending on the protected data itself is quite common both for classical methods and algorithms of digital watermarks, and for the considered problematic area of data protection of wireless sensor networks and the Internet of things.

So the work [9] is devoted to the problem of authentication of data coming from Internet of things devices. It is proposed to extract the stochastic characteristics of data streams and form digital watermarks on their basis. As a method of embedding digital watermarks in a data stream, the spread spectrum method is used. An important distinguishing feature of this work is the attempt to solve the problem of data authentication in the Internet of things systems with limited resources. It is solved using the mathematical apparatus of game theory, supported by deep learning methods.

[10] also discusses the use of various characteristics of captured data in the formation of a digital watermark including data length, occurrence frequency, and capturing time. A distinctive feature of the digital watermark embedding algorithm proposed in this study is encryption. The generated digital watermark is encrypted using the secret key before the embedding. Another distinctive feature of the work as a whole is the consideration of two scenarios for protecting a wireless sensor network data using digital watermarks. In the first case, one-to-one based scenario is considered where each sensor node is directly linked to base station in data communication. In the second case, we are talking about cluster based scenario, when cluster node collects data from authorized nodes and sends it to base station through intermediate nodes.

The study [11] focuses on the problem of energy conservation in wireless sensor networks. The authors propose a hybrid scheme for embedding digital watermarks in the data of wireless sensor networks, combining irreversible and reversible digital watermarks. This hybrid scheme includes three stages. They are the intra-cluster authentication (for protecting the sensed data transmission between the sensor nodes and the cluster head), the data aggregation in the cluster head, and the inter-cluster authentication (for protecting the aggregated data transmission between the cluster head and the sink). The task of energy saving is solved at the stage of intra-cluster authentication. The incorporation of reversible digital watermarks consists in a simple hashing of the sensor values generated by the sensors and then attaching the resulting hash values to the data packets.

The considered articles offer methods and algorithms designed to work primarily with abstract sensory data without specifying their origin. There are also studies that deal with sensory data of a particular type.

For example, the article [12] describes the method of embedding digital watermarks into LiDAR data, which represent information about the position of points on the Earth's surface, stored in a standardized format. The watermark scheme proposed in this paper can be used to protect copyrights and track the source of the data. In order to embed a digital watermark, the vector of marker positions is first determined, i.e. those positions in which the bits of the digital watermark will be embedded using a pseudo-random generator. Embedding is performed in a circular area around the marker positions, which is divided into smaller areas evenly distributed in the circle. The distance vector is calculated on the base of points in the obtained areas. Then the discrete cosine transform is applied to the vector. The direct implementation of a digital watermark is accomplished by modulating the last coefficient of the discrete cosine transform. The experimental results demonstrate resistance to the most likely attacks, such as cropping or accidentally deletion of points.

A more detailed and systematic description of the methods of embedding digital watermarks in the data of the Internet of things systems can be found in the review paper [13] and other publications in the field of information security [14, 15].

The present research proposes a new algorithm for embedding digital watermarks in the data of wireless sensor networks, featuring the ability to control the level of distortions introduced into the original data as a result of embedding.

3 Proposed Algorithm

The main elements of wireless sensor networks are sensors that measure the values of various physical quantities. This may be temperature, humidity, pressure, vibration level, wind speed, etc. Each sensor with a certain periodicity captures the measured value and transmits it to the head node.

In general, these numbers are real with an accuracy of 3–5 decimal places. In the present work it is proposed to use the fractional part of these values for embedding elements of a digital watermark. Since the fractional part contains, among other things, the noise component, a potential intruder will not be able to restore the digital watermark from the intercepted values while they are transmitted to the head node. However, it is necessary to provide the ability to control the level of distortions of embedding, since the significance of the fractional part of the measured values for physical quantities of different nature may be different.

The description of the proposed algorithm is given below. It operates with digital watermarks, which are binary sequences of some finite length. In accordance with this algorithm, the bits of a digital watermark are sequentially embedded into each L values of a physical quantity captured by the sensor.

Input: a group of real sensory data $S = s_1 s_2 \ldots s_L$; scaling factor $\lambda \geq 1$; the number of embedding intervals n; digital watermark $W = w_1 w_2 \ldots w_N$, $w_i \in \{0, 1\}$, $i = \overline{1, N}$.

Output: a group of sensory data $D = d_1 d_2 \ldots d_L$ with a recorded digital watermark.

Step 1. For $i = \overline{1, L}$ do the following:

Step 1.1. Calculate $j = (i \bmod N) + 1$.

Step 1.2. Calculate $x = \{\lambda s_i\}$.

Step 1.3. If $w_j = 0$, then generate a random value r that falls into the closest to x semi-interval from the partition $\left[0; \frac{1}{n}\right) \cup \left[\frac{2}{n}; \frac{3}{n}\right) \cup \cdots \cup \left[\frac{n-2}{n}; \frac{n-1}{n}\right)$. Otherwise, generate a random value r that falls in the closest to x semi-interval from the partition $\left[\frac{1}{n}; \frac{2}{n}\right) \cup \left[\frac{3}{n}; \frac{4}{n}\right) \cup \cdots \cup \left[\frac{n-1}{n}; 1\right)$.

Step 1.4. Calculate $d_i = \frac{\lambda s_i - \{\lambda s_i\} + r}{\lambda}$.

Step 2. Return $D = d_1 d_2 \ldots d_L$ and end the algorithm.

Let us explain the presented formal description of the proposed algorithm.

The scaling factor λ allows us to select the decimal digits of the fractional part of the sensor data element s, in which the digital watermark bit will be recorded. Thereto, the data element s is multiplied by λ, and then the fractional part x is separated from the obtained auxiliary value.

Let $s = 12.3685$. Then, for the scaling factor $\lambda = 1$, the value of x will be 0.3685, for the scaling factor $\lambda = 10$, the value of x will be 0.685, etc.

Thus, an increase in the coefficient λ makes it possible to reduce the number of digits of the fractional part of the sensor data element used to record a digital watermark bit. This reduces the level of embedding distortion.

Embedding directly consists in the fact that the value of the fractional part of the data element (after scaling) is replaced by a randomly generated value that encodes the corresponding bit of a digital watermark. Thereto, the interval of possible values of the fractional part of the real number $[0, 1)$ is divided into n consecutive disjoint semi-intervals of equal length. The values falling into the odd semi-intervals will be regarded as corresponding to the zero bit, and the values falling into the even semi-intervals will be regarded as corresponding to the one bit.

Let $n = 4$. Then the following semi-intervals $\left[0; \frac{1}{4}\right)$, $\left[\frac{2}{4}; \frac{3}{4}\right)$ correspond to the zero bit of a digital watermark. In its turn, the following semi-intervals $\left[\frac{1}{4}; \frac{2}{4}\right)$, $\left[\frac{3}{4}; 1\right)$ correspond to the one bit of a digital watermark.

Let the digital watermark bit be $w = 1$. Let us take the value $x = 0.685$ obtained earlier for the case $\lambda = 10$. We can see from Fig. 2 that the closest suitable semi-interval with respect to this value is $\left[\frac{3}{4}; 1\right)$.

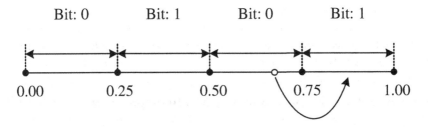

Fig. 2. Partitioning of the interval $[0, 1)$ into semi-intervals when embedding a bit of a digital watermark.

We shall take the value $r = 0.832$ as a random variable falling in this interval. Then, in accordance with step 1.4 of the algorithm, the new value of the sensory data element will be equal to 12.3832.

For simplicity, the presented algorithm is formulated in such a way as if it works with groups of simultaneously generated data. However, the values generated by the sensor and protected by a digital watermark are not transmitted simultaneously in groups, but sequentially; therefore, the head node checks the authenticity of each individual value based on one bit of the digital watermark extracted from it.

When using this algorithm, it is recommended to record in each sensor, included in the wireless sensor network, a unique digital watermark that does not repeat the digital watermarks of other sensors. The fact of the appearance of two sensors that produce a sequence of values containing the same digital watermark will indicate an intruder's attack on the protected infrastructure by reproducing previously intercepted values.

The scaling factor allows you to control the level of distortions of embedding, therefore, the proposed algorithm is applicable to work with physical quantities of various nature.

The algorithm for extracting a digital watermark from a group of sensory data is organized as follows.

Input: a group of sensory data $D = d_1 d_2 \ldots d_L$ with recorded digital watermark; scaling factor $\lambda \geq 1$; the number of embed intervals n.

Output: a digital watermark $W = w_1 w_2 \ldots w_N$, $w_i \in \{0, 1\}$, $i = \overline{1, N}$.

Step 1. For $i = \overline{1, L}$ do the following:

Step 1.1. Calculate $x = \{\lambda s_i\}$.

Step 1.2. If x falls into any of the semi-intervals in the partition $\left[0; \frac{1}{n}\right) \cup \left[\frac{2}{n}; \frac{3}{n}\right) \cup \cdots \cup \left[\frac{n-2}{n}; \frac{n-1}{n}\right)$, then assign $w_i = 0$. Otherwise assign $w_i = 1$.

Step 2. For $i = \overline{N + 1, L}$ do the following:

Step 2.1. Calculate $j = (i \bmod N) + 1$.

Step 2.2. If $w_j \neq w_i$, return the message "digital watermark is incorrect".

Step 3. Return a digital watermark $W = w_1 w_2 \ldots w_N$ and end the algorithm.

The extraction of a digital watermark occurs on the side of the head node, which thus verifies the authenticity of the data received from the sensors.

The presented digital watermark extraction algorithm contains a check that the bits of the digital watermark are cyclically repeated in N steps. If this condition is not fulfilled, this means that the digital watermark has been damaged and the sensory data is distorted.

4 Computing Experiments and Discussion of Results

4.1 Wireless Sensor Network Model

To carry out computing experiments, a simple wireless sensor network model was used, constructed using the OMNeT++ network simulator (Fig. 3).

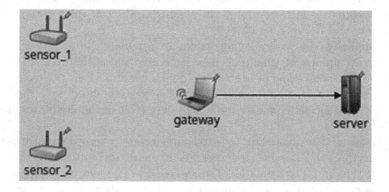

Fig. 3. Wireless sensor network model.

This model includes two temperature sensors, an intermediate gateway and a server. Information is embedded on the side of temperature sensors in the data they generate. The modified temperature values get to the intermediate gateway via the wireless network. Then the gateway carries out the authentication and sends the data to the server via a wired connection.

4.2 Results of the Experiments

The data for the experiments was obtained by means of simulation modeling in the OMNeT++ network simulator. For each sensor, the initial temperature values obtained from the sensor and modified values containing the bits of the digital watermark were recorded.

Examples of the obtained data sequences for different values of the scaling factor are shown in Table 1. In all cases, a random binary sequence was taken as a digital watermark. To assess the effectiveness of the proposed algorithm according to the criterion of invisibility of embedding, we use the methods of mathematical statistics.

Table 1. The results of the algorithm operation.

Scaling factor λ	Initial data	Data with a digital watermark
1	8.8803	8.8803
1	11.6848	11.952
1	8.1888	8.1888
1	12.1891	12.2964
1	8.9417	8.9417
1	11.5443	11.5443
1	9.1568	9.4994
1	11.3317	11.2756
1	9.5142	9.5142
1	10.642	10.9827
1	9.6357	9.2837
1	10.7926	10.7926
10	8.8803	8.8803
10	11.6848	11.6848
10	8.1888	8.1511
10	12.1891	12.1891
10	8.9417	8.9417
10	11.5443	11.5443
10	9.1568	9.1362
10	11.3317	11.3431
10	9.5142	9.5142
10	10.642	10.642
10	9.6357	9.6357
10	10.7926	10.7926
100	8.8803	8.884
100	11.6848	11.6848
100	8.1888	8.1856
100	12.1891	12.1891
100	8.9417	8.94
100	11.5443	11.5443
100	9.1568	9.1587
100	11.3317	11.3317
100	9.5142	9.5183
100	10.642	10.6432
100	9.6357	9.6335
100	10.7926	10.7926

Let us suggest a hypothesis that there is no significant statistical difference between the initial data and the modified data containing a digital watermark.

Let's check this hypothesis using Wilcoxon T-test. This statistical criterion is used to assess the differences between two rows of measurements made for the same feature under investigation, but in different conditions or at different time.

The results of the Wilcoxon T-test for the studied model sensory data samples are presented in Table 2. Critical values were obtained from the table of critical values of the Wilcoxon T-test at a significance level of 0.01 and 0.05.

Table 2. Wilcoxon T-test calculation results.

λ	T_{emp}	$T_{crit}(0.01)$	$T_{crit}(0.05)$
1	26	9	17
10	55	9	17
100	26	9	17

Since the empirical value in all cases exceeds the critical value, the hypothesis of the absence of statistical distinguishability is accepted.

5   Conclusion

This paper presents a new algorithm for embedding digital watermarks in the data of wireless sensor networks. The purpose of the proposed algorithm is to protect against the substitution of a data source in a wireless sensor network.

An important distinctive feature of the proposed algorithm is the ability to control the level of distortions introduced as a result of embedding. This allows us to recommend this algorithm for use in wireless sensor networks, including sensor nodes of various types, designed to measure physical quantities of different nature.

The development of this work will consist in designing a set of algorithms for embedding digital watermarks in the data of wireless sensor networks. These algorithms will differ from each other in the provided ratio between the indicators of the embedding effectiveness and will be intended to protect against various types of attacks.

References

1. Hu, F., Cao, X.: Wireless Sensor Networks: Principles and Practice. Auerbach Publications, New York (2010)
2. Fridrich, J.: Steganography in Digital Media: Principles, Algorithms, and Applications. Cambridge University Press, Cambridge (2009)
3. Zhang, W., Liu, Y., Das, S.K., De, P.: Secure data aggregation in wireless sensor networks: a watermark based authentication supportive approach. Pervasive Mob. Comput. 4(5), 658–680 (2008)

4. Panah, A.S., Van Schyndel, R., Sellis, T., Bertino, E.: In the shadows we trust: a secure aggregation tolerant watermark for data streams. In: Proceedings of the WoWMoM 2015: A World of Wireless Mobile and Multimedia Networks, Boston, MA, USA, pp. 1–9. IEEE (2015)
5. Wang, C., Bai, Y., Mo, X.: Data secure transmission model based on compressed sensing and digital watermarking technology. Wuhan Univ. J. Nat. Sci. **19**(6), 505–511 (2014)
6. Alromih, A., Al-Rodhaan, M., Tian, Y.: A randomized watermarking technique for detecting malicious data injection attacks in heterogeneous wireless sensor networks for Internet of Things applications. Sensors (Switzerland) **18**(12), 1–19 (2018)
7. Rubio-Hernan, J., De Cicco, L., Garcia-Alfaro, J.: Adaptive control-theoretic detection of integrity attacks against cyber-physical industrial systems. Trans. Emerg. Telecommun. Technol. **29**(7), 1–17 (2018)
8. Wang, B., Kong, W., Li, W., Xiong, N.N.: A dual-chaining watermark scheme for data integrity protection in Internet of Things. Comput. Mater. Continua **58**(3), 679–695 (2019)
9. Ferdowsi, A., Saad, W.: Deep learning for signal authentication and security in massive Internet-of-Things systems. IEEE Trans. Commun. **67**(2), 1371–1387 (2019)
10. Hameed, K., Khan, A., Ahmed, M., Goutham Reddy, A., Rathore, M.M.: Towards a formally verified zero watermarking scheme for data integrity in the Internet of Things based-wireless sensor networks. Future Gener. Comput. Syst. **82**, 274–289 (2018)
11. Wang, C.F., Wu, A.T., Huang, S.C.: An energy conserving reversible and irreversible digital watermarking hybrid scheme for cluster-based wireless sensor networks. J. Internet Technol. **19**(1), 105–114 (2018)
12. Lipuš, B., Žalik, B.: Robust watermarking of airborne LiDAR data. Multimedia Tools Appl. **77**(21), 29077–29097 (2018)
13. Evsutin, O.O., Kokurina, A.S., Meshcheryakov, R.V.: A review of methods of embedding information in digital objects for security in the Internet of Things. Comput. Opt. **43**(1), 137–154 (2019)
14. Iskhakov, A.: Adaptive authentication technologies in behavioral analysis solutions of robotic systems. In: The VIth International Workshop "Critical Infrastructures: Contingency Management, Intelligent, Agent-Based, Cloud Computing and Cyber Security" (IWCI 2019), Irkutsk, Baikalsk, Russia, pp. 49–54. Atlantis Press (2019)
15. Shumskaya, O., Iskhakova, A.: Application of digital watermarks in the problem of operating signal hidden transfer in multi-agent robotic system. In: Proceedings of 2019 International Siberian Conference on Control and Communications (SIBCON), Tomsk, Russia, pp. 1–5. IEEE (2019)

Investigation of Video Traffic Transmission via Augmented Reality Devices in a Mesh Network

V. Blagodarova[1] and M. Makolkina[1,2(✉)]

[1] The Bonch-Bruevich Saint-Petersburg State University of Telecommunications,
22 Prospekt Bolshevikov, St. Petersburg, Russian Federation
blagodarova@gmail.com
[2] Peoples Friendship University of Russia (RUDN University),
6 Miklukho-Maklaya St., Moscow, Russian Federation
makolkina@list.ru

Abstract. Increasing attention in research in the field of networks and communication systems attracts the creation and development of communication networks of the fifth generation. An important role is given to new applications that determine the capabilities of these networks in the provision of services. Augmented Reality (AR) is one of the most promising applications. Requirements for the provision of Augmented Reality services are quite stringent, especially in terms of the circular delay, the permissible value of which is determined by the value of 5 ms. This requires the development of adequate solutions for the network structure, which in the fifth generation communications network are based on mobile edge computing technology. The article is devoted to the actual topic of using Augmented Reality technology and provides an investigation of the transmission of video data for Augmented Reality applications through a mesh network. The basis of the material presented is the solution of the problem of simulation modeling of the use of Augmented Reality in a mesh network, made using the COOJA simulator.

Keywords: Augmented Reality · Mesh networks · Video traffic · Quality of service · Simulation · Simulator COOJA

1 Introduction

Over the past few years, Augmented Reality has undergone significant changes and has become an integral part of the lives of many people. This technology provides services of the most diverse types: from gaming and entertainment functions to assistance in training and executing work in almost all areas of activity [14,16]. Currently, users with the help of Augmented Reality applications can create content themselves, interact with it, send it to other people, etc., that is, the implementation of the technology itself is simplified more and more [12,13]. Nowadays, almost smart phone and glasses with high quality camera are used

© Springer Nature Switzerland AG 2019
V. M. Vishnevskiy et al. (Eds.): DCCN 2019, CCIS 1141, pp. 586–596, 2019.
https://doi.org/10.1007/978-3-030-36625-4_47

for AR applications. When the phone camera recognizes the object, the information of this object will be shown in the phone display. In recent years, AR technology has been growing up quickly that provide users different experiences about everything. In the fashion market, consumers can visualize products and imagine what it might feel like to own the product or experience the service before actually purchasing it. During driving, AR application can support the car or driver with augmented information about traffic, weather, etc.

At the same time, WiFi networks are replenished every few years with new IEEE 802.11 standards, which is associated with an increase in the number of consumers of wireless Internet access, as well as an increase in the amount of information several hundred and thousand times. This led to the emergence of high-speed mesh networks (IEEE 802.11s), which allow to transmit traffic quickly, efficiently, safely and with the best quality of service (QoS) performance [10]. Augmented Reality technology and mesh networks are both promising and in demand among users. In the mesh network devices can share AR content to each other by transmitting data hop-by-hop. Therefore, no need to send requests to cloud server if the AR content can be get information from the other device in the network. Besides WiFi, The Bluetooth 5.0 (BLE) technology is coming with supporting mesh connection. A study in [9] was investigated in transmission of AR contents over BLE mesh network. Devices connect and share data with each other in the same Bluetooth network. When smartphones are provided with new chip BLE 5.0, it could be benefit for AR application developers that there are different choices to provide AR applications to users.

Since augmented reality improves every year and spreads to various areas of activity, its applications are used not only by end users, but also organizations to provide assistance to others (for example, in medicine) or in the service sector (advertising and marketing, construction, education etc.). It should be mentioned that most aspects of human life are currently related to information technology. The wider use of the Internet of Things has made possible the concept of a smart city. It covers many areas, from environmental protection to automated traffic management. The smart city includes a combination of advanced technologies, including the Internet of things, artificial intelligence and smart sensors. One of the main components of the smart city concept is the task of creating an urban environment that is more people-oriented. And it is here that Augmented Reality plays an important role.

In addition, Augmented Reality applications allow users to visually interact with the system, which greatly simplifies the implementation of various tasks and functions. During deploying AR application, the users are interested in Quality of Experience that gives them different feeling about using this application. Moreover, QoS ensures the network requirement provided to consumers.

Augmented Reality attracts more and more users to its market every year due to a number of advantages over other technologies. These include: interactivity, accessibility of information, interaction with reality, new ways of using in a particular field of activity. Therefore, Augmented Reality is quite profitable to use for various companies to promote their products and services.

Consequently, the number of AR devices will grow every year, as will the profit from them. Taking advantage of AR devices, their connection in the mesh network can be considered to transmit video in particular cases.

The AR technology involves the transmission of different types of traffic. In each AR application there are several types of data transmitted and shared between devices. The data type can be sound, voice, text, image or video, etc. Among these, video traffic is the most time-consuming type of data, requiring high quality network and its characteristics. In order to analyze the transmitted video traffic between AR-devices in a WiFi mesh network, it is necessary to conduct a study of the dependence of the network characteristics for different formats of transmitted video. As a method, this article uses simulation modeling in the COOJA simulator.

2   Related Works

Currently, one of the driving factors in the modernization of modern communication networks is the transition to interaction scenarios within the fifth generation 5G/IMT-2020 communication networks [4,17]. The future network continues considering issues to ensure the high QoS, low latency, ubiquitous coverage as AR applications and services require. In these articles, the authors analyzed the requirements and challenges of the new network that will support several applications with different requirements, where the AR application is a part of this network.

Fifth-generation networks assume scenarios on the basis of which a seamless connection should be provided between devices and applications of the Internet of Things [1,17]. One of these scenarios is the interaction of Internet of Things devices in the structure of a self-organizing network. Self-organizing networks typically have a mesh topology and allow for the delivery of data from node to node by building a route to the destination. Due to the fact that according to the reference architecture of the construction of Smart Cities, continuous monitoring of various urban infrastructure facilities with the possibility of obtaining data from sensors arises, it becomes possible to use a cellular topology to organize a network for collecting data from such facilities, as well as delivering them to Augmented Reality applications for greater informativity [2,3].

The paper [8] discusses different approaches to the resource allocation in the provision of augmented reality services, and also proposes new architectures for distributing data on servers. A method was considered for selecting the structural parameters of the service system when providing AR service. The resource allocation goal was a task of data clustering and localization of data processing. The AR data or content is allocated and processed by optimal way.

In this article [7], the authors explore the different traffic patterns of Augmented Reality and the requirements that the Augmented Reality applications make to network characteristics. The investigation of interaction between AR and flying ubiquitous sensor networks was considered. Indeed such modern applications as AR require development of the new traffic patterns that can ensure the further Quality of Experience and estimation.

3 Simulation in the Application Package COOJA

Working Principle of AR Devices in the Mesh Network. One of the smartphones is connected to the server via a local WiFi network (LAN), so it acts as a gateway. The server through the Ethernet port (WAN) has access to the Internet. Through the gateway, the other devices included in the mesh network are connected to the server and exchange data with each other. Mesh topology allows for the implementation of unique municipal capabilities networks focused on rapid response services. It is necessary to take into account when creating mesh networks, the following problems may arise:

- limited frequency resource (frequency ranges 802.11 in the largest cities),
- the need to confirm the results of radio frequency planning with practical studies of the state of the radio environment in the network deployment area (presence of unregistered users),
- organization of the placement of access points in the maximum proximity from subscribers, providing round-the-clock power supply, etc.

Currently, there are various application packages for simulation of communication networks. These systems are focused on modeling various systems and processes. Among the large number of application packages of simulation in the field of communication, the COOJA system was chosen. COOJA is a simulator of network processes in the operating system Contiki, designed to simulate devices and networks. It simplifies the development and debugging of software in wireless networks [11,15]. At the network layer, work with routing protocols is carried out, hardware reporting is omitted. This layer controls the radio part of the sensor node and its specific properties. Modeling the execution of native code is implemented at the operating system level. At the level of machine code instructions, it is possible in nodes with different basic structures to simulate them in java based on microcontrollers instead of the compiled system Contiki.

The COOJA package was chosen for the study, since it is most suitable for modeling wireless nodes and terminals. Also this simulator meets all the requirements for analyzing the data that will be obtained.

The simulated node in COOJA has three main properties: data in memory, node type and its components. The type of the node determines their common properties. The main distinguishing feature of this simulator is that it allows to control three different levels simultaneously – the network level, the operating system level, and the level of machine code [15]. The COOJA configuration is flexible, as many parts of the model can be replaced or extended with additional functionality [5,6,11].

COOJA allows to simulate four types of wireless channels: Unit Disk Graph Medium (UDGM)-Constant Loss – channel with constant losses in the area bounded by a circle, UDGM–Distance Loss with losses depending on the distance in the circle and Directed Graph Radio Medium (DGRM) – loss depending on the direction.

The UDGM constant loss model is a wireless channel model in which the communication area looks like a perfect circle. Outside this circle, nodes do not receive packets, but within the circle they receive.

The UDGM model with losses depending on distance is the same as the previous one, but it additionally takes into account two more factors:

– interference (in which case packets are lost at the distance of interference),
– The coefficient of successful transmissions and receptions can be set.

DGRM – a channel in which the network topology is limited. It is mainly used to determine the transmission success rate and propagation delay in asymmetric channels.

Formulation of the Problem. For the experiment, the network simulator COOJA is used, in which the data stream is specified in three variations and has the following dimensions: 39 Mbs, 173 Mbs and 1.56 Gbs (depending on the traffic format). Augmented Reality devices are a set of smartphones on which an AR-application is installed. A total of 50 devices are used, and they are formed into a single Wi-Fi mesh network, where device number 51 act as a gateway for accessing the Internet, which is necessary for data exchange between the network in question and the Augmented Reality server. A TCP/UDP server is used as a base station, and TCP/UDP clients are used as mobile stations. The main task of simulation is to study the video stream in the Wi-Fi mesh-network of Augmented Reality devices, comparing the basic network characteristics (route length, delay and loss) when transmitting different video formats. In this simulation, IPv6 protocol was used for implementation.

The Composition of the Model. The network model includes the Cooja network simulator, which implements the IPv6 protocol, as well as the following parameters:

– number of TCP/UDP clients: 50 (nodes 1–50),
– number of TCP/UDP servers: 1 (node 51),
– room size: 100×100 m.

Figure 1 shows the constructed scheme in the COOJA system with the types of nodes used on it.

Figure 2 shows the process of transmitting video traffic in a mesh network.

The node at number 51 is the gateway that allows smartphones to generate a video stream by accessing the server of Augmented Reality via the Internet. At different points in time, the direction of movement of traffic between devices changes, some nodes cease to participate in its transmission, while others, conversely, are activated. Some part of the device acts as the head transit node, which at a certain time t_n sends information simultaneously to other nodes close to it.

Analysis of Simulation Results. The gateway is the node number 51, which allows the considered mesh network to access the Internet to communicate with the server of augmented reality. At different times tn in the network, head nodes

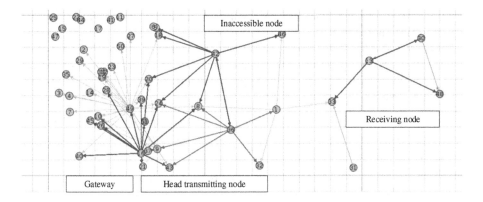

Fig. 1. Types of used nodes

are formed, which transmit video traffic simultaneously to several devices located closest to it. In order to evaluate the network performance, it is necessary to analyze the main network characteristics (delay, loss, and route length), as well as to compare how they vary for different video traffic formats (576i, 720p and 2160p). Table 1 displays the parameters specified in the settings of the simulator.

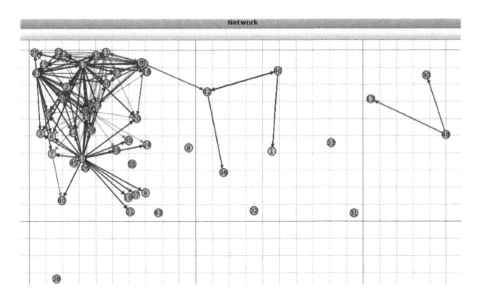

Fig. 2. Traffic transmission in the mesh-network at time t_n

The symbols "i" and "p" in video formats mean interlaced and progressive modes. The mode "p" is better, since each frame of a progressive video has a full size. In an interlaced video, each frame is divided into two half-frames, consisting of lines selected through one. In fact, a single frame of interlaced video has half the resolution in height.

It is also necessary to consider an important parameter as the frame rate per second. One of the most common frequencies are: 25, 30, 50, 60 frames per second (FPS). The higher the frequency, the more smooth (natural) the video looks. An example of a short and meaningful video designation, including a frame rate of 2160p60s.

Table 1. Parameters for modeling

The duration of the experiment	15 min
Content size 576i	39 Mb
Content size 720p	172 Mb
Content size 2160p	1.56 Gb
Frame rate	60 FPS

After the parameters in the COOJA simulator were set and the network conditions were set, the experiment resulted in a duration of 15 min and numerical results were obtained, downloaded in Excel. Figure 3 shows the dependence of the end-to-end delay on the number of passable hops (route length) by traffic during video transmission of 39 Mb in the 576i format.

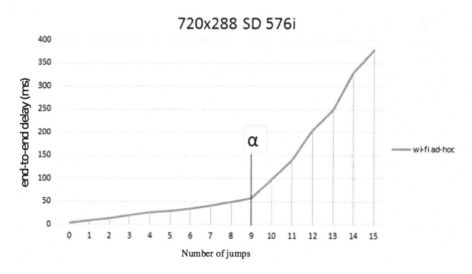

Fig. 3. Dependence of the delay on the length of the route when transmitting video format 576i

In Fig. 3 it can be seen that the limiting value of the route length α is equal to 9, that is, if this video is transmitted through more than 9 devices, the delay increases dramatically and the video quality starts to drop dramatically. In order for the delay to be almost imperceptible, it is necessary not to exceed 100 ms, therefore, to prevent the transmission of video through more than 10 devices.

Figure 4 shows the dependence of the delay on the length of the route already for transmitting a 720p video stream.

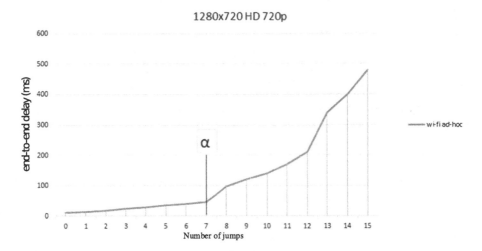

Fig. 4. Dependence of the delay on the length of the route when transmitting video format 720p

According to the results of the experiment, it is clear that the maximum length of the route for the optimal delay is 7 nodes. If the traffic transmission value exceeds more than 8 nodes, the delay will exceed its limit value, which can lead to data packet losses.

Next, the simulation result is obtained for a 2160p video format, shown in Fig. 5, by which we can conclude that the maximum allowable value of traffic passing through the nodes α is 5.

In Figs. 3, 4 and 5, it can be seen that with increasing video resolution and quality, network requirements become more stringent, that is, the maximum allowable number of nodes through which video is transmitted decreases by several devices. Further, in Fig. 6, all three graphs are described, as described above. The parameters λ, β, and α are 5, 7, and 9, respectively. Consequently, when transmitting more information, it is necessary to adhere to a smaller number of hopes during data transmission.

In order to estimate the losses occurring during the operation of a mesh network from AR devices, it is necessary to trace the dependence of the decimal logarithm of the number of transmitted packets on the length of the route. Figure 7 for 576i, 720p, and 2160p formats shows the loss as a function of the number of hops.

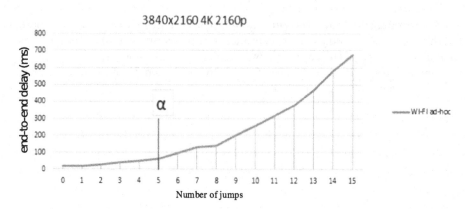

Fig. 5. Dependence of the delay on the length of the route in the transmission of video format 2160p

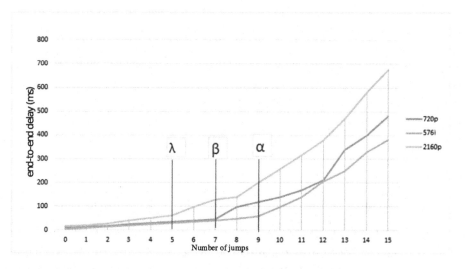

Fig. 6. General graph of the dependence of the delay on the length of the route

It can be concluded that for the 576i and 720p format the losses differ by less than one value (<1), since the difference in the amount of transmitted video is not significant, namely 134 Mb (with 39 Mb and 173 Mb, respectively). Even with an increase in the length of the route by 5–10 devices, the losses are very small. For the format of 2160p, the results are opposite: since the volume of transmitted video is quite large – 1.56 Gb – when the allowed value of the number of hops is 5, the probability of loss increases dramatically. To avoid this, it is necessary to choose the minimum route for transmitting traffic.

Simulation is a method of knowledge, consisting in the creation and study of models. Simulation models, as one of the types, make it possible not only to display reality with varying degrees of accuracy, but also to simulate it.

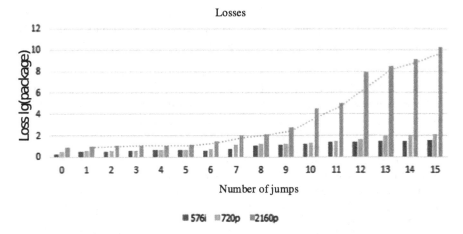

Fig. 7. The dependence of losses on the length of the route for different video formats

This article used the COOJA simulator program, which can simulate any network processes, and in this case, the study of video traffic between AR devices in a mesh network of 51 nodes.

4 Conclusion

As a result of the experiment, it is concluded that the main network characteristic affecting the quality of data transmission in a mesh network is the route length. Dependencies of delays and losses on this characteristic showed that with a larger video format, it is optimal to transmit traffic through a smaller number of nodes. This will avoid losses and minimize delays. If the required quality of service is observed, the augmented reality devices will be able to exchange video traffic with each other in the best quality.

Acknowledgment. The publication has been prepared with the support of the "RUDN University Program 5-100" and funded by RFBR according to the research projects No. 12-34-56789 and No. 12-34-56789.

References

1. Asadi, A., Wang, Q., Mancuso, V.: A survey on device-to-device communication in cellular networks. IEEE Commun. Surv. Tutor. **16**(4), 1801–1819 (2014)
2. Ateya, A., Muthanna, A., Gudkova, I., Abuarqoub, A., Vybornova, A., Koucheryavy, A.: Development of intelligent core network for tactile internet and future smart systems. J. Sens. Actuator Netw. **7**(1), 1 (2018)
3. Ateya, A.A., Muthanna, A., Gudkova, I., Vybornova, A., Koucheryavy, A.: Intelligent core network for tactile internet system. In: Proceedings of the International Conference on Future Networks and Distributed Systems, p. 22. ACM (2017)

4. Ateya, A.A., Muthanna, A., Koucheryavy, A.: 5G framework based on multi-level edge computing with D2D enabled communication. In: 2018 20th International Conference on Advanced Communication Technology (ICACT), pp. 507–512. IEEE (2018)

5. Jadhao, A.R., Solapure, S.S.: Analysis of routing protocol for low power and lossy networks (RPL) using Cooja simulator. In: 2017 International Conference on Wireless Communications, Signal Processing and Networking (WiSPNET), pp. 2364–2368. IEEE (2017)

6. Kugler, P., Nordhus, P., Eskofier, B.: Shimmer, Cooja and Contiki: a new toolset for the simulation of on-node signal processing algorithms. In: 2013 IEEE International Conference on Body Sensor Networks, pp. 1–6. IEEE (2013)

7. Makolkina, M., Koucheryavy, A., Paramonov, A.: Investigation of traffic pattern for the augmented reality applications. In: Koucheryavy, Y., Mamatas, L., Matta, I., Ometov, A., Papadimitriou, P. (eds.) WWIC 2017. LNCS, vol. 10372, pp. 233–246. Springer, Cham (2017). https://doi.org/10.1007/978-3-319-61382-6_19

8. Makolkina, M., Paramonov, A., Koucheryavy, A.: Resource allocation for the provision of augmented reality service. In: Galinina, O., Andreev, S., Balandin, S., Koucheryavy, Y. (eds.) NEW2AN/ruSMART -2018. LNCS, vol. 11118, pp. 441–455. Springer, Cham (2018). https://doi.org/10.1007/978-3-030-01168-0_40

9. Makolkina, M., Pham, V.D., Dinh, T.D., Ryzhkov, A., Kirichek, R.: Transmission of augmented reality contents based on BLE 5.0 mesh network. In: Galinina, O., Andreev, S., Balandin, S., Koucheryavy, Y. (eds.) NEW2AN/ruSMART -2018. LNCS, vol. 11118, pp. 394–404. Springer, Cham (2018). https://doi.org/10.1007/978-3-030-01168-0_36

10. Morais, A., Cavalli, A.: A quality of experience based approach for wireless mesh networks. In: Masip-Bruin, X., Verchere, D., Tsaoussidis, V., Yannuzzi, M. (eds.) WWIC 2011. LNCS, vol. 6649, pp. 162–173. Springer, Heidelberg (2011). https://doi.org/10.1007/978-3-642-21560-5_14

11. Naik, K.P., Joshi, U.R.: Performance analysis of constrained application protocol using Cooja simulator in Contiki OS. In: 2017 International Conference on Intelligent Computing, Instrumentation and Control Technologies (ICICICT), pp. 547–550. IEEE (2017)

12. Pankratz, F., Dippon, A., Coskun, T., Klinker, G.: User awareness of tracking uncertainties in AR navigation scenarios. In: 2013 IEEE International Symposium on Mixed and Augmented Reality (ISMAR), pp. 285–286. IEEE (2013)

13. Papagiannis, H.: Augmented reality (AR) joiners, a novel expanded cinematic form. In: 2009 IEEE International Symposium on Mixed and Augmented Reality-Arts, Media and Humanities, pp. 39–42. IEEE (2009)

14. Park, M.K., Lim, K.J., Seo, M.K., Jung, S.J., Lee, K.H.: Spatial augmented reality for product appearance design evaluation. J. Comput. Des. Eng. 2(1), 38–46 (2015)

15. Romdhani, I., Qasem, M., Al-Dubai, A.Y., Ghaleb, B.: Cooja simulator manual. Edinb. Napier Univ. 1 (2016)

16. Underwood, J.: Conduct-AR: desktop micromanaging at its finest. https://www.macstories.net

17. Yastrebova, A., Kirichek, R., Koucheryavy, Y., Borodin, A., Koucheryavy, A.: Future networks 2030: architecture & requirements. In: 2018 10th International Congress on Ultra Modern Telecommunications and Control Systems and Workshops (ICUMT), pp. 1–8. IEEE (2018)

Digital Object Architecture as an Approach to Identifying Internet of Things Devices

Dmitriy Sazonov$^{(\boxtimes)}$ and Ruslan Kirichek

The Bonch-Bruevich Saint-Petersburg State University of Telecommunications,
22 Prospekt Bolshevikov, 193232 St. Petersburg, Russian Federation
`dim-saz@yandex.ru`

Abstract. The analysis of the possibility of building a system for identifying Internet of things devices based on the Digital Objects Architecture is given. The model of the Handle resolution system for identifiers of digital objects as a queuing system is proposed. The optimization experiment was performed and a configuration to reduce the time for identifier resolution was obtained. The ways of possible improvement of algorithms with the aim of reducing the time for identifier resolution are proposed. A block diagram of a prototype authentication and identification system for Internet of things devices based on the Handle resolution system is proposed. The administration server for working with Handle system and the mobile client application for the administrator of the descriptors, which allows to manage handle entries was implemented.

Keywords: Internet of Things · Digital Object Architecture · Handle System · Identification

1 Introduction

In modern society a considerable part of the market of technical systems is occupied by the Internet of Things. These devices find place in many areas, beginning from simple household use, medicine and finishing with application in the military purposes. By rough estimates, the number of IoT devices reaches about 28 billion and the digit grows every year. To ensure correct and fast work with a huge information flow from such devices, a reliable addressing and identification system is required. The key features of identification for the Internet of Things [1, 4, 9]:

- Different life cycle of devices (some devices may work for a rather long time, while others - vice versa);
- relationship of objects of the Internet of things with other entities which are not included in this system (owners and administrators of devices can change during "life", which affects identification processes);

The publication has been prepared with the support of the "RUDN University Program 5–100"

© Springer Nature Switzerland AG 2019
V. M. Vishnevskiy et al. (Eds.): DCCN 2019, CCIS 1141, pp. 597–611, 2019.
https://doi.org/10.1007/978-3-030-36625-4_48

- special requirements for the context in which the devices operate (in certain cases, access to objects for the same data may be allowed or limited depending on the situation);
- requirements for the provision of protection mechanisms (when designing these mechanisms, it is worth considering the limited resources of the Internet of things in terms of resources and performance);
- the ability to expand the identification system to a huge number of devices (over a billion devices);
- the ability to work effectively for a wide variety of devices (devices on the network can be extremely heterogeneous in their resources and performance);
- security requirements (in certain cases, the access of objects to the same data may be allowed or limited depending on the situation);
- transparency of the addressing system and independence from the network (in contrast to the classical addressing systems used, for example, in the Internet, the identification of Internet of things devices should be independent from the devices network location or belonging to the user; in addition, it should be have in view that devices can change their location during lifetime);
- a flexible and efficient mechanism for resolving identifiers (Internet of things devices must be precisely defined regardless of their location; in addition, there must be ease in connecting and configuring a new object to an existing network);
- safety and security of user data (do not forget that devices often work with a huge amount of personal data, which requires additional protection).

Today there are several approaches for creation of an identification system for the Internet of things [4,8,9]. One of possible solutions is creation of a system of identification on the basis of the Digital Objects Architecture (DOA).

2   Handle Resolution System

The main structural elements of DOA are a digital object, a Handle System, and a repository and registry of digital objects. The digital object in this architecture is characterized not only by information about location of the object. In addition, it is possible to obtain various information about the object itself: access requirements, authentication, information about the author, etc. [7]. All this information is entered by the administrator of the digital object. For this, a special infrastructure is integrated into the DOA architecture, providing the necessary encryption and access verification. Important part of the DOA architecture is the resolution system called Handle System. For each digital object in the described architecture the unique identifier of digital objects - DOI (Digital Object Identifier) is connected. This identifier is similar to the URL on which the modern Internet is built. However, unlike the latter, the assigned identifiers remain constant and do not depend on the state of the digital object. It is the resolution system that connects the identifier with information about the current status of a digital object (location, access, information about authenticity) [2,5,7]. The classical Handle system has two-level architecture [5–7]. The first level of the resolution system is the global register (GHR). The second level is a set of local registers (LHR). To resolve the identifier in this subsystem, first a call is made to the global GHR registry, which reports information about the local LHR registry, which contains the necessary information about the digital object. Schematically this architecture is presented in Fig. 1.

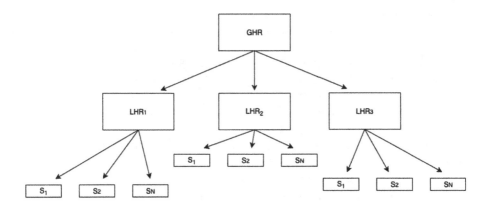

Fig. 1. Handle system infrastructure

The DOI identifier structure itself also corresponds to a two-level system [7]. For example, consider an identifier: 77.TEST/0001. The first part, located before the "/", is named after the prefix; the second part is suffix. The prefix allows you to set information about the local registry of the digital LHR object. This correspondence between the prefix and the information about the administrator is stored in the global GHR registry. The suffix unambiguously identifies a specific object, and this information linking the suffix to a specific object is stored in the local LHR registry.

3   Handle System as a Queuing System

In order to characterize the effectiveness of the identifier resolution system in the DOA architecture, we modeled the Handle system as a queuing system (QS). It was decided to take a model with an exponential distribution of the service time of requests and an exponential distribution of the time between the receipt of applications. The simulation time interval was chosen equal to 200 s. This queuing system model was prototype based on the analysis of the existing implementation of the resolution system [3,7,10]. In this implementation of the Handle System, not one GHR server is used, but several servers belonging to the so-called top-level administrators MPA (Multi-Primary Administrators) controlled by DONA Foundation [11] are used. Infrastructure of the GHR servers was obtained and average delay for the request resolution is defined. All MPA servers are equivalent among themselves, a request for resolution is received sequentially on all servers and the first response received is analyzed [11,12]. There is no analysis of the delay time from client to the server in existing infrastructure. In fact, the resolution system guarantees that if a request for resolution arrives in the system, it will certainly be fulfilled, however, the time that may be required for this is clearly not regulated. Table 1 presents the characteristics of the MPA servers used as GHRs.

Table 1. MPA servers specifications

MPA	IP address	Average resolution delay, ms
America	132.151.20.9; 38.100.138.153; 38.100.138.153; 38.100.138.153; 132.151.20.9; 2001:550:100:6::138:153; 2001:550:100:6::4; 132.151.1.179	243.548
Switzerland	156.106.193.160	71.33
China 1	119.90.34.34	473.583
China 2	47.90.103.77	410.693
Tunisia	41.231.118.2	82.510
Germany	134.76.30.197	44.356
Saudi Arabia	86.111.195.107	318.450
Kenya	196.12.152.22	258.450

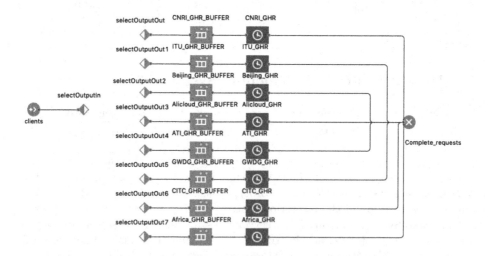

Fig. 2. QS model

Figure 2 is showed the model of the QS developed in the package Anylogic. The *clients* block on the Fig. 2 corresponds to the source of requests from devices. Then there is a branching into 8 channels, each of which corresponds to the infrastructure of a specific MPA. The probability of choosing each of the channels in the existing system is equal. Each MPA server consist of a receiving requests buffer and an identifier processing server. The number of channels in the processing server corresponds to the number of the servers of each specific MPA given in Table 1. The average resolution delay in the system is modeled using the *Delay* object (in the Fig. 2 these are blocks with the GHR postfix). The delay value in this blocks corresponds to the values given in Table 1 for each server. The buffer of each MPA server is unlimited, which guarantees the processing of all requests. It should also be noted that the model presents only the upper level of the resolution system - interaction with the GHR and the next level interaction with LHR was not analyzed. Interaction with local servers and analysis of their configuration should be considered separately as part of the specific problem.

The key parameters affecting the performance of the Handle resolution subsystem are the network delay time for an incoming request, the speed of request processing and the number of processing channels. The main characteristic of such system is the average time to resolve a single request. This time will depend on the system configuration and the intensity of the requests. In Fig. 3 shows the dependence of the average identifier resolution time on the intensity of incoming requests for the current system configuration from Table 1. The parameter λ is a parameter of the exponential distribution of time between incoming requests.

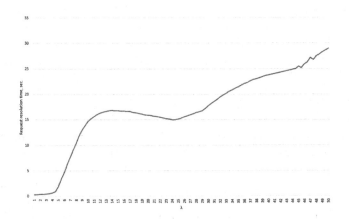

Fig. 3. Dependence of resolution time on the intensity of requests

To obtain this dependence, 100 test runs of the created model in the Anylogic package were performed. The simulation time in each run was 200 s. At each run, the exponential distribution parameter λ increased with a uniform step from 1 to 50. As can be seen from Fig. 3, with an increase of λ, the average time for resolving one identifier also increases, and for large volumes of requests, this time reaches 30 s, which is a lot for real-time applications.

In order to improve the query resolution time indicator in the system, a modification of the resolution system is required. In classic client-server systems, which include the Handle resolution system, there are two most obvious ways to increase the performance of the server side processing in order to increase the speed of processing incoming requests. The first approach, which is the most correct from the point of view of long-term prospects, but also the most complex and time-consuming in terms of labor and material costs, is to optimize server software, modify it, and search for the so-called bottle necks in the program code - the places where the largest percentage is lost performance. This approach to the modification of the server system guarantees long-term stable operation and increase in productivity. However, as has already been emphasized, this approach requires resource costs, since it implies a deep analysis of program code with a further expenditure on the work of programmers in order to make optimization corrections into the code.

There is a second approach to increase performance and increase the speed of processing of the server-side infrastructure. This approach is called horizontal scaling of the server-side infrastructure of the system and consists in increasing the number of physical devices that process requests. The increase in performance is achieved due to the fact that the total number of requests is now processed on more servers. In addition, this solution is extremely simple and does not take much time, as modern server architectures are adapted for fast horizontal scaling. Also, this solution is beneficial from an economic point of view, since it implies costs only for new server equipment, which is becoming cheaper every year. However, despite the fact that this approach is simple and understandable from the point of view of business and obtaining quick benefits, it implies solving the problem in the moment. The benefit that will be achieved by horizontal scaling is extremely short-term and after a certain period of time it may be necessary to continue to build up the server infrastructure. This approach does not solve the problem, but only gives a postponement. For a longer-term perspective, the first approach is most preferable, although it requires more labor both in terms of resources and in terms of the time required to make improvement.

We will perform an optimization experiment aimed at establishing the most suitable infrastructure for GHR servers with the current configuration of time delays in order to reduce the average time for identifier resolution. The number of the GHR servers used by each of MPA will be a key parameter for optimization. Resolving time will be set to 1 s. The results of the optimization experiment are presented in Table 2, where: alfa - load intensity parameter; d1 ... d8 - the number of servers of each MPA. The optimization process consists in sequentially launching the model with varying optimization parameters (number of GHR servers) to achieve the set goal (identifier resolution time less than 1 s). The *Current* column presents the parameters of the optimized model at the current iteration step. The *Functional parameter* row shows the value of the optimization function at the current step. At the end of the optimization process, we get a

Table 2. Optimization experiment results

Current		Best
Iteration step	500	60
Functional parameter	3.947	0.878
Optimization parameters		
λ	50	50
d1	7	7
d2	9	10
d3	4	1
d4	9	10
d5	8	10
d6	8	10
d7	10	10
d8	8	10

set of parameters (the number of GHR servers) that most closely provide an identifier resolution time of no more than 1 s. For the current configuration of the model, the number of servers is 7, 10, 1, 10, 10, 10, 10, 10 for each MPA from Table 1, respectively, as shown in the "Best" column. As you can see from the *Functional parameter* row for the best iteration, the value of the identifier resolution time was 0.878 s.

Figure 4 is showed the dependence of the resolution time on the intensity when configuration for servers taken from the optimization experiment results.

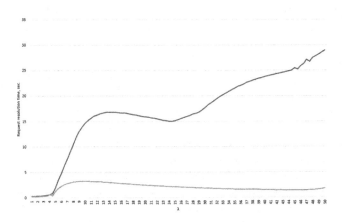

Fig. 4. Request resolution time from requests intensity at optimal configuration

According to the dependence in Fig. 4, it can be seen that with this configuration of the GHR servers, identifiers resolution in the system is much faster. The increase in speed reaches 15 times at maximum load intensity.

4   Authentication and Identification System

A feature of Internet of Things devices in terms of authentication and authorization is that devices operate autonomously without external access. During the life cycle of such devices, the direct access of the administrator to the main processes is not implied. Such devices often also do not have a user interface. Accordingly, such devices should support the automatic authorization and authentication process in the networks in which the work is carried out. To implement this functionality, it is proposed to use the Handle resolution system and DOI identifiers. Considering the features of the functioning of the Handle resolution system, considered in the proposed model described earlier, we will offer a structural diagram of the system of identification and authentication of Internet of Things devices based on client software provided by Handling.net [8]. The interaction scheme is shown in Fig. 5.

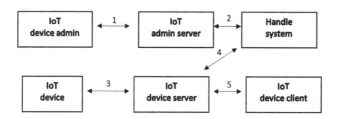

Fig. 5. Diagram of the identification and authentication system for the Internet of Things devices

This diagram shows the main participants in the automated device identification and authentication process based on the Handle system, and also shows the phased interaction of various components in this system. Key components of this system:

- IoT device admin – the administrator of the Internet of Things device. This part is responsible for registering the device in the system and further control;
- IoT admin server – administration server implemented on the basis of the Handling.net client software and providing interaction between the device administrator and the Handle system;
- Handle System – Handle descriptor resolution system, responsible for information management and resolution identifiers;
- IoT device – the device of the Internet of Things with the integrated Handle identifier;

- IoT device server – the main server that provides client interaction with the Internet of Things device, as well as responsible for the identification and authentication process;
- IoT device client – client application that receives information from the device or interacts with the device in any other way.

The main processes in this system are divided into stages and can be divided into two main groups: processes of administration and registration and processes of client interaction. Stages 1 and 2 belong to processes of administration. At 1 stage the administrator sends a request for registration of the IoT device to the administration server of the system. This request contains the main information about the device and the descriptor for the Handle system. After that on the step 2 record with the received information is created in the Handle resolution system and connected with the Handle descriptor. After this stages registration of the device is complete and the administrator can control and change this information.

Process of the client interaction consists of steps 3, 4, 5. On the step 3 there is primary authentication of the IoT device for connection it to the system. After the device is initialized, an authentication request is sent to the IoT device server. In this request the device send own Handle descriptor and additional information, for example, a secret string which can be integrated in the device too and is added by the administrator to the special hidden field in the Handle system record. Based on this secret string more secure authentication process can be implemented. After receiving of the request the client server start the process of the resolution of the device descriptor using the Handle system (step 4). If the descriptor is not found in the Handle system the device authentication is refused and all further requests from this device are ignored by the server. If the descriptor is found in the Handle System the additional authentication based on the secret string can be performed if such option is implemented in a system. After successful authentication the client server adds the device descriptor to the list of authenticated devices with the necessary information obtained from the Handle system. It is necessary for reduction of number of requests to the resolution system. Further, at each request from the device to the client server, for example, for data transmission from the device, these request is identified on the server according to the data obtained at an authentication stage. On a step 5, the client application interacts with the server, and get the information from all authenticated devices to which this application has an access. The offered authentication and identification system is a prototype and requires further implementation in the form of a test system. Additionally, it is necessary to solve the issue of storing private keys on the administration server. In addition, the development of an interaction scheme with additional authentication based on a secret string embedded in the device itself is required. Existing client software provided by Handle.Net implements functionality only for the administration of descriptors and manual management of them. This process does not involve any automation. To implement an automatic system, the development of own server for interaction with the Handle system is required. This functionality can be obtained by integrating the client software libraries provided by Handle.Net into your program code. As part of this work, the IoT admin software server that is responsible for the administration of descriptors was implemented. As described earlier, the administration server in the diagram shown in Fig. 5 is responsible for the main interaction between the administrator of the devices and Handle System. Also, as part of the preparation of the test bed, a client mobile application was developed for interaction with the administration server. The server functionality is based on the freely distributed client library from Handle.Net. This library is implemented in the Java programming language and contains all the necessary entities and logic for the working with the Handle system infrastructure. Detailed documentation for this library is provided in Javadoc format on Handle.Net site [13]. To implement the administration web server, it was decided to use the Java language for better integration with the Handle.Net client library. The server is implemented using REST API technology [14]. This technology is the modern standard for developing client-server applications, allowing you to implement communication with the server using the HTTP protocol and JSON format of requests and responses.

Now let's look at the client application that implements an interface for managing handles. The client mobile application implements requests to the administration server and does not contain any additional logic. The application consists of several screens. Figure 6 shows the first screen that appears immediately after entering the application.

Fig. 6. Application start screen

This screen provides a list of all devices that are registered by the administrator in the system. This information is requested from the administration server during application initialization. To add a new device to the system, the administrator clicks the *Add new device* button. Figure 7 shows the interface for adding a new device.

Fig. 7. Add new device screen

On the add new device screen, the administrator indicates basic information about the new device to authenticate it in the system, as described in Fig. 5. On this screen, you must specify the device name, its id (may not be specified), IP address allocated for the device in the system, a Handle descriptor that is associated with this device and also a description of the device to be added.

After entering the necessary data, the administrator clicks on the Add Device button, the server is accessed, and if the entry in the descriptor with the new device is successfully added, the inscription shown in Fig. 8 appears.

Fig. 8. Successful device registration status

After successful registration, the new device will appear in the list of devices on the initial screen (Fig. 9).

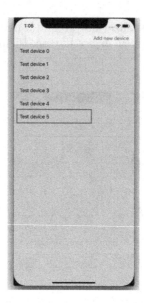

Fig. 9. The added new device is displayed in the list of devices

Clicking on the name of the device in the list opens a screen with detailed information for each device (Fig. 10).

Fig. 10. Detailed device information

As shown in Fig. 10, the bottom of the screen shows the status of the device in the system (the device is authenticated or not). This status is requested by the application from the server. The process of authenticating a device in the system is implemented according to the interaction scheme shown in Fig. 5. The new device goes through the authentication and authorization process in the system after adding an entry by the administrator. If the administrator added a new device, but it has not yet passed the authorization and authentication process on the server, the status will appear on the detailed information screen (Fig. 11).

Fig. 11. Failed device authorization on the server

As mentioned earlier, after the device is added by the administrator, a new value rows is created in the specified handle for the registered device in the Handle system. Record data is presented in Fig. 12.

Handle.Net®

Handle Values for: 77.TEST/0001			
Index	Type	Timestamp	Data
1	Device name	2019-09-15 08:15:41Z	Test device 5
2	URL	2019-09-15 08:15:10Z	172.17.0.1
3	Description	2019-09-15 08:16:12Z	Test device #5
100	HS_ADMIN	2019-03-14 21:04:20Z	handle=77.TEST/0001; index=200; [read val,modify val,del val,add val]
200	HS_VLIST	2019-03-14 21:16:04Z	300:77.TEST/0001

Fig. 12. Handle System records for registered device

Figure 12 shows the client web interface for working with the proxy resolution server provided by Handle.Net. This server provides only descriptor resolution and not allowing to manage them. This server and web client are suitable for quickly checking the status of handle records during the debugging process while working on application. As can be seen from the Fig. 12, as a result of the operation of the administration server, a request that was sent from the client mobile application to register a new device was successfully perform on Handle system side. All the necessary data that was provided by the client was saved which will allow to authorize the registered device or access it by using provided IP in URL row.

5   Conclusion

In this work, an assessment was made of the possibility of using the Digital Object Architecture technology to solve the problem of identification and authentication of Internet of Things devices. An analysis of the existing infrastructure and software of the Handle descriptor resolution system was made. Based on the analysis, a simulation model was developed in the Anylogic program and the results of modeling the system's operation with a large number of requests were obtained, which is important for Internet of Things applications. Based on the results of system simulation, it was concluded that the current infrastructure of the Handle system requires scaling and distribution. An optimization experiment was provided aimed at obtaining one of the possible configurations of the resolution system infrastructure in order to accelerate the processing of incoming requests. It was shown that the optimization of Handle system infrastructure is extremely important for the further development of various user solutions based on this architecture. After the description of the model and the results of its optimization, one of the possible implementations of the authentication and identification system for Internet of Things devices built on the basis of the Digital Object Architecture and using the existing handle resolution system was proposed. The diagram of the possible interaction of all participants in the authentication and identification process was presented and after that described a typical scenario for the operation of such a system. In addition, as part of this work, the process of implementing a test bed of the described authentication and identification system was started. The administration server that integrates the application logic into the Handle system infrastructure was implemented as well as the simple mobile client application for the administrator of the descriptors, which allows you to manage handle entries. In the future, it is expected to continue work on the implementation of the proposed test scheme and its subsequent evaluation when working with Internet of Things devices and a comparison of the test bed with the simulation model.

References

1. Al-Bahri, M., Yankovsky, A., Borodin, A., Kirichek, R.: Testbed for identify IoT-devices based on digital object architecture. In: Galinina, O., Andreev, S., Balandin, S., Koucheryavy, Y. (eds.) NEW2AN/ruSMART -2018. LNCS, vol. 11118, pp. 129–137. Springer, Cham (2018). https://doi.org/10.1007/978-3-030-01168-0_12

2. Al-Bahri, M., Yankovsky, A., Kirichek, R., Borodin, A.: Smart system based on DOA & IoT for products monitoring & anti-counterfeiting. In: 2019 4th MEC International Conference on Big Data and Smart City (ICBDSC), pp. 1–5. IEEE (2019)

3. Albahri, M., Kirichek, R., Ateya, A.A., Muthanna, A., Borodin, A.: Combating counterfeit for IoT system based on DOA. In: 2018 10th International Congress on Ultra Modern Telecommunications and Control Systems and Workshops (ICUMT), pp. 1–5. IEEE (2018)

4. Da, B., Esnault, P.P., Hu, S., Wang, C.: Identity/identifier-enabled networks (ideas) for Internet of Things (IoT). In: 2018 IEEE 4th World Forum on Internet of Things (WF-IoT), pp. 412–415. IEEE (2018)

5. ISO, I.: 26324: 2012 information and documentation-digital object identifier system. BSI British Standards (2012)

6. ITU-T: An architecture for IoT interoperability. Recommendation ITU-T T. 181203 (2018)

7. Kahn, R., Wilensky, R.: A framework for distributed digital object services. Int. J. Digit. Libr. **6**(2), 115–123 (2006)

8. Khan, R., Kristin, M.L., Sharp, T.: The role of architecture in internet defense. Americas Cyber Future: Security and Prosperity in the Information Age, Center for a New American Security, Washington, DC (2011)

9. Lam, K.-Y., Chi, C.-H.: Identity in the Internet-of-Things (IoT): new challenges and opportunities. In: Lam, K.-Y., Chi, C.-H., Qing, S. (eds.) ICICS 2016. LNCS, vol. 9977, pp. 18–26. Springer, Cham (2016). https://doi.org/10.1007/978-3-319-50011-9_2

10. Handle.Net Registry. http://www.handle.net/index.html

11. DONA Foundation. https://www.dona.net/handle-system

12. Handle.Net Software. http://www.handle.net/download.html

13. Handle.Net Software Javadoc. http://handle.net/hnr-source/api-javadoc-v9.2.0/index.html

14. What is REST. https://restfulapi.net

Potential Applications of Smart Contract Technology in Corporate Business Processes

O. V. Boychenko and I. V. Gavrikov(✉)

Institute for Economics and Management, Crimean Federal University,
Simferopol, Russia
bolek61@mail.ru, painttool@gmail.com

Abstract. There is more to blockchain technology than just cryptocurrencies – its more interesting and promising applications include powering so-called "smart contracts". Smart contracts are special programs running on the blockchain that define relations (usually monetary) between parties and enforce them without needing to involve a third party. However, designing smart contracts can be difficult due to a dearth of experienced specialists and tools that enable their rapid development. In this paper, a prototype of one such tool is proposed and described – a tool designed to enable the development of smart contracts by non-experts in the field.

Keywords: Smart contracts · Digital economy · Blockchain

1 Introduction

Today the world is entering a new era – the era of digital economy, where information is the principal resource and product. Businesses that aim to thrive in this economy must leverage new technology and stay on the cutting edge to remain competitive – this means constantly evolving their IT infrastructure by integrating new IT developments. There is a number of new IT phenomena that aim to disrupt the global economy, but few have generated as much discourse as the *blockchain*, and one intriguing application of blockchain technology is the creation and use of so-called smart contracts.

An analysis of related work shows that a significant portion of the research done on blockchain technology has concentrated on its applications for cryptocurrencies, especially Bitcoin, and the implications of their existence and use.

For instance, [19] provides an overview of the state of Bitcoin – in technical, legal and other aspects. Similarly but more narrowly, [4] performs an in-depth technical analysis of Bitcoin's distributed network.

Research has also been done on applications of blockchain technology narrowly specific to its technical aspects, or to a certain field or use case. [12], for example, explores economic aspects of blockchain consensus protocols. [20] provides a model of business blockchain application coupled with Internet of Things

© Springer Nature Switzerland AG 2019
V. M. Vishnevskiy et al. (Eds.): DCCN 2019, CCIS 1141, pp. 612–624, 2019.
https://doi.org/10.1007/978-3-030-36625-4_49

technologies, and [11] provides a case study of applying blockchain to knowledge sharing in industry. A general examination of the applications of blockchain for business, and the advantages and issues associated with its use, has been done in [21].

Regarding smart contracts themselves, there have been many demonstrations and case studies of applying smart contracts to a specific problem, or a single field or industry – for example, [7] describes the application of smart contracts to energy exchange through auctions, and [8] presents a design and a use case for smart contracts in real estate.

In spite of all this, there have been few systematic explorations of applying blockchain and smart contracts to business processes. [3] and [16] have studied smart contract development methodology through the prism of security, however their proposals are the further development of such a methodology. [5] has explored semi-automated creation of smart contracts by means of a domain-specific language, relatively human-readable so as to facilitate the translation of real-world contracts into smart contracts. However, visual approaches to smart contract creation remain scarce; further study in this area is warranted.

2 Blockchain: Theoretical Principles

A blockchain is essentially a distributed system of storing and exchanging information. Its principal distinguishing feature is how it stores data – the system has no trust mechanism, and any participant can add information to it. Blocks of information then become an integral part of the system, being included in an unbroken chain from the first block to the latest (hence the name of the technology). New transactions are verified using a "consensus algorithm", which is an algorithm for solving the Byzantine fault problem, also formulated as the "Byzantine generals problem". The theoretical basis for the problem (lemmas, theorems and detailed descriptions of special cases) are available in [10]. A review of a solution for a special case of the problem follows below.

Let there be four "generals", who are passing information to each other about the numbers of their armies. One of the generals (general #4) is a traitor ($n = 4, m = 1$).

1. Each general sends a message to the others that contains the number of troops in his army. "Loyal" generals (i.e. ones who are not traitors) send the true number of troops in thousands, whereas traitors send random numbers. Thus, the first general sends 1, the second 2, the third 3, and the fourth (the traitor) sends x, y, z to the first, second and third generals, respectively.
2. Each general forms a vector from the data he received. As a result, there are four vectors:

$$v_1 = (1, 2, 3, x)$$
$$v_2 = (1, 2, 3, y)$$
$$v_3 = (1, 2, 3, z)$$
$$v_4 = (1, 2, 3, 4)$$

3. Loyal generals send their vectors to the others. The traitor general sends a vector with random values. Table 1 contains the final results of this data exchange, where each cell represents the data received by the general represented by the row from the general represented by the column.

Table 1. Data after exchange

Recipient	General 1	General 2	General 3	General 4
General 1	$(\mathbf{1,2,3,x})$	$(1,2,3,y)$	$(1,2,3,z)$	(a,b,c,d)
General 2	$(1,2,3,x)$	$(\mathbf{1,2,3,y})$	$(1,2,3,z)$	(e,f,g,h)
General 3	$(1,2,3,x)$	$(1,2,3,y)$	$(\mathbf{1,2,3,z})$	(i,j,k,l)
General 4	$(1,2,3,x)$	$(1,2,3,y)$	$(1,2,3,z)$	$(\mathbf{1,2,3,4})$

4. Each general determines the sizes of the other armies for himself. To determine the size of army i, each general takes three values – the size of the army received from every general except the commander of army i itself. According to [10], consensus in a system may only be reached with $2m + 1$ loyal generals – this means that in this case, if a value is repeated at least twice in the set of three values, it will be placed in the resulting vector as the size of army i; otherwise the respective element of the vector is marked as unknown. Consequently, the random values $(a, b, \ldots, l)$ are discarded, as they only ever appear once, and the loyal generals have a vector of the form $(1, 2, 3, f(x, y, z))$, where $f(x, y, z)$ may be either a number repeated twice or an unknown. Since the values x, y, z and the function f are the same for all loyal generals, consensus is reached.

Nakamoto's implementation of the Bitcoin blockchain formulated a more general solution for this problem, where the number of "generals" (participating nodes) is unlimited and variable, which allowed for decentralized confirmation of new transactions in the blockchain. This solution is the so-called "proof-of-work" (PoW) consensus algorithm, where participating nodes compete in solving a cryptography problem, which, when solved, automatically confirms the block, and where the longest chain of blocks is considered to be the "authoritative" one.

Such a system is guaranteed to work correctly when more than half of its computational power is distributed among "loyal" participants (an improvement over Lamport's two-thirds as demonstrated in [10]). Nakamoto in [13] provides a Poisson distribution to determine the probability of an attacker catching up to the "honest" chain and overtaking it. For a given p equal to the probability of an honest node discovering the next block, q equal to the probability of the attacker discovering it, and z equal to the number of blocks in the honest chain after the initial block selected by the attacker as the origin of their malicious chain, the expected value of the Poisson distribution λ becomes

$$\lambda = z\frac{q}{p} \tag{1}$$

Therefore, the probability P of the attacker catching up to the chain may be determined as follows:

$$P = \sum_{k=0}^{\infty} \frac{\lambda^k e^{-\lambda}}{k!} \cdot \left\{ \begin{matrix} (\frac{q}{p})^{z-k} & k < z \\ 1 & k > z \end{matrix} \right\} \tag{2}$$

where k is the number of blocks in the chain after block z.

Nakamoto's computations in [13] show that as z increases, the probability P decreases exponentially.

3 Smart Contracts

3.1 Definition

Although the blockchain is most famous as the technology backing crypto-currencies, being a system developed around requirements for decentralization and trustless communications allows it to find varied applications outside of finance [3].

1. Ensuring data integrity. Although the blockchain is not the most optimal solution for data storage as such, it has potential as a component of a data security solution that emphasises transparency of data access. Combined with a decentralized data storage system, this allows to securely and transparently store data of varied formats.
2. Reducing overhead costs. Traditionally, record keeping systems use a special paper document as proof of participation (ID card, certificate, etc.), and any changes to stored data also require paper confirmation. Expenses related to these documents (printing, storage and destruction) may be eliminated completely by implementing blockchain technology without a negative impact on data security and integrity.
3. Another promising application of blockchain technology today is its use for "smart contracts". A smart contract is an algorithm designed to automate the process of fulfilling contractual obligations. The body of a smart contract contains conditions and actions executed as a consequence of meeting said conditions. Smart contracts do not require intermediaries or third parties by design, as all conditions and actions are checked and effected automatically.

This definition of a smart contract was first given by Szabo in [18]. According to Szabo's definition, there are four goals that must be met in the process of developing a smart contract:

1. *Observability.* Contract principals must have a means of observing each other's performance of the contract and proving their performance to each other.
2. *Verifiability.* Contract principals must have a means of proving a breach of contract.
3. *Privity.* Information about the contents and performance of a contract must be distributed among parties only inasmuch as is necessary for the performance of the contract.
4. *Enforceability.* Contract principals must have a means of enforcing contract performance by other parties.

Within the context of today's blockchain development, smart contracts are usually implemented as general purpose computations performed on the blockchain. Because of this, current implementations of smart contract protocols mean that smart contracts today are less contracts in the traditional sense and more computer programs developed with the purpose of regulating relations between parties [1,2]. As such, smart contracts are subject to many of the same concerns as software development in general, compounded by concerns unique to, or at least more prominent in, smart contract platforms.

3.2   Corporate Smart Contracts

The main benefit of smart contracts in corporate business processes is the streamlining of certain processes through removing the human component from them. Although not as well-suited for complex contracts and corporate processes at the current stage of development, the codification of some relatively simple contractual relations as smart contracts has the potential to eliminate many errors caused by the human component – therefore, as long as parties to a contract agree to use a single platform or a set of compatible tools, contractual relations codified as smart contracts become in effect self-enforcing, removing the need for human employees for each party (and potentially one or several third-parties facilitating the operation) checking and performing actions as part of the contractual relationship.

The codification of contractual relations as smart contracts also results in the added benefit of transparency, as automation of business processes eliminates many avenues of corruption, and the transparency of a public blockchain and smart contract platform such as Ethereum enables any interested parties to audit any transactions on the part of a smart contract.

The downside to the use of code for enforcing contractual obligations is the complexity of its development and implementation.

Development of smart contracts requires qualified specialists, which translates to significant expenses on the part of the company that wants to integrate smart contracts into its business processes. Blockchain technology and smart contracts both require specific knowledge and skills, which most IT specialists on the market do not yet possess. Because of this, companies have to make a choice between hiring blockchain specialists, purchasing freelance services, or paying for training programs for employees already with the company.

Each of these approaches has their own disadvantages:

- hiring a specialist requires expending money and time for the on boarding process and later for their labor;
- purchasing freelance services does not guarantee their reliability and quality, especially when multiple freelance specialists are involved;
- training employees also requires significant expenses, as well as finding reputable and high-quality providers of educational programs.

Additionally, code quality concerns are especially pertinent for smart contracts. The reason is two-fold. Firstly, smart contracts may not be altered or patched after being published on the blockchain – the altered smart contract would have to be republished under a new address, and the old contract retired or turned into a delegate contract for the new version, which requires additional development work. This is compounded by the fact that the smart contract ecosystem is still relatively young and developing – patterns, libraries, utilities, and best practices are not yet fully established and tested, which means many components of smart contracts are routinely "reinvented".

There are differing definitions for the term "code quality". Hereafter, code quality is defined as a combination of security (lack of vulnerabilities), maintainability (amount of time and labour required to fix or continue evolving the code), correctness (lack of erroneous results for given inputs), and efficiency (computational resources expended to achieve a result).

Blockchain security concerns were well-illustrated by the DAO hack; an exploitation of a vulnerability in the code of the DAO (Decentralised Autonomous Organisation), a set of programs on the Ethereum smart contract platform designed as a form of capital venture fund without central control or management. In June 2016, a bug in the smart contract allowed a hacker (or a hacker group) to siphon some 70 million USD in Ethereum currency to their own account, and the sheer scale of the attack prompted a "hard fork" (a bifurcation of the blockchain) of the entire Ethereum chain that reversed the effects of the hack. The reasons, specifics and consequences of "forking" in blockchain systems, as well as certain details of the DAO hack itself, are further examined in [9].

An additional illustrative example of the need for code quality assurance in smart contracts was the hacks of the Parity software. Parity is an Ethereum client designed for storing tokens on the Ethereum blockchain; one of the features it offers is multi-signature ("multisig") wallets – cryptocurrency storage requiring authentication by two or more private keys, as opposed to the standard single-key authentication. In July 2017, a flaw in the smart contract powering Parity's multisig wallets allowed hackers to steal approximately 30 million USD in Ethereum tokens [17]. In November 2017, another bug was used (apparently accidentally) to freeze over 100 million dollars in assets [15].

Security remains a problem even today - [14] shows that as of Q1 2018, 89% out of the sampled 3759 smart contracts on the Ethereum blockchain have exploitable flaws and vulnerabilities. The high incidence of flaws in smart contract code and low levels of standardisation across an industry already creating significant value and having the potential to create even more has repeatedly prompted calls for better code quality. [3] and [16], for example, call for the development of a discipline they call "blockchain-oriented software engineering", which would concentrate on raising smart contract code quality through the development and definition of best practices and smart contract development methodology, design patterns, and testing techniques and tools.

Promoting such a discipline and adopting a well-defined smart contract development methodology, coupled with choosing the right approach to managing personnel and expenses associated with smart contract development (as described above), would also address maintainability concerns, as code developed using standardised tools according to well-known best practices and established development processes within a department or team would be significantly easier to maintain and evolve along with a company's business processes.

Even if a smart contract has no immediately apparent vulnerabilities, smart contract code must provision for so-called "edge" and "corner cases" – actions handled by the code for which inputs are within expected parameters, but close to their limits. While provisioning for edge cases is an important component in software development and quality assurance regardless of field or application, it becomes especially important in smart contract code, which directly handles the storage and transfer of assets, and for which an extensive body of experience has not yet been accrued.

The specific requirements for smart contract efficiency differ between implementations of the smart contract platform. It is obvious that a business process codified as a smart contract should aim to be at least as efficient as, and ideally more efficient than, the same business process performed by humans. However, in an Ethereum context, a smart contract's efficiency directly correlates with the costs of its execution and maintenance – the Ethereum's concept of "gas" means that any action performed on the Ethereum blockchain leads to a certain amount of ether (Ethereum's currency) being charged to the transaction sender's account. More computationally expensive operations are therefore also more financially expensive. Furthermore, an inefficient contract that is also insecure may present unique vulnerabilities – for example, in Ethereum, a computation of a transaction that costs more than the transaction's pre-defined gas limit would be terminated by the system. If a contract recursively attempts to execute such a transaction (for example, in the case of a flawed loop in the code), the contract's assets would be perpetually frozen and unable to be accessed by any party. This kind of vulnerability caused the November 2017 Parity asset freeze described in [15]. [6] presents a more in-depth study of smart contract efficiency concerns and out-of-gas conditions in the Ethereum platform.

4   A Visual Approach to Building Smart Contracts

Considering the above specifics, it becomes more pertinent than ever to formulate a balanced approach to smart contract development, which would manage an equilibrium between avoiding significant expenses associated with educating existing personnel or hiring specialists as employees for smart contract development projects, and producing quality code protected from attacks.

As a solution to this problem, a system for smart contract development is proposed, powered by a visual interface. The system is a visual "constructor" of smart contracts, comprised of a client app and a serverside API. The client app allows a user to formulate conditions and actions within a smart contract by visually connecting functional elements with connections of various types. The structure the client app yields is then sent to the API, which processes it and translates it into blockchain-ready code.

The code produced by the system is built from blocks of code associated with the functional elements assembled by the user. The code blocks are designed specifically to maximise standardisation and security, by drawing from opensource experience and best practices, and to interlock with minimal potential for conflict.

The current prototype supports two principal classes of connections:

- *Attribute connections.* Components have attributes, which are used by the component to function, and exports, which are values returned by the component as it performs its function. Attribute connections allow the user to connect attributes to use exported values of other components.
- *Trigger connections.* Components that correspond to events or functions must have conditions for their execution or creation. Trigger connections allow the user to specify the order of procedure execution and event creation.

Attributes are strongly typed, where the most important are the `money` and `address` types – as an example, in Solidity, the language powering smart contracts on the Ethereum platform, these correspond to `uint256` and `address`, respectively.

The prototype supports three component categories:

- *Storage* components store values in contract state, usually as an array or map.
- *Event* components include external and internal events. External events occur "unconditionally" as far as the contract is concerned, i.e. they are triggered externally relative to the contract. Internal events occur after being triggered from within the contract, provided certain conditions are met.
- *Utility* components are required to bridge the gap between program code and full visual abstraction. The current prototype supports several internal utility components, which correspond to various programming patterns and functions.

The current version of the prototype contains several component classes available for use. Henceforth, any values returned as exports for the system's use are considered exported, and any values open for connections or input as attributes are considered accepted.

- *Payment.* Event component (external). Represents the transfer of funds to the smart contract address. Exports the payer's address and the paid amount.
- *Transfer.* Event component (internal). Represents the transfer of funds from the smart contract address to another address. Accepts recipient address and amount paid.
- *Request.* Event component (external). Represents externally calling a public function on the contract. Exports the caller's address. Component name is used as the function name.
- *Balance.* Storage component. Represents a map that naively stores money values under address keys. Designed to work with utility components, does not accept or export values.
- *Balance mutator.* Utility component. Unconditionally changes the value of an entry in the balance storage. Accepts an address for entry lookup and amount.
- *Balance accessor.* Utility component. Accepts address, exports stored balance for provided address as money.
- *Balance adder.* Utility component. Adds amount to balance. Accepts address and amount.
- *Balance subtractor.* Utility component. Unconditionally subtracts amount from balance. Accepts address and amount.
- *Validator.* Utility component. Validates whether a given value is within the given bounds. Accepts value and lower and upper bounds as money.

The system interface is split into three parts (see Fig. 1):

- The left pane contains all components that may be used to construct a smart contract.
- The middle zone, or "canvas", contains all components currently added to the smart contract and displays their connections.
- The right pane contains information about the component currently selected on the canvas.

After adding a component to the canvas, it becomes part of the smart contract. Right-clicking the component on the canvas opens a context menu which allows the user to delete the component, clone it, or connect it with another component on the canvas. When connecting two components, their properties become available to each other for attribute and trigger connections.

Selecting an element in the canvas displays its properties on the right pane (see Fig. 2). These are divided into three principal categories:

- General properties contain information about the element and certain shared properties that are independent of its type: name, connections, and triggers, if applicable.

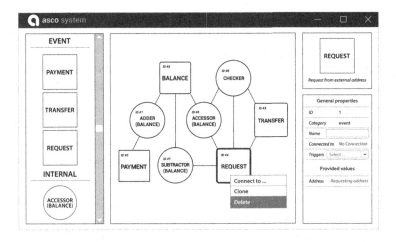

Fig. 1. Schematic of app UI

- Exported values (rendered in the info pane as "provided values") list properties that are available as sources for connections to this element.
- Object properties list properties that are available as receiving ends of connections to this element.

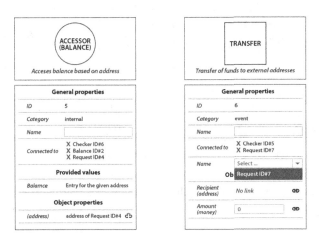

Fig. 2. Information pane for a selected element with various properties

Different possible values of an object property depend on its type. Properties of type `address` only support connections to exported values of other elements; properties of other types (e.g. `money`) support manual input.

Event objects may also have "triggers" – an event with a trigger will only be executed if the trigger condition is satisfied. For instance, the Request event is

622   O. V. Boychenko and I. V. Gavrikov

an external event that occurs when a request to the smart contract is sent from the outside. Thus, using it as a trigger for a Transfer action would execute the transfer only after the request arrives.

A naïve smart contract example is provided to demonstrate the process of creating a smart contract using the constructor. The example contract is a simple "piggy bank" algorithm:

- Each user has an account in the piggy bank associated with their address.
- The user transfers tokens to their account in the piggy bank.
- At any moment, the user may withdraw all tokens stored in their piggy bank account.

Figure 3 displays a schematic representation of the smart contract described above, using elements available in the constructor to build it. The displayed smart contract contains two external events: Payment and Request.

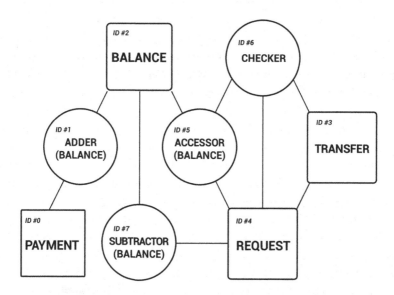

Fig. 3. Schematic representation of the algorithm, built using elements from the constructor

The Payment event adds funds to the payer's account balance in the contract. To achieve this, the Adder element adds the funds transferred via the Payment event to the account associated with the payer in the Balance element – the Balance element handles safely creating and using separate balances for addresses in the contract.

The Request event transfers all stored funds to the address requesting withdrawal. To do this, the Request event acts as a trigger for the Transfer internal event. A key element of this functionality is the Accessor element, which returns funds stored under an address. Here, the address used for the Accessor is the address of the party sending a request to the smart contract, and the funds returned by the accessor are used by the Subtractor element to change the associated balance.

5   Conclusion

The blockchain has made possible the creation of various decentralised and automated systems very closely intertwined with economics on a scale never seen before. While cryptocurrencies in and of themselves are an oft-cited result of the development of blockchain technology (to the point of often being conflated with the blockchain itself), an even more interesting and potentially promising consequence of the blockchain's emergence is smart contract technology.

Smart contracts are a promising technology with a number of advantages and intriguing applications. However, a significant obstacle to their widespread integration in corporate business processes is the difficulty of developing them.

The difficulty lies in the complexity of balancing costs of smart contract development, which arise from hiring specialists or purchasing consulting services for a smart contract development project, and the quality of smart contract code produced as a result. Smart contracts, while being an attractive technology for creating value, must be very carefully designed and implemented, since flaws and vulnerabilities may result in disastrous consequences, as experience has shown.

In this paper, a prototype solution was designed for simple creation of smart contracts without special programming skills – one based on visual schematic building using pre-defined functional components and connections between them. In the future, the solution may be improved by extending its functionality (e.g. by adding import and export functionality to store created schematics), as well as by expanding the variety of pre-defined components for building smart contracts and creating templates for quick creation of common smart contracts and smart contract patterns, thereby enabling more widespread adoption of smart contracts in business.

References

1. Buterin, V.: Ethereum: a next-generation smart contract and decentralized application platform (2013). https://github.com/ethereum/wiki/wiki/White-Paper
2. Cachin, C.: Architecture of the hyperledger blockchain fabric. In: Workshop on Distributed Cryptocurrencies and Consensus Ledgers, vol. 310, p. 4 (2016)
3. Destefanis, G., Marchesi, M., Ortu, M., Tonelli, R., Bracciali, A., Hierons, R.: Smart contracts vulnerabilities: a call for blockchain software engineering? In: 2018 International Workshop on Blockchain Oriented Software Engineering (IWBOSE), pp. 19–25. IEEE (2018)
4. Feld, S., Schönfeld, M., Werner, M.: Analyzing the deployment of bitcoin's p2p network under an as-level perspective. Proc. Comput. Sci. **32**, 1121–1126 (2014)

5. Frantz, C.K., Nowostawski, M.: From institutions to code: towards automated generation of smart contracts. In: 2016 IEEE 1st International Workshops on Foundations and Applications of Self* Systems (FAS* W), pp. 210–215. IEEE (2016)
6. Grech, N., Kong, M., Jurisevic, A., Brent, L., Scholz, B., Smaragdakis, Y.: MadMax: surviving out-of-gas conditions in ethereum smart contracts. Proc. ACM Program. Lang. 2(OOPSLA), 116:1–116:27 (2018). https://doi.org/10.1145/3276486
7. Hahn, A., Singh, R., Liu, C.C., Chen, S.: Smart contract-based campus demonstration of decentralized transactive energy auctions. In: 2017 IEEE Power and Energy Society Innovative Smart Grid Technologies Conference (ISGT), pp. 1–5. IEEE (2017)
8. Karamitsos, I., Papadaki, M., Al Barghuthi, N.B.: Design of the blockchain smart contract: a use case for real estate. J. Inf. Secur. 9(03), 177 (2018)
9. Kiffer, L., Levin, D., Mislove, A.: Stick a fork in it: analyzing the Ethereum network partition. In: Proceedings of the 16th ACM Workshop on Hot Topics in Networks, HotNets-XVI, pp. 94–100. ACM, New York (2017). https://doi.org/10.1145/3152434.3152449
10. Lamport, L., Shostak, R., Pease, M.: The byzantine generals problem. ACM Trans. Program. Lang. Syst. (TOPLAS) 4(3), 382–401 (1982)
11. Li, Z., Liu, L., Barenji, A.V., Wang, W.: Cloud-based manufacturing blockchain: secure knowledge sharing for injection mould redesign. Proc. CIRP 72, 961–966 (2018)
12. Luu, L., Teutsch, J., Kulkarni, R., Saxena, P.: Demystifying incentives in the consensus computer. In: Proceedings of the 22nd ACM SIGSAC Conference on Computer and Communications Security, pp. 706–719. ACM (2015)
13. Nakamoto, S., et al.: Bitcoin: a peer-to-peer electronic cash system (2008)
14. Nikolić, I., Kolluri, A., Sergey, I., Saxena, P., Hobor, A.: Finding the greedy, prodigal, and suicidal contracts at scale. In: Proceedings of the 34th Annual Computer Security Applications Conference, pp. 653–663. ACM (2018)
15. Palladino, S.: The $280m ethereum's parity bug (2017). https://blog.comae.io/the-280m-ethereums-bug-f28e5de43513. Accessed 09 Sept 2019
16. Porru, S., Pinna, A., Marchesi, M., Tonelli, R.: Blockchain-oriented software engineering: challenges and new directions. In: 2017 IEEE/ACM 39th International Conference on Software Engineering Companion (ICSE-C), pp. 169–171. IEEE (2017)
17. Suiche, M.: The parity wallet hack explained (2017). https://blog.openzeppelin.com/on-the-parity-wallet-multisig-hack-405a8c12e8f7/. Accessed: 09 Sept 2019
18. Szabo, N.: Formalizing and securing relationships on public networks. First Monday 2(9) (1997)
19. Tsukerman, M.: The block is hot: a survey of the state of bitcoin regulation and suggestions for the future. Berkeley Technol. Law J. 30(4), 1127–1170 (2015)
20. Zhang, Y., Wen, J.: An IoT electric business model based on the protocol of bitcoin. In: 2015 18th International Conference on Intelligence in Next Generation Networks, pp. 184–191. IEEE (2015)
21. Zhao, J.L., Fan, S., Yan, J.: Overview of business innovations and research opportunities in blockchain and introduction to the special issue (2016)

Heterogeneous Network Security Effective Monitoring Method

A. O. Kalashnikov$^{(\boxtimes)}$ and E. V. Anikina$^{(\boxtimes)}$

V.A. Trapeznikov Institute of Control Sciences Russian Academy of Sciences,
Moscow, Russia
aokalash@ipu.ru, janet0584@mail.ru

Abstract. This paper considers one of the methods of efficient allocation of limited resources in special-purpose devices (sensors) to monitor
heterogeneous network unit cybersecurity.

Keywords: Heterogeneous network · Security monitoring · Security
sensor · Resource allocation · Bipartite graph

1 Introduction

A more rapid transition to digital economy is characterized by intense implementation and use of cybertechnologies in the areas of economy and finances, industry and energy, transport and communication, federal and municipal administration, defense and security, science and culture, education and medical care and
many others. Such extensive use of information technologies is impossible without security. First of all, it is related to the issues of cybersecurity management
of Russian critical IT infrastructure (hereinafter 'RCITI') facilities and RCITI in
general. The fact that the importance of this problem is clearly realized, including at the level of the President and the Government of the Russian Federation,
is confirmed by Federal Law dated 26.07.2017 No 187-FZ On Security of the
Critical IT Infrastructure of the Russian Federation [1]. In accordance with [1]
(Article 7), categorization of RCITI facilities includes setting up compliance of
an RCITI facility with significance criteria, assigning one of three significance
categories and checking data on the results of such assigning.

In order to successfully implement security measures for RCITI facilities and
RCITI in general, it is required to solve a whole number of complex research
and engineering problems with RCITI cybersecurity monitoring, including using
special-purpose devices, being one of the key problems.

It is the solution of the problem of efficient allocation of limited resources in
special-purpose devices (sensors) to monitor complex network, such as RCITI or
internet, unit cybersecurity that this paper deals with.

It should be noted that for the first time a similarly stated problem was considered in [2] (see detailed analysis and reference list there), where the problem of
computer system limited resource allocation represented by a certain number of
various devices at a variety of users belonging to various classes, was considered.
A key variance of the problem considered herein from the problem presented in

© Springer Nature Switzerland AG 2019
V. M. Vishnevskiy et al. (Eds.): DCCN 2019, CCIS 1141, pp. 625–635, 2019.
https://doi.org/10.1007/978-3-030-36625-4_50

[2] is the transition from considering a variety of unique devices to considering a variety of groups of similar devices which is a substantial generalization and complication of the original problem. Nevertheless, certain results obtained in [2] are presented in this paper.

Let's consider a formal setting of the problem.

Supposedly, there is a complex network consisting of units belonging to various classes $\mathcal{K} = \{1, \ldots, K\}$. Examples of such networks are a total of networks that are constituent elements of RCITI or internet. Let us denote a number of network units belonging to $k \in \mathcal{K}$ class as X_k and a total number of network units as $X = \sum_{k=1}^{K} X_k$.

Further, supposedly, there is a certain number of special-purpose devices – sensors – for monitoring network unit cybersecurity (hereinafter 'security sensor'), also belonging to various $\mathcal{M} = \{1, \ldots, M\}$ classes. Let us denote a number of sensors belonging to $m \in \mathcal{M}$ class as Y_m and a total number of sensors as $Y = \sum_{m=1}^{M} Y_m$. We shall assume that security sensors of various classes have, in general, various efficiency of cybersecurity monitoring in respect of network units, also belonging to various classes.

Let's assume that the efficiency (utility value) of an $m \in \mathcal{M}$ security sensor is a function of a number of units that are monitored by the above sensor: $\sigma_m(x), m \in \mathcal{M}, x \in \mathbb{N}_0 = \{0, 1, 2, \ldots\}$. Within the framework of this paper, for the sake of simplicity, we shall assume that $\sigma_m(x)$ is a concave function that doesn't depend on specific unit classes but is defined by a total number of network units which are monitored by this sensor.

Let us denote a maximum number of units of $k \in \mathcal{K}$ class that can be monitored by a security sensor of $m \in \mathcal{M}$ class as $N_{m,k} \geq 0$ and a total maximum number of units, which can be monitored by a security sensor of $m \in \mathcal{M}$ class as $N_m \geq 0$. This constraint is quite natural as, on one side, within a residual period of time, any security sensor can check only a limited number of units and, on the other side, it may turn out that an $m \in \mathcal{M}$ class security sensor cannot be used for monitoring $k \in \mathcal{K}$ class units.

Thus, our task is to find such an allocation of network units across security sensors that will maximize the total efficiency (utility value) of all the devices as long as the above constraints are satisfied.

Supposedly, $x_{m,k}^i$ - is a total number of units of $k \in \mathcal{K}$ class which are monitored by i-th sensor belonging to $m \in \mathcal{M}$ class, $i \in \{1, 2, \ldots, Y_m\}$. Let us denote: a total number of units belonging to variety of groups as $x_m^i = \sum_{k=1}^{K} x_{m,k}^i$, which are monitored by i-th sensor belonging to $m \in \mathcal{M}$ class, a total number of units belonging to variety of groups as $x_m = \sum_{i=1}^{Y_m} x_m^i$, which are monitored by sensors belonging to $m \in \mathcal{M}$ class, and a total number of units belonging to $k \in \mathcal{K}$ as $z_k = \sum_{m=1}^{M} \sum_{i=1}^{Y_m} x_{m,k}^i$ which are monitored by sensors all of classes. Then our problem can to formally written as:

$$\sum_{m=1}^{M} \sum_{i=1}^{Y_M} \sigma_m(x_{m,k}^i) \to max \text{ (Task)} \tag{1}$$

with the following constraints:
$x_{m,k}^i \in \{0, 1, \ldots, N_{m,k}\},$

$x_m^i \leq N_m,$
$z_k = X_k,$
$k \in \mathcal{K}, m \in \mathcal{M}, i \in \{1, 2, \ldots, Y_m\}.$

2 Algorithm Description

Let us consider, by analogy with [3], a sequence of intermediate optimization problems of the following form:

$$\sum_{m=1}^{M} \sum_{i=1}^{Y_M} \sigma_m(x_{m,k}^i) \to max \ (\text{Task}(j)) \tag{2}$$

with the following constraints:
$\sum_{m=1}^{M} x_m(j) = j,$
$x_{m,k}^i(j) \in \{0, 1, \ldots, N_{m,k}\},$
$x_m^i(j) \leq N_m,$
$z_k(j) \leq X_k,$
$k \in \mathcal{K}, m \in \mathcal{M}, i \in \{1, 2, \ldots, Y_m\},$
　　where:
$x_m^i(j) = \sum_{k=1}^{K} x_{m,k}^i(j),$
$x_m(j) = \sum_{i=1}^{Y_m} x_m^i(j),$
$z_k(j) = \sum_{m=1}^{M} \sum_{i=1}^{Y_m} x_{m,k}^i(j).$
The problem Task handling algorithm will be continuous handling of Tasks (j) at $j = 1, 2, \ldots, X$. Obviously, the problem Task (X) will be identical to the original Task.

　　Let us call $\overline{X} = [x_{m,k}^i]$ the assignment for the Task problem, if the constraints are fulfilled:
$x_{m,k}^i \in \{0, 1, \ldots, N_{m,k}\},$
$z_k \leq X_k,$
$k \in \mathcal{K}, m \in \mathcal{M}, i \in \{1, 2, \ldots, Y_m\}.$
The allocation $\overline{X} = [x_{m,k}^i]$ is called an admissible assignment for a problem Task, if and only if $\overline{X} = [x_{m,k}^i]$ is an assignment for a problem Task and the constraints are fulfilled: $x_m^i \leq N_m, m \in \mathcal{M}.$
Let us denote $\overline{X} = [x_{m,k}^i]$ is a partially admissible assignment, if not all network units are distributed by security sensors. We will call a completely admissible assignment as $\overline{X} = [x_{m,k}^i]$ if all the constraints of the problem Task are fully satisfied.

　　Consider a partially admissible assignment $\overline{X} = [x_{m,k}^i]$ for which we define the non-empty set generated by it not of yet-distributed network units $\overline{X}_0(\overline{X})$. We will introduce a "dummy" class of security sensors with the label "0" which consist of one element (that is, $m = 0, i = 1$) and allocate all network units belonging to the set $\overline{X}_0(\overline{X})$ to it. Obviously, $\overline{X}_0(\overline{X}) = [x_{0,k}^1]$ is an assignment for which the constraints are satisfied:
$x_{0,k}^1 = X_k - z_k \leq X_k,$

$x_0^1 = \sum_{k=1}^{K} x_{0,k}^1$,
$k \in \mathcal{K}$.

Let us denote the set of all security sensors as $S = \{s_m^i\}, m \in \mathcal{M} = \{1, \ldots, M\}, i \in \{1, 2, \ldots, Y_m\}$ and "dummy" sensor as s_0^1.

Definition 1. *Supposedly $\overline{X} = [x_{m,k}^i]$ is a admissible assignment, then $G(\overline{X})$ is a admissible bipartite graph with variety of vertices $V = \{s_0^1\} \cup S$ and a variety of edges $E = \{(e_1, e_2)\}, if : e_1 \in V, e_2 \in S$ and there is at least one class $k \in \mathcal{K} = \{1, \ldots, K\}$ such that $x_{e_1,k}^i > 0$ and $x_{e_2,k}^i < N_{e_2,k}$.*

Admissible bipartite graph $G(\overline{X})$, implements all potential allocations of each network unit across security which are allowed by admissible assignment $\overline{X} = [x_{m,k}^i]$.

The following should be noted that the graph $G(\overline{X})$ has no edges leading to the "dummy" sensor s_0^1 and if there is a finite sequence of edges $(e_1, e_{i_1}), \ldots, (e_{i_n}, e_2)$ in the graph $G(\overline{X})$ (in other words, there is a path from vertex e_1 to vertex e_2 in the graph $G(\overline{X})$), then it is possible to generate a new assignment $\widetilde{X} = [\widetilde{x}_{m,k}^i]$, which differs from $\overline{X} = [x_{m,k}^i]$ in the number of network nodes distributed between e_1 and e_2. The new assignment has exactly one node less on sensor e_1 and exactly one node more on sensor e_2. If $\widetilde{x}_{e_2}^i < N_{e_2}$, then the new assignment will be admissible.

Let us denote a path in the graph $G(\overline{X})$ as $= (e_1, e_2, \ldots, e_n)$, if $(e_i, e_{i+1}) \in E$ for all of $i \in \{1, 2, \ldots, n-1\}$.

Definition 2. *Supposedly $\overline{X} = [x_{m,k}^i]$ - is a admissible assignment for a problem Task and $L = (e_1, e_2, \ldots, e_n)$ a path without cycles in the graph $G(\overline{X})$, then we will define $T(\overline{X}, L)$ as: $T(\overline{X}, L) = T(\overline{X}, (e_1, \ldots, e_n)) = T(T(\overline{X}, (e_1, e_2)), (e_2, \ldots, e_n)), n > 2$ where assignment $\widetilde{X} = T(\overline{X}, (e_1, e_2))$ have the form:*

$$\widetilde{x}_{m,k}^i == \begin{cases} x_{e_2,k}^i + 1, & m = e_2; \\ & k = \min\left\{j | x_{e_1,j}^i > 0 \text{ and } x_{e_2,j}^i < N_{e_2,j}\right\}; \\ x_{e_2,j}^i - 1, & m = e_1; \\ & k = \min\left\{j | x_{e_1,j}^i > 0 \text{ and } x_{e_2,j}^i < N_{e_2,j}\right\}; \\ x_{m,k}^i, & else. \end{cases} \quad (3)$$

It should be noted that if $\overline{X} = [x_{m,k}^i]$ – is a admissible assignment and $\widetilde{x}_{e_2}^i < N_{e_2}$, then the new assignment $T(\overline{X}, L)$ will be admissible.

Let us denote $\Delta_m(x) = \sigma_m(x) - \sigma_m(x-1), m \in \mathcal{M}, x \in \{1, 2, \ldots\}$ – is the increment of efficiency (utility value), due to the increase in the number of network nodes per unit distributed to the $m \in \mathcal{M}$ class safety sensor.

Let us now proceed to the description of the algorithm for solving the problem Task (j), provided that the optimal assignment $\overline{X}(j-1) = [x_{m,k}^i(j-1)]$, for the Task(j-1) has already been found.

Algorithm 0.

0. Supposedly $S_{Task} := \{s_m^i \in S \mid x_m^i < N_m\}$;

1. Search $s_m^i \in S_{Task}$, such that $\Delta_m(x_m^i(j-1)+1)$ maximally.

If there is a path without cycles $L = (s_0^1, \ldots, s_m^i)$ in the graph $G(\overline{X}(j-1))$ from vertex s_0^1 to vertex s_m^i,

then $\overline{X}(j) := T(\overline{X}(j-1), L)$,

else $S_{Task} := S_{Task} - \{s_m^i\}$;

if $S_{Task} = \{\varnothing\}$, then end,

else go to step 1.

Recall that the general algorithm for solving the problem Task is a continuous handling of the problems Tasks (j) at $j = 1, 2, \ldots, X$ using the Algorithm 0. Now let's go to the description of the General algorithm for handling the problem Task.

Algorithm 1.

0. Supposedly $S_{Task} := S$;

1. For k := 1 to K do:

$x_{0,k}^1 := X_k$

For m := 1 to M do:

For i := 1 to Y_m do:

$x_{m,k}^i(0) := 0$;

2. Initialize graph $G(\overline{X}(0))$;

3. For j := 1 to X do:

4. Search $s_m^i \in S_{Task}$, such that $\Delta_m(x_m^i(j-1)+1)$ maximally.

If there is a path without cycles $L = (s_0^1, \ldots, s_m^i)$ in the graph $G(\overline{X}(j-1))$ from vertex s_0^1 to vertex s_m^i,

then $\overline{X}(j) := T(\overline{X}(j-1), L)$,

else $S_{Task} := S_{Task} - \{s_m^i\}$;

if $S_{Task} := \{\varnothing\}$, then end,

else go to step 4.

The above-defined function $T(\overline{X}, L)$ is an additive function that allows an efficient modification of the graph $G(\overline{X})$ by translating it into a graph $G(T(\overline{X}, L))$.

It should be noted that Algorithm 1 goes into a "greedy" algorithm in the absence of any restrictions on the assignment of network nodes to security sensors (see, for example, [4,5]).

Also, it should be noted that the correctness of the algorithms described above is significantly determined by the fact that if some security sensor $s_m^i, m \in \mathcal{M} = \{1, \ldots, M\}, i \in \{1, 2, \ldots, Y_m\}$ was removed from the S_{Task} set when solving the problem Task (j), then the specified sensor will also not be used to solve the problems Task (j + 1), Task (j + 2) and so on further, up to Task (X). Let us give are the main points of proof this fact. The detailed proof of this fact is presented in [3].

3   The Proof of the Correctness Algorithm

In this section, we will show the Algorithm 0 works correctly. Then, based on the proven, we show that if some security sensor is not be used in step j, then it will not be used in the future.

Let us prove first three lemmas, which are generalized analogues of Lemmas 1–3, is presented in [3].

Lemma 1. *Let set:*

- *a variety of network units belonging to various classes* $\mathcal{K} = \{1, \dots, K\}$ *where* X_k *- a number of network units belonging to* $k \in \mathcal{K}$ *class and* $X = \sum_{k=1}^{K} X_k$ *- a total number of network units;*
- *a variety of sensors* $S = \{s_m^i\}, m \in \mathcal{M}, i \in \{1, 2, \dots, Y_m\}$ *belonging to various* $\mathcal{M} = \{1, \dots, M\}$ *classes, where* Y_m *- a number of sensors belonging to* $m \in \mathcal{M}$ *class, and* $Y = \sum_{m=1}^{M} Y_m$ *- a total number of sensors;*
- s_0^1 *– "dummy" sensor;*
- $\overline{X} = [x_{m,k}^i]$ *and* $\overline{\widetilde{X}} = [\widetilde{x}_{m,k}^i]$ *– are a admissible assignments such that for some* $s_{m_0}^{i_0} \in S : \widetilde{x}_{m_0}^{i_0} = x_{m_0}^{i_0} + 1$ *and* $\widetilde{x}_m^i = x_m^i$ *in other cases.*

Then: there is a direct path without cycles $L = (s_0^1, \dots, s_{m_0}^{i_0})$ *in the graph* $G(\overline{X})$.

The proof:
Supposedly, all of network units not distributed across security sensor from the $S = \{s_m^i\}$ set are assigned on a "dummy" sensor s_0^1. Let us denote $z_k = \sum_{m=1}^{M} \sum_{i=1}^{Y_m} x_{m,k}^i$ and $z_k = \sum_{m=1}^{M} \sum_{i=1}^{Y_m} \widetilde{x}_{m,k}^i$. Let us construct an algorithm that allows us to calculate the path $L = (s_0^1, \dots, s_{m_0}^{i_0})$ in the graph $G(\overline{X})$.

Algorithm 2.

0. Initialize the path $L := ()$;

1. $L := L^\circ(s_0^1)$ *(add a "dummy" sensor* s_0^1 *to the path L):*
find $1 \leq k \leq K$ such that $\widetilde{z}_k > z_k$
find $s_{m_1}^{i_1} \in S$ such that $\widetilde{x}_{m_1,k}^{i_1} > x_{m_1,k}^{i_1}$
Then $x_{m_1,k}^{i_1} = x_{m_1,k}^{i_1} + 1, x_{0,k}^1 := x_{0,k}^1 - 1$;

2. $L := L^\circ(s_{m_1}^{i_1})$ *(add a sensor* $s_{m_1}^{i_1} \in S$ *to the path L):*
If $s_{m_1}^{i_1}$ matches $s_{m_0}^{i_0}$, then end,
find $1 \leq k \leq K$ such that $\widetilde{x}_{m_1,k}^{i_1} < x_{m_1,k}^{i_1}$,
Then $x_{m_1,k}^{i_1} = x_{m_1,k}^{i_1} - 1$
find $s_{m_2}^{i_2} \in S$ such that $\widetilde{x}_{m_2,k}^{i_2} > x_{m_2,k}^{i_2}$,
if such $s_{m_2}^{i_2} \in S$ does not exist,
then $x_{0,k}^1 := x_{0,k}^1 + 1$ and go to step 1,
else $x_{m_1,k}^{i_1} = x_{m_1,k}^{i_1} + 1$ and go to step 2.

The above algorithm has the following properties [3]:

1. A network unit is an assigned to the security sensor only once.

2. The number of assignments (transitions) of the presented algorithm is estimated from above as: $\sum_{s_m^i \in S \cup \{s_0^1\}} \sum_{k=1}^{K} | x_{m,k}^i - \widetilde{x}_{m,k}^i | < \infty$.

Thus, this algorithm always terminates according by property 2 and thus generates a direct path L, which represents all possible assignments of network units to security sensors.

Path L may contain a cycle. It is possible to delete all cycles in the path L and get a new acyclic path $\widehat{L}(s_0^1, \ldots, s_{m_0}^{i_0})$, which has the desired properties applying the procedure described in [3]. Each pair (e_i, e_{i+1}), belonging to path $\widehat{L} = (s_0^1, \ldots, s_{m_0}^{i_0})$, is associated with the network unit that was moved from vertex e_i to vertex e_{i+1} in the original path. Due to property 1, this network unit assigned to the sensor corresponding to vertex e_i before the algorithm used. Hence, the edge (e_i, e_{i+1}), is represented in $G(X)$.

The proof is complete.

Lemma 2. *Let $\overline{X}(1) = [x_{m,k}^i(1)]$ and $\overline{X}(2) = [x_{m,k}^i(2)]$ are admissible assignments such that:*
$C_1 = \{s_m^i \in S : x_m^i(1) > x_m^i(2)\} \neq \varnothing;$
$C_2 = \{s_m^i \in S : x_m^i(2) > x_m^i(1)\} \neq \varnothing;$
then for anyone $s_{m_2}^{i_2} \in C_2$ exists $s_{m_1}^{i_1} \in C_1$ such that assignment $\overline{X}(3) = [x_{m,k}^i(3)]$, such that:

$$x_m^i(3) == \begin{cases} x_m^i(3) = x_m^i(1), \text{ if } i \neq i_1, i_2 \text{ and } m \neq m_1, m_2; \\ x_{m_1}^{i_1}(3) = x_{m_1}^{i_1}(1) - 1; \\ x_{m_2}^{i_2}(3) = x_{m_2}^{i_2}(1) + 1. \end{cases}$$

is a admissible.

The proof:
Let us construct a new partially assignment $\overline{X}(4) = [x_{m,k}^i(4)]$ for the selected $s_{m_2}^{i_2} \in C_2$:

$$x_m^i(4) = \begin{cases} x_m^i(1), \text{ if } i = i_2 \text{ and } m = m_2; \\ x_m^i(1), \text{ if } s_m^i \notin C_1 \cup \{s_{m_2}^{i_2}\}; \\ x_m^i(2), \text{ if } s_m^i \in C_1. \end{cases}$$

The assignment $\overline{X}(4) = [x_{m,k}^i(4)]$ is also admissible.

Let us construct a new total assignment $\overline{X}(5) = [x_{m,k}^i(5)]$ from $\overline{X}(1) = [x_{m,k}^i(1)]$, by removing all units assigned to sensors belonging to the set C_1. This assignment is also admissible.

There is a path $\widehat{L} = (s_0^1, \ldots, s_{m_2}^{i_2})$ in the graph $G(\overline{X}(5))$, by Lemma 1, which can be used to determine the units that need to be reassigned to convert the assignment $\overline{X}(5)$ to $\overline{X}(4)$, that is $\overline{X}(4) = T(X(5), \widehat{L})$. Network units can be reassigned to $\overline{X}(1)$ by method which used for construct $\overline{X}(5)$.

Let us consider the first few steps to move units in this transformative sequence. The resulting assignment $T(\overline{X}(5), \widehat{L})$ is equivalent to $\overline{X}(3)$ and $i = i_1$ and $m = m_1$ if some unit is assigned to the sensor $s_m^i \in C_1$. The unit can be removed from the given set and assigned to the "dummy" sensor s_0^1 and $s_{m_1}^{i_1} \in C_1$, if the first unit in the sequence is assigned to the "dummy" sensor. In other words, we generate the assignment $\overline{X}(3)$ which we were looking for by reassigning a single unit from sensors $s_m^i \in C_1$ to a "dummy" sensor s_0^1 in assignment $T(\overline{X}(1), \widehat{L})$.

In this case $i = i_1$ and $m = m_1$. Therefore, $\overline{X}(3)$ is an admissible assignment. The proof is complete.

Let us introduce the following definition.

Definition 3. Supposedly $\overline{X}(1) = [x_{m,k}^i(1)]$ and $\overline{X}(2) = [x_{m,k}^i(2)]$ – are admissible assignments for a problem Task then:
$$\Delta(\overline{X}(1), \overline{X}(2)) = \sum_{s_m^i \in S} |\, x_m^i(1) - x_m^i(2) \,|.$$

Lemma 3. There are optimal assignments $\overline{X}(j) = [x_{m,k}^i(j)]$ and $\overline{X}(j-1) = [x_{m,k}^i(j-1)]$ for sequential problems Task (j) and Task (j-1), respectively, between which the difference is the number of network units assigned to a single security sensor. In other words, there exists $s_{m_0}^{i_0} \in S$ such that:

$$x_m^i(j) = \begin{cases} x_m^i(j-1), & \text{if } i \neq i_0, m \neq m_0; \\ x_m^i(j-1) + 1, & \text{if } i = i_0, m = m_0. \end{cases}$$

The proof: Let us hold the proof by contradiction. Let our statement is false. Let us define two optimal assignments $\overline{X}(j) = [x_{m,k}^i(j)]$ and $\overline{X}(j-1) = [x_{m,k}^i(j-1)]$ for sequential problems Task (j) and Task (j-1), respectively, which minimize: $\Delta(\overline{X}(j), \overline{X}(j-1)) = \sum_{s_m^i \in S} |\, x_m^i(j) - x_m^i(j-1) \,|.$
Due to our assumption the following must be executed: $\Delta(\overline{X}(j), \overline{X}(j-1)) > 1.$
Let us define the following three sets of security sensors:
$C_1 = \{s_m^i \in S : x_m^i(j) > x_m^i(j-1)\}; C_2 = \{s_m^i \in S : x_m^i(j) < x_m^i(j-1)\}; C_3 = \{s_m^i \in S : x_m^i(j) = x_m^i(j-1)\}.$
It should be noted that neither C_1 nor C_2 can be empty.
The first of all, let us show there exists $s_{m_2}^{i_2} \in C_2$, such that $\Delta_m(x_m^i(j)) < \Delta_{m_2}(x_{m_2}^i(j) + 1)$ for all $s_m^i \in C_1$. Let us select $s_{m_1}^{i_1} \in C_1$. Let us construct a new assignment $\widetilde{\overline{X}}(j-1) = [\widetilde{x}_{m,k}^i(j-1)]$ by starting with $\overline{X}(j)$ and removing the network unit previously assigned to the sensor other than $s_{m_1}^{i_1} \in C_1$. Let us remove the user from $s_{m_1}^{i_1} \in C_1$ if $C_1 = \{s_{m_1}^{i_1}\}$. The key note is that: $\widetilde{x}_{m_1}^{i_1}(j-1) > x_{m_1}^{i_1}(j-1)$. The above assignment is admissible but not optimal for problem Task (j-1), because otherwise it contradicted the statement that $\Delta(\overline{X}(j), \overline{X}(j-1))$ is minimal.

Let us apply Lemma 2 to assignments $\widetilde{\overline{X}}(j-1)$ and $\overline{X}(j)$ and obtain another admissible assignment $\widetilde{\widetilde{X}}(j-1)$, which differs from $\overline{X}(j)$ in that:
$\widetilde{\widetilde{x}}_{m_1}^{i_1}(j-1) = x_{m_1}^{i_1}(j-1) + 1$
and there exists such $s_{m_2}^{i_2} \in C_2$ that:
$\widetilde{\widetilde{x}}_{m_2}^{i_2}(j-1) = x_{m_2}^{i_2}(j-1) - 1$. The number of units assigned to all other sensors does not change. Again, this assignment $\widetilde{\widetilde{X}}(j-1)$ not to be optimal.

Therefore, for all $s_m^i \in C_1$ is true:
$\Delta_m(x_m^i(j)) \leq \Delta_m(x_m^i(j-1) + 1) < \Delta_{m_2}(x_{m_2}^{i_2}(j-1)) \leq \Delta_{m_2}(x_{m_2}^{i_2}(j) + 1).$

Let us apply Lemma 2 again to assignments $\overline{X}(j)$ and $\overline{X}(j-1)$ and obtain a new admissible assignment $\widetilde{\widetilde{X}}(j)$, such that: $\widetilde{\widetilde{x}}_{m_2}^{i_2}(j) = x_{m_2}^{i_2}(j) + 1$ and $\widetilde{\widetilde{x}}_{m_1}^{i_1}(j) = x_{m_1}^{i_1}(j) - 1$ for some $s_{m_1}^{i_1} \in C_1$.

But then the assignment of $\widetilde{\widetilde{X}}(j)$ must have greater total efficiency than $\overline{X}(j)$, which contradicts our assumption of the optimality of $\overline{X}(j)$.

Therefore, the optimal assignments $\overline{X}(j)$ and $\overline{X}(j-1)$ for the successive problems Task (j) and Task (j-1) differ only in the number of users on a single element $s_{m_0}^{i_0} \in S$.

The proof is complete.

Let us turn to the proof of the main result of this paper, which is a generalization of Theorem 1 in [3].

Statement 1. *The assignment $\overline{X}(j)$ is obtained as a result of applying Algorithm 0 for the problem Task (j) and is optimal for all $j = 1, 2, \ldots, X$.*

The proof:

Let us hold the proof out by induction on j.

Reference step.

Algorithm 0 generates optimal assignment $\overline{X}(1)$ in a trivial way for $j = 1$.

Induction step.

Let us assume the statement is true for all $j = 1, 2, \ldots i$, such that $i < X$ and let us prove it for $j = i + 1$.

The assignments $\overline{X}(j)$ and $\overline{X}(j-1)$, according by Lemma 3, differ only in the number of units on a single dedicated sensor, let us say $s_{m_0}^{i_0} \in S$.

There is at least one straight path without cycles $\widehat{L} = (s_0^1, \ldots, s_{m_0}^{i_0})$ in the graph $G(\overline{X}(j-1))$ from s_0^1 to $s_{m_0}^{i_0} \in S$ according by Lemma 1. Algorithm 0 uses one of these paths.

Therefore, Algorithm 0 generates an optimal assignment.

The proof is complete.

Let us show any network unit is assigned to the security sensor only once in the process of Algorithm 1.

Statement 2. *Supposedly, j_0 is a minimum number of the problem Task (j) for which some sensor $s_{m_0}^{i_0} \in S$ is selected and there is no path $\widehat{L} = (s_0^1, \ldots, s_{m_0}^{i_0})$ in the graph $G(\overline{X}(j_0 - 1))$ during the operation of Algorithm 1. Then, sensor $s_{m_0}^{i_0} \in S$ will not be used to assign network nodes to it for the rest of the algorithm operation time.*

The proof of this statement coincides with the proof of a similar Theorem 2 in [3], so let us omit it.

Thus, the correctness of the General algorithm for solving the problem Task (Algorithm 1) is a fully proved.

4 Algorithm Complexity Evaluation

Paper [2] gives a complexity evaluation for the algorithm that solves the problem of finding computer network limited resource allocation represented by a certain number of various devices across a variety of users belonging to various classes and being in the following form:

$$O\left(M\left(LM + M^2 + LK\right)\right) \tag{4}$$

where M is a number of devices, L is a number of users and K is a number of user classes.

A key variance of the problem under consideration herein, as it has already been said above, is the transition from considering a variety of individual devices to considering a variety of groups of similar devices which is a substantial generalization and complication of the original problem. Nevertheless, as, like in [2], Task handling is a continuous handling of Tasks (j) at $j = 1,2,\ldots,X$, then, in this case, the Task handling algorithm complexity will be in the following form:

$$O\left(Y\left(XY + Y^2 + XK\right)\right) \tag{5}$$

where Y is a total number of security sensors, X is a total number of network units and K is a number of unit classes.

5 Conclusion

This paper considers the problem of finding of such an allocation of complex network units, belonging to various classes, across security sensors, also belonging to various classes, which would maximize the total efficiency (utility value) of functioning of all the sensors considering the constraints defined for them. A general algorithm for solving the above-mentioned problem has been recommended, its correctness has been proved and its complexity has been evaluated.

It should be noted that the recommended method of efficient allocation of cybersecurity monitoring facilities within the framework of a complex network can be successfully used within the framework of solution of other problems. For example, when assessing RCITI security, including based on a wavelet analysis [6], RCITI cybersecurity control based on identifying its abnormal conditions using comprehensive evaluation [7] and cluster analysis [8,9] mechanisms.

References

1. Federal Law dated 26.07.2017 No 187-FZ On Security of the Critical IT Infrastructure of the Russian Federation
2. Tantawi, A.N., Towsley, G., Wolf, J.: Optimal allocation of multiple class resources in computer systems. ACM SIGMETRICS Perform. Eval. Rev. **16**(1), 253–260 (1988)

3. Kalashnikov, A.O., Anikina, E.V.: A method of efficient allocation of scanners for information security monitoring of nodes in a heterogeneous network. Inf. Secur. **21**(4), 455–464 (2018)
4. Korte, B. Kombinatornaya optimizaciya. Teoriya i algoritmi / B. Korte, Y. Figen - M.: MCNMO, p. 720 (2015)
5. Novikov, F.A.: Diskretnaya matematika: Uchebnik. / F.A. Novikov - SPb.: Piter, p. 432 (2013)
6. Kalashnikov, A.O., Sakrutina, E.A.: A model of the critical information infrastructure security assessment on the wavelet analysis basis. Inf. Secur. **20**(4), 478–491 (2017)
7. Kalashnikov, A.O.: Information risk management of organizational system: integrated rating mechanisms. Inf. Secur. **19**(3(4)), 315–322 (2016)
8. Kalashnikov, A.O., Anikina, E.V.: A model of the information security management of critical information infrastructure based on the anomaly detection (Part 1). Inf. Secur. **21**(2(4)), 145–154 (2018)
9. Kalashnikov, A.O., Anikina, E.V.: A model of the information security management of critical information infrastructure based on the anomaly detection (Part 2). Inf. Secur. **21**(2(4)), 155–164 (2018)

Development of Models and Methods for Using Heterogeneous Gateways in 5G/IMT-2020 Network Infrastructure

Lidiia Vlasenko[1], Vyacheslav Kulik[1(✉)], Ruslan Kirichek[1,2],
and Andrey Koucheryavy[1]

[1] The Bonch-Bruevich Saint-Petersburg State University of Telecommunications,
22 Prospekt Bolshevikov, Saint-Petersburg 193232, Russia
vslav.kulik@gmail.com
[2] Peoples' Friendship University of Russia (RUDN University), 6 Miklukho-Maklaya
Street, 117198 Moscow, Russian Federation
http://www.sut.ru,
http://www.rudn.ru

Abstract. In the course of the presented work, the types of traffic expected to be used in the infrastructure of 5G/IMT-2020 networks were investigated. Video services of 4K, virtual reality services, 360-degree broadcasting services, work in cloud storage were analyzed. As a result of the analysis, mathematical distributions were formed to build a model network. A model network was built. The model network is scaled up. Recommendations on the organization of 5G/IMT-2020 networks were made.

Keywords: 5G/IMT-2020 · Testing · Heterogeneous gateways · Model network · AnyLogic

1 Introduction

Machine-to-machine interaction is increasingly predominant over man-to-machine interaction every year. There are already several solutions available for interconnecting mobile devices. But with the advent of new services, the limited frequency spectrum, the increase in the number of mobile terminal devices, the requirements for communication channels and gateways at network nodes exceed the possibilities of the technologies used [1]. However, it must be remembered that network gateways must also be able to communicate between subnetworks operating on different technologies. It is to solve the problems with high latency, low bandwidth, narrow coverage area it is proposed to use mobile networks 5G/IMT-2020 [2,3].

One of the challenges in organizing 5G/IMT-2020 networks is to meet the requirements for different services. For example, for the implementation of the Smart Home network and the unmanned car network, the minimum level of

© Springer Nature Switzerland AG 2019
V. M. Vishnevskiy et al. (Eds.): DCCN 2019, CCIS 1141, pp. 636–645, 2019.
https://doi.org/10.1007/978-3-030-36625-4_51

delay, the number of devices to be connected, and the coverage width will be different [4,5]. If in the first case we need to connect millions of mobile devices that are at a short distance from each other, communicate with the Internet as needed, do not require a low level of delay, in the second case the terminals will travel long distances, requiring a very low delay and constant communication with the coordinator/sensors.

To solve these problems, the network will be divided into virtual "network layers", each of which is optimized for different requirements of data transmission speeds, delays, bandwidth and so on. To separate the "layers", but to combine technologies for interaction between subnetworks, it is necessary to use heterogeneous gateways, capable not only to carry out data transmission and transformation according to the standards of mobile networks of previous generations but also to implement new technologies of 5G/IMT-2020 networks [6].

Modeling and testing 5G/IMT-2020 mobile networks are key ways to predict and require intermediate network nodes, links and heterogeneous gateways. They allow determining the minimum level of delay, channel capacity, power consumption requirements of intermediate network nodes, method of load distribution, coverage width, etc.

2   5G/IMT-2020 Traffic Study

The following services from the immersive category were selected for the study, namely:

- Video services 4K;
- 360-degree video broadcasting;
- Virtual reality services;
- Working in a cloud storage facility.

Between 70,000 and 100,000 packages for each of the selected services were captured for analysis. The packets were filtered by IP address, protocol, and port. As a result of the study of the obtained data, the requirements for the communication channels were defined, which are presented in Table 1.

Table 1. Requirements for 5G/IMT-2020 network services

Network parameters	360-degree video (Kb)	4K video (MBs)	Virtual reality services (ms)	Cloud services (ms)
Average packet size	1.4	1.4	0.3	0.4
Average throughput	0.6	1.6	2.2	0.4
Delay	4.0	0.0	33.0	4.0
Jitter	2.0	0.0	5.8	2.0

When building a 5G/IMT-2020 based on these requirements network that will provide 4K video services, 360-degree video broadcasting services, virtual reality services and will allow to work in cloud storage, it is required that the average bandwidth of the communication channel is at least 2.2 Mbit/s with a delay of no more than 33 ms. Estimated values of delay and capacity of 5G/IMT-2020 networks certainly in many times exceed the received indicators, but they are theoretical and are not realized in practice yet. It can be assumed that the first mass implementations of 5G/IMT-2020 networks will not be in line with the declared indicators [7].

To determine the network node requirements for 5G/IMT-2020 services, it is necessary to examine the traffic for the respective services in the time interval between the sending of the request from the processor and receiving the response to this request from the server [8].

3 Package Arrival Rate Models

To create a traffic generator it is necessary to define the distribution law, with the help of which it would be possible to generate traffic. For the types of traffic described above, a two-parameter gamma, exponential and two-parameter beta distributions that described by the next functions [9, 10]:

Gamma probability distribution:

$$F(x) = \frac{1}{G(\beta)} \int_0^{\alpha x} t^{\alpha-1} e^{-t} dt, \tag{1}$$

where α - shape parameter, β - scale parameter ($\alpha > 0$, $\beta > 0$),

$$G(\beta) = \int_0^\infty t^{\beta-1} e^{-t} dt, \tag{2}$$

gamma function.

Exponential probability distribution:

$$F(x) = 1 - e^{-\alpha x}, \tag{3}$$

where α - scale parameter ($\alpha > 0$).

Beta probability distribution:

$$F(x) = \frac{B_x(\alpha, \beta)}{B(\alpha, \beta)}, \tag{4}$$

where α, β - shapes properties ($\alpha > 0, \beta > 0$).

$$B_x(\alpha, \beta) = \int_0^x t^\alpha (1 - t)^{\beta-1} dt, \tag{5}$$

incomplete beta function.

$$B(\alpha, \beta) = \int_0^\infty \frac{t^{\alpha-1}}{(1 + t)^{\alpha+\beta}} dt, \tag{6}$$

beta function.

It is selected following objective function to choose the function coefficient that based on the generalized reduced gradient method by the least square summary coefficient:

$$K_{lsm}(t) = \sum_{1}^{n} [P(t_i) - P(t_i, t_{i+1}]^2, \tag{7}$$

where $P(t)$ is the probability of hitting a random value of the time interval between the receipt of messages in the interval from t_i prior to t_{i+1} according to experimental data,

$$P(t_i, t_{i+1}) = |F(t_{i+1}) - F(t_i)| \tag{8}$$

is the probability of hitting a random value of the time interval between message receipts, according to the chosen distribution law $F(t)$.

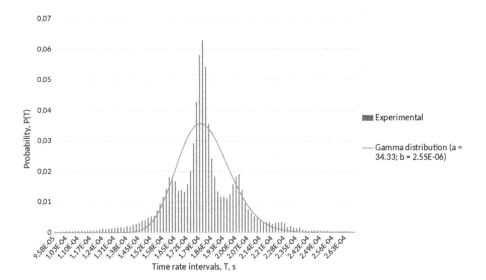

Fig. 1. Histogram of the ratio of experimental data to the chosen probability distribution: (a) for 360-degree video

The similarity of theoretical and practical distributions is checked using the Kolmogorov–Smirnov test at 95% confidence probability [11]:

$$D_n = \max[F_n(x) - F(x)], \tag{9}$$

If the value of D_n^* does not exceed the Kolmogorov's distribution quantile K_n, the distributions are considered identical:

$$D_n^* <= K_n, \tag{10}$$

where D_n^* is practical quantile of the distribution:

$$D_n^* = \sqrt{\frac{n * m}{n + m}}, \tag{11}$$

where n is practical distribution size and m is a theoretical distribution size (for this case n = m).

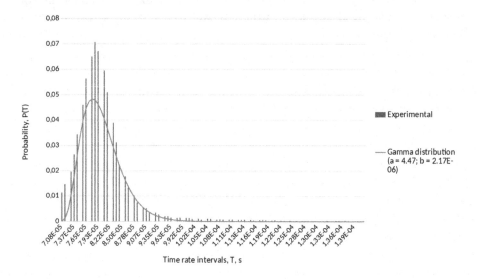

Fig. 2. (b) for 4k video

A graph showing the ratio of the probability of a random value of the time interval between messages and observations for the 360-degree video broadcasting service is shown in Fig. 1. As a result, the following form values were obtained: $\alpha = 34.33$, $\beta = 2.55\text{E}{-}06$.

A graph showing the ratio of the probability of a random value of the time interval between message arrivals and observations for 4K video broadcasting service is shown in Fig. 2. As a result, the following form values were obtained: $\alpha = 4.47$, $\beta = 2.17\text{E}{-}06$.

A graph showing the ratio of the probability of a random value of the time interval between message arrivals and observations for virtual reality services is shown in Fig. 3. As a result, the following form values were obtained: $\alpha = 62.91$.

A graph showing the ratio of the probability of occurrence of a random value of the time interval between the receipt of messages and observations to work with cloud storage is shown in Fig. 4. As a result, the following form values were obtained: $\alpha = 67.40$, $\beta = 1348376$.

Also, two-parameter gamma-distributions were built to simulate delays during packet processing at a heterogeneous gateway in the 5G/IMT-2020 model network.

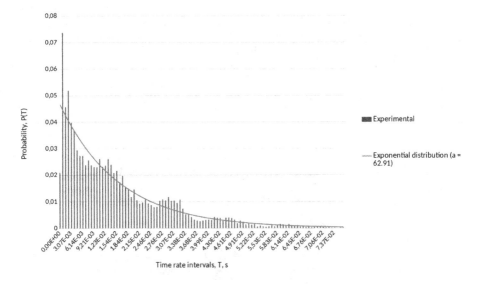

Fig. 3. (c) for virtual reality services

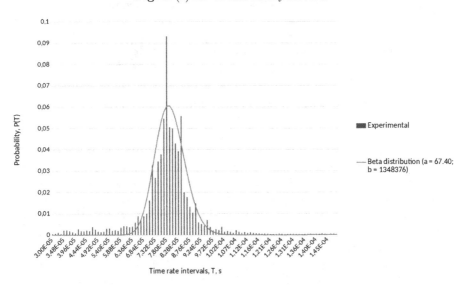

Fig. 4. (d) for remote cloud services

4 Development and Testing of a 5G/IMT-2020 Model Network

To determine the methods of heterogeneous gateway application in 5G/IMT-2020 networks, it is necessary to develop a model complex simulating 5G/IMT-2020 network operation, as well as to conduct load testing. According to ITU-T

Q.3900, "Testing methods and architecture of model networks for testing IPS technical means used in public telecommunication networks", a model network is a network that simulates the possibilities similar to those existing in existing telecommunication networks, has a similar architecture and has the same functionality, and uses the same technical means of telecommunication [12].

The model network was developed in AnyLogic. The architecture of this network is shown in Fig. 5.

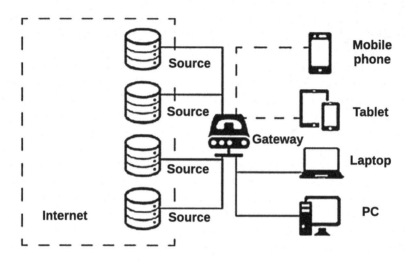

Fig. 5. 5G/IMT-2020 model network architecture

The source of traffic for this system is video resources placed on the Internet, as well as a VR-video game. Depending on features of construction of a network and tasks in view all specified above elements and subsystems can be realized both together, and separately in various combinations.

The source elements used in AnyLogic were source elements, the packages in which arrived according to the distribution law defined in Sect. 2. Also, the packages were of the specified size as defined in Sect. 2. To simulate a heterogeneous gateway, the elements queue, delay and sink were taken. The queue element emitted a heterogeneous gateway buffer with a capacity of 100 packets. An additional sink element was connected to the queue element to allow packages that did not have time to be processed to be supplanted. The delay element was chosen as the element that simulates the processing of the packet at the gateway, where the distribution law was set according to the previously defined one.

As a result of this testing, the capacity requirements for the selected services were obtained and presented in Table 2.

To determine the throughput of a heterogeneous gateway with a greater number of traffic sources, each of the obtained models was scaled up to 10, 50 and 100 sources.

Table 2. Minimum capacity requirements for 5G/IMT-2020 services

Services	Minimum throughput requirements
360-degree video broadcasting	967 kBs
Video services 4K	16 MB/s
Virtual reality services	16 Kb/s
Working with cloud storage	204 kB/s

When scaling the model network with the use of 360-degree video broadcasting service with the use of 10 sources, the minimum value of bandwidth was 9 kB/s, with the use of 50 sources - 47 MB/s, with the use of 100 sources, the network fails.

At scaling of a model network with the use of video services 4K at the use of 10 sources the minimum value of throughput has made 166 MB/s, at the use of 50 sources the network refuses.

At the scaling of a model network with the use of services of virtual reality at the use of 10 sources the minimum value of throughput has made 36 Kb/s, at the use of 50 sources - 99 kB/s, at the use of 100 sources the network refuses.

At scaling of a model network with use of services of a virtual reality at use of 10 sources the minimum value of throughput has made 20 MB/s, at use of 50 sources - 99 MB/s, at use of 100 sources - 197 MB/s, at use of 1000 sources the network refuses.

5   Conclusion

In this article, we studied the types of 5G/IMT-2020 traffic, the services of this network, and the requirements of the respective services. Certain types of traffic have been selected for the experiment and model network construction. Also, the traffic of 360-degree video broadcasting, 4K video services, virtual reality services, and cloud storage was studied. Heterogeneous gateway requirements to provide these services, namely average throughput, are obtained. The constants for building the corresponding types of traffic in the model network are defined.

A 5G/IMT-2020 model network was developed to study the heterogeneous gateway in it. The network is tested with each of the selected services. Heterogeneous gateway requirements corresponding to certain distributions are obtained. The data obtained can be used to build a model network with 360-degree video broadcasting traffic, 4K video services, virtual reality services, traffic when working with cloud storage, as well as the data obtained can be used to identify possible difficulties in working with the heterogeneous gateway.

Each of the networks has been scaled up. The gateway bandwidth required to process traffic coming from specified distributions is determined. It was determined that among the selected services the most resistant to scaling was a model network simulating the work of 5G/IMT-2020 network when working with cloud storage. The advantages of using heterogeneous gateways are listed and their necessity in 5G/IMT-2020 networks is confirmed.

This work allows you to build a private network using a heterogeneous gateway for 360-degree video broadcasting services, 4K video services, virtual reality services when working with cloud storage. If the requirements defined in the course of work are met, with a specified number of traffic sources, it is possible to determine the required performance of the gateway in advance. The methods and models of heterogeneous gateways application studied in the work also allow to determine the necessity of using heterogeneous gateways in each specific 5G/IMT- 2020 technology.

Acknowledgements. The publication has been prepared with the support of the "RUDN University Program 5-100" and funded by RFBR according to the research projects No. 12-34-56789 and No. 12-34-56789.

References

1. Paramonov, A., Koucheryavy, A.: M2M traffic models and flow types in case of mass event detection. In: Balandin, S., Andreev, S., Koucheryavy, Y. (eds.) NEW2AN 2014. LNCS, vol. 8638, pp. 294–300. Springer, Cham (2014). https://doi.org/10.1007/978-3-319-10353-2_25
2. Mahjoubi, A.E., Mazri, T., Hmina, N.: M2M and eMTC communications via NB-IoT, Morocco first testbed experimental results and RF deployment scenario: new approach to improve main 5G KPIs and performances. In: 2017 International Conference on Wireless Networks and Mobile Communications (WINCOM) (2017). https://doi.org/10.1109/WINCOM.2017.8238156
3. Beshley, H., Kyryk, M., Beshley, M., Panchenko, O.: Method of information flows engineering and resource distribution in 4G/5G heterogeneous network for M2M service provisioning. In: 2018 IEEE 4th International Symposium on Wireless Systems within the International Conferences on Intelligent Data Acquisition and Advanced Computing Systems (IDAACS-SWS), pp. 229–233 (2018). https://doi.org/10.1109/IDAACS-SWS.2018.8525680
4. Rhee, S.: Catalyzing the Internet of Things and smart cities: global city teams challenge. In: 2016 1st International Workshop on Science of Smart City Operations and Platforms Engineering (SCOPE) in partnership with Global City Teams Challenge (GCTC) (SCOPE - GCTC) (2016). https://doi.org/10.1109/SCOPE.2016.7515058
5. Giyenko, A., Cho, Y.I.: Intelligent UAV in smart cities using IoT. In: 2016 16th International Conference on Control, Automation and Systems (ICCAS-2016), pp. 207–210 (2016). https://doi.org/10.1109/ICCAS.2016.7832322
6. Ignatova, L., Khakimov, A., Mahmood, A., Muthanna, A.: Analysis of the Internet of Things devices integration in 5G networks. In: Conference: Systems of Signal Synchronization 2017 (2017). https://doi.org/10.1109/SINKHROINFO.2017.7997524
7. Zikria, Y.B., Kim, S.W., Afzal, M.K., Wang, H., Rehmani, M.H.: 5G mobile services, and scenarios: challenges and solutions. Sustainability **3626**, 1–9 (2018). https://doi.org/10.3390/su10103626

8. Kostopoulos, A., Agapiou, G., Kuo, F.-C., Pentikousis, K., et al.: Scenarios for 5G networks. In: Conference: 2016 23rd International Conference on Telecommunications (ICT) (2017). https://doi.org/10.1109/ICT.2016.7500421

9. Ross, S.M.: Introduction to Probability Models, 11th edn, p. 784. Academic Press, Norwell (2014). eBook ISBN: 9780124081215

10. Hogg, R.V., Craig, A.T.: Introduction to Mathematical Statistics, 4th edn. Macmillan, New York (1978). ISBN 0-02-978990-7

11. Sheskin, D.J.: Handbook of Parametric and Nonparametric Statistical Procedures. CRC Press, Boca Raton (2003). ISBN 1-58488-440-1

12. ITU-T Q.3900 : Methods of testing and model network architecture for NGN technical means testing as applied to public telecommunication networks. International Telecommunication Union. Telecommunication Standardization Sector (ITU-T) (2006)

Reliability Evaluation of a Hexacopter-Based Flight Module of a Tethered Unmanned High-Altitude Platform

D. V. Kozyrev[1,2(⊠)] , Nguyen Duy Phuong[3], H. G. K. Houankpo[2] ,
and Alexander Sokolov[1]

[1] V. A. Trapeznikov Institute of Control Sciences of Russian Academy of Sciences,
65 Profsoyuznaya street, Moscow 117997, Russia
[2] Department of Applied Probability and Informatics, Peoples' Friendship University
of Russia (RUDN University), 6 Miklukho-Maklaya Street,
Moscow 117198, Russian Federation
kozyrev-dv@rudn.ru, gibsonhouankpo@yahoo.fr
[3] Moscow Institute of Physics and Technology (National Research University),
9 Institutskiy per., Dolgoprudny, Moscow Region 141701, Russia
ndphuong2207@gmail.com

Abstract. This article discusses a model on the basis of a multidimensional Markov process applied for evaluation of the reliability characteristics of a tethered multirotor high-altitude platform based on a hexacopter. The proposed model takes into account the increase in the functional load after the failure of an element on the remaining operating ones, and also takes into account the location of the failed elements.

Keywords: UAV · High-altitude platform · Hexacopter · Reliability · k-out-of-n system

1 Introduction and Motivation

One of the promising directions in the framework of the concept of creating the next-generation 5G/IMT-2020 networks is the development of broadband wireless networks based on autonomous and tethered unmanned aerial vehicles (UAVs). The advantage of such networks is their fast and flexible deployment, a wider area of telecommunication coverage and enhanced reliability of wireless communications, controllable mobility, reduced operating costs, etc., which ensures their effective application in both civil and defense industries [1–4].

At present time, tethered high-altitude unmanned telecommunication platforms, whose long-term operation is ensured by transmission of electric energy from ground to board via a thin cable rope [6], have received widespread development. The tethered high-altitude platforms fall in-between satellite systems and terrestrial systems whose equipment (cellular base stations, radio-relay and radar equipment, etc.) is deployed at high-rise structures. The tethered high-altitude platforms, as compared with expensive satellite systems, are highly cost efficient.

© Springer Nature Switzerland AG 2019
V. M. Vishnevskiy et al. (Eds.): DCCN 2019, CCIS 1141, pp. 646–656, 2019.
https://doi.org/10.1007/978-3-030-36625-4_52

Diagnostics of the performance of unmanned aerial vehicles (UAVs) is becoming a new trend in the scientific work of many researchers [5]. In addition to the interest in high-altitude platforms implemented on autonomous UAVs, research centers of advanced countries of the world are currently carrying out intensive scientific work on the design and implementation of tethered unmanned high-altitude platforms, given the vastness of their practical application [7,8]. The possibility of long-term operation of tethered unmanned high-altitude platforms, which is one of the main advantages compared to autonomous UAVs, puts forward a number of new requirements on the reliability of both the individual components and the high-altitude platform as a whole [4].

In this regard, the subject of the current paper, aimed at developing and studying methods for reliability modelling and reliability enhancement of tethered high-altitude platforms, is relevant.

2   Model Description

A tethered multi-rotor unmanned high-altitude platform is a hexacopter system consisting of 6 identical rotors (elements) arranged uniformly in a circle and in pairs symmetrically with respect to the center of the circle. The structural diagram of the system is shown in Fig. 1, and an image of an example of such a system is shown in Fig. 2.

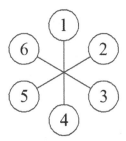

Fig. 1. Block diagram of a hexacopter

Fig. 2. Hexacopter

Assume that the system is operational while at least 4 rotors are operating, and the failed 2 rotors are not located next to each other, i.e. the system is inoperative if either 2 adjacent rotors or any 3 rotors fail. We suggest to denote this system as a "(2, 3)-out-of-6: F" system. We consider a repairable multiple cold-standby system with one repair device, with an exponential distribution law for the uptime of its elements with a parameter λ, and an arbitrary distribution law for their repair time as a mathematical model of a multi-rotor flight module. Due to the fact that failures of some elements increase the functional load on the remaining workable elements, and their reliability decreases, the failure rate of each of the elements after the failure of the first one increases and will be equal to $\lambda_1 > \lambda$ and the failure rate of each of the remaining elements after failure the second one will be equal to $\lambda_2 > \lambda_1$.

To describe the functioning of the system, we consider the space of its possible states $E = \{0, 1, 2, 3\}$, where

(0) – initial state of the system when all rotors are operational.

(1) – system state when one rotor is in failure mode.

(2) – state of the system when two non-adjacent rotors are in the failure mode.

(3) – state of the system when either 2 adjacent rotors or any 3 rotors fail.

We introduce a stochastic process $v(t)$ on the state space E—the number of elements that are in a failure state at time t.

To markovize this process, that is, to describe the behavior of the system using the Markov process (MP) [9,10], we introduce an additional variable $x(t) \in R_+$ – the time spent at time t to repair the failed element and use the expanded state space $\varepsilon = E \times R_+$. As a result, we obtain a two-dimensional process $(v(t), x(t))$ with an expanded state space $\varepsilon = \{(0), (1), (2), (3, x)\}$.

3   Calculation of the Stationary Probability Distribution of System States

We introduce the following notations:

A – random variable (r.v.), time to failure of the main element,

B – r.v., recovery time of a failed element,

$B(x)$ – cumulative distribution function (CDF) of the r.v. B,

$b(x)$ – probability density function (PDF) of the r.v. B,

$\beta(x) = \frac{b(x)}{1-B(x)}$ – conditional distribution density of the residual duration of repair of an element under repair time t.

Denote by $p_k(t)$ the probability of the system being at state k at time t, $k = \overline{0,2}$, and by $p_3(t, x)$ the PDF of the probabilities that the system is at state 3 at time t. The rate transition diagram is shown in Fig. 3.

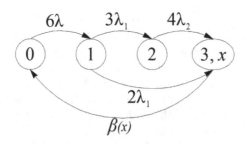

Fig. 3. The rate transition diagram

We compose the equations for the state probabilities of the system using the formula of total probability. As we pass to the limit as $\Delta \to 0$, we derive the Kolmogorov system of differential equations, which allows us to find the probabilities of the states of the considered system.

$$\frac{\partial p_0(t)}{\partial t} = -6\lambda p_0(t) + \int_0^t p_3(t,x)\beta(x)dx \tag{1.1}$$

$$\frac{\partial p_1(t)}{\partial t} = 6\lambda p_0(t) - 5\lambda_1 p_1(t) \tag{1.2}$$

$$\frac{\partial p_2(t)}{\partial t} = 3\lambda_1 p_1(t) - 4\lambda_2 p_2(t) \tag{1.3}$$

$$\frac{\partial p_3(t,x)}{\partial t} + \frac{\partial p_3(t,x)}{\partial x} = -\beta(x)p_3(t,x) \tag{1.4}$$

The boundary condition has the following form:

$$p_3(t,0) = 2\lambda_1 p_1(t) + 4\lambda_2 p_2(t) \tag{1.5}$$

Assume that for the described process, there exists a stationary probability distribution as $t \to \infty$. Then the transformed equations will take the following form (balance equation system):

$$-6\lambda\, p_0 + \int_0^\infty p_3(x)\,\beta(x)dx = 0 \tag{2.1}$$

$$6\lambda\, p_0 - 5\lambda_1\, p_1 = 0 \tag{2.2}$$

$$3\lambda_1\, p_1 - 4\lambda_2\, p_2 = 0 \tag{2.3}$$

$$\frac{\partial p_3(x)}{\partial x} = -\beta(x)p_3(x) \tag{2.4}$$

with the boundary condition:

$$p_3(0) = 2\lambda_1\, p_1 + 4\lambda_2\, p_2 \tag{2.5}$$

We proceed to the solution of the obtained equation system.
As a result, we obtain the macro-state probabilities of the system:

$$p_0 = \frac{10\lambda_1\lambda_2}{10\lambda_1\lambda_2 + 9\lambda\lambda_1 + 12\lambda\lambda_2 + 60\lambda\lambda_1\lambda_2 b}$$

$$p_1 = \frac{12\lambda\lambda_2}{10\lambda_1\lambda_2 + 9\lambda\lambda_1 + 12\lambda\lambda_2 + 60\lambda\lambda_1\lambda_2 b}$$

$$p_2 = \frac{9\lambda\lambda_1}{10\lambda_1\lambda_2 + 9\lambda\lambda_1 + 12\lambda\lambda_2 + 60\lambda\lambda_1\lambda_2 b}$$

$$p_3 = \frac{60\lambda\lambda_1\lambda_2 b}{10\lambda_1\lambda_2 + 9\lambda\lambda_1 + 12\lambda\lambda_2 + 60\lambda\lambda_1\lambda_2 b}$$

4   Special Case. Numerical Example

In this section, we consider a special case of the model of the repairable "(2, 3)-out-of-6: F" system with the exponential distribution of the repair time of the failed components with a parameter μ ($\beta(x) = 1 - e^{-\mu x},\ x \geq 0$).

Thus, the mean repair time of the failed system's components is $b = \mu^{-1}$ and we obtain the following stationary probabilities of the states of the considered redundant system:

$$p_0 = \frac{10\mu\lambda_1\lambda_2}{\mu(10\lambda_1\lambda_2 + 12\lambda\lambda_2 + 9\lambda\lambda_1) + 60\lambda\lambda_1\lambda_2};$$

$$p_1 = \frac{12\mu\lambda\lambda_2}{\mu(10\lambda_1\lambda_2 + 12\lambda\lambda_2 + 9\lambda\lambda_1) + 60\lambda\lambda_1\lambda_2};$$

$$p_2 = \frac{9\mu\lambda\lambda_1}{\mu(10\lambda_1\lambda_2 + 12\lambda\lambda_2 + 9\lambda\lambda_1) + 60\lambda\lambda_1\lambda_2};$$

$$p_3 = \frac{60\lambda\lambda_1\lambda_2}{\mu(10\lambda_1\lambda_2 + 12\lambda\lambda_2 + 9\lambda\lambda_1) + 60\lambda\lambda_1\lambda_2}.$$

State 3 coincides with the system's failure state, i.e. the probability of failure of the system $p_{[system failure]}$ is

$$p_{[system failure]} = p_3 = 1 - K = \frac{60\lambda\lambda_1\lambda_2}{\mu(10\lambda_1\lambda_2 + 12\lambda\lambda_2 + 9\lambda\lambda_1) + 60\lambda\lambda_1\lambda_2}.$$

From this formula, we find the availability factor

$$K = \frac{\mu(10\lambda_1\lambda_2 + 12\lambda\lambda_2 + 9\lambda\lambda_1)}{\mu(10\lambda_1\lambda_2 + 12\lambda\lambda_2 + 9\lambda\lambda_1) + 60\lambda\lambda_1\lambda_2}.$$

To plot the dependence of the system availability factor K on the relative recovery rate $\rho = \frac{\mu}{\lambda}$, we use the following numerical values for the failure rates of the elements: $\lambda = 1$, $\lambda_1 = 1.1$, $\lambda_2 = 1.4$ (Fig. 4). Then the availability factor has the form

$$K = \frac{42,1\mu}{42,1\mu + 92,4}$$

Table 1 gives some values of μ and the corresponding values of ρ and K.

Table 1. Parameter values μ, ρ and K

μ	0	1	2	3	4	5
ρ	0	1	2	3	4	5
K	0	0,313	0,4768	0,5775	0,6457	0,6949

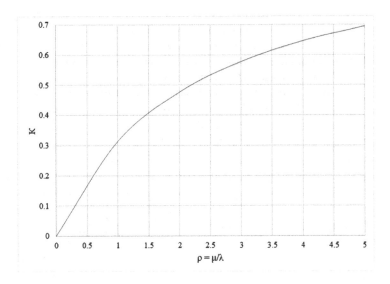

Fig. 4. System availability factor K versus the relative recovery rate $\rho = \mu/\lambda$

5 Simulation Model

It is not always possible to obtain the explicit analytical expressions for the steady-state distribution of the considered system. For this case we developed a simulation model based on the discrete-event approach. We introduce the following variables to describe the algorithm for simulation of the $\langle M_{3<6}/GI/1 \rangle$ system reliability:

- double t—modelling time (hours), change in case of failure or restoration of the system's elements;
- int i, j—system state variables; when an event occurs, the transition from i to j takes place;
- double $t_{nextfail}$—service variable, that stores the time till the next element's failure;
- double $t_{nextrepair}$—service variable, that stores the time till the next repair of a failed element;
- int k—counter of iterations of the main loop.

For the sake of clarity, the simulation model is presented graphically in Fig. 5 in the form of a flowchart.

The criterion for stopping the main cycle of the simulation model is reaching the maximum model execution time T.

The algorithm of the discrete-event process of simulation modeling is also provided in the form of a pseudo-code with comments (Algorithm 1).

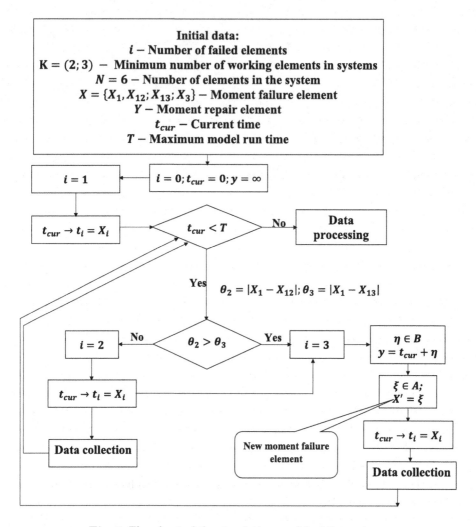

Fig. 5. Flowchart of the simulation model of the system

Algorithm 1. Pseudo-code of the $\langle M_{3<6}/GI/1 \rangle$ system simulation process
Input: a1, b1, N,T, NG, "GI".

$a = \{3, 30, 300\}$ - Mean failure time of the first element
$a1 = \{2, 20, 200\}$ - Mean failure time of the second element
$a2 = \{1, 10, 100\}$ - Mean failure time of the third element
$b1 = 3$ - Average repair time
$k = (2, 3)$ - Maximum number of failed elements
$N = 6$ - The number of elements in the system
$T = 1000$ - Maximum model run time
$NG = 500$ - Number of trajectories

"*GI*" - Distribution function.

Output: stationary state probabilities P_0, P_1, P_2, P_3.

Begin

array $r[\ ] := [0, 0, 0]$; // multidimensional array containing results, k-steps of the main loop

double $t := 0.0$; // time clock initialization

int $i := 0$; $j := 0$; // system state variables

double $t_{nextfail} := 0.0$; // time till the next element's failure

double $t_{nextrepair} := 0.0$; // time till the next repair is completed

int $k := 1$; // initialization of the counter of iterations of the main loop

$s := rf\_exp(6, 1/a)$; // generation of an exponential random variable s – time to the first event (failure)

$sr := rf\_GI(\delta(x))$; // generation of an arbitrary random variable sr – repair time of the failed element)

$t_{nextfail} := t + s$;

$t_{nextrepair} := t + sr$;

while $t < \infty$ **do**

    if $i = 0$ **then**

        $t_{nextrepair} := \infty$; $j := j + 1$; $t := t_{nextfail}$;

    else if

        if $i = 1$ **then**

            $s_{12} := rf\_exp(5, 1/a1)$; $s_{13} := rf\_exp(5, 1/a1)$;

            $t_{nextfail12} := t + S_{12}$; $t_{nextfail13} := t + S_{13}$;

            if $abs(s - s_{12}) < abs(s - s_{13})$ **then**

                $j := j + 1$; $t := t_{nextfail12}$;

            else

                $j := j + 2$; $t := t_{nextfail13}$;

            end

        else

            $i = 2$; $t_{nextfail2} := t + s2$; $j := j + 1$; $t := t_{nextfail2}$;

        end

    else

        $i = 3$; $t_{nextfail} := \infty$; $j := j - 3$; $t := t_{nextrepair}$;

        if $t > T$ **then**

            $t = T$

        end

    $r[,, k] := [t, i, j]$; $i := j$; $k := k + 1$;

end do

Estimated duration of stay in each state $i, i = 0, 1, 2, 3.$; calculation of stationary probabilities

$$\widehat{P}_i = \frac{1}{NG} \sum_{j=1}^{NG} (\text{ length of stay } i/T)_j$$

end

Software implementation of Algorithm 1 was performed in the programming language R.

Table 2 shows the values of stationary state probabilities, calculated via the simulation approach. For the analysis of the results, the following distribution of repair times of the system's elements were chosen: exponential, Weibull-Gnedenko, Pareto and lognormal. However, the developed simulation model is not limited to the choice of the repair time distribution. As a model parameter the value $\rho = \frac{a}{b_1}$ is considered.

For analyzing the sensitivity of the model to the form of the distribution function, the magnitude values increase $\rho = 1, 10, 100$, where $b_1 = 3$ in all the considered cases, the distribution parameters are chosen in such a way as to correspond to the model parameter value.

Table 2. Simulation results for the values of stationary state probabilities p_i; $i = \{0, 1, 2, 3\}$ of the system $\langle M_{3<6}/GI/1\rangle$ calculated for different values of the model parameter $\rho = 1, 10, 100$.

ρ; p_i		M	$WB(W = 1/2)$	$PAR(P = 3)$	$LN(sig = 1)$
$\rho = 1$	p_0	0.20435677	0.2836192	0.13484044	0.18822212
	p_1	0.13259454	0.2001307	0.09383081	0.12818292
	p_2	0.04856519	0.0758115	0.03422376	0.04699674
	p_3	0.61119058	0.4411852	0.73613228	0.62652607
$\rho = 10$	p_0	0.4102325	0.4336593	0.39863381	0.4158587
	p_1	0.2836765	0.3035402	0.27556396	0.2847193
	p_2	0.1019977	0.1081576	0.09877994	0.1042716
	p_3	0.2168704	0.1296582	0.22537897	0.1956799
$\rho = 100$	p_0	0.50902727	0.51475970	0.50738918	0.5154946
	p_1	0.34215798	0.33933409	0.33800919	0.3410063
	p_2	0.11827470	0.11959156	0.12034636	0.1253436
	p_3	0.02828614	0.02859619	0.02925021	0.0296693

Numerical results show that for $\rho = 100$ the stationary probabilities of the state of the system for the distribution under consideration decrease $p_0 > p_1 > p_2 > p_3$, that is, the faster the restoration of system elements, the more reliable the system

Plots in Fig. 6 show the dependence of the failure-free probability of the system versus the model parameter ρ.

Where $a = 1$; $a1 = 0.91$; $a2 = 0.71$; $b1 = 0.04$; $k = (2, 3)$; $N = 6$; $T = 1000$; $NG = 500$;

The graphical results show that in this case, the Pareto distribution is the most reliable repair time model.

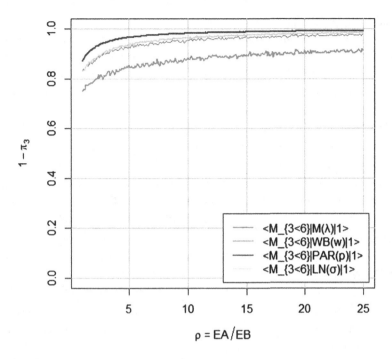

Fig. 6. Plots of the failure-free probability of the system versus the model parameter $\rho = \frac{a}{b_1}$

6   Conclusions and Future Work

In the present work, a reliability model for the flight module of a tethered multi-rotor high-altitude platform based on a hexacopter is constructed in the form of a model of the "(2, 3)-out-of-6"-type system. The proposed analytical model is a generalization of classical models of the "k-out-of-n"-type redundant systems since it takes into account the heterogeneity of the structure of failures, i.e. the increase in the functional load after the failure of an element to the remaining ones, and the location of the failed elements.

The results of calculating the stationary state probabilities and stationary reliability characteristics of the considered system were obtained. It was assumed that the repair time of the failed elements was generally distributed. A discrete-event simulation model was constructed for the study of the general case. The results obtained with both analytical and simulation approaches are in close agreement.

The proposed reliability model of multi-rotor high-altitude aerial modules allows a number of further generalizations. Future work concerns deeper reliability analysis, calculation of non-stationary reliability characteristics of the system, as well as reliability analysis of the multi-rotor flight module in a random environment.

Acknowledgments. The publication has been prepared with the support of the "RUDN University Program 5-100" (D.V. Kozyrev, mathematical model development, H.G.K. Houankpo, simulation model development). The reported study was funded by RFBR, project number 19-29-06043 and 17-07-00142 (D.V. Kozyrev, numerical analysis).

References

1. Chandrasekharan, S., et al.: Designing and implementing future aerial communication networks. IEEE Commun. Mag. **54**(5), 26–34 (2016)
2. Morales-Perryman, Q., Lee, D.D.: Tethering system for unmanned aerial vehicles, pp. 1–7. Hampton University, Electrical Engineering (2018)
3. Kiribayashi, S., Yakushigawa, K., Nagatani, K.: Design and development of tether-powered multirotor micro unmanned aerial vehicle system for remote-controlled construction machine. In: Hutter, M., Siegwart, R. (eds.) Field and Service Robotics. SPAR, vol. 5, pp. 637–648. Springer, Cham (2018). https://doi.org/10.1007/978-3-319-67361-5_41
4. Vishnevsky, V.M., Efrosinin, D.V., Krishnamoorthy, A.: Principles of construction of mobile and stationary tethered high-altitude unmanned telecommunication platforms of long-term operation. In: Vishnevskiy, V.M., Kozyrev, D.V. (eds.) DCCN 2018. CCIS, vol. 919, pp. 561–569. Springer, Cham (2018). https://doi.org/10.1007/978-3-319-99447-5_48
5. Khan, M.A., Hamila, R., Kiranyaz, M.S., Gabbou, A.M.: A novel UAV aided network architecture using WiFi direct. IEEE Access, **7**, 67305–67318 (2019)
6. Vishnevsky, V., Tereschenko, B., Tumchenok, D., Shirvanyan, A.: Optimal method for uplink transfer of power and the design of high-voltage cable for tethered high-altitude unmanned telecommunication platforms. In: Vishnevskiy, V.M., Samouylov, K.E., Kozyrev, D.V. (eds.) DCCN 2017. CCIS, vol. 700, pp. 240–247. Springer, Cham (2017). https://doi.org/10.1007/978-3-319-66836-9_20
7. Perelomov, V.N., Myrova, L.O., Aminev, D.A., Kozyrev, D.V.: Efficiency enhancement of tethered high altitude communication platforms based on their hardware-software unification. In: Vishnevskiy, V.M., Kozyrev, D.V. (eds.) DCCN 2018. CCIS, vol. 919, pp. 184–200. Springer, Cham (2018). https://doi.org/10.1007/978-3-319-99447-5_16
8. Vishnevskiy, V.M., Shirvanyan, A.M., Tumchenok, D.A.: Mathematical model of the dynamics of operation of the tethered high-altitude telecommunication platform in the turbulent atmosphere. In: Proceedings of International Scientific Conference 2019 Systems of Signals Generating and Processing in the Field of on Board Communications IEEE Conference No. 46544, Moscow, pp. 1–7. IEEE, Moscow (2019)
9. Rykov, V., Kozyrev, D.: Reliability model for hierarchical systems: regenerative approach. Autom. Remote Control **71**(7), 1325–1336 (2010). https://doi.org/10.1134/S0005117910070064
10. Rykov, V.V., Kozyrev, D.V.: Analysis of renewable reliability systems by Markovization method. In: Rykov, V.V., Singpurwalla, N.D., Zubkov, A.M. (eds.) ACMPT 2017. LNCS, vol. 10684, pp. 210–220. Springer, Cham (2017). https://doi.org/10.1007/978-3-319-71504-9_19

Author Index

Printed in the United States
By Bookmasters